Y0-BSD-767

LITHIUM CHEMISTRY

LITHIUM CHEMISTRY

A Theoretical and Experimental Overview

Edited by
ANNE-MARIE SAPSE
New York, New York

and

PAUL VON RAGUÉ SCHLEYER
Erlangen, Germany

A Wiley-Interscience Publication
JOHN WILEY & SONS, INC.
New York • Chichester • Brisbane • Toronto • Singapore

This text is printed on acid-free paper.

Library of Congress Cataloging in Publication Data:

Lithium chemistry: a theoretical and experimental overview / edited by Anne-Marie Sapse and Paul von Ragué Schleyer.

p. cm.

"A Wiley-Interscience publication."

Includes index.

ISBN 0-471-54930-4 (acid-free)

1. Lithium. 2. Organolithium compounds. I. Sapse, Anne-Marie. II. Schleyer, Paul von R., 1930–

QD181.L5L487 1994

546'.381—dc20 94-20176

Printed in the United States of America

10 9 8 7 6 5 4 3 2 1

CONTRIBUTORS

Andrews, Lester, Department of Chemistry, University of Virginia, Charlottesville, Virginia

Bach, Ricardo, Lake Wylie, South Carolina

Bachrach, Steven M., Department of Chemistry, Northern Illinois University, DeKalb, Illinois

Bartsch, Richard A., Department of Chemistry, Texas Tech University, Lubbock, Texas

Bauer, Walter, Institut fur Organische Chemie 1, Universität Erlangen-Nürnberg, Erlangen, Germany

Boche, Gernot, Fachbereich Chemie, Philipps Universität, Marburg, Germany

Dorigo, Andrea, Institut fur Organische Chemie I, Universität Erlangen-Nürnberg, Erlangen, Germany

Jain, Duli C., Natural Sciences Department, York College of the City University of New York, Jamaica, New York

Kimura, Keiki, Chemical Process Engineering, Faculty of Engineering, Osaka University, Osaka, Japan

Liebman, Joel F., Department of Chemistry and Biochemistry, University of Maryland, Baltimore Country Campus, Baltimore, Maryland

Lohrenz, John C.W., Fachbereich Chemie, Philipps Universität, Marburg, Germany

Maercker, A., Institute of Organic Chemistry, University of Siegen, Siegen, Germany

Manceron, Laurent, Laboratoire de Spectrochimie Moleculaire, Universite Pierre et Marie Curie, Paris, France

Opel, Achim, An der Linde 8, Schoffengrund, Germany

Pauer, Frank, Department of Chemistry, University of California, Davis, Davis, California

Power, Philip P., Department of Chemistry, University of California, Davis, Davis, California

Raghavachari, Krishnan, AT&T Laboratories, Murray Hill, New Jersey

Ramesh, Visvanathan, Materials Technology Group, TATA Research Development Design Center, Pune, India

Sapse, Anne-Marie, Graduate School and John Jay College, City University of New York, New York, New York; and Rockefeller University, New York, New York

Scheiner, Steve, Department of Chemistry, Southern Illinois University at Carbondale, Carbondale, Illinois

Schleyer, Paul von Ragué, Institut fur Organische Chemie, Universität Erlangen-Nürnberg, Erlangen, Germany

Shono, Toshiyuki, Department of Applied Chemistry, Faculty of Engineering, Osaka University, Osaka, Japan

Simoes, Jose A. M., Departamento de Quimica, Faculdade de Ciencias, Universidade de Lisboa, Lisbon, Portugal

Slayden, Suzanne W., Department of Chemistry, George Mason University, Fairfax, Virginia

Snaith, Ralph, University Chemical Laboratory, Cambridge, Great Britain

Streitwieser, Andrew, Jr., Department of Chemistry, University of California, Berkeley, Berkeley, California

Wright, Dominic, University Chemical Laboratory, Cambridge, Great Britain

CONTENTS

PREFACE

Lithium, the lightest of alkali metals and the least reactive, has formed the object of numerous experimental and theoretical studies. One might say that lithium chemistry is a branch of chemistry in itself. Anomalous properties of lithium, because of its small size, include such phenomena as the lithium bond (similar to the hydrogen bond), formation of aggregates, and complexation by crown ethers.

This book presents the work of a number of researchers in the lithium field and tries to include most of the aspects of this multifaced subject.

Chapters 1 and 2 as well as part of Chapter 7 discuss the applications of quantum chemical methods (ab initio) to the study of lithium compounds and their complexes. Chapter 7 presents also X-ray crystallography data.

Chapter 3 compares the lithium and the hydrogen bonds from a theoretical point of view.

Chapter 4 studies lithium matrix reactions with small molecules such as NO, N2, and others, via infrared techniques.

Thermochemistry aspects of lithium are presented in Chapter 6.

NMR techniques (HOESY), applied to organolithium compounds, are discussed in Chapter 5. Lithium salts of heteroatom compounds of main groups 3, 4, 5, and 6 are subjected to a comprehensive study in Chapter 9. Their description is undertaken through the use of low-temperature X-ray data collection and crystal mounting techniques.

Inorganic lithium salt complexes are discussed in Chapter 8. Synthetic methods are presented, as well as structural and bonding studies.

An important aspect of lithium chemistry, the methodology of separation and identification of the ion via the use of synthetic multidentate ligands, is presented in Chapter 10.

The preparation and reactions of organic compounds containing two or more lithium atoms are discussed in Chapter 11.

This book is for researchers in the domain of lithium studies and requires some background at graduate level in quantum chemistry and experimental physical chemistry.

Anne-Marie Sapse
Paul von Ragué Schleyer

New York, New York
Erlangen, Germany
February 1995

LITHIUM CHEMISTRY

1

BONDING, STRUCTURES AND ENERGIES IN ORGANOLITHIUM COMPOUNDS

ANDREW STREITWIESER
*Department of Chemistry, University of California
Berkeley, California*

STEVEN M. BACHRACH
*Department of Chemistry, Northern Illinois University
De Kalb, Illinois*

ANDREA DORIGO AND PAUL VON RAGUÉ SCHLEYER
*Computer Chemistry Center, Institute für Organische Chemie, Universität Erlangen-Nürnberg
Erlangen, Germany*

Lithium Chemistry: A Theoretical and Experimental Overview, Edited by Anne-Marie Sapse and Paul von Ragué Schleyer.
ISBN 0-471-54930-4

1 INTRODUCTION

The nature of the carbon—lithium bond has been a subject of considerable controversy in recent years. At issue is the degree of covalent or ionic character of the bond. Until this controversy arose, the carbon—lithium bond was considered to be highly covalent. Typical opinions may be quoted from textbooks published a few years ago. For example, C. D. Ritchie stated didactically, "the carbon—lithium bond in simple alkyllithiums is now known to be highly covalent."[1] In their survey of main group organometallic compounds, Coates et al. considered alkyllithiums as "probably essentially covalent."[2] This view stemmed largely from the properties of alkyllithium compounds and their differences from other alkali metal organometallic compounds. Alkylsodium compounds are nonvolatile solids insoluble in benzene, typical properties expected for ionic structures, $R^- M^+$. Alkyllithium compounds, however, frequently have appreciable volatility and are soluble in benzene, properties that have led to the statements, "It is clear that the bonding in such compounds is not ionic"[3] and "Organolithium compounds have properties which are typical of 'covalent' substances."[4]

Attempts have been made to quantify ionic and covalent character. Such attempts are impossible in a strict sense since "covalent character" and "ionic character" are not physical observables and are not eigenfunctions of quantum mechanical operators. Accordingly, any quantitative definition is necessarily ad hoc and has merit only to the degree that the results are useful in thinking about chemical properties. The relative electronegativity of the bonded elements has been used to estimate ionic character. An approach suggested by Pauling gives the C—Li bond an ionic character of 45%.[5] Ebel[6] considered this high value to be inconsistent with the solubility of alkyllithiums in petroleum ether. Using a principle of electronegativity equalization he derived a lower value of 27% for the ionic character of the C—Li bond in methyllithium. But the corresponding value in methylsodium is 29%, only slightly higher. Methods based on electronegativity differences in fact generally show only small differences among the alkali metals. For example, another definition based on electronegativity differences gives an estimate of 43% for the ionic character of the C—Li bond.[7] The same approach assigns an ionic character of only 47% to the C—Na bond, which qualitatively would seem to be much more ionic. Even the C—Cs bond is assigned an ionic character of only 57%. For comparison, the structure of solid methylpotassium is that of an ionic crystal.[8] Despite these problems in definition, as qualitative concepts, covalent and ionic character are widely used in chemical discussions and attempts to provide useful definitions persist. Recent approaches have been based on *ab initio* quantum mechanical calculations.

In 1976, Streitwieser et al.[9] called attention to the low magnitude of the computed electron density function at its minimum between carbon and lithium in methyllithium. They pointed out that covalent character requires shared electron density and the low magnitude of such density in methyllithium is indicative of high ionic character. In their paper they showed how the properties and previous computations of methyllithium are consistent with high ionic character. Many studies have been added since then and are discussed in the following sections.

2 COMPUTATIONAL APPROACH

2.1 The Covalent Case

The arguments supporting an essentially covalent C—Li bond are based on population analysis and the orientation of atomic orbitals. Population analysis partitions the electrons in a molecule into atomic regions, making use of the contribution of the atomic orbital basis functions in the occupied orbitals. The difference between the electron population assigned to an atom and the nuclear charge then gives the net atomic charge. An ionic bond would thus have highly (and oppositely) charged neighboring atoms, whereas neighboring atoms would be nearly neutral in a covalent bond. One must keep in mind, however, that the partitioning scheme used in the population analysis is completely arbitrary since orbitals have no real existence—they are not physical observables.

The most common population method is the Mulliken population.[10] Mulliken populations are determined by simply summing the number of electrons that occupy the atomic orbitals centered on the atom of interest. We start with molecular orbitals ψ_i expanded as a linear combination of atomic orbitals ϕ_μ (Eq. 1);

$$\psi_i = \sum_{\mu} c_{i\mu} \phi_\mu \tag{1}$$

If n_i is the number of electrons in ψ_i and $S_{\mu\nu}$ is the overlap between atomic orbitals ϕ_μ and ϕ_ν, then the atomic population of atom k is

$$N(k) \sum_{i}^{mos} n_i \sum_{\mu}^{aos} \left[c_{i\mu}^2 + \sum_{\nu}^{aos} c_{i\mu} c_{i\nu} S_{\mu\nu} \right] \tag{2}$$

Here, the sum over μ involves only atomic orbitals on atom k and the sum over ν involves atomic orbitals on all other atoms. The second term in the expression above is the overlap population. A sizable overlap population between neighboring atoms has been taken as evidence for a covalent bond.

Early *ab initio* studies of methyllithium supported the concept of a relatively covalent C—Li bond. Baird et al.[11] reported a charge distribution of $C^{-0.58}$ and $Li^{+0.27}$ with an overlap population of 0.556 while Hinchliffe[12] found the charges on carbon and lithium to be −0.22 and +0.64, respectively, with an overlap population of 0.638. Guest, Hillier, and Saunders[13] reported a lithium charge of +0.35 in methyllithium monomer, but a higher charge of +0.61 in the tetramer. Studies of other organolithium compounds are similar to these findings. For example, the Mulliken charge on carbon in CLi_4, CLi_5, and CLi_6 is −0.81, −0.81, and −0.93, respectively.[14] Table 1.1 lists the Mulliken charge on carbon and lithium for several lithiated ethylenes.[15] The charges

TABLE 1.1 Mulliken Charges[15] for Lithiated Ethenes at HF/6-31G*

Compound	$q(C_\alpha)$	$q(C_\beta)$	$q(Li)$
1,1-Dilithioethene (twisted)	−0.286	−0.362	0.192
1,1-Dilithioethene (planar)	−0.368	−0.355	0.284
Lithioethene	−0.397	−0.361	0.429
cis-1,2-Dilithioethene	−0.392	−0.392	0.327
trans-1,2-Dilithioethene	−0.324	−0.324	0.253
1,2-Dilithioethene ($C_{\infty v}$)	−0.489	−0.489	0.191

have relatively low magnitude. The overlap population between carbon and lithium is large. For example, the overlap populations in ethyllithium[16] and allyllithium[17] are 0.618 and 0.300, respectively. In dilithioethane, where the Li—C—C angle is only 71°, the C_α—Li and C_β—Li overlap populations are 0.547 and 0.247, respectively.[18] The sharing of electrons between carbon and lithium, as indicated by these large overlap populations, along with small charge transfer would suggest a predominantly covalent bond.

The covalent view demands orbital overlap between the bonded carbon and lithium atoms. Formally, lithium is $1s^22s^1$ and would be able to form just one bond using its 2*s* orbital. Lithium, however, is frequently found in a bridging environment which requires participation of its 2*p* orbitals in bonding. With both the 2*s* and 2*p* orbitals active in bonding, most organolithium structures can be rationalized in terms of covalent interactions. Bonding to lithium usually involves the Li 2*s* orbital donating electrons and the Li 2*p* acting as an electron acceptor. Two examples are described below.

The minimum energy structure of dilithioacetylene[19] has a bridged geometry, shown in **1**. This structure is stabilized by the interaction of the C—C bond perpendicular to the molecular plane overlapping with the empty Li *p* orbitals, schematically represented in **2**. This four-center two-electron ring is

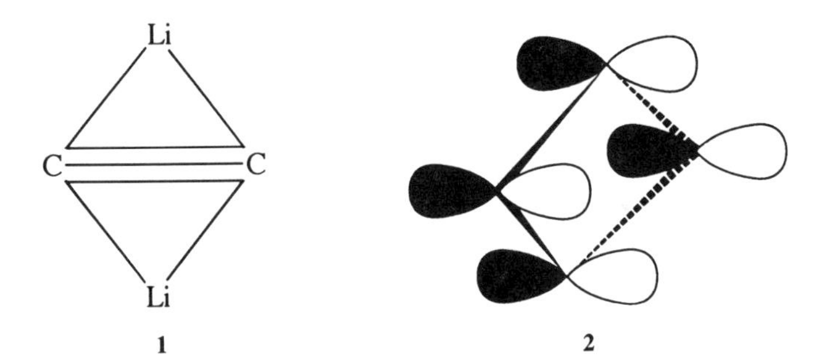

1 **2**

formally aromatic. This type of stabilization of lithiated rings is invoked for many organolithium structures, such as planar dilithiomethane[18,20] and 1.4-dilithio-1,3-butadiene.[21] The minimum energy structure **3** of allyllithium[17,22] has the lithium atom above the carbon plane, bridging the terminal carbon atoms **3**. The bridging lithium atom stabilizes the allyl HOMO through the 2*p*

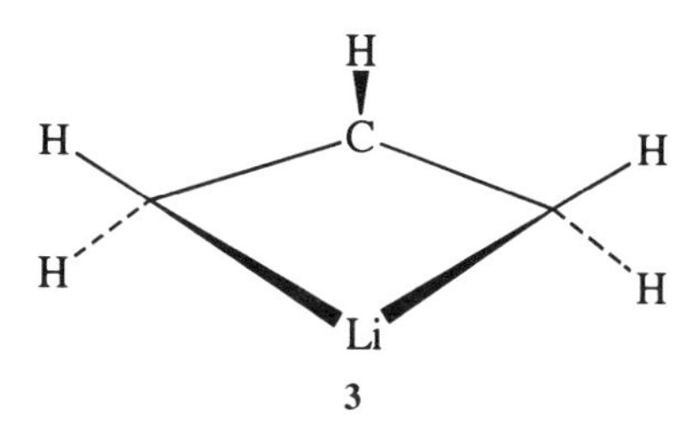

3

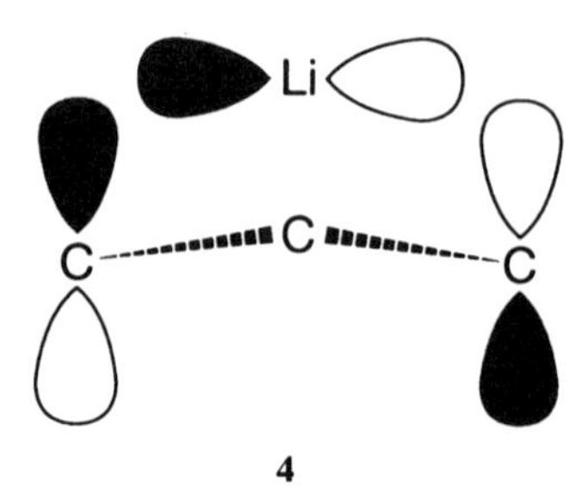

4

orbital, shown in **4.** The hydrogen on C1 and C3 bend out of plane to distort the *p* orbital toward lithium for better overlap.

Schiffer and Ahlrichs[23] focused attention on the HOMO of methyllithium, an orbital dominated by C—Li bonding but with a small contribution from the hydrogens. The electron density of this orbital includes contours about lithium and does not correspond to that of a purely ionic system. To get a better measure of the role of covalency, they used a so-called "shared electron number," SEN, a number derived from population differences compared to those of modified orbitals of the constituent atoms.[24] Some SEN values of highly covalent compounds for comparison are: H_2, 1.46; CH_4, 1.43; C_2H_6 (C—C) 1.39; Cl_2, 0.86. The value for C—Li in CH_3Li is 0.96, a result that would make the C—Li bond rather covalent. Some compounds known to be essentially ionic have relatively low SEN values: Na—F, 0.17 and Na—Cl, 0.34. However, the high SEN for other highly ionic compounds (Li—Cl, 0.84 and Li—F, 0.70) are almost as high as for C—Li and suggest that SEN values are not really chemically useful.

2.2 The Ionic Case

Streitwieser and co-workers first argued for an ionic C—Li bond with their study of methyllithium.[7] Finding extremely small values of the electron density between the carbon and lithium atoms, particularly when examining the valence space, they concluded that little sharing of electrons could take place. The large Mulliken overlap populations, they argued, arise from an integration over all space as opposed to just the "bonding region." They concluded that "the electron density criterion leads to the conclusion that the C—Li bond has essentially nil covalent character." This conclusion was criticized by Graham, Marynick, and Lipscomb,[25] who emphasized the effect of the rather long C—Li bond, 2.02 Å. The minimum value of the electron density along the C—Li line, 0.04 e au^{-3}, is low compared to that of a C—H bond, 0.27 e au^{-3}, but they pointed out that this electron density is generally low for long bonds. They gave as examples Li_2, presumably a covalent molecule, whose bond length is 2.67 Å and whose minimum density is only 0.013 e au^{-3}. For LiF the minimum density is 0.08 e au^{-3} at the normal bond length, 1.56 Å, but only 0.25 e au^{-3}, lower than that for CH_3Li, when the F—Li bond is stretched to the

C—Li distance. They considered that the value of the minimum electron density along a bond is not a universally valid measure of covalency (or ionicity), since this minimum value is so dependent on the atoms involved and the bond distance. Another measure of ionic character is the degree of charge transfer from one atom to another. Based on Mulliken populations and dipole moments, and estimating polarization effects, they deduced the charge carried by lithium to be about +0.55 to +0.60 and considered this to be a reasonable measure of the degree of ionic character of the C—Li bond.

Instead of partitioning the molecular electron density via an orbital occupation (Hilbert space) scheme, as in the Mulliken procedure, the Streitwieser group argued that density should be divided among the atoms based on the spatial distribution of the electrons. Electrons close to a nucleus should be assigned to that atom; orbital occupancy is irrelevant. Given the volume an atom occupies in space Ω_k, integration of the electron density r within this volume gives the atomic population

$$N(k) = \int_{\Omega_k} \rho(r)\, dr \tag{3}$$

The key is the definition of the atomic volume. This can be rigorously defined using the topological electron density method developed by Bader.[26] In the topological method, the volume of an atom is defined by the interior of all zero-flux surfaces about the nucleus. A zero-flux surface is defined as the union of all points such that $\nabla\rho \cdot \mathbf{r} = 0$, in which $\mathbf{r}$ is the normal vector to the zero-flux surface; that is, the electron density is a minimum normal to the zero-flux surface. An analogy is that of a mountain range and a pass. Just as the pass separates adjacent mountains, the zero-flux surface separates atoms. A line of greatest electron density between two bonded atoms is the "bond path." The electron density ρ falls off in directions perpendicular to the bond path. The minimum electron density along the bond path occurs at the "critical point," the analogy of the pass between mountains, and the electron density at the critical point is ρ_c. The zero-flux surface between the two atoms includes the critical point. Figure 1.1 shows the electron density contours of one plane in methyllithium.[27] Shown in this figure are the bond paths between Li and C and between C and one of the H's, and the corresponding zero-flux surfaces. The bond critical point occurs where the zero-flux surface intersects the bond path.

Shared electron density is necessary for covalency; accordingly, having significant electron density at the bond critical point is a necessary condition for covalent bonding, but it is not a sufficient condition. It has been pointed out, for example, that dihelium, He_2, an unstable species at normal bonding distances, has significant electron density at the bond critical point, a point intermediate between the two nuclei.[28] This arises because of the doubly occupied molecular orbital that results from the plus combination of the two He 1s atomic orbitals, $\phi_1 + \phi_2$. The doubly occupied antibonding molecular orbital, $\phi_1 - \phi_2$, has a node between the two atoms and contributes no electron density

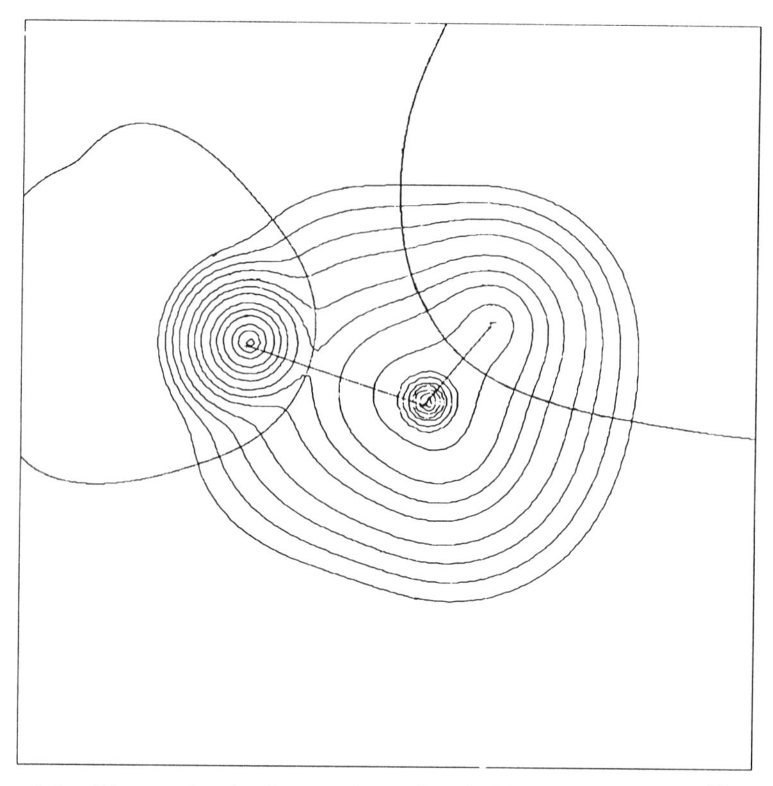

Figure 1.1. Electron density for one plane of methyllithium at 6-31++G**. Lithium is on the right. The outermost level is 0.001 e au^{-3} and the others increase progressively by factors of 2. The ρ_c values are: C—Li, 0.041 e au^{-3}; C—H, 0.268 e au^{-3}.

to the bond critical point. In this case, of course, the covalent bonding is not enough to overcome the repulsion of the two nuclei.

The foregoing discussion points up another aspect of the argument presented by Graham, Marynick, and Lipscomb[25] discussed above. The critical density of the C—Li bond was presented as low because the bond is long. Let us consider what happens when the bond is shrunk to lengths typical of other first row atoms. The critical density is increased accordingly, but this increased density does not result in net bonding. The increased density comes largely from the 1*s* core and is insufficient to overcome the increased nuclear repulsion—the compressed bond has high total energy.

Bonds to alkali metals have low covalency because there remains only a single positive charge for the valence shell and the single electron in this shell in the neutral atom is consequently only weakly held. Other atoms in the same row have larger effective nuclear charges and the valence electrons, which incompletely screen each other from this charge, are more tightly held. In bonds between such atoms and alkali metals, these higher positive charges also provide greater attraction for the single valence electron of the alkali metal and bonds with high "ionic character" result.

A "pure" ionic bond would result from juxtaposition of two oppositely charged ions with no electronic orbital overlap. Such a situation is, or course, impossible since wave functions diminish exponentially but never reach zero. Thus, all real "ionic" bonds must have some shared electron density and therefore some covalency. Bonds to alkali metals that we think of as ionic are just dominantly so; they are characterized by relatively low critical densities combined with large amounts of charge transfer as given by some reasonable measure of "atomic charge."

The use of the topological properties of the electron density provides an approach to such charges that has important advantages. The electron density is real in the sense that it is a physical observable. In principle, the electron density can be determined experimentally or by quantum mechanical calculations that make no use of "orbitals." The electron population of an atom in a molecule is given by integration of the electron density ρ within the zero-flux surface enclosing that nucleus, the "atomic basin."[26] In their original work, the Streitwieser group employed an approximation of the topological definition.[29] They made use of a "projected electron density" in which the electron density was projected onto a plane. Integration within atomic volumes defined by the two-dimensional analogues of the zero-flux surfaces gives integrated spatial electron population (ISEP), later renamed the integrated projected population (IPP). The charge derived in this way for lithium in methyllithium[9] (+0.88), planar CH_2Li_2[30] (0.79), and tetrahedral CH_2Li_2[28] (+0.82) indicates high ionic character for the C—Li bond.

The projection method was used because integrations within zero-flux surfaces are computationally intensive. However, the projection method gives integrated charges that have no fundamental significance and which are at best only approximations to charges that correspond to zero-flux surfaces. The method finds use now primarily for aiding visualization of the electron density. For example, the molecular plane of a conjugated system is a convenient plane for showing σ-electrons, but it is the nodal plane for the π-electrons; the projection method is convenient for representing both types of electron in the same plane.[13] With modern computers it is now, however, routine to accomplish complete integrations within atomic basins using Bader's PROAIM program.[32]

Ritchie and Bachrach[33] examined a series of organolithium compounds and determined the charge on lithium using the topological analysis. The topological analysis makes no assumptions concerning the nature of the orbitals or which plane to project upon, and is therefore free of many of the biases found in other population schemes. Some of their results are listed in Table 1.2. For all cases, the charge on lithium is nearly +0.9, indicating nearly complete transfer of an electron to carbon. Another tool in the topological method is the value of the Laplacian of the density at the bond critical point. The Laplacian measures the concentration of the density relative to its neighbors. A positive value for the Laplacian indicates that charge is concentrated in the atomic bas-

TABLE 1.2 Topological Charge on Li in some Organolithium Compounds at HF/3-21G[33]

Compound	q(Li)
Methyllithium	0.91
Ethyllithium	0.91
Vinyllithium	0.92
Cyclopropyllithium	0.92
Dilithiomethane	0.88
Tetralithiomethane	0.86
Hexalithiomethane	0.73
Dilithioacetylene	0.91
1,3-Dilithiopropane	0.89
Allyllithium	0.90
1,4-Dilithiobutadiene	0.90

ins, away from the internuclear region, while a negative value indicates charge buildup along the bond. Bader et al.[34] suggested that an ionic bond will have positive Laplacian values. Ritchie and Bachrach[33] found only positive Laplacian values for all C—Li critical points.

Other population methods also support a large ionic character in the C—Li bond. Cioslowski[35] proposed a population method based on atomic polar tensors. He found the charge distribution in methyllithium at HF/6-31G** as $C^{-0.388}$ and $Li^{+0.667}$. Natural population analysis[36] (NPA) assigns electrons to orbitals that attempt to define individual atoms in a molecule. The NPA charge on lithium in methyllithium is 0.88 at a high basis level, 6-311+G*.[37] Other charges at a somewhat lower basis set level are +0.811, +0.844, and +0.884, for ethyllithium, vinyllithium, and ethynyllithium, respectively.[38] Note that the NPA method is an orbital-based procedure, but still leads to high lithium charges. An alternative approach for projecting molecular orbitals onto atomic orbitals suggested by Jug et al.[39] gives a somewhat lower lithium charge for methyllithium, 0.68, and greater variation in different lithium compounds.

Horn and Ahlrichs[40] have proposed measuring ionic character by comparing the energy of an enforced ionic electronic structure with the actual energy. Methyllithium is found by this criterion to be highly ionic.

Valence bond calculations of methyllithium have also indicated a highly ionic C—Li bond with an estimated weighting of the C^- Li^+ contribution of close of 80%.[41] A later application of the closely related spin-coupled valence bond theory to methyllithium showed the C—H bonds to be similar to those in methane, essentially C9(sp^3)—H($1s$), but the C—Li bond to be largely ionic with an estimated contribution of ionic structures to be again about 80%.[42]

How are we to reconcile this consensus picture of a dominantly ionic C—Li bond with populations given by Mulliken-type procedures? The central problem is that mathematical functions used to describe electrons anywhere in a molecule are assigned by the Mulliken method to the point or nucleus on which the function is centered. For example, it is possible to describe methane using orbitals centered only on carbon.[43] All of the electrons in such a calculation would then be assigned to carbon and the hydrogen populations would be zero! This type of problem is particularly troublesome with diffuse orbitals since they may be used to describe regions closer to other atoms than the atom upon which they are centered. This is exactly the situation that pertains to lithium.[28] The outer Gaussian 2*p* orbital of lithium is very diffuse, having an exponent of 0.0286, for example, in the 3-21G basis set. This Gaussian function has its maximum value at a distance 2.21 Å away from lithium.[28] The typical C—Li bond length is about 2.0 Å. Electrons in this orbital are actually located more at carbon than at lithium. It is clearly inappropriate to assign electrons in these diffuse orbitals to lithium, and it leads to a greater Li population (a smaller positive charge) than would be expected based on the spatial distribution. This is a kind of basis set superposition error (BSSE)[44] in which the lithium *p* functions are acting as superposition functions, used to describe density at carbon, not at lithium. This problem is especially acute for minimal basis sets, including most semiempirical methods, because for lithium compounds the same number of basis orbitals are used to describe the small valence electron density on lithium cation and the much larger valence density of the associated anion.

Several studies of basis set effects have confirmed the notion of the superposition effect. Streitwieser et al. examined the C 2*p* coefficients in 1,3-dilithiopropene, $CH_2CHCHLi_2$, the corresponding propenylidene dianion, and a model formed from the propenylidene dianion with point charges replacing the lithium atoms.[45] The outer C 2*p* coefficient in the HOMO of propenylidene dianion is grater than in the HOMO of dilithiopropene itself. For covalent bonding between C and Li, one might expect this C orbital contribution to be larger than in the dianion. Instead, the Li *p* orbitals are aiding in the description of the carbanion, reducing the need of the C *p* orbitals. Similarly, in a study of allyllithium, Clark et al. noted that the addition of diffuse functions to carbons, to aid in the description of the allyl anion, greatly reduces the Li 2*p* contribution to the wavefunction.[22] Furthermore, the relative energy ordering of three allyllithium structures was well reproduced when the 3-21G basis set was used with the Li 2*p* or the Li 2*sp* shells deleted. Bachrach and Streitwieser studied the energy decomposition of ethynyllithium into acetylide anion and lithium cation.[46] This decomposition scheme shows an extremely large basis set superposition error with standard basis sets (3-21G, 4-31G, and 6-31G), but addition of diffuse functions to carbon reduces the superposition error to essentially nil. With the addition of the diffuse functions, the coefficients of the outer Li 2*p* orbitals become insignificant in the description of the HOMO and the subjacent orbital. These studies clearly indicate that the Li outer 2*p* orbitals serve only as superposition functions, helping to describe the carbanion, and do not play any part in covalent interaction. There appears, however, to be a small covalent contribution played by the inner parts of the Li 2*p* orbitals.

Population analysis based on orbital occupations will be completely distorted by the superposition effects and must be used with great care in interpreting the density distribution, and thus the bonding, in organolithium systems. As indicated above, this caveat holds for all orbital-based population methods, and particularly for minimal basis set methods, which includes most semiempirical methods.

If the C—Li bond is predominantly ionic, one should be able to reproduce the structures and energies of these molecules with a simple point charge model. Using the simplest type of such a model, the structure of the methyllithium tetramer, which can be thought of as two intersecting tetrahedra of different size formed of either the four lithium atoms or the four methyl groups, was shown to be understandable as the minimum energy arrangement of four positive charges and four negative charges, each restricted to form a tetrahedron.[47] Baird et al. had previously suggested that methyllithium is associated in the dimer by electrostatic attractions.[11]

Somewhat more sophisticated approaches have been applied to a number of organolithium compounds. In their study of 1,3-dilithiopropene, Streitwieser et al.[45] modeled the lithiums as point charges associated with the propenylidene dianion. This model reproduced the energy ordering of four structures of dilithiopropene. Interestingly, the energies of the bare dianions of the four structures have dramatically different energy relationships. Some structural features in the dilithium compound were shown to be well modeled by this use of point charges.

Clark, Rhode, and Schleyer[22] compared the optimized structure of allyllithium **3** with the structure obtained when a point positive charge replaced the lithium. The agreement was not exact, but the point charge model does reproduce all major structural features, particularly the small C—C—C angle and the out-of-plane deviations of the hydrogen atoms. Note that in this model orbital interactions of the type shown in **4** cannot occur. Moreover, a geometry optimization of allyl anion with ghost lithium orbitals (i.e., the lithium basis set without the lithium nucleus) resulted in distortions of the allyl framework opposite those found in allyllithium. Thus, electrostatic interactions alone account for the structure of allyllithium.

A related approach was applied to 1,3-dilithiopropane. Schleyer et al. found four structures of 1,2-dilithiopropane at 3-21G.[48] Replacing the lithium atoms with point positive charges placed at the position of the lithium nuclei reproduced the relative energetic ordering of these structures, though the energy differences are quantitatively different. They also reported a simple electrostatic calculation assuming point positive charges at lithium and point negative charges at the terminal carbon atoms. These calculations fail to predict the correct energy ordering. However, placing the point negative charge at the carbon nucleus is a gross approximation of an anion in an sp^3 orbital. This points out the error in assigning positions of anionic charge. One must keep in mind that the carbanion is not spherically symmetric and cannot (usually) be considered as residing at the nucleus.

The electrostatic model accurately predicts other minimum energy struc-

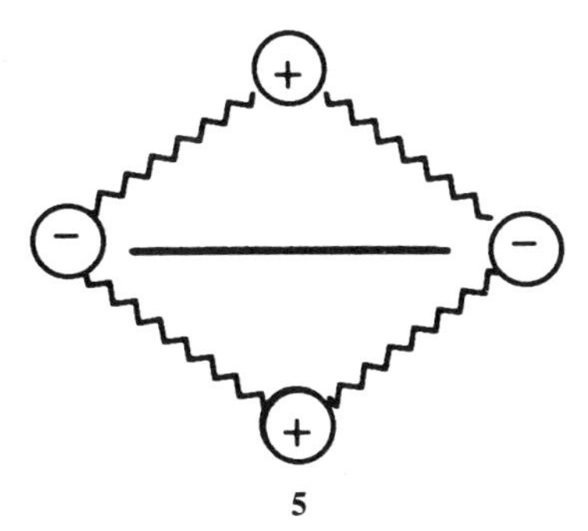

5

tures. Dilithioacetylide can be dissected into the triple ion formed by two lithium cations and the acetylide dianion. The minimum energy structure would have the geometry of **5.**

Cyclopentadienyllithium can be treated as a contact ion pair of lithium cation and cyclopentadienyl anion. Electrostatics alone would predict that the cation sits above the center of the cyclopentadienyl ring, bridging all five partially negatively charged carbon atoms. The cyclopentadienyl anion is planar. However, to maximize the electrostatic interaction with lithium, one might expect the hydrogens to bend away from the lithium cation, thereby polarizing the carbon *p* orbitals so as to bring more negative charge onto the side facing the lithium. The optimized structure of cyclopentadienyllithium **6** shows just this nonplanar distortion.[35,49]

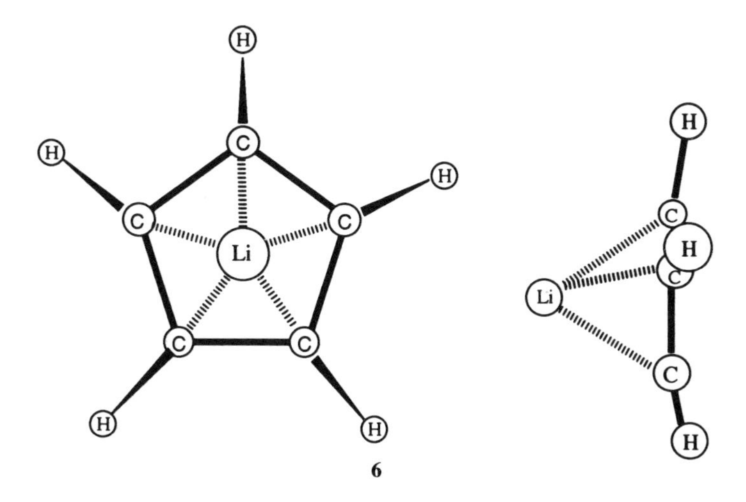

6

1,4-Dilithiobutadiene has the doubly bridging structure shown in **7.** It was originally argued[21] that this arrangement allows for both Hückel- and Möbius-type stabilization as seen in the schematic diagrams of the two highest MOs in **8.** One can also, and more simply, argue that **7** is stable due to the formation of an ideal triple ion.[50] The two lithium cations bridge the terminal sp^2 anions

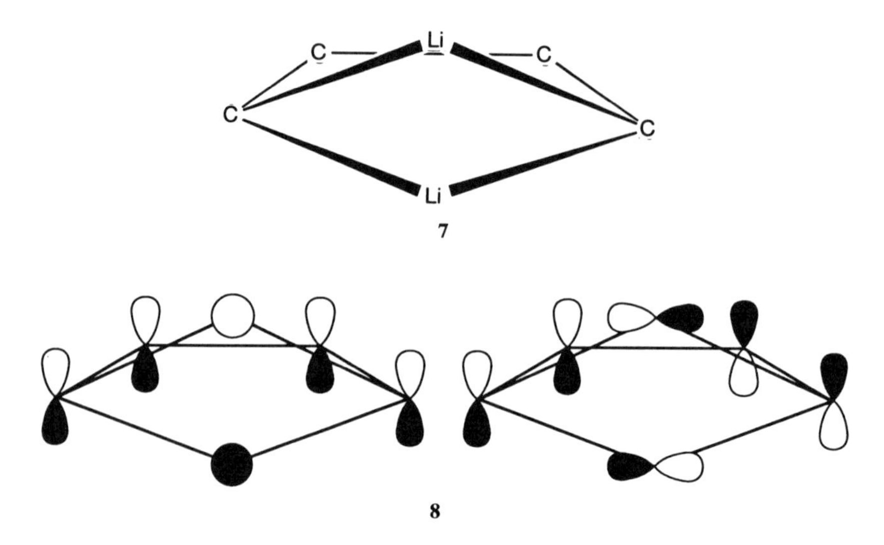

directed inside the ring. Other possible geometries lack this electrostatically favorable triple ion structure.

The structure of benzyllithium points up possible limitations of a simple electrostatic model. A simple ionic model gives an η^1 structure with the lithium associated only with a somewhat pyramidal α-carbon.[51] Ab initio calculations give as the lowest energy structure an η^3 structure with the lithium bridged between the $\alpha-$ and ortho-carbons.[51-53] The potential surface, however, is rather flat and the different structures are separated by only a kilocalorie per mol or so.[53]

One final organolithium compound deserves some attention. Hexalithiomethane is a hypercoordinate carbon system that is formally hypervalent. The Schleyer group originally proposed that carbon bears the usual eight electrons and is not hypervalent.[13] The extra two electrons reside in a molecular orbital that is spherical about carbon but describes a bonding interaction among the lithium atoms only. Ritchie and Bachrach offered an alternative view.[33] The topological analysis indicates that there is no bonding interaction between the lithium atoms, and that bonds exist only between the carbon and each lithium. The integrated charge on carbon is -4.38, which would appear to exceed the octet on carbon. Jug et al.[39] rejected the topological charge analysis out of hand "because there is an accumulation of more than 10 electrons on carbon." However, in the electrostatic field generated by the six octahedrally placed lithium cations, the carbon 3*s* orbital is sufficiently lowered in energy so that it may be partially occupied. Simple electrostatic models explain the stability and geometry of the hyperlithiated species, without resorting to unusual bonding, strange orbitals, or various definitions of populations.[54]

Monomeric methyllithium computationally has a high dipole moment of 5.7–6.0 D.[9,24,37] The experimental infrared spectrum indicates a high dipole moment.[55] Two electronic charges separated by 2.0 Å correspond to a dipole moment of 9.6 D. The experimental dipole moment is only about 60% of this value but, as mentioned above, the carbon lone pair in the methyl anion would be concentrated in front of the carbon such that the distance between the center of charges is less than the internuclear distance and would produce a smaller dipole moment. Polarization effects would further reduce the dipole moment. Thus, a μ value of about 6 D is not inconsistent with an ionic character on the order of 80–90% in the C—Li bond.

The Raman spectrum of the tetramer of *t*-butyllithium has been interpreted as indicating only minute Li—Li bonding,[56] a result consistent with association via electrostatic interactions. The absence of observable $^6Li-^7Li$ coupling in the methyllithium tetramer also indicates little valence electron density in the Li—Li region.[57] For aggregates of alkyllithiums, the ionic portions—the lithium carbamide units—are in the core of the aggregate; the external surface is hydrocarbon. This morphology readily accounts for the hydrocarbon solubility of these reagents and their volatility.

Scalar $^7Li-^{13}C$ coupling is significant in methyllithium tetramer, 15 Hz.[58] The $^6Li-^{13}C$ coupling constants for a number of compounds have been tabulated by Seebach et al.[59] and range up to 18 Hz ($^7Li-^{13}C$ are higher than $^6Li-^{13}C$ by a factor of 2.64) for some compounds at low temperatures. The listings include monomeric organolithium compounds with relatively localized carbanions. Bridged organolithium compounds show no coupling. Bauer et al.[60] added additional coupling constants for monovalent organolithium compounds and summarized much data by the simple formula given in Eq. 3.

$$J(^{13}C, ^6Li) = (17 \pm 2)/n \text{ Hz} \tag{4}$$

where n is the degree of aggregation.

Carbon—lithium coupling may be indicative of some covalency,[61] but it can also result[9] from incomplete charge transfer with spin polarization (through-space coupling); that is, the contributions of the two structures C↑Li↓ and C↓Li↑ differ depending on the nuclear spin states.[62] Bauer et al.[60] point out that the validity of Eq. 3 independently of the nature of the carbon, whether *sp*, sp^2 of sp^3, strongly suggests the dominance of the through-space mechanism.

Thus far in this chapter we have emphasized that the carbon—lithium bond in theory and in chemical properties can be modeled as an essentially ionic bond. The chemical usefulness of this model has been further demonstrated in convincing fashion by Schleyer et al.[63] who showed that the relative energies of a number of organolithium compounds parallel those of the corresponding carbanions. In the following sections we give several examples of systems that illustrate the importance of ionic structures in organolithium chemistry.

3 STRUCTURES AND ENERGIES

Insights into the structure of isolated organolithium monomers were first gained computationally, as experimental studies are complicated by aggregation and solvation. Some of the initial discoveries were startling. Thus CH_2Li_2 had nearly the same energy in planar as in tetrahedral geometries, both for singlet and triplet states. Similarly, $H_2C{=}CLi_2$ was computed to prefer a perpendicular over the usual planar geometry. Figure 1.2, reproduced from a 1984 review,[64] emphasized some of the peculiar organolithium structures which were discovered computationally. Additional reviews of this early work are available.

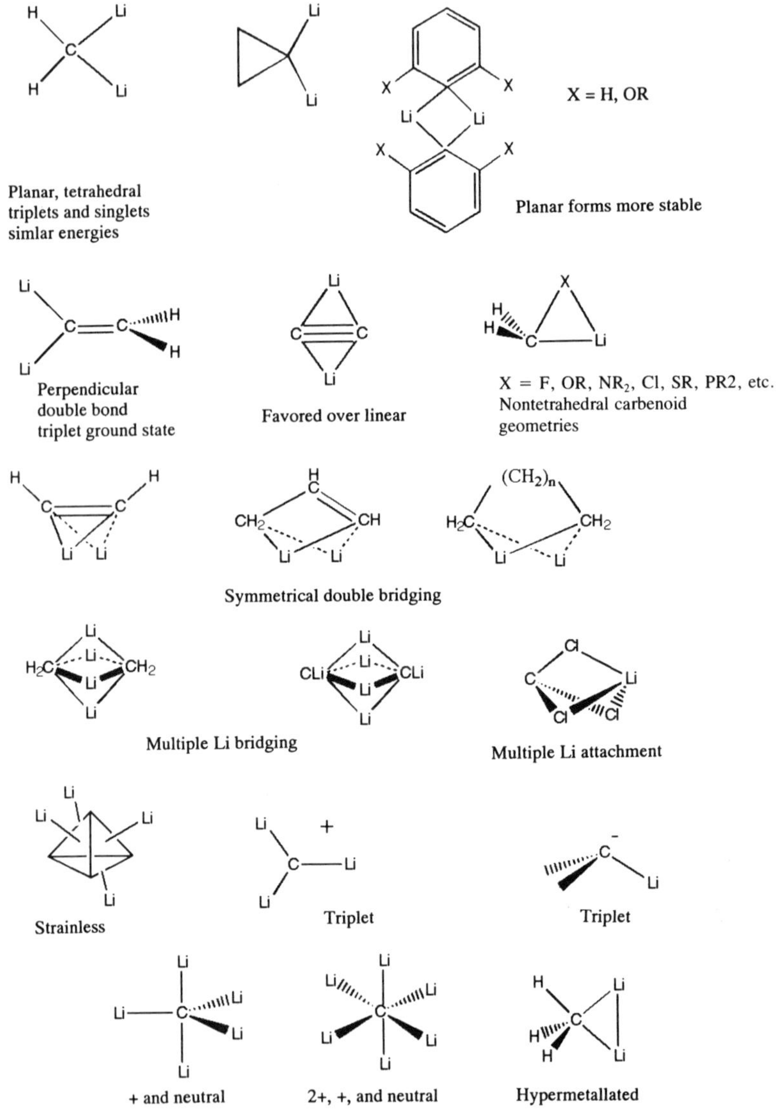

Figure 1.2. Some published rule-violating lithium structures discovered computationally.

The main purpose of this section is not to review the earlier contributions, but rather to bring them up to date. Most of the computational results, such as those on the molecules illustrated in Figure 1.2, were carried out many years ago at what now seems to be an unsophisticated level of theory. Small basis sets were employed, geometries were not optimized at electron correlated levels, and frequency analyses (to ascertain the nature of stationary points, minima, transition structures, etc.) could not be carried out. For the most part, potential energy surfaces were not searched widely in the earlier work, which concentrated on a few possible geometries of higher symmetry.

The enormous progress in computational methodology since the 1970s, both with respect to the much faster performance of computers and to the greater degree of sophistication of the quantum chemical programs, now facilitates the reexamination of the species at modern levels. This has been a continuing process over the years, although updated results have not always been published. While the essential features of much of the earlier work is confirmed, there are many surprises and new findings. For example, some of the aesthetically pleasing higher-symmetry structures proved to have several imaginary frequencies, that is, they do not have any practical validity. We have drawn the newer structures together in this chapter. Bonding principles of organolithium compounds emerge and these help us to recognize structural relationships among the various systems.

4 METHANE ANALOGUES

4.1 Methyllithium, CH_3Li

Methyllithium is the simplest organolithium compound, and consequently many theoretical investigations on methyllithium have been reported.[65,66] The geometry of the monomer is straightforward. The carbon—lithium bond distance is near 2.02 Å and the carbon atom is pyramidal, with the HCH angle somewhat smaller than in methane.[66] The slightly large Li—C—H angle is not unexpected for electropositive substituents and may be explained by electrostatic repulsion between the positively charged lithium and hydrogen atoms, and by attraction to the carbanion lone pair.

The essentially ionic nature of the C—Li bond in this species and in related molecules has been discussed in detail above. The methyllithium dimer structure can likewise be explained in terms of electrostatics alone,[67] even though an interpretation of the structure in covalent terms has also been proposed.[16,68,69]

4.2 Dilithiomethane, CH_2Li_2

Earlier calculations showed the lowest energy structure of dilithiomethane to be the tetrahedral singlet;[70] however, without inclusion of CI, the tetrahedral triplet was the energy minimum.[20,30] Since it is well known[71] that triplets are

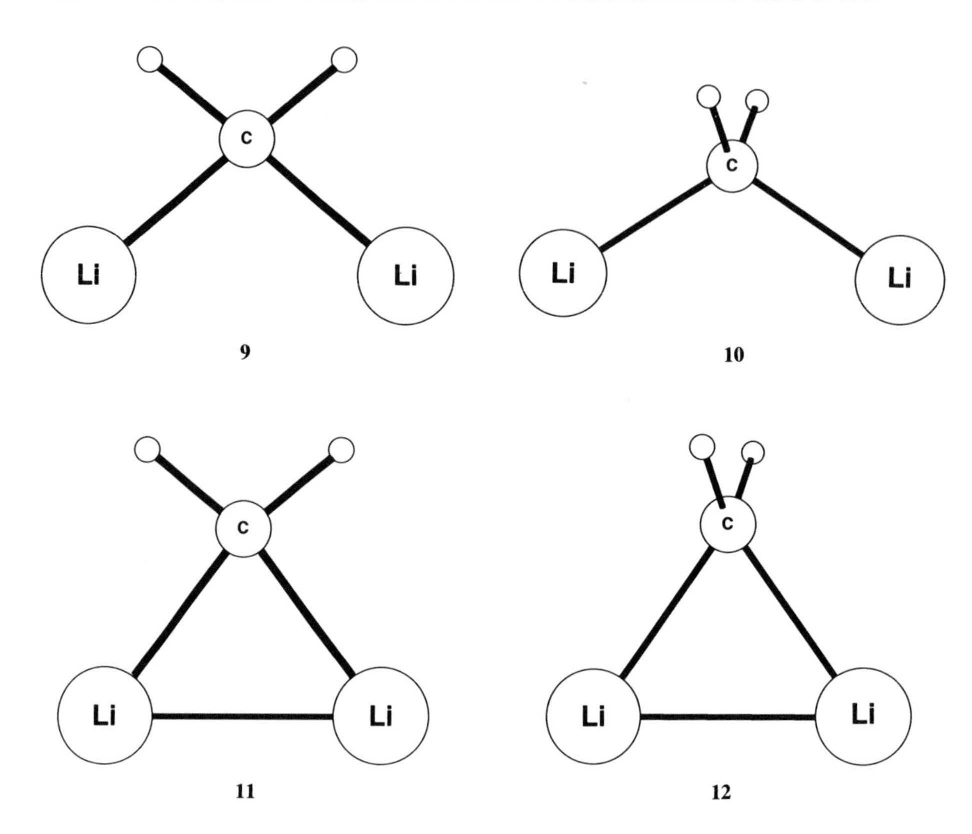

artificially favored over singlets at the SCF level, we have reoptimized[72] the structures of four isomers—singlet and triplet planar and tetrahedral (structures **9–12**)—at QCISD/6-31G**. The results of these calculations (Table 1.3) confirm the preference for a tetrahedral singlet. The planar singlet, the tetrahedral triplet, and the planar triplet are 5.5, 6.5, and 8.4 kcal/mol higher than the tetrahedral singlet, respectively. The narrow range of the four minimum energy structures is of interest. The stabilization of the planar structures is due to a small amount of three-center π-bonding between the carbon and the two lithium atoms.[20] However, the structures are still dominated by electrostatics, as the ISEP charges for lithium are +0.8 and +0.6 in the singlets and triplets,

TABLE 1.3 Relative Energies (kcal/mol) of the Dilithiomethane Isomers 9–12

	HF/3-21G //HF/3-21G	HF/6-31G* //HF/6-31G*	QCISD/6-31G** //QCISD/6-31G**
9 planar singlet	26.0	25.1	5.5
10 Tetrahedral singlet	18.5	19.1	0.0
11 Planar triplet	3.2	2.4	8.4
12 Tetrahedral triplet	0.0	0.0	6.5

respectively. Electrostatics alone predict the tetrahedral structures to be lower in energy by placing the positively charged lithiums in a plane perpendicular to the positively charged hydrogens. The singlets may be described as CH_2^{2-} $(Li)_2^+$ species, while the triplets can be viewed in terms of an ion–radical pair, $CH_2^{-\bullet}Li_2^{+\bullet}$. The latter description is justified in terms of both the calculated charges and the short distance between the two lithium atoms in the triplets, similar to that in $Li_2^{+\bullet}$.[72]

Jemmis, Schleyer, and Pople examined the dimer of dilithiomethane.[73] The lowest energy structure according to the authors, **13a,** has all four lithium atoms lying nearly in the same plane bridging the two carbon atoms. This arrangement is described in the paper as twisted ethylene with a ring of lithium atoms about its middle. The dimerization energy for forming this compound is 35.1 kcal/mol at 4-31G//STO-3G. This structure is quite stable, which is remarkable considering that the wrong ends of the dipoles are brought together. However, we have found[72] the C_{2h}-symmetric dimer **13b** to be more stable by 12.3 kcal/mol at MP2/6-31G*//HF/3-21G.

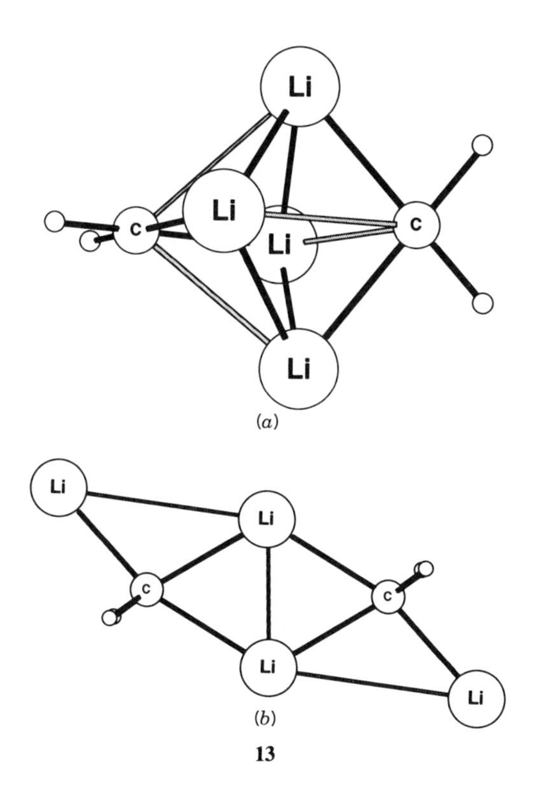

13

4.3 Tri- and Tetralithiomethane, $CHLi_3$ and CLi_4

Only two references to trilithiomethane and tetralithiomethane have been reported.[20, 74] In an effort to obtain a stable planar four-coordinate carbon atom, Schleyer et al.[20] optimized tri- and tetralithiomethane in the planar and tetrahedral configurations. Tetrahedral trilithiomethane, at the 4-31G//4-31G level,[14] is approximately 8.7 kcal/mol lower in energy than the planar form. The preference for the tetrahedral form is most likely due to decreased electrostatic repulsions between the positively charged lithium atoms. Tetralithiomethane is 16.4 kcal/mol more stable in the tetrahedral form than in the planar form at HF/4-31G//4-31G.[20] We have reinvestigated[72] this species at HF/6-31G* and at MP4SDTQ/6-31G*//HF/6-31G*, but the preference for the tetrahedral form remains similar (15.4 and 17.1 kcal/mol, respectively). This molecule should be viewed as C^{4-} surrounded by four lithium cations, which would then be placed in a tetrahedral structure to achieve minimum energy. Placing all four lithiums in one plane dramatically increases the electrostatic repulsion between the lithium atoms, much more so than having just two or even three lithiums in one plane. This added repulsion explains both why tetrahedral tetralithiomethane is lower in energy than the planar structure and why the energy difference between the two structures is much larger in tetralithiomethane than in trilithiomethane.

In a communication that generated much notice, Schleyer, Pople and coworkers presented a series of lithiomethanes that contain a formally hypervalent carbon atom.[14] Pentalithiomethane **14** and hexalithiomethane **15** are both quite stable to any decomposition route. For example, loss of Li_2 from CLi_6 is 65.2 kcal/mol endothermic at the 3-21G level. We have reoptimized both **14** and **15** at HF/6-31G*.[72] Even at this level, both are thermodynamically stable, **14** being 58.9 kcal/mol lower in energy than $CLi_4 + Li_2$.

These hyperlithiated compounds can be viewed as C^{4-} surrounded by Li_n^{4+} with the lithium atoms placed to minimize their mutual repulsion. For pentalithiomethane and hexalithiomethane electrostatics would predict a trigonal bipyramid (D_{3h}) and octahedral structure, respectively. This is in agree-

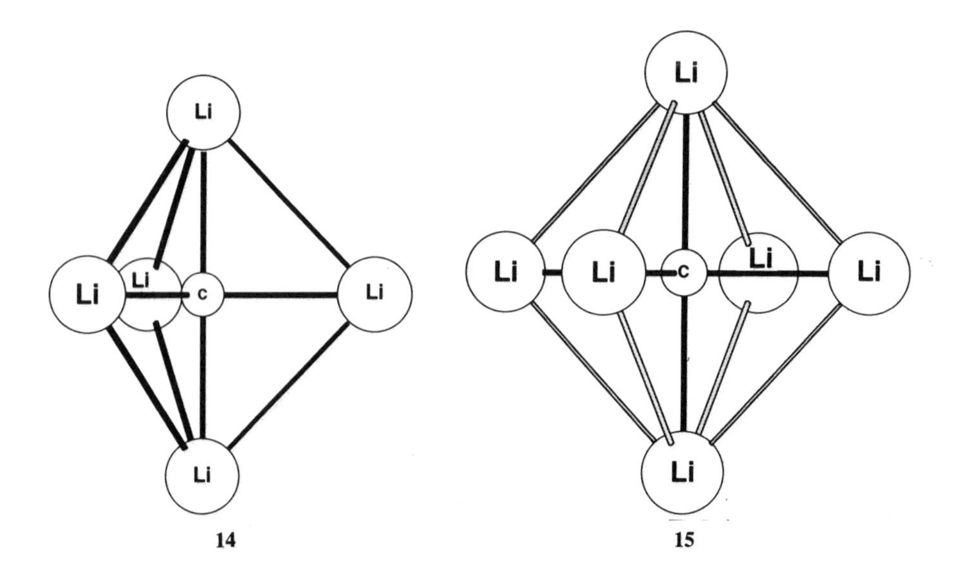

ment with the calculated geometries, although the C_{3v} CLi_5 isomer is only 0.1 kcal/mol higher in energy than the D_{3h} form at HF/6-31G*.[72] The extra lithium electrons are placed in an orbital which is completely symmetric, being a Li—Li bonding orbital among all lithium atoms (with a small contribution from the 2*s* carbon orbital). These compounds do not violate the octet rule since carbon has eight valence electrons (recall the carbon should be considered as C^{4-}); the additional electrons populate the all-Li bonding orbital which is not associated with the central carbon atom (see above).

5 ETHANE ANALOGUES

5.1 Ethyllithium, C_2H_5Li

The minimum energy structure of ethyllithium is the staggered conformation,[9,16] which is completely analogous to ethane itself. The electrostatic picture is that of a combination of ethyl anion and lithium cation. Lithium positions itself so as to minimize repulsions with the hydrogens on C_2 and maximize the electrostatic interaction with the anion in the sp^3 orbital on C_1.

5.2 1,2-Dilithioethane, $C_2H_4Li_2$

The ionic model portrays 1,2-dilithioethane as a C_2H_4 dianion interacting with two lithium cations. The lithium cations form a bridge between the carbons to enhance the stabilization afforded by this ion triplet structure. The optimized HF/3-21G structure[18] **16** has the lithiums bridging, but not in a symmetric fashion, and has therefore only C_{2h} symmetry. The symmetrically bridged (D_{2h}) structure is only 1.9 kcal/mol less stable at this level, but becomes disfavored

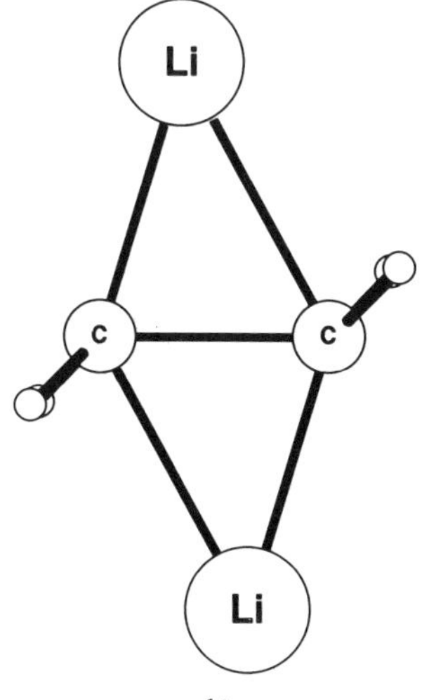

16

by 4.8 kcal/mol at HF/6-31G*.[72] Electrostatic calculations (using point charges) predict the D_{2h} structure to be the minimum, so that covalent interactions may cause the C_{2h} structure to be favored. Schleyer pointed out that the D_{2h} HOMO is the π^*_{CC} orbital, which has large C—C antibonding character. By allowing the carbons to hybridize, the π^* orbital distorts to decrease its antibonding nature, and thereby stabilize the structure. Some lithium p orbital interaction is seen in these HOMOs, but it is necessary to go to very small contour levels before the effects of lithium orbitals are seen in the density plots.

A convincing argument that organolithium compounds are dramatically different from hydrocarbons is seen with dilithioethane. The bridging D_{2h} structure for ethane lies approximately 200 kcal/mol above the minimum energy structure, staggered ethane. For 1,2-dilithioethane, the bridged structure is the minimum.

5.3 1,1-Dilithioethane, $C_2H_4Li_2$

This molecule, having two lithium atoms on the same carbon atom, also is a candidate for a structure having a planar four-coordinate carbon center (see our previous discussion on CH_nLi_{4-n} isomers). Indeed, the C_s-symmetric singlet having a planar HC(Li_2)C configuration is a minimum (structure **17**), as is also

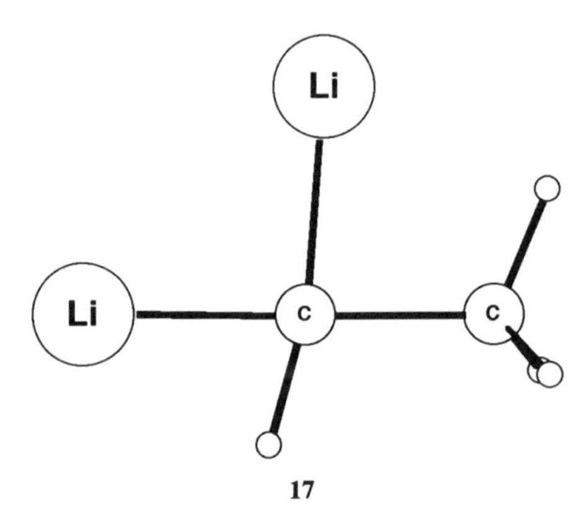

17

the C_1-symmetric structure **(18)** having both carbon atoms essentially tetrahedral and lying only 1.3 (0.8) kcal/mol lower at MP4SDTQ/6-31G*//HF/6-31G*

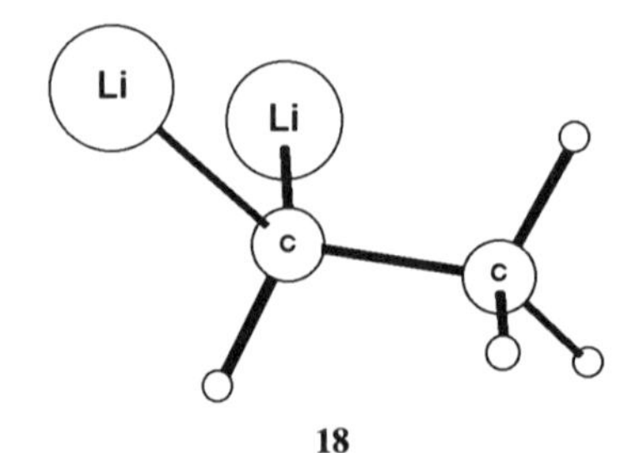

18

(HF/6-31G*).[72] Note that the global $C_2H_4Li_2$ minimum is far more stable than both of the structures described above (being 35.8 kcal/mol lower in energy than the C_1-symmetric species at MP4SDTQ/6-31G*//HF/6-31G*), and it is not an *ethane* derivative. Its structure, **19,** can be described as a mixed dimer of vinyl lithium and lithium hydride, or, alternatively, as the complex between the vinyl anion and the $[Li_2H]^+$ cation. Structure **19,** which would be obtained

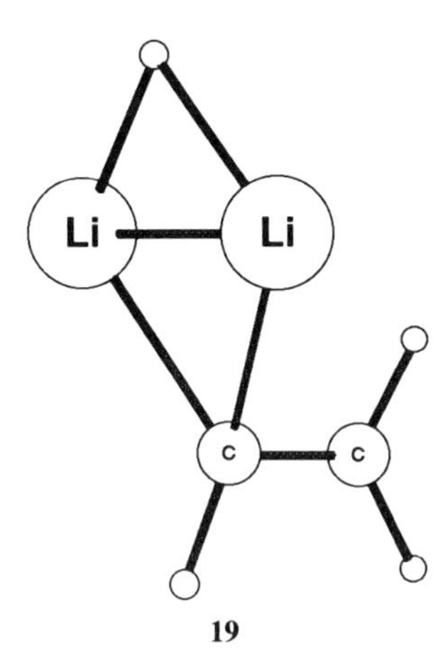

19

through a hydride shift from **18,** is characterized by a short C—C bond (1.344 Å, essentially a double bond). The presence of stable ionic fragments, such as $[Li_2H]^+$ in this case, is a feature that recurs in organolithium compounds and underscores the difference in bonding in these species with respect to the parent alkanes. The next "ethane" derivative in this discussion (C_2Li_6) offers an impressive example of this feature.

5.4 Hexalithioethane, C_2Li_6

Early studies determined the structure of the molecule (HF/STO-3G[75]) to be the C_{2h}-symmetric **20.** Electrostatic point-charge calculations would predict the D_{4h} structure **21** to be the minimum, yet it lies 7.0 kcal/mol higher in energy. However, both structures turn out to be far higher in energy than those corresponding to acetylide "salts." Thus the C_2-symmetric **22,** which can be described as $Li_3^+C_2^{-2}Li_3^+$, is 43.5 kcal/mol more stable than **21** at HF/6-31G*.[72] In addition to the natural charges on the carbon atoms—both equal to −0.95—the CC bond length of 1.27 Å in **22** is indicative of an acetylide derivative (this bond length is essentially the same as that of Li_2C_2, dilithium acetylide, described later in this chapter). The stability of the acetylide dianion and of the Li_3^+ cation are at the origin of the low energy of this structure. Note that the peripheral lithium atom in each Li_3^+ fragment—that is, the one which is not bound to either carbon—has a negative charge; the Li_3^+ cation is thus entirely analogous to the $[Li_2H]^+$ ionic fragment discussed in the previous section. The

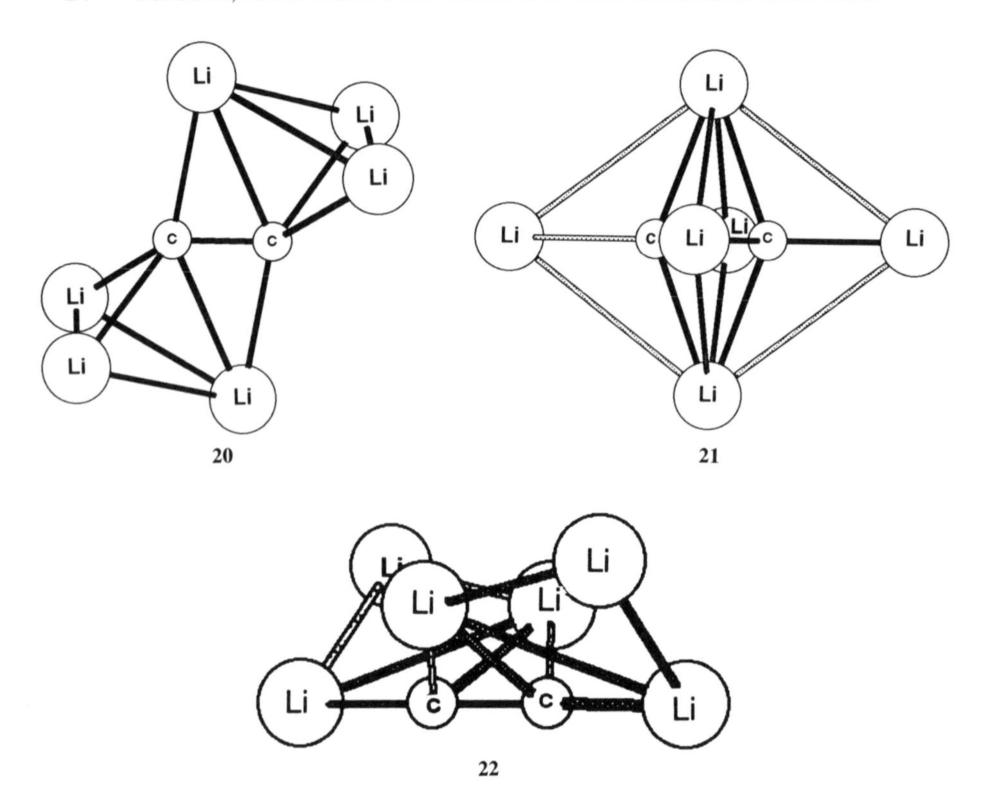

20

21

22

acetylide dianion polarizes the cationic Li_3^+ aggregate so as to enhance the positive charge on the nearer two lithium atoms while inducing a negative charge on the third. This quadrupole-like stabilization of **22** is more advantageous than the alternative approach in which all three lithium atoms interact with the negative charged carbon atoms, as in **20.** The Li_2 edge-complexation in **22** is favored over the Li_3 face-complexation in **20,** and this turns out to be generally true, as we shall see in later examples. The picture in C_2Li_6 is even more complex however. Structure **22** is a local minimum, but the global minimum at MP2/6-31G* is **23,**[72] which lies 1.0 kcal/mol lower in energy than

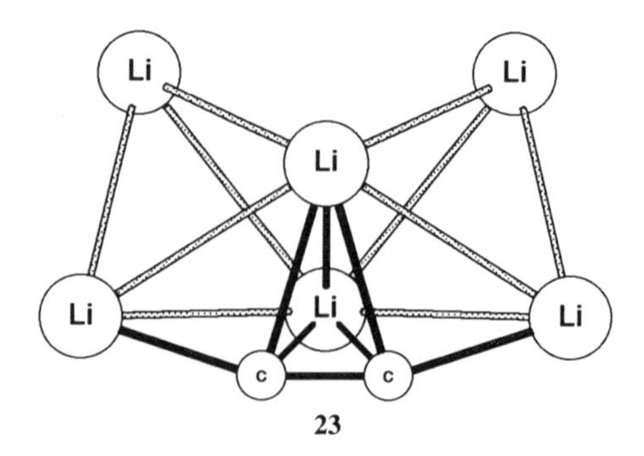

23

22 at this level. Structure **23,** which has C_{2v} symmetry, has all six lithium atoms on the same side of the C—C bond—a long way from its "parent" hydrocarbon ethane! This remarkable structure is best described as the "salt" of the acetylide dianion with the Li_6^{2+} cation, which also has C_{2v} symmetry. More work may be necessary to determine the relative stability of **22** and **23,** but there is no doubt that both of these *acetylene* derivatives are far more stable than ethane-like alternatives.

6 ETHYLENE ANALOGUES

6.1 Vinyllithium, C_2H_3Li

The structure of vinyllithium is similar to ethylene,[15] as the H_2CCH fragment is nearly identical in the two molecules. The C—Li bond length is about 1.97 Å with most basis sets, which is consistent with other organolithium compounds. The Li—C—C angle is basis-set dependent, being 78.6° at STO-3G and a more reasonable 121.3° at 4-31G; MP2/6-31G* also gives a "normal" value of 116.5°. Here too, there has been some discussion about the degree of covalency of the C—Li bond, but calculations by both Streitwieser[9] and Hinchliffe[76] suggest a predominantly ionic bond, similar to that in methyl- and ethyllithium.

6.2 1,1-Dilithioethylene, CH_2CLi_2

This molecule was originally examined in depth by Apeloig and Schleyer,[15] and subsequently by Laidig and Schaefer.[77] Using a near DZ+P basis set, they optimized four structures of 1,1-dilithioethylene—the singlet and triplet planar and twisted conformations **24–27.** Energies were then calculated with a DZ+P basis set with CI and Davidson's correction. We have recently reoptimized[72b] the structures at MP2/6-31G*, followed by QCISD(T) single points (Table

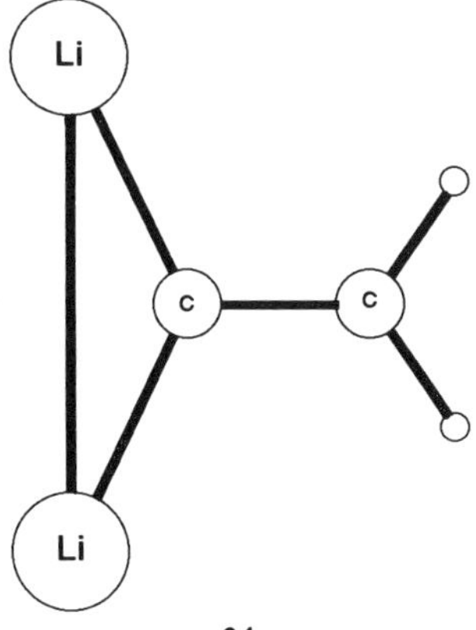

24

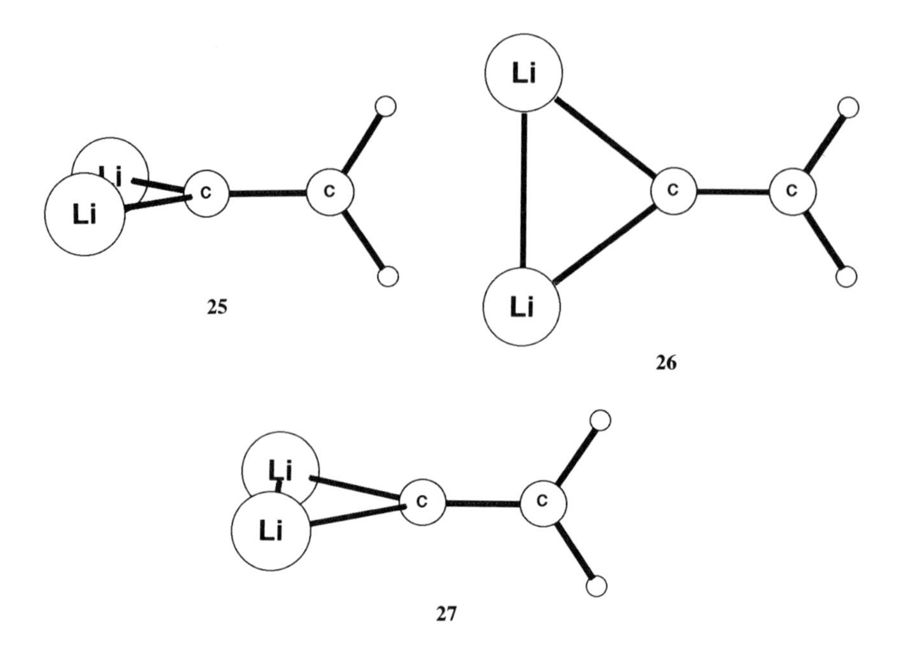

1.4). At all levels except MP2/6-31G*, the twisted triplet **27** is the minimum energy structure. At the highest level of theory, the planar triplet **26** lies 1.62 kcal/mol above **27.** The singlets are substantially higher in energy.

The stability of the twisted structure has been interpreted in two ways. The orbital argument[78, 79] claims that a three-center C—Li—Li cyclic π bond, involving the lithiums and the C_2 p orbital coplanar with them, leads to the stability of the twisted states. The ionic argument is that a twisted arrangement will minimize the repulsions between the lithiums and hydrogens.

The geometries of these structures are quite remarkable. For all four states the C—C distance indicates a full double bond, which led Schleyer to conclude that 1,1-dilithioethylene has nearly free rotation about a C=C bond. The C—Li

TABLE 1.4 Relative Energies in kcal/mol of the 1,1-Dilithioethylene Structures

	HF/6-31G*	MP2(FU) /6-31G*	MP4SDTQ/ 6-31G*// MP2(FU)/6-31G*	QCISD(T)/6-31G* //MP2(FU)/6-31G*
		Singlets		
Planar (**24**)	34.3	2.7	16.7	19.5
Twisted (**25**)	32.7	0.0	9.4	13.0
		Triplets		
Planar (**26**)	1.3	3.9	1.9	1.6
Twisted (**27**)	0.0	1.8	0.0	0.0

bond is longer in the triplets than in the singlets; the singlets have the expected C—Li bond lengths. The Li—C—Li angle is considerably wider in the singlets than in the triplets. In fact, the triplets have an acute Li—C—Li angle of 73°–74° with most basis sets. In addition, the direction of the dipole moment is reversed for the triplets.

The unusual geometries for the 1,1-dilithioethylenes are similar to the structures found for dilithiomethane. The interpretation of these structures was presented by Schaefer[77] and Streitwieser.[30] The singlets are formed from the ethylene dianion and two lithium cations. Fairly normal C—Li distances result and the large Li—C—Li angles are due to strong Li—Li electrostatic repulsion. To describe the triplets, one must start with the singlet geometry and then excite one electron from the HOMO into the next highest orbital ($b_2 \rightarrow ba$). The *b* symmetry orbital (the singlet HOMO) represents the lone pair on C_1 with some small contribution from the lithium atoms. The *a* symmetry orbital (the singlet LUMO) is a Li—Li bonding orbital with slight C—Li antibonding character. Removing one electron from *b* reduces the C—Li attraction, lengthening the C—Li bond. This electron is then placed in a Li—Li bonding orbital, bringing the lithiums closer together and reducing the positive charge on each lithium. This reduction in charge is actually enough to flip the dipole moment. Triplet 1,1-dilithioethylene may perhaps be described as triplet $H_2C{=}C$: associated with Li_2.

6.3 1,2-Dilithioethylene, $C_2H_2Li_2$

1,2-Dilithioethylene presents yet another example of organolithium compounds having structures quite different from their hydrocarbon analogues. Several structures were considered in an earlier work by Schleyer.[72b,80] At the MP2/6-31G*//HF/6-31G* level, the *trans* isomer **29** and the doubly bridged **30** are the two most stable structures; **30** is lower in energy by 1.0 kcal/mol. The planar *cis* structure **28** is 21.9 kcal/mol higher than **30,** and is a saddle point.[80] On the other hand, geometry optimization at MP2/6-31G* gives **29** as the lowest-energy isomer,[72] with **30** lying 1.3 kcal/mol higher in energy. The *trans* structure **29** has two lithium atoms bridging the carbons, but not symmetrically. This is clearly reminiscent of 1,2-dilithioethane (described above).

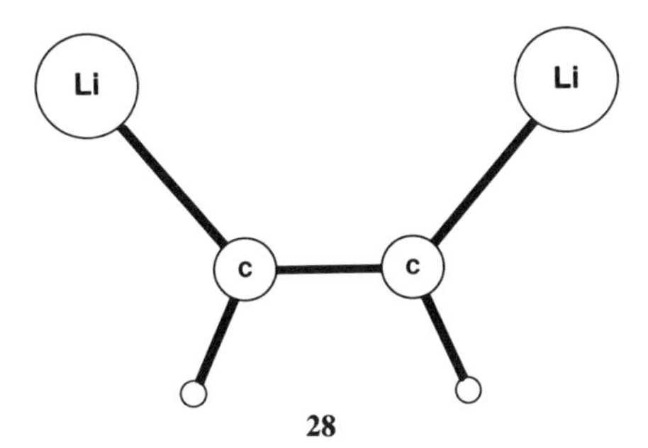

28

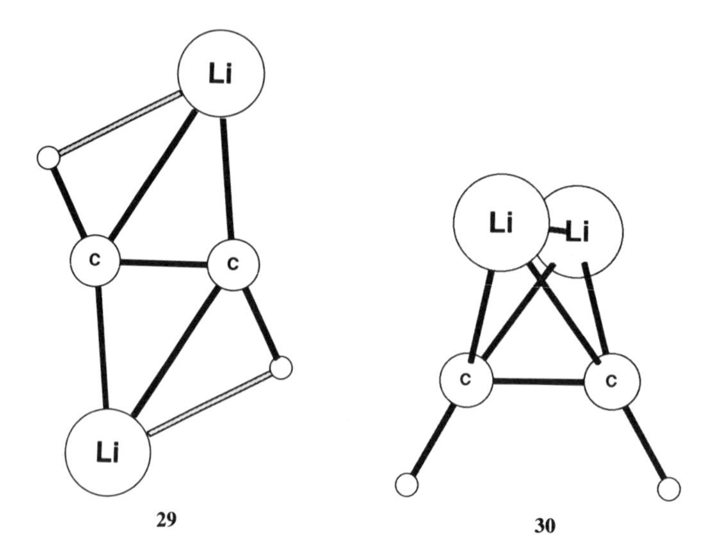

29 **30**

Schleyer[80] contended that this partially bridging structure is stable due to strong Li—H electrostatic interaction. The Li—H distance is quite short, the Li—H overlap population significant, and the energy of the HOMO highly dependent on the C—C—Li angle. Additionally, structure **29** allows for delocalization of the π bond into the empty lithium p orbitals to obtain a four-center two-electron system.

The *cis* structure **28** is higher in energy for a number of reasons. Unlike **29**, **28** does not have any Li—H interactions; moreover, lithium coulombic repulsion destabilizes the structure, and is responsible for the wide C—C—Li angle. Schleyer suggested that the doubly bridged structure **30** is stabilized by delocalization of the dianion into lithium p orbitals (see **31**).

Electrostatics can be used to interpret these structures. Schleyer pointed out that the relative stability of ethylene dianions follows the order **32** (most stable) > **33** > **34.** Based on anion stabilities alone, one would expect the *trans*-1,2-dilithioethylene to be the most stable, which is the case at MP2/6-31G*.[72] By having the lithium partially bridge, however, some ion triplet stabilization is

31

32 > **33** > **34**

gained. Structure **30,** even though it involves the unstable *cis*-ethylene dianion, is dramatically stabilized by the strong ion triplet structure where both lithium atoms can fully interact with each carbon anion orbital.

The third dianion (1,1-ethylidene dianion, **34**) is quite unstable, but placing two lithium cations near the anionic orbitals is a favorable energetic conformation. The difference in energy between the planar 1,1-dilithioethylene, where each lithium interacts primarily with one anion orbital, and the twisted form, where each lithium bridges the two anion orbitals, is relatively minor since each arrangement has considerable coulombic stabilization of comparable magnitude. Still, planar 1,1-dilithioethylene singlet **24** is 16.2 kcal/mol less stable than **29,** reflecting the lower stability of the 1,1-dianion of ethylene relative to the 1,2-*trans* dianion.

Finally, we note that the global minimum $C_2Li_2H_2$ structure is not one of the ethylene derivatives **22–30,** but the acetylene derivative **35,** which is a

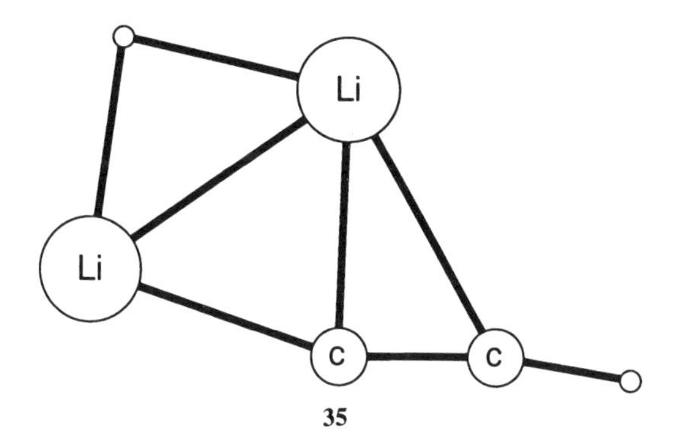

35

complex between lithium hydride and lithium acetylide. This is analogous to what we have already seen in the case of $C_2Li_2H_4$.

6.4 Tetralithioethylene, C_2Li_4

The first and, until recently, the only *ab initio* calculation on only tetralithioethylene in the literature was in a paper by Morokuma,[79] who reported the twisted structure to be more favorable than the planar form by 11.2 kcal/mol at STO-3G. This result can be explained in terms of minimization of electrostatic repulsion of the four lithium cations. However, Schleyer has recently shown[81] that even this species is not an ethylene derivative, but rather an acetylene "salt" having the Li^+ and Li_3^+ counterions, respectively. The global minimum **(36)** has both ions bridged, but the potential energy surface is quite flat and two other structures (**37** and **38**) are also minima. Structure **36** is only 1.0 kcal/mol more stable than **37** at MP4SDTQ/6-31+G*//MP2/6-31G*. The presence of the Li_3^+ fragment in **36–38** confirms the stability of this unit, previously seen in the discussion above on C_2Li_6.

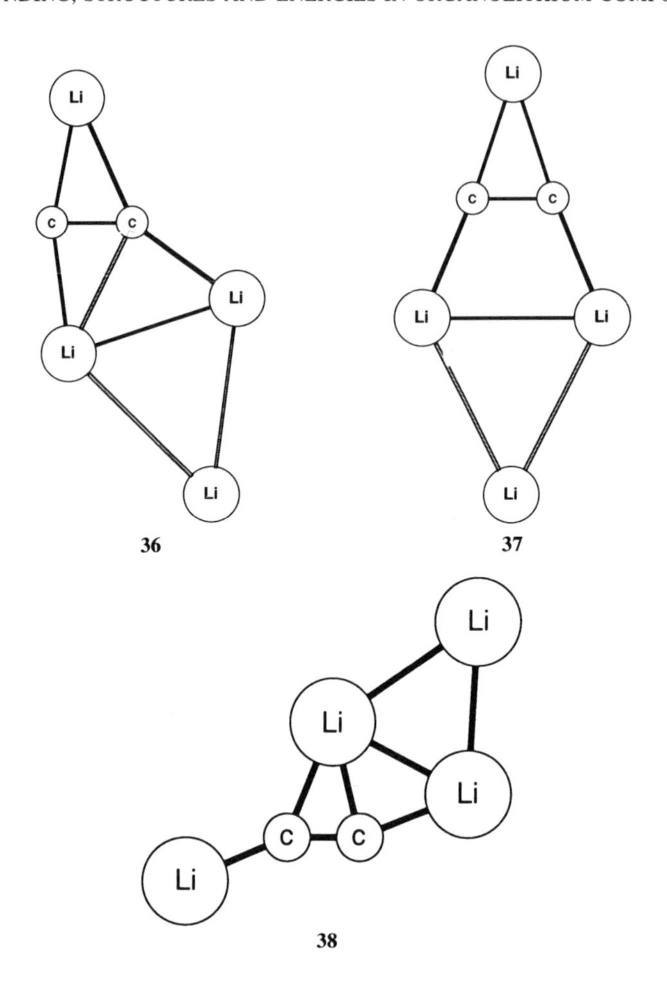

7 ACETYLENE ANALOGUES

7.1 Ethynyllithium, C_2HLi

Hinchliffe has examined ethynyllithium using a DZ+P basis set on an INDO optimized structure.[82] The structure is mundane: linear, with a carbon—lithium distance of 1.9 Å. Both Hinchliffe and Streitwieser's[46] analyses indicate that the bond is very ionic—the MO for the "C—Li" bond was shown by Streitwieser to be only the *sp* anion orbital.

7.2 Dilithioacetylene

The structure of this molecule is quite basis-set dependent.[83,84] For small basis sets without *d* functions (STO-3G, 4-31G, 6-31G) the linear structure is the minimum even with correlation through MP2. Inclusion of *d* functions, however, makes the bridged D_{2h} structure **39** more stable than the linear form. At MP2/6-31+G*,[84] the bridged structure is 3.9 kcal/mol more stable than the

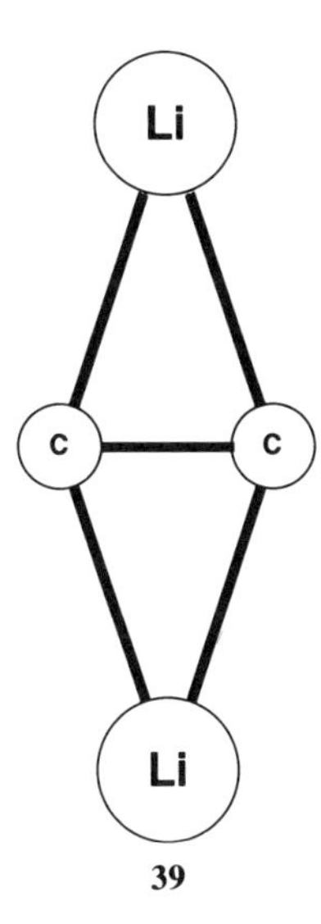

39

linear structure. This difference is 3.6 kcal/mol at the BECKE3LYP/6-3HG* density functional level. Ritchie[19] has pointed out that force constant analysis on the bridged D_{2h} structure shows it to be a transition state, not a minimum. Distortion of the structure by moving one Li atom out of the plane by about 10° provides the true minimum, having C_{2v} symmetry. This minimum is only on the order of *calories*/mol lower in energy than the D_{2h} form.

8 PROPANE ANALOGUES

8.1 2,2-Dilithiopropane, $C_3H_6Li_2$

2,2-Dilithiopropane is another molecule which might be envisioned to have a planar tetracoordinate carbon. Just as in the case of dilithiomethane, four isomers were analyzed:[72] the singlet and triplet planar and tetrahedral geometries **(40–43)**. MP2fc/6-31G* optimization gives the perpendicular triplet form **43** as the most stable isomer, but the planar triplet **42** is only 0.1 kcal/mol higher in energy, with the singlets being less stable (Table 1.5). With respect to dilithiomethane, both planar forms gain substantially in relative stability. The reason for this can be ascribed to the greater electron-donating tendency of the methyl groups, which enhance the contribution of three-center two-electrons bonding[20] to the stability of the planar form. An additional factor may be an ''agostic'' interaction (possible only in the planar form) between one of the CH bonds and the Li center.[72c]

9 PROPENE ANALOGUES

9.1 Allyllithium, C_3H_5Li

Allyllithium is at least 17 kcal/mol more stable in the bridged form **44** than in the classical form.[17,85,86] At first, the explanation for the stability of the bridged

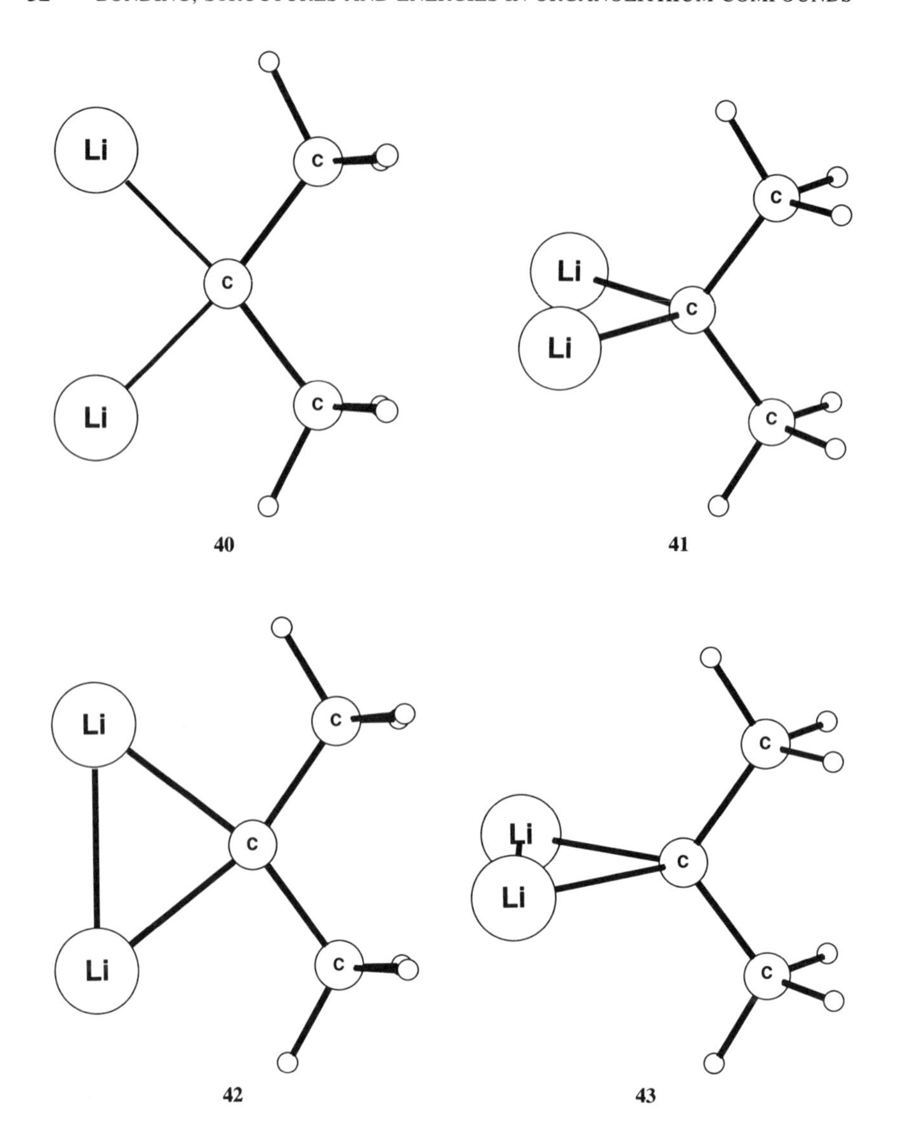

TABLE 1.5 Relative Energies (kcal/mol) of 2,2-Dilithiopropane Isomers 40–43

	HF/3-21G //HF/3-21G	HF/6-31G* //HF/6-31G*	MP2fc/6-31G* //MP2fc/6-31G*
40 Planar singlet	35.7	34.5	1.8
41 Tetrahedral singlet	38.2	38.0	5.7
42 Planar triplet	1.9	1.5	0.1
43 Tetrahedral triplet	0.0	0.0	0.0

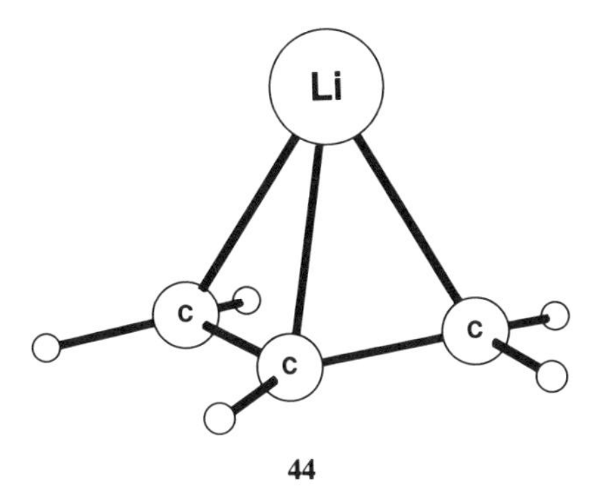

44

structure involved covalent overlap of the lithium 2*p* orbital with the allyl HOMO.[17,87] However, point-charge models and basis-set variations have recently implied that the interaction in allyllithium is mainly ionic.[86] Thus, calculations of allyllithium without selected Li 2*p* functions indicate that lithium 2*p* inner functions stabilize allyllithium by about 4–12 kcal/mol.[86] The lithium outer 2*p* functions serve only as superposition functions. The out-of-plane bending of the hydrogens results from hybridization of the carbon *p* orbitals so as to orient them toward the lithium. This hybridization brings the electron density of the anion closer to the lithium cation for stronger coulombic interactions.

10 CYCLOPROPANE AND CYCLOPROPENE ANALOGUES

10.1 Cyclopropyllithium, C_3H_5Li

The structure of cyclopropyllithium is analogous to cyclopropane.[20,88] Boggs[88] argued that bonding involved back-donation into the empty lithium 2*p* orbitals. However, the ionic model predicts the same structure, with the lithium cation attached to the sp^3 anion.

10.2 1,1-Dilithiocyclopropane, $C_3H_4Li_2$

This molecule is predicted to have a planar tetracoordinated carbon atom.[20] Structure **45,** which has the three carbon atoms and both lithium atoms in the same plane is favored at every level of theory up to and including MP4SDTQ/6-31G**.[72] Relative energies of structures **45–50** are given in Table 1.6. Note that the singlet twisted form **47** is C_s-symmetric—the two lithium atoms occupy a pseudo-axial and pseudo-equatorial position, respectively. The C_{2v}-symmetric **46** is a second-order stationary point. The stability of the planar structure involves the same arguments as for dilithiomethane and 1,1-dilithioethylene. In addition, cyclopropane offers less of a steric problem in achieving planarity since the carbon atoms are tied back in the ring.

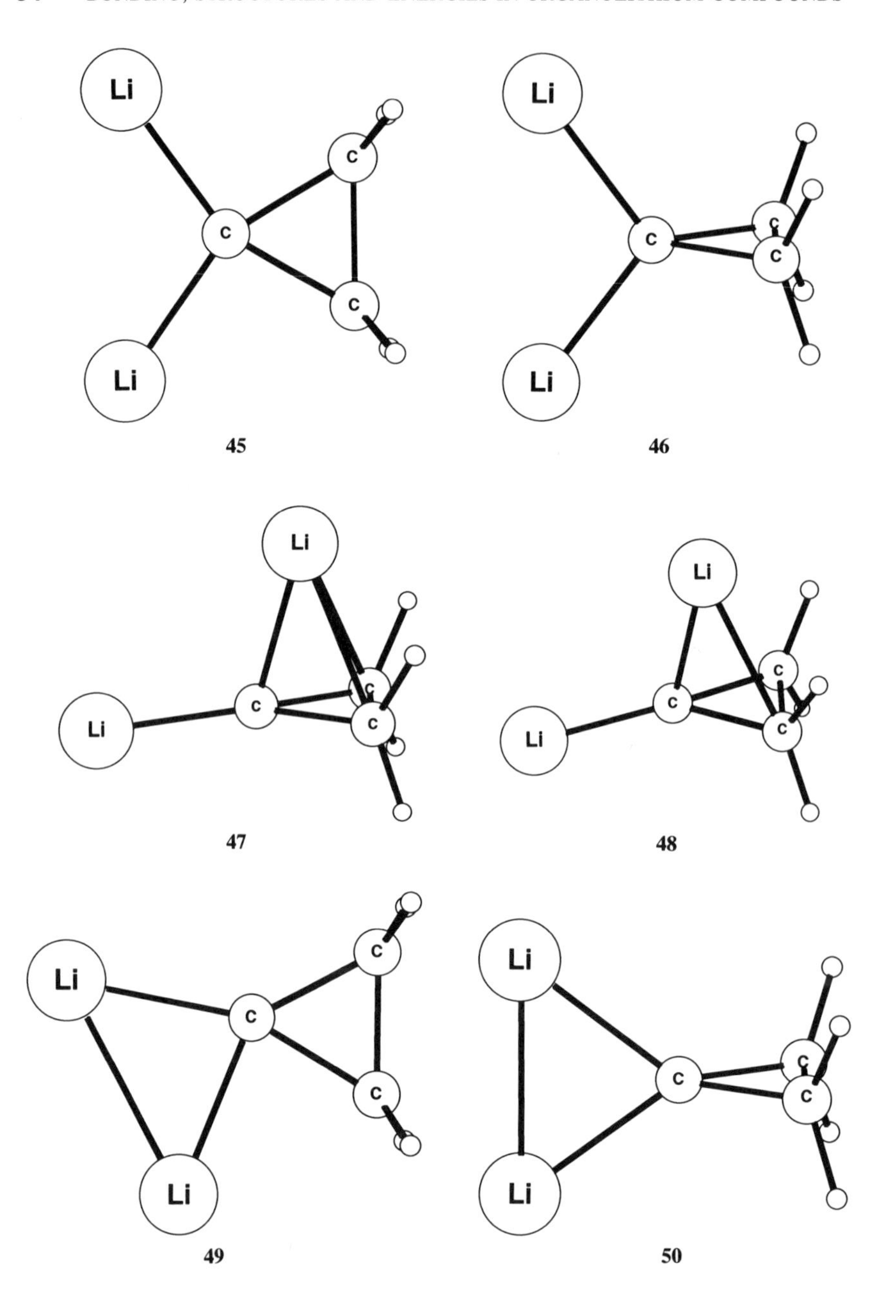

All six cyclopropane derivatives **45–50** are higher in energy than the isomeric 1-cyclopropyllithium—lithium hydride complex **51,** which is more stable than **45** by 28.6 kcal/mol at HF/6-31G*.[70] Even **51** contains a planar carbon; the twisted isomer **52** lies 1.2 kcal/mol higher in energy at HF/6-31G*.[70] This suggests that the dimer of 1-cyclopropyllithium should also contain a

TABLE 1.6 Relative Energies (kcal/mol) of 1,1-Dilithiocyclopropane Isomers 45–50

	HF/3-21G //HF/3-21G	HF/6-31G* //HF/6-31G*	MP2fc/6-31G* //MP2fc/6-31G*	MP4sdtq/6-31G** //MP2fc/6-31G*
45	27.0	27.5	0.0	0.0
46	34.7	35.7	9.1	7.7
47	33.9	33.0	7.2	6.2
48	32.3	29.4	7.3	6.4
49	0.0	0.0	1.8	0.4
50	2.4	2.3	6.2	4.5

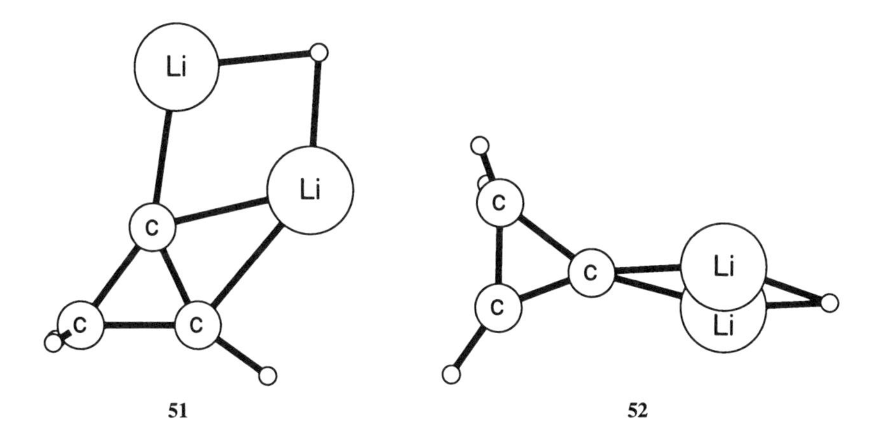

51 **52**

planar tetracoordinate carbon atom per monomeric unit. We will return to this point in the following section.

10.3 Lithiocyclopropene, C_3H_3Li

1-Lithiocyclopropene is approximately 27 kcal/mol more stable than 3-lithiocyclopropene.[89] Since vinylic hydrogens are more acidic than methylenic ones, the relative energies of the two lithium derivatives are in the expected order.

The dimer of 1-lithiocyclopropene, as discussed in the previous section, is of interest since it is a molecule which might have two planar carbon atoms.

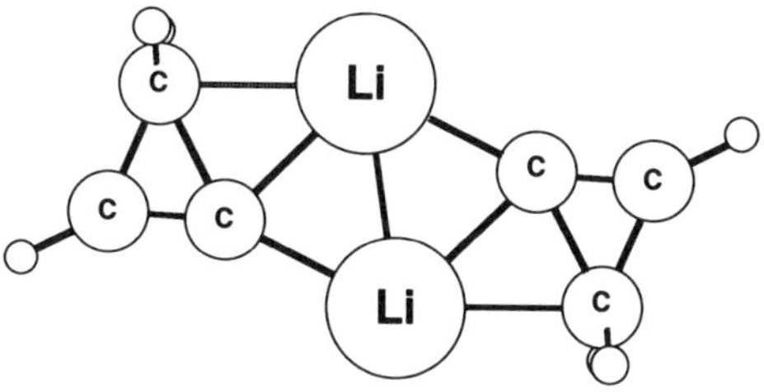

53

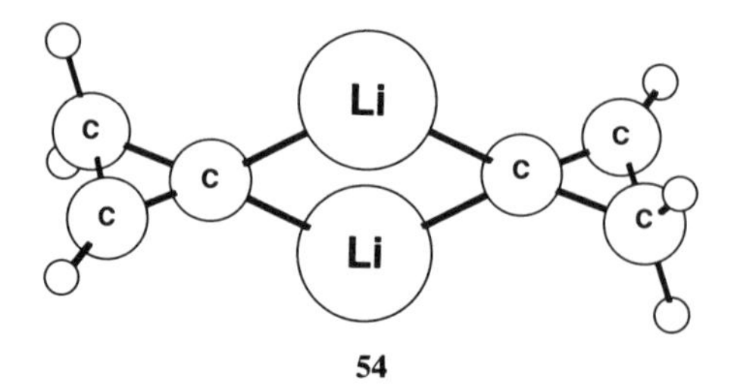

54

Geometry optimizations[72] at the HF/6-31G* level confirm that the C_{2h} dimer **53** is more stable than the twisted counterpart **54** by 9.25 kcal/mol. Attempts at preparing **53** experimentally are now in progress.

11 ALLENE ANALOGUES

11.1 C_3Li_4

The lowest energy structure reported for the C_3Li_4 molecule is **55.**[90] Schleyer[90] calculated this to be the lowest-energy species among several isomers considered at HF/4-31G//STO-3G. The next most stable isomer was calculated to be

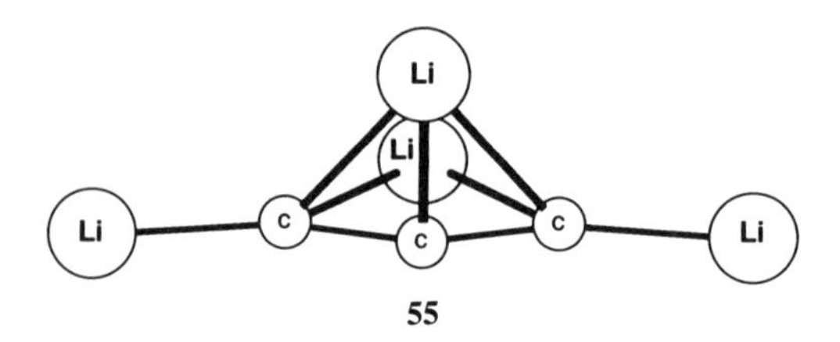

55

the C_{3v}-symmetric tetralithioallene form **56.** Our more recent MP2/6-31G* geometry optimizations[72] show that **55** remains favored over **56,** but only by 2.7 kcal/mol. Both are minima at HF/6-31G*. In **55,** the C_3Li_2 dianion is

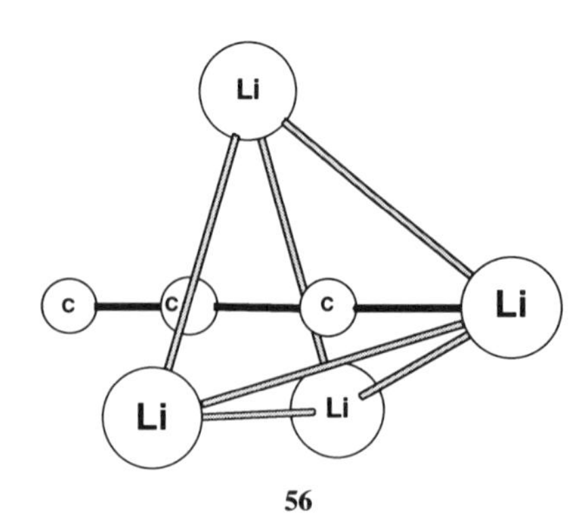

56

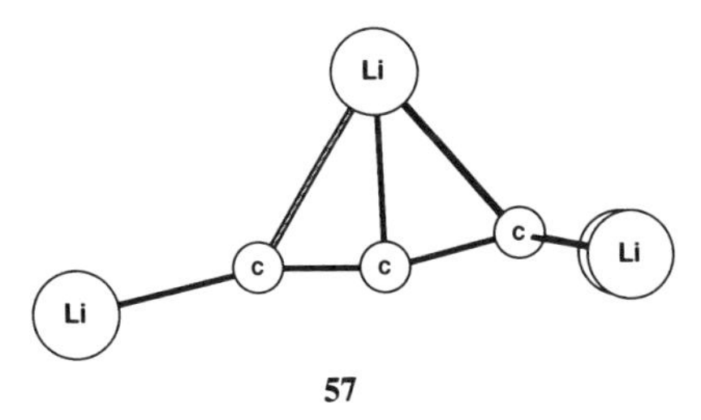

57

doubly bridged by two lithium cations. The C_s-symmetric **57,** in which only one lithium atom is bridging the terminal carbons, is less stable than **55** by 12.4 kcal/mol at HF/6-31G*,[72] and is a saddle point at this level. Note that the D_{2d}- and D_{2h}-symmetric structures are high-energy species, and each is a third-order stationary point.

12 HALOGENATED ORGANOLITHIUM SPECIES

12.1 Trifluoromethyllithium, CF_3Li

This molecule[91] represents another striking example of how the geometries of organolithium molecules differ from the "classical" structures of the parent, nonmetallated species. The C_{3v} structure **58** is a second-order stationary point at HF/6-31G* and is 9.4 kcal/mol higher in energy than the two minima on the PES: the C_s-symmetric $LiF{-}CF_2$ complex **59** and the "inverted" C_{3v}-symmetric structure **60,** the latter being the global minimum at HF/6-31G*. The preference for **60** over the classical species **58** is a consequence, once again, of the primarily electrostatic nature of bonding to lithium. The negative charge in the CF_3^- anion sits primarily on the very electronegative fluorine atoms, so that lithium prefers to be coordinated to them rather than to the carbon atom. A higher level of theory may be required to determine the relative stability of **60** and **59.** Single-point calculations indicate **60** to be more stable by 2.1 kcal/mol at (MP4SDTQ/6-31G*//HF/6-31G*), but the influence of geometry optimization at correlated levels has not been explored.

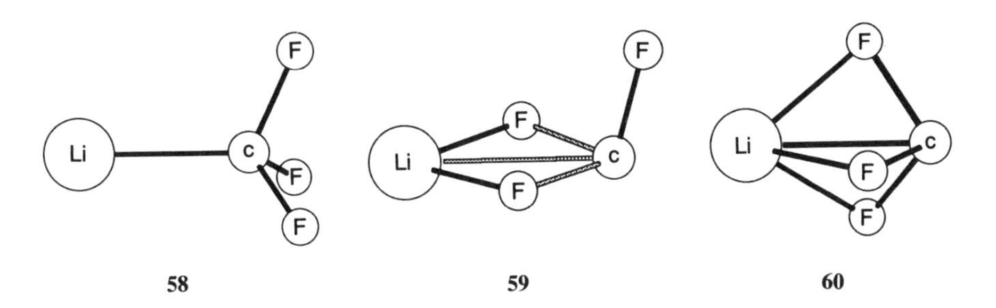

58 59 60

12.2 Trichloromethyllithium, CCl_3Li

Schleyer analyzed this molecule at HF/4-31G,[91] and again the "classical" C_{3v} form **61** was found to be disfavored. At this level the global minimum is the "inverted" C_{3v} isomer **63,** which is analogous to **60,** and which is more stable than **61** and LiCl:Cl_2 complex **62** by 6.5 and 1.5 kcal/mol, respectively. Our more recent calculations at MP4/6-31G*//HF/6-31G* give however a 7.3 kcal/mol preference for **62** over **63.** Here, too, the influence of correlation on geometry optimization must still be examined.

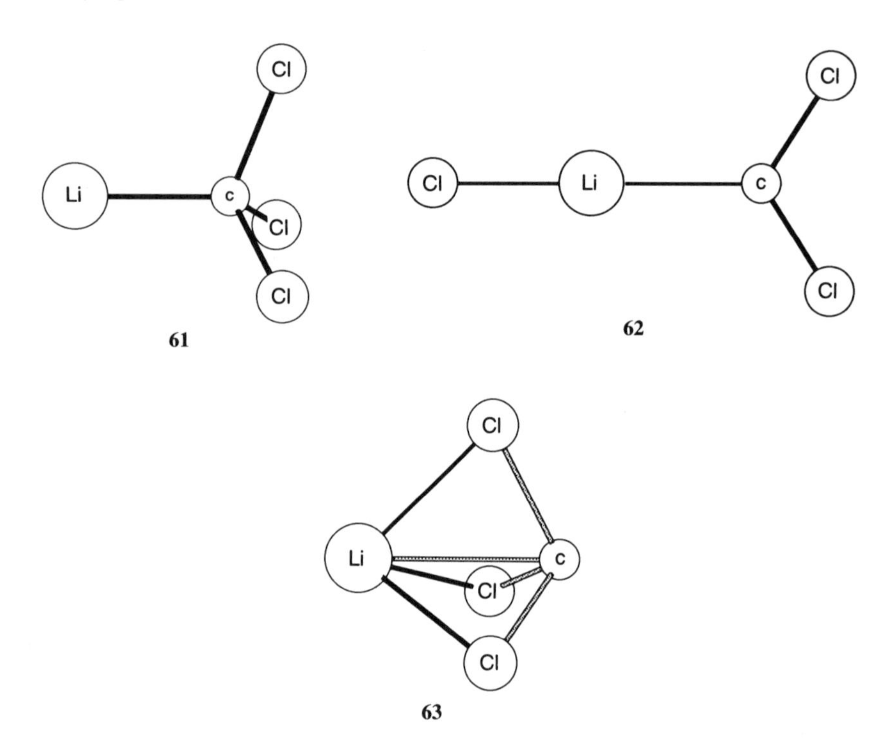

13 CHARGED SYSTEMS

13.1 CLi_6^{2+} and CLi_5^+

These cations have been examined by Schleyer[20] and by Weinhold.[92] Both are stable minima at HF/6-31G*. CLi_6^{2+} **(64),** which has octahedral symmetry, loses Li^+ to give CLi_5^+ **(65).** The transition structure[72] **66** has C_{4v} symmetry and lies 33.9 kcal/mol above **64.** The CLi_5^+ species **65** is a trigonal bipyramid, having D_{3h} symmetry; the C_{4v} structure lies only 0.8 kcal/mol above **65** at both HF/6-31G* and MP4SDTQ/6-31G*//HF/6-31G*. This "flat" potential surface is found also for the neutral CLi_5 species.

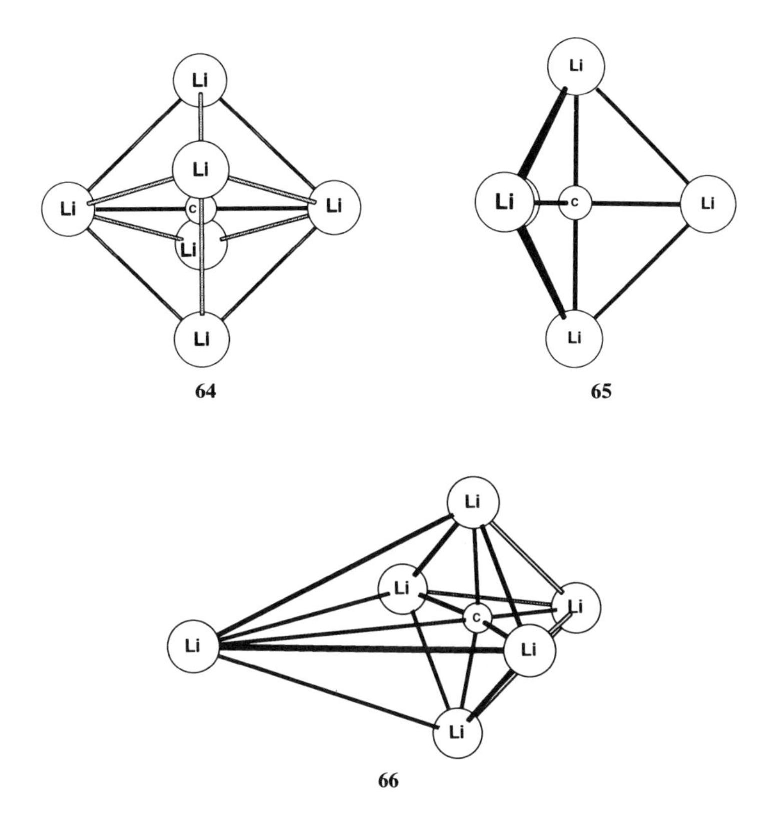

14 CONCLUSION

In this part we have discussed some of the more representative organolithium structures published in the literature. We have also presented some updated, higher-level calculations which in some cases change the stability order of the isomers involved. Most stable geometries are remarkably different from the structures of the corresponding "parent" hydrocarbons. The main bonding interaction in organolithium compounds is electrostatic. The best structures are those in which stable anions and cations (or Li_n^{m+} cationic "fragments") are present. This is an important guideline which will continue to serve in the elucidation of more complex carbolithiated compounds in the future.

ACKNOWLEDGMENT

This work was supported in part by the National Science Foundation, the Convex Computer Corporation, and the Fonds der Chemischen Industrie.

REFERENCES

1. Ritchie, C. D. *Physical Organic Chemistry—The Fundamental Concepts*; New York: Marcel Dekker, 1975; p. 169.
2. Coates, G. E.; Green, M. L. H.; Wade, K. *Organometallic Compounds*, Vol. 1, 3rd ed.; Methuen; London, 1967; p. 32.
3. Cartmell, E.; Fowles, G. W. A. *Valency and Molecular Structure*, 3rd ed.; Van Nostrand: New York, 1966; p. 157.
4. Armstrong, D. R.; Perkins, P. G. *Coord. Chem. Rev.* **1981,** *38*, 139–275.
5. Pauling, L. *The Nature of the Chemical Bond*, 3rd ed.; Cornell University Press: Ithaca, NY, 1960; p. 96.
6. Ebel, H. F. *Tetrahedron* **1965,** *21*, 699–707.
7. Rochow, E. G.; Hurd, D. T.; Lewis, R. N. *The Chemistry of Organometallic Compounds*; Wiley: New York, 1957; p. 18.
8. Weiss, E.; Sauerman, G. *Chem. Ber.* **1970,** *103*, 265; *J. Organomet. Chem.* **1970,** *21,* 1.
9. Streitwieser, A., Jr.; Williams, J. W.; Alexandratos, S.; McKelvey, J. M. *J. Am. Chem. Soc.* **1976,** *98*, 4778.
10. Mulliken, R. S. *J. Chem. Phys.* **1955,** *23*, 1833, 1841, 2338, 2343.
11. Baird, N. C.; Barr, R. F.; Datta, R. K. *J. Organomet. Chem.* **1973,** *59*, 65.
12. Hinchliffe, A.; Saunders, E. J. *J. Mol. Struct.* **1976,** *31*, 283.
13. Guest, M. F.; Hillier, I. H.; Saunders, V. R. *J. Organomet. Chem.* **1972,** *44*, 59.
14. Schleyer, P. v. R.; Würthwein, E. U.; Kaufmann, E.; Clark, T.; Pople, J. A. *J. Am. Chem. Soc.* **1983,** *105*, 5930.
15. Apeloig, Y.; Clark, T.; Kos, A. J.; Jemmis, E. D.; Schleyer, P. v. R. *Isr. J. Chem.* **1980,** *20*, 43.
16. Graham, G.; Richtsmeier, S.; Dixon, D. *J. Am. Chem. Soc.* **1980,** *102*, 5759.
17. Clark, T.; Jemmis, E. D.; Schleyer, P. v. R.; Binkley, J. S.; Pople, J. A. *J. Organomet. Chem.* **1978,** *150*, 1.
18. Kos, A. J.; Jemmis, E. D.; Schleyer, P. v. R.; Gleiter, R.; Fischbach, U.; Pople, J. A. *J. Am. Chem. Soc.* **1981,** *103*, 4996.
19. (a) Apeloig, Y.; Schleyer, P. v. R.; Binkley, J. S.; Pople, J. A.; Jorgenson, W. L. *Tetrahedron Lett.* **1976,** 3923. (b) Ritchie, J. P. *Tetrahedron Lett.* **1982,** *23*, 4999.
20. Collins, J. B.; Dill, J. D.; Jemmis, E. D.; Apeloig, Y.; Schleyer, P. v. R.; Seeger, R.; Pople, J. A. *J. Am. Chem. Soc.* **1976,** *98*, 5419.
21. Kos, A. J.; Schleyer, P. v. R. *J. Am. Chem. Soc.* **1980,** *102*, 7928.
22. Clark, T.; Rhode, C.; Schleyer, P. v. R. *Organometallics* **1983,** *2*, 1344.
23. Schiffer, H.; Ahlrichs, R. *Chem. Phys. Lett.* **1986,** *124*, 172.
24. Ehrhardt, C.; Ahlrichs, R. *Theor. Chim. Acta* **1985,** *68*, 231.
25. Graham, G.D.; Marynick, D. S.; Lipscomb, W. N. *J. Am. Chem. Soc.* **1980,** *102*, 4572.
26. Bader, R. F. W. *Atoms in Molecules—A Quantum Theory*; Clarendon Press: Oxford, 1990. See also Bader, R. F. W. *Acc. Chem. Res.* **1985,** *18*, 9.

27. Laidig, K. E.; Streitwieser, A., unpublished results.
28. Streitwieser, A.; Grier, D. L.; Kohler, B. A. B.; Vorpagel, E. R.; Schriver, G. W. in *Electron Distribution and the Chemical Bond*; Coppens, P.; Hall, M. B.; Eds.; Plenum: New York, 1982; p. 447.
29. (a) Collins, J. B.; Streitwieser, A.; McKelvey, J. M. *Comp. Chem.* **1979,** *3*, 79. (b) Streitwieser, A.; Collins, J. B.; McKelvey, J. M.; Grier, D.; Sender, J.; Toczko, G. *Proc. Natl. Acad. Sci. USA* **1979,** *76*, 2499.
30. Bachrach, S. M.; Streitwieser, A. *J. Am. Chem. Soc.* **1984,** *106*, 5818.
31. Agrafiotis, D. K.; Tansy, B.; Streitwieser, A. *J. Comput. Chem.* **1989,** *11*, 1101-1110; Agrafiotis, D. K.; Tansy, B.; Streitwieser, A. *QCPE Bull.* **1991,** *11*, 13.
32. Biegler-König, F. W.; Bader, R. F. W.; Tang, T.-H. *J. Comput. Chem.* **1982,** *3*, 317.
33. Ritchie, J. P.; Bachrach, S. M. *J. Am. Chem. Soc.* **1987,** *109*, 5909.
34. (a) Bader, R. F. W.; MacDougall, P. J.; Lau, C. D. H. *J. Am. Chem. Soc.* **1984,** *106*, 1594. (b) Bader, R. F. W.; Essen, H. *J. Chem. Phys.* **1984,** *80*, 1943.
35. Cioslowski, J. *J. Am. Chem. Soc.* **1989,** *111*, 8333.
36. Reed, A. E.; Weinstock, R. B.; Weinhold, F. *J. Chem. Phys.* **1985,** *83*, 735.
37. Kaufmann, E.; Raghavachari, K.; Reed, A. E.; Schleyer, P. v. R. *Organometallics* **1988,** *7*, 1597.
38. Bachrach, S. M., unpublished results.
39. Jug, K.; Fasold, E.; Gopinathan, M. S. *J. Comp. Chem.* **1989,** *10*, 965.
40. Horn, H.; Ahlrichs, R. *J. Am. Chem. Soc.* **1990,** *112*, 2121.
41. Hiberty, P. C.; Cooper, D. L. *J. Mol. Struct.* (*Theochem.*) **1988,** *169*, 437.
42. Penotti, F. E. G.; Gerratt, J.; Cooper, D. L.; Raimondi, M. *J. Chem. Soc., Faraday Trans. 2* **1989,** *85*, 151.
43. Saturno, A. F.; Parr, R. G. *J. Chem. Phys.* **1960,** *33*, 22.
44. Kolos, W. *Theor. Chim. Acta* **1979,** *51*, 219.
45. Kost, D.; Klein, J.; Streitwieser, A.; Schriver, G. W. *Proc. Natl. Acad. Sci. USA* **1982,** *79*, 3922.
46. Bachrach, S. M.; Streitwieser, A. *J. Am. Chem. Soc.* **1984,** *106*, 2283.
47. Streitwieser, A. *J. Organomet. Chem.* **1978,** *156*, 1. See also Bushby, R. J.; Steel, H. L. *ibid.* **1987,** *336*, C25.
48. Schleyer, P. v. R.; Kos, A. J.; Kaufmann, E. J. *J. Am. Chem. Soc.* **1983,** *105*, 7617.
49. Streitwieser, A.; Waterman, K. C. *J. Am. Chem. Soc.*, **1984,** *106*, 3138.
50. Streitwieser, A., Jr. *Acc. Chem. Res.* **1984,** *10*, 353–357.
51. Sygula, A.; Rabideau, P. W. *J. Am. Chem. Soc.* **1992,** *114*, 821.
52. Buhl, M.; van Eikema Hommes, N. J. R.; Schleyer, P. v. R.; Fleischer, U.; Kutzelnigg, W. *J. Am. Chem. Soc.* **1991,** *113*, 2459.
53. Anders, E.; Opitz, A.; van Eikema Hommes, N.; Hampel, F. *J. Org. Chem.* **1993,** *58*, 4424.
54. An NPA analysis shows a nearly complete octet on carbon with the additional electrons assigned to lithium (Reed, A. E.; Weinhold, F. *J. Am. Chem. Soc.* **1985,** *107*, 1919).
55. Andrews, L. J. *J. Chem. Phys.* **1967,** *47*, 4834.

56. Scovell, W. M.; Kimura, B. Y.; Spiro, T. G. *J. Coord. Chem.* **1971,** *1*, 107.
57. Brown, T. L.; Kimura, B. Y.; Seitz, L. M. *J. Am. Chem. Soc.* **1968,** *90*, 3245.
58. McKeever, L. D.; Waack, R.; Doran, M. A.; Baker, E. B. *J. Am. Chem. Soc.* **1969,** *91*, 1057.
59. Seebach, D.; Hässig, R.; Gabriel, J. *Helv. Chim. Acta* **1983,** *66*, 308; Seebach, D.; Gabriel, J.; Hässig, R. *ibid.* **1984,** *67*, 1083.
60. Bauer, W.; Winchester, W. R.; Schleyer, P. v. R. *Organometallics,* **1987,** *6*, 2371.
61. Brown, T. L. *Pure & Appl. Chem.* **1970,** *23*, 447.
62. See the review and discussion of Günther, H.; Moskau, D.; Bast, P.; Schmalz, D. *Angew. Chem. Int. Ed. Engl.* **1987,** *26*, 1212.
63. Schleyer, P. v. R.; Chandrasekhar, J.; Kos, A. J.; Clark, T.; Spitznagel, G. W. *J. Chem. Soc. Chem. Commun.* **1981,** 882.
64. Schleyer, P. v. R. *Pure & Appl. Chem.* **1984,** *56*, 151.
65. Baird, N. C.; Barr, R. F.; Datta, R. K. *J. Organomet. Chem.* **1973,** *59*, 65.
66. Graham, G. D.; Marynick, D. S.; Lipscomb, W. N. *J. Am. Chem. Soc.* **1980,** *102*, 4572.
67. Kaufmann, E.; Clark, T.; Schleyer, P. v. R. *J. Am. Chem. Soc.* **1984,** *106*, 1856.
68. Clark, T.; Schleyer, P. v. R.; Pople, J. A. *J. Chem. Soc. Chem. Commun.* **1978,** 137.
69. Herzig, L.; Howell, J. M.; Sapse, A. M.; Singman, E.; Snyder, G. *J. Chem. Phys.* **1982,** *77*, 429.
70. Laidig, W. D.; Schaefer, H. F., III *J. Am. Chem. Soc.* **1978,** *100*, 5972.
71. Bachrach, S. M.; Streitwieser, A., Jr. *J. Am. Chem. Soc.* **1984,** *106*, 5818.
72. a) Schleyer, P. v. R., unpublished results. b) Bolton, E. E.; Schaefer, H. F. III; Laidig, W. D.; Schleyer, P. v. R. *J. Am. Chem. Soc.* **1994,** *116*, 9602. c) Sorger, U.; Kottke, T.; Hommes, N. J. R. v. E.; Schleyer, P. v. R. *J. Am. Chem. Soc.*, submitted for publication.
73. Jemmis, E. D.; Schleyer, P. v. R.; Pople, J. A. *J. Organomet. Chem.* **1978,** *154*, 327.
74. Wurthwein, E. U.; Sen, K. D.; Pople, J. A.; Schleyer, P. v. R. *Inorg. Chem.* **1983,** *22*, 496.
75. Kos, A.; Poppinger, D.; Schleyer, P. v.R.; Thiel, W. *Tetrahedron Lett.* **1980,** *20*, 2515.
76. Hinchliffe, A. *J. Mol. Struct.* **1977,** *37*, 289.
77. Laidig, W. D.; Schaefer, H. F., III *J. Am. Chem. Soc.* **1979,** *101*, 7184.
78. Apeloig, Y.; Schleyer, P. v. R.; Binkley, J. S.; Pople, J. A. *J. Am. Chem. Soc.* **1976,** *98*, 4332.
79. Nagase, S.; Morokuma, F. *J. Am. Chem. Soc.* **1978,** *100*, 1661.
80. Schleyer, P. v. R.; Kaufmann, E.; Kos, A.; Clark, T.; Pople, J. A. *Angew. Chem.* **1986,** *25*, 169.
81. Dorigo, A. E.; Hommes, N. J. v. E.; Krogh-Jespersen, K.; Schleyer, P. v. R. *Angew. Chem.*, in press.
82. Hinchliffe, A. *J. Mol. Struct.* **1977,** *37*, 145.

83. Disch, R. L.; Schulman, J. M.; Ritchie, J. P. *J. Am. Chem. Soc.* **1984,** *106,* 6246.

84. a) Schleyer, P. v. R. *Pure & Appl. Chem.* **1983,** *55*, 355. b) Schleyer, P. v. R. *J. Phys. Chem.* **1990,** *94*, 5560.

85. Lambert, C.; Schleyer, P. v. R. *Angew. Chemie Int. Ed. Engl.* **1994,** *33*, 1129, and references therein. See also: Lambert, C.; Schleyer, P. v. R., *Houben-Weyl Methoden der Organischen Chemie, Band E 19d*, Georg Thiene Verlag; Stuttgart, 1993: p. 1–99.

86. Clark, T.; Rohde, C.; Schleyer, P. v. R. *Organometallics* **1983,** *103*, 7617.

87. Bingini, A.; Cianelli, G.; Cardillo, G.; Palmieri, P.; Umani-Ronchi, A. *J. Organomet. Chem.* **1976,** *110*, 1.

88. Skancke A.; Boggs, J. E. *J. Mol. Struct.* **1978,** *50*, 173.

89. Jemmis, E. D.; Chandrasekhar, J.; Schleyer, P. v. R. *J. Am. Chem. Soc.* **1979,** *101*, 2848.

90. Jemmis, E. D.; Poppinger, D.; Schleyer, P. v. R.; Pople, J. A. *J. Am. Chem. Soc.* **1977,** *99*, 5796.

91. a) Clark, T.; Schleyer, P. v. R. *J. Am. Chem. Soc.* **1979,** *101*, 7747. b) For related calculations on CH_2FLi, see: Mareda, J.; Rondan, N.; Houk, K. N.; Clark, T.; Schleyer, P. v. R. *J. Am. Chem. Soc.* **1983,** *105*, 6997. Luke, B. T.; Pople, J. A.; Schleyer, P. v. R.; Clark, T. *Chem. Phys. Lett.* **1984,** *102*, 148. c) See also: Hehre, W. J.; Radom, L.; Schleyer, P. v. R.; Pople, J. A. *Ab initio molecular orbital theory*, John Wiley & Sons; New York, 1985: p. 417–419. At the level quoted here (HF/6-31G*//HF/3-21G*) the LiCl:CCl_2 complex is even more strongly favored, being 12 kcal/mol more stable than the next most stable isomer (the "classical" C_{3v} form **61**).

92. Reed, A. E.; Weinhold, A. *J. Am. Chem. Soc.* **1985,** *107*, 1919.

2

THEORETICAL STUDIES OF AGGREGATES OF LITHIUM COMPOUNDS

ANNE-MARIE SAPSE
City University of New York, Graduate Center and John Jay College, New York, New York

DULI C. JAIN
York College of the City University of New York, Jamaica, New York

KRISHNAN RAGHAVACHARI
AT&T Bell Laboratories, Murray Hill, New Jersey

1 INTRODUCTION

Lithium compounds are known to associate in solvents, in the crystal, and even in the gas phase.[1] It is doubtful that the monomeric species can exist except

---

Lithium Chemistry: A Theoretical and Experimental Overview, Edited by Anne-Marie Sapse and Paul von Ragué Schleyer.
ISBN 0-471-54930-4

under very unusual conditions. Experiments consisting of colligative measurements, NMR investigations, X-ray crystal structure determinations, mass spectroscopic observations, and other methods have shown that aggregation is typical in these compounds. Cryoscopic techniques[2–5] have shown, for instance, that the ratio of the molar weight to formula weight of ethyllithium is close to 6 in benzene and cyclohexane. The same ratio was observed for *n*-butyllithium in the same solvents by cryoscopic[3] and isopiestic[6] methods. Other organolithium compounds, such as $(CH_3)_3CLi$ and $(CH_3)_2CHCH_2Li$, are present as tetramers in hexane and benzene as determined by cryoscopic and ebulioscopic methods.[7,8]

In most cases, different hydrocarbon solvents lead to the formation of the same oligomers for the same organolithium compound. An exception is $(CH_3)_3SiCH_2Li$, which exists as a tetramer in benzene but adopts a hexameric form in cyclohexane.[3] In general, most alkyllithium compounds are hexamers in hydrocarbon solvents, except when α or β positions feature branching, which leads to tetrameric structures.[9] However, when the alkyl group is very large, as for example in menthyllithium, benzyllithium, or allylic polymer lithium compounds such as (polyisopren)yl lithium, dimers are present, as shown for the latter by light-scattering and viscosity measurements.[10]

In the crystal phase, methyllithium and ethyllithium adopt tetrameric structures, as shown by X-ray studies.[11–13] In the gas-phase, mass spectrometric studies of ethyllithium, *t*-butyllithium and trimethylsilylmethyllithium indicate the presence of tetrameric and hexameric structures,[14,15] while methyllithium is found to exist as a tetramer.[16–18]

The experimental evidence for the existence of oligomers of lithium compounds as well as investigations of their structures can be complemented by theoretical studies. Ab initio and semiempirical calculations can elucidate the nature of binding in the clusters, determine their preferential conformations, and provide detailed information about their geometries. In addition, binding energies as well as the energies of disproportionation reactions can be obtained providing a more complete description of these clusters. In this work, we review some of the recent theoretical studies on the oligomers of lithium compounds.

2 Li_nH_m THEORETICAL INVESTIGATIONS

The simplest lithium compound that forms aggregates is LiH. These clusters have aroused considerable interest because they may provide simple models for hydrogen adsorption on bulk metals, using them as hydrogen-storing mediums.[19] As shown by empirical studies,[20,21] the interatomic distance between two hydrogen atoms has to be larger than about 2.1 Å for the hydrogen to be retained by the metal, so the available interstitial space is of great importance for the stability of the hydrides.[22]

Some of the first calculations on LiH clusters were performed by Dill et

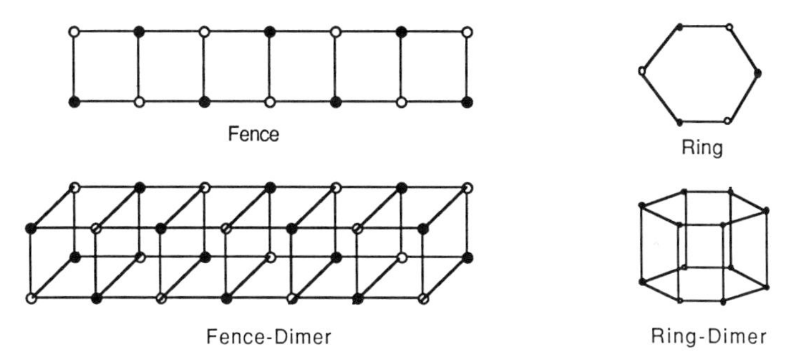

Figure 2.1

al.[23] and by Kato et al.[24] The latter found that $(LiH)_n$ is the most stable species in the Li_nH_m series. For stoichiometric $(LiH)_n$ clusters, they investigated such structures as fence, ring, fence-dimer, and ring-dimer, as shown in Figure 2.1. Using the Hartree–Fock method with a double-zeta basis set, they found that the most stable structure of $(LiH)_6$ is the ring-dimer, which is formed from two rings parallel to each other.

To obtain information about hydrogen binding to lithium clusters, Rao et al.[25] investigated the energetics and equilibrium geometries of lithium clusters with varying hydrogen content and cluster size. The Li_nH_m clusters were examined with the unrestricted Hartree–Fock method using a 6-31G basis set, with the correlation energy evaluated by the Moller–Plesset perturbation procedure up to fourth order.

Rao et al.[25] studied Li_nH_m clusters ($n \leq 4$, $m < n$) and compared them with valence isoelectronic lithium clusters featuring $m + n$ atoms. They observed that while Li_nH_m clusters are planar for $n < 3$ and $m < 3$, lithium clusters with more than five atoms are three dimensional, so even though the two species contain the same number of valence electrons, their geometries are different. For instance, Li_4H_2 is three dimensional while Li_3H_3 is planar and Li_6 is a pentagonal pyramid. It is clear that the *s* electron of the hydrogen behaves very differently from those of the alkali metals, which can easily be explained by its lack of shielding by core electrons.[25]

However, in some of the smaller systems, such as Li_4 and Li_2H_2, the geometries are fairly similar. Both structures are rhombic, but the Li—Li distance of 3.1 Å is much larger than the Li—H distance of 1.81 Å. This causes the Li_3H cluster to feature a distorted rhombic geometry, as seen in Figure 2.2. The value of 1.81 Å for the Li—H distance is somewhat smaller than the value of 2.04 Å found in solid lithium hydrides[26] and greater than the value of 1.60 Å present in gas phase LiH.[27] It is interesting to notice that a short Li—H distance is also found in organolithium compounds, as mentioned by Novoa et al.,[28] which contributes to the ease of LiH removal from these compounds.[29] In the dimer, Rao et al.[25] found a value of 1.64 Å for the Li—H distance, and the bond length remained nearly 1.7 Å for all the Li_nH_m clusters.

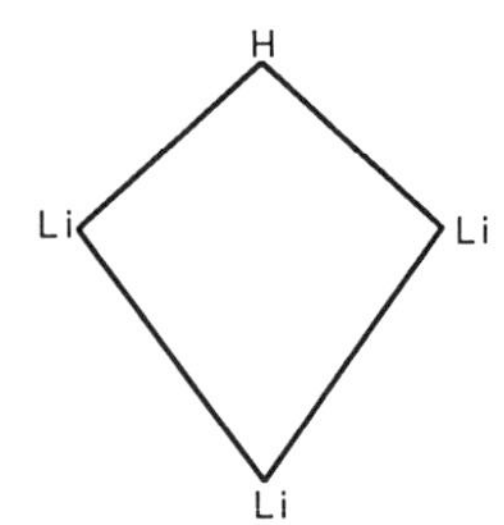

Figure 2.2

The binding energies of the clusters, defined as the energy of the cluster minus the sum of the energies of the individual atoms, depends on cluster size for Li_nH and Li_nH_3 but not for Li_nH_2. The LiH_2 is not bound because the strong covalent energy of the H_2 molecule is not compensated by the bond formation with Li when the H_2 bond is broken. Rao et al.[25] found the binding energy of LiH itself to be 2 eV, somewhat smaller than the experimental value 2.43 eV and the value of 2.35 eV obtained by Talbi and Saxon,[30] who used a large basis set, including *d* and *f* functions.

The binding energies of the clusters were used by Rao et al.[22] to follow the increase of the values of the heats of solution with the size of the cluster. These are defined as

$$-\Delta E_{\mathrm{H}} = E(\mathrm{M}_n\mathrm{H}_{2n}) - nE(\mathrm{H}_2)$$

Indeed, they rapidly approach the heat of solution of crystal LiH which is 0.93 eV per hydrogen atom. They find that the clusters with an even number of hydrogens are more stable than those with an odd number for Li_2H, Li_2H_2, and Li_2H_3. The reverse is true for Li_3H, Li_2H_2, and Li_3H_3 species, showing that the stability of the clusters as a function of hydrogen content depends on the metal content. It is clear that clusters containing an even number of electrons are more stable due to electron pairing. The effect can also be understood by examining the HOMO levels induced by each hydrogen addition. For instance, in Li_3H_2, four of the five valence electrons occupy the two lower H levels but the fifth must occupy the higher Li 2*s* level. In Li_3H_3, there are three low hydrogen levels and the six electrons can occupy them, affording more stability to the molecule.

With the exception of the Li_3H_2 cluster, the Mulliken population analysis shows that the hydrogen features a net negative charge of 0.4 eu. When charge density contours of the Li_3H_2 and Li_3H_3 species are examined, Rao et al. conclude that the Li—H bond has a strong ionic character and there is practically no bonding between Li atoms or between the H atoms. The strong ionic character of the Li—H bond thus resembles the totally ionic character proposed by Streitwieser et al.[31] for the C—Li bond in CH_3Li.

The energetics and geometries of Li_nH_m clusters were also examined by Michels and Wadehra.[32] Their interest in this topic springs from a desire to better understand the role of Li_nH_m (Cs_nH_m) molecules formed by seeding an alkali metal in a hydrogen plasma. The formation of such ions can be of importance for their application in gaseous discharges, fusion plasmas, and gas lasers.[33] They have investigated Li_xH_y clusters and their role in dissociative attachment reactions (e.g., $Li_2H_2 + e \rightarrow LiH + LiH^-$) leading to negative ion production. In particular, Li_2H, Li_3H, Li_2H_2, and their negative ions were studied by Michels and Wadehra with ab initio methods, using the triple-zeta 6-311G basis set which includes *d* functions on the lithium atoms and *p* functions on the hydrogen atoms. To better describe the negative species, diffuse functions were added both to lithium and to hydrogen. The correlation energy effects were evaluated by the second-order Moller–Plesset method (MP2).

The Li_2H exhibited a C_{2v} geometry and was stable when compared to $Li_2 + H_2$. The corresponding anion is slightly bound in comparison to the neutral species. These results, obtained via MP2 calculations, confirm the Hartree–Fock results of Cardelino et al.[34] The optimum geometry for the Li_3H cluster also corresponds to a C_{2v} symmetry (a kite-like structure) in agreement with Rao et al.[25] In addition, a higher-energy minimum with trigonal pyramidal structure is also found. The anions in both conformations are unbound. Higher-level MCSCF/Cl studies on the singlet and triplet forms of Li_3H were carried out by Talbi and Saxon.[30] They found that the planar configuration is the global minimum for the ground state and for the first excited state. For the ground state a pyramidal local minimum was also found, while no such form seems to exist for the excited state. Calculations on the $(LiH)_2$ dimer by Michels and Wadehra[32] agree with the results of Rao et al.[25] that its most stable conformation is the rhombus (D_{2h}) form. The linear conformation is a saddle point, with one imaginary frequency. The negative ion is unstable.

3 METHYLLITHIUM CLUSTERS

The transition from LiH clusters to the more complicated organolithium clusters can be made by examining CH_3Li clusters. First prepared by Schlenk and Holtz[35] in 1917, methyllithium is of prime importance in organic synthesis and in the preparation of transition metal complexes. Infrared studies indicated as early as in 1957 that methyllithium does not exist as a monomer even in the gas phase, though the detailed assignments were not definitive. In 1964, Weiss and Lucken[13] deduced that methyllithium exists as a tetramer from its X-ray powder diffraction pattern. The methyl groups in the tetramer bridge the faces of a lithium tetrahedron, and the hydrogen atoms are staggered with respect to the corresponding Li site. Similar structures have been seen for ethyllithium also. The tetramers persist in solution and are present even in the gas phase, as shown using a flash vaporization technique.[36] Equilibrium between tetramers and dimers has been seen in ether solution. The infrared spectrum of the monomer was measured in an argon matrix in 1967 and indicated a low H—C—Li bending force constant and a high dipole moment (6 D), suggesting that the C—Li bond has a high degree of ionic character.[11,37]

There have been a wide variety of theoretical studies on the oligomers of methyllithium. It has been suggested that the association in the oligomers occurs simply by the electrostatic attraction between the positively charged lithiums and the negatively charged carbons.[38] The extent of ionicity of the C—Li bond is of central interest and has been debated extensively. Streitwieser et al.[31] claimed essentially ionic bonding in the monomer of methyllithium based on electron density projection functions, and rationalized the tetrameric structure of solid CH_3Li by a totally ionic model. However, Lipscomb et al.[39] suggested that the oligomers exhibit multicenter, electron-deficient bonding with a charge separation of only about 0.6 electrons. Based on large basis-set configuration interaction calculations, Schiffer and Ahlrichs[40] concluded that the C—Li bond in CH_3Li has nonnegligible covalent contributions. A detailed analysis of the different electron distribution contributions to the dipole moment of CH_3Li was carried out by Kaufmann et al.[41] They concluded that while the stabilization is primarily electrostatic (88% ionic), a small but important covalent component is present as well.

Herzig et al.[42] carried out Hartree–Fock studies using the 4-31G basis set on several oligomers of methyllithium. Among other findings, they reported the tetrahedral structure of the tetramer as seen experimentally. Kaufmann et al.[41] performed systematic optimizations on all the oligomers up to the tetramer with a variety of basis sets (up to 6-31G*). They also included the effects of electron correlation by second-order Moller–Plesset perturbation theory (MP2). The cyclic C_{2h} structure (Fig. 2.3) for the dimer is consistent with the results from previous studies. To obtain more definitive results, we carried out additional optimizations using the MP2 method and the larger 6-311G** basis set. These calculations indicate that the dimer is almost symmetric with the Li—C distances, differing by only 0.01 Å. Such a structure, with all four Li—C bonds having nearly equal lengths, implies the absence of a "lithium bond" similar to the hydrogen bond in these compounds. The presence of a lithium bond would have led to a structure with two short polar covalent C—Li bonds and two long C—Li lithium bonds. The trimer has a planar cyclic arrangement with each methyl group again being shared by two lithiums. The tetramer is tetrahedral, with an eclipsed methyl group orientation. The staggered conformation found experimentally has been attributed to the packing effect involving interaggregate interactions.

The association energies of the dimer, trimer, and tetramer were calculated by Kaufmann et al. to be −44.3, −79.0, and −122.9 kcal/mol. The bond dissociation energy in the monomer derived by these authors is 46.4 kcal/mol. The calculated oligomerization energies are relatively independent of the basis set employed. As is typical for ionic systems, the effect of electron correlation is small, ranging from 2% for the dimer and trimer to 5% for the tetramer. The largest binding energy (per monomer) found for the tetramer is consistent with its dominant presence in different media. No trimers have been observed experimentally, even though the calculations indicate a higher binding energy for the trimers than for the dimers. This appears to be due to entropy effects

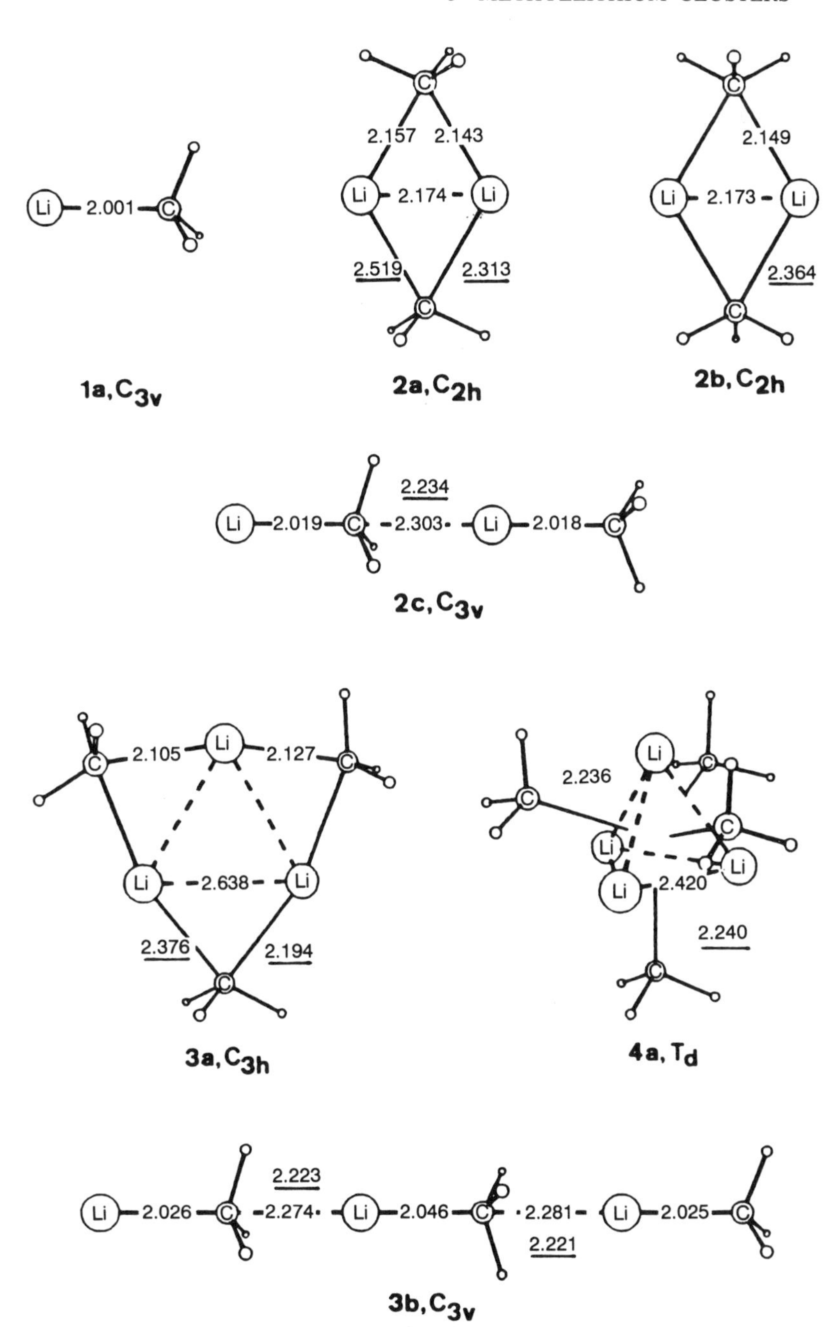
Li
2.001
C
1a, C_{3v}
2.157
2.143
2.174
2.519
2.313
2a, C_{2h}
2.149
2.173
2.364
2b, C_{2h}
2.234
2.019
2.303
2.018
2c, C_{3v}
2.105
2.127
2.638
2.376
2.194
3a, C_{3h}
2.236
2.420
2.240
4a, T_d
2.223
2.026
2.274
2.046
2.281
2.025
2.221
3b, C_{3v}

Figure 2.3

associated with solvation. Semiempirical MNDO schemes used to model solvation effects indicate that indeed the dimers are favored in solution. In the presence of the solvent, the dimer to trimer reaction is strongly endothermic.

4 INORGANIC LITHIUM COMPOUND CLUSTERS

Inorganic compounds of lithium are also found experimentally to associate in solution and in the gas phase. The compound $LiN(SiMe_3)_2$ forms oligomers in solution and the degree of association varies considerably with the conditions. In tetrahydrofuran, a dimer–monomer equilibrium is present, while in hydrocarbon solvents, a dimer–tetramer equilibrium is mostly found. The trimer could be crystallized from petroleum ether or benzene[43,44] and shows a planar six-membered ring with alternating N and Li. Theoretically, the $LiNH_2$ monomer was studied by Wurthwein et al.[45] and the dimer and trimer formed the subject of an ad initio study by Sapse et al.[46] who compared the theoretically obtained geometries with the X-ray results. The study used standard basis sets, such as split-valence sets (3-21G, 6-21G, 6-31G), polarized basis sets (6-31G*) that include *d*-functions on nonhydrogen atoms, and diffuse-function augmented basis sets (3-21+G, 6-31+G). Single-point calculations with the MP2 method were performed using the Hartree–Fock geometries. In addition, MNDO calculations were used to obtain heats of formation, based on Thiel and Clark's parameterization for lithium,[47] which is quite successful for illustrating the energetics of nitrogen–lithium interactions.

The monomer $LiNH_2$ is found to be planar at high levels of theory,[48] even though NH_3 is pyramidal. Indeed, electropositive substituents induce planarity, which is further favored by the *p-pi* delocalization. Sapse et al. found that both $(LiNH_2)_2$ and $(LiNH_2)_3$ form planar cyclic arrangements with alternating Li and N atoms. The D_{2h} and D_{3h} forms of the dimer and trimer are shown in Figure 2.4. The experimental work on $[LiN(SiMe_3)_2]_3$[49] shows it to be planar and this result is confirmed theoretically. It is also of interest to investigate the

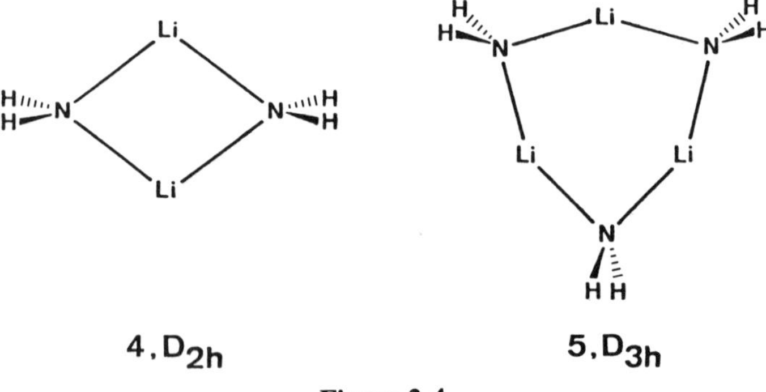

Figure 2.4

energy cost involved if the oligomers stray from planarity. Sapse et al. considered the "boat" and "chair" conformations by displacing one N atom and its opposite Li atom by 10° from the plane, and found that the "chair" distortion is only 0.4 kcal/mol higher than the planar conformation, indicating that large amplitude motion should be possible.

The geometrical parameters obtained with different basis sets are within ±0.02 Å of each other, while the angle variation is within 1°. Upon oligomerization, the N—Li bond length increases by 0.2 Å. In general, the calculated values are shorter than the X-ray results, but this might be due to the difference between $LiNH_2$ and the heavily substituted species studied experimentally. In both the dimer and the trimer, the NH_2 groups are perpendicular to the NLi ring (Figure 2.4), which allows the nitrogen lone pair to lie in the ring plane and to interact with the lithium atoms. If the NH_2 groups are constrained to lie in the plane of the ring, the energies increase significantly (22–60 kcal/mol, depending on the oligomer and on the basis set used). The greater stability of the perpendicular over the planar arrangements of the NH_2 groups is thus clearly due to the more favorable orientation of the nitrogen "lone pair" orbital toward the two adjacent lithium atoms. The total N—Li overlap population is much higher in the perpendicular than in the planar form. The overlap population in both species shows the Li· · ·Li interaction to be antibonding.

The energies of dimerization and trimerization are sensitive to the choice of the basis set, and are overestimated considerably by inadequate basis sets (superposition error[50]). Split-valence basis sets do not describe the lone pair on the nitrogen accurately[51]—diffuse functions improve the description and are more effective than polarization functions. Since these molecules have a high degree of ionic character, the correlation energy corrections are small.

Since the lone pair of electrons on the nitrogen atoms of the $LiNH_2$ molecule play an important role in the structure of the oligomers, it is interesting to examine the effect of the perturbation of the electronic cloud by attaching substituent atoms. Sapse and Jain[52] performed Hartree–Fock calculations on the dimers of NF_2Li and $NHLi_2$ species in order to compare the effects of electronegative and electron-donating atoms on the dimerization. Optimizations were performed with the 3-21G basis set followed by single-point energy evaluations with the larger 6-31G* basis. It was found that the NF_2Li dimer features an asymmetric geometry, whereas the $NHLi_2$ dimer is symmetric. Indeed, the NF_2Li dimer optimized by constraining the N—Li bonds to be equal is 6 kcal/mol higher in energy than the complex where these bonds were unconstrained. The long and short bonds exhibit values of 2.40 and 2.01 Å in the optimized structure. The binding energy of the $NHLi_2$ dimer is much larger than that of the NF_2Li dimer, 67.9 versus 49.1 kcal/mol (3-21G level). This loss of binding is attributed to the loss of negative charge on the nitrogen.

Another inorganic lithium compound whose dimer has been investigated in detail is LiOH. Before exploring its capacity for dimerization, Raghavachari[48] considered the hydration of the Li_2O molecule, which eventually leads to the dimer of LiOH. The Li_2O molecule is difficult to study with both experimental methods (since in the gas phase at high temperatures it dissociates into elements) and theoretical methods (since it is linear, with a low bending frequency and its surface of interaction with other molecules is flat). Only by using a large basis set such as 6-31G** with diffuse functions and correlation energy terms could reliable results be obtained. It was found that the interaction between Li_2O and H_2O initially proceeds via a lithium bond by forming the complex shown in Figure 2.5*b*. However, it rearranges with a very low activation barrier (2 kcal/mol) to the more stable D_{2h} structure shown in Figure 2.5*a*. The rearrangement potential energy surface was found to be extremely flat. It was concluded that the reaction between Li_2O and H_2O proceeds without any overall activation energy, leading to the formation of the LiOH dimer. The overall energy of reaction is -84.1 kcal/mol, which agrees very well with the experimental value of -83.6 kcal/mol.[53] Vibrational analysis carried out for all the investigated structures shows that both Li_2O and LiOH have low bending frequencies (134 and 467 cm^{-1}). Indeed, both these species were incorrectly considered bent in the early literature.[53,54]

The LiF dimer was examined recently[55] and the geometries of two configurations were optimized with the MP2/6-311+G* method, which used diffuse functions and polarization functions. The linear configuration shown in Figure 2.6*a* is considerably (28 kcal/mol) less stable than the ring structure shown in Figure 2.6*b*. However, both structures were found to be local minima by vibrational analysis. The bond between Li_1F_2 (1.784 Å) is longer than the other LiF (1.639 Å) bonds in Figure 2.6*a* and, therefore, it is a lithium bond. Conversely, in the D_{2h} dimer shown in Figure 2.6*b*, all the LiF bonds are equal, as is typical in a multicenter bonding situation. The binding energy defined as $[E(\text{complex}) - 2E(\text{LiF})]$ is 34.6 kcal/mol for structure *a* and 62.9 kcal/mol for structure *b*. It should be concluded that the dimer is present as a ring, similar to other small dimers such as $(LiOH)_2$ and $(LiNH_2)_2$.

Kaufmann et al.[56] systematically calculated the dimerization energies of all $LiXH_n$ dimers (XH_n = Li, BeH, BH_2, CH_3, NH_2, OH, F). They suggested that electrostatic interactions are dominant in determining the dimerization energy. They used a simple electrostatic model where Li and XH_n are replaced by point positive and negative charges at the optimized positions of the heavy atoms. The dimerization energy is then the increase in electrostatic energy on going from two monomers to a dimer. The overall excellent agreement of this

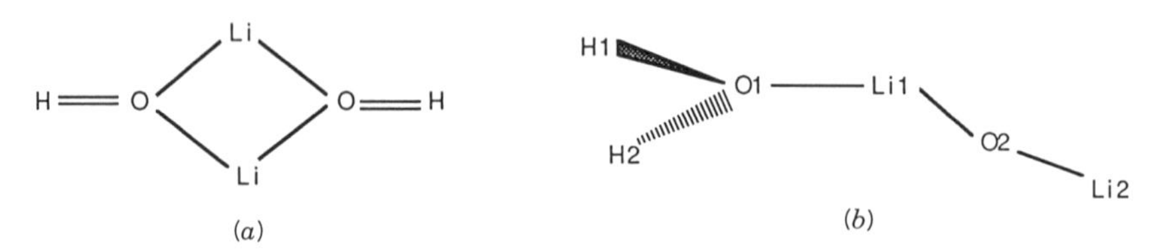

Figure 2.5

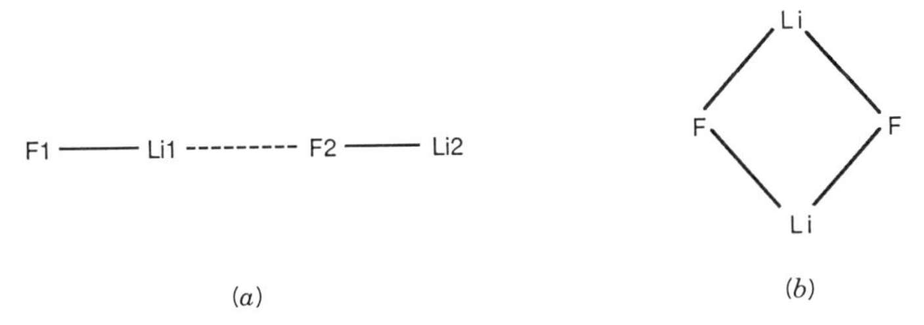

Figure 2.6

model suggested the dominance of the electrostatic interaction. Deviations seen for the dimers $LiNH_2$ and LiOH were explained by the participation of the lone pair on nitrogen or oxygen (covalent contributions) as discussed earlier.

Higher oligomers such as tetramers and hexamers of LiF, $LiNH_2$, and LiOH were studied by Raghavachari et al.[57] and by Sapse et al.[58] As mentioned earlier, organic tetramers of lithium compounds such as methyllithium feature a cubic arrangement of alternating lithium atoms and more electronegative atoms, a structure similar to the one found for methylsodium.[59,60] It has also been found, though in rarer occurrences, that metallic tetramers may adopt planar, eight-membered ring arrangements. Examples of such structures can be found in coinage metal[61] and in some mixed-metal derivatives involving lithium.[62-64] Lappert et al.[65] found a planar Li—N framework for the tetramer of $Li[NCMe_2(CH_2)_3CMe_2]$, while the tetramer of $LiN{=}C(C_6H_5)_2$ exhibits a cubic arrangement in the crystal.[66]

To investigate the structural preference of LiF, LiOH, and $LiNH_2$ tetramers, Sapse et al.[58] used a number of Gaussian basis sets to perform Hartree–Fock calculations. The geometries of LiF, LiOH, and $LiNH_2$ tetramers in tetrahedral and planar forms were optimized using the 3-21G basis set. The geometries obtained are shown in Figure 2.7. Some other lower-symmetry structures were also investigated but were found to be higher in energy. The effect of increasing the basis-set size was estimated by performing a 6-31G geometry optimization on the LiF tetramer followed by single-point computations with the 6-31G + *sp* + *d* basis set, which also includes diffuse *sp* functions and a set of *d* functions on F, O, and N. Diffuse functions are usually important in systems with unshared lone pairs of electrons. To estimate the effect of the inclusion of the correlation energy on the relative stability of different conformations, Moller–Plesset (MP2) calculations were also performed.

The study concluded that at all theory levels the most stable isomers are the T_d forms for the tetramers of LiF and LiOH and the D_{4h} tetramer for $LiNH_2$. The planarity of the Li—N framework is in agreement with the experimentally found structure of the tetramer of $Li[NCMe_2(CH_3)_3CMe_2]$. The computed geometrical parameters, such as the Li—N distance of 1.91 Å, the Li—N—Li angle of 104.0°, and the N—Li—N angle of 166° are in good agreement with the corresponding parameters of the larger compound (Li—N of 2 Å, Li—N—Li of 101.5°, and N—Li—N of 168.5°).

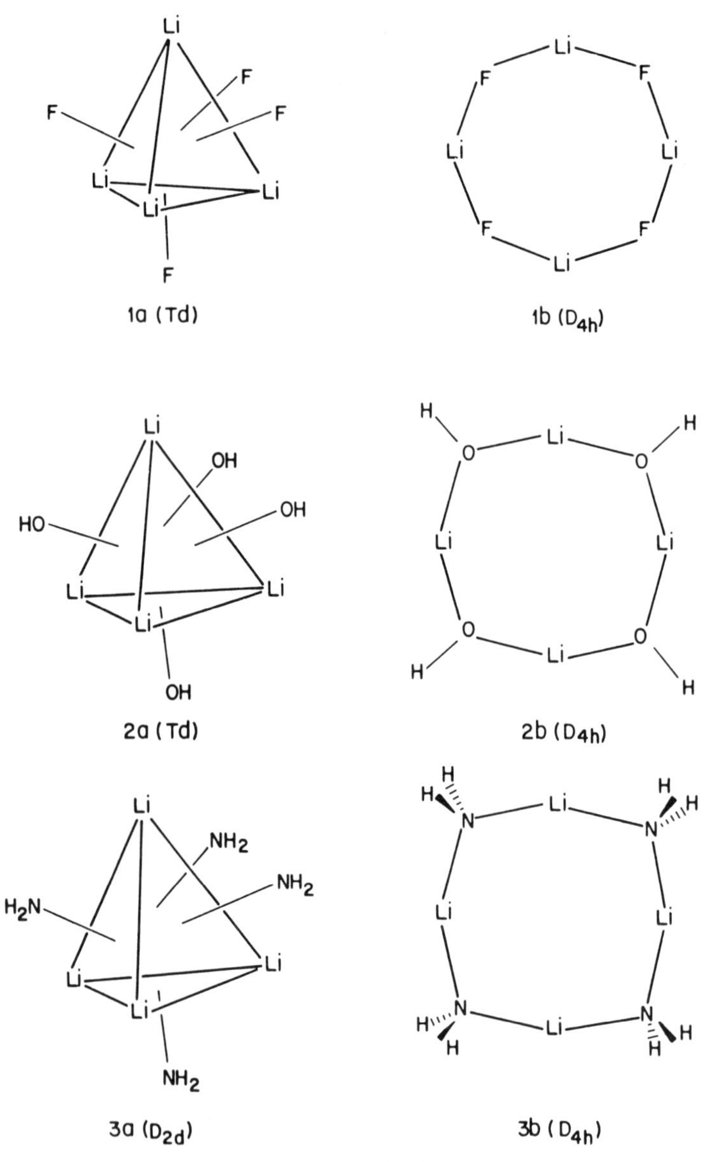

Figure 2.7

It was also found that the choice of basis set influences the magnitude of the differences between the energies of the T_d and D_{4h} structures. For instance, this difference for $(LiF)_4$ is 27.9 kcal/mol at the 3-21G level and only 8.6 kcal/mol at the 6-31G + *sp* + *d* level. This is due in part to the inadequacy of the small 3-21G basis set, where the tighter tetrahedral structure benefits from the basis-set superposition error. A smaller but similar effect is seen in the case of

Table 2.1 Binding Energies (kcal/mol) of the Tetramers at Different Levels of Theory

	$(LiF)_4$		$(LiOH)_4$		$(LiNH_2)_4$	
Symmetry	T_d	D_{4h}	T_d	D_{4h}	D_{2h}	D_{4h}
HF/3-21G	275.0	247.1	252.1	218.2	199.9	211.5
HF/6-31G	224.9	211.8	211.8	187.1	172.2	182.0
MP2/6-31G	221.1	207.4	206.6	181.6	172.3	181.6
MP3/6-31G	224.7	210.5	210.8	185.2	175.1	183.7
HF/6-31G + *sp* + *d*	187.8	179.2	189.8	173.6	158.4	167.3

LiOH. These effects are small for $(LiNH_2)_4$, showing that the 3-21G basis set is better for N. The correlation energy effects are not significant, as expected in such highly ionic compounds. Some of the results are summarized in Table 2.1.

Although the simple electrostatic model discussed earlier reproduced the dimerization energies of Li_xH_y compounds, it does not account for the energies of tetrameric structures. In such a model, the D_{4h} structure is found to be more stable than the T_d form in all cases, and the binding energies are overestimated. This appears to result because although the interactions between lithium and the electronegative atoms are mostly ionic, they also exhibit multicenter covalent character, which is neglected in the purely electrostatic model.

It is interesting to examine the degree of association in a highly ionic compound such as LiF and a lesser one such as $LiNH_2$. In LiF the disproportionation reaction between the trimer and tetramer is highly exothermic, and much less so in $LiNH_2$. This result agrees with the experimental findings on trimers of $LiNH_2$[46] in solution.

Higher oligomers, such as hexamers, have also been investigated with theoretical as well as experimental methods. For example, Frankel et al.[67] established on the basis of NMR line shape analysis that the state of aggregation of 2-methylbutyl lithium in hydrocarbon solvents is six and they suggested an octahedral geometry for it. Barr et al.[66] found hexameric forms for several iminolithium compounds and also suggested octahedral geometries. To determine the relative stabilities of octahedral versus planar conformations for gas phase oligomers of small lithium compounds, Raghavachari et al.[57] performed calculations on the hexameric structures of LiF, LiOH, and $LiNH_2$. Only the two basic structures, namely a distorted octahedral form (D_{3d}) and a hexagonal planar arrangement (D_{6h}), were considered.

Raghavachari et al. used the minimal STO-3G and the split-valence 3-21G basis sets to perform geometry optimizations for the two main structures of each of the three hexamers considered. The harmonic force constants and the associated vibrational frequencies were then evaluated, confirming that both structures were real minima. The frequency calculations showed that the D_{3d} structure has higher frequencies than the more loosely bound planar D_{6h} form.

Also, the zero-point energy of the former is 5 kcal/mol larger than that of the latter, leading us to conclude that the inclusion of zero-point energies modifies the D_{3d}–D_{6h} energy difference. Single-point calculations using larger basis sets such as 6-31 + *sp* + *d* were also performed.

It was found at all levels of theory and for all three compounds that the D_{3d}

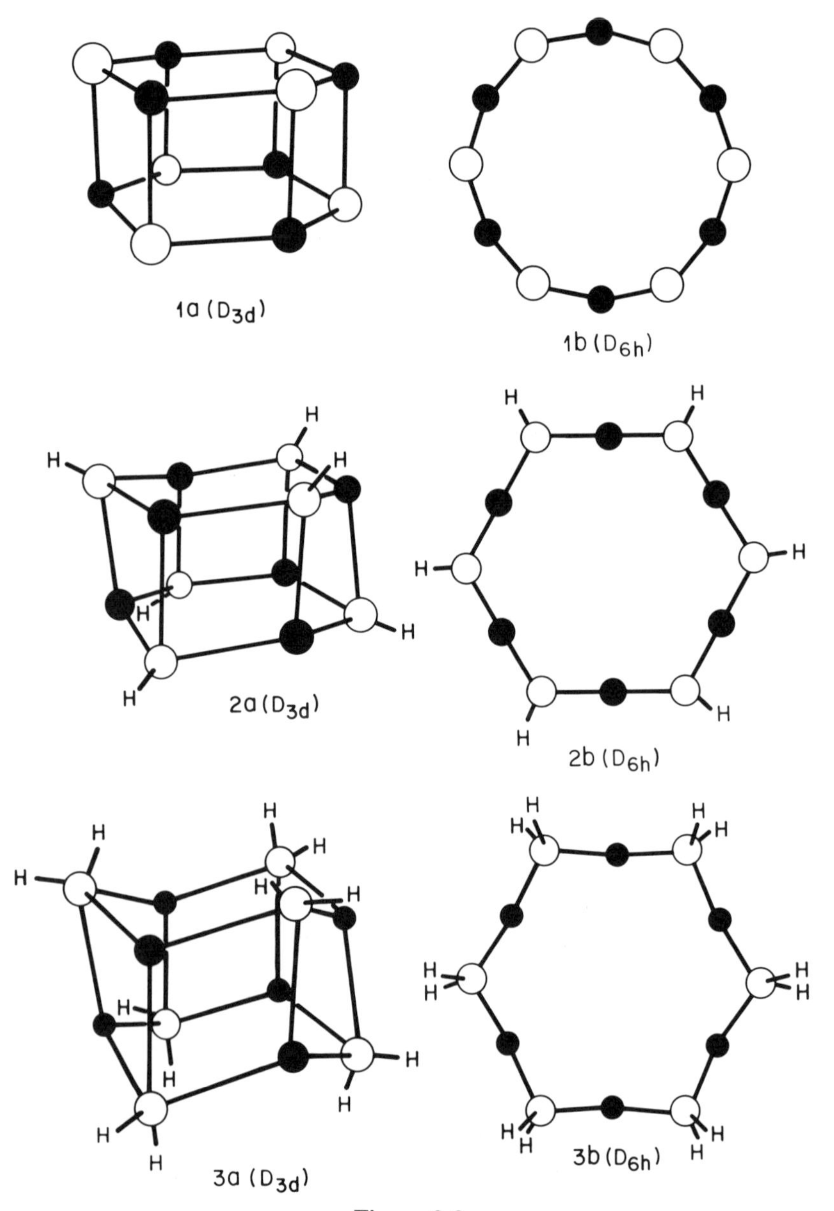

Figure 2.8

Table 2.2 Binding Energies (kcal/mol) of Different Oligomers at the HF/6-31G + *sp* + *d* Level of Theory

	Binding Energy per Monomer				
Structure	Dimer	Trimer	Tetramer	Hexamer	Solid[a]
$(LiF)_n$	32.5	42.0	47.0	52.1	65
$(LiOH)_n$	33.6	41.5	47.5	51.2	59
$(LiNH_2)_n$	32.8	40.3	41.8	45.4	

[a]Experimental values from reference 76. These values probably have an uncertainty of about 3–5 kcal/mol.

structures are more stable. These structures, which can be described as two interpenetrating octahedrons, are similar to the experimentally found structure of the LiN=C(Ph)NMe_2 hexamer investigated by Barr et al.[66] In the $LiNH_2$ hexamer, the theoretical calculations indicate the presence of two short (1.99 Å) and one long (2.06 Å) Li—N bonds, in agreement with the experimental values of 1.98, 2.01, and 2.05 Å. Calculated Li—Li distances are similar to the ones obtained experimentally by Zerger et al.[68] for the hexamer of cyclohexyllithium. The structures are illustrated in Figure 2.8.

As in the case of the tetramers, the relative energies are basis-set dependent, with the smaller basis sets overestimating the stability of the D_{3d} structure. The binding energies of LiF and LiOH species are similar (about 310 kcal/mol). Comparing these results to the binding energies of smaller oligomers, it can be seen that the binding energy per monomer increases as the size of the cluster increases. Table 2.2 shows these values and it is clear that there is quite a large difference between the energies of the hexamers and the experimental energy[69] of the solid, which is to be expected for these ionic molecules.

5 SOLVENT EFFECT

The calculations described above have been performed in the gas phase without taking the solvent effect into account. To estimate the effect of solvent on the dimerization energy of some lithium compounds, Sapse and Jain[70] use a modified Born model[71] which considers a molecule of solute placed in a cavity formed in the solvent. The solvent is treated as a continuum, characterized by its dielectric constant. The interaction between the solute and the solvent provides a stabilization energy. When the energy of stabilization for the dimer is compared to that for the monomer, taken twice, one can determine whether the solvent will favor or disfavor the dimerization. The stabilization or destabilization energy is defined as the difference between solvent–oligomer electrostatic energy and solvent–monomer interaction energies.

The compounds examined are the monomers and dimers of $LiNH_2$, $LiCH_3$, $LiNF_2$, Li_2NH, LiF, and $LiN(SiH_3)_2$. The description of these compounds is obtained from Hartree–Fock calculations, using Gaussian basis sets. In these calculations, a water molecule is used to model an ether, which is a typical solvent used in experimental studies.

Since the compounds investigated are not completely ionic, the Born equation cannot be used. Instead, a modified Born equation which takes into account dipoles and higher multipoles is applied. This equation writes the energy of interaction as

$$U = \frac{1}{2} \sum_{n=0}^{\infty} \frac{(n-1)(1-\epsilon)}{n+(n+1)\epsilon} \sum_{j=1}^{N} \sum_{k=1}^{N} Q_j Q_k \frac{(r_j r_k)^n}{a^{2n+1}} P_n(r_j r_k)$$

Here Q_j denotes the net charge on the jth atom of the solute, N is the total number of atoms, $P_n(r_j r_k)$ is the Legendre polynomial of degree n, r_j locates the jth atom in space, ϵ is the dielectric constant of the solvent, and a is the radius of the spherical cavity in which the solute is placed.

The positions of atoms were determined by geometry optimization employing the 6-31G* basis set except for Li_2NH and $LiNF_2$ for which the 3-21G basis set was used. The minimal basis sets underestimate charge separation and split-valence basis sets overestimate it; the 6-31G* basis set, which includes polarization functions on nonhydrogen atoms, is generally more reliable. Therefore, the net atomic charges obtained from the Mulliken population analysis with the 6-31G* basis set was used. The net atomic charges for Li_2NH and $LiNF_2$ were obtained by single-point 6-31G* calculations at 3-21G optimized geometry. The LiF monomers and dimers were optimized at the 6-31G* level. The hydrated $LiNH_2$ as well as the $LiN(SiH_3)_2$ monomer and dimer were optimized with the 6-31G basis set.

An examination of the geometries of the monomers and dimers under investigation revealed that a prolate spheroidal cavity in the solvent was the most appropriate one. If the molecule is encapsulated in a prolate spheroidal cavity, the interaction energy assumes the form

$$U = \frac{1}{2}\left[\frac{1}{\epsilon} - 1\right] \sum_{n=0}^{\infty} \sum_{m=0}^{n} \frac{F_n^m}{\Delta_n^m} [(\gamma_n^m)^2 + (\sigma_n^m)^2]$$

Here the semimajor and semiminor axes are denoted by a and b, respectively, $f = (a^2 + b^2)^{1/2}$, and $\xi_0 = a/f$. The other symbols are

$$\gamma_n^m = \sum_{j=1}^{N} Q_j P_n^m(\xi_j) P_n^m(\eta_j) \cos(m\phi_j)$$

$$\sigma_n^m = \sum_{j=1}^{N} Q_j P_n^m(\xi_j) P_n^m(\eta_j) \sin(m\phi_j)$$

$$F_n^m = \frac{1}{f}[2 - \delta_{m,o}](-1)^m(2n+1)\left[\frac{(n-m)!}{(n+m)!}\right]^2$$

$$\Delta_n^m = \frac{P_n^m(\xi_0)}{Q_n^m(\xi_0)} - \frac{1}{\epsilon}\frac{[P_n^m(\xi_0)]'}{[Q_n^m(\xi_0)]'}$$

In these equations, ξ_j, η_j, and ϕ_j are the prolate spheroidal coordinates of the jth ion, and $P_n^m(\xi_0)$ and $Q_n^m(\xi_0)$ are associated Legendre functions of the first and second kind, respectively.

The dimensions of the cavity were selected in the following manner: Values for the semimajor and semiminor axes a and b as well as the Euler angles defining the orientation of the cavity relative to the molecule are randomly selected. If the molecule fits the cavity, a search is made for the minimum surface area that would encapsulate the molecule. This procedure is carried out for a total of approximately 50,000 sets of randomly selected values of a and b, and the values that give the smallest surface area are obtained. This corresponds to minimizing the surface energy associated with the cavity formation. The distance from atoms to the walls of the cavity is kept at least equal to the value of their van der Waals radii. A general value of 1.5 Å has been used for the van der Waals radius. After the size of the cavity is determined, a summation over contribution from individual azimuthal indices m is carried out, since now the symmetry is axial and the energy of interaction will depend sensitively on the orientation of the molecule relative to the axis of the cavity.

Since monomers occupy a smaller cavity than dimers, the solvation energies for the monomers are almost always larger than those for the dimers. Two other reasons for this result are that the monomers come into closer contact with the solvent and that the dimers feature negligible dipole moments due to their symmetry. So their energy of interaction results mainly from higher multipoles. For less polar monomers such as $LiCH_3$ and $LiNF_2$, this effect is smaller. Therefore, the destabilization will occur primarily for polar solutes. Obviously, for less polar solvents, the destabilization of dimers versus monomers is much smaller. A solvent such as pentane will provide no destabilization. Thus it is expected that the monomer–dimer equilibrium will be shifted toward the dimer in less polar solvents. Kimura and Brown[43] found, through NMR studies, that lithium bis(trimethylsilyl)amide, for which $LiN(SiH_3)_2$ has been used as a model, in tetrahydrofuran exhibits an equilibrium constant

$$K = [\mathrm{M}]^2/[\mathrm{D}]$$

of 1.1 at a temperature of $-41°$C in dilute solution. In methylcyclohexane, at a similar temperature ($-45°$C), K decreases to 0.28. When the ratio of the equilibrium constant for $LiN(SiH_3)_2$ in furan and benzene is compared, it is found that

$$K_\mathrm{F}/K_\mathrm{B} = \exp(-\Delta H/RT)$$

where $\Delta H = H_{\mathrm{furan}} - H_{\mathrm{benzene}} = 1.83$ kcal/mol. Using 230 K as typical of the temperatures used by Kimura and Brown yields

$$K_\mathrm{F}/K_\mathrm{B} \approx 0.2$$

This value is similar to $0.28/1.1 = 0.25$, which was found by Kimura and Brown for similar solvents at similar temperatures. This calculation assumes no difference in entropy between the two solvents. Freezing point lowering and isopiestic methods used by Kimura and Brown show that lithium bis(trimethylsilyl)amide exists as a dimer in benzene at 23°C and exhibits a degree of dimerization of about 80% in ether, again showing the decrease of the dimerization by a more polar solvent.

The calculated gas-phase energies of dimerization are quite large (between 35 and 70 kcal/mol), indicating total dimerization. Several factors can explain why the monomer is also present in certain solvents:

1. Neglect of the entropy of solvation.
2. Neglect of the gas-phase entropy. Indeed, the entropy stabilizes the monomers versus the dimers, though the effect is likely to be large only at high temperatures.
3. The basis-set superposition error in the calculations which yields binding energies that are too large.

Another possible explanation for the difference between the gas phase and solvent when the solvent is able to bind through hydrogen bonds or "lithium bonds" (similar to hydrogen bonds) is the preferential stabilization of the monomer versus the dimer by this type of bond. For example, the $LiNH_2$–water complex $[LiNH_2(H_2O)]$ features a dimerization energy of 57 kcal/mol at the 6-31G level, which is smaller than the 72 kcal/mol obtained for $LiNH_2$. When the already hydrated complexes are placed in a cavity in water, the dimer is destabilized by an additional 8 kcal/mol, indicating that in solvents capable of direct bonding to the monomers and dimers, the continuum model is improved by considering the local interactions. Indeed, the 8 kcal/mol due to the continuum model, added to the 15 kcal/mol which is the difference between the dimerization energy of the hydrated and nonhydrated species, provides a destabilization energy of the dimer versus the monomer of 23 kcal/mol. The destabilization of the nonhydrated dimer due to the continuum model is only 16 kcal/mol. Therefore, no numerical predictions for the monomer–dimer equilibrium in solvents should be made using the gas phase values for the monomer–dimer equilibrium. However, comparing different solvents is possible, as shown by the results concerning $LiN(SiH_3)_2$ in furan and benzene.

When spherical instead of prolate spheroid cavities are used the following pattern emerges: For the monomers which can be adequately described in spherical terms (such as $LiCH_3$ and $LiNF_2$), there is no significant difference between the results obtained by using spherical or prolate cavities. For $LiNH_2$ and Li_2NH, which are planar, the spherical cavity does not describe the monomers adequately and thus the dimerization energies are overestimated in comparison with the ones obtained using prolate cavities. In the case of the LiF molecule, the results obtained with spherical and prolate cavities do not have any significant difference. When trends are examined, the results obtained with spherical and prolate cavities are in general agreement.

6 CONCLUSION

This chapter provided a brief overview of some of the recent theoretical studies on the oligomers of lithium compounds. It is clear that the geometrical parameters and binding energies of the gas phase oligomers agree quite well with the experimental results. However, solvation effects are very important for describing equilibria between oligomers in solution.

REFERENCES

1. Wakefield, B. J. *The Chemistry of Organometallic Compounds*; Pergamon Press: Oxford, 1974.
2. Brown, T. L.; Gerteis, R. L.; Bafus, D. A.; Ladd, J. A. *J. Am. Chem. Soc.* **1964,** *86*, 2135.
3. Brown, T. L. *Acc. Chem. Res.* **1968,** *1*, 23.
4. Brown, T. L. *Adv. Organomet. Chem.* **1965,** *3*, 365.
5. Brown, T. L.; Ladd, J. A.; Newman, G. N. *Adv. Organomet. Chem.* **1965,** *3*, 1.
6. Margerison, D.; Newport, J. P. *Trans. Faraday Soc.* **1963,** *59*, 2058.
7. Bywater, S.; Worsfold, D. J. *Adv. Organomet. Chem.* **1967,** *10*, 1.
8. Weiner, M.; Vogel, C.; West R. *Inorg. Chem.* **1962,** *2*, 654.
9. Lewis, H. L.; Brown, T. L. *J. Am. Chem. Soc.* **1970,** *92*, 4664.
10. Morton, M.; Fetters, L. J.; Pett, R. A.; Meier, J. F. *Macromolecules* **1970,** *3*, 327.
11. Dietrich, H. *Acta Cryst.* **1963,** *16*, 681.
12. Weiss, E.; Hencken, G. *J. Organomet. Chem.* **1970,** *21*, 265.
13. Weiss, E.; Lucken, E. A. C. *J. Organomet. Chem.* **1964,** *2*, 197.
14. Darensbourg, M. Y.; Kimura, B. Y.; Hartwell, G. E.; Brown, T. L. *J. Am. Chem. Soc.* **1970,** *92*, 1236.
15. Hartwell, G. E.; Brown, T. L. *Inorg. Chem.* **1966,** *5*, 1257.
16. Berkowitz, J.; Bafus, D. A.; Brown, T. L. *J. Phys. Chem.* **1961,** *65*, 1380.
17. Craubnerm, I. *Z. Phys. Chem.* **1966,** *51*, 15.
18. Chinn, Jr., J. W.; Lagow, R. J. *J. Organomet.* **1984,** *3*, 75.
19. *Hydrogen Storage Materials*; Barnes, R. G. Ed.; Trans Tech: Zurich, 1988.
20. Switendick, A. C. *Z. Phys. Chem. Neue Folge* **1979,** *117*, 89.
21. Westlake, D. G.; Shaked, H.; Mason, P. R.; McCart, B. R.; Mueller, M. H.; Matsumoto, T.; Amano, M. *J. Less Common Met.* **1982,** *88*, 17.
22. Rao, B. K.; Jena, P. *Phys. Rev. B*, **1985,** *31*, 6726.
23. Dill, J. D.; v. R. Schleyer, P.; Binkley, J. S.; Pople, J. A. *J. Am. Chem. Soc.* **1977,** *99*, 6159.
24. Kato, H.; Hirao, K.; Akagi, K. *Inorg. Chem.* **1981,** *20*, 3659.
25. Rao, B. K.; Khanna, S. N.; Jena, P. *Phys. Rev. B* **1991,** *43*, 1416.
26. Zintl, E.; Harder, A. *Z. Phys. Chem. Abt. B* **1975,** *28*, 478.

27. Herzberg, G. *Molecular Spectra and Molecular Structure. I. Diatomic Molecules*; Prentice Hall: New York, 1939; p. 489.
28. Novoa, J. J.; Whangbo, M. H.; Stucky, G. D. *J. Org. Chem.* **1991,** *56*, 3181.
29. (a) Rhine, W. E.; Stucky, G. D.; Pertersen, S. W. *J. Am. Chem. Soc.* **1975,** *97*, 6401. (b) Stucky, G. D.; Eddy, M. M.; Harrison, W. H.; Lagow, R.; Cox, D. E. *J. Am. Chem. Soc.* **1990,** *112*, 2425.
30. Talbi, D.; Saxon, R. P. *Chem. Phys. Lett.* **1989,** *157*, 419.
31. Streitwieser, Jr., A.; Williams, J. E.; Alexandrotos, S.; McKelvey, J. M. *J. Am. Chem. Soc.* **1976,** *98*, 4778.
32. Michels, H. H.; Wadehra, J. M. *Int. J. Quantum Chem.: Quantum Chem. Symp.* **1990,** *24*, 521.
33. Prelec, K., Ed. *Proceedings of the Third International Symposium on the Production and Neutralization of Negative Ions and Beams*, AIP Conference Proceedings 111; AIP: New York, 1984.
34. Cardelino, B. H.; Eberhardt, W. H.; Borkman, R. F. *J. Chem. Phys.* **1986,** *84*, 3230.
35. Schlenk, W.; Holtz, J. *J. Ber. Dtsch. Chem. Ges.* **1917,** *50*, 262.
36. Landro, F. J.; Gurak, J. A.; Chinn, Jr., J. W.; Lagow, R. J. *J. Organomet. Chem.* **1983,** *249*, 1.
37. Dietrich, H. *J. Organomet. Chem.* **1981,** *205*, 291.
38. Baird, N. C.; Barr, R. F.; Datta, R. K. *J. Organomet. Chem.* **1973,** *59*, 65.
39. Graham, G. D.; Marynick, D. S.; Lipscomb, W. N. *J. Am. Chem. Soc.* **1980,** *102*, 4572.
40. Schiffer, H.; Ahlrichs, R. *Chem. Phys. Lett.* **1986,** *124*, 172.
41. Kaufmann, E.; Raghavachari, K.; Reed, A. E.; v. R. Schleyer, P. *Organometallics* **1988,** *7*, 1597.
42. Herzig, L.; Howell, J. M.; Sapse, A. M.; Singman, E.; Snyder, G. *J. Chem. Phys.* **1982,** *77*, 429.
43. Kimura, B. Y.; Brown, T. L. *J. Organomet. Chem.* **1971,** *26*, 57.
44. Mootz, D.; Zinnius, A.; Bottcher, B. *Angew. Chem.* **1969,** *81*, 398.
45. Wurthwein, E.-U.; Sen, K. D.; Pople, J. A.; v. R. Schleyer, P. *Inorg. Chem.* **1983,** *22*, 496.
46. Sapse, A. M.; Kaufman, E.; v. R. Schleyer, P.; Gleiter, R. *Inorg. Chem.* **1984,** *23*, 1569.
47. MNDO; Dewer, M. J. S.; Thiel, W. *J. Am. Chem. Soc.* **1977,** *99*, 4899, 4907.
48. Raghavachari, K. *J. Chem. Phys.* **1982,** *76*, 5421.
49. (a) Rogers, R. D.; Atwood, J. L.; Gruning, R. J. *J. Organomet. Chem.* **1978,** *157*, 229. (b) For the X-ray structure of $[(PhCH_2)_2NLi]_3$ see Barr, D.; Clegg, W.; Mulvey, R. E.; Snaith, R. *J. Chem. Soc., Chem. Commun.* **1984,** 285.
50. Boys, S. F.; Bernardi, F. *Mol. Phys.* **1970,** *19*, 553.
51. (a) Collins, J. B.; Streitwieser, Jr., A. *J. Comput. Chem.* **1987,** *1*, 81. (b) Klein, J.; Kost, D.; Schriver, W. G.; Streitwieser, Jr., A. *Proc. Natl. Acad. Sci. USA* **1982,** *79*, 3922.
52. Sapse, A. M.; Jain, D. C. *Chem. Phys. Lett.* **1984,** *110*, 251.

53. Berkowitz, J.; Meschi, D. J.; Chupka, W. A. *J. Chem. Phys.* **1960,** *33*, 533.
54. Berkowitz, J.; Chupka, W. A.; Blue, G. D.; Margrave, J. L. *J. Phys. Chem.* **1959,** *63*, 644.
55. Sapse, A. M., unpublished results.
56. Kaufman, E.; Clark, T.; v. R. Schleyer, P. *J. Am. Chem. Soc.* **1984,** *106*, 1856.
57. Raghavachari, K.; Sapse, A. M.; Jain, D. C. *Inorg. Chem.* **1987,** *26*, 2585.
58. Sapse, A. M.; Raghavachari, K.; v. R. Schleyer, P.; Kaufman, E. *J. Am. Chem. Soc.* **1985,** *107*, 6483.
59. Review of the X-ray structures of lithium compounds: Setzer, W.; v. R. Schleyer, P. *Adv. Organomet. Chem.* **1985,** *24*, 353.
60. Weiss, E.; Sauermann, G.; Thirase, G. *Chem. Ber.* **1983,** *116*, 74.
61. Gambarotta, S.; Floriani, C.; Chiesi-Villa, A.; Gaustini, G. *J. Chem. Soc., Chem. Commun.* **1983,** 1087–1089, 1156–1158 and literature cited therein.
62. (a) van Koten, G.; Noltes, J. G. *Comp. Organomet. Chem.* **1982,** *2*, 709. (b) van Koten, G.; Jastrzebski, D. B. H.; Stamm, C. H.; Neimann, N. C. *J. Am. Chem. Soc.* **1984,** *106*, 1880.
63. van Koten, G.; Jastrzebski, D. B. H.; Muller, F.; Stamm, C. H. *J. Am. Chem. Soc.* **1985,** *107*, 697.
64. Stewart, K. R.; Lever, J. R.; Whangbo, M.-H. *J. Org. Chem.* **1982,** *47*, 1472.
65. Lappert, M. F.; Slade, M. J.; Singh, A.; Atwood, J. L.; Rogers, R. D.; Shakir, R. *J. Am. Chem. Soc.* **1983,** *105*, 302.
66. Barr, D.; Clegg, W.; Mulvey, R. E.; Snaith, R. *J. Chem. Soc., Chem. Commun.* **1984,** 79–80.
67. Frankel, G.; Beckenbaugh, W. E.; Yang, P. P. *J. Am. Chem. Soc.* **1976,** *98*, 6879.
68. Zerger, R.; Rhine, W.; Stucky, G. *J. Am. Chem. Soc.* **1974,** *96*, 6048.
69. Karapetyants, M. K.; Karapetyants, M. L. *Thermodynamic Constants of Inorganic and Organic Compounds*; Ann Arbor: London, 1970.
70. Sapse, A. M.; Jain, D. C. *J. Phys. Chem.* **1987,** *91*, 3923.
71. Gersten, J. I.; Sapse, A. M. *J. Am. Chem. Soc.* **1985,** *107*, 3786.

3

COMPARISON OF LITHIUM AND HYDROGEN BONDS

STEVE SCHEINER
Department of Chemistry & Biochemistry
Southern Illinois University
Carbondale, Illinois

1 INTRODUCTION

Lithium is similar to hydrogen in that it has one valence electron, but differs in that it contains an inner core of two electrons while hydrogen does not. Their valence electronic structure permits either atom to form a single covalent bond to a neighboring atom as in H—F or Li—F. Hydrogen has been known for decades to participate in a secondary form of interaction known as hydro-

Lithium Chemistry: A Theoretical and Experimental Overview, Edited by Anne-Marie Sapse and Paul von Ragué Schleyer.
ISBN 0-471-54930-4

gen bonding.[1-3] A hydrogen bond is formed when the hydrogen is covalently bonded to an electronegative atom such as O or F. When approached by another molecule in which an electronegative atom contains a free pair of electrons, the H atom acts as a sort of bridge, as in X—H· · ·X—H. The hydrogen atom usually positions itself right along the direction of the electron donor's lone pair in such a way that it also lies directly along the X· · ·X axis.

While this sort of interaction has been well known for some time,[1-3] the analogous situation wherein Li takes the place of H is much less common. The first suggestion of this idea appeared in the literature shortly before 1960.[4] Following a theoretical inquiry by Kollman, Liebman, and Allen in 1970,[5] the first unambiguous observation of a Li bond was made in 1975 by Ault and Pimentel[6] using spectroscopic measurements of LiCl and LiBr combined with nitrogen bases in matrix isolation. The Li—X stretching frequency was found to shift in a direction similar to that observed in H bonds, but by a much smaller amount. Another parallel was drawn to the possibility that the bridging Li, like the proton in a H bond, is capable of transferring from one group to the other. This possibility was indicated by a minimum in the curve which plots the relative frequency change versus a normalized difference in proton affinity of the two groups.[6]

Information concerning the structural and energetic aspects of the lithium analogue to the H bond has been sparse. For this reason, the major fraction of currently available data has been derived from quantum chemical calculations. Many of these calculations have focused on binary complexes in which each monomer contains a Li atom. In such situations, it is quite common to see a bridge formed of both Li atoms like

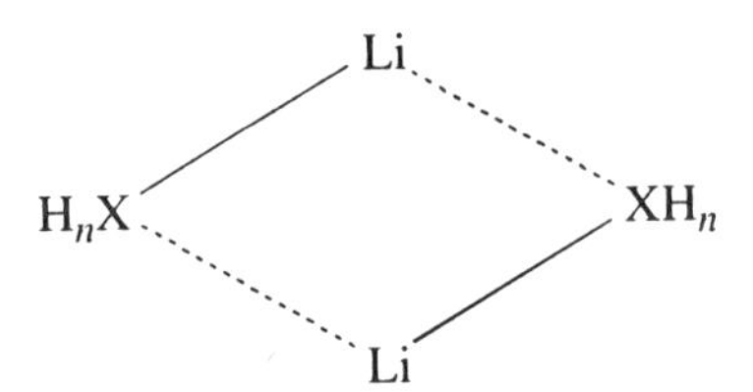

Examples of systems where this type of geometry is preferred include the dimers of LiH,[7] $LiBH_2$,[8] $LiCH_3$,[8] $LiNH_2$,[7b-9] LiOH,[8,10] LiF,[7b,8,11,12] LiCN,[13] and $LiNF_2$.[14] Whereas these geometries are predicted by ab initio methods, the experimental information[15,16] cannot conclusively confirm this sort of bonding, but is instead limited to interaction energies which may be compared to the calculated data. Analogous "cyclic" H bonds are comparatively rare. A cyclic geometry is likely for the ammonia dimer,[17] but it is questionable whether one would refer to this structure as a true H bond. Moreover, the cyclic geometry appears not to represent a true minimum on the potential energy surface of dimers of other simple molecules such as H_2O or HF.[18,19]

When only one of the two molecules involved in the interaction contains a Li atom, a symmetric cyclic arrangement is not possible. In such cases the Li tends to locate itself directly along the line connecting the donor and acceptor atoms, forming a structure akin to the more traditional linear H bonds.[20–26] Examples include complexes of NH_3 with LiH, LiBeH, $LiCH_3$, $LiNH_2$, LiOH, LiF, and LiCl as well as a number of similar complexes with NH_3 replaced by H_2O.

This sort of Li bond, analogous to the conventional linear H bond, represents the subject of this chapter. We first compare in some detail the geometric and energetic aspects of the two types of interactions, focusing our attention on a small set of complexes. The next point of comparison refers to the spectroscopic characteristics of hydrogen and lithium bonds. Not only do the vibrational spectra provide a useful point of contact with experiment, but perturbations caused by the complexation allow one to probe the accompanying electronic redistributions. One of the more fascinating aspects of hydrogen bonding is the possibility of transferring the bridging hydrogen from the donor molecule to the acceptor.[27–41] A similar possibility was raised in the original paper by Ault and Pimentel.[6] For this reason, the last section of this chapter considers the possibility of transferring the Li center from one molecule to the other and contrasts this behavior with what is known of proton transfer.

2 GEOMETRIES AND ENERGETICS

Kollman, Liebman, and Allen[5] were the first to examine the possibility of Li bonding from a theoretical perspective. Computer limitations in 1970 forced them to restrict themselves to the Hartree–Fock level with only a modest basis set. Nonetheless, they were able to arrive at a number of qualitative conclusions that remain valid. By examining the various conformations possible for the pair HF and LiF, they were able to contrast the H bonding in LiF· · ·HF with the Li bond in HF· · ·LiF. Also examined were cyclic geometries wherein both types of bonds are present, although they are nonlinear. The interfluorine distance was computed to be some 1.2 Å longer in the Li-bridged structure, which was more weakly bound by a factor of about 2. The cyclic geometry was found to be similar in stability to the Li-bonded structure. When they examined the LiF dimer, the cyclic geometry was clearly more stable than the linear arrangement containing a single Li bond.

A survey of small complexes containing either a hydrogen or lithium as bridging atom is compiled in Table 3.1. Perusal of the data makes it immediately apparent that the two types of bonds follow very different patterns. The strength of each H bond is highly sensitive to the nature of the proton donor, increasing as the donor becomes more acidic. Taking the water molecule as proton acceptor, the strength of the interaction varies from less than 1 kcal/mol when methane is donor to 11 kcal/mol for FH. At the same time, the separation between the bridging proton and the acceptor O atom diminishes from 2.67 to 1.72 Å. The contrast with the Li bonds in the lower half of Table 3.1 is striking. The binding shows almost no sensitivity to the nature of the Li

Table 3.1 Comparison of Calculated Properties of H and Li Bonds

Complex	R(H· ·N) Å	$-\Delta E$ (kcal/mol)	Method	Ref.
$H_3CH\cdot\cdot NH_3$	2.9	0.6	MP4/[532/32]	42
$H_2NH\cdot\cdot NH_3$	2.33	4.0	MP2/6-31G(2d1p)	17
$HOH\cdot\cdot NH_3$	1.973	7.0	MP2/+VP^M	17
$FH\cdot\cdot NH_3$	1.743	15.1	MP2/6-31G(2d1p)	23
$ClH\cdot\cdot NH_3$	1.827	11.0	MP2/6-31G(2d1p)	23
	R(H· ·O)			
$H_3CH\cdot\cdot OH_2$	2.67	0.6	MP4/[532/32]	43
	2.672	0.5	MP2/6-31++G(2d2p)	44
$H_2NH\cdot\cdot OH_2$	2.208	4.1	SCF/4-31G	45
$HOH\cdot\cdot OH_2$	2.083	4.4	MP2/DZP′	46
$FH\cdot\cdot OH_2$	1.716	11.0	MP2/6-311G**	47
	R(Li· ·N)			
$H_3CLi\cdot\cdot NH_3$	2.076[a]	23.6	MP2/6-31 + G*	22
$H_2NLi\cdot\cdot NH_3$	2.083	23.3	MP2/6/31 + G*	22
$HOLi\cdot\cdot NH_3$	2.087	23.7	MP2/6-31 + G*	22
$FLi\cdot\cdot NH_3$	2.073	23.5	MP2/6-31G(2d1p)	23
$ClLi\cdot\cdot NH_3$	2.036	26.7	MP2/6-31G(2d1p)	23
	R(Li· ·O)			
$H_3CLi\cdot\cdot OH_2$	1.949[b]	19.8	MP2/6-31 + G*	22
$H_2NLi\cdot\cdot OH_2$	1.941	19.6	MP2/6-31 + G*	22
$HOLi\cdot\cdot OH_2$	1.952	19.5	MP2/6-31 + G*	22
$FLi\cdot\cdot OH_2$	1.954	20.9	MP2/6-31 + G*	22

[a] Li is nearly midway between C and N : R(C· ·Li) = 2.026 Å.
[b] Li is nearly midway between C and O : R(C· ·Li) = 2.020 Å.

donor. For example, the interaction energy with water is right around 20 kcal/mol for all donors varying from H_3CLi to FLi. The intermolecular distance is quite constant as well, remaining at 1.94–1.95 Å for all four complexes with water.

Inspection of the data also reveals that the Li bonds are considerably stronger than the hydrogen analogues as a general rule. This discrepancy is most obvious with regard to the weak donors. Whereas the interaction energies involving H_2NH as proton donor are some 4 kcal/mol, this property increases to 20 kcal/mol or higher for H_2NLi. The greater strength of the Li bonds was confirmed in an earlier study by Kulkarni and Rao, who limited themselves to a minimal basis set.[48] Their calculations further suggested that Na bonds are of intermediate strength.

Another interesting difference involves the geometry of the weakest donor of a bridging lithium atom, H_3CLi. In its complex with either NH_3 or OH_2, the Li atom is situated very nearly midway between the C atom and the N or O center. In contrast, the proton of H_3CH is hardly stretched at all from its position close to the C, lying nearly 3 Å from the oxygen of OH_2.

One caveat should be issued at this point. The flatness of the potential energy surfaces for complexes of this type can cloud the identification of the true minima. For example, Sapse and Raghavachari[49] recently observed that a nearly linear arrangement of $H_2NLi\cdot\cdot OH_2$ is indeed a minimum on the surface. However, even more stable by several kilocalories per mole is a cyclic geometry containing both a bent H and Li bond. In fact, the latter complex can transfer nearly simultaneously the H and Li centers to form the even more stable $H_3N\cdot\cdot LiOH$ geometry. A similar picture emerges for $H_2O\cdot\cdot LiF$, where the competitive stabilities of the cyclic and linear arrangements can depend upon whether basis set superposition is corrected and the particular level of electron correlation applied to the problem.[26] A cyclic geometry appears to be a minimum on the surface of $H_2O\cdot\cdot LiOH$ as well.[26]

Table 3.2 provides more detailed information about the geometric, energetic, and electronic aspects of a particular set of complexes chosen to highlight the fundamental differences between hydrogen and lithium bonds. Ammonia, as a prototype base, is combined with HF or HCl and with LiF or LiCl. Calculations are performed both at the SCF and MP2 levels so as to gauge the contribution of electron correlation to these two types of bonds. The first row of data in Table 3.2 refers to the distance between heavy atoms N and X (F or Cl). The Li bonds are uniformly 1 Å longer than the H bonds, consistent with the presence of a 1s core of two electrons on the Li center. Correlation reduces the length of the H bond in either complex whereas the MP2 Li bond is longer than the SCF value.

Of course the LiF or LiCl bond distance is longer than is HF or HCl. What is more interesting is the change in this length upon formation of the complex with NH_3. As illustrated by the values reported in Table 3.2 for Δr, formation of the complex causes the bond to elongate by 0.02–0.04 Å. As in the case of $R(N\cdot\cdot X)$, correlation has the opposite effect upon the H and Li bonds. The MP2 bond elongations are considerably larger for the H bonds than the SCF stretches; the Li bond stretch is smaller when correlated.

The next rows of Table 3.2 report the calculated energetics of formation of the various complexes. The Li bonds are evidently much stronger than the H bonds, with ΔE values exceeding 20 kcal/mol, as compared to 10–15 for the two H bonds. Whereas the H bonds are weakened upon changing the acid from HF to HCl, the opposite is noted for the Li bonds, where LiCl forms a stronger complex than does LiF. While correlation is responsible for a substantial increase in the binding energy of the H bonds, its contribution to the Li bond is minimal.

The final two rows in Table 3.2 illustrate the much larger dipole moments of the Li-bonded complexes, as compared to the H bonds. On the other hand,

Table 3.2 Calculated Properties of H and Li Bonds with NH_3[a]

	H_3N-HF		H_3N-HCl		H_3N-LiF		$H_3N-LiCl$
	SCF	MP2	SCF	MP2	SCF	MP2	SCF
R(N· ·X) (Å)	2.728	2.693	3.297	3.144	3.652	3.665	4.118
r(XZ) (Å)	0.922	0.950	1.293	1.317	1.582	1.592	2.082
Δr(XZ) (Å)[b]	0.022	0.028	0.023	0.040	0.018	0.014	0.020
ΔE^{SCF} (kcal/mol)	−11.84	−11.80	−9.29	−6.59	−22.89	−22.81	−25.46
ΔE^{MP2} (kcal/mol)	−14.86	−15.09	−10.42	−11.03	−23.48	−23.52	−26.65
μ^{SCF} (D)	4.39	4.56	3.90	4.30	8.74	8.79	10.16
$\Delta\mu^{SCF}$ (D)[c]	0.97	1.12	1.13	1.52	0.91	0.90	1.14

[a]X=F, Cl; Z=H, Li. Data from reference 23.
[b]$\Delta r = r(\text{complex}) - r(\text{isolated subunit})$. [c]$\Delta\mu = \mu(\text{complex}) - \Sigma\mu(\text{isolated subunits})$.

much of this greater moment is a simple result of the higher dipole of LiX in comparison to the HX molecule. The last row shows that the *change* in moment caused by complexation, that is, the difference between the moment of the complex and the sum of dipoles of the two isolated monomers, is quite comparable for the two types of complexes.

In capsule form, Li bonds are stronger than H bonds. Part of this greater strength derives from the larger dipole moment of the LiF and LiCl monomers, which acts to increase the electrostatic component of the interaction energy. Indeed, the ultimate source of the Li bond is largely electrostatic in nature. While the Coulombic energy of H bonds is important, there are other terms that make very significant contributions as well. One of these auxiliary terms is dispersion, absent at the SCF level, and one of the forces responsible for the contraction of the R(N· ·X) H-bond length. A second factor is that correlation tends to reduce the dipole moment of most molecules.[50–52] The resulting lowered electrostatic attraction shows up in the calculations as a correlation-induced "repulsion." Although the latter effect is outweighed by dispersion attraction for H bonds, there is a more even balance for the Li bonds, explaining the near absence of correlation effects on the latter interaction. Another point of distinction emphasized by Sannigrahi et al.[26] is that whereas the R(X· ·Y) distance in H bonds of the general type XH· ·Y is typically shorter than the sum of van der Waals radii of X and Y, the reverse is true in Li bonds.

3 SPECTROSCOPIC FEATURES

Another format for examining some of the fundamental properties of the Li bond is through infrared spectroscopic data. Earlier work highlighted several glaring distinctions with the properties of the H bond.[6] Whereas the ν_s stretching mode is red shifted and greatly intensified when a hydrogen bond is formed, little of this behavior is seen when Li is substituted for the bridging hydrogen. To determine the reasons for these differences, and to probe more deeply other features of the electronic rearrangements that accompany motion of the nuclei, ab initio calculations were performed on the H_3N· ·LiCl system, which was then compared to H_3N· ·HCl.[53]

Table 3.3 lists the vibrational spectral data computed using the 6-31G** basis set and assuming a purely harmonic vibrational force field. Conforming to common notation, ν_σ refers to oscillations in the intermolecular separation, approximately R(N· ·Cl). These frequencies are larger for the Li- than for the H bonds at either the SCF or MP2 level, consonant with the stronger nature of the Li bond. Also consistent with the previous finding of a much larger contribution by correlation to the H bond, the MP2 value of ν_σ is 16% larger than the SCF result for H_3N· ·HCl, whereas the SCF and MP2 frequencies are very similar for the Li bond.

The rocking of the H_3N molecule corresponds to the ν_β motion. This fre-

Table 3.3 Calculated Intermolecular Vibrational Frequencies and Intensities of H and Li Bonds[a]

	$H_3N\cdot\cdot HCl$		$H_3N\cdot\cdot LiCl$	
	SCF	MP2	SCF	MP2
	Frequencies (cm^{-1})			
ν_σ	173	200	248	238
ν_β	239	276	525	499
ν_b	700	912	56	70
	Intensities (km/mol)			
ν_σ	10	43	24	10
ν_β	46	44	550	208
ν_b	150	114	80	22

[a]Data from Reference 53.

quency is nearly twice as large for the Li bond as compared to $H_3N\cdot\cdot HCl$, again in agreement with the stronger nature of the former. Bending about the central (H or Li) nucleus is associated with ν_b. The much larger frequency for the H bond is connected with the smaller mass of a proton as compared to the Li nucleus. Inclusion of correlation has a profound effect upon the value computed for this frequency.

Similar comparisons can be drawn between the two types of bonds with respect to intramolecular frequencies, listed in Table 3.4. The frequencies of the unperturbed monomers are presented in the first pair of columns, followed in order by the H and Li bonds. The first row illustrates the strong red shift of the HCl stretching frequency resulting from formation of the H bond. The ν_s frequency drops by some 400 cm^{-1} at the SCF level and 900 cm^{-1} at MP2. In striking contrast, the LiCl frequency follows an opposite trend and increases by a small amount upon complexation with H_3N. With regard to the vibrations within the H_3N molecule, the a_1 and e stretches are virtually unaffected by formation of the H bond but undergo small decreases when combined with LiCl. The a_1 bending frequency climbs a little upon complexation with HCl, and even more in the Li bond. Small decreases are noted upon formation of either bond in the $\nu_b(e)$ mode.

Vibrational intensities offer a valuable opportunity to examine the way in which the electronic distributions of each molecule are affected by formation of a complex. Considering first the intermolecular modes, inspection of Table 3.3 indicates that correlation affects these intensities a great deal. For example, whereas the MP2 value of the intensity of the ν_σ band is four times larger than the SCF value for the H bond, the reverse is true in $H_3N\cdot\cdot LiCl$, where correlation reduces the intensity to less than half. Hence, SCF and MP2 would lead to opposite conclusions as to whether the ν_σ band is more intense in the

Table 3.4 Calculated Intramolecular Vibrational Spectra of $H_3N\cdot\cdot HCl$ and $H_3N\cdot\cdot LiCl$[a]

	$H_3N + ZCl$		$H_3N\cdot\cdot HCl$		$H_3N\cdot\cdot LiCl$	
	SCF	MP2	SCF	MP2	SCF	MP2
	Frequencies (cm^{-1})					
ν_s (Z=H)	3164	3084	2740	2186	—	—
ν_s (Z=Li)	637	623	—	—	710	716
ν_{st} (a_1)	3706	3568	3706	3565	3693	3558
ν_{st} (e)	3843	3726	3843	3723	3817	3700
ν_b (a_1)	1143	1116	1217	1187	1319	1247
ν_b (e)	1812	1727	1805	1713	1800	1712
	Intensities (km/mol)					
ν_s (Z=H)	120	46	1087	2283	—	—
ν_s (Z=Li)	135	123	—	—	331	111
ν_{st} (a_1)	0.15	0.3	4	6	26	11
ν_{st} (e)	1.5	1.0	22	36	166	60
ν_b (a_1)	217	170	200	119	574	191
ν_b (e)	42	32	56	52	156	50

[a]Data from reference 53.

H or Li bond. At the correlated (or SCF) level, it seems clear that the ν_β vibration is more intense in the Li bond and ν_b in the H bond.

Turning now to the intramolecular vibrations, it is well known that the ν_s band is intensified by formation of a hydrogen bond, which is amply reproduced by the intensities reported in the lower half of Table 3.4. At the SCF level, ν_s is increased by a factor of 9 upon H-bond formation; a factor of 50 is noted at the MP2 level. In contrast, the same band in the Li bond is strengthened by only 2.5 at the SCF level and is slightly weakened after inclusion of correlation. Other intramolecular modes are also heavily influenced by the complexation. The very weak $\nu_{st}(a_1)$ stretch in isolated NH_3 is strongly enhanced in the H bond and even more so in $H_3N\cdot\cdot LiCl$, as is the other stretching motion. On the other hand, the two bending modes in NH_3 do not seem much affected by the complexation.

3.1 Electronic Redistributions

To make optimal use of these intensification patterns, it is helpful to translate the data into electronic redistribution information. This alternate interpretation can arise by means of atomic polar tensors which partition the total intensity into contributions from each individual atom.[54–58] Within the context of these tensors, the equilibrium charge on any atom is related to the change in the dipole moment of the entire system with respect to the motion of that atom.

Specifically, this charge q is defined as $\partial\mu_x/\partial x = P_{xx}$ where the x direction is perpendicular to the bond connecting this atom to the remainder of the molecule. When the latter bond is stretched, its polarity is altered, which can be represented as charge flux, that is, electronic redistribution, toward or away from the particular atom. This flux, CF, can be extracted from the charge since $P_{zz} = \partial\mu_z/\partial z = q + z(\partial q/\partial z)$.

The charges and charge flux evaluated in this manner are reported for the bridging H or Li atoms in Table 3.5. At either the SCF or MP2 level, the charge on the hydrogen is +0.21 in neutral HCl. This charge is reduced by a small amount when the complex is formed with NH_3. Whether in the complex or not, the stretching motion of the H away from the Cl will change the dipole moment of the system as would motion of any charged unit. Of particular significance here is the charge flux, which measures the susceptibility of the atomic charge to bond lengthening. In isolated HCl, the CF is quite small, which suggests that the charge on the hydrogen will remain nearly constant as this bond is stretched. Since vibrational intensities are related to the effect of a given motion upon the dipole moment, essentially all of the intensity of the ν_s band in HCl is due to the changing position of the fixed charge of the hydrogen. The situation is dramatically different, however, when this molecule is engaged in a hydrogen bond with NH_3. The very large CF here adds another component to the intensity. The dipole increases when the H moves, not only because of the motion of its fixed charge, but also because its charge increases sharply as it moves. It is thus no wonder that the appropriate intensity, also listed in Table 3.5, shows such a dramatic increase when the H bond is formed.

Turning as a comparison now to the Li bond, the charge of the Li atom is quite large in LiCl, near unity. Its motion will thus result in a substantial change in moment of the molecule. The charge flux is in the 0.13–0.15 range, which translates into a further increase in charge when the Li atom moves. Together, these two factors yield a significantly intense vibration in LiCl. Complexation with NH_3 lowers the charge of the Li somewhat, thereby adding a negative contribution to the intensity. Countering this effect is the increase in CF, which tends to increase the intensity by allowing the Li atom to become more positively charged as it moves away from the Cl. The net result is a fairly insignificant change in the intensity of the LiCl stretch in the Li-bonded system.

Table 3.5 Atomic Charge (q) and Charge Flux (CF) of Bridging H and Li Atoms Derived from Atomic Polar Tensors[a]

	SCF			MP2		
	q	CF	A_s	q	CF	A_s
HCl	0.210	0.008	120	0.208	0.006	46
$H_3N\cdot\cdot HCl$	0.190	0.859	1087	0.163	1.356	2186
LiCl	0.750	0.149	135	0.729	0.129	123
$H_3N\cdot\cdot LiCl$	0.628	0.288	331	0.593	0.262	111

[a]Data from reference 53.

In a fundamental sense, these patterns can be understood in terms of the polarity of the molecules involved. The Li—Cl bond is highly polar in the isolated monomer. The low electron density surrounding the Li nucleus has difficulty in following it, thereby raising its positive charge as it is stretched away from Cl, which manifests itself in a significant value of CF. Interaction with NH_3 adds electron density to Li, lowering q, but leads to a small increase in CF. These two effects nearly cancel each other leading to no substantial increase in $\partial\mu/\partial z$ upon complexation. The intensity of ν_s is thus not very sensitive to formation of a Li bond.

The atomic polar tensors also shed light on the intensity patterns for intramolecular modes within the NH_3 subunit. As indicated in Table 3.6, the hydrogen atoms are again of positive charge and their stretching motion away from the N atom raises the dipole moment and results in infrared intensity. Working in the other direction, however, is the charge flux, which is negative. The stretching motion therefore acts to lower the hydrogen's positive charge and thereby detracts from the intensity. Hence it is not surprising that the simple $A_{st}(a_1)$ stretching mode is quite weak in NH_3. We now turn our attention to the complexes. For either the H or Li bond, the charge on the H atom suffers a small drop in magnitude. More significant, however, is the strong drop in the negative value of the charge flux. In the complexes, therefore, the hydrogens do not lose so much of their positive charge as they stretch away from N, and the intensity is thus considerably stronger. This effect is somewhat stronger for the Li than for the H bond.

Numerous attempts have been made over the years to portray the charge rearrangements that accompany the formation of a hydrogen bond. These approaches commonly resort to partitioning of the total electron density into contributions from each atom and thence to "atomic charges." Since these charges do not correspond to a quantum mechanical operator and are arbitrary in the sense of where one chooses to draw the dividing lines between atoms, they lack rigorous physical meaning. In contrast, vibrational intensities are indeed observable, as are the atomic polar tensors derived from them. It then becomes feasible to talk about atomic charges which are real, at least in the sense that they reproduce the experimentally observed intensities.

Table 3.7 lists the *changes* in the charge of each atom that result from the formation of the complex, relative to the isolated monomers. For example, the N—H hydrogens become slightly more negative when NH_3 combines with

Table 3.6 Atomic Charges and Charge Fluxes of Hydrogen Atoms of NH_3[a]

	SCF			MP2		
	q	CF	$A_{st}(a_1)$	q	CF	$A_{st}(a_1)$
NH_3	0.351	−0.346	0.15	0.335	−0.339	0.3
$H_3N\cdot\cdot HCl$	0.312	−0.232	4	0.288	−0.177	6
$H_3N\cdot\cdot LiCl$	0.324	−0.183	26	0.313	−0.172	11

[a]Data from reference 53.

Table 3.7 Changes in Atomic Charge that Accompany Formation of H or Li Bonds[a]

	SCF		MP2	
	$H_3N\cdot\cdot HCl$	$H_3N\cdot\cdot LiCl$	$H_3N\cdot\cdot HCl$	$H_3N\cdot\cdot LiCl$
H_N	−0.039	−0.027	−0.047	−0.022
N	0.223	0.199	0.320	0.197
H or Li	−0.020	−0.122	−0.046	−0.136
Cl	−0.085	0.003	−0.134	0.005
CT[b]	0.105	0.119	0.180	0.131

[a]Data from reference 53; positive quantities refer to greater positive charge.
[b]Charge transferred from NH_3 to HCl or LiCl.

either HCl or LiCl, whereas the N atom becomes more positive. In total, the NH_3 molecule donates between 0.1 and 0.2 electrons to the HCl or LiCl molecule, as indicated by the charge transfer (CT) contained in the last row of the table. The H and Li bonds differ most in where this donated charge winds up. In the former case, the bulk of the negative charge, some 20–25%, bypasses the hydrogen and is added to the more remote Cl. The situation is quite the opposite for the Li bond, where all of the charge donated from NH_3 ends up on the Li atom, with essentially no change in Cl charge at all.

3.2 Anharmonic Effects

The previous analyses of the vibrations in the hydrogen- and lithium-bonded systems assumed a purely harmonic force field. But the weak nature of these bonds leads to the expectation of anharmonicity which has been verified experimentally in a number of cases. An attempt to study this anharmonicity was made recently for the $FZ\cdot\cdot\cdot NH_3$ complexes where again Z=H or Li.[59] Ab initio methods were used to generate a potential energy surface as a function of both the intermolecular $R(F\cdot\cdot\cdot N)$ distance and the r(FZ) bond length. This grid of points was fit to a fourth-order polynomial which determines the nuclear Schrödinger equation and can be solved by a variational procedure.[60] One can obtain in this manner a set of vibrational wave functions which correspond to the ν_{FZ} and $\nu_{F\cdot\cdot N}$ stretching fundamentals and overtones.

The anharmonic fundamentals reported in Table 3.8 reaffirm some of the principles elucidated above under the assumption of harmonic force fields. Specifically, correlation greatly reduces the ν_{FH} stretch but has very little effect upon the same mode in the Li-bonded complex. With respect to the intermolecular $\nu_{F\cdot\cdot N}$ (ν_σ) stretch, correlation increases this frequency by a significant amount for the H bond and produces a small reduction for the Li bond. With or without correlation, the frequency of this stretch is larger for the latter interaction than for the former. The MP2 harmonic frequency of the ν_{FH} stretch is intermediate between the anharmonic SCF and MP2 values for the H bond,

Table 3.8 Anharmonic and Harmonic Vibrational Frequencies (cm^{-1}) Calculated for $H_3N\cdot\cdot HF$ and $H_3N\cdot\cdot LiF$[a]

	$H_3N\cdot\cdot HF$				$H_3N\cdot\cdot LiF$			
	SCF		MP2		SCF		MP2	
	Anh	Harmonic	Anh	Harmonic	Anh	Harmonic	Anh	Harmonic
ν_{FZ}	3773	4011	3331	3514	932	953	947	1010
$\nu_{F\cdot\cdot N}$	240	230	263	265	292	286	281	272

[a]Data from reference 59.

but is greater than either for the Li bond. The harmonic frequency is surprisingly accurate for the ν_σ stretch, especially for the H bond.

This method of analyzing the frequencies permits calculation of overtone spectra. These overtones are reported in Table 3.9 for both the hydrogen- and lithium-bonded complexes. Most notable is the spacing between the various bands. For example, the 00→01 excitation of $H_3N\cdot\cdot HF$, corresponding to its ν_σ stretch, requires 263 cm^{-1} while 00→02 requires 509, or 246 more than the former. The step in exciting to the third level is still smaller, only 230 cm^{-1}. These decreases result from the anharmonicity, and continue up to higher levels as well. The lower grouping of data in Table 3.9 illustrates a similar decrement in the steps when the excitation in $\nu_{F\cdot\cdot N}$ is accompanied by a jump by 1 of the vibrational quantum number characterizing the ν_{FZ} band. Very similar results are noted for the $H_3N\cdot\cdot LiF$ complex as well.

Table 3.9 Overtones and Combination Bands (cm^{-1}) Calculated at the MP2 Level for $H_3N\cdot\cdot HF$ and $H_3N\cdot\cdot LiF$[a]

	$H_3N\cdot\cdot HF$	$H_3N\cdot\cdot LiF$
00 → 01	263	281
00 → 02	509	557
00 → 03	739	829
00 → 04	958	1096
00 → 05	1172	
00 → 10	3331	947
00 → 11	3644	1272
00 → 12	3921	1551
00 → 13	4180	1822
00 → 14	4425	
00 → 15	4658	

[a]Data from reference 59.

[b]Excitations 00 → *mn* where *m* refers to ν_{FZ} and *n* to $\nu_{F\cdot\cdot N}$.

4 LITHIUM ION TRANSFERS

As noted previously, one of the more interesting aspects of the hydrogen bond is the possibility of the bridging proton transferring across from the donor molecule to the acceptor. Although a great deal is understood about the proton transfer process from both experimental and theoretical perspectives, the corresponding lithium transfer is comparatively unexplored, although two recent studies[61,62] did examine the possibility of simultaneous exchange of H and Li.

The system chosen for study, based on its simplicity and likelihood of occurrence, is that in which an ion, H^+ or Li^+, acts as the glue holding together a pair of water molecules. It had been determined in a number of previous calculations, and supported by experimental observations, that the equilibrium geometry of $(H_2O\cdot\cdot H\cdot\cdot OH_2)^+$ contains a symmetric H bond wherein the bridging proton is located precisely midway between the two oxygens.[63] The same conclusion was reached for $(H_2O\cdot\cdot Li\cdot\cdot OH_2)^+$ with a number of different basis sets.[64] Indeed, when combined with a pair of anions such as OH^- or NH_2^-, the Li^+ ion in complexes such as $(HO\cdot\cdot Li\cdot\cdot NH_2)^-$ also adopts a position approximately midway between the first-row atoms.[62]

On the other hand, there are a number of differences between the equilibrium structures of these two complexes. $R(O\cdot\cdot O)$ is quite short in the H-bonded species, around 2.4 Å, whereas it is 1.3 Å longer in $(H_2O\cdot\cdot Li\cdot\cdot OH_2)^+$. Another minor distinction is the orientation of the two water molecules. In the case of the short H bond, the HOH bisector makes an angle of some 120° with the $O\cdot\cdot O$ axis,[65] but this angle is 180° in $(H_2O\cdot\cdot Li\cdot\cdot OH_2)^+$. In other words, the oxygen atoms are pyramidal in $(H_2O\cdot\cdot H\cdot\cdot OH_2)^+$ but planar for the Li bond.

Just as in the case of a neutral complex, the Li center is more positively charged than the bridging hydrogen in an ionic complex such as $(H_2O\cdot\cdot Z\cdot\cdot OH_2)^+$. The Mulliken charge on lithium is 0.90, as compared to 0.66 for H.[64] Another distinction between the two types of complexes has to do with their binding energies. Interaction of the Z^+ ion with the first water molecule is referred to as the proton (lithium) affinity of water. This quantity is calculated to be around 180 kcal/mol in the case of H^+ but only about 40 kcal/mol for Li^+.[64] The large difference has been attributed to the much stronger covalent nature of the O—H bond as compared to the ionic O— —Li^+ interaction. Comparison of the hydration energy for adding a second water to $H_2O\cdot\cdot Z^+$ underscores this conclusion. In the case of either H or Li, this second hydration energy is around 40 kcal/mol, quite close to the value obtained for the first hydration energy of Li^+.[64] This finding is consistent with the highly ionic electrostatic nature of each interaction in $(H_2O\cdot\cdot Li\cdot\cdot OH_2)^+$.

As mentioned above, the proton transfer potential in fully relaxed $(H_2O\cdot\cdot H\cdot\cdot OH_2)^+$ is of symmetric single-well character. However, the potential acquires a second well if the intermolecular $R(O\cdot\cdot O)$ distance is stretched to values longer than its equilibrium length. The two wells correspond to the $(H_2OH\cdot\cdot OH_2)^+$ and $(H_2O\cdot\cdot HOH_2)^+$ configurations and the fully

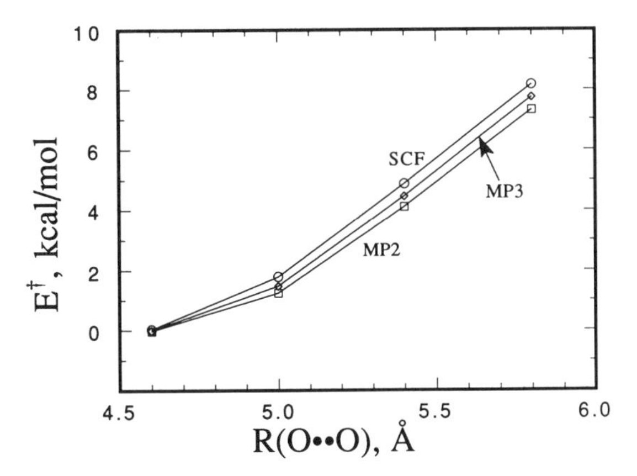

Figure 3.1 Energy barries for lithium ion transfer in $(H_2O\cdot\cdot Li\cdot\cdot OH_2)^+$ for a range of R(OO) distances, computed with the 6-311+G** basis set.

symmetric $(H_2O\cdot\ \cdot H\cdot\ \cdot OH_2)^+$ geometry represents the transition state separating these two wells. The height of this energy barrier rises swiftly as the interoxygen distance is increased.

Certain similarities were noted in the behavior of the Li-bonded analogue in that the transfer barrier is related to the intermolecular separation, as illustrated in Figure 3.1. The second well appears in the Li transfer potential for R(O· · ·O) distances longer than 4.6 Å, nearly 1 Å longer than the equilibrium intermolecular separation. As in the case of proton transfer, the barrier height rises with increasing R. However, the Li transfer barrier is considerably less sensitive to the intermolecular separation. For example, the proton transfer barrier rises from 1.7 to 17.5 kcal/mol as a result of a 0.4-Å stretch in the H bond length from 2.55 to 2.95 Å. In comparison, the lithium transfer barrier rises by only 4 kcal/mol when R(O· ·O) is stretched by twice this amount from 4.6 to 5.4.[64] More details concerning the more gradual change of barrier with increasing intermolecular distance for the Li bond are presented in Table 3.10. The values of ΔR listed there refer to the stretch of the interoxygen separation from its equilibrium value. For example, despite being elongated

Table 3.10 Computed Intermolecular Bond Lengths and Stretches (Å) and Transfer Barriers (kcal/mol)

$(H_2O-H-H_2O)^+$			$(H_2O-Li-H_2O)^+$		
R(O· ·O)	ΔR(O· ·O)[a]	$E^\dagger$	R(O· ·O)	ΔR(O· ·O)[a]	$E^\dagger$
2.55	0.19	1.7	5.0	0.4	2.7
2.75	0.39	8.0	5.4	0.8	6.9
2.95	0.59	17.5	5.8	1.2	11.2

[a]Stretch from optimized value.

by 0.8 Å, the Li transfer barrier is only 6.9 kcal/mol, as compared to a higher barrier of 8.0 kcal/mol for an elongation of only half this amount in $(H_2O—H—H_2O)^+$.

The data in Figure 3.1 suggest that correlation lowers the Li transfer barrier a small amount, at least for the 6-311+G** basis set. A more complete elucidation of the sensitivity of the Li^+ transfer barrier to basis set and correlation is summarized in Figure 3.2, which focuses on the particular R(O· ·O) distance of 5.4 Å. With any basis set, the SCF barrier is lowered when correlation is included. The MP3 barriers are slightly higher than MP2. This trend in correlation-induced lowering is consistent with trends noted earlier for proton transfers.[65-67] On the other hand, the behavior of the barrier with respect to basis set is distinctly different. Proton transfer barriers generally increase with each increment in the basis set that makes it more flexible. The Li^+ transfer barriers behave less regularly. There is a marked drop when the 4-31G basis set is improved to 6-31G* by the addition of polarization functions, whereas the barriers rise again upon adding such functions to the hydrogens and enlarging the valence core. However, the barrier is lowered again by the presence of the diffuse functions in 6-311+G**.

This seemingly erratic behavior of the Li transfer barrier is further evidence that the Li bond is highly ionic in character. Also included in Figure 3.2 as the broken curve is the dipole moment computed for the water monomer with each of the basis sets. There is an obvious correlation between this property and the computed transfer barrier. This relationship can easily be rationalized on the basis of an electrostatic interaction between the Li^+ cation and each of the water monomers. Motion of the ion away from the water is opposed by an electrostatic interaction with the dipole moment of the neutral molecule. Large dipole moments will more vigorously oppose the removal of the Li^+ and thereby raise the transfer barrier.

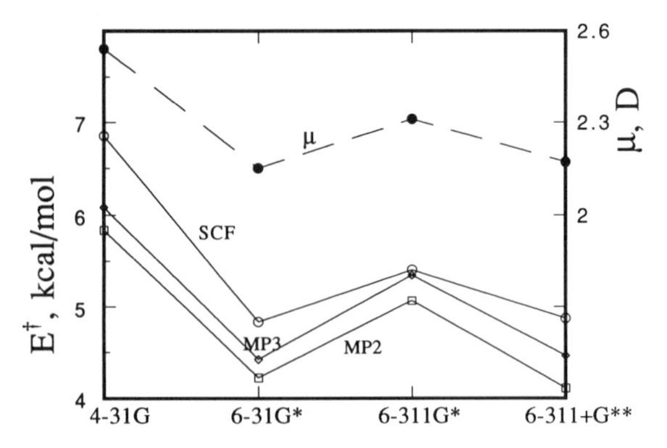

Figure 3.2 Lithium ion transfer barriers shown as solid lines for various basis sets using the scale on the left. Barriers are all computed for R(OO) = 5.4 Å. Dipole moments for the isolated water monomer refer to the right-hand scale and are shown as a broken line.

4.1 Angular Deformations

Previous calculations had revealed that the energy barrier for proton transfer is highly sensitive to angular distortions of the H bond.[63,68–71] The types of distortions considered are described in Figure 3.3, where either or both of the two water molecules may be turned such that their HOH bisectors make angles of α_1 and α_2 with the O· ·O internuclear axis. As was done previously for proton transfer, the Li was allowed to follow its lowest energy path between the two HOH molecules, keeping frozen the remainder of the geometry. Three modes of distortion were investigated. In the first case, the donor molecule was not distorted at all (α_1 = 0). A second mode is conrotatory in the sense that both molecules are turned by equal amounts in the same direction ($\alpha_1 = \alpha_2$). Misorientation in opposing directions comprises the disrotatory mode ($\alpha_1 = -\alpha_2$).

The effects of each sort of misorientation are exhibited in Figure 3.4, where it is clear that both the SCF and correlated MP2 barriers obey similar trends. In either case, the solid curves make it evident that the barrier rises gradually if only one of the molecules is misoriented. The broken curves indicate that

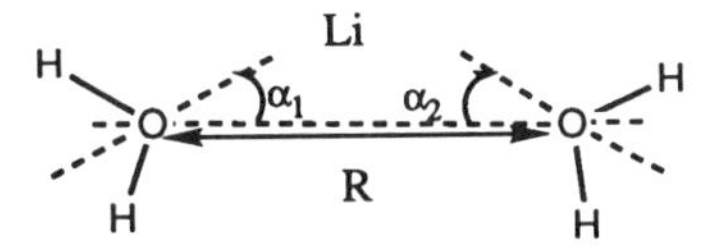

Figure 3.3 Parameters used to define angular distortions in $(H_2O-Li-OH_2)^+$. α_1 and α_2 refer to the angle between the O· ·O axis and the HOH bisector of the Li donor and acceptor molecules, respectively. Both of these angles are positive in the example shown.

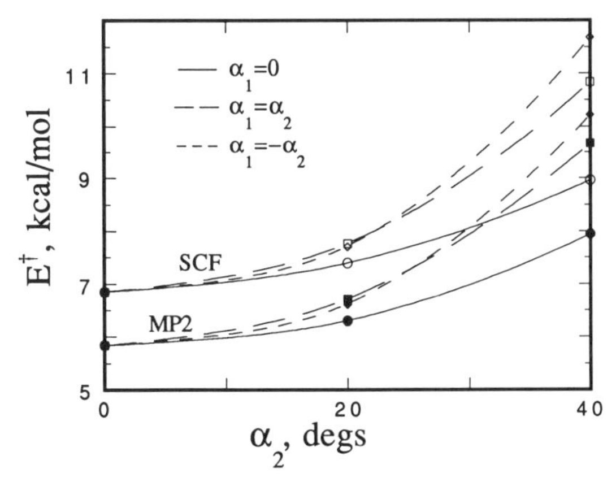

Figure 3.4 Dependence of lithium transfer barrier in $(H_2O\cdot\cdot Li\cdot\cdot OH_2)^+$ upon angular distortions of the Li bond, calculated with a 6-31+G** basis set. α_1 and α_2 refer to deformations of lithium donor and acceptor molecules respectively (see Fig. 3.3). The interoxygen distance is held fixed at 5.4 Å.

rotation of both molecules causes a somewhat larger barrier increase, especially the disrotatory mode. These trends are consistent with what was noted earlier for proton transfers.

What is different between the Li and H transfers is the rapidity with which the barriers rise with increasing angular distortion. Note, for example, from Figure 3.4 that the Li transfer barrier increases by 4 kcal/mol from 6.9 in the undistorted case to 10.8 when $\alpha_1 = \alpha_2 = 40°$ at the SCF level. In contrast, the proton transfer barrier increases by 13, from 17 to 30 kcal/mol, upon undergoing a like distortion [for R(OO) = 2.95 Å].[63]

The much smaller sensitivity of the Li^+ transfer barrier to angular distortion is directly linked to the longer distance separating this ion from the water, as compared to the proton. When R(OO) = 5.4 Å, the lithium center is nearly 3 Å from each water molecule at the transfer transition state. Since it was shown earlier[34,70] that the behavior of the proton transfer barrier with respect to angular deformations could be rationalized on the basis of electrostatic interactions, it is likely that such interactions play an even larger role in the case of lithium transfer, in light of the more ionic character of the Li bond. At the longer distance of 3 Å, misorientation of the dipole of the neutral water will have a less dramatic effect upon the electrostatic energy of the system, explaining the lower sensitivity of the Li^+ transfer barrier to angular distortion.

5 SUMMARY

Lithium bonds are generally considerably stronger than their proton-centered counterparts. While electrostatic interactions are important to both Li and H bonds, they dominate in the former. For this reason, the strength of the Li bond is nearly independent of the nature of the Li donor molecule, as compared to H bonds, where the identity of this molecule is of overriding importance.

Calculated intermolecular vibrational frequencies are consistent with the stronger nature of the Li bond and reproduce the strong red shift of ν_s in the H bond and its weak blue shift in the Li bond. Patterns observed in the approximation of a harmonic force field are largely confirmed when anharmonicity is accounted for.

The changes induced in the intensities of the infrared bands can be rationalized on the basis of the atomic charges and charge flux within each covalent bond. Both the H and Li centers bear a fractional positive charge in a molecule such as HCl or LiCl. Formation of a hydrogen bond with a proton acceptor such as NH_3 causes the charge on the H center to increase as the HCl bond is elongated. Even in isolated LiCl, the charge on the Li is near unity, imparting significant intensity to the bond stretch. Formation of the Li bond produces little net change since the Li center becomes slightly less positive, but this effect is countered by an increase in the susceptibility of the Li charge to Li—Cl bond length. Another fundamental distinction between the H and Li bond is the final destination of charge transferred from the electron donor molecule upon formation of the bond. This electron density bypasses the bridging hydrogen in the H bond, but resides fully on the Li center.

Just as in a hydrogen bond where the bridging proton may be transferred from the donor to the acceptor molecule, Li^+ transfers are also possible. In either case, the height of the energy barrier rises as the donor and acceptor are separated from one another. However, the barrier to Li^+ transfer is considerably less sensitive to intermolecular distance than is the proton transfer barrier. The propensities of the barriers to rise upon imposing angular distortions into the bond are likewise reduced in the case of Li^+ as compared to H^+. Where the two types of transfer most differ is in their dependence upon subtle aspects of the basis set being used to study them. These trends reinforce the dominating influence of coulombic interactions in the case of Li bonds.

ACKNOWLEDGMENT

I am grateful to my co-workers, who brought to the group their interest and insights into Li bonding. This work was supported by the National Institutes of Health (GM29391 and GM36912).

REFERENCES

1. *The Hydrogen Bond, Recent Developments in Theory and Experiments*; Schuster, P.; Zundel, G.; Sandorfy, C., Eds.; North-Holland: Amsterdam, 1976.
2. Jeffrey, G. A.; Saenger, W. *Hydrogen Bonding in Biological Structures*; Springer-Verlag: Berlin, 1991.
3. Scheiner, S. In *Theoretical Models of Chemical Bonding*; Maksic, Z. B. Ed.; Springer-Verlag: Berlin, 1991; Part 4, pp. 171–227.
4. Shigorin, D. N. *Spectrochim. Acta* **1969,** *14*, 198.
5. Kollman, P. A.; Liebman, J. F.; Allen, L. C. *J. Am. Chem. Soc.* **1970,** *92*, 1142.
6. Ault, B. S.; Pimentel, G. C. *J. Phys. Chem.* **1975,** *79*, 621.
7. (a) Cardelino, B. H.; Eberhardt, W. H.; Borkman, R. F. *J. Chem. Phys.* **1986,** *84*, 3230. (b) Hodoscek, M.; Solmajer, T. *J. Am. Chem. Soc.* **1984,** *106*, 1854.
8. Kaufmann, E.; Clark, T.; v. R. Schleyer, P. *J. Am. Chem. Soc.* **1984,** *106*, 1856.
9. Sapse, A. M.; Kaufmann, E.; v. R. Schleyer, P.; Geiter, R. *Inorg. Chem.* **1984,** *23*, 1969.
10. Raghavachari, K. *J. Chem. Phys.* **1982,** *76*, 5421.
11. Rupp, M.; Ahlrichs, R. *Theor. Chim. Acta* **1977,** *46*, 117. Kollman, P. *J. Am. Chem. Soc.* **1977,** *99*, 4875.
12. Boldyrev, A. I.; Solomonik, V. G.; Zakzhevskii, V. G.; Charkin, O. P. *Chem. Phys. Lett.* **1980,** *73*, 58.
13. Marsden, C. J. *J. Chem. Soc., Dalton Trans.* **1984,** 1279.
14. Sapse, A. M.; Jain, D. C. *Chem. Phys. Lett.* **1984,** *110*, 251.

15. Ihle, H. R.; Wu, C. H. *Adv. Mass Spectrosc.* **1978,** *7a*, 636.
16. Snelson, A. *J. Chem. Phys.* **1967,** *46*, 3652.
17. Latajka, Z.; Scheiner, S. *J. Chem. Phys.* **1986,** *84*, 341.
18. Scheiner, S. In *Reviews in Computational Chemistry*; Lipkowitz, K. B.; Boyd, D. B., Eds.; VCH: New York, 1991; Vol. 2, pp. 165–218.
19. Smith, B. J.; Swanton, D. J.; Pople, J. A.; Schaefer, III, H. F.; Radom, L. *J. Chem. Phys.* **1990,** *92*, 1240.
20. Sannigrahi, A. B.; Kar, T.; Niyogi, B. G. *J. Sci. Ind. Res.* **1989,** *48*, 428.
21. Szczesniak, M. M.; Ratajczak, H. *Chem. Phys. Lett.* **1980,** *74*, 243.
22. Kaufmann, E.; Tidor, B.; v. R. Schleyer, P. *J. Comput. Chem.* **1986,** *7*, 334.
23. Latajka, Z.; Scheiner, S. *J. Chem. Phys.* **1984,** *81*, 4014.
24. Szczesniak, M. M.; Ratajczak, H.; Agarwal, U. P.; Rao, C. N. R. *Chem. Phys. Lett.* **1976,** *44*, 465.
25. Szczesniak, M. M.; Latajka, Z.; Piecuch, P.; Ratajczak, H.; Orville-Thomas, W. J.; Rao, C. N. R. *Chem. Phys.* **1985,** *94*, 55.
26. Sannigrahi, A. B.; Kar, T.; Niyogi, B. G.; Hobza, P.; v. R. Schleyer, P. *Chem. Rev.* **1990,** *90*, 1061.
27. Bell, R. P. *The Proton in Chemistry*; Cornell University Press, Ithaca, NY, 1973.
28. Kresge, A. J. *Acc. Chem. Res.* **1975,** *8*, 354.
29. Bordwell, R. G.; Boyle, W. J., Jr. *J. Am. Chem. Soc.* **1975,** *97*, 3447.
30. Caldin, E. F.; Gold, V. *Proton Transfer Reactions*; Halsted: New York, 1975.
31. Stewart, R. *The Proton: Applications to Organic Chemistry*; Academic: Orlando, FL, 1985.
32. Dodd, J. A.; Baer, S.; Moylan, C. R.; Brauman, J. I. *J. Am. Chem. Soc.* **1991,** *113*, 5942.
33. Meot-Ner, M.; Smith, S. C. *J. Am. Chem. Soc.* **1991,** *113*, 862.
34. Scheiner, S. *Acc. Chem. Res.* **1985,** *18*, 174.
35. Truong, T. N.; McCammon, J. A. *J. Am. Chem. Soc.* **1991,** *113*, 7504.
36. Bosch, E.; Moreno, M.; Lluch, J. M.; Bertran, J. *Chem. Phys.* **1990,** *148*, 77.
37. Shida, N.; Barbara, P. F.; Almlöf, J. E. *J. Chem. Phys.* **1989,** *91*, 4061.
38. Hodoscek, M.; Hadzi, D. *J. Mol. Struct. (Theochem)* **1990,** *209*, 411, 421.
39. Howard, N. W.; Legon, A. C. *J. Chem. Phys.* **1988,** *88*, 4694; Legon, A. C.; Wallwork, A. L.; Rego, C. A. *J. Chem. Phys.* **1990,** *92*, 6397.
40. Brciz, A.; Karpfen, A.; Lischka, H.; Schuster, P. *Chem. Phys.* **1984,** *89*, 337.
41. Jasien, P. G.; Stevens, W. J. *Chem. Phys. Lett.* **1986,** *130*, 127.
42. Szczesniak, M. M. (unpublished).
43. Szczesniak, M. M.; Chalasinski, G.; Cybulski, S. M. *J. Chem. Phys.* **1993,** *98*, 3078.
44. Novoa, J. J.; Tarron, B.; Whangbo, M.-H.; Williams, J. M. *J. Chem. Phys.* **1991,** *95*, 5179.
45. Umeyama, H.; Morokuma, K. *J. Am. Chem. Soc.* **1977,** *99*, 1316.
46. Vos, R. J.; Hendriks, R.; van Duijneveldt, F. B. *J. Comput. Chem.* **1990,** *11*, 1.
47. Szczesniak, M. M.; Scheiner, S.; Bouteiller, Y. *J. Chem. Phys.* **1984,** *81*, 5024.

48. Kulkarni, G. V.; Rao, C. N. R. *J. Mol. Struct.* **1983,** *100*, 531.
49. Sapse, A. M., Raghavachari, K. *Chem. Phys. Lett.* **1989,** *158*, 213.
50. Amos, R. D. *Chem. Phys. Lett.* **1980,** *73*, 602.
51. Diercksen, G. H. F.; Sadlej, A. J. *J. Chem. Phys.* **1981,** *75*, 1253.
52. Diercksen, G. H. F.; Kellö, V.; Sadlej, A. J. *J. Chem. Phys.* **1983,** *79*, 2918.
53. Szczesniak, M. M.; Kurnig, I. J.; Scheiner, S. *J. Chem. Phys.* **1988,** *89*, 3131.
54. Person, W. B. In *Vibrational Intensities in Infrared and Raman Spectroscopy*, Person, W. B.; Zerbi, G. Eds.; Elsevier: Amsterdam, 1982; Chapter 4.
55. Zilles, B. A.; Person, W. B. *J. Chem. Phys.* **1983,** *79*, 65.
56. Amos, R. D. *Chem. Phys.* **1986,** *104*, 145.
57. Swanton, D. J.; Bacskay, G. B.; Hush, N. S. *Chem. Phys.* **1983,** *83*, 303.
58. Kurnig, I. J.; Szczesniak, M. M.; Scheiner, S. *J. Chem. Phys.* **1987,** *87*, 2214.
59. Bouteiller, Y.; Latajka, Z.; Ratajczak, H.; Scheiner, S. *J. Chem. Phys.* **1991,** *94*, 2956.
60. Mijoule, C.; Allavena, M.; Leclerq, J. M.; Bouteiller, Y. *Chem. Phys.* **1986,** *109*, 207.
61. Sudhakar, P. V.; Lammertsma, K.; v. R. Schleyer, P. *J. Mol. Struct. (Theochem.)* **1992,** *255*, 309.
62. Sapse, A. M.; Jain, D. C. *Chem. Phys. Lett.* **1990,** *171*, 480.
63. Scheiner, S. *J. Am. Chem. Soc.* **1981,** *103*, 315.
64. Duan, X.; Scheiner, S. *J. Phys. Chem.* **1992,** *96*, 7971.
65. Latajka, Z.; Scheiner, S. *J. Mol. Struct. (Theochem.)* **1991,** *234*, 373.
66. Scheiner, S.; Szczesniak, M. M.; Bigham, L. D. *Int. J. Quantum Chem.* **1983,** *23*, 739.
67. Szczesniak, M. M.; Scheiner, S. *J. Chem. Phys.* **1982,** *77*, 4586.
68. Scheiner, S. *J. Phys. Chem.* **1982,** *86*, 376.
69. Scheiner, S. *J. Chem. Phys.* **1982,** *77*, 4039.
70. Hillenbrand, E. A.; Scheiner, S. *J. Am. Chem. Soc.* **1984,** *106*, 6266.
71. Scheiner, S.; Bigham, L. D. *J. Chem. Phys.* **1985,** *82*, 3316.

4

LITHIUM ATOM MATRIX REACTIONS WITH SMALL MOLECULES

LAURENT MANCERON

Laboratoire de Spectrochimie Moléculaire, Université Pierre et Marie Curie, Paris, France

LESTER ANDREWS

Chemistry Department, University of Virginia Charlottesville, Virginia

---

Lithium Chemistry: A Theoretical and Experimental Overview, Edited by Anne-Marie Sapse and Paul von Ragué Schleyer.
ISBN 0-471-54930-4

1 INTRODUCTION

Alkali metal atom reactions have been used to produce reactive chemical species since the early work of Polanyi and his co-workers.[1] Applications of the "sodium flame" technique to the study of free radical reactions have been discussed by Steacie.[2] In more recent work, crossed-molecular-beam reactions of alkali metal atoms and halogen compounds have provided a means of studying reaction dynamics; the early crossed-beam work has been reviewed by Herschbach.[3] Alkali metal atom reactions and product characterization were the focus of a matrix reaction technique first developed by Andrews and Pimentel[4] in a study of the lithium atom–nitric matrix codeposition reaction. Subsequent use of alkali metal atom matrix reactions by Andrews and co-workers produced a series of free radicals[5] by halogen abstraction,

$$M + R{-}X \longrightarrow MX + R \tag{1}$$

and new alkali metal species[6] by direct synthesis,

$$M + O_2 \longrightarrow M^+O_2^- \tag{2}$$

which were trapped in the matrix spectroscopic study. In first part of this chapter we review where the lithium atom serves as an electron donor to generate anionic or quasi-ionic species to be studied spectroscopically. These were observed with molecules having favorable electron affinities such as O_2 O_3, X_2 ($X = F$, Cl, Br, I), NO, NO_2, SO_2, and others. These species are characterized either by near complete charge transfer, as deduced from ESR data, and/or by the vibrational properties of the molecular anion being fairly invariant with the natura of the alkali counterion.

A second family of complexes was studied more recently. These involve complexes of lithium with electron-rich molecules having unfavorable electron affinities in which the Li atom plays a far less predictable role. These complexes have variable strengths and stoichiometries (multiple ligand uptake becomes possible) according to the nature of the interaction in play. These can belong to two different classes: (1) polar molecules that have electron lone pairs (H_2O, NH_3) and poor or negative electron affinities and (2) simple unsaturated organic molecules such CO, C_2H_2, C_2H_4, and C_6H_6 which also have unfavorable electron affinities, but are nevertheless notoriously good π^* electron acceptor and π or σ electron donors. The second part of the chapter is devoted to characterizing some of the properties of these complexes.

2 EXPERIMENTAL

The experimental goal is to bring lithium atoms and reactive molecules together long enough for primary reactions to take place and then to quickly trap the primary reaction products for spectroscopic study. This is accomplished by depositing a vapor stream of alkali metal atoms from a Knudsen cell along with some reactive molecules at high dilution in argon together on a substrate cooled to 10–15 K. Figure 4.1 illustrates the reaction geometry. During the condensation process, collisions between lithium atoms and reactive molecules take place, which in many cases produce reaction products that are quickly trapped and prevented from further reaction by the matrix host after sample solidification. In the infrared transmission experimental apparatus shown in Figure 4.1, the substrate is a CsI plate. Sample deposition is governed by the

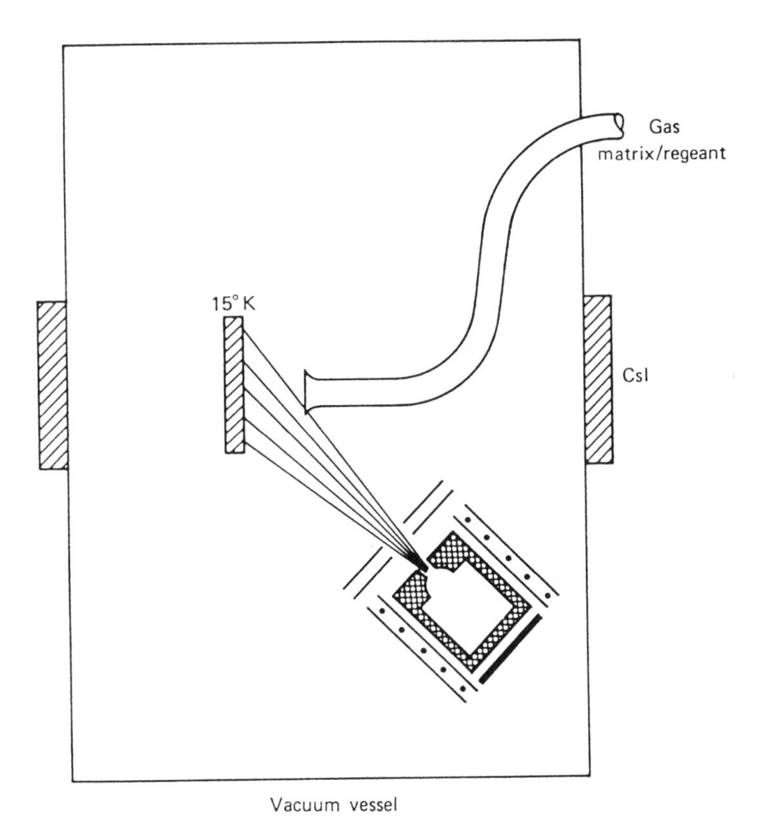

Figure 4.1 Cross-section of the vacuum vessel base showing reaction geometry, Knudsen cell alkali–metal–atom source, and gas-deposition line.

desire to maximize the yield of reaction products, while at the same time maintain infrared transmission and good spectral conditions of the sample. For the Raman scattering experiment, a tilted metal wedge is used to condense the gas sample streams, and for ESR studies a sapphire rod or a CaF_2 slab is employed. The maximum yield of products is counterbalanced by the need to keep the sample light in color to maximize light scattering and minimize light absorption by the sample. The most critical aspect of both infrared and Raman matrix experiments is sample preparation.

3 MATRIX REACTION WITH NO

The first new species synthesized using matrix reactions of alkali metal atoms was LiON.[4]

$$\mathrm{Li} + \mathrm{NO} \longrightarrow \mathrm{LiON} \tag{3}$$

In this pioneering work, Andrews and Pimentel assigned intense bands at 651 and 1352 cm^{-1} to the Li—O and the N—O stretches of a presumably bent species LiON; the use of ^6Li and ^7Li atoms and ^{15}N and ^{18}O enriched NO reagents as important for characterizing the vibrational modes. Subsequently Milligan and Jacox also observed the 651- and 1352-cm^{-1} bands; mercury arc photolysis markedly decreased the 651-cm^{-1} feature while the 1352-cm^{-1} signal remained.[7] The workers attributed the 1352-cm^{-1} band to $(NO)^-$ and the 651-cm^{-1} feature to some other lithium–nitric oxide species. In the most recent study, aimed at comparison of the $M^+O_2^-$ and M^+NO^- species, Tevault and Andrews[8] prepared matrix samples of LiON and studied their mercury arc photolysis behavior. The decrease in intensity of the Li—O mode at 651 cm^{-1} was accompanied by the growth of a new Li—O mode at 447 cm^{-1}, while the $(NO)^-$ mode at 1352 cm^{-1} remained. The photolysis behavior was explained as the photoisomerism of the triangular ionic species $Li^+(ON)^-$ to a bent form. Here the frequency of the interionic $Li^+ \leftrightarrow NO^-$ mode was changed by a rearrangement in cation–anion structure, whereas the intraionic mode was not measurably affected.

$$(\mathrm{Li})^+ \left(\begin{array}{c} \mathrm{O} \\ | \\ \mathrm{N} \end{array} \right)^- \xrightarrow{h\nu} (\mathrm{Li})^+ \left(\begin{array}{cc} \mathrm{O} & \\ & \diagdown \\ & \mathrm{N} \end{array} \right)^- \tag{4}$$

Tevault and Andrews also observed three bands that showed isotopic splittings appropriate to a species containing two equivalent Li atoms and one NO molecule.[8] These bands were assigned to the two antisymmetric interionic modes and the intraionic $(N—O)^{2-}$ mode in the secondary reaction product $Li^+(NO)^{2-}Li^+$.

$$(\mathrm{Li})^+(\mathrm{ON})^- + \mathrm{Li} \longrightarrow \mathrm{Li}^+(\mathrm{NO})^{2-}\mathrm{Li}^+ \tag{5}$$

Table 4.1 Fundamental Frequencies (cm^{-1}) Assigned to the ν_1 Intraionic and the ν_2 Interionic Modes of the $M^+(NO)^-$ Species

Metal	ν_1	ν_2
^6Li	1353.5	692.3
^7Li	1352.5	651.8
Na	1358.0	361.0
K	1372.0	280.4
Rb	1373.0	235.0
Cs	1374.0	219.0

In studies involving all of the heavier alkali metal reagents, Tevault and Andrews observed the M^+NO^- species for all of the alkali metal reagents.[9] The M^+NO^- spectra were characterized by two bands, an $(N \leftrightarrow O)^-$ intraionic mode in the 1352–1374 cm^{-1} region and a $M^+ \leftrightarrow (NO)^-$ interionic mode between 692 and 219 cm^{-1}, depending upon the alkali counterion. Table 4.1 lists the M^+NO^- frequencies. Note the significant alkali metal mass effect on ν_2, and $M^+ \leftrightarrow NO^-$ mode. Note also the small reverse effect on ν_1, the $(N{-}O)^-$ mode. Here the larger, more polarizable alkali cation accommodates more of the antibonding anion electron which results in the removal of a small amount of antibonding charge density from NO^- and a slight increase in the $(N{-}O)^-$ frequency. The M^+ effect on the $(N{-}O)^-$ frequency is relatively small compared to the difference between $(N{-}O)^-$ frequencies as a function of M^+ (1352–1374 cm^{-1}) and NO (1875 cm^{-1}).

4 MATRIX REACTIONS WITH O_2

Alkali metal atom–oxygen molecule matrix reactions have been extensively studied in this laboratory using infrared and Raman techniques. The primary reaction yields the superoxide species (see Eq. 2), which will be discussed in detail. Two secondary reactions yield the peroxide species

$$M^+O_2^- + M \longrightarrow M^+O_2^{2-}M^+ \tag{6}$$

and the disuperoxide species

$$M^+O_2^- + O_2 \longrightarrow M^+O_4^- \tag{7}$$

In the first study of $M{-}O_2$ matrix reactions, Andrews reported a new chemical species, lithium superoxide, $Li^+O_2^-$, with ionic binding between Li^+ and O_2^- and an isosceles triangular structure.[6] These conclusions were based upon near agreement between the O—O mode for LiO_2 and the O_2^- fundamental and the observation of sharp triplet bands in the reaction of $^{16}O_2/^{16}O^{18}O/^{18}O_2$ with

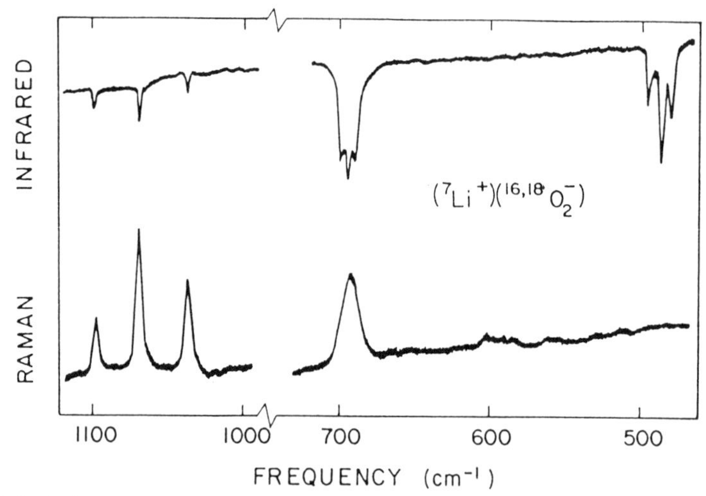

Figure 4.2 Infrared and Raman spectra of lithium superoxide, $Li^+O_2^-$, using lithium −7 and 30% $^{18}O_2$, 50%, $^{16}O^{18}O$, and 20% $^{16}O_2$ (Ar/O_2 = 100). Raman spectrum recorded using 200 mW of 4800-Å excitation and a long-wavelength-pass dielectric filter in the 1000-cm^{-1} region.

^{7}Li atoms at high dilution in argon. Figure 4.2 contrasts the infrared spectrum in this critical experiment with the Raman spectrum of a similar argon matrix sample.[10]

First, examine the O_2^- region near 1100 cm^{-1}. The 1096.1-cm^{-1} Raman band observed at 1096.6 cm^{-1} in the infrared is due to $^7Li^{+16}O_2^-$; the 1065.7-cm^{-1} Raman band recorded at 1066.5 cm^{-1} in the IR is due to $^7Li^{+16}O^{18}O^-$; the 1034.6-cm^{-1} Raman signal measured at 1035.2 cm^{-1} in the infrared is due to $^7Li^{+18}O_2^-$. The large 61-cm^{-1} oxygen isotopic shift indicates the pure oxygen character of this vibration, which is identified as ν_1, the intraionic $(O \leftrightarrow O)^-$ mode. The central feature reveals the molecular structure. In both infrared and Raman experiments, the single 16–18 isotopic band was sharp, just as sharp as the 16–16 and 18–18 isotopic bands. This indicates that the 16–18 isotope produces a single molecular arrangement $M\langle^{16}_{18}$ with equivalent oxygen atoms, in contrast to two possible M-16–18, M-18–16 arrangements which would be expected to produce a split central band, as was observed for $H^{16}O^{18}O$ and $H^{18}O^{16}O$.[11] In the infrared experiments, this band was very sharp, 1.0 cm^{-1} half-width. It appears that 0.2 cm^{-1} is a reasonable upper limit on the inequivalence of the 0 atoms in $Li^+O_2^-$ in the $Li^{+16}O^{18}O^-$ isotopic band.

Second, in the Li—O stretching region, two intense resolved triplet bands were observed in the infrared; a single broader, weaker feature was observed in the Raman spectrum. The large lithium isotopic shift and smaller oxygen isotopic shift indicate that the band near 700 cm^{-1} is the symmetric interionic mode ν_2, $Li^+ \leftrightarrow O_2^-$, and the smaller lithium isotopic shift and large oxygen isotopic shift show that the band near 500 cm^{-1} is ν_3, the antisymmetric interionic mode. The G-matrix elements for these normal modes weight the participation of Li and O atoms differently.[6]

Comparison of band intensities in Figure 4.2 also reflects on the bonding in the $Li^+O_2^-$ species. The infrared spectrum shows a weak intraionic mode, ν_1, and very intense interionic modes, ν_2 and ν_3, as would be expected for an ionic model. The Raman spectrum contains a very intense intraionic mode ν_1, a moderately intense symmetric interionic mode ν_2, and the antisymmetric interionic mode ν_3 is absent. The ionic model for LiO_2 is supported by the complementary Raman spectrum. Finally the LiO_2 molecule has been examined using a number of theoretical calculations[12] which have verified the charge-transfer model and isosceles triangular geometry for LiO_2 deduced from the matrix infrared spectrum.[6]

The heavier alkali metals Na, K, Rb, and Cs react with oxygen to produce the superoxide $M^+O_2^-$ species.[13–15] Raman spectra for the Na, K, Rb, and Cs atom–O_2 reactions show a strong band near 1110 cm^{-1} for the O_2^- vibration.[16,17] The $Cs^+O_2^-$ vibrational data are contrasted in Table 4.2 with frequencies for the other $M^+O_2^-$ molecules. Note the increase in ν_1 with increasing cation size. This trend has been rationalized[10] as follows: increasing the cation polarizability increases the induced dipole moment on the cation, which is opposite in sense to the ion dipole: accordingly, a small amount of antibonding electron density is removed from O_2^-, increasing the fundamental in the direction of the O_2 value (1552 cm^{-1}). Also note the larger decrease in ν_2 and smaller decline in ν_3 with increasing alkali atomic weight, which is required by the G-matrix elements for these normal modes.

The peroxide species, $M^+O_2^{2-}M^+$, has been observed for the five alkali metals. Intense antisymmetric Li—O stretching bands at 796 and 446 cm^{-1} for the ^7Li species showed isotopic splittings appropriate for a species containing *two* equivalent lithium atoms and *two* equivalent oxygen atoms. A rhombus structure with an O—O bond was suggested for the $M^+O_2^{2-}M^+$ molecule.[8] In the heavier species, the sodium frequency counterparts were observed at 525 and 254 cm^{-1}.[13] The upper band only was observed for the K, Rb, and Cs species at 433, 389, and 357 cm^{-1}.[14,15]

Table 4.2 Fundamental Frequencies (cm^{-1}) Assigned to the ν_1 Intraionic and ν_2 and ν_3 Interionic Modes of the C_{2v} Alkali Metal Superoxide Molecules in Solid Argon at 15 K

Molecule	ν_1	ν_2	ν_3
$^6LiO_2$	1097.4	743.8	507.3
$^7LiO_2$	1096.9	698.8	492.4
NaO_2	1094	390.7	332.8
KO_2	1108	307.5	—
RbO_2	1111.3	255.0	282.5
CsO_2	1115.6	236.5	268.6

The disuperoxide species, $M^+O_4^-$, have been observed for the heaviest alkali metals. The small lithium ion apparently is not large enough to stabilize the large O_4^- anion. The new MO_4 species was first observed in K and Rb studies which produced extraordinarily intense, very sharp, bands at 993.4 and 991.7 cm^{-1}, respectively.[14] The stoichiometry of the new species is revealed by the mixed isotopic reactions; the $^{16}O_2 + {}^{18}O_2$ sample produced a symmetric triplet, which indicates the presence of two equivalent O_2 molecules in the new species, and a well-resolved sextet was produced in the scrambled oxygen isotopic experiment. The explicit interpretation of the sextet indicates a species with equivalent O_2 units with equivalent O atoms in each unit. Subsequently, Jacox and Milligan pointed out the existence of the O_4^- ion in ion–molecule reactions and suggested the O_4^- arrangement.[18] The O_4^- anion was suggested to contain two "superoxide" bonds and a weak intermolecular bond between the two O_2 units. A very strong Raman band near 300 cm^{-1} for the $K^+O_4^-$, $Rb^+O_4^-$, and $Cs^+O_4^-$ species was assigned to this symmetric intermolecular mode $(O_2 \leftrightarrow O_2)^-$; interpretation of this low-frequency Raman band requires a very weak oxygen bond which is provided by the O_4^- anion in the $M^+O_4^-$ species.[17] More recent studies on $K^+O_4^-$ led to the observation of new spectral features confirming that the O_4^- anion has then a chained, *trans*-structure with O_2 subunits substantially less perturbed than in O_2^-.[19]

5 MATRIX REACTIONS WITH OZONE

Extensive infrared and Raman studies of matrix-isolated isotopic ozone molecules were conducted by Andrews and Spiker.[20] More recently, ozone complexes with such molecules were prepared and their photochemistry studied.[21,22] In this work isotopic molecules of ozone ($^{16}O_3$, $^{18}O_3$, $^{16,18}O_3$) were synthesized by Tesla coil discharge of O_2 gas in a pyrex finger immersed in liquid nitrogen. Oxygen was outgassed from the blue, liquid ozone sample by evacuating the sample at 77 K. Argon–ozone samples were prepared using standard nanometric techniques in a stainless steel vacuum system.

Matrix reactions of alkali metal atoms and ozone have also been studied using infrared and Raman techniques. The infrared spectra[23] were characterized by very intense bands near 800 cm^{-1} depending upon the alkali atom and weaker bands near 600 cm^{-1} for the heavier alkali metal reactions. The intense 800 cm^{-1} bands were assigned to ν_3 and the weak 600 cm^{-1} bands were attributed to ν_2 of the O_3^- in the $M^+O_3^-$ species. Raman spectra[24] revealed a very strong signal at 1010 $\pm$ 10 cm^{-1} depending on the alkali metal; the $Li^+O_3^-$ signal at 1012 cm^{-1} lacked the site splitting of the heavier alkali ozonide species and was more sensitive to laser photolysis. Resonance enhancement was observed as the exciting laser wavelength decreased from yellow to green to blue to violet. The absorption spectrum revealed a strong absorption at 440 nm with resolved vibronic structure (900 $\pm$ 10 cm^{-1} fundamental) for the alkali ozonides.[25] Again $Li^+O_3^-$ gave the sharpest vibronic structure and was the most photosensitive.

6 MATRIX REACTIONS WITH Cl_2

The dihalide anion Cl_2^- is another species of limited stability which requires matrix synthesis and stabilization. Hass and Griscom attributed a Raman band at 265 cm^{-1} in γ-irradiated alkali halide–alkali borate glasses to the Cl_2^- fundamental.[26] Howard and Andrews extensively studied matrix reactions of Cl_2 and alkali atoms using Raman and infrared techniques.[27] The Raman spectra were characterized by intense bands near 250 cm^{-1} depending upon the alkali reagent; the fundamental was observed at 246 cm^{-1} with Li, 225 cm^{-1} with Na, 264 cm^{-1} with K, 260 cm^{-1} with Rb, and 259 $\pm$ 1 cm^{-1} with Cs. These frequency differences indicate the interaction of the interionic mode ν_2 $M^+ \leftrightarrow Cl_2^-$, with the observed frequencies assigned to ν_1, the symmetric intraionic vibration $(Cl \rightarrow Cl)^-$, in the $M^+Cl_2^-$ species, reaction (8).

$$M + Cl_2 \longrightarrow M^+Cl_2^- \tag{8}$$

Owing to the yellowish-orange color of the $M^+Cl_2^-$ sample and high Raman intensity of the fundamental Cl_2^- band, the higher-frequency region was searched for overtones. The optical absorption of Cl_2^- hole-type centers tails out above 500 nm,[26] and the matrix-isolated $M^+Cl_2^-$ species gives rise to a very strong 350-nm absorption which tails into the visible region.[28] Argon ion excitation falls within the tail of the electronic band. Figure 4.3 shows the regular progression of overtones observed out to 9ν for Cl_2^- in $Li^+Cl_2^-$ using either 4880 or 5145 Å excitation. A similar regular progression of overtones was observed for the $Cs^+Cl_2^-$ species using 4579 Å excitation. The fundamental and overtone bands are nicely fit by the well-known equation: $\nu(v) = v\omega_e - \omega_e x_e(v + 1)(v)$. The regularly decreasing intensity of the overtones with increasing vibrational quantum number suggests that these spectra are due to the resonance Raman effect.[27]

Infrared spectra of lithium-6 and -7 isotopic reactions with Cl_2 in this laboratory produced intense absorptions at 616.9 and 579.5 cm^{-1} which are due to ^6LiCl and ^7LiCl and bands at 519.6, 511.4, and 375.7 cm^{-1} and 489.2, 481.1, and 352.6 cm^{-1} which are due, respectively, to $(^6LiCl)_2$ and $(^7LiCl)_2$. The LiCl monomer bands have an isotopic frequency ratio of 0.9394 in excellent agreement with the calculated 0.9372 ratio with the expected small anharmonic deviation; the dimer assignments also produce the correct isotopic frequency ratio. In addition, infrared bands were observed at 552.4 cm^{-1} in the lithium-6 reaction and at 517.8 cm^{-1} with lithium-7, which are appropriate for assignment to the ν_2 interionic mode $Li^+ \leftrightarrow Cl_2^-$. The LiCl monomer and dimer bands and the ν_2 $Li^+Cl_2^-$ band were not observed in the Raman spectrum because of the low polarizability change for interionic modes.

7 MATRIX REACTIONS WITH N_2

There has been a lot of recent research activity on metal atom–dinitrogen matrix reactions with lithium atoms and transition metal atoms. Spiker and co-

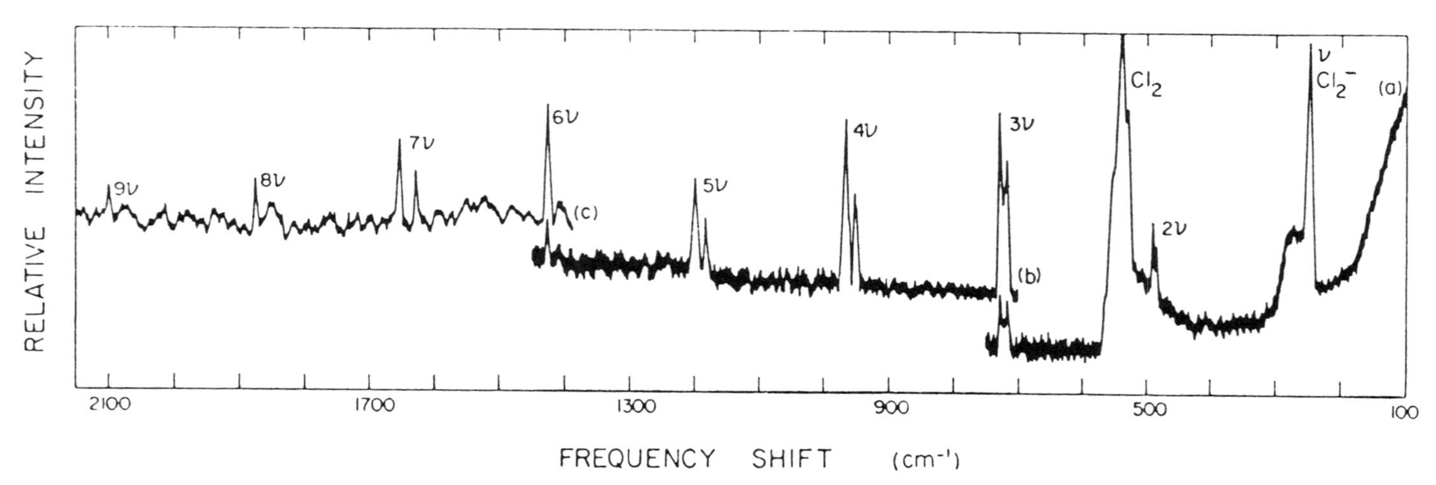

Figure 4.3 Resonance Raman spectrum of Cl_2^- in the matrix-isolated $Li^+Cl_2^-$ species using approximately 50 mW of 5145-Å excitation (Ar/Cl_2 = 100). Amplification range (*a*) 1 × 10^{-9} A, (*b*) 0.3 × 10^{-9} A, (*c*) 0.1 × 10^{-9} A (from ref. 27, used with permission).

workers observed sharp bands at 1800 and 1535 cm^{-1} and a broad band at 2300 cm^{-1} following the deposition of lithium atoms into pure nitrogen matrices.[29] The intense 1535-cm^{-1} band is illustrated in Figure 4.4 along with spectra using nitrogen-15 enriched matrices. With a $^{15}N_2$ matrix the broad band shifted to 2222 cm^{-1}, and the sharp bands shifted respectively to 1740 and 1487 cm^{-1}; these large isotopic shifts indicate pure nitrogen stretching modes.

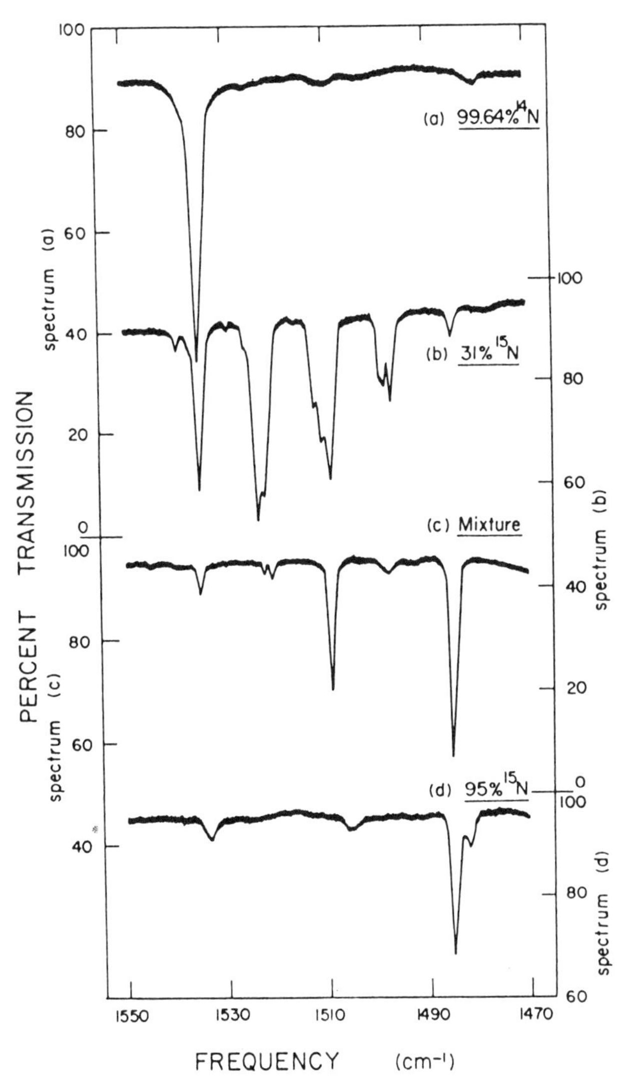

Figure 4.4 Infrared spectra of lithium–atom–nitrogen matrix cocondensation products. (*a*) Natural isotopic N_2; (*b*) 9% $^{15}N_2$, 43% $^{15}N^{14}N$, 48% $^{14}N_2$; (*c*) 72% $^{15}N_2$, 8% $^{15}N^{14}N$, 18% $^{15}N^{14}N$, 18% $^{14}N_2$; (*d*) 90% $^{15}N_2$, 10% $^{15}N^{14}$ $^{15}N^{14}N$ (from ref. 29, used with permission).

A mixed $^{14}N_2/^{15}N_2$ matrix sample produced an intense central component at 1509 cm^{-1} in addition to the 1535- and 1487-cm^{-1} bands, indicating the presence of two equivalent N_2 units in this molecular species. Using a scrambled $^{14,15}N_2$ isotopic matrix, the 1537-cm^{-1} band was split into a multiplet of nine components which showed that the N_2 molecules were bonded "end on" (i.e., inequivalent N atoms) to the lithium reagent. It was suggested that this new species, with the lowest reported N—N frequency, has the structure N=N(Li)$_2$N=N. The spectra certainly show that two dinitrogen molecules complex with a lithium species. Owing to the proximity of the 1800-cm^{-1} nitrogen frequency to N_2^- Raman bands in alkali halide lattices,[30] the 1800-cm^{-1} band was assigned to the intraionic (N—N) mode in $Li^+N_2^-$. The broad feature at 2300 cm^{-1} just below the fundamental of diatomic N_2 was attributed to a nitrogen molecular mode perturbed by lithium in solid nitrogen.[29]

8 INTERACTION WITH ELECTRON LONE-PAIR CONTAINING MOLECULES (H_2O, NH_3)

The existence of this class of complexes was first predicted in 1978 by Trenary et al.[31] They correspond to complexes in which the alkali atom was predicted to play the unusual role of electron acceptor. Calculations have predicted equilibrium geometries, dissociation energies (on the order of 10–15 kcal/mole), and large dipole moment enhancements. Experimentally, the Li—OH_2 and Li—NH_3 have been studied using ESR[32] and IR spectroscopies.[33-35]

All the experimental data are consistent with the maximum symmetry structures proposed by the various calculations:[31,36,37]

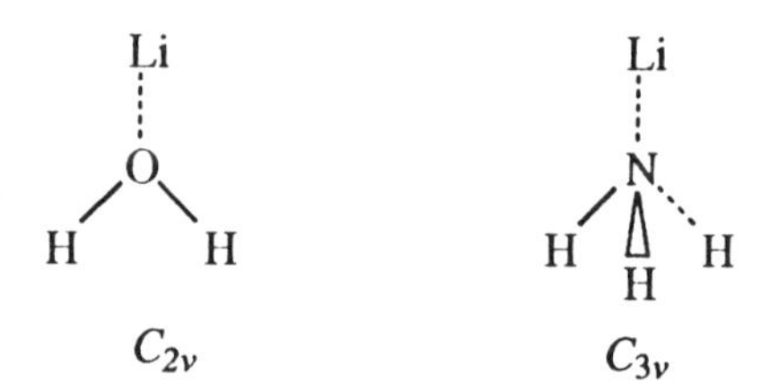

The absence of isotopic splitting using LiOHD, $LiNH_2D$, and $LiND_2H$ could be shown spectroscopically, thus confirming the C_{2v} and C_{3v} symmetries for $LiOH_2$ and $LiNH_3$, respectively. The observation of the Li—N stretching mode around 320 cm^{-1} confirmed the nitrogen coordination in the case of $LiNH_3$,[35] but the Li—O stretching motion could not be observed.

Unfortunately, in their ESR study Meier et al.[32] could not infer, from the observed reduction of unpaired spin density on the Li atom, the direction of the charge transfer. However, the steady decrease of the perturbation of the molecular moiety when going from Li—OH_2 to Cs—OH_2 and from Li—NH_3 to Cs—NH_3 evidenced in the infrared studies is compatible with the molecule-to-metal donation scheme proposed by theoretical calculations.

Although the perturbation of the water and ammonia molecules are relatively the largest with lithium, they remain overall very small (1–3% on the N—H or O—H stretching motions). The large blue shift of the ν_2 "umbrella" bending mode motion (1147 cm^{-1} in Li—NH_3 vs. 974 cm^{-1} in free NH_3) seems at first view to create an exception. But it was shown to be due to the appearance of a potential constant involving the HNLi bending in the normal coordinate associated with the ν_2 mode, and does not reflect a stiffening of the HNH bond angle coordinate.[35] Indeed, these species correspond to weak complexes in which the submolecules retain their individuality. It was not possible, on the basis of the isotopic studies (^{6}Li/^{7}Li, H/D) and subsequent vibrational analysis, to detect noticeable changes in the H$\hat{N}$H or H$\hat{O}$H band angles.

The most remarkable property in the infrared spectrum of these complexes lies in the intensity inversion of the symmetrical and antisymmetrical stretching modes. The symmetrical modes are roughly one order of magnitude weaker in the free molecules, but become one order of magnitude stronger in the complex. This effect has been discussed in reference 33 within the framework of the theory of infrared intensities of charge transfer complexes developed by Friedrich and Person.[38] The mechanism required for increased variation of dipole moment with nuclear vibration (and hence infrared enhancement) is the modulation of electronic wavefunctions with explicit nuclear dependence. It was shown that vibrations with the same symmetry properties as that of the charge transfer can modulate these electronic wave functions, with, as a consequence, a modulation of the overall dipole moment of the complex.

9 LITHIUM-UNSATURATED HYDROCARBON COMPLEXES

9.1 The LiC_2H_2 Molecule

9.1.1 IR Studies. In contrast with group IB metals,[39] the reaction of one Li atom with one acetylene molecule did not produce an ethyl radical-like adduct. Infrared studies demonstrated that the LiC_2H_2 species is likely to be planar, and has lithium bridging the Π-system and a *cis*-bent C_2H_2 group.[40] An $Li_2C_2H_2$ species was also detected (Fig. 4.5) and both LiC_2H_2 and $Li_2C_2H_2$ have equivalent C—H bonds. Figure 4.2 compares the IR spectra in the C—H and C=C stretching regions for the LiC_2H_2, $Li^{13}C_2H_2$, LiC_2D_2, and LiC_2HD isotopic species. The unicity of the LiC_2HD signals (open arrows) in each C=C, C—D, and C—H stretching region demonstrates the equivalence of the two C—D(D) bonds, as the existence of an ethyl radical-like adduct would have led to two sets of bands for each isotopomer:

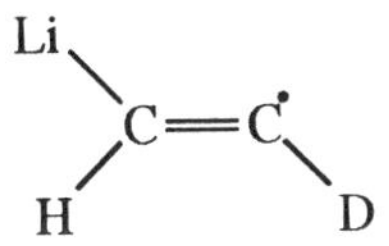

Li C=Ċ H D and Li C=Ċ D H

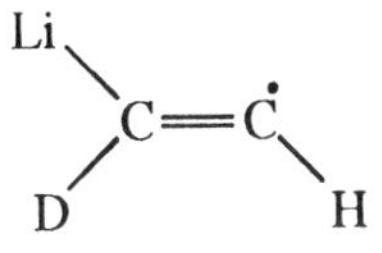

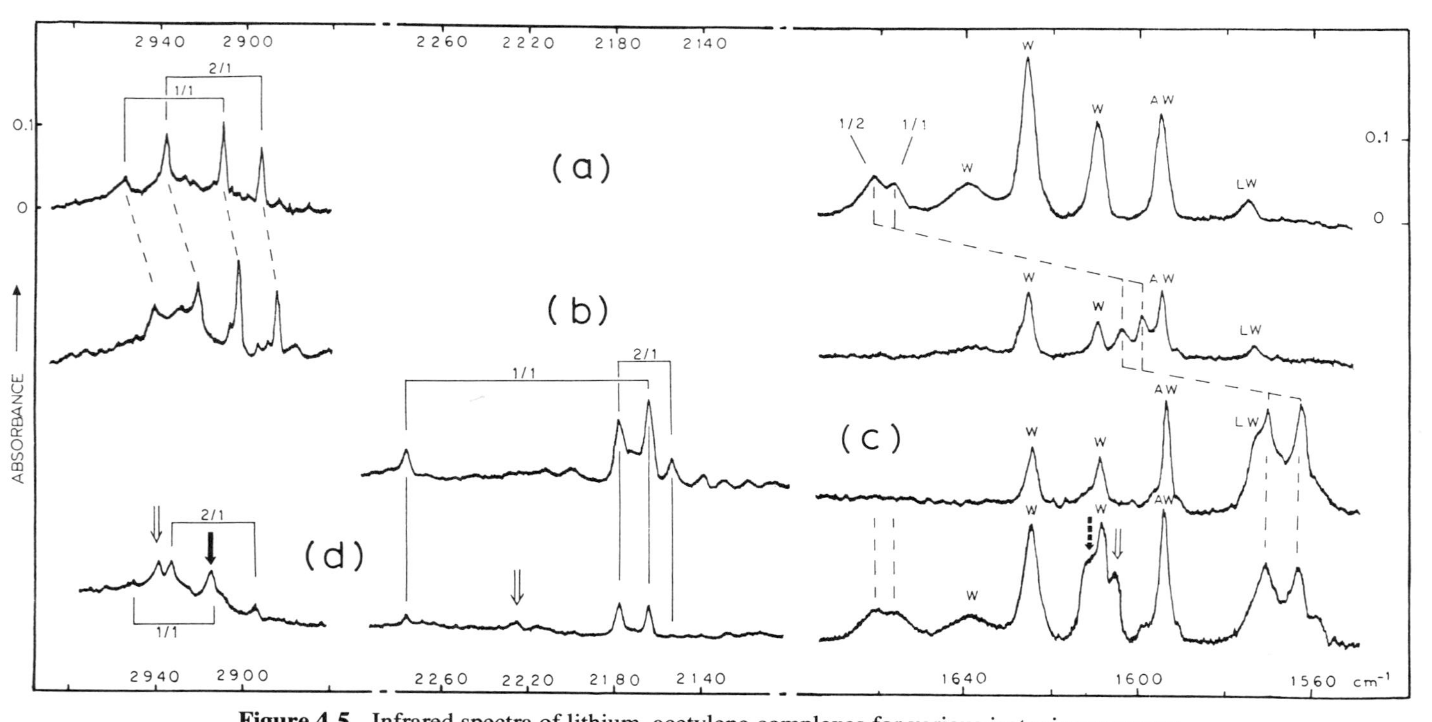

Figure 4.5 Infrared spectra of lithium–acetylene complexes for various isotopic precursors. (*a*) Li + C_2H_2; (*b*) Li + $^{13}C_2H_2$; (*c*) Li + C_2D_2; (*d*) Li + C_2H_2, C_2HD, C_2D_2 mixture. Open arrows designate absorptions due to the Li acetylene complex (1/1), bold arrows designate the dilithium–acetylene complex (2/1). The signals marked by 1/2 are attributable to lithium–acetylene solvated by another acetylene molecule. Bands marked with W are due to water impurity (from ref. 40, used with permission).

The two C—H symmetric and antisymmetric stretching motions are shifted a long way down from their values for free acetylene itself (2952 and 2907 vs. 3374 and ≅3290 cm^{-1}), indicating substantial rehybridization of the acetylene unit. The symmetrical C—H motion comes at higher frequency than the antisymmetrical one as a result of the residual coupling with the C=C stretching motion of same symmetry. Interestingly, the coupling between the carbon–carbon and symmetrical carbon—hydrogen stretching coordinates is much reduced compared with free acetylene. The "ν_{CC}" motion, for instance, experiences only a 95-cm^{-1} shift in LiC_2H_2 with respect to LiC_2D_2, compared to about 210 cm^{-1} for free acetylene. This reduction also contains structural information, as it results, on the one hand, from the weakening of the carbon–carbon bond, and, on the other hand, from the bending of the C_2H_2 group.

Normal coordinate analysis using an harmonic model and the isotopic data for LiC_2H_2, $Li^{13}C_2H_2$, LiC_2D_2, LiC_2HD and their ^6Li and ^7Li isotopic counterparts were best fit with the following C_{2v} structure

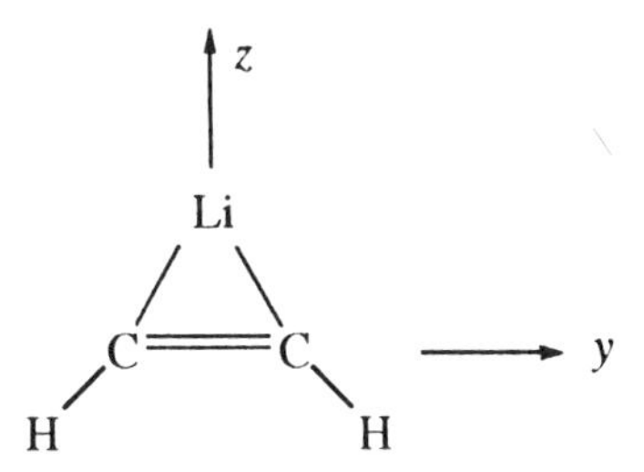

with $C\hat{C}H$ and $C\hat{L}iCH$ bond angles of 140 ± 10° and 40 ± 5°, respectively.

As can be suspected from the large frequency shifts, the carbon–carbon bond is clearly the most affected by the metal fixation: the F_{CC} force constant is decreased by 40% from 16 to about 10.5 mdyne/Å. Empirical relationships between carbon force constants are useful to translate the extent of the corresponding geometry change (Fig. 4.6). A C=C distance in the vicinity of 1.25–1.3 Å can be inferred, intermediate between a double and a triple bond value.

9.1.2 ESR Studies. Although vibrational spectroscopy provided information regarding bond strengths and geometrical structure of the LiC_2H_2 species, it cannot distinguish between two possible bonding schemes for the lithium/Π-system interaction. Both the Li s and p_z atomic and the acetylene Π_{zy} molecular orbitals, as well as the Li p_y and acetylene Π^*_{zy} orbitals, have the correct symmetry to interact. Depending on the strengths of these interactions, there are two possible ground states for LiC_2H_2, corresponding to schemes A and B in Figure 4.7.

Placing the unpaired electron in the Li 2s-Π antibonding combination before the Li 2p-Π* bonding combination would result in a $^2A_1$ ground state with the unpaired electron in a molecular orbital having still substantial 2s character, while the reverse possibility would correspond to a $^2B_2$ ground state with un-

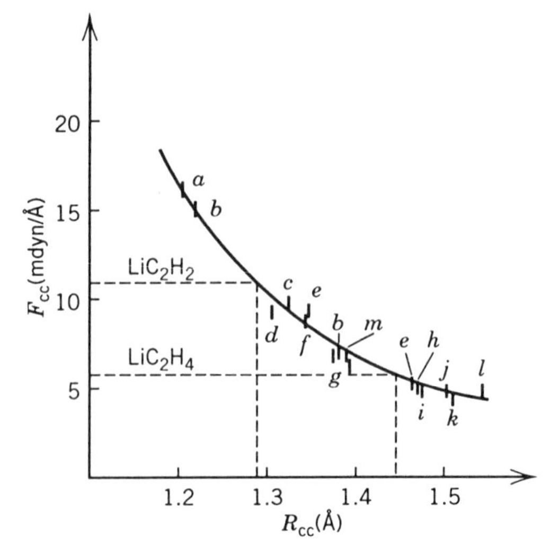

Figure 4.6 Comparison of carbon–carbon bond force constant for LiC_2H_4 and Li-C_2H_2 complexes with known force constants and C–C bond distances for some stable hydrocarbons. (*a*) C_2H_2; (*b*) diacetylene; (*c*) C_2Cl_4; (*d*) cyclopropene; (*e*) butadiene; (*f*) C_2H_4; (*g*) C_6H_6; (*h*) sytrene; (*i*) ethylene oxide; (*j*) toluene; (*k*) cyclopropane; (*l*) C_2H_6; (*m*) Zeise's salt.

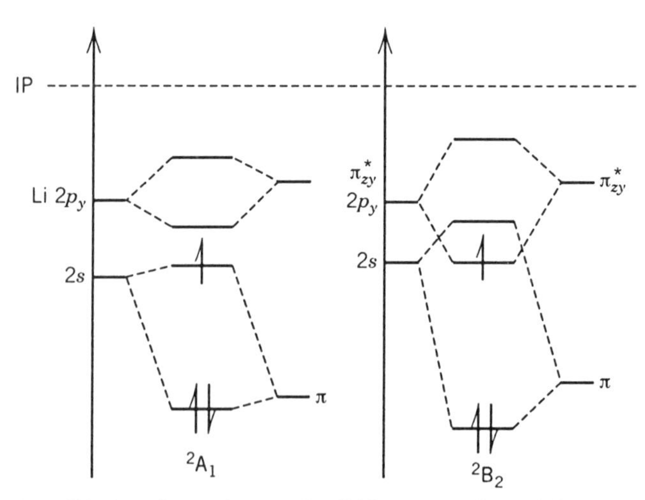

Figure 4.7 Possible bonding schemes for lithium atom $2s$ and $2p_x$ levels with a π system.

paired electron density in an orbital with large Li $2p$ partial character and virtually no $2s$.

Figure 4.8 presents the ESR spectra of the LiC_2H_2 molecule for the two isotopes of lithium (^6Li, I = 1, μ_I = 0.822 nm and ^7Li, I = $\frac{3}{2}$, μ_I = 3.257 nm) superimposed to the unreacted atom signals.[42] As LiC_2H_2 contains two magnetic nuclei Li and H, one observes a double hyperfine structure (triplet of triplet for $^6LiC_2H_2$, triplet of quadruplet for $^7LiC_2H_2$) complicated by spectral anisotropy. The change in hyperfine structure when going from $^6LiC_2H_2$ to $^7LiC_2H_2$ indicates that the larger hyperfine structure ($\simeq$ 170 MHz) is due to interaction with the two equivalent H atoms), while the Li hyperfine structure is the smallest (25–18 MHz). Such values demonstrate a 93–96% reduction in the Li hyperfine splitting.[41]

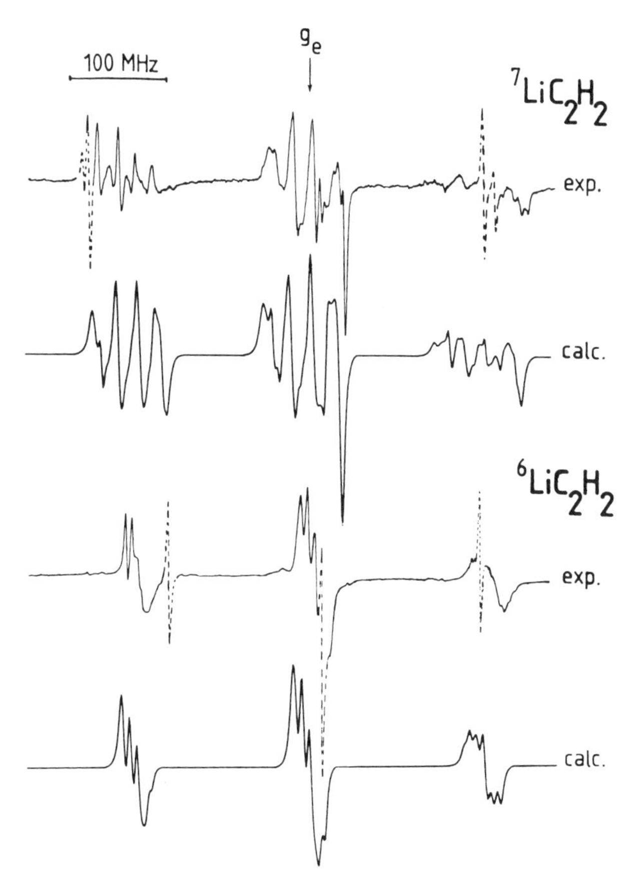

Figure 4.8 Comparison of experimental and calculated X-band ESR spectra of LiC_2H_2 in solid argon. Top spectrum: $^7Li + C_2H_2$; bottom spectrum: following parameters: $g_x = g_y = 2.0039$, $g_z = 2.0006$; $A_{Li_x} = 26$ MH_2; $A_{Li_z} = A_{Li_y} = 17$ MHz for ^7Li and corresponding values for ^6Li; $A_{H_x} = 167$, $A_{H_y} = 185$, $A_{H_z} = 180$ MHz. G_e designates the free electron resonance position.

More quantitatively using standard powder spectra simulation procedures, as presented in Figure 4.8, it is possible to estimate the absolute value of the elements of the hyperfine interaction tensor A_{Li}. Assuming that the singly occupied molecular orbital (SOMO) of the complex can be approximated in the vicinity of the Li nucleus by a combination of Li 2s or 2p atomic orbitals, one can evaluate the Li 2s or 2p character in the electronic wavefunction of the unpaired electron. The A_{Li} can be written as the sum if isotropic and anisotropic contributions:

$$A_{Li} = \begin{vmatrix} A_{\perp} & 0 & 0 \\ 0 & A_{\perp} & 0 \\ 0 & 0 & A_{\parallel} \end{vmatrix} = \begin{vmatrix} |A_{iso} - A_{dip}| & 0 & 0 \\ 0 & |A_{iso} - A_{dip}| & 0 \\ 0 & 0 & |A_{iso} + 2A_{dip}| \end{vmatrix}$$

where $A_{iso} = 800\,(\mu_I/I)\,|\psi_{(0)}|^2$ (in MHz) represents the Fermi contact term, proportional to the Li 2s character and $A_{dip} = 3.2\,(\mu_I/I)\,\langle (a_0/r)^3 \rangle_{av}$ (in MHz) for the magnetic dipolar interaction due to Li 2p character.[43] An A_{iso} value of 20.7 ± 0.5 MH_2 corresponding to at most 5% Li 2s character and a small but nonnegligible A_{dip} term ($\simeq$2.5 MHz) can be deduced, which proves that the LiC_2H_2 has a $^2B_2$ ground state (corresponding to scheme B). The substantial unpaired electron density at the H atoms (10–15% of the free H atom hyperfine splitting) also demonstrates a large amount of metal-to-acetylene charge transfer.

Two ab initio studies lend further support to these conclusions. A first study calculated that the formation of *cis*-bent LiC_2H_2 was energetically most favored, the Li-vinylidene rearrangement, calculated only 2 kcal/mole more stable, being prohibited by a high energy barrier.[44] The predicted geometry is here in excellent agreement with the experimental estimation (R_{CC} = 1.267 Å, CCH and CLiC bond angles 138.9 and 36.6°, respectively). The second study predicted that the LiC_2H_2 molecule should have a $^2B_2$ ground state rather than $^2A_1$.[45]

9.2 The $Li(C_2H_4)_{1,2,3}$ and $Li_2C_2H_4$ Molecules

The chemistry of lithium atoms with ethylene is surprisingly richer than that with acetylene.[46] At least five different products can be evidenced which demonstrate different Li/C_2H_4 stoichiometries. As for LiC_2H_2, it can be shown spectroscopically by the use of partial asymmetrical deuteration [Li_n-$(CH_2CD_2)_m$] that the lithium bridges symmetrically the carbon–carbon bond. Such structures are compatible with the structures presented in Figure 4.9. The bridging position is further confirmed by the $^6Li/^7Li$ or $^{12}C/^{13}C$ isotopic shifts on the 400-cm^{-1} Li—C_2H_4 symmetrical stretching mode which would be too large and too small, respectively, for a simple C—Li oscillator.

The LiC_2H_4, $Li(C_2H_4)_2$, and $Li(C_2H_4)_3$ species are mainly characterized by

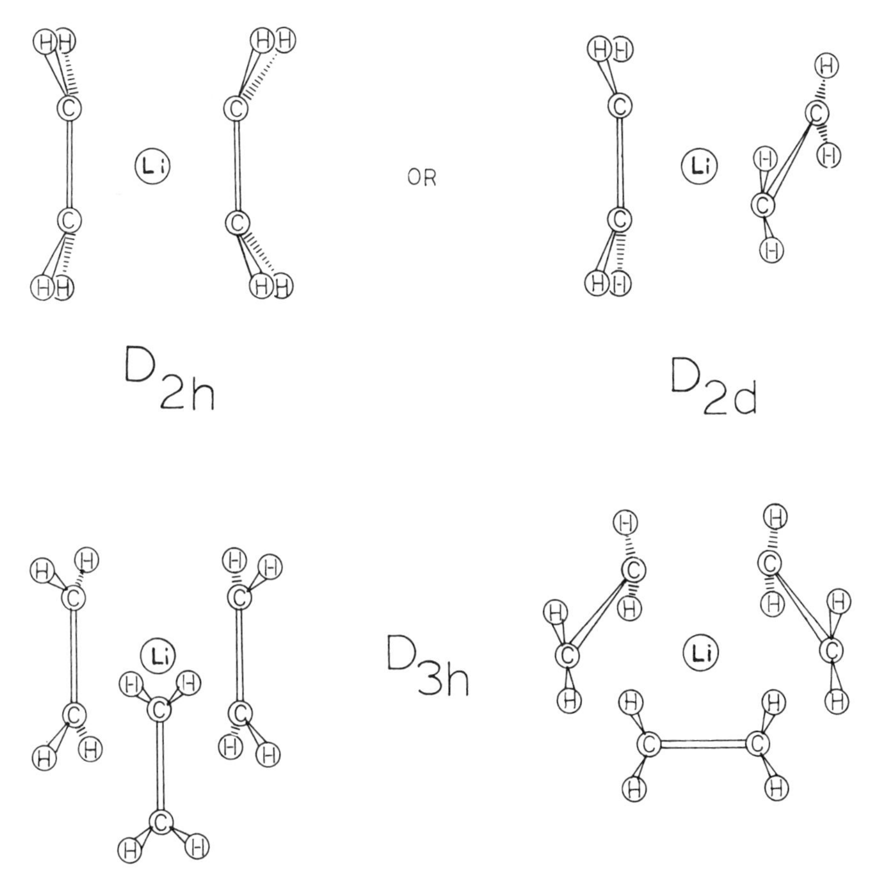

Figure 4.9 Possible structure for $Li(C_2H_4)_n$, $n = 1, 2, 3$ complexes.

intense IR absorptions in the C=C and CH_2 scissoring motions range, showing large H/D shifts but small or nonexistent $^6Li/^7Li$ isotropic shifts and absorptions in the low frequency region, demonstrating more substantial $^6Li/^7Li$ isotopic shifts.

9.2.1 LiC_2H_4. The vibrational spectrum of LiC_2H_4 bears some striking similarities to that of ethylene bonded to other metals.[46–48] The modes involving the C—C stretching and CH_2 in-phase scissoring motions are very strongly IR-activated; also seen were the other symmetrical vibrations—CH_2 wagging and Li—C symmetrical stretch at about 700 and 370 cm^{-1}.

The normal coordinate analysis could quantify:

1. The extent of the perturbation of the C=C bond, again by far the most affected by the complexation. The C=C bond force constant is lowered by about 30%, suggesting a R_{CC} distance in the vicinity of 1.4 Å (see Fig. 4.3).

2. The relative weakness of the metal–carbon force constant in comparison with other metal–ethylene complexes (0.33 vs. 0.58 mdyne/Å).

3. The importance of the charge transfer effects on the spectroscopic properties of the complex. On the one hand, one observes a substantial vibrational coupling between the C—C and the C—Li motions (presence of a small $^{6}Li/^{7}Li$ shift on the modes involving the C—C coordinate). On the other hand an analysis of the infrared intensities shows that the main contribution to the infrared activity comes from the carbon–carbon oscillator itself. That is, fluctuation of the C—C distance produces a very large variation of the overall dipole moment of the complex, in the direction of the metal, perpendicularly to the C—C bond itself. This conclusion shows that it is not necessary to invoke large puckering distortion of the ethylene group to explain the large infrared intensification of the C—C coordinate.

Figure 4.10 presents the ESR Spectrum of Li + C_2D_4 deposited in solid argon. In the central part of the spectrum, which corresponds to the lithium ethylene complex, the small lithium-7 hyperfine structure of the complex is still clearly recognizable, though much reduced in comparison with the free Li atoms. As for LiC_2H_2, a drastic reduction ($\simeq$ −90%) of Li 2*s* character has taken place in the molecular orbital occupied by the unpaired electron.

This means that the energetic cost of promoting the Li valance electron into the 2*p* level is, here again, compensated for by the stabilization resulting from

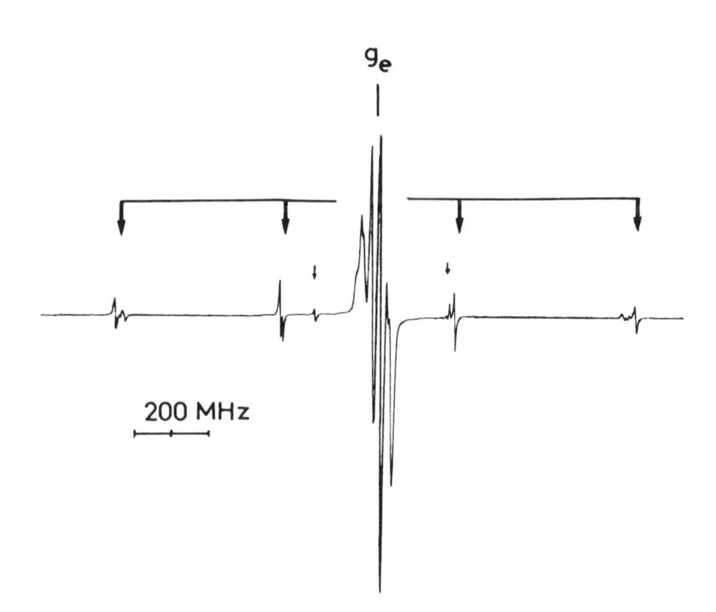

Figure 4.10 X-Band ESR spectrum for Li + C_2D_4 deposited in solid argon. The arrows designate unreacted atomic lithium lines and g_e represents the three-electron resonance position.

the good overlap between the Li $2p$ and ethylene Π^* lowest unoccupied molecular orbital (LUMO). It results in a substantial charge transfer of the Li valence electron in the π^* LUMO of ethylene, leading to the lengthening of the C—C distance and to the appearance of a large permanent dipole moment, oriented perpendicularly to the C—C bond.

9.2.2 $Li(C_2H_4)_2$ and $Li(C_2H_4)_3$. Likewise, the use of isotopic mixtures enabled us to demonstrate that for di- and triethylene lithium, the $Li(C_2H_4)(C_2D_4)$ and $Li(C_2H_4)_2(C_2D_4)$ species have only one isotopomer and therefore two and three equivalent ligand molecules (see Fig. 4.9).

These isotopic studies also showed that the ethylene molecules do not vibrate independently one from another when bonded on the same lithium. All in-phase C—C motions are at higher energy than out-of-phase ones, showing that stretching of one of the C—C bonds within a complex favors shortening of the other(s). This observation is coherent with the calculated increase of the F_{CC} force constant toward the free ethylene value when successively two and three molecules are bonded to the metal [-30% in LiC_2H_4, -21% in $Li(C_2H_4)_2$, and -16% in $Li(C_2H_4)_3$].

9.2.3 $Li_2C_2H_4$. A fourth species was identified which contained more than one Li atom, with at least one of them bridging the C—C bond. This species however, did not contain equivalent Li atoms nor did it present an inversion center. Such findings are compatible with a structure of the type

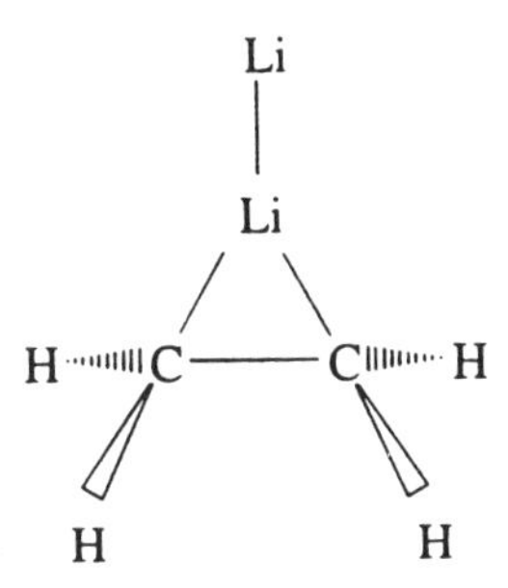

with a C—C bond slightly more perturbed and Li—C interactions somewhat stiffer, as in the monolithium complex LiC_2H_4. A normal coordinate analysis showed that such a model is capable of reproducing satisfactorily the measured isotopic shifts on the 12 observed fundamentals (out of 18). However the Li—Li vibration, lying presumably in the far-infrared range, but crucial to the positive identification of such a species, was not observed.

An ab initio study of the isomers of 1,2 dilithioethane[51] pointed to more symmetrical structures and did not consider the one suggested by our experimental findings. Also, this structure may be only an energy subminimum kinetically stabilized in our experimental conditions by a sizable barrier to rearrangement. However, the fact that addition of a second metal atom takes place preferably on the first metal center, rather than on the other side of the C—C bond, could also indicate why an α,β-dilithioalkane reagent has so far, to our knowledge, not been synthetized.

9.3 Ternary $Li/C_2H_4/N_2$ Complexes

It was observed that complexes of Li with ethylene or other alkenes have the ability, unique among the unsaturated hydrocarbon series, to fix one to several nitrogen molecules.[52] Let us recall that at the Li + N_2 isolated-pair level, no strong interaction was observed in an inert rare gas matrix (Fig. 4.11*a*). The LiN_2 and $N_2Li_2N_2$ species with strongly perturbed N_2 subunits were only observed in neat nitrogen.[29]

With ternary mixtures $Li/C_2H_4/N_2$ one observes a number of $Li(C_2H_4)_x(N_2)_y$ mixed complexes which are formed spontaneously even in inert rare gas matrices (see Fig. 4.11*b–d*). These complexes are characterized by very intense N≡N stretching signals in the 2200–2300 cm^{-1} range and by other IR signals

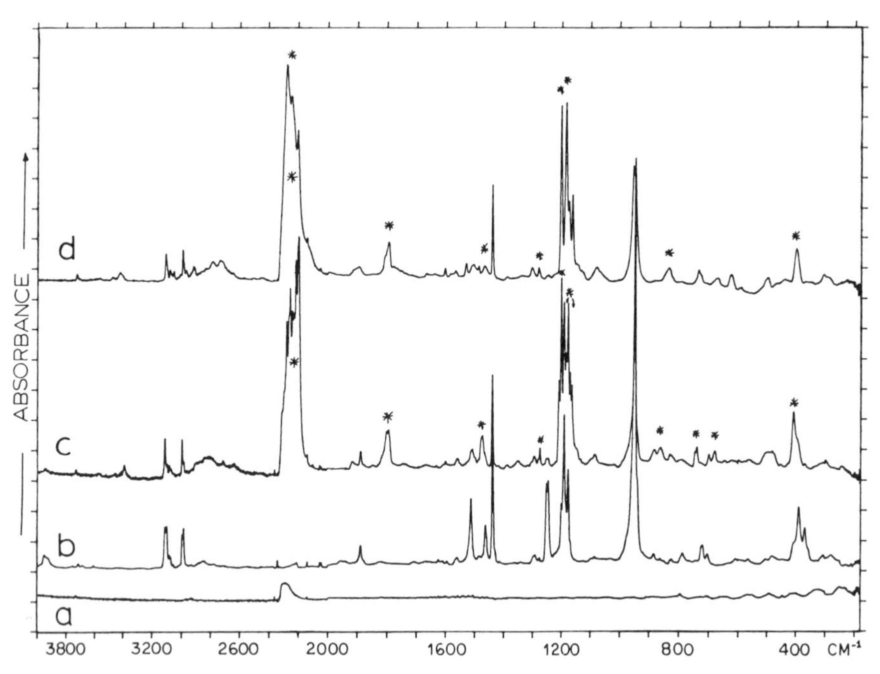

Figure 4.11 Coordination of molecular nitrogen by atomic lithium assisted by ethylene. Infrared spectra observed for a mixture of ethylene and/or nitrogen in argon deposited with Li atoms at 15–16 K; (*a*) $Li/N_2/Ar \cong 0.8/4/400$ ($T_{Li} = 445°C$); (*b*) $Li/C_2H_4/Ar \cong 0.3/3/800$ ($T_{Li} = 430°C$); (*c*) Li/C_2H_4/N/i2/Ar ≅ 1/1/8/800 ($T_{Li} = 445°C$); (*d*) $Li/C_2H_4/N_2/Ar \cong 1/1/270/800$ ($T_{Li} = 445°C$). The asterisks designate products involving lithium, ethylene, and nitrogen (from ref. 52, used with permission).

corresponding to the Li-bonded C_2H_4 subunit falling in the close vicinity of the LiC_2H_4 and $Li(C_2H_4)_2$ signals. Species corresponding to $Li(C_2H_4)(N_2)_y$ with $y = 1$–3 and N≡N stretching modes ranging from 2195 to 2275 cm^{-1} respectively are very instable and require large excesses of nitrogen to be observed. The observed ν_{NN} stretching modes are about 100 wavenumbers higher than with transition metal complexes. Considering also the very small perturbations of the LiC_2H_4 subunit with respect to uncoordinated LiC_2H_4, the nitrogen molecules must hence be very weakly interacting. The stability of the mixed complexes increases noticeably with the number of ethylene ligands: the $Li(C_2H_4)_2(N_2)$ complex is stable up to the temperature at which the rare gas matrix sublimates. It contains two equivalent ethylene ligands and a single nitrogen molecule characterized by an intense ν_{NN} structure of C_{2v} symmetry (Fig. 4.12) with a value of $140 \pm 15°$ for the angle θ between ethylene ligand axes, as deduced from the relative intensity ratios of the A_1 and B_2 symmetry motions of the coupled ethylene subunits.

A fifth species was observed, favored by high lithium concentrations, which displays a N=N stretching mode at 1795 cm^{-1} with equivalent nitrogen atoms, characteristic of the $Li^+N_2^-$ ion pair with side-on bonding of the lithium atom.[29] It must, however, be emphasized that this species was *not* observed in solid argon *without* ethylene added to the sample (see Fig. 4.11*a*). The catalytic role played by ethylene remains mysterious, but both the nonlinear lithium dependence of $Li^+N_2^-$ and the fact that $Li(C_2H_4)_{1,2}$ complexes add N_2 more readily than Li itself, suggest a two-step mechanism of the type

$$LiC_2H_4 \xrightarrow[+N_2]{} (N_2)Li(C_2H_4) \xrightarrow[+Li]{} LiC_2H_4 + Li^+N_2^- \quad (9)$$

It should be noted at this point that the assisted complexation of nitrogen to lithium (observed also with methyl-substituted ethylene and butadiene) is a characteristic property of lithium–alkene complexes, as experiments performed

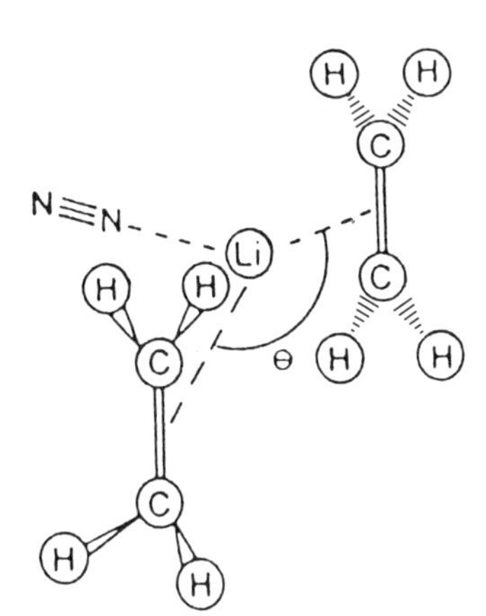

Figure 4.12 Structure of the lithium diethylene mononitrogen complex.

with other lithium complexes did not yield, so far, such ternary complexes. If one can easily envision that the fractional positive charge on the lithium in LiC_2H_4 and $Li(C_2H_4)_2$ facilitates the coordination of N_2 with, presumably, a σ-donation to lithium, and possibly, to a weaker extent, p-donation from the metal, it is difficult to rationalize why for instance LiC_2H_2 and LiC_2H_4 (in which the SOMO have very similar Li s and p characters, as deduced from ESR studies) behave so differently with respect to nitrogen.

9.4 Li–Benzene Complexes

The reaction of lithium atoms with benzene molecules provides an interesting model for the interaction of lithium atoms with larger aromatic systems, such as encountered in intercalation compounds with aromatic systems, where intercalation compounds with graphite, LiC_6, correspond to the highest stoichiometry.[53,54] Two ab initio studies have been published[55,56] which rely on calculations made with unrelaxed benzene geometries and the comparison with experimental data is needed. Two complexes corresponding to LiC_6H_6 and $Li(C_6H_6)_2$ were identified in the vibrational spectra of Li + benzene deposits in solid argon[57] (see Fig. 4.13).

With the help of isotopic substitution and mixtures (^{6}Li, ^{7}Li, C_6H_6, C_6D_6), some important structural information was derived from the following observations.

1. The strong infrared activation of some of the IR-inactive ring-stretching motions (ring breathing, ring deformation) demonstrates the loss of the sixfold symmetry axis.
2. The magnitude of the ^{6}Li/^{7}Li isotopic shift on the observed metal-ring stretching motion indicates that the Li atom occupies an axial position. These two observations combined together limit the overall symmetry to, at most, C_{3v} or C_{2v} for the "half-sandwich" compound LiC_6H_6.
3. The position of distinct signals corresponding to $Li(C_6H_6)(C_6D_6)$ in isotopic mixtures indicates that the complex contains two equivalent benzene ligands, whose ring-stretching motions are in fact coupled into two sets of vibrations (in-phase and out-of-phase), the higher set of which is either not IR-active or is very weak. Given the axial position proposed for the LiC_6H_6 complex, mutual exclusion due to the presence of an inversion center in a sandwich structure seems very likely.

The observation of a vibrational coupling between the ring-breathing and a former E_{2g} symmetry C—H deformation further indicates that, for the $Li(C_6H_6)_2$ sandwich compound, the overall symmetry is at most D_{2h} (Fig. 4.14). For both mono- and dibenzene lithium, the new IR-active vibrations are, in addition to metal–carbon stretches, skeletal vibrations corresponding to ν_1, ν_8, or ν_9 of free benzene, formerly of A_1 and E_{2g} symmetry. In other words, strong variations of global dipole moment occur in the complexes for normal

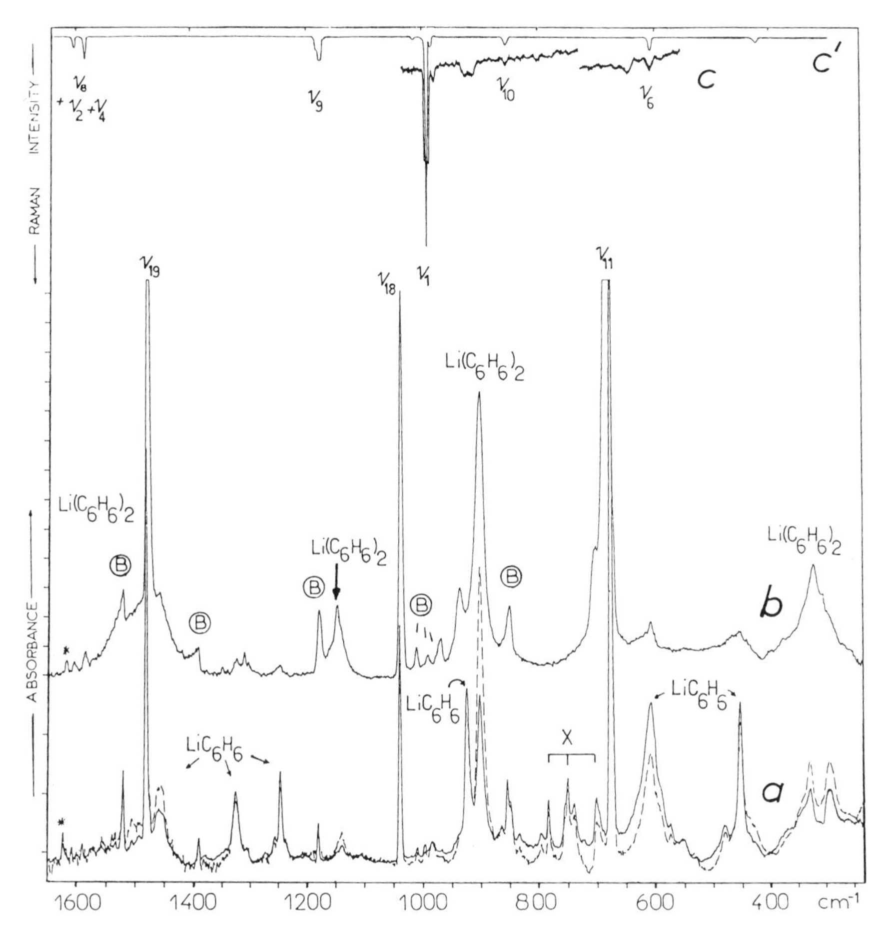

Figure 4.13 Comparison of IR and Raman spectra for various lithium benzene–argon samples in the 1600- to 250-cm^{-1} region. (*a*) $Li/C_6H_6/Ar$ (1.2/1/1000) (plain line after deposition, dotted line after annealing to 30–35 K). (*b*) $Li/C_6H_6/Ar$ (0.8/5/1000). (*c*) $Li/C_6H_6/Ar$ (2.4/10/1000) (exc 514.4 nm, 40 mW power). (*c'*) C_6H_6/Ar (0.0/10/1000), same as *a* but without metal. B designes parent molecule combinations; (*) represents water impurity absorptions (from ref. 57, used with permission).

modes where all C—C distances increase or corresponding to a D_{2h} distortion of the ring. Thus metal-to-ligand charge flows are not only related to C—C or Li—C distance variations, but also to C_{2v} or D_{2h} distortions of the lithium-capped benzene.

Unfortunately, too few modes were observed to yield a complete normal coordinate analysis, as was prepared for the other hydrocarbon complexes. A reduction of at least 10% on the *average* carbon–carbon force constant can however be estimated. This is substantially more than the 4% observed for $Cr(C_6H_6)_2$,[58] which keeps a D_{6h} symmetry and is bound according to a different mechanism.

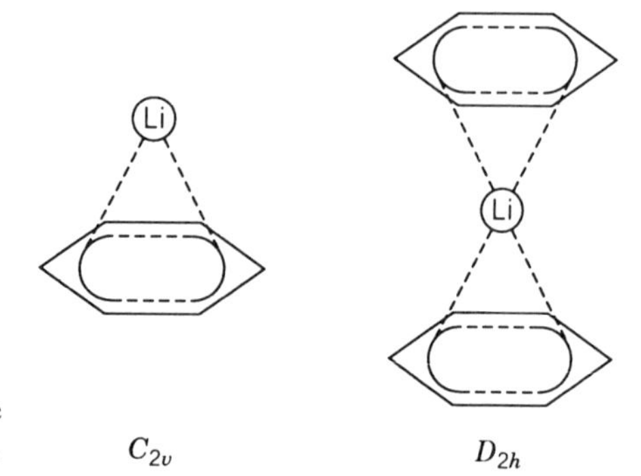

Figure 4.14 Most likely structures of the lithium mono- and dibenzene complexes.

In the ESR spectra of the lithium mono- and dibenzene complexes the hyperfine structure due to the lithium nucleus has completely disappeared.[41] The prominent triplet of quintets hyperfine structure observed with LiC_6H_6 disappears completely using LiC_6D_6 (Fig. 4.15). Hence, the unpaired electron is situated in a molecular orbital with no measurable Li 2*s* or 2*p* character, that is, it is practically completely transferred to the benzene ligand. In a Jahn–Teller situation such as the benzenide anion, addition of an extra electron to one of the two degenerate LUMOs of E_{2g} symmetry will lead to a molecular distortion with the degeneracy removed. The possible carbon skeleton distor-

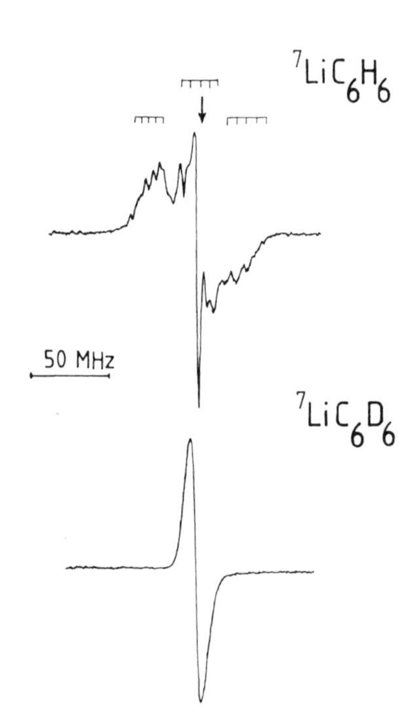

Figure 4.15 X-Band ESR spectra of $^7LiC_6H_6$ and $^7LiC_6D_6$ complexes isolated in solid argon at about 10 K. The arrow designates the free electron resonance position.

tions should belong to the same E_{2g} representation. We believe that such a distortion is identified here for LiC_6H_6 and $Li(C_6H_6)_2$. This mechanism is conceivably backed by electron back-donation from the filled E_{1g} HOMO of benzene into the empty Li $2p_x$ and $2p_y$ levels (if z is taken along the C_2 axis). This could explain why in our experiment, as well as in others,[59,60] no interaction was observed at the atom + molecule pair level between benzene and Na, where the $3p$ orbitals are more diffuse and further apart in energy from the $3s$ level than $2p$ from $2s$ in Li and, therefore, a bonding interaction is less likely to occur in spite of a slightly smaller ionization potential. It is interesting to note that $Li(C_6H_6)_2$ can also be synthesized by lithium vapor deposition in neat benzene and decomposes slowly in vacuo at temperatures between 150 and 200 K, which would correspond to binding energies of a few kilocalorics per mole.

9.5 Li/Phenylacetylene and Styrene Complexes

After studying the structure and bonding of Li atoms to some model unsaturated systems, it was desired to compare, at least qualitatively, the binding energies of lithium when CC triple or double bonds coexist on the same ligand as a benzene ring (phenylacetylene or styrene).

As deduced from the infrared spectra,[61] two different isomers of the lithium complexes are formed by reaction of ground state Li atoms with either phenylacetylene or styrene.

With the help of isotopic studies, it was possible to show that each complex has one isomer corresponding to complexation of the lithium atom on the arene ring and the other corresponding to either a triple or double carbon–carbon bond. It was also shown that each isomer has distinct absorptions in the UV and visible range. For phenyl acetylene, selective electronic excitation in the visible range leads to interconversion between the two isomers, red-yellow light converts the ''ethynyl-group-complex'' into the ''ring-complexed'' structure, but blue light only achieves partial reconversion. Excitation with large excess energy in the UV range (3.8 eV) or temperature effects produce only photoisomerization into the more stable Li–arene complex. For the Li–styrene complex, both isomers also coexist, but differences in stability could be clearly evidenced. Both diphenylacetylene and distyrene complexes are also observed, and have structures comparable to the dibenzene species.

10 LITHIUM CARBONYL COMPLEXES

The products observed in the reactions of lithium with carbon monoxide molecules in various rare gas matrices[62] can be divided into three groups: (1) $Li(CO)_n$ monometallic species with $n = 1$–4 or more, with ν_{co} stretching frequencies in the 2060- to 1700-cm^{-1} range (Fig. 4.16): (2) polymetallic $Li_m(CO)_n$ species where $m = 2, 3$ and $n = 1, 2$ with CO stretching frequencies

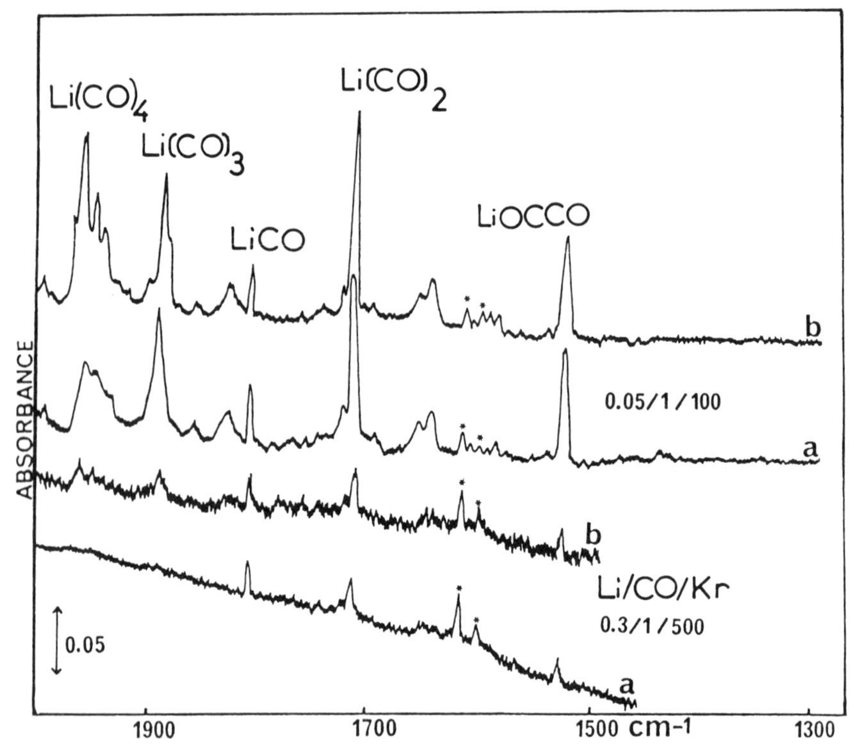

Figure 4.16 Infrared spectra of various monolithium carbonyl complexes isolated in solid krypton. (*a*) After deposition at 10 K; (*b*) after warmup to enable molecular diffusion (from ref. 62, used with permission).

ranging from 1600 to 1470 cm^{-1}, in which each carbon monoxide molecule can be considerated as a separately coordinated carbonyl group; and (3) a class of compounds in which true chemical bonds between carbonyl groups have been formed. For the heavy alkali atoms, only species of the second and third groups can be observed.[63,64]

10.1 Monolithium Carbonyl Complexes

Four monolithium species can be identified using vibrational spectroscopy based on CO concentration dependencies and isotopic mixtures. Although the carbonyl groups remain separately bonded on the metal center, delocalized electronic structures cause strong vibrational coupling between the carbonyl entities, which thus do not vibrate independently but see their motion coupled into in-phase and out-of-phase vibrations. Multiplet patterns result, using $^{12}CO + ^{13}CO$ isotopic mixtures which constitute characteristic signatures of the cluster stoichiometry: doublets for Li_mCO single carbonyl species (Fig. 4.17), triplet and quadruplet patterns for di- and tricarbonyls (Fig. 4.18), and so on (see references 65 and 66 for a complete discussion).

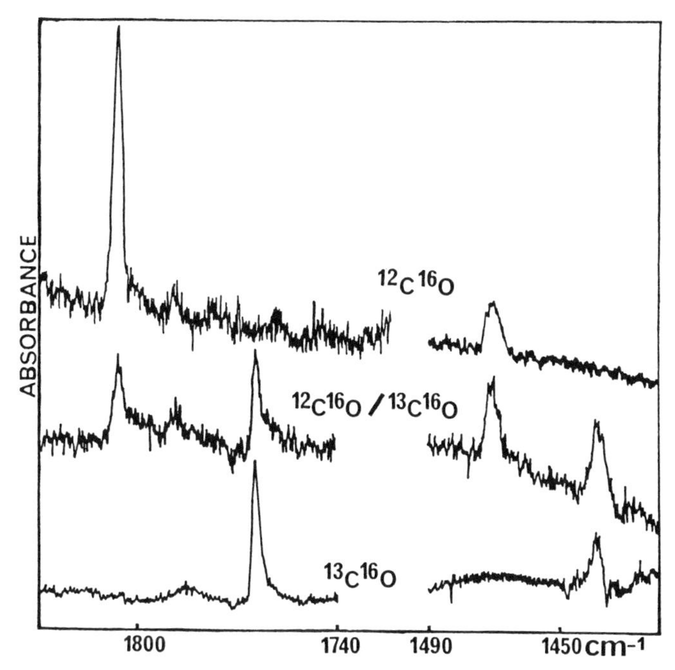

Figure 4.17 Infrared spectra of the CO stretching region for LiCO and Li_3CO complexes for ^{12}CO (top), ^{12}CO + ^{13}CO (middle), and ^{13}CO (bottom) precursors.

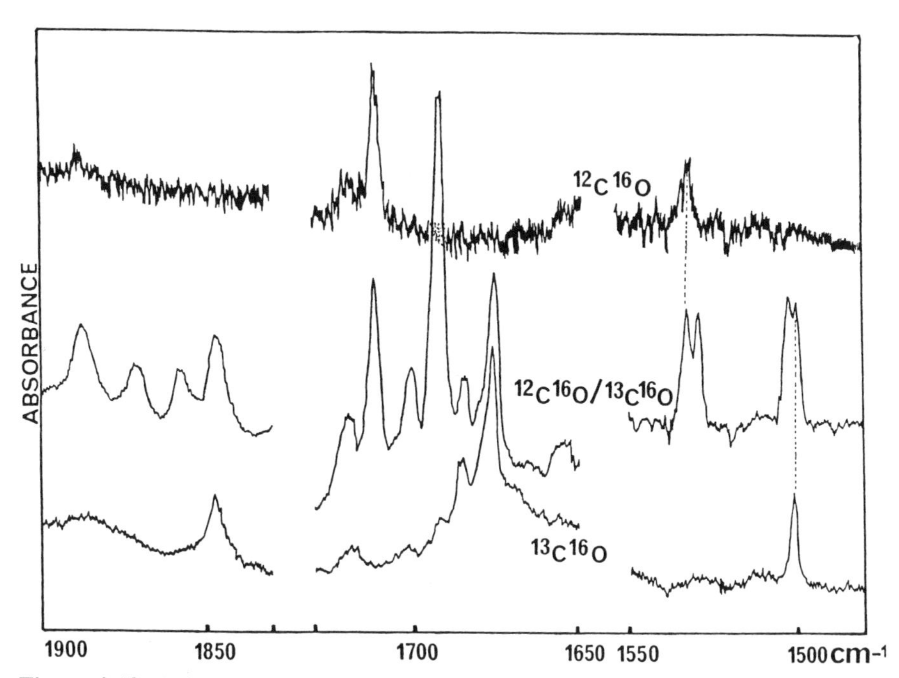

Figure 4.18 Infrared spectra of the CO stretching region for $Li(CO)_3$, $Li(CO)_2$; and LiOCCO species using various isotopic precursors.

The Li—CO complex is characterized by a CO stretching frequency at about 1800 cm^{-1} and a lithium stretching mode around 620 cm^{-1}. The third motion (bending mode) was not observed and it was not possible to deduce from IR data alone the structure of this species. The 340-cm^{-1} reduction of the CO stretching frequency is however remarkable, more than twice as much as for other known metal–carbonyl complexes of the same stoichiometry (e.g., 1995 cm^{-1} for NiCO).

A subsequent ab initio study[67] concluded that LiCO has a linear geometry and a $^2\Pi$ ground state correlating with the Li $^2P_{3/2}$ first excited state. No substantial binding interaction was found for the $^2\Sigma$ state of the complex (corresponding with the Li $^2S_{1/2}$ ground state). The Li—C and CO distances of 1.930 and 1.143 Å, respectively, were calculated at the 6-31G* level, and both calculated frequencies and intensities were in qualitative agreement with the experimental values, which provide a crucial test for the identification of the ground state electronic structure. Only the $^2\Pi$ ground state is characterized by a large red shift in the C—O stretching vibration (−32.7% calculated at the 6-31G* level vs. −29.5% experimentally), while the $^2\Sigma$ state of the same molecule, as well as all calculated states of the Li− −OC configuration, are characterized by frequencies that are almost unaffected by complexation. The following diagram indicates the charge transfer processes of Li—CO, as they can be deduced from the Mulliken populations:

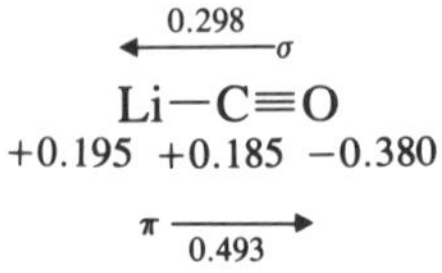

$$\overset{\xleftarrow{0.298}\ \sigma}{\mathrm{Li{-}C{\equiv}O}} \qquad {}^2\Pi \text{ ground state}$$

$$+0.195 \quad +0.185 \quad -0.380$$

$$\pi \xrightarrow[0.493]{}$$

Here the numbers under the atoms give the Mulliken net charges in electrons, while the numbers near the arrows give the numbers of electrons transferred from the σ and π systems, respectively, in the directions indicated by the arrows. In the $^2\Pi$ state, about 0.5 e is transferred from the Li atom to the antibonding $*\pi$ orbital of CO, and 0.3 e is back-donated in the empty $2s$ orbital of Li. The net charge transfer (0.2 $\bar{e}$) is substantial, but by no means complete, and should produce a large dipole moment (on the order of 5 D). Both charge transfers weaken the CO bond and have an additive effect on the red shift of the C—O stretching frequency, which has been analyzed in detail.

The $Li(CO)_2$ complex gives rise to two sets of C—O stretching bands in the IR spectrum situated around 1715 and 1995 cm^{-1} which originate from slightly different conformations due to different trapping sites in the matrix. This and the isotopic pattern prove that the $Li(CO)_2$ complex has a C_{2v} symmetry structure, which contradicts the ab initio calculations that predicted a linear $D_{\infty h}$ structure but, nevertheless, a very "soft" C—Li—C coordinate. Making the common (but probably erroneous) assumption that the total dipole moment derivative $\partial\vec{\mu}/\partial r_{co}$ is directed exactly along the C—O bond, the C—Li—C angle θ can be deduced from the measured intensity ratio between in-phase and out-of-phase modes using the relationship

$$I_s/I_a = (1 + \cos\theta)/(1 - \cos\theta)$$

The value found varies between 124° and 152° according to the matrix site, which corroborates the suggestion that the Θ coordinate is quite floppy.

For the third species absorbing at 1888 cm^{-1}, quartet patterns typical of symmetrical tricarbonyl species are observed in isotopic mixture experiments (Fig. 4.18). A second C—O stretch at 2065 cm^{-1} indicates a nonplanar C_{3v} symmetry. With the results of normal coordinate analysis the angle θ between the C—O axis is determined from the relationship

$$I_s/I_d = (1 + 2\cos\theta)/(1 - \cos\theta)$$

using the experimental intensity ratio of the symmetrical and degenerate C—O modes, and a value of 114° is obtained, close to the value of 120° for a planar D_{3h} structure.

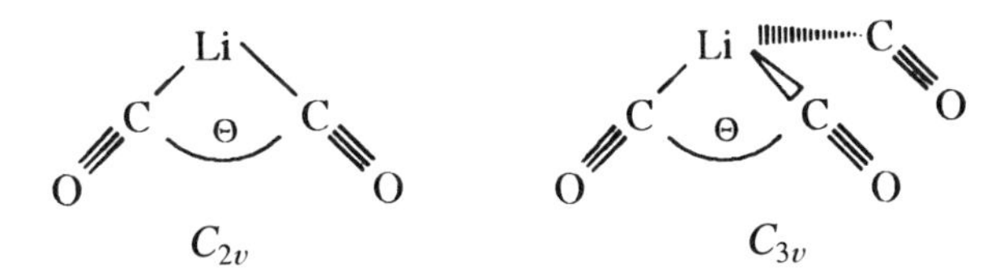

A fourth monometallic species was identified by a C—O stretching mode at 1965 cm^{-1} and was shown to contain more CO than $Li(CO)_3$. However, a certain identification could not be obtained from concentration studies, and its poorly resolved isotopic pattern is compatible with that of a $Li(CO)_4$ species with D_{4h} symmetry as well as with that of a hexacarbonyl complex.

10.2 Polymetallic Complexes

Both Li_2CO and Li_3CO complexes are identified. These have Li dependence markedly different from the Li—CO complex and are characterized by very low frequency ν_{co} at 1596 and 1470 cm^{-1} respectively. Another species having a C—O stretch in this frequency range absorbs at 1613 cm^{-1} but contains two equivalent carbonyl groups, as inferred from its isotopic pattern obtained with isotopic mixtures. Its relative metal dependence also indicates that it contains two lithium atoms, therefore identifying this species as $Li_2(CO)_2$ with moderately coupled CO molecules. Since the other in-phase stretching motion was not observed, this species is presumably of D_{2h} symmetry.

Li
O=C C=O
Li

10.3 Oxocarbon Species

Several species were observed which certainly involve chemically bonded carbonyl groups. These species necessitated that photoexcitation be formed at a low temperature from the abovementioned carbonyl complexes, and were observed with lithium as well as with other, heavier alkali atoms. The simplest and best characterized of these species is identified through the observation of a C—O stretching mode at about 1250 cm^{-1}, with a triplet structure using $^{12}CO + ^{13}CO$ isotopic mixtures, signaling a species containing two equivalent C—O bonds. Their associated force constant is nevertheless quite low ($\cong 9$ mdyne/Å) and closer to a single C—O bond value than to a double C=O bond value.

Observing the isomerization process from the dilithium dicarbonyl complex into this species confirms its identification as dilithium acetylene diolate:[64]

$$Li^+ \quad ^-O-C\equiv C-O^- \quad Li^+$$

This species was first isolated as a white pyrophoric compound in 1933 by Pearson.[68] The structure of the potassium analogue was later determined by Weiss and Büchner by X-ray diffraction with the following internuclear distances:[69] C—O = 1.28 Å and C—C = 1.21 Å. In an ab initio study at the 6-31G level,[70] the electronic structure of the molecule was calculated to be a $^1\Sigma_g$ state, with an almost ionic description, as given by the Mulliken net charges.

	Li	O	C	C	O	Li
Mulliken net charges (e)	+0.72	−1.02	+0.3	+0.3	−1.02	+0.72

$$Li-O-C\equiv C-O-Li$$

Internuclear distances (Å): 1.307 (C—O), 1.197 (C≡C)

Other species containing two nonequivalent molecules and one or two lithium atoms were also evidenced through relatively low-frequency C—O stretching modes at 1530 and 1130 cm^{-1}. Structures of the type Li O—C=C=O and Li OLiC=C=O have been suggested and discussed, which could possibly account for the observed isotopic patterns, but they remain speculative in the absence of further information.

11 CONCLUSION

Alkali metal atoms and small molecules can form complexes with CO and unsaturated hydrocarbon molecules with unfavorable electron affinities. Large differences of reactivities are found, however, between Li and the heavier alkali metals. These should prevent hasty systematic rationalization using simplistic ionic models. Obviously, other mechanisms are at work ($\Pi^* \rightarrow p$ back bonding, stabilization of the alkali metal p level), first, to make the coordination of single atoms possible only for lithium and, second, for species seen with either lithium or the heavier alkalis, to induce systematically larger perturbations of the complexed molecules for the lithium species.

REFERENCES

1. Beutler, H.; Bogdandy, S. V.; Polanyi, M. *Naturwiss.* **1925,** *13*, 711; *14*, 164.
2. Steacie, E. W. R. *Atomic and Free Radical Reactions*, 2nd ed.; Reinhold: New York, 1954.
3. Herschbach, D. R. *Appl. Opt., Supp. 2 Chem. Lasers* **1965,** 128.
4. Andrews, W. L. S.; Pimentel, G. C. *J. Chem. Phys.* **1966,** *44*, 2361.
5. Andrews, L. *J. Chem. Phys.* **1968,** *48*, 972; Andrews, L. *Ann. Rev. Phys. Chem.* **1971,** *22*, 109.
6. Andrews, L.; *J. Chem. Phys.* **1969,** *50*, 4288.
7. Milligan, D. E.; Jacox, M. E. *J. Chem. Phys.* **1971,** *55*, 3404.
8. Tevault, D. E.; Andrews, L. *J. Phys. Chem.* **1973,** *77*, 1640.
9. Tevault, D. E.; Andrews, L. *J. Phys. Chem.* **1973,** *77*, 1646.
10. Andrews, L.; R. R. Smardzewski, *J. Chem. Phys.* **1973,** *58*, 2258.
11. Smith, D. W.; Andrews, L. *J. Chem. Phys.* **1974,** *60*, 81.
12. Alexander, M. H. *J. Chem. Phys.* **1978,** *69*, 3502 and references therein.
13. Andrews, L. *J. Phys. Chem.* **1969,** *73*, 3922.
14. Andrews, L. *J. Chem. Phys.* **1971,** *54*, 4935.
15. Andrews, L.; Hwang, J. T.; Trindle, C. *J. Phys. Chem.* **1973,** *77*, 1065.
16. Smardzewski, R. R.; Andrews, L. *J. Chem. Phys.* **1972,** *57*, 1327.
17. Smardzewski, R. R.; Andrews, L. *J. Phys. Chem.* **1973,** *77*, 801.
18. Jacox, M. E.; Milligan, D. E. *Chem. Phys. Lett.* **1972,** *14*, 518.
19. Manceron, L.; Le Quéré, A. M.; Perchard, J. P. *J. Phys. Chem.* **1989,** *93*, 2960.
20. Andrews, L.; Spiker, Jr., R. C. *J. Phys. Chem.* **1972,** *76*, 3208.
21. Withnall, R.; Hawkins, M.; Andrews, L. *J. Phys. Chem.* **1986,** *90*, 575.
22. Withnall, R.; Andrews, L. *J. Phys. Chem.* **1987,** *91*, 784.
23. Spiker, Jr., R. C.; Andrews, L. *J. Chem. Phys.* **1973,** *59*, 1851.
24. Andrews, L.; Spiker, Jr., R. C. *J. Chem. Phys.* **1973,** *59*, 1863.
25. Andrews, L. *J. Chem. Phys.* **1975,** *63*, 4465.
26. Hass, M.; Griscom, D. L. *J. Chem. Phys.* **1969,** *51*, 5185.
27. Howard, Jr., W. F.; Andrews, L. *J. Am. Chem. Soc.* **1973,** *95*, 2056.
28. Andrews, L. *J. Am. Chem. Soc.* **1976,** *98*, 2147.
29. Spiker, Jr., R. C.; Andrews, L.; Trindle, C. *J. Am. Chem. Soc.* **1972,** *94*, 2401.
30. Holzer, W.; Murphy, W. F.; Bernstein, H. J. *J. Mol. Spectrosc.* **1969,** *32*, 13.
31. Trenary, M.; Schaefer, H. F.; Kollman, P. *J. Am. Chem. Soc.* **1977,** *99*, 3885; *J. Chem. Phys.* **1978,** *68*, 4047.
32. Meier, P. F.; Hauge, R. H.; Margrave, J. L. *J. Am. Chem. Soc.* **1978,** *100*, 2108.

33. Manceron, L.; Loutellier A.; Perchard, J. P. *Chem. Phys.* **1985,** *92*, 75.
34. Suzer, S.; Andrews, L. *Chem. Phys. Lett.* **1987,** *140*, 300.
35. Loutellier, A.; Manceron, L.; Perchard, J. P. *Chem. Phys.* **1990,** *146*, 179.
36. Nicely, V. A.; Dye, J. L. *J. Chem. Phys.* **1970,** *52*, 4795.
37. Curtis, L. A.; Pople, J. A. *J. Chem. Phys.* **1985,** *82*, 4230.
38. Friedrich, H. B.; Person, W. B. *J. Chem. Phys.* **1966,** *44*, 2161, and references therein.
39. Kasaï, P. H.; McLeod D.; Watanabe, T. *J. Am. Chem. Soc.* **1983,** *105*, 6704.
40. Manceron, L.; Andrews, L. *J. Am. Chem. Soc.* **1985,** *107*, 563.
41. Manceron, L.; Schrimpf, A.; Bornemann, T.; Rosendahl, R.; Stöckmann, H. J. *Chem. Phys.*, submitted for publication.
42. Jen, C. K.; Bowers, V. A.; Cochran, E. L.; Foners, S. N. *Phys. Rev.* **1962,** *126*, 1749.
43. Weltner, W. *Magnetic Atoms and Molecules*; Van Nostrand Rheinhold: New York, 1983.
44. Pouchan, C. *Chem. Phys.* **1987,** *111*, 87.
45. Nguyen, M. T. *J. Phys. Chem.* **1988,** *92*, 1426.
46. Manceron, L.; Andrews, L. *J. Phys. Chem.* **1986,** *90*, 4514.
47. Heberhold, M. *Metal π-Complexes*, Vol. II; Elsevier: Amsterdam, 1974.
48. Hiraishi, J. *Spectrochim. Acta Part A* **1969,** *25A*, 749.
49. Manceron, L.; Andrews, L. *J. Phys. Chem.* **1989,** *93*, 2964.
50. Andrews, L. *J. Chem. Phys.* **1967,** *47*, 4834.
51. Kos, A.; Jemmis, E. D.; Schleyer, P. R.; Gleiter, R.; Fischbach, U.; Pople, J. *J. Am. Chem. Soc* **1981,** *103*, 4996.
52. Manceron, L.; Hawkins, M.; Andrews, L. *J. Phys. Chem.* **1986,** *90*, 4987.
53. Brooks, J. J.; Rhine, W. E.; Stucky, G. D. *J. Am. Chem. Soc.* **1972,** *94*, 7339; **1975,** *97*, 737.
54. Rhine, W. E.; Davis, J.; Stucky, G. D. *J. Am. Chem. Soc.* **1975,** *97*, 2079.
55. Jebli, R.; Volpilhac, G.; Hoarau, J.; Achard, F. *J. Chim. Phys.* **1984,** *81*, 13.
56. Morton Blake, D. A.; Corish, J.; Beniere, F. *Theor. Chim. Acta* **1985,** *68*, 389.
57. Manceron, L.; Andrews, L. *J. Am. Chem. Soc.* **1988,** *110*, 3840.
58. Cyvin, S. J.; Brunvoll, J.; Schafer, L. *J. Chem. Phys.* **1971,** *54*, 1517.
59. McCullough, J. D.; Duley, W. W. *Chem. Phys. Lett.* **1972,** *15*, 240.
60. Moore, J.; Thornton, C.; Coller, W.; Delvin, J. P. *J. Phys. Chem.* **1981,** *85*, 350.
61. Manceron L.; Andrews, L. *J. Phys. Chem.* **1988,** *92*, 2150.
62. Ayed, O.; Loutellier, A.; Manceron, L.; Perchard, J. P. *J. Am. Chem. Soc.* **1986,** *108*, 8138.
63. Ayed, O.; Manceron, L.; Silvi, B. *J. Phys. Chem.* **1988,** *92*, 37.
64. Manceron, L.; Perchard, J. P. *Surface Sci* **1989,** *220*, 230.
65. Cotton, F. A.; Kraihanzel, C. S. *J. Am. Chem. Soc.* **1962,** *84*, 4432.
66. Moskovits, M.; Ozin, G. A. In *Vibrational Spectra and Structure*, Vol. 4; During, J., Ed.; Elsevier: Amsterdam, 1975.

67. Silvi, B.; Ayed, O.; Person, W. *J. Am. Chem. Soc.* **1986,** *108*, 8148.
68. Pearson, T. G. *nature (London)* **1933,** *131*, 166.
69. Weiss, E.; Büchner, W. *Helv. Chim. Acta* **1963,** *46*, 1121; *Z. Anorg. Allg. Chem.* **1964,** *330*, 251.
70. Silvi, B.; Manceron, L.; Ayed, O.; Person, W. *J. Mol. Struct. (Theochem.)* **1988,** *181*, 325.

5

NMR OF ORGANOLITHIUM COMPOUNDS: GENERAL ASPECTS AND APPLICATION OF TWO-DIMENSIONAL HETERONUCLEAR OVERHAUSER EFFECT SPECTROSCOPY (HOESY)

WALTER BAUER

Institut für Organische Chemie der Friedrich-Alexander-Universität Erlangen-Nürnberg, Erlangen, Germany

Lithium Chemistry: A Theoretical and Experimental Overview, Edited by Anne-Marie Sapse and Paul von Ragué Schleyer.
ISBN 0-471-54930-4

1 INTRODUCTION

The structural analysis of alkali metal organic compounds, mainly organolithium compounds, has benefitted during the last two decades from the tremendous evolution of NMR spectroscopy. Along with the development of new NMR techniques, many "direct" insights into the solution structures of this class of compounds have been achieved. A particularly fruitful technique has turned out to be ^{6}Li, ^{1}H heteronuclear Overhauser effect spectroscopy (HOESY). Here, the merits of two-dimensional (2D) NMR coincide with the favorable properties of the ^{6}Li nucleus. In the first part of this chapter we give a brief description of the basic principles of HOESY. The second part describes fundamental properties and NMR aspects of the alkali metal nuclei. The third part presents examples obtained in our group by using the ^{6}Li ^{1}H HOESY technique. In addition to the description of lithium NMR, and for comparison purposes, some aspects of ^{133}Cs NMR are treated as well.

2 GENERAL NMR ASPECTS OF HOESY

2.1 The Nuclear Overhauser Effect (NOE)

In its original presentation, the "Overhauser effect" referred to dynamic polarization of nuclei in metals. It was predicted that the intensity of an NMR resonance line would increase when the ESR transition of an unpaired electron in the vicinity of the observed nucleus was irradiated.[1] The prediction was verified for solutions of sodium metal in liquid ammonia: a proton signal increase by a factor of -658 was obtained. More usefully, however, the Overhauser effect has been applied since its discovery to pairs of *nuclei* (nuclear Overhauser effect, NOE), both with spin quantum numbers $I > 0$.[2]

Different purposes are traced out by the application of the NOE in NMR spectroscopy: (1) the study of nuclear relaxation; (2) chemical exchange studies; (3) intensity enhancement of weak signals (e.g., in ^{13}C NMR spectroscopy); (4) determination of internuclear distances. An excellent and comprehensive description of all aspects of the NOE is given in the book by Neuhaus and Williamson.[3] We will focus here only briefly on some fundamentals of NOE.

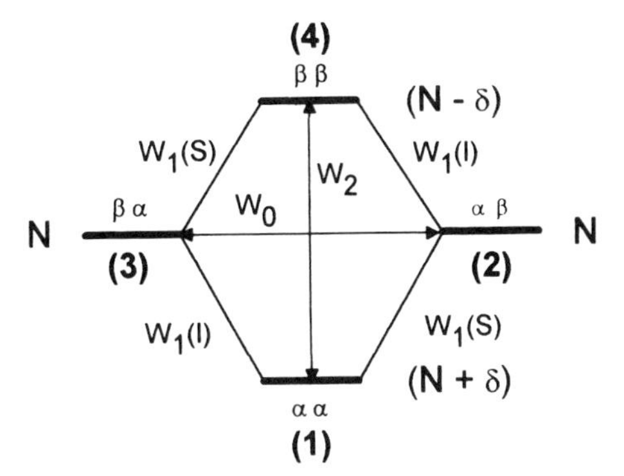

Figure 5.1 Energy level diagram (Solomon Scheme) of a homonuclear I, S spin system with indicated transition probabilities W and populations N (adapted from ref. 3).

The NOE is based on dipolar relaxation. Consider the energy level diagram of a two-spin I,S system (Solomon Scheme,[4] Fig. 5.1). It is assumed that there is dipolar but not scalar (J) coupling between I and S. In Figure 5.1, W denotes the individual transition probabilities; W_1 denotes single quantum transitions. These are "allowed" transitions in the sense that they may be excited by radio frequency (RF) pulses. Likewise, spin lattice relaxation may occur via W_1. The symbols W_0 and W_2 denote zero and double quantum transitions, respectively. These are "forbidden" with respect to their direct excitation by RF pulses. However, the spin system may *relax* across these transitions ("cross relaxation"). This is the exclusive mechanism that manifests in NOE enhancements.

In Figure 5.1 the Boltzmann equilibrium populations of spin states 1–4 are denoted by $(N + \delta)$, N, and $(N - \delta)$. Continuous irradiation (saturation) of the I resonance line results in pairwise population equilibration of spin states 1 and 3 ($= N + \frac{1}{2}\delta$), and 2 and 4 ($= N - \frac{1}{2}\delta$). The system reestablishes a new, perturbed equilibrium by relaxation across $W_1(S)$, W_0, and W_2. Relaxation across $W_1(S)$ does not change the involved resonance line intensities. Relaxation by the zero quantum mechanism, W_0, *decreases* the S-line intensity, whereas relaxation by the double quantum mechanism, W_2, *increases* the S-line intensity. The amount of intensity increase for the S line (the nuclear Overhauser enhancement factor η) is given by Eq. 1 for the general (homo- and heteronuclear) case; note that in the homonuclear case the term γ_I/γ_S vanishes.

$$\eta_S(I) = \frac{S - S_0}{S_0} = \frac{\gamma_I}{\gamma_S} \frac{W_2 - W_0}{2W_1(S) + W_2 + W_0} \tag{1}$$

In Eq. 1, S and S_0 are the S-line signal intensities with and without irradiation of I, respectively; γ is the magnetogyric ratio. The cross relaxation rate, σ_{IS}, for the general case is defined by Eq. 2.

$$\sigma_{IS} = W_2 - W_0 = \left(\frac{\mu_0}{4\pi}\right)^2 \frac{\gamma_I^2 \gamma_S^2 \hbar^2 \tau_c}{10 r^6} \cdot \left(\frac{6}{1 + (\omega_I + \omega_S)^2 \tau_c^2} - \frac{1}{1 + (\omega_I - \omega_S)^2 \tau_c^2}\right) \tag{2}$$

In Eq. 2, μ_0 is the permeability constant in a vacuum, r is the internuclear distance, ω is the Larmor frequency, and τ_c is the molecular rotational correlation time. For the homonuclear case and for $\omega\tau_c << 1$, Eq. 2 simplifies to Eq. 3.

$$\sigma_{\text{homo}} = W_2 - W_0 = \frac{1}{2}\left(\frac{\mu_0}{4\pi}\right)^2 \frac{\gamma^4 \hbar^2 \tau_c}{r^6} \tag{3}$$

The W_2 relaxation mechanism is predominant under conditions of the "extreme narrowing limit,"[3] that is, for small, rapidly tumbling molecules in nonviscous solvents ($\omega\tau_c < 1$, $\tau_c \approx 10^{-11}$ sec). By contrast, W_0 is the prevailing cross-relaxation mechanism in the "black spectral limit" (i.e., for large, slowly tumbling molecules and/or viscous solvents; $\omega\tau_c >> 1$, $\tau_c \approx 10^{-7}$ sec). It can be shown[3] that for

$$\omega\tau_c = \left(\frac{5}{4}\right)^{1/2} = 1.12 \tag{4}$$

the homonuclear NOE passes zero. Provided that relaxation of the involved spins is entirely due to the dipolar mechanism, under extreme narrowing conditions ($\omega\tau_c << 1$) the maximum NOE, $\eta_{\max}$, is

$$\eta_{\max} = \frac{1}{2}\frac{\gamma_I}{\gamma_S} \tag{5}$$

In Eq. 5, γ_I and γ_S are the magnetogyric ratios of the irradiated and the observed nucleus, respectively. Equation 5 holds both for the homo- and the heteronuclear case. In the homonuclear case (e.g., ^{1}H, ^{1}H), the maximum intensity enhancement due to the NOE is 50%. Likewise, a ^{13}C NMR signal may increase by a maximum of about 200% upon irradiation of ^{1}H (symbolized ^{13}C{^{1}H}). If γ_I and γ_S have the same sign, the NOE is positive for $\omega\tau_c <$ 1.12, and negative for $\omega\tau_c > 1.12$.

As is evident from Eq. 4, the cross-relaxation rate, σ, is proportional to the inverse sixth power of the internuclear distance. Hence the NOE drops rapidly with increasing internuclear separation. In the ^{1}H, ^{1}H homonuclear case, a distance of approximately 4.5 Å is considered to be the upper limit for the detection of an NOE.

Equation 2 is the fundamental basis for the quantitative estimation of internuclear distances using the NOE. This has led to the usage of NOE methods as a very powerful tool in the structure determination of large molecules like proteins and nucleic acids.[3,5,6] With respect to today's instrumental and methodological facilities, the statement "NOE = X-ray in solution" is not very far from reality. We will return to quantitative internuclear distance determination in Section 4.6.

2.2 Steady State and Transient NOEs

The description of the origin of the NOE to this point was based on a steady state system. Therein, one particular transient in a two-spin system is irradiated continuously, and the distortions in the equilibrium populations manifest in the observed (positive or negative) signal enhancements. By using this method, one-dimensional homonuclear difference NOE spectroscopy is frequently carried out for structure analysis.[3,7] The basic scheme is shown in Figure 5.2*a*.

There is, however, a further method for the generation of an NOE. A selective 180° RF pulse on a single *I* resonance line may be employed to invert the spin populations of nucleus *I* (shown for the homonuclear case in Fig. 5.2*b*). The spin system is then left without perturbation for a "mixing time," τ_m. During τ_m, a "transient" NOE builds up. A nonselective 90° pulse sam-

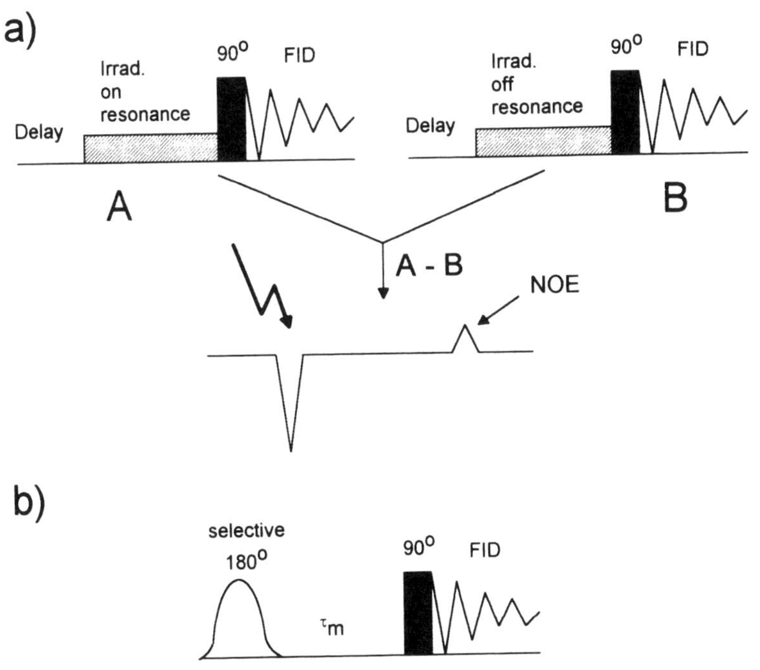

Figure 5.2 Pulse schemes for (*a*) one-dimensional steady state difference NOE spectroscopy and (*b*) one-dimensional transient NOE experiments.

ples the populations of nuclei *I* and *S*. Since the *I* nucleus is not irradiated constantly, the relaxation probabilities (cf. Fig. 5.1) differ for the steady state and the transient NOE; in the case of the transient NOE, the direct relaxation rate, R_I, of spin *I* must be taken into account as well. Generally, in the extreme narrowing limit, the transient NOE is smaller than the steady state NOE: for a homonuclear spin system, η_{max} (transient) $= 38.5\%$ η_{max} (steady state) $= 50\%$.[3] In the slow motion limit, the maximum NOEs are identical for both methods [η_{max} (transient) $= \eta_{max}$ (steady state) $= -100\%$ in the homonuclear case]. Both homonuclear (NOESY) and heteronuclear (HOESY) two-dimensional NOE spectra are based on *transient* NOEs.

The buildup kinetics of the steady state and the transient NOE differ (Fig. 5.3). Whereas in the steady state NOE the achieved signal enhancements remain constant with irradiation time after the initial buildup, the transient NOE reaches a maximum at a certain mixing time and then drops to zero.

Steady state experiments that use an irradiation period shorter than the time required to gain the maximum attainable NOE are termed "truncated driven NOE" (TOE). The buildup kinetics of both transient NOE and TOE experiments may be employed to derive quantitative internuclear distances.[3]

2.3 The Heteronuclear NOE

From a sensitivity point of view, Eq 5 dictates that in heteronuclear NOE experiments the *sensitive* nucleus (e.g., ^{1}H) should be *irradiated*, whereas the *insensitive* nucleus (e.g., ^{13}C) should be *detected*. Figure 5.4 shows the dependence of the magnitude of the maximum theoretical heteronuclear NOE on the molecular correlation time.

Whereas for ^{19}F the NOE changes its sign, the NOE is positive for ^{13}C even in the slow motion limit. It can be shown[3] that for $\gamma_I/\gamma_S > 2.38$, the heteronuclear NOE remains positive, irrespective of the tumbling speed of the molecule. Except for ^{19}F, this condition is met in the case of X{^{1}H} for all

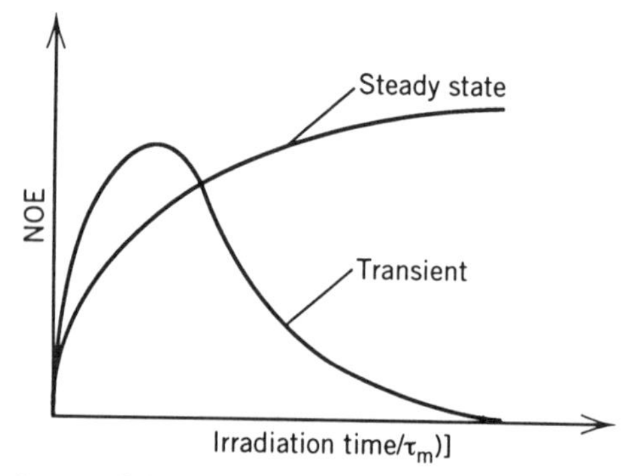

Figure 5.3 Dependence of the NOE on the irradiation time in a steady state NOE experiment, and on the mixing time in a transient NOE experiment.

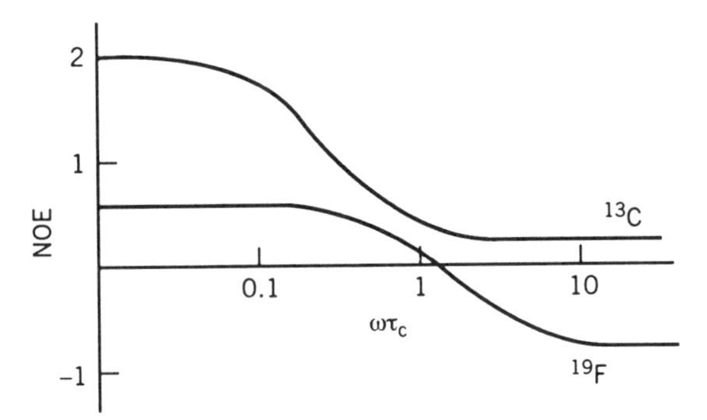

Figure 5.4 Dependence of the maximum steady state NOE, η_{max}, on $\omega\tau_c$ for heteronuclear spin systems X{^{1}H}, X = ^{19}F, ^{13}C (reproduced with permission from ref. 3).

stable isotopes X with spin quantum number $I_X > 0$ and with positive magnetogyric ratios ($\gamma_X > 0$). In Figure 5.4, the graph of ^{7}Li would be located beneath, whereas the graphs of ^{6}Li and ^{133}Cs would be located above the graph of ^{13}C.

If the ratio γ_I/γ_S is negative, the NOE is negative as well. This is the case, for example, for ^{1}H, ^{15}N, or ^{1}H, ^{29}Si pairs of nuclei. In these cases, the NMR signal of the heteronucleus may vanish if the NOE is −100%. The isotopes considered in this paper (^{6}Li, ^{7}Li, and ^{133}Cs) have positive magnetogyric ratios. Thus, as long as no "indirect" three-spin effects are involved,[3] NOEs between these alkali metal isotopes and ^{1}H are positive throughout, irrespective of the molecular correlation time.

2.4 The 2D HOESY Experiment

In 1983, Rinaldi[8] and (independently) Yu and Levy[9] reported a 2D NMR pulse sequence suited for the detection of spatially nearby heteronuclei. The method was given the acronym "HOESY" (= *h*eteronuclear *O*verhauser *e*ffect *s*pectroscop*y*). This pulse sequence is derived from the homonuclear 2D NOESY sequence ($90° - t_1 - 90° - \tau_m - 90° - t_2$). Figure 5.5*a* shows the HOESY pulse sequence which may be analyzed by using the classical vector picture (Fig. 5.5*c*). This will be outlined briefly; note that Figure 5.5*c* uses the "left-handed notation" which has conventionally been employed for the description of spin dynamics in the Bloch vector picture.[10]

For convenience, we designate the "irradiated" nucleus I (usually ^{1}H) and the detected heteronucleus S (e.g., ^{13}C). The experiment starts with a 90° pulse on the I nucleus, thus creating I transverse magnetization (state 1 in Fig. 5.5*c*). During the first half of the evolution period, $t_1/2$, I precesses in the rotating frame of reference with its chemical shift frequency. Additionally, the I magnetization vector will split due to heteronuclear I,S scalar coupling (state 2). A 180° pulse on the heteronucleus inverts the S nuclear spin states. As a con-

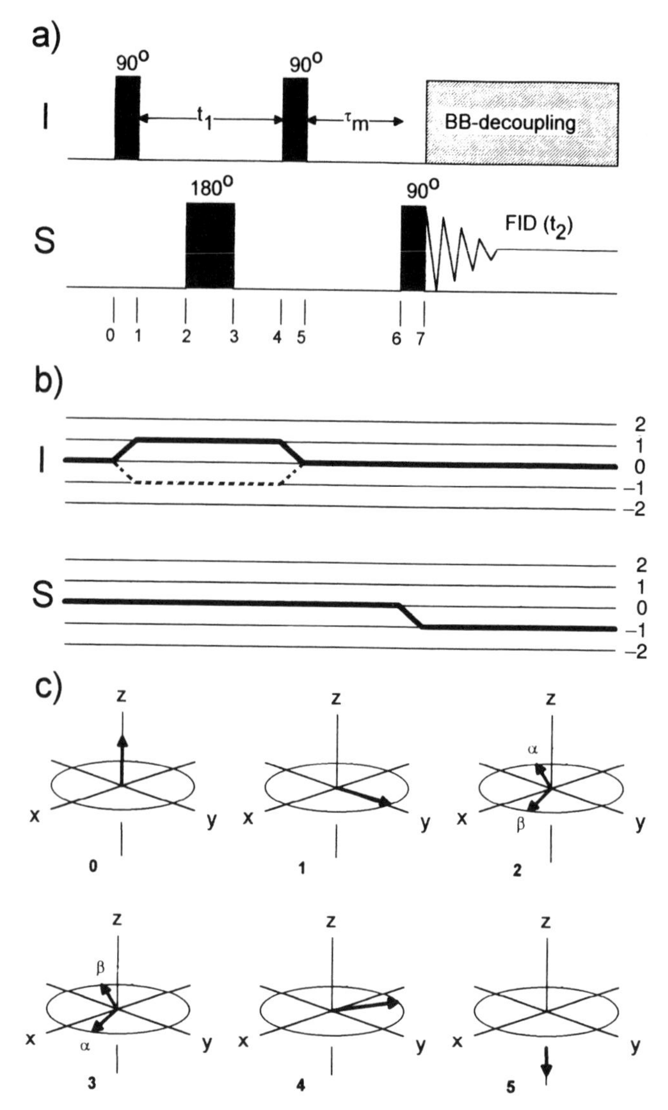

Figure 5.5 The 2D HOESY experiment. (*a*) Pulse sequence. (*b*) Coherence transfer pathways. (*c*) Vector description for the *I* nucleus magnetization (S = spin 1/2 nucleus).

sequence, the individual *I* magnetization vectors change their precession velocities (state 3). After another period $t_1/2$, splitting due to *I*,*S* coupling has refocussed (state 4). The 180° pulse in the middle of the evolution period thus eliminates scalar heteronuclear coupling in the t_1 domain. A 90°_x pulse on the *I* nucleus creates longitudinal (*z*) magnetization (state 5). The magnitude of I_z at state 5 is equal to the *y* component of the vector at state 4.

With increasing periods of t_1 during a 2D HOESY experiment, the magnitude of the I_z magnetization vector will oscillate. In other words, I_z magnetization becomes modulated by the I nucleus chemical shift. Since I_z magnetization represents I state population, any cross relaxation between I and a spatially close S nucleus will likewise be modulated by the I chemical shift in the course of the 2D HOESY experiment. Exchange of magnetization due to dipolar relaxation may occur during the mixing period, τ_m. Thus, if I and S are sufficiently close, the S nucleus magnetization is labelled by the I-chemical shift. A 90°_x pulse at the end of the mixing period (state 6) converts longitudinal (S_z) into observable transverse (S_y) magnetization (state 7). The FID is acquired on the S nucleus. Noise decoupling on the I nucleus removes scalar I,S coupling during t_2. As usual in 2D experiments, double Fourier transformation yields the 2D spectrum in its two frequency domains.

Figure 5.5*b* shows the coherence transfer pathways[11] of the 2D HOESY experiment. The initial 90° pulse creates I coherence of order ± 1 which evolves during $t_1/2$. Since the 180° pulse on the S nucleus inverts the S populations only, no change in the S coherence level is introduced. A 90° (I) pulse transfers I spin coherence back to level 0. However, owing to the second 90° pulse on the I nucleus, coherence orders -2 and $+2$ are populated as well. Phase cycling ensures that both homonuclear I single and double quantum coherences are eliminated; only longitudinal I magnetization (coherence order 0) is selected. Note that because of the 180° pulse in the middle of the evolution period no heteronuclear antiphase terms may evolve during t_1. This prevents the evolution of heteronuclear multiple quantum coherence during the mixing period.[3] Subsequent to the incoherent $I_z \rightarrow S_z$ magnetization transfer during τ_m, the final 90° (S) pulse creates detectable single quantum S coherence.

The description of the HOESY experiment in the product operator formalism[12] is given in Eqs. 6–8 in the right-handed notation.[13] Note that by using this notation the sense of rotation is opposite to that shown in Figure 5.5*c*. The superscripts at the operator symbols denote the assigned time states in the pulse sequence (Fig. 5.5*a*).

$$\begin{aligned} I_z^0 + S_z^0 &\xrightarrow{90^\circ_x(I)} -I_y^1 + S_z^1 \xrightarrow{t_1,\ 180^\circ_x(S)} \\ &-I_y^4 \cos\omega_I t_1 + I_x^4 \sin\omega_I t_1 - S_z^4 \xrightarrow{90^\circ_x(I)} \\ &-I_z^5 \cos\omega_I t_1 + I_x^5 \sin\omega_I t_1 - S_z^5 \end{aligned} \tag{6}$$

During the mixing period, τ_m, longitudinal magnetization is transferred incoherently from the I nucleus to the S nucleus as a result of cross relaxation.

$$-I_z^5 \xrightarrow{\tau_m(\text{NOE})} S_z^6 \tag{7}$$

The perturbed S_z population is transferred into observable magnetization by the final 90° "read" pulse:

$$S_z^6 \xrightarrow{90^\circ_x(S)} -S_y^7 \tag{8}$$

Since no scalar heteronuclear couplings between I and S evolve in the course of the HOESY pulse sequence, operator products, IS do not appear in Eqs. 6–9. For this reason, the description of the HOESY experiment in the classical vector picture is possible.

Both the basic HOESY pulse sequence as well as variants have been extensively studied and reviewed by Kövér and Batta.[14]

3 NMR OF ALKALI METAL ORGANIC COMPOUNDS AND NUCLEAR PROPERTIES OF THE ALKALI METAL ISOTOPES

All stable alkali metal isotopes are quadrupolar nuclei (spin quantum number $I > 1/2$; see Table 5.1). With respect to NOE measurements, only the elements lithium and cesium have received attention so far. Other isotopes exhibit high quadrupole moments and/or short spin lattice relaxation times (T_1) and must therefore considered to be nonsuitable for heteronuclear NOE studies.

3.1 Natural Abundance, Magnetogyric Ratios, and Receptivity

Lithium consists of two stable isotopes: ^{6}Li, abundance 7.4%, and ^{7}Li, 92.6%. Cesium is a pure isotope element (^{133}Cs, 100% abundance). The magnetogyric ratio, γ, of ^{7}Li is comparatively high. At a magnetic field strength of 9.4 T (^{1}H: 400 MHz), the correlated Larmor frequency of ^{7}Li is 155.45 MHz. Because of its high γ number and its high natural abundance. ^{7}Li is a sensitive and easy to detect nucleus (relative receptivity, D^c, as referenced to ^{13}C; $D^c(^7\mathrm{Li}) = 1540$.[15-17]) The magnetogyric ratios, γ, of both ^{6}Li and ^{133}Cs are considerably smaller than those of ^{7}Li. The involved Larmor frequencies at 9.4 are 58.86 MHz (^{6}Li) and 52.47 MHz (^{133}Cs). Because of its 100% natural abundance, ^{133}Cs is an easily detectable nucleus [$D^c(^{133}\mathrm{Cs}) = 269$]. By contrast, ^{6}Li in natural abundance is more difficult to observe [$D^c(^6\mathrm{Li}) = 3.58$], and measurements involving the ^{6}Li isotope in natural abundance became possible only with the advent of pulse Fourier transform spectrometers. However, owing to other favorable physical properties (see below), many NMR studies on organolithium compounds are now frequently carried out by observing the ^{6}Li nucleus, preferably in ^{6}Li isotope-enriched samples.

Isotopic labelling with ^{6}Li may be carried out comparatively easy. Introduction of ^{6}Li is usually achieved by reaction of the precursor with ^{6}Li metal[18-20] or in a secondary reaction with an already ^{6}Li-enriched organolithium compound. (^{6}Li) *n*-Butyllithium[21] is often employed for this purpose. The ^{6}Li metal itself is comparatively inexpensive. This is obviously because of the use of ^{6}Li in nuclear weapons.

3.2 Chemical Shifts

The shielding constant, σ, of a nucleus is commonly defined by Eq. 9.[22]

$$\sigma = \sigma_{\mathrm{dia}} + \sigma_{\mathrm{para}} \tag{9}$$

Table 5.1 Nuclear Properties of Stable Alkali Metal Isotopes[a]

Isotope	Natural Abundance (%)	Spin Quantum Number I	υ_0 at 9.4 T (^1H = 400 MHz)	Quadrupole Moment Q (10^{-28} m^2)	Receptivity D^c (^{13}C = 1.00)	Sternheimer Antishielding Factor ($1 + \gamma_\infty$)	Width Factor WF (^7Li = 1.0)
^6Li	7.42	1	58.862	-8×10^{-4}	3.58	0.74	2.0×10^{-3}
^7Li	92.58	3/2	155.454	-4.5×10^{-2}	1540	0.74	1.00
^{23}Na	100.0	3/2	105.805	0.12	525	5.1	343
^{39}K	93.1	3/2	18.666	5.5×10^{-2}	2.69	18.3	1359
^{41}K	6.88	3/2	10.245	6.7×10^{-2}	0.0328	18.3	—
^{85}Rb	72.15	5/2	38.620	0.25	43	48.2	54200
^{87}Rb	27.85	3/2	130.885	0.12	277	48.2	—
^{133}C	100.0	7/2	52.468	-3×10^{-3}	269	111	15

[a]Adapted from references 16, 17, and 25.

The diamagnetic term, σ_{dia}, describes the free rotation of electrons about the nucleus, whereas the paramagnetic term, σ_{para}, refers to hindrance of this rotation caused by other electrons and nuclei in the molecule.[23-25] Whereas for the higher alkali metals the paramagnetic term is dominant, σ_{dia} and σ_{para} are of the same order of magnitude for ^{6}Li and ^{7}Li. Since σ_{dia} and σ_{para} are of opposite effect, the chemical shift range of Li usually is quite small (ca. -3 to $+3$ ppm).[16] However, exceptional upfield chemical shifts are found when lithium is located in the shielding cone of an aromatic ring. Thus, in cyclopentadienyl lithium (CpLi) the lithium resonance line is considerably shifted diamagnetically ($\delta = -8.37$ ppm in THF at room temperature[26,27]). Lithium sandwiched between two Cp rings experiences even higher upfield chemical shifts (up to -13 ppm, see Section 4.4). By analogy, the lithium counter ions of the 8π-antiaromatic substituted benzene dianion in **1** are extremely deshielded ($\delta = +10.7$ ppm[28]). In the X-ray structure of **1**, the lithium cations are located above and below the six-membered ring.

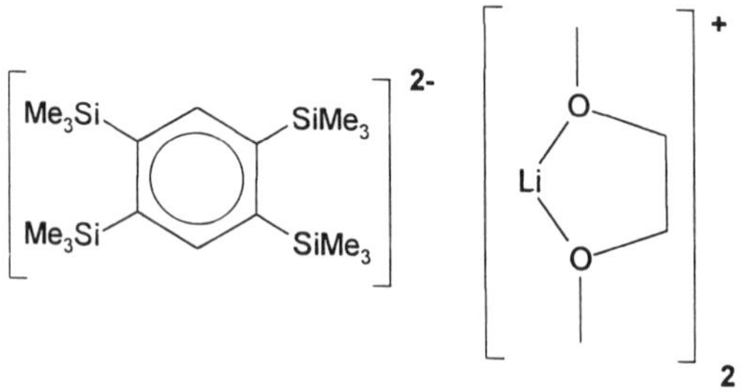

Me$_3$Si SiMe$_3$ Me$_3$Si SiMe$_3$ 2- Li O O + 2

1

A major drawback with respect to the relatively small chemical shift range of lithium is the lack of a generally agreed NMR reference standard. To circumvent chemical exchange, external standards must be employed for referencing Li–NMR spectra of organolithium compounds. These are, for example, LiCl in H_2O or LiBr in THF. Since bulk susceptibility corrections are rarely carried out, and owing to variations in the concentration of the standards, large percentage uncertainties of reported Li chemical shifts are quite common.

A more general method of referencing ^{6}Li/^{7}Li NMR chemical shifts has been proposed recently by Jackman.[29] The lithium NMR chemical shifts obtained by this method are free of errors due to different bulk magnetic susceptibilities. A compilation of Li chemical shifts obtained under identical experimental conditions is found in reference 30.

Theoretical calculations of lithium NMR chemical shifts have made progress because of the application of the IGLO (individual gauge for localized orbitals) method.[31] The strong upfield chemical shifts in CpLi and in the "lithiocene" anion $[(C_5H_5)_2Li]^-$ are reproduced by IGLO with remarkable accuracy.[32]

By contrast to lithium, the chemical shift range of cesium is very large, as is generally expected for heavier elements. The chemical shift dispersion of cesium salts and cesium salt complexes reaches from δ at about +240 ppm to −80 ppm.[16]

3.3 Spin-Spin Couplings

The pioneering work of Fraenkel[33] and Seebach[21] demonstrated the merits of ^{6}Li isotope enrichment. Owing to the well-known multiplicity rule ($M = 2nI + 1$; M = number of resonance lines, n = number of coupled nuclei, I = spin quantum number), the splitting pattern of a ^{13}C NMR signal due to resolved $^1J(^{13}C, ^6Li)$ coupling is an invaluable tool for establishing the solution structures of organolithium compounds. The observation of similar coupling patterns between ^{13}C and the major abundant ^{7}Li nucleus is often obscured: since dynamic processes are usually encountered with organolithium compounds, low temperatures must often be maintained to resolve scalar C,Li coupling. In the extreme narrowing limit, the quadrupole relaxation rates, T_{1Q}^{-1} and T_{2Q}^{-1}, of a quadrupolar nucleus are defined by Eq. 10.[34]

$$T_{1Q}^{-1} = T_{2Q}^{-1} = \frac{3}{10}\pi^2 \frac{2I+3}{I^2(2I-1)} \chi^2 \tau_c \tag{10}$$

Here I is the spin quantum number, χ is the nuclear quadrupole coupling constant (e^2qQ/h), and τ_c is the molecular correlation time. Hence at lower temperatures (i.e., for increasing τ_c) quadrupole relaxation of ^{7}Li becomes more efficient and scrambling of resolved ^{13}C, ^{7}Li coupling multiplets may be observed.[35,36] Isotopic enrichment with the ^{6}Li nucleus circumvents this problem owing to the small quadrupole moment of ^{6}Li.

Empirically it was found[36–38] that one-bond ^{13}C, ^{6}Li scalar coupling constants follow the rule given in Eq. 11.

$$^1J(^{13}C, ^6Li) = 17/n \text{ (Hz)} \tag{11}$$

Here, n is the number of ^{6}Li nuclei bonded to ^{13}C in terms of the NMR time scale. Hence, apart from the multiplicity pattern, the magnitude of $^1J(^{13}C, ^6Li)$ is indicative of the aggregate size of an organolithium compound. Surprisingly, the magnitude of $^1J(^{13}C, ^6Li)$ is virtually independent of the type of the metallated carbon atom, that is, aliphatic, olefinic, aromatic, and acetylenic lithiated carbon atoms all show approximately identical numbers of $^1J(^{13}C, ^6Li)$ as long as the aggregate size remains constant. This is entirely different from the well-known dependence of carbon, proton coupling constants on carbon atom hybridization [$\%s = 0.2\ ^1J(^{13}C, ^1H)$].[39] Table 5.2 exemplifies this phenomenon.

The insensitivity of ^{13}C, ^{6}Li coupling constants on the hybridization of ^{13}C is presumably due to two opposing effects: in going from sp^3 to sp, the C—Li covalent bond order decreases. By contrast, the s character of the C,Li bond

Table 5.2 $^{1}J(^{13}C, ^{6}Li)$ Coupling Constants of Various Organolithium Aggregates with Differently Hybridized Carbon Atoms

Aggregate	$^{1}J(^{13}C, ^{6}Li)$ (Hz)	Ref.
	Monomers	
^{t}BuLi	11.9	36
PhLi	15.6	36
1-Lithio-1-(2-methylphenyl)ethene	12.0	40
	Dimers	
$(^{t}BuLi)_2$	7.6	36
$(PhLi)_2$	8.0	21, 41
$(Ph\text{-}C{\equiv}C\text{-}Li)_2$	8.2	42
	Static Tetramers	
$(^{t}BuLi)_4$	5.4	43
$(H_2C{=}CHLi)_4$	5.9	44
$(^{t}Bu\text{-}C{\equiv}C\text{-}Li)_4$	6.0	45, 46

increases in the same direction. The product (bond order x percent *s*) is approximately constant.[47]

Scalar couplings to lithium have also been observed for nitrogen,[48] silicon,[49] and phosphorus.[50] Resolved scalar ^{1}J-coupling between lithium and hydrogen was reported for a transition metal hydride.[51] Scalar $^{6}Li, ^{1}H$ coupling across two and three bonds ($\leq$ 0.2 Hz) has been detected recently in our group for vinyl lithium by using polarization transfer methods.[52] Coupling between chemically nonequivalent pairs of ^{6}Li nuclei has been observed by the application of $^{6}Li, ^{6}Li$–INADEQUATE,[53,54] and $^{6}Li, ^{6}Li$–COSY.[55,56]

No scalar couplings have been reported as yet which involve alkali metals other than lithium, apparently owing to the larger quadrupole relaxation rates of these isotopes (see below).

3.4 Quadrupole Moments and Relaxation

All alkali metal isotopes listed in Table 5.1 are quadrupolar nuclei, that is, there is nonspherical distribution of the electric charge at the site of the nucleus. Instead, because of the electric quadrupole moment of these nuclei, the nuclear charge adopts a spheroidal shape. The interaction of the quadrupole moment with an external electric field gradient is an efficient means for relaxation of spin $I > 1/2$ nuclei. Except for symmetrical surroundings (e.g., in the $[Li(H_2O)_4]^+$ ion), electric field gradients at the site of the nuclei are ubiquitous. Hence nuclei with moderate to high quadrupole moment (^{23}Na, ^{39}K, etc.) usually relax entirely owing to the quadrupole mechanism.[25] Exceptions are the isotopes ^{6}Li, ^{7}Li, and ^{133}Cs (cf. Table 5.1). Actually, ^{6}Li has the small-

Table 5.3 Percentage Contributions of Relaxation Mechanisms to $^6Li^+$ in H_2O (as 3.9 M LiCl) at +40°C[a]

Relaxation Mechanism	Contribution (%)
Dipole, dipole (^{6}Li, ^{1}H)	84
Spin rotation	8
Dipole, dipole (^{6}Li, ^{7}Li, intermolecular)	<1
Quadrupolar	<0.1
Unidentified	7

[a]From reference 59.

est quadrupole moment of all stable spin $I > 1/2$ isotopes. As a result, relaxation of ^{6}Li is only partly of quadrupolar origin.

Wehrli studied the relaxation mechanisms of ^{6}Li in aqueous LiCl solution[57] as well as in some organolithium compounds.[58] For a 3.9 M solution of LiCl in H_2O at + 40°C (natural abundance lithium), the contributions to spin-lattice relaxation of ^{6}Li were identified (%) as listed in Table 5.3.[59]

The predominantly dipolar relaxation mechanism of $^6Li^+$ in H_2O was deduced from a heteronuclear steady state NOE experiment: irradiation of the solvent protons resulted in an NOE enhancement of $\eta = 2.61$ (theoretical maximum: $\eta = 3.40$; cf. Eq. 5). For MeLi in Et_2O, *n*BuLi in hexane, and PhLi in C_6D_6/Et_2O (70:30), the dipole, dipole contributions to spin lattice relaxation of ^{6}Li were 20, 35, and 17%, respectively.[58]

In the absence of surrounding ^{1}H nuclei with their high magnetic moments, spin lattice relaxation times of ^{6}Li may be extraordinarily long: $T_1(^6\text{Li}) = 1040$ sec for LiCl in D_2O at + 80°C.[57] Even in unsymmetrical surroundings, ^{6}Li relaxes unexpectedly slowly: $T_1 = 72$ sec (MeLi in Et_2O[58]) and 81 sec (vinyllithium in THF-d_8[44]). Wehrli concludes that "... ^{6}Li possesses most virtues of a spin $-1/2$ nucleus ..."[57] The high dipolar contribution to spin-lattice relaxation of ^{6}Li is of fundamental importance for the successful application of ^{6}Li,^{1}H heteronuclear Overhauser experiments.

For the ^{7}Li isotope, a HOESY experiment was described only recently.[60] Because of the much larger quadrupole moment of ^{7}Li as compared to ^{6}Li and, hence, the smaller contribution of dipolar relaxation, the application of NOE experiments to ^{7}Li appears to be restricted to favorable cases. In our group, ^{7}Li,^{1}H HOESY has been successfully applied to vinyl lithium.[61]

For the ^{133}Cs nucleus, the maximum theoretical NOE is $\eta_{max} = 381\%$. Spin lattice relaxation times of ^{133}Cs may be quite long in some cases ($T_1 = 13.3$ sec for CsCl in H_2O at ±25°C).[62] However, despite its comparatively low quadrupole moment, spin lattice relaxation of the ^{133}Cs isotope is nearly completely of quadrupolar origin.[63] In ^{133}Cs{^{1}H} experiments on CsCl in H_2O, no NOE was observed by Wehrli.[63] We have detected a very weak (0.2%) NOE enhancement of the ^{133}Cs NMR signal in steady state ^{133}Cs{^{1}H} NOE experiments on cesium 3-methylpentyl-3-oxide **2**. Two dimensional ^{133}Cs,^{1}H HOESY experiments on $CsBPh_4$ in pyridine revealed the ion pair character in

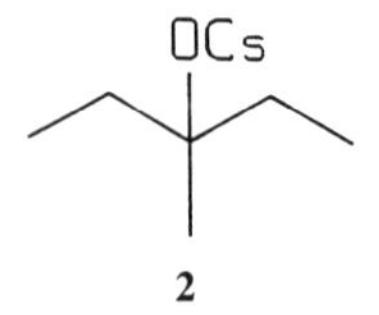

2

the presence and the absence of 18-crown-6.[64] For reasons not yet completely understood, the 2D HOESY experiment appears to work better for ^{133}Cs than its 1D steady state analogue.

3.5 The Sternheimer Antishielding Factor

When placed into an electric field gradient, the electron shell of a nucleus is polarized. For an alkali metal ion this results in a distortion of its otherwise spherical electron environment. This distortion causes an additional electric field gradient at the site of the nucleus. These additionally induced field gradients may strongly influence the relaxation behavior of the involved quadrupolar nuclei.

The Sternheimer antishielding factor $(1 + \gamma_\infty)$ is a measure of this effect. The numbers listed in Table 5.1 are calculated for free ions. Whereas for Li^+ the antishielding effect is negligible, a drastic influence may be anticipated for Cs^+ in nonsymmetrical environments. For alkali metals involved in bonds with partly covalent character, the antishielding effect is smaller than for the free ions.[16]

3.6 Spin Quantum Numbers and Line Width Factors

The ^6Li-isotope is one of the very rare "u,u" nuclei (uneven number of protons, uneven number of neutrons); there are only five stable u,u isotopes in the periodic table.[65] For this reason, ^6Li has an integer spin quantum number $(I = 1)$. By contrast, ^7Li $(I = 3/2)$ and ^{133}Cs $(I = 7/2)$ are "u,e" nuclei (uneven number of protons, even number of neutrons; 49 stable, u,e-isotopes in the periodic table) with consequently half-integer spin quantum numbers.

The line width factors, WF, listed in Table 5.1 are derived from Eq. 12. Therein, I is the spin quantum number, Q is the quadrupole moment, and $(1 + \gamma_\infty)$ is the Sternheimer antishielding factor.

$$WF = \frac{2I + 3}{I^2(2I - 1)} Q^2(1 + \gamma_\infty)^2 \tag{12}$$

The spin term of Eq. 12 decreases rapidly on going from $I = 1$ to $9/2$. Consequently, provided that Q and $(1 + \gamma_\infty)$ are constant, quadrupolar nuclei with higher spin quantum numbers show smaller line widths. In Table 5.1, WF of ^7Li is arbitrarily set to 1. Owing to its very small quadrupole moment,

WF (^{6}Li) is smaller than WF (^{7}Li) by three orders of magnitude. By contrast, WF (^{133}Cs) is 15 times larger than WF (^{7}Li), despite the smaller quadrupole moment of ^{133}Cs as compared to ^{7}Li. This is chiefly due to the large Sternheimer antishielding factor of ^{133}Cs.

4 APPLICATION OF ^{6}Li,^{1}H HOESY AND ^{133}Cs,^{1}H HOESY TO ALKALI METAL ORGANIC COMPOUNDS

4.1 Ortho-Directed Lithiations

The pioneering work of Gilman[66] and Wittig[67] demonstrated that aromatic rings substituted with heteroatom bearing groups undergo specific lithiation in the ortho position (Scheme 1). This synthetically important type of reaction was termed "ortho lithiation," "heteroatom facilitated lithiation," and so on.[68] We have studied the prototype reaction, the lithiation reaction of anisole **3** with *n*BuLi (Scheme 2) by NMR.[69]

Figure 5.6*a* shows the one-dimensional ^{1}H NMR spectrum of anisole, dissolved in toluene-d_8, at −64°C. The addition of 1 equiv. of *n*Bu^{6}Li leads to significant changes (Fig. 5.6*b*): the signals of the ortho protons H2,6 and of the OCH_3 protons are shifted downfield, whereas H3,4,5 appear as upfield shifted signals. Note that anisole is not yet lithiated under these conditions. The observed shift phenomena are indicative of the formation of a tight complex between anisole and *n*BuLi. Subsequent addition of 1 equiv. of TMEDA (TMEDA = tetramethylethylenediamine) completely reverses the changes brought about by *n*BuLi (Fig. 5.6*c*): the spectrum of the anisole protons in the 1 : 1 : 1 anisole : *n*BuLi : TMEDA mixture is completely identical to the spectrum of anisole without any additives. This indicates that anisole is released from the anisole–*n*BuLi complex upon the addition of TMEDA.

The ^{13}C-NMR chemical shifts of the anisole : *n*BuLi = 1 : 1 mixture indicate

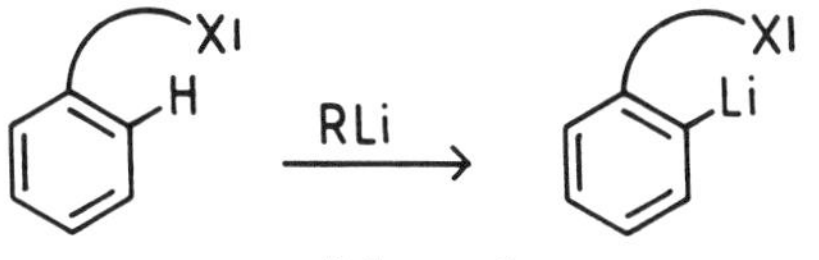

Scheme 1

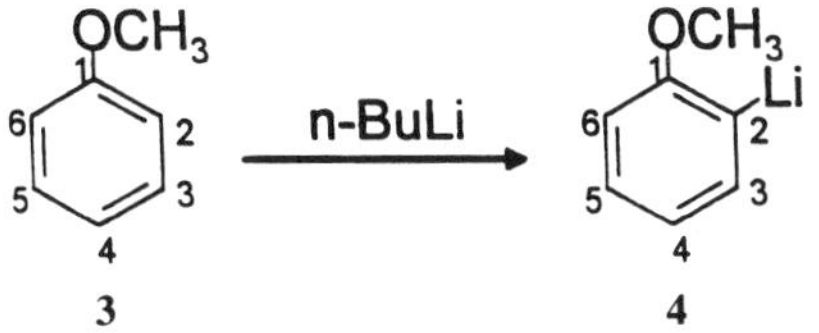

Scheme 2

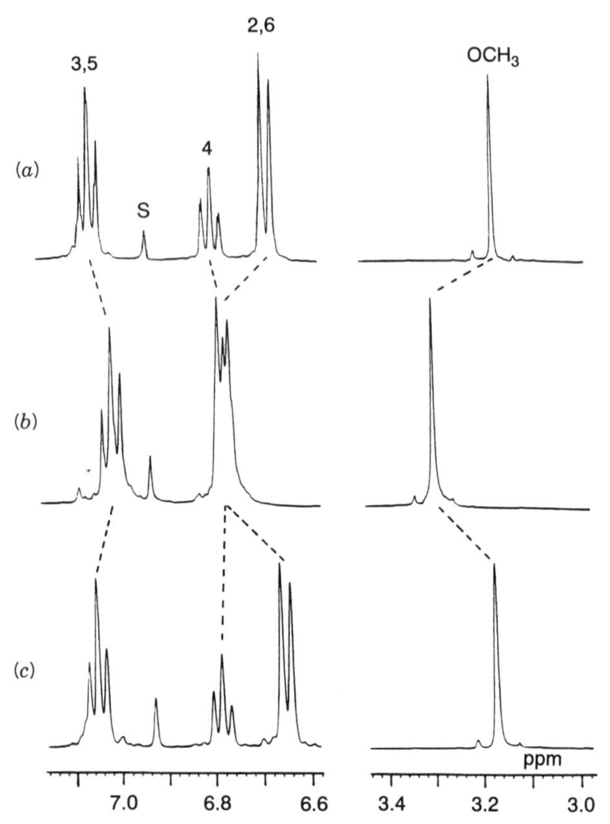

Figure 5.6 ^{1}H NMR spectrum (toluene-d$_8$, −64°C) of (*a*) anisole (ca. 1.6 M), (*b*) anisole + 1 equiv. of *n*BuLi (ca. 0.8 M anisole), (*c*) anisole + 1 equiv. of *n*BuLi + 1 equiv. of TMEDA (ca. 0.8 M anisole); *s* = solvent signal.

that *n*BuLi is tetrameric under these conditions, and that *n*BuLi gets dimeric when 1 equiv. of TMEDA is added.

The ^{6}Li,^{1}H HOESY spectra nicely support the findings discussed above. Figure 5.7*a* shows the ^{6}Li,^{1}H HOESY spectrum of anisole:*n*Bu^{6}Li = 1:1 in toluene-d$_8$ at −64°C. Apart from the expected cross peaks between lithium and the *n*BuLi–^{1}H resonances, cross peaks are detected which involve the OCH_3—and the ortho protons of the aromatic ring. This indicates close proximity of the involved ^{1}H positions to lithium, in agreement with a tight anisole–*n*BuLi complex. After addition of 1 equiv. of TMEDA (Fig. 5.7*b*), any correlations between lithium and the anisole ^{1}H NMR resonances have vanished in the HOESY spectrum. Instead, new cross peaks between *n*BuLi and the TMEDA resonance lines are observed. This indicates that anisole is "free" under these conditions. Scheme 3 summarizes the NMR findings.

Strikingly, the chemical behavior of the binary anisole-*n*BuLi mixture and the ternary anisole-*n*BuLi/TMEDA mixture is in contrast to what might be

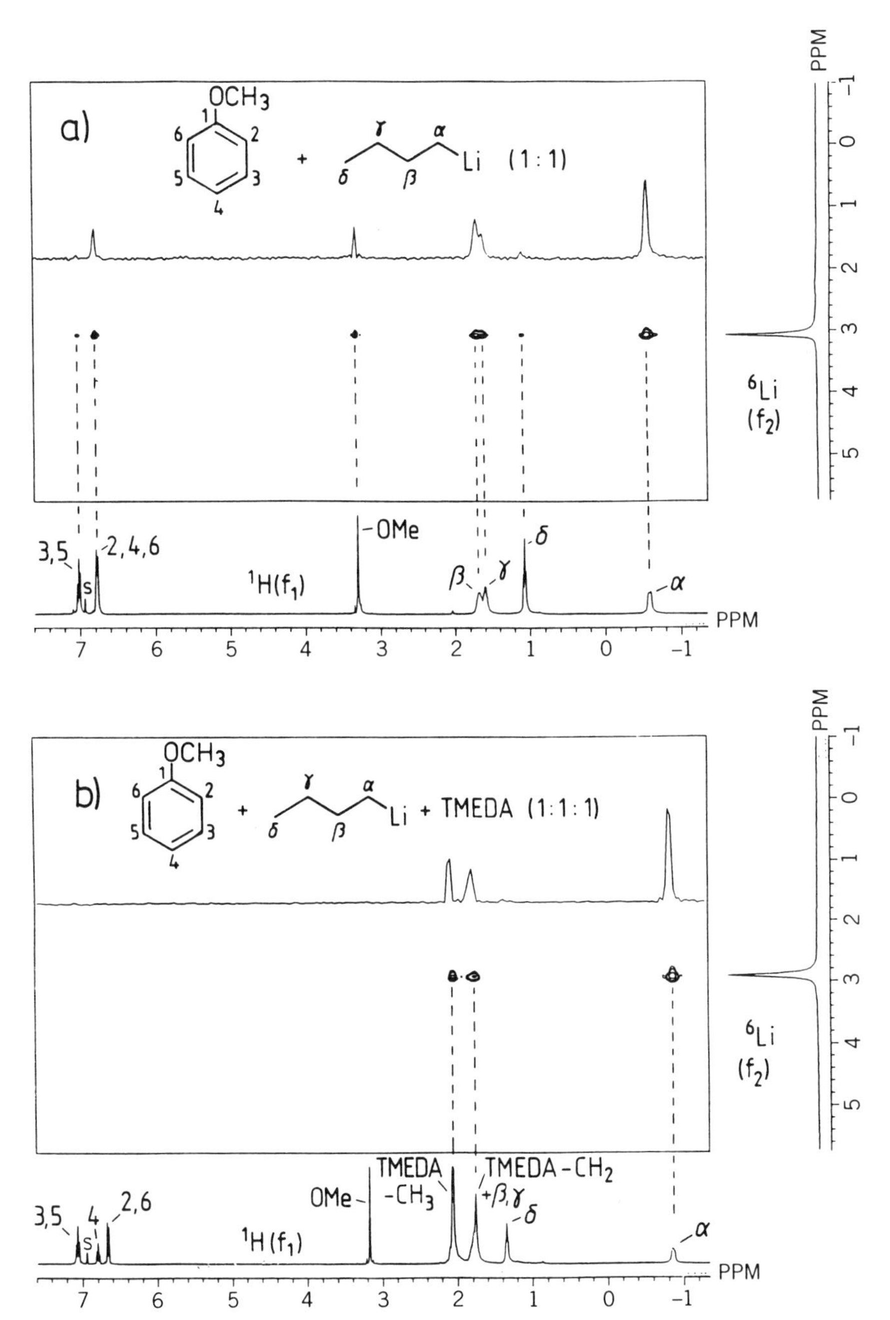

Figure 5.7 Phase-sensitive ^{6}Li,^{1}H HOESY spectrum (toluene-d$_8$, −64 °C) a) of a 1 : 1 anisole **3** – *n*Bu^{6}Li mixture; b) of a 1 : 1 : 1 anisole **3** – *n*Bu^{6}Li – TMEDA mixture (ca. 0.8 M **3**). Insets: f$_1$-cross sections cut at the chemical shift of the Li signal.; s = solvent signal.

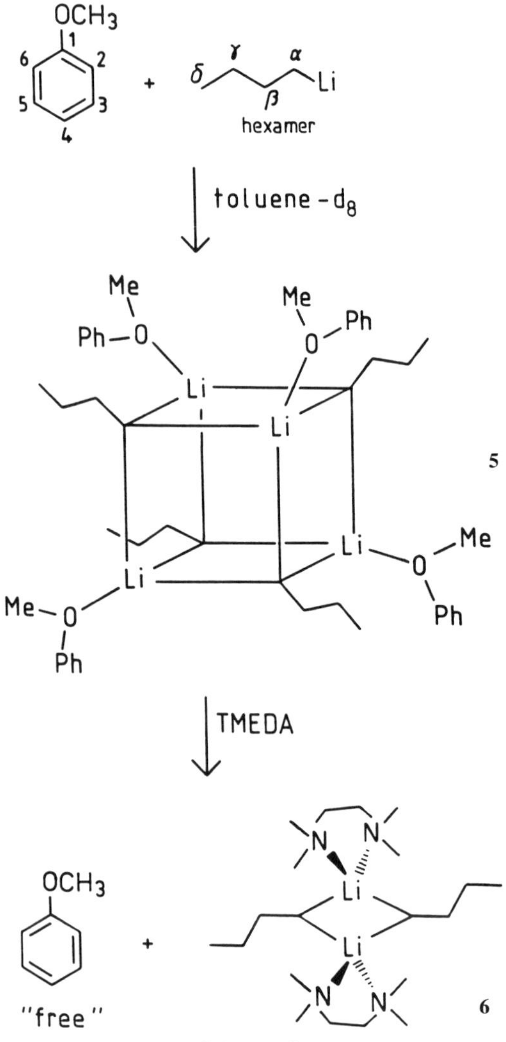

Scheme 3

expected: under the conditions of Figure 5.7*a* (where anisole is permanently close to the organolithium compound), no metallation is observed even at −10°C over several hours. By contrast, under the conditions of Figure 5.7*b* (with "free" anisole), ortho lithiation occurs moderately even at −70°C. We have tentatively ascribed this reaction behavior to the aggregation state of *n*BuLi: in a tetrameric *n*BuLi aggregate, only *one* coordination site at Li is accessible to an external ligand. Thus, occupation of this lithium site by the anisole oxygen atom leaves no more residual space for the attachment of an additional ligand (e.g., an ortho aromatic proton). By contrast, in dimeric *n*BuLi, one TMEDA chelate ligand might dissociate completely. This would allow coordination by anisole both with the oxygen atom *and* with one of the ortho aromatic protons (Scheme 4).

The MNDO calculations were in agreement with this assumption: in the MNDO optimized geometry of **9**, one ortho hydrogen atom exhibits agostic interaction with lithium and is indicated to undergo subsequent metal–hydrogen exchange. However, the hypothetical reactive intermediate **9** does not account for the phenomenon of facilitated ortho lithiation *as such*. Instead, it has been shown by recent ab initio calculations[70] that ortho-directing substituents effectively stabilize the *transition state* of a lithiation reaction. Thus the activation energy for ortho lithiation of phenol by LiH is 22.6 kcal/mol when starting from the initial addition complex of the reactants. By contrast, the activation barrier for the analogous metallation of benzene by LiH is higher by 11.9 kcal/mol. Thus, facile ortho lithiation of aromatic rings in the presence

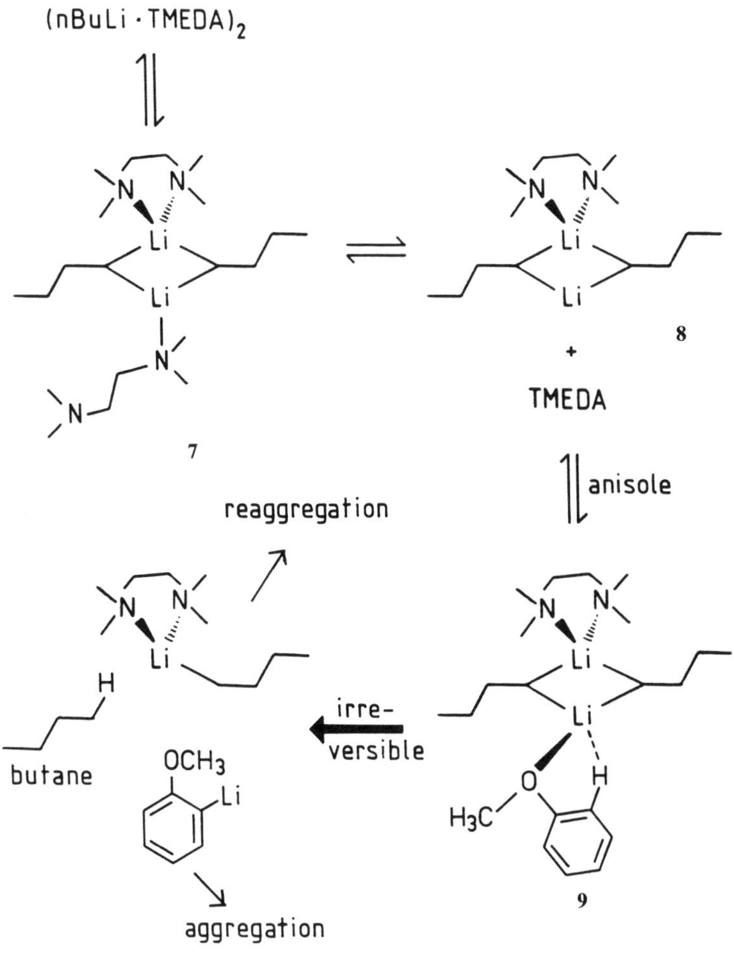

Scheme 4

of ortho-directing groups may be ascribed to "kinetic enhancement" by these substituents.

The claimed calculations[70] also solved the enigma of a further experimental observation:[69] fluorobenzene is known to undergo facile ortho lithiation.[68,71,72] However, by using the same NMR methods as described above for the study of anisole, no complex formation was observed between fluorobenzene and *n*BuLi in toluene-d_8 at low temperatures. This agrees with ab initio results,[70] according to which the formation of a fluorobenzene–LiH complex is less favorable than the analogous formation of phenol–LiH and benzene–LiH complexes. However, the activation barrier for the ortho lithiation of PhF is even *lower* than for phenol–LiH, in agreement with the experimentally observed reactivity of PhF toward organolithium reagents.

With respect to the complexation behavior toward *n*BuLi, the compounds thioanisole, 1,3-dimethoxybenzene, and *N,N*-dimethylaniline were found to behave very similar to anisole.[69]

4.2 Directed Second Lithiations

Monolithiated organic compounds may be converted to dilithiated derivatives in certain cases. The first lithium substituent often determines the position of the second metallation.[73,74]

4.2.1 **N-*Phenylpyrrole*.** One of the first compounds studied by ^{6}Li,^{1}H HOESY was 2-lithio-1-phenylpyrrole **11**.

Monolithio compound **11** may be generated by treatment of *N*-phenylpyrrole **10** with *n*BuLi (Scheme 5). Crystals of **11** were isolated and subjected to X-ray analysis.[75] In the solid state, **11** is a dimer with a central four-membered C-Li-C-Li ring and with the phenyl rings in syn arrangement. Addition of one further equivalent of *n*BuLi to **11** leads to introduction of a second lithium atom specifically in the ortho position of the phenyl ring.[76]

The ^{6}Li,^{1}H HOESY spectrum of crystals of (**11** × TMEDA)$_2$ in THF-d_8 at −70°C reveals an intense cross peak for the ortho protons H7,11.[75] Less intense cross peaks are detected which involve H3 and the TMEDA—CH_3 protons. This indicates the proximities between lithium and these H-positions and suggests "agostic" interactions between lithium and the ortho hydrogen atoms of the phenyl ring. In agreement with the HOESY results, during the course of the chemical reaction the second lithium atom is introduced into the 7/11

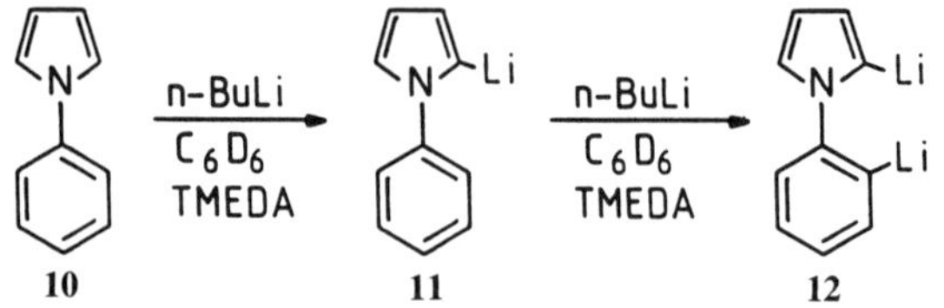

N
10
n-BuLi
C_6D_6
TMEDA
N
Li
11
n-BuLi
C_6D_6
TMEDA
N
Li
Li
12

Scheme 5

position of **11**. Thus, ^{6}Li,^{1}H HOESY appears to be a suitable tool for predicting the position of second metallations.

However, it should be emphasized that the occurrence of Li,H HOESY cross peaks is indicative only of close proximities between the involved positions. Any conclusions about "activation" of H positions by lithium must be treated with care. Under the conditions of kinetic control, the actual position of the introduction of a lithium atom is determined by the activation barrier of a specific reaction route. As described above for the mechanism of ortho-directed metallations, stabilization of the transition state[70] by functional groups (including lithium itself as a "functional group") must be considered to be the driving force in (second) lithiations.

4.2.2 Naphthalene. 1-Lithionaphthalene may be conveniently prepared from 1-bromonaphthalene by halogen/metal exchange (Scheme 6).

The ^{1}H NMR spectrum of isolated monolithio compound **14** (with 1 equiv. of TMEDA) exhibits two low-field doublets at δ = 8.57 and 8.47 ppm (cf. Fig. 5.8).[77] By using standard 2D NMR techniques these may be assigned to H2 and H8, respectively.

The deshielding of these ^{1}H positions may be attributed to the proximity of lithium and has been found in other cases as well.[36] This effect may be due to the electric field produced by lithium.

The ^{6}Li,^{1}H HOESY spectrum of (**14** $\times$ TMEDA) (Fig. 5.9; recorded with natural abundance lithium) shows a cross peak which involves the TMEDA–CH_3 proton signal. This indicates the (expected) chelation of lithium under these conditions. A further cross peak is detected in the low-field region. This may be interpreted to involve H2 exclusively or, owing to its elongated shape, both H2 and H8. It has been shown by one-dimensional ^{6}Li{^{1}H} experiments on the same compound that both H2 and H8 are involved in creating a HOE at the lithium site.[78] This agrees with the MNDO calculated[77] proximities between lithium and H2/H8 in the dimer[79] of **14**: the mean Li,H distances are 3.26 Å (Li,H8) and 2.89 Å (Li,H2), respectively.

The second lithiation of **14** is a stepwise process. Addition of a further organolithium reagent, RLi (R = Me, *n*Bu), to **14** in benzene results in an initial exothermic formation of a mixed aggregate between **14** and RLi. Although **14** is not yet reacted chemically at this stage, its ^{1}H NMR spectrum has changed significantly (Fig. 5.8*b*). In particular, the chemical shifts of H2 and H8 have crossed. The ^{6}Li,^{1}H HOESY spectrum of this mixed complex

Br → (nBuLi, TMEDA, −100 °C) → Li → (nBuLi, TMEDA, Δ) → Li Li

13 **14** **15**

Scheme 6

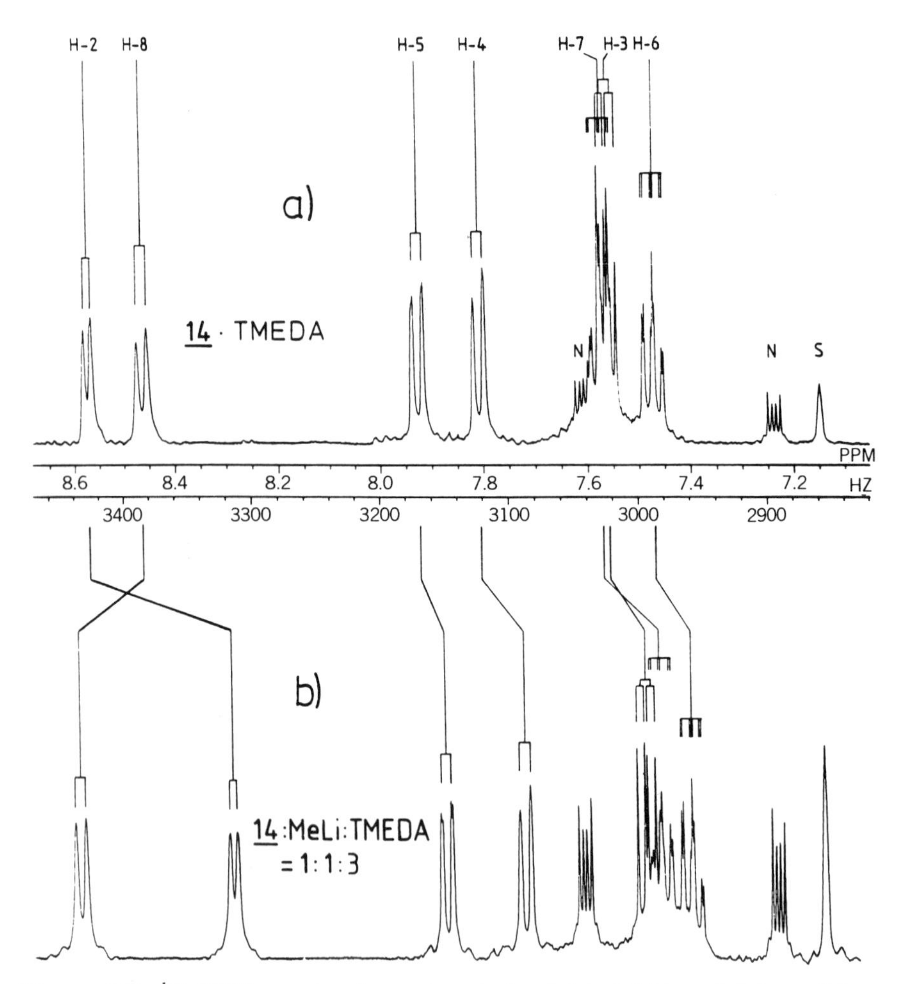

Figure 5.8 ^{1}H NMR spectra (aromat region) of (*a*) 1-lithionaphthalene (**14** × TMEDA) and (*b*) the mixed aggregate (**14** × CH_3Li × *n*TMEDA) in C_6D_6 at +23°C; N = naphthalene; S = solvent.

shows (apart from two other expected cross peaks) a cross signal which involves H8 of **14** (Fig. 5.10). No cross peaks are detected which involve H2. This may be taken as indicative that H8 is the "activated" position under these conditions. Actually, prolonged heating of the **14**:RLi = 1 : 1 complex in benzene leads to 1,8-dilithionaphthalene in moderate yields.[74] The MNDO calculations on the mixed **14**–CH_3Li complex agree qualitatively: the mean Li, H8 distance (2.99 Å) is now shorter than the mean Li,H2 separation (3.10 Å).

4.2.3 1-Lithio-1,2-diphenylhex-1-ene. The addition reaction of *n*BuLi to diphenylacetylene (tolane) was studied two decades ago by Mulvaney[80] (Scheme

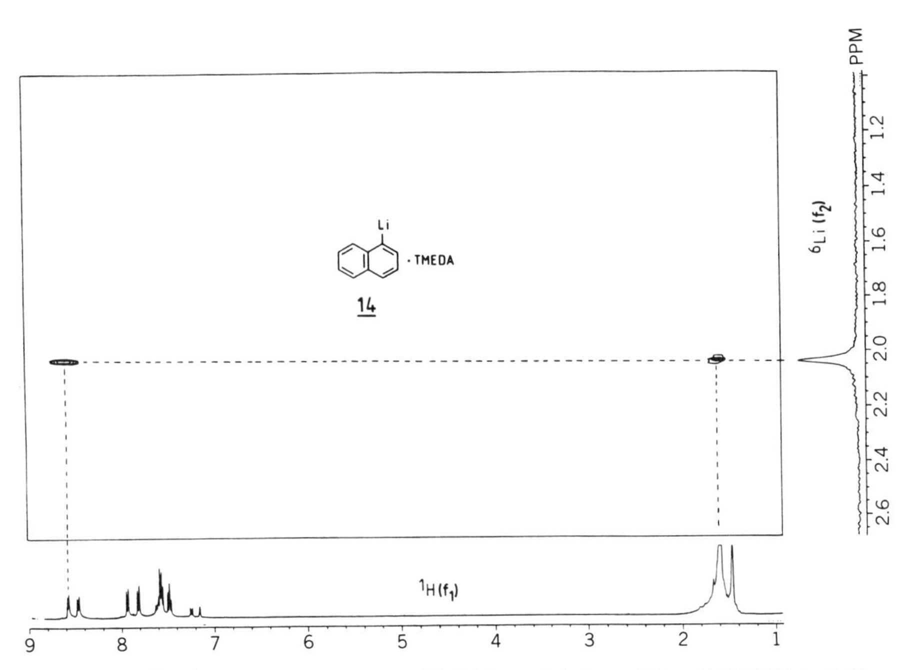

Figure 5.9 $^{6}Li,^{1}H$ HOESY spectrum of 1-lithionaphthalene (**14** × TMEDA) in C_6D_6, + 23°C. Mixing time, 2.1 sec. For the ^{1}H NMR assignments, see Figure 5.8.

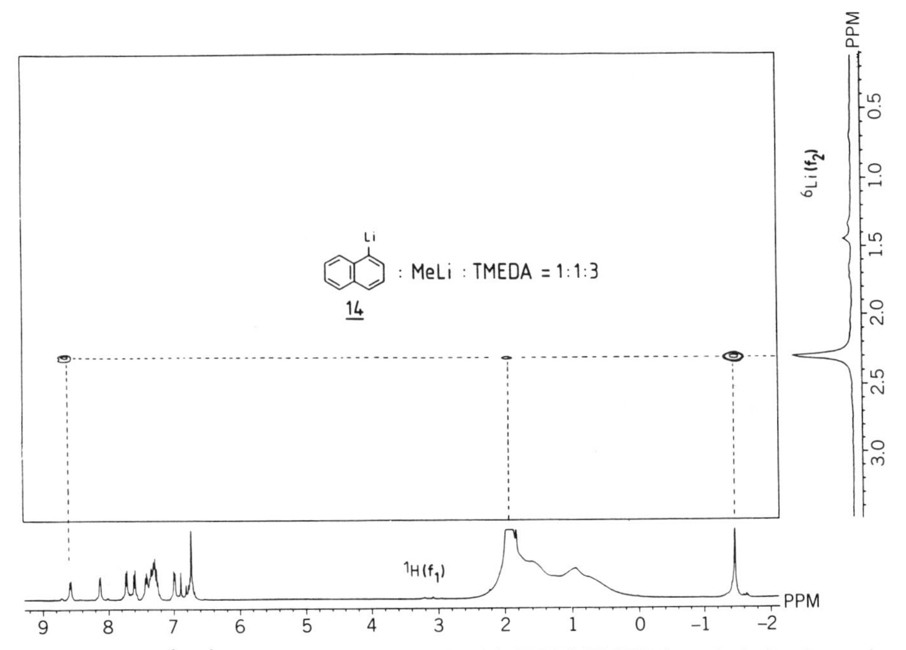

Figure 5.10 $^{6}Li,^{1}H$ HOESY spectrum of a **14**:CH_3Li:TMEDA = 1 : 1 : 1 mixture in toluene-d_8 at −85°C; mixing time, 1.6 sec. For the ^{1}H NMR assignments, see Figure 5.8.

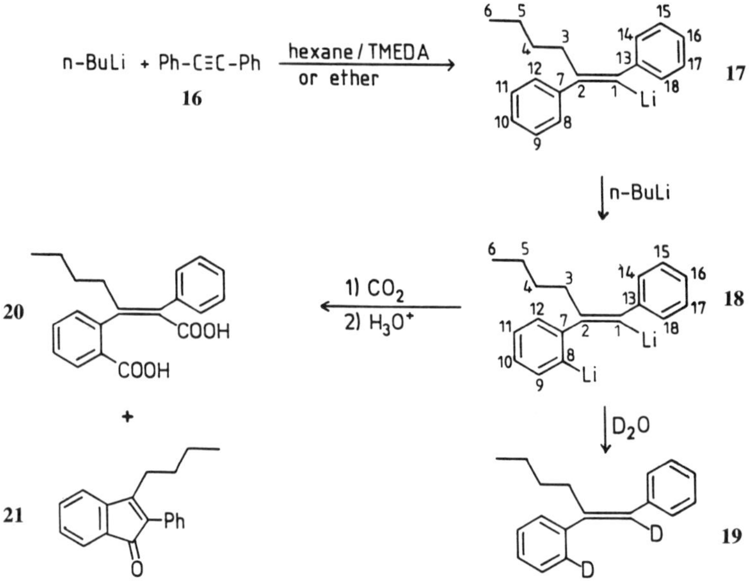

Scheme 7

7). The primary addition product **17** may undergo further lithiation to give dilithio product **18**. Intermediate **18** may be reacted with D_2O or CO_2 to yield products **19**, **20**, and **21**.

We have studied monolithio compound **17** and dilithio species **18** by NMR. In addition, an X-ray analysis was carried out for **18**.[56]

When the addition of *n*BuLi to tolane is carried out in THF, the reaction may be stopped at the stage of the primary product **17**. It may be shown by standard 1D and 2D NMR methods that **17** adopts *trans*-geometry of the phenyl rings with respect to the double bond. Furthermore, according to the $^{13}C,^{6}Li$ and $^{13}C,^{7}Li$ coupling patterns, **17** is monomeric in THF at −90°C. The $^{6}Li,^{1}H$ HOESY spectrum of **17** in THF at −30°C reveals an intense cross peak which involves H8/12, that is, the ortho positions of the phenyl ring which is remote from the metallated position in **17**. A further, much less intense cross peak is detected for the second pair of ortho protons H14/18. Rotation of the two phenyl rings in **17** is rapid on the NMR time scale under these conditions.

The MNDO calculations are in agreement with these findings: the distance Li,H8 is 2.79 Å as compared to 3.86 Å for Li,H18 in trisolvated **17** (H_2O ligands). For **17** without any ligand the involved MNDO distances are 2.34 and 3.98 Å, respectively.

Dilithio product **18**, obtained by reacting tolane with 2 equiv. of *n*BuLi/TMEDA in hexane, is a monomer in the crystalline state. Figure 5.11 shows the X-ray crystal structure of (**18** × 2TMEDA). Similar to dilithio biphenyl[73] and related species,[81] double bridging of lithium is observed in (**18** × 2TMEDA). This structural feature may be considered as the intramolecular

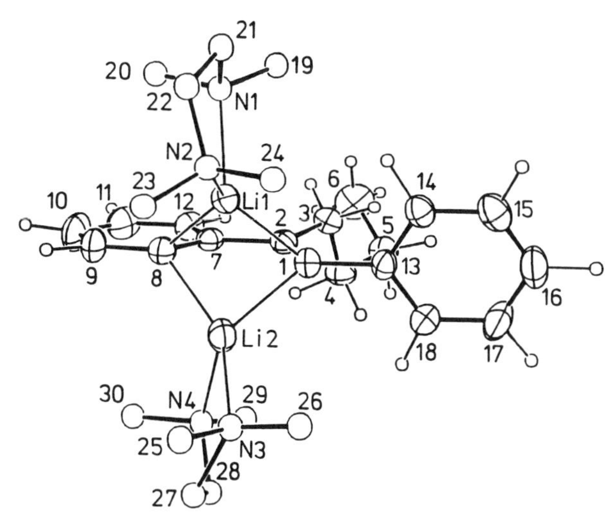

Figure 5.11 X-Ray structure of (**18** × 2TMEDA). The TMEDA hydrogen atoms have been omitted for clarity.

equivalent of dimerization.[82] The dilithio-bridged structure of (**18** × 2TMEDA) as determined by X-ray analysis is retained in benzene solution. This has been confirmed by solid state and solution NMR studies: the ^{13}C NMR chemical shifts of (**18** × 2TMEDA) obtained in C_6D_6 solution are identical to those recorded for solid (**18** × 2TMEDA) by using CP/MAS.[83]

The ^{1}H and ^{13}C NMR spectra of a THF-d_8 solution of dissolved crystals of (**18** × 2TMEDA) reveal the presence of both monomeric and dimeric **18** at −70°C. The dimer has been deduced to have a cube shape as in **22**.

The ^{13}C NMR signals of the lithiated carbon atoms in **22** show complex splitting patterns due to coupling with chemically nonequivalent ^{6}Li atoms. The phase-sensitive ^{6}Li,^{1}H HOESY spectrum of **18** in THF-d_8 is shown in Figure 5.12.[84]

In the one-dimensional ^{6}Li-spectrum (f_2 axis), the two peripheral signals (δ = 1.21 and 0.52 ppm) are assigned to dimer **22**, whereas the peak at δ = 0.73 ppm results from monomeric **18**. In the 2D HOESY contour plot, cross peaks between the ^{6}Li signal of the monomer and hydrogen atoms H9,14/18 are observed. In addition, intense cross peaks are detected which involve the TMEDA protons. Thus, even in presence of excess THF-d_8, TMEDA still chelates the Li atoms of monomeric **18** under these conditions. By contrast, no cross peaks are found which involve the TMEDA ^{1}H-signals and the ^{6}Li-resonances of dimer **22**. In structure **22** only one coordination site is available at each lithium atom. Since TMEDA is a powerful chelate ligand, however, a weaker monodentate ligand, THF effectively competes with TMEDA for Li-solvation in the dimer of **18**.

The assignment of the ^{6}Li NMR signals of the dimer of **18** may be achieved

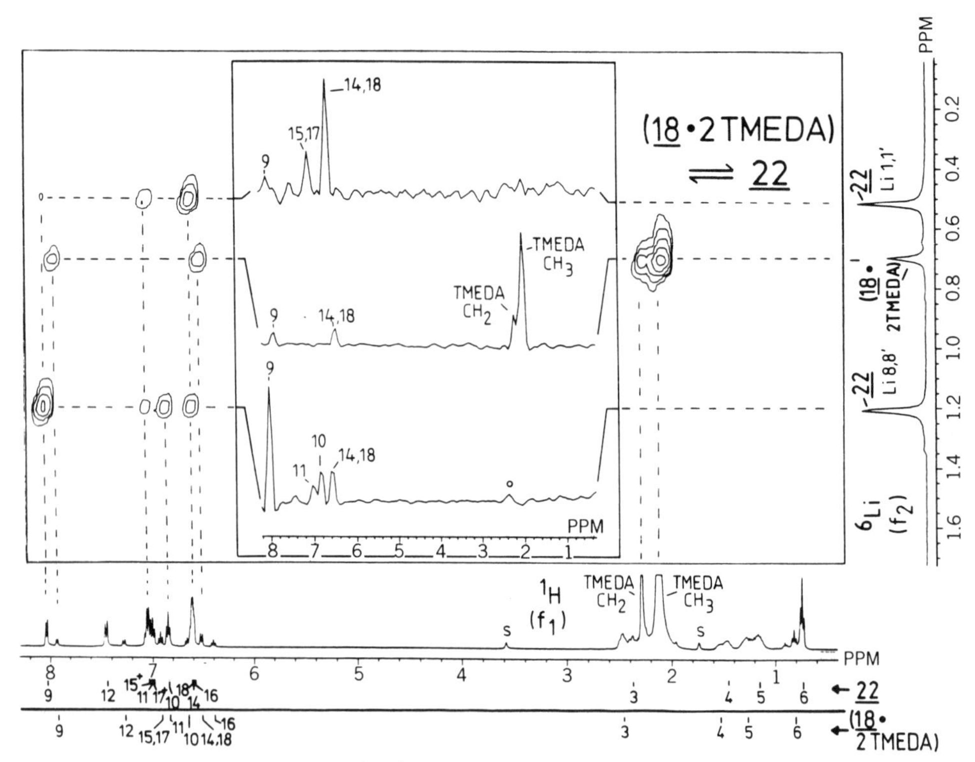

Figure 5.12 Phase-sensitive ^{6}Li,^{1}H HOESY spectrum of **18** at −64°C; crystals of (**18** × 2TMEDA), enriched with ^{6}Li, were dissolved in THF-d$_8$. For numbering, see formula **18** in Scheme 7. The inset shows f_1 cross sections cut at the chemical shifts of the three ^{6}Li signals; s = solvent; ° cross peak between Li8,8′ and H3,3′ in dimer **22**; + assignments may be interchanged.

by the HOESY spectrum depicted in Figure 5.12: the low-field ^{6}Li-signal shows an intense cross peak with H9 and is therefore assigned to Li(8,8′) in **22**. By contrast, the corresponding high-field ^{6}Li NMR signal has cross peaks in common with the ortho and meta positions of the nonmetallated aromatic rings. This signal is therefore assigned to Li(1,1′) in **22**.

By means of variable temperature measurements (van't Hoff plot) the thermodynamic parameters of the monomer–dimer equilibrium of **18** (Eq. 13) were determined:

$$\underset{\text{Monomer}}{2(\mathbf{18} \times 2\text{TMEDA})} + 4\ \text{THF} \rightleftarrows \underset{\text{Dimer}}{\mathbf{22}} + 4\ \text{TMEDA} \qquad (13)$$

$\Delta H^{\circ} = 9.4 \pm 0.6$ kcal/mol and $\Delta S^{\circ} = 22.2 \pm 2.1$ eu. Line shape analysis of the temperature-dependent ^{6}Li NMR spectra of **18** in THF-d$_8$ yielded the activation parameters: the dimer–monomer conversion proceeds with $\Delta H^{\ddagger} =$

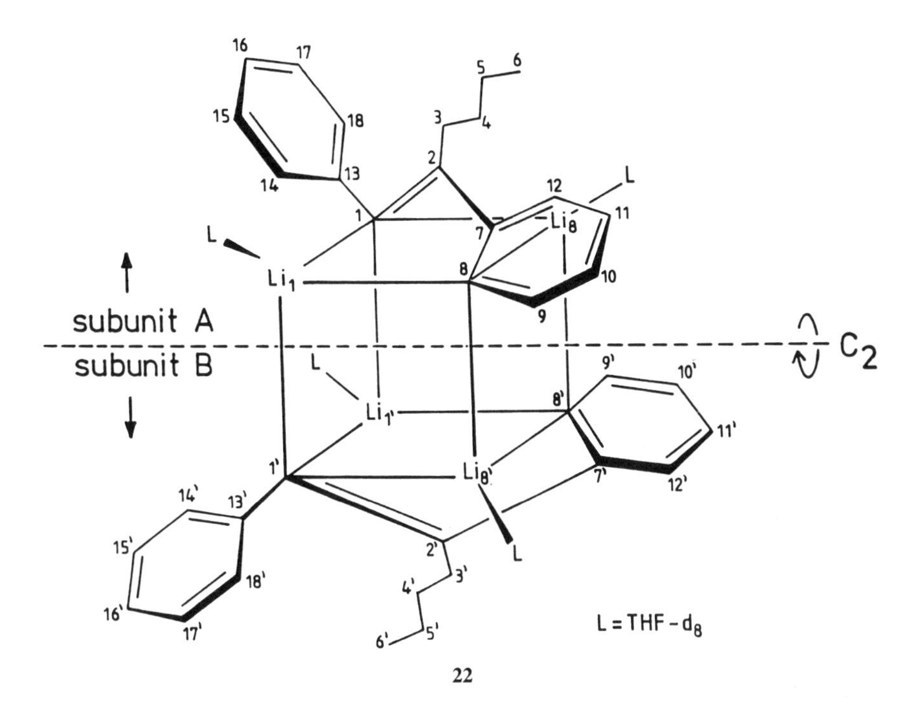

22

9.4 ± 0.7 kcal/mol, and $\Delta S^{\ddagger} = -23.5 \pm 2.5$ eu. A second dynamic process, the exchange of the chemically nonequivalent lithium positions Li(1,1′) and Li(8,8′) in dimer **22**, is related to $\Delta H^{\ddagger} = 16.1 \pm 0.4$ kcal/mol, and $\Delta S^{\ddagger} = 1.2 \pm 1.1$ eu. The latter process may be considered similar to a rotation in Rubik's cube.

Small scalar coupling between the two chemically nonequivalent lithium positions in dimer **22**, though not detectable in 1D NMR spectra, is indicated by a ^{6}Li,^{6}Li COSY spectrum with fixed delays.[85,86] Mutual exchange of lithium positions between monomer (**18** × 2TMEDA) and dimer **22**, as well as exchange of the different lithium sites within dimer **22**, may be demonstrated by ^{6}Li,^{6}Li-EXSY spectroscopy[37,78,87] (Fig. 5.13). The cross peaks in Figure 5.13 are due to magnetization transfer based on chemical exchange during the mixing period of the EXSY pulse sequence.

4.3 Characterization of Ion Pairs

According to earlier spectroscopic studies, mainly UV measurements, it has been postulated that alkali metal organic compounds may coexist as contact ion pairs (CIPs) and solvent separated ion pairs (SSIPs).[88] Fluorenyl lithium (FlLi) **23** has been extensively studied with respect to ion pair phenomena.

In the solid state, **23** has been characterized as CIP **24** (with two quinuclidine ligands)[89] and as SSIP **25** (with four THF ligands around lithium[90]). Ultraviolet spectroscopy measurements revealed that **23** consists predominantly

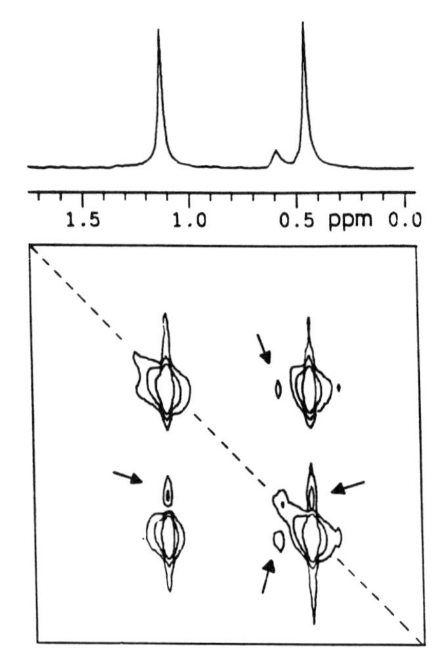

Figure 5.13 Two-dimensional $^{6}Li,^{6}Li$ exchange (EXSY) spectrum of **18** in THF-d_8, 0.2 M, −20°C, mixing time, 2.4 sec. Arrows denote cross peaks due to chemical exchange between monomer (**18** × 2TMEDA) and dimer **22**. For assignment of the ^{6}Li resonances, see Figure 5.12.

of SSIPs in THF solution at −30°C.[91] The SSIP–CIP equilibria in fluorenyl lithium derivatives are temperature dependent; the amount of CIPs was generally found to increase with increasing temperature.[88]

We applied $^{6}Li,^{1}H$ HOESY to FlLi **23**.[92] Figure 5.14 shows the HOESY contour plot of a solution of **24** in THF-d_8 at −90°C (*a*) and at −10°C (*b*). Crystalline material of (FlLi × 4THF), **25**, was employed to prepare the NMR

Li^+

23

24

THF
Li
THF
THF
THF

25

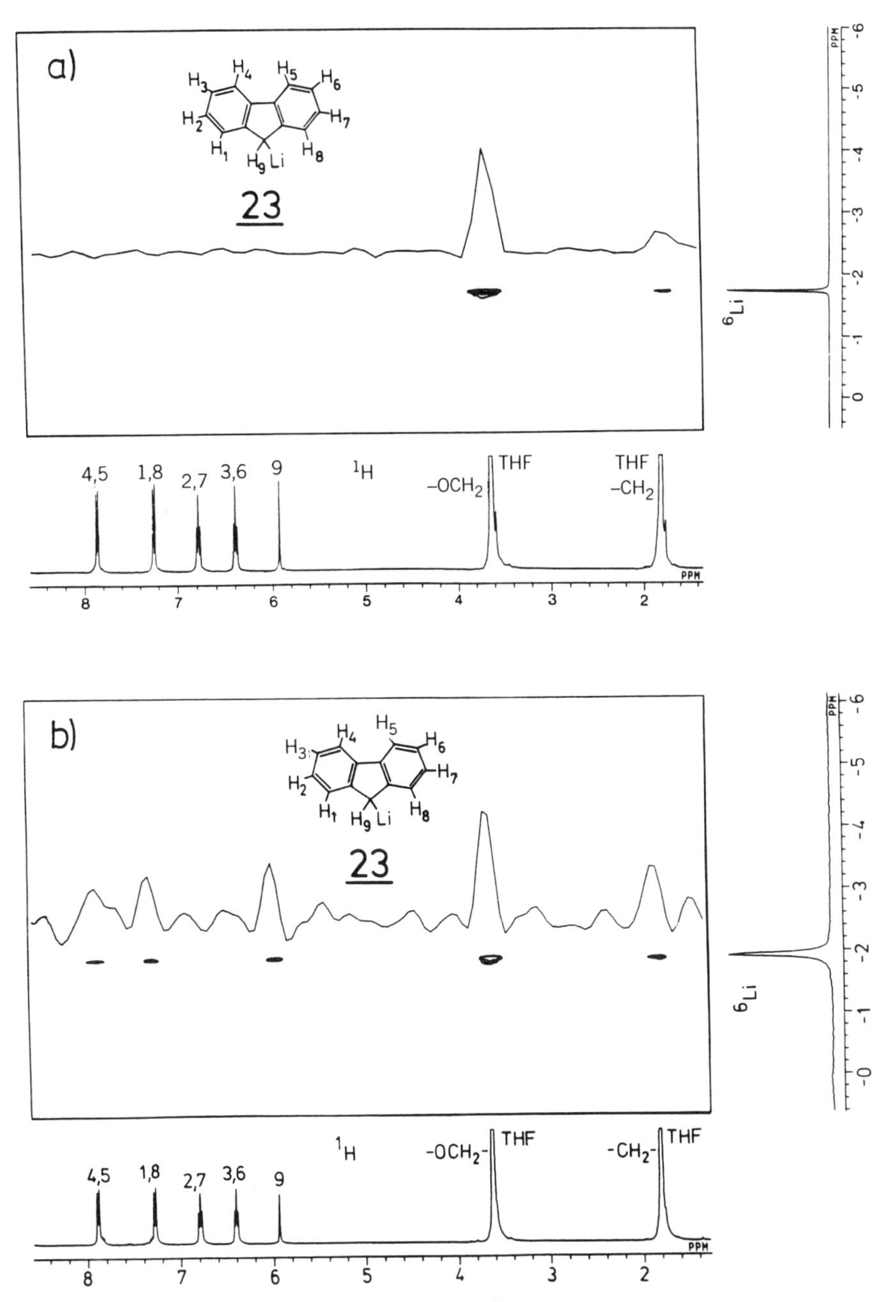

Figure 5.14 ^{6}Li,^{1}H HOESY contour plot of ^{6}Li-fluorenyl **22** in THF-d_8, 0.8 M, at $-90°$C (*a*) and at $-10°$C (*b*). Crystals of **23** with 4 equiv. of nondeuterated THF were employed. Mixing time, 2.0 sec. Insets: f_1 cross section cut at the chemical shift of the ^{6}Li resonance line.

sample. The one-dimensional ^{1}H- and ^{6}Li NMR spectra show only one set of signals. Since the involved CIP–SSIP equilibria are rapid on the NMR time scale, averaged signals are observed.

The HOESY plot in Figure 5.14*a* shows only cross peaks which involve the ^{1}H NMR signals of nondeuterated THF, introduced by the crystals. No cross peaks appear which include the signals of the fluorenyl moiety. This is indicative of a large SSIP/CIP ratio under these conditions. When the sample of Figure 5.14 is warmed to $-10°C$, the HOESY spectrum (Fig. 5.14*b*) now shows additional cross peaks. These include positions 4/5, 1/8, and 9 of the fluorenyl moiety. Thus it may be concluded that a larger amount of the CIP is present at higher temperatures, consistent with earlier findings.[88] Furthermore, as deduced from the cross peak positions in Figure 5.14*b*, lithium seems to adopt a structure in the CIP which must be attributed to η^5 bonding to the central Cp ring.

A somewhat different structure appears to be present in solution when crystals of (**23** × 4THF) are dissolved in diethyl ether-d_{10}: an intense cross peak is detected which involves H9. Hence, in the corresponding CIP, lithium appears to be located at the periphery of the anion.

With respect to conclusions drawn about CIP/SSIP equilibria, the HOESY results obtained on FlLi **23** are in complete agreement with data obtained earlier by using different spectroscopic methods.

4.4 Stereochemical Assignments in Norbornane and Camphor-Fused Lithio Cyclopentadienide Derivatives

At low temperatures ($< -80°C$), the ^{1}H and ^{13}C NMR spectra of lithium isodicyclopentadienide (isodiCpLi) **26** in THF-d_8 indicate the presence of two species.[32]

According to cryoscopic measurements, a monomer–dimer equilibrium is indicated. The ^{6}Li NMR spectrum of **26** shows three signals at $\delta = -1.10$,

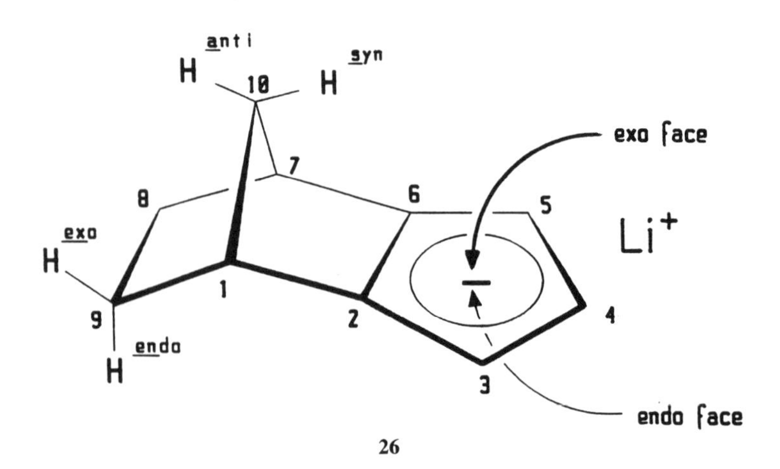

26

−7.64, and −12.78 ppm under these conditions. The exceptional high-field shifts of two of these signals may be attributed to ring current phenomena, that is, location of lithium in the magnetically anisotropic environment of one or two Cp rings.[26,27,93,94] An increase in the concentration of **26** leads to increase of the two "outer" ^{6}Li resonance lines at the expense of the "inner" line. Since the intensity changes of the "outer" lines are simultaneous, it is concluded that these two lines are due to one and the same species. This agrees with an ion pair **27**: therein, one lithium atom is sandwiched between two Cp^- rings and experiences an exceptional upfield shift ($\delta = -12.78$ ppm). The ^{6}Li-

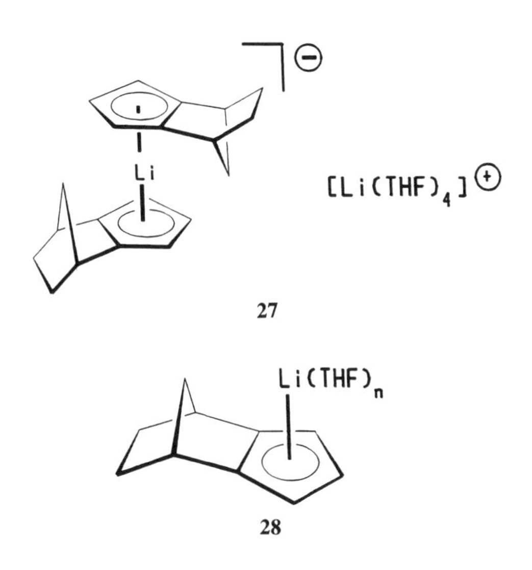

resonance of the counter ion, $[Li(THF)_4]^+$, lies in the "normal" range ($\delta = -1.10$ ppm).

The ^{6}Li signal at $\delta = -7.64$ ppm may be attributed to lithium located tightly above a single Cp ring as in **28**.

The stereochemical assignment of **27** as an *exo,exo* sandwich dimer and of **28** as an *exo* monomer results from ^{6}Li,^{1}H HOESY. Figure 5.15 shows the HOESY contour plot of **26** in THF-d_8 at −108°C.

Intense cross peaks appear in Figure 5.15 between the Cp ring proton ^{1}H signals and all three ^{6}Li signals. Although $[Li(THF)_4]^+$, assigned to the low-field ^{6}Li signal, has no direct H contacts with the anion, rapid exchange with lithium located in monomer **28** transfers magnetization between the involved H-positions. Hence, the information contents of slices *c* and *d* in Figure 5.15 are identical.

In slice *c*, a cross peak at the *syn*-H of the methano bridge protons H10 indicates close proximity between lithium and this H-position in monomer **28**. In addition, since a cross peak at the chemical shift of H8,9 (*endo*) is missing, strong evidence for *exo*-located lithium is provided. Trace *b* in Figure 5.15 has the same information contents as trace *c* owing to rapid chemical exchange.

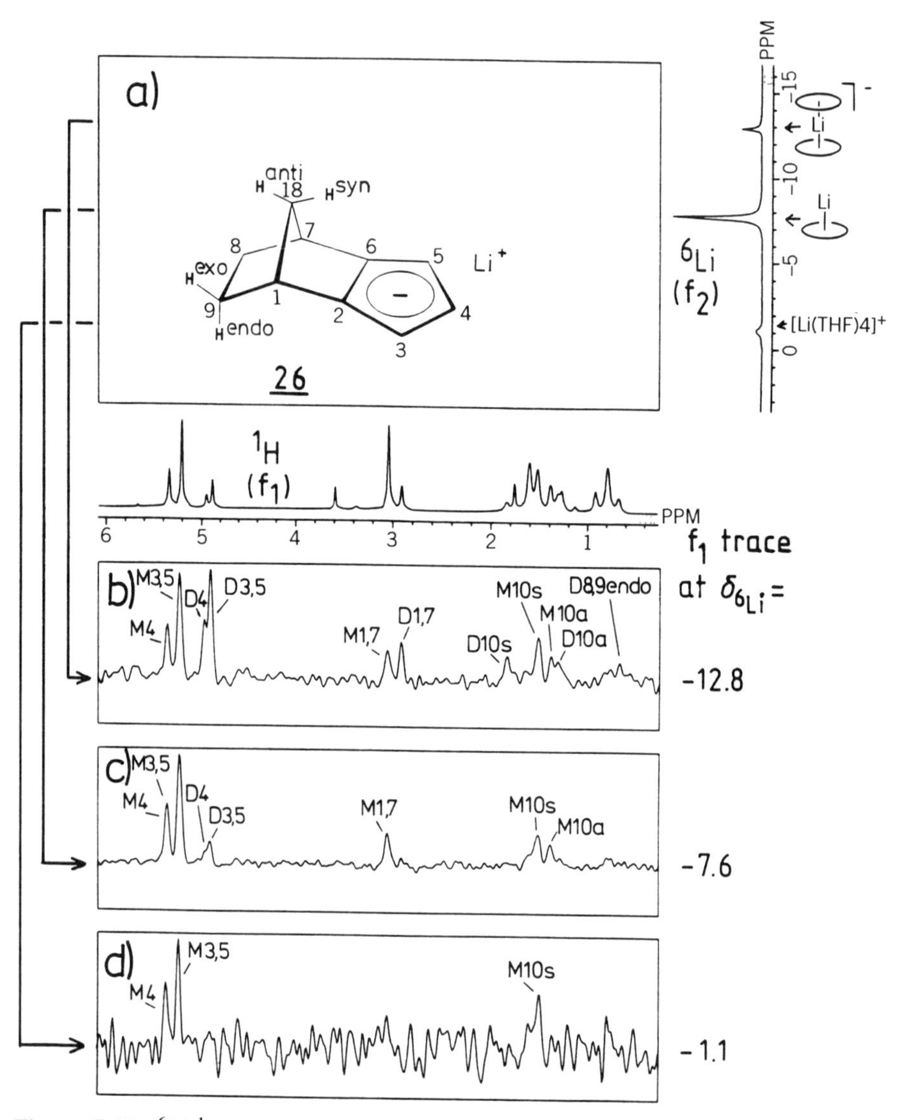

Figure 5.15 ^{6}Li,^{1}H HOESY spectrum of isodiCpLi **26** in THF-d$_8$, -108°C, 0.52 M, mixing time 1.1 sec. (*a*) contour plot; (*b*)–(*d*) f_1 cross sections of the ^{6}Li signals at $\delta = -12.78$, -7.64, and -1.10 ppm, respectively; D = dimer **27**, M = monomer **28.**

In addition, a significant cross peak at the chemical shift of H10 (*syn*) in dimer **27** is indicative of the *exo,exo* location of lithium in the sandwich dimer anion of **27**. Thus, alternative structures like *endo* monomer **29** and *endo,endo* dimer **30** may be ruled out. A mixed *exo,endo* dimer **31** should inherently exhibit two ^{1}H and ^{13}C NMR signal sets in 1 : 1 intensity ratio; this is not observed.

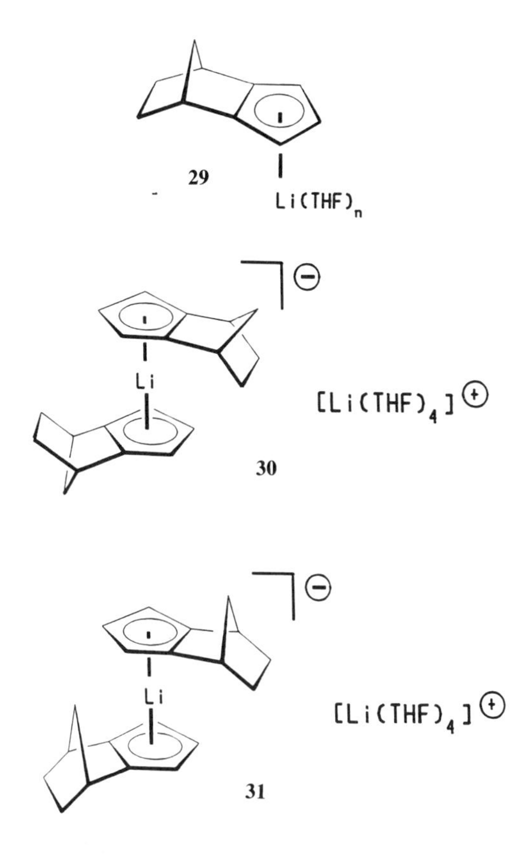

Line-shape analysis of the temperature-dependent ^{6}Li spectra of **26** in THF-d_8 reveals two independent kinetic processes: rapid exchange of "external" $[Li(THF)_4]^+$ with monomer-bound lithium, and slower exchange between monomer-bound and sandwiched lithium. No direct exchange between $[Li(THF)_4]^+$ and sandwiched lithium occurs. At high temperatures (+25°C), the monomer **28**/dimer **27** equilibrium is shifted toward the side of monomer **28**.

The MNDO calculations on various isodiCpLi isomers agree with the relative stabilities deduced from ^{6}Li,^{1}H HOESY. On the basis of the NMR results, a mechanistic scheme was proposed for the reaction of **26** with electrophiles.[32] It was found that **26** reacts mainly at the endo face at −80°C, whereas predominant exo attack is observed at +25°C.[95] At low temperatures, in dimer **27** (assumed to be the more reactive species), only *endo* faces are available for electrophilic attack (Scheme 8). By contrast, at +25°C, in the prevailing monomer **28** the electrophile is conducted by the lithium cation to the *exo* face (Scheme 9).

Contrary to isodiCpLi **26**, the related camphor-derived compound CCpLi **32** does not react with electrophiles at −80°C for steric reasons. At +25°C, preference for endo attack is observed. An NMR investigation on **32** re-

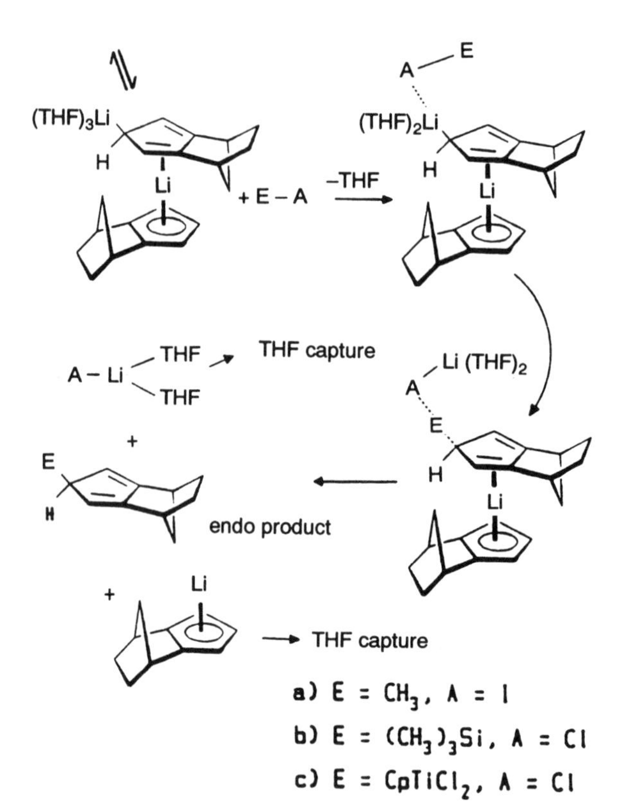

Scheme 8

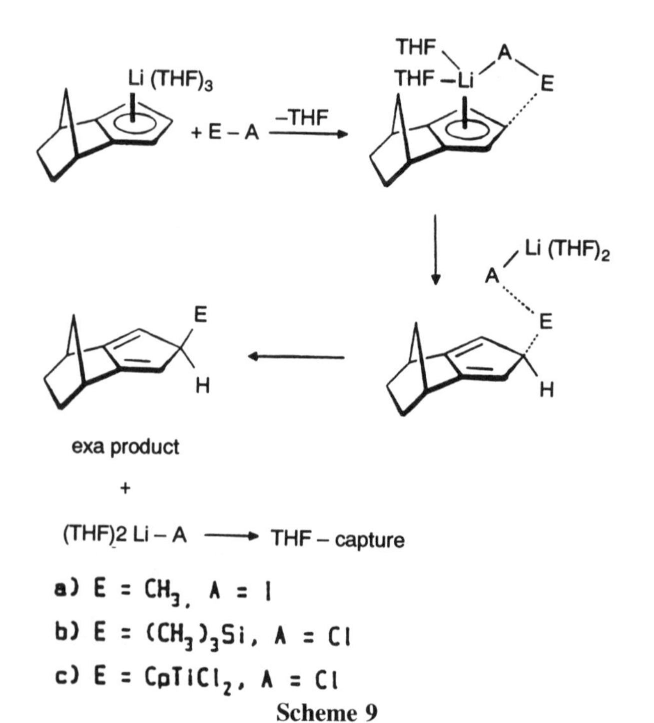

Scheme 9

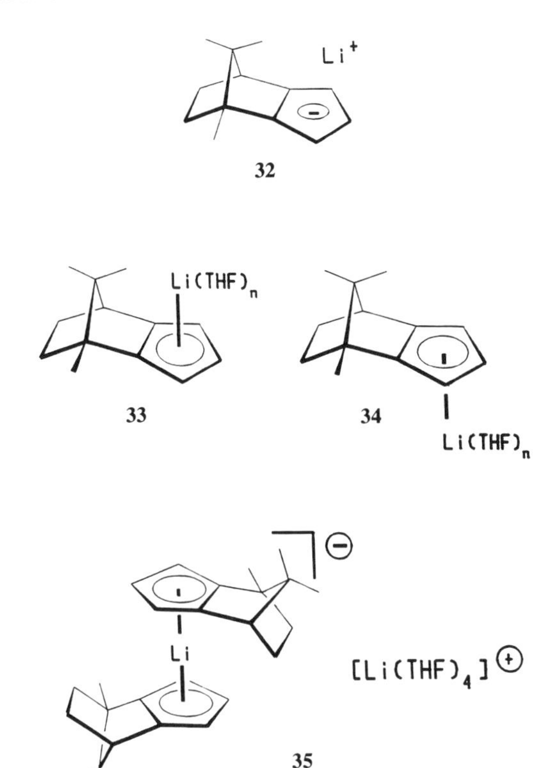

vealed—for the first time—a *ternary* equilibrium in THF solution.[96] Both at +25°C and at −107°C, **32** exists as a mixture of *exo* lithio monomer **33**, *endo* lithio monomer **34**, and *endo,endo* lithio dimer **35** in approximately 57 : 21 : 22 molar ratio. The identification of these stereoisomers is based on one-dimensional ^{1}H and ^{13}C spectra, as well as on ^{6}Li,^{1}H HOESY and ^{1}H ROESY[97] experiments. Similar to isodiCpLi **26**, unusual upfield ^{6}Li NMR chemical shifts are observed for CCpLi **32** (δ = −7.83 ppm for monomers **33** and **34**; δ = −12.45 ppm for sandwiched lithium in dimer **35**).

4.5 Studies on "Super Bases"

More than two decades ago, Lochmann[98] and, independently, Schlosser[99] realized that the efficiency of organolithium compounds as polymerization initiating and deprotonating reagents may be largely enhanced upon the addition of a potassium alkoxide. The prototype reagent, a 1 : 1 mixture of *n*BuLi and potassium tert-butoxide (KO*t*Bu) is now widely employed as a "super base."[18–20] These types of mixtures (also termed "LICKOR" reagents) may be employed either as a slurry in hydrocarbon solvents (e.g., hexane), or as homogeneous solution in THF.[100] It has been known that transmetallation occurs upon mixing *n*BuLi and KO*t*Bu in heptane: pure *n*BuK precipitates.[101] However, the nature of the prevailing species in solution remained unknown.

Both separate components (*n*BuK and LiO*t*Bu)[102] as well as mixed aggregates, (e.g. **36**[99]), have been discussed as the predominant species.

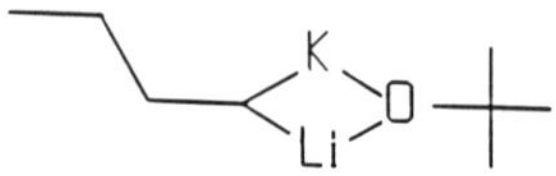

36

Since no stable potassium isotope is amenable to NOE measurements due to the high quadrupole moments (see Table 5.1 and Section 3.1.4), we employed cesium instead. The ^{133}Cs-isotope may be used for ^{133}Cs,^{1}H HOESY experiments in favorable cases,[64] though with considerably reduced sensitivity as compared to ^{6}Li. With respect to good solubility, the branched cesium alkoxide **37** was employed in our studies.[103] Likewise, with respect to reduced reactivity toward the solvent, THF-d_8, trityl lithium **38** was used as a model organolithium compound.

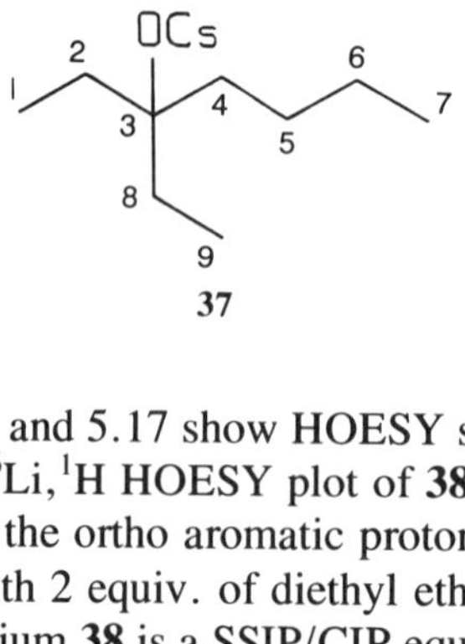

37 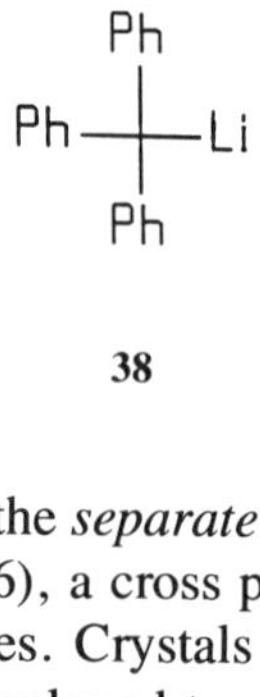 **38**

Figures 5.16 and 5.17 show HOESY studies on the *separate* compounds **37** and **38**. In the ^{6}Li,^{1}H HOESY plot of **38** (Fig. 5.16), a cross peak is detected which involves the ortho aromatic proton resonances. Crystals of **38** enriched with ^{6}Li and with 2 equiv. of diethyl ether were employed to generate Figure 5.16. Trityl lithium **38** is a SSIP/CIP equilibrium mixture under the conditions of Figure 5.16, with the SSIP strongly dominating (SSIP : CIP = 39 : 1).[104] Since only the CIP may lead to observable cross peaks between lithium and the protons of the aromatic ring, the signal to noise ratio in Figure 5.16 is comparatively low.

Figure 5.17 shows the ^{133}Cs,^{1}H HOESY spectrum of cesium alkoxide **37**. Though **37** is a pure CIP under these conditions (as deduced from the extremely low field ^{133}Cs chemical shift), the cross peaks of Figure 5.17 are weak and reflect the insensitivity of ^{133}Cs,^{1}H HOESY. Cross peaks in Figure 5.17 appear at the expected ^{1}H positions near the alkoxide oxygen. Remote ^{1}H positions such as protons 6 and 7 in **37** show no cross peaks.

The ^{6}Li,^{1}H HOESY spectrum of an approximately equimolar mixture of trityl–^{6}Li **38** and cesium alkoxide **37** is depicted in Figure 5.18. Intense cross peaks are now found between lithium and the alkoxide protons. This indicates that quantitative transmetallation must have occurred in homogeneous THF solution. Furthermore, the lack of cross peaks which involve the protons of

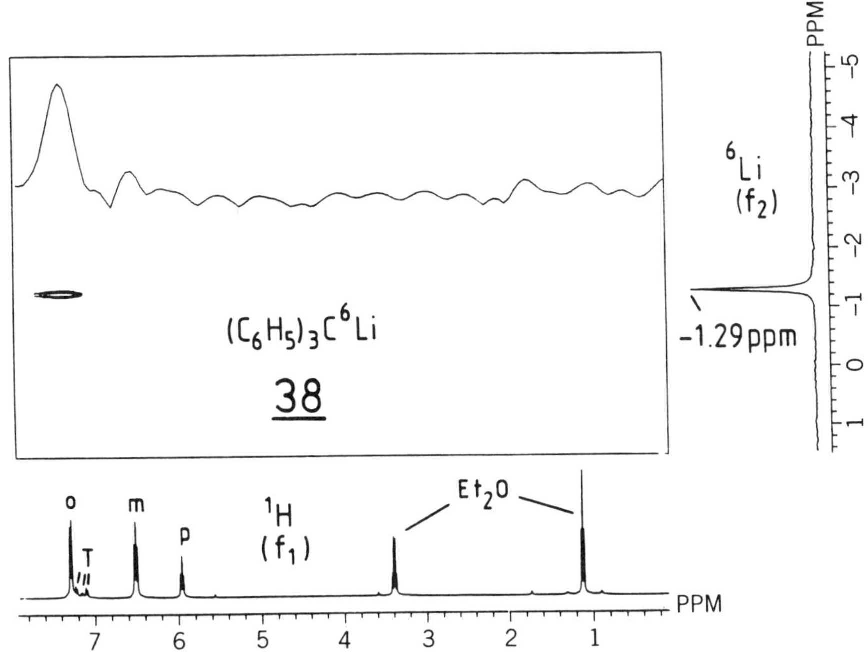

Figure 5.16 $^6Li,^1H$ HOESY contour plot of lithium trityl **38**; crystals of **38** (with 2 equivs. of diethyl ether and enriched 96% with ^6Li) were dissolved in THF-d_8 (0.35 M, +26°C, mixing time 2.0 sec.) Inset: f_1 cross section at δ (^6Li) = −1.29 ppm; T = triphenylmethane.

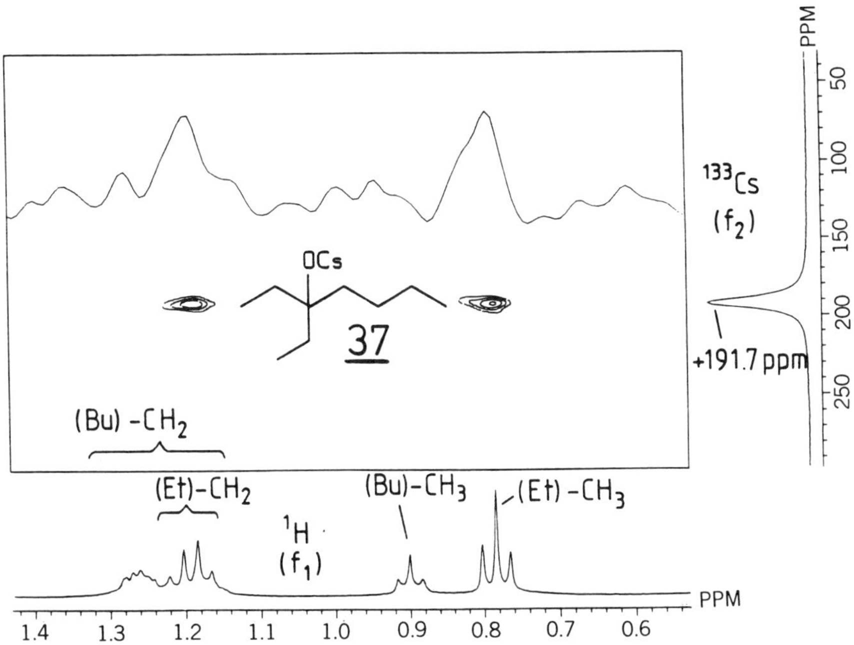

Figure 5.17 $^{133}Cs,^1H$ HOESY contour plot of cesium 3-ethyl-3-heptoxide **37**; crystals of **37** were dissolved in THF-d_8 (1.2 M, +40°C, mixing time 0.3 sec). Inset: f_1 cross section at δ (^{133}Cs) = +191.7 ppm.

OLi

39

Ph
Ph—C—Cs
Ph

40

the phenyl rings suggests that the newly formed lithium alkoxide **39** and cesium trityl **40** are *separate* species under these conditions.

This is confirmed by inspection of the complementary ^{133}Cs,^{1}H HOESY spectrum shown in Figure 5.19. An intense cross peak is found which involves the ortho aromatic protons. This is indicative of cesium being located above the central trityl C-atom, at least averaged in terms of the NMR time scale. However, in agreement with the conclusions drawn from Figure 5.18, no cross peaks are found in Figure 5.19 between the cesium and the alkoxide resonance lines. This confirms that lithium alkoxide **39** and cesium trityl **40** do *not* inter-

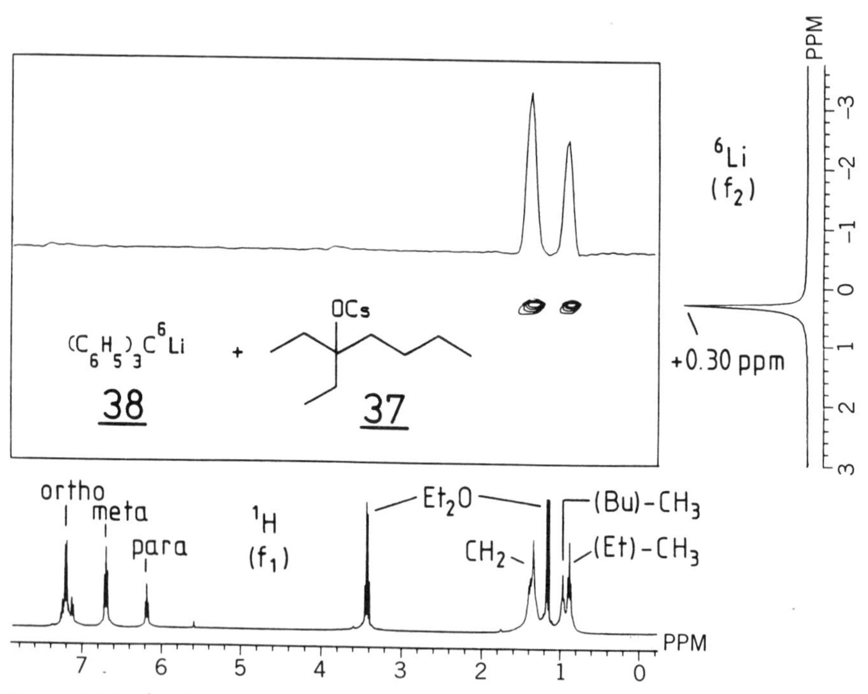

Figure 5.18 ^{6}Li,^{1}H HOESY contour plot of a mixture of lithium trityl **38** and cesium 3-ethyl-3-heptoxide **37** (1 : 1.5 molar ratio); crystals of **38** (with 2 equiv. of diethyl ether and enriched 96% with ^{6}Li) and crystals of **37** were dissolved in THF-d_8 (0.6 M in **38**, +26°C, mixing time 2.0 sec). Inset: f_1 cross section at δ (^{6}Li) = +0.30 ppm.

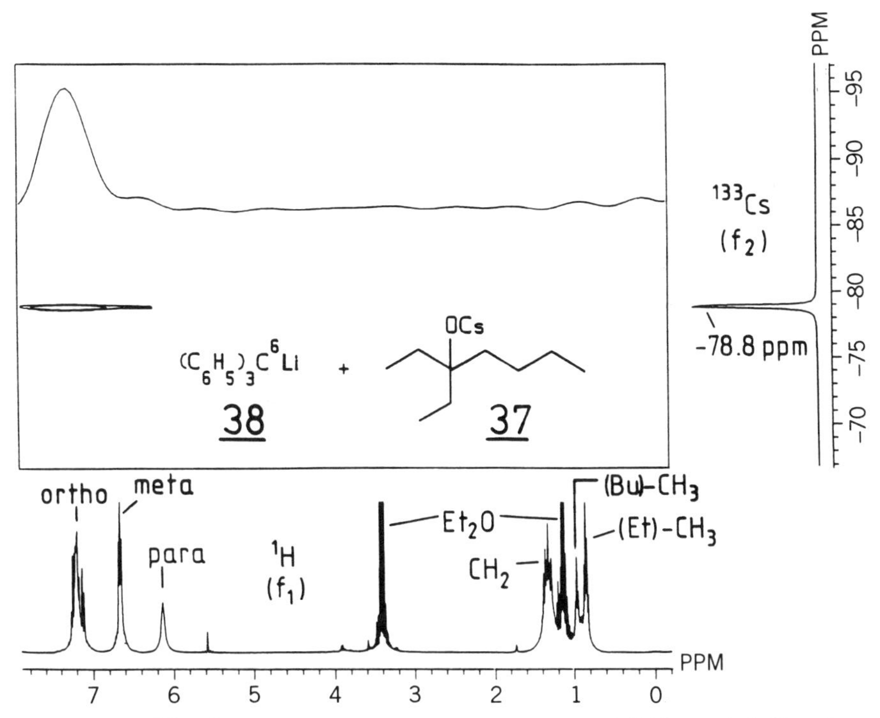

Figure 5.19 ^{133}Cs,^{1}H HOESY contour plot of a mixture of lithium trityl **38** and cesium 3-ethyl-3-heptoxide **37** (1.2 : 1 molar ratio); crystals of **38** (with 2 equiv. of diethyl ether and enriched 96% with ^{6}Li) and crystals of **37** were dissolved in THF-d_8 (0.6 M in **38**, +26°C, mixing time 0.35 sec.). Inset: f_1 cross section at δ (^{133}Cs) = −78.8 ppm.

act in THF-d_8 solution. Thus the model "super base" system gives no hints for the presence of a mixed aggregate, for example, **41**.

The MNDO calculations on Ph_3CLi and KOtBu are in agreement with the NMR findings: the transmetallation reaction is calculated to be highly exo-

Cs
O
Li

41

thermic. By contrast, the formation of a mixed Ph_3CK/LiOtBu aggregate is an endothermic process.

It should be emphasized that lithium trityl, employed for the described NMR and MNDO studies, may not be typical of real "super bases" which consist of an alkyl lithium compound. The existence of mixed aggregates for the latter compound with KOtBu may not be ruled out at present.

4.6 Quantitative Determination of Li,H Distances

The ^{6}Li,^{1}H HOESY spectra described so far were indicative only qualitatively of close internuclear Li,H contacts. The cross peak intensities in a *single* ^{6}Li,^{1}H HOESY spectrum may be treated only as a crude measure for Li,H separations. However, similar to the determination of homonuclear ^{1}H,^{1}H distances by 2D NOESY,[3] *quantitative* internuclear Li,H distances may be obtained from a series of ^{6}Li,^{1}H HOESY spectra. The initial buildup rates, f, of the transient heteronuclear NOEs are related to the corresponding distances, r, by Eq. 14.

$$[f(^{6}\text{Li},^{1}\text{H}^{\text{A}})/f(^{6}\text{Li},^{1}\text{H}^{\text{X}})]^{1/6} = r(^{6}\text{Li},^{1}\text{H}^{\text{X}})/r(^{6}\text{Li},^{1}\text{H}^{\text{A}}) \quad (14)$$

Here H^A denotes a proton whose distance from the lithium nucleus is known (e.g., from X-ray data), whereas H^X is a proton of unknown lithium separation. For our quantitative ^{6}Li,^{1}H HOESY studies[44] we employed vinyl lithium **42**.

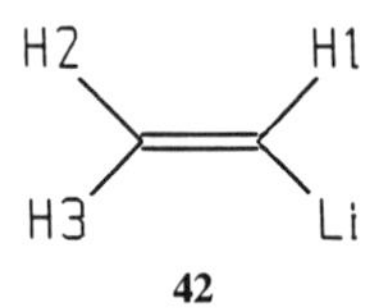

42

The X-ray crystal structure of **42** revealed a tetramer as shown in Figure 5.20. In THF solution at +26°C, **42** is tetrameric as well. In the ^{6}Li,^{1}H HOESY spectrum of **42** in THF-d_8, the order of the cross peak intensities is (as expected): H1 > H3 > H2. Figure 5.21*a* shows the dependence of cross peak intensities in a series of ^{6}Li,^{1}H HOESY spectra on **42** on the mixing time. Owing to the long spin-lattice relaxation time of ^{6}Li under the conditions of Figure 5.21 [$T_1(^{6}\text{Li}) = 81$ sec], the heteronuclear NOE builds up much slower than is usually found in ^{1}H,^{1}H NOESY.

The initial HOE buildup rates may be derived from HOESY spectra with short mixing times (Figure 5.21*b*). The ratio of the slopes H1:H2:H3 is 7.96:1.00:3.12. By using Eq. 14, the involved Li—H distance ratios may be derived: r(Li,H1)/r(Li,H2) = 1.41; r(Li,H1)/r(Li,H3) = 1.17; r(Li,H2)/r(Li,H3) = 0.83. By choosing a particular Li,H distance from the X-ray analysis of **42** (Figure 5.20) as a reference distance, the remaining Li,H separations may be calculated. In matrix representation, Table 5.4 shows the $(\langle r^{-6}\rangle^{-1/6})$

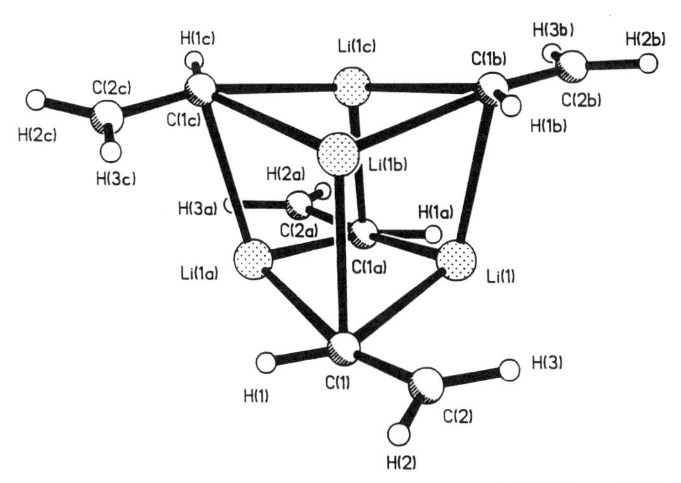

Figure 5.20 Molecular structure of vinyllithium–THF, $[C_2H_3Li \times C_4H_8O]_4$; THF ligands have been omitted for clarity.

weighted internuclear Li,H distances from the X-ray analysis of **42** (underlined values) as well as the Li,H distances calculated from the HOESY data.

There is good agreement between the X-ray structural distances and the ^{6}Li,^{1}H HOESY data. Ideally, the numbers within one vertical column in Table 5.4 should be identical. The mean deviation within all columns of Table 5.4

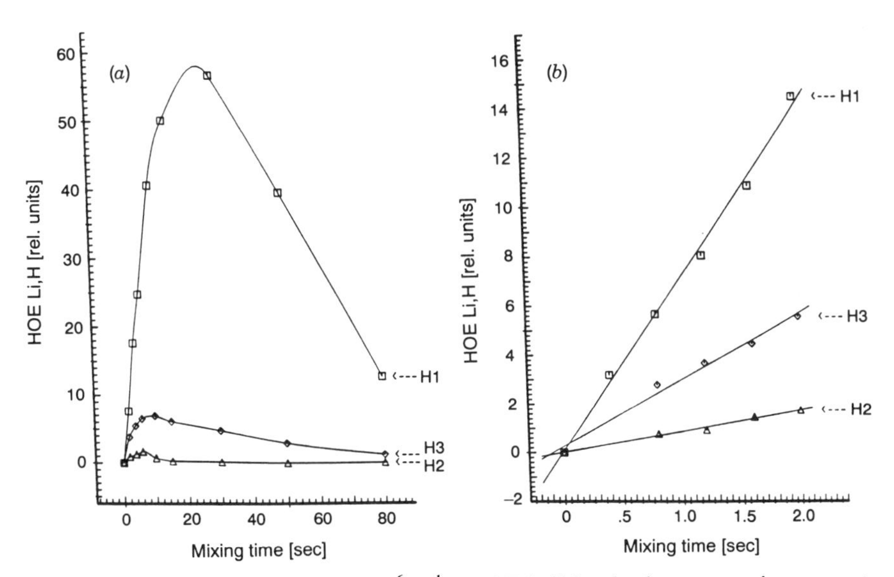

Figure 5.21 (*a*) Dependence of the ^{6}Li,^{1}H HOE buildup in the tetrameric aggregate of **42** on the mixing time, τ_m (THF-d$_8$, + 26°C, c = 2.0 M); (*b*) initial ^{6}Li,^{1}H HOE buildup in the tetrameric aggregate of **42**, determined from cross-peak volume integrals for short mixing times.

Table 5.4 Intramolecular Li, H Distances (Å) in the Tetrameric Aggregate of Vinyllithium 42[a]

	Li, H1	Li, H2	Li, H3
Li, H1	2.49	3.52	2.91
Li, H2	2.73	3.86	3.19
Li, H3	2.68	3.80	3.14

[a]Underlined values are weighted data ($(\langle r^{-6}\rangle^{-1/6})$) from the X-ray analysis (Fig. 5.20). Other data in horizontal rows represent distances derived from ^{6}Li, ^{1}H HOESY by using the underlined reference distance of the same row.

is 0.2 Å. Thus, internuclear Li,H distances may be determined by using ^{6}Li,^{1}H HOESY with reasonable accuracy. However, a suitable calibration distance must be given in this approach.

REFERENCES

1. Overhauser, A. W. *Phys. Rev.* **1953,** *89,* 689. *Ibid.* **1955,** *92*, 411.
2. Bloembergen, N.; Solomon, I. *J. Chem. Phys.* **1956,** *25*, 261.
3. Neuhaus, D.; Williamson, M. P. *The Nuclear Overhauser Effect in Structural and Conformational Analysis*; VCH: New York, 1989.
4. Solomon, I. *Phys. Rev.* **1955,** *99*, 559.
5. Wüthrich, *NMR of Proteins and Nucleic Acids*; Wiley: New York, 1986.
6. Bertini, I.; Molinari, H.; Niccolai, N., Eds.; *NMR and Biomolecular Structure*; VCH: Weinheim, Germany, 1991.
7. Derome, A. *Modern NMR Techniques for Chemistry Research*; Pergamon: Oxford, 1987.
8. Rinaldi, P. *J. Am. Chem. Soc.* **1983,** *105*, 5167.
9. Yu, C.; Levy, G. C. *J. Am. Chem. Soc.* **1983,** *105*, 6994. *Ibid.*, **1984,** *106*, 6533.
10. Freeman, R. *A Handbook of Nuclear Magnetic Resonance*; Longman Scientific & Technical: Essex, England, 1988 p. 164.
11. Bodenhausen, G.; Kogler, H.; Ernst, R. R. *J. Magn. Reson.* **1984,** *58*, 370.
12. Sørensen, O. W.; Eich, G. W.; Levitt, M. H.; Bodenhausen, G.; Ernst, R. R. *Progr. Nucl. Magn. Reson. Spectrosc.* **1983,** *16*, 163.
13. Ernst, R. R.; Bodenhausen, G.; Wokaun, A. *Principles of Nuclear Magnetic Resonance in One and Two Dimensions*; Clarendon: Oxford, 1987.
14. Kövér, K. E.; Batta, G. *Progr. NMR Spectroscopy* **1987,** *19*, 223.
15. Brevard, C.; Granger, P. *Handbook of High Resolution Multinuclear NMR*; Wiley: New York, 1981.
16. Akitt, J. W. In *Multinuclear NMR*; Mason, J., Ed.; Plenum Press: New York, 1987.

17. Lambert, J. B.; Riddell, F. G. *The Multinuclear Approach to NMR Spectroscopy*; Reidel: Dordrecht, Holland, 1983.
18. Wakefield, B. J. *Organolithium Methods*; Academic: London, 1988.
19. Brandsma, L.; Verkruijsse, H. *Preparative Polar Organometallic Chemistry I*; Springer: Berlin, 1987.
20. Brandsma, L. *Preparative Organometallic Chemistry II*; Springer: Berlin, 1990.
21. Seebach, D.: Hässig, R.; Gabriel, J. *Helv. Chim. Acta* **1983,** *66*, 308.
22. Ramsey, N. F. *Phys. Rev.* **1950,** *78*, 689.
23. Webb, G. A. In *NMR and the Periodic Table*; Harris, R. K.; Mann, B. E., Eds.; Academic: London, 1978.
24. Webb, G. A. In *NMR of Newly Accessible Nuclei, Vol. 1*; Laszlo, P., Ed.; Academic: New York, 1983.
25. Detellier, C. In *NMR of Newly Accessible Nuclei, Vol. 1*; Laszlo, P., Ed.; Academic: New York, 1983.
26. Cox, R. H.; Terry, H. W., Jr.; Harrison, L. W. *J. Am. Chem. Soc.* **1971,** *93*, 3297.
27. Cox, R. H.; Terry, H. W. *J. Magn. Reson.* **1974,** *14*, 317.
28. Sekiguchi, A.; Ebata, K.; Kabuto, C.; Sakurai, H. *J. Am. Chem. Soc.* **1991,** *113*, 7081.
29. Jackman, L. M.; Rakiewicz, E. F.; Benesi, A. J. *J. Chem. Soc.* **1991,** *113*, 4101.
30. Scherr, P. A.; Hogan, R. J.; Oliver, J. P. *J. Am. Chem. Soc.* **1974,** *96*, 6055.
31. Kutzelnigg, W.; Fleischer, U.; Schindler, M. In *NMR—Basic Principles and Progress, Vol. 23*; Diehl, P.; Fluck, E.; Günther, H.; Kosfeld, R.; Seelig, J., Eds.; Springer: Berlin, 1991.
32. Paquette, L. A.; Bauer, W.; Sivik, M. R.; Bühl, M.; Feigel, M.; Schleyer, P. v. R. *J. Am. Chem. Soc.* **1990,** *112*, 8776.
33. Fraenkel, G.; Fraenkel, A. M.; Geckle, M. J.; Schloss, F. *J. Am. Chem. Soc.* **1979,** *101*, 4745.
34. Harris, R. K. *Nuclear Magnetic Resonance Spectroscopy*; Longman Scientific & Technical: Essex, England, **1987,** p. 134.
35. Seebach, D.; Siegel, H.; Gabriel, J.; Hässig, R. *Helv. Chim. Acta* **1980,** *63*, 2046.
36. Bauer, W.; Winchester, W. R.; Schleyer, P. v. R. *Organometallics* **1987,** *6*, 2371.
37. Bauer, W.; Schleyer, P. v. R. In *Advances in Carbanion Chemistry, Vol. 1*, Snieckus, V., Ed.; Jai Press: Greenwich, CT, 1992.
38. Fraenkel, G.; Henrichs, M.; Hewitt, J. M.; Su, B. M.; Geckle, M. *J. Am. Chem. Soc.* **1980,** *102*, 3345.
39. Kalinowski, H.-O.; Berger, S.; Braun, S. *Carbon-13 NMR Spectroscopy*; Wiley: Chichester, England, 1988.
40. Knorr, R.; von Roman, T.; von Roman, U. Private communication.
41. Fraenkel, G.; Hsu, H.; Su, B. M. In *Lithium: Current Applications in Science, Medicine, and Technology*; Bach, R. O., Ed.; Wiley: New York, 1985.
42. Hässig, R.; Seebach, D. *Helv. Chim. Acta* **1983,** *66*, 2269.

43. Thomas, R. D.; Clarke, M. T.; Jensen, R. M.; Young, T. C. *Organometallics* **1986,** *5*, 1851.
44. Bauer, W.; Hampel, F. *J. Chem. Soc. Chem. Commun.* **1992,** 903.
45. Fraenkel, G.; Stier, M. *Symposium on Advances in Carbanion Chemistry presented before the Division of Petroleum Chemistry, INC.* ACS Chicago Meeting, September 8–13, 1985.
46. Fraenkel, G.; Pramanik, P. *J. Chem. Soc. Chem. Commun.* **1983,** 1527.
47. van Eikema Hommes, N. R. J.; Schleyer, P. v. R. Unpublished results.
48. Kallmann, N.; Collum, D. B. *J. Am. Chem. Soc.* **1987,** *109*, 7466. Galiano-Roth, A. S.; Michaelides, E. M.; Collum, D. B. *J. Am. Chem. Soc.* **1988,** *110*, 2658. De Pue, J. S.; Collum, D. B. *J. Am. Chem. Soc.* **1988,** *110*, 5518. Galiano-Roth, A. S.; Collum, D. B. *J. Am. Chem. Soc.* **1989,** *111*, 6772. Jackman, L. M.; Scarmoutzos, L. M.; Porter, W. *J. Am. Chem. Soc.* **1987,** *109*, 6524. Jackman, L. M.; Scarmoutzos, L. M. *J. Am. Chem. Soc.* **1987,** *109*, 5348. Jackman, L. M.; Scarmoutzos, L. M.; Smith, B. D.; Willard, P. G. *J. Am. Chem. Soc.* **1988,** *110*, 6058.
49. Edlund, U.; Lejon, T.; Venkatachalam, T. K.; Buncel, E. *J. Am. Chem. Soc.* **1985,** *107*, 6408.
50. Anderson, D. M.; Hitchcock, P. B.; Lappert, M.; Leung, W.-P.; Zora, J. A. *J. Organomet. Chem.* **1987,** *333*, C13.
51. Gilbert, T. M.; Bergman, R. G. *J. Am. Chem. Soc.* **1985,** *107*, 6391.
52. Bauer, W.; Griesinger, C. *J. Am. Chem. Soc.* **1993,** *115*, 10871.
53. Eppers, O.; Günther, H.; Klein, K.-O.; Maercker, A. *Magn. Reson. Chem.* **1991,** *29*, 1065.
54. Eppers, O.; Fox, T.; Günther, H. *Helv. Chim. Acta* **1992,** *75*, 883.
55. Günther, H.; Moskau, D.; Dujardin, R.; Maercker, A. *Tetrahedron Lett.* **1986,** *27*, 2251.
56. Bauer, W.; Feigel, M.: Müller, G.; Schleyer, P. v. R. *J. Am. Chem. Soc.* **1988,** *110*, 6033.
57. Wehrli, F. W. *J. Magn. Reson.* **1976,** *23*, 527.
58. Wehrli, F. W. *Org. Magn. Reson.* **1978,** *11*, 106.
59. Wehrli, F. W. *J. Magn. Reson.* **1978,** *30*, 193.
60. Field, L. D.; Gardiner, M. G.; Kennard, C. H. L.; Messerle, B. A.; Raston, C. *Organometallics* **1991,** *10*, 3167.
61. Bauer, W. Unpublished results.
62. Hertz, H. G.; Holz, M.; Klute, R.; Stadilis, G.; Versmold, H. *Ber. Bunsenges. Phys. Chem.* **1974,** *78*, 24. Hertz, H. G. Holz, M.; Keller G.; Versmold, H.; Yoon, C. *Ibid.* **1974,** *78*, 493.
63. Wehrli, F. W. *J. Magn. Reson.* **1977,** *25*, 575.
64. Bauer, W. *Magn. Reson. Chem.* **1991,** *29*, 494.
65. Lieser, K. H. *Einführung in die Kernchemie*; VCH: Weinheim, Germany, 1969.
66. Gilman, H.; Young, R. V. *J. Am. Chem. Soc.* **1934,** *56*, 1415.
67. Wittig, G.; Pockels, U.; Dröge, H. *Chem. Ber.* **1938,** *71*, 1903.
68. Gilman, H.; Morton, J. W. *Org. React.* **1954,** *8*, 258. Gschwend, H. W.; Rodriguez, H. R. *Org. React.* **1979,** *26*, 1. Wardell, J. L. In *Comprehensive Or-*

ganometallic Chemistry; Wilkinson, G.; Stone, F. G. A.; Abel, E. W., Eds.; Pergamon: Oxford, 1982; Vol. 1, p. 57. Beak, P.; Snieckus, V. *Acc. Chem. Res.* **1982,** *15*, 306. Narasimhan, N. S.; Mali, R. S. *Top. Curr.* Chem. **1987,** *138*, 63. Snieckus, V. *Bull. Soc. Chem. Fr.* **1988,** 67.

69. Bauer, W.; Schleyer, P. v. R. *J. Am. Chem. Soc.* **1989,** *111*, 7191.
70. van Eikema Hommes, N. J. R.; Schleyer, P. v. R. *Angew. Chem.* **1992,** *104*, 768. *Angew. Chem. Int. Ed. Engl.* **1992,** *33*, 755.
71. Gilman, H.; Soddy, T. *J. Org. Chem.* **1957,** *22*, 1715.
72. Wittig, G.; Pieper, G.; Fuhrmann, G. *Chem. Ber.* **1940,** *73*, 1193.
73. Neugebauer, W.; Kos, A. J.; Schleyer, P. v. R. *J. Organomet. Chem.* **1982,** *228*, 107. Schubert, U.; Neugebauer, W.; Schleyer, P. v. R. *J. Chem. Soc. Chem. Commun.* **1982,** 1184.
74. Neugebauer, W.; Clark, T.; Schleyer, P. v. R. *Chem. Ber.* **1983,** *116*, 3283.
75. Bauer, W.; Müller, G.; Pi, R.; Schleyer, P. v. R. *Angew. Chem.* **1986,** *98*, 1130. *Angew. Chem. Int. Ed. Engl.* **1986,** *25*, 1103.
76. Cheeseman, G. W. H.; Greenberg, S. G. *J. Organomet. Chem.* **1979,** *166*, 139.
77. Bauer, W.; Clark, T.; Schleyer, P. v. R. *J. Am. Chem. Soc.* **1987,** *109*, 970.
78. Günther, H.; Moskau, D.; Schmalz, D.; *Angew. Chem.* **1987,** *99*, 1242. *Angew. Chem. Int. Ed. Engl.* **1987,** *26*, 1212.
79. Rodionov, A. N.; Shigorin, D. N.; Talalaeva, T. V.; Tsareva, G. V.; Kocheshkov, K. A. *Zh. Fiz. Khim.* **1966,** *40*, 2265; *Chem. Abstr.* **1967,** *66*, 28268e.
80. Mulvaney, J. E.; Garlund, Z. G.; Garlund, S. L. *J. Am. Chem. Soc.* **1963,** *85*, 3897. Mulvaney, J. E.; Garlund, Z. G.; Garlund, S. L.; Newton, D. J. *J. Am. Chem. Soc.* **1966,** *88*, 476. Mulvaney, J. E.; Carr, L. J. *J. Org. Chem.* **1968,** *33*, 3286. Mulvaney, J. E.; Newton, D. J. *J. Org. Chem.* **1969,** *34*, 1936.
81. Setzer, W. N.; Schleyer, P. v. R. *Adv. Organomet. Chem.* **1985,** *24*, 353.
82. Schleyer, P. v. R. *Pure Appl. Chem.* **1984,** *56*, 151.
83. Bauer, W.; Sebald, A.; Schleyer, P. v. R. Unpublished results.
84. Bauer, W.; Schleyer, P. v. R. *Magn. Reson. Chem.* **1988,** *26*, 827.
85. Morris, G. A. *Magn. Reson. Chem.* **1986,** *24*, 371.
86. Günther, H.; Moskau, D.; Dujardin, R.; Maercker, A. *Tetrahedron Lett.* **1986,** *27*, 2251. Moskau, D.; Brauers, F.; Günther, H.; Maercker, A. *J. Am. Chem. Soc.* **1987,** *109*, 5532.
87. Meier, B. H.; Ernst, R. R. *J. Am. Chem. Soc.* **1979,** *101*, 6441.
88. Szwarc M., Ed.; *Ions and Ion Pairs in Organic Reactions*; Wiley: New York, 1972; Vol. 1; 1974, Vol. 2. J. Smid, *Ibid.*; 1972, Vol. 1, p. 85. Hogen-Esch, T. E. In *Advances in Physical Organic Chemistry*; Gold, V.; Bethell, D., Eds.; Academic: London, 1977; Vol. 15, p. 154.
89. Zerger, R.; Rhine, W.; Stucky, G. D. *J. Am. Chem. Soc.* **1974,** *96*, 5441.
90. Hoffmann, D.; Stalke, D.; Schleyer, P. v. R. Unpublished results.
91. Hogen-Esch, T. E.; Smid, J. *J. Am. Chem. Soc.* **1966,** *88*, 307, 318.
92. Hoffmann, D.; Bauer, W.; Schleyer, P. v. R. *J. Chem. Soc. Chem. Commun.* **1990,** 208.
93. Exner, M. M.; Waack, R.; Steiner, E. C. *J. Am. Chem. Soc.* **1973,** *95*, 7009.
94. Jutzi, P.; Leffers, W.; Pohl, S.; Saak, W. *Chem. Ber.* **1989,** *122*, 1449.

95. Paquette, L. A.; Charumilind, P.; Kravetz, T. M.; Böhm, M. C.; Gleiter, R. *J. Am. Chem. Soc.* **1983,** *105*, 3126. Paquette, L. A.; Charumilind, P.; Gallucci, J. C. *J. Am. Chem. Soc.* **1983,** *105*, 7364.
96. Bauer, W.; O'Doherty, G. A.; Schleyer, P. v. R.; Paquette, L. A. *J. Am. Chem. Soc.* **1991,** *113*, 7093.
97. Kessler, H.; Gehrke, M.; Griesinger, C. *Angew. Chem.* **1988,** *100*, 507. *Angew. Chem. Int. Ed. Engl.* **1988,** *27*, 490.
98. Lochmann, L.; Pospísil, J.; Lim, D. *Tetrahedron Lett.* **1966,** 257.
99. Schlosser, M. *J. Organomet. Chem.* **1967,** *8*, 9.
100. Schlosser, M.; Hartmann, J. *J. Am. Chem. Soc.* **1976,** *98*, 4674. Stähle, M; Hartmann, J.; Schlosser, M. *Helv. Chem. Acta* **1977,** *60*, 1730.
101. Pi, R.; Bauer, W.; Brix, B.; Schade, C.; Schleyer, P. v. R. *J. Organomet. Chem.* **1986,** *306*, C1. Lochmann, L.; Lím, D. *J. Organomet. Chem.* **1971,** *28*, 153.
102. Lochmann, L. *Collect. Czech. Chem. Commun.* **1987,** *52*, 2710.
103. Bauer, W.; Lochmann, L. *J. Am. Chem. Soc.* **1992,** *114*, 7482.
104. Buncel, E.; Menon, B. *J. Org. Chem.* **1979,** *44*, 317. O'Brien, D. H.; Russell, C. R.; Hart, A. J. *J. Am. Chem. Soc.* **1979,** *101*, 633.

6

ASPECTS OF THE THERMOCHEMISTRY OF LITHIUM COMPOUNDS

JOEL F. LIEBMAN
Department of Chemistry and Biochemistry
University of Maryland, Baltimore Country Campus
Baltimore, Maryland

JOSÉ ARTUR MARTINHO SIMÕES
Departamento de Química
Universedada de Lisboa
Portugal

SUZANNE W. SLAYDEN
Department of Chemistry
George Mason University
Fairfax, Virginia

Thermochemical data for lithium compounds are very scarce, despite the importance of these substances in areas ranging from synthetic to theoretical chemistry. The available experimental enthalpies of formation[1-14] for lithium hydrocarbyls (alkyls, aryls, etc.), halides, and alkoxides are collected in Table 6.1. A recent paper by Leroy and co-workers[15] reports high-level ab initio derived enthalpies of formation of several alkyls in the gas phase. These data are collected in Table 6.2.

Most of the experimental data for the hydrocarbyl-substituted lithiums were derived from results reported by Holm[1], who used a steady state heat flow

Lithium Chemistry: A Theoretical and Experimental Overview, Edited by Anne-Marie Sapse and Paul von Ragué Schleyer.
ISBN 0-471-54930-4

Table 6.1 Standard Enthalpies of Formation of Lithium Compounds (kJ/mol)

Compound	ΔH_f^0(1/c)	Method[a]	Ref.	$\Delta H_v^0/\Delta H_s^0$	Ref.	ΔH_f^0(g)
LiMe,c	−74.9	RSC	1			
	−71.1 ± 2.1	EVAL	2			
LiEt,c	−55.2	RSC	1	116.7 ± 0.8	4	
	−53.0 ± 2.2	EVAL	2			
	−58.7 ± 5.4	SB	3,4			
Li allyl,c	−5.9	RSC	1			
	−5.0 ± 2.2	EVAL	2			
Li(*i*-Pr),l	−57.7	RSC	1			
	−74.5 ± 2.1	EVAL	2			
LiBu,l	−109.2	RSC	1	107.1 ± 2.9	4	
	−108.7 ± 2.1	EVAL	2			
	−133.1 ± 7.1	SB	3,5			
	−132.2 ± 3.3	RSC	5,6			
Li(*s*-Bu),l	−87.9	RSC	1			
	−87.4 ± 2.1	EVAL	2			
Li(*t*-Bu),l	−92.0	RSC	1			
	−91.1 ± 2.1	EVAL	2			
LiPh,c	51.5	RSC	1			
	51.4 ± 2.1	EVAL	2			
$LiCH_2Ph$,c	17.6	RSC	1			
	17.4 ± 2.1	EVAL	2			
Li(C_6H_4-*p*-Me),c	5.9	RSC	1			
	5.7 ± 2.1	EVAL	2			
Li(C_6H_4-*p*-Cl),c	7.5	RSC	1			
	7.6 ± 2.2	EVAL	2			
Li_2cot,c[b]	−140.9 ± 8.5	RSC	7			
Li_2C_2,c	−67.7 ± 4.3	OTH	8			
Li tcnq,c[b]	340.8 ± 8.7	OTH	9			
LiH,c	−90.6 ± 0.1	EVAL	10	231.2	10	140.62 ± 0.04
LiF,c	−616.9 ± 0.8	EVAL	10	276.1	10	−340.8 ± 8.4
LiCl,c	−408.3 ± 1.1	EVAL	10	212.6	10	−195.7 ± 12.6
LiBr,c	−350.9 ± 0.4	EVAL	10	196.9	10	−154.0 ± 13
LiI,c	−270.1 ± 0.4	EVAL	10	179.1	10	−91.0 ± 8.4
LiOH,c	−487.5 ± 0.1	RSC	11	250.6	10	−236.9
LiOMe,c	−433.0 ± 2.4	RSC	12			
LiOEt,c	−473.0 ± 2.5	RSC	12			
LiOPr-*i*,c	−499.2 ± 1.4	RSC	12			
LiOBu,c	−512.3 ± 0.9	RSC	12			
LiOBu-*t*,c	−508.6 ± 2.2	RSC	12			
LiNHMe,c	−95.9 ± 1.7	RSC	4,13			

[a]RSC = reaction-solution calorimetry; EVAL = literature evaluation from composite experimental data; SB = static-bomb combustion calorimetry; OTH = other methods; see text discussion.
[b]cot = cyclooctatetraene; tcnq = 7,7,8,8-tetracyano-*p*-quinodimethane.

Table 6.2 Quantum Chemically Derived Standard Enthalpies of Formation and Bond Dissociation Enthalpies of Gaseous Lithium Alkyls[15] (kJ/mol)[a]

LiR	ΔH_f^0(g)	D(Li-R)
LiMe	105.6	201
LiEt	109.5	169
LiPr	88.5	167
Li(*i*-Pr)	104.0	144
LiBu	67.5	166
Li(*s*-Bu)	78.5	152
Li(*t*-Bu)	89.4	119

[a]The bond dissociation enthalpies rely on the ΔH_f^0(R, g) data selected in Table 6.3

calorimeter to measure the enthalpies of reactions (Eq. 1) where the lithium alkyls, aryls, and allyl were handled as solutions or suspensions in ether or in petroleum ether.

$$\text{LiR(soln/susp)} + \text{HBr(soln)} \longrightarrow \text{LiBr(c)} + \text{RH(g/soln)} \quad (1)$$

Although the enthalpies of formation of LiR were calculated in Holm's paper, no details are given regarding the auxiliary data (enthalpies of solution) used to derive the enthalpies of formation of the crystalline compounds. Pedley and Rylance[2] have quoted Holm's enthalpy of reaction data as it refers to reaction 2.

$$\text{LiR(c/l)} + \text{HBr(g)} \longrightarrow \text{LiBr(c)} + \text{RH(g/l)} \quad (2)$$

Although most of the $\Delta H_r^o(2)$ values in the Sussex–NPL Tables are identical to Holm's, those for R = Me, Et, and *i*-Pr differ by 5–16 kJ/mol (1 kcal/mol = 4.184 kJ/mol) from the original numbers. These discrepancies are probably due to corrections introduced by Pedley and Rylance, for example, the enthalpies of solution of RH in the calorimetric mixture.

The assessment of the hydrocarbyl-substituted lithium data in Table 6.1 is not simple. First, as seen in this table, there is another reaction–solution calorimetry value for R = Bu, obtained from the enthalpy of hydrolysis of LiBu,[5,6] which differs by more than 20 kJ/mol from the one based on Holm's data. That value, −132.2 kJ/mol, is in excellent agreement with a static bomb combustion calorimetry result, −133.1 kJ/mol,[3,5] a technique which is considered satisfactory for probing the energetics of organolithium compounds.[16] Yet, another static bomb combustion result, ΔH_f^o(LiEt,c) = −58.7 kJ/mol,[3,4] is in only fair agreement with ΔH_f^o(LiEt,c) = −53.0 kJ/mol derived from Holm's data.

When plotted against the number of carbon atoms in the alkyl group, the points for the quantum mechanically derived ethyl, propyl, and butyl lithiums define a perfect straight line with a slope of −21.00 (very close to the "universal" slope of −20.6 kJ/mol [5,17]) and a constant equal to 151.0 kJ/mol. This profound linearity is not unexpected since the authors have used an additivity scheme to predict the latter two values from the calculated ΔH_f^o(LiEt,g). (The additivity scheme can be used to estimate the energetics of other lithium alkyls since a set of bond enthalpy terms is provided in the paper by Leroy and coworkers). As is the case for the methyl-substituted members in other homologous series,[5,17] the methyl lithium enthalpy deviates from the line; however, relative to other reported series, the methyl lithium deviation from linearity is in the opposite direction. Indeed, the enthalpy of formation of methyl lithium is more negative than the enthalpy of formation of ethyl lithium. This is qualitatively in accord with Montgomery and Rossini's[17] excellent correlation between the value (and necessarily the direction) of the methyl deviation and the Pauling electronegativity of the attached group. However, using the Montgomery and Rossini equation with an electronegativity value of 1.0 for lithium, one obtains a value for the methyl lithium deviation which does not closely correspond to that observed from this plot (−51.1 kJ/mol versus −24.1 kJ/mol).[18] Using a different approach, Luo and Benson[19] present "universal" equations for bond energies and enthalpies of formation in terms of an alternative electronegativity scale, V_χ, which they call the "covalent potential." Using their V_χ for lithium, we find enthalpy of formation differences [ΔH_f^o(RLi,g) − ΔH_f^o(MeLi,g)] of 0.2, −14.5, and −41.1 kJ/mol for R = ethyl, isopropyl, and *tert*-butyl respectively. These may be compared with the differences of 3.9, −1.6, and −16.2 kJ/mol calculated from the enthalpy values in Table 6.2. Should one accept Luo and Benson's expression and the enthalpies of formation of the alkyl lithiums suggested by Leroy and his coworkers, a negative value of V_χ would be obtained. This is very different from the value suggested by the former authors, as well as seemingly unphysical.

A method that seems to be valid for assessing the data for many families of compounds, say ML_n, consists of plotting $\Delta H_f^o(ML_n)$ versus ΔH_f^o(LH), with ML_n and LH in either their standard reference states (their stable physical states at 298.15 K and 1 bar, ca. the more "conventional" 25°C and 1 atm) or in the gas phase.[20-25] It has been observed that many, if not all, of the plots discussed above which involve reliable thermochemical data define excellent straight lines. This empirical linear relationship may be expressed as Eq. 3.

$$\Delta H_f^o(ML_n) = a[\Delta H_f^o(LH)] + b \qquad (3)$$

The meaning of this observation is seen by considering Scheme 1. Both $\Delta H^o(4)$ and $\Delta H^o(5)$ are the enthalpies of hypothetical reactions of a "family" of compounds ML_n, where reactants and products are in the standard reference states and in the gas phase, respectively, and ΔH_V^o are vaporization or sublimation enthalpies.

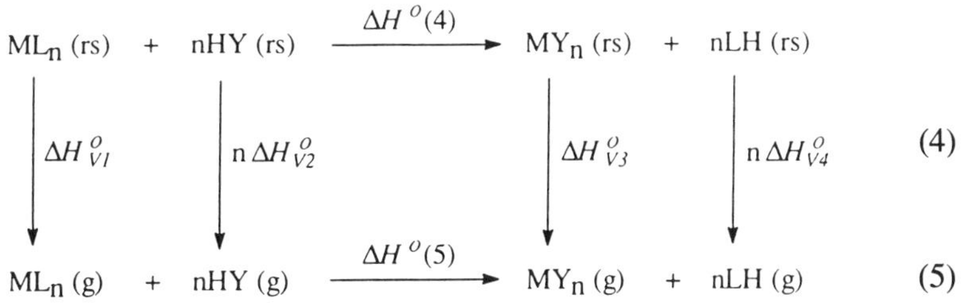

$$\begin{array}{ccccccc} ML_n\,(rs) & + & nHY\,(rs) & \xrightarrow{\Delta H^o(4)} & MY_n\,(rs) & + & nLH\,(rs) \\ \downarrow \Delta H^o_{V1} & & \downarrow n\Delta H^o_{V2} & & \downarrow \Delta H^o_{V3} & & \downarrow n\Delta H^o_{V4} \\ ML_n\,(g) & + & nHY\,(g) & \xrightarrow{\Delta H^o(5)} & MY_n\,(g) & + & nLH\,(g) \end{array} \quad \begin{array}{c}(4)\\ \\ (5)\end{array}$$

Scheme 1

Observation of an empirically linear equation (Eq. 3) for the gas phase enthalpy of formation data implies that $\Delta H^o(4)$ is constant for the series of compounds ML_n.

Also, $\Delta H^o(4)$ can be expressed in terms of the bond dissociation enthalpies (Eq. 6) by again using Scheme 1.

$$\Delta H^o(4) = n[\overline{D}(M{-}L) - D(L{-}H)] + [\Delta H^o_{V1} - n\Delta H^o_{V4}] + n\Delta H^o_{V2} + [nD(H{-}Y) - n\overline{D}(M{-}Y) - \Delta H^o_{V3}] \quad (6)$$

The third bracketed term in Eq. 6 is constant for a given Y, implying that $n[\overline{D}(M{-}L) - D(L{-}H)] + [\Delta H^o_{V1} - n\Delta H^o_{V4}]\}$ is also constant for the series of compounds that obey the linear correlation.[26] It is therefore very likely that $\overline{D}(M{-}L)$ and $D(L{-}H)$ follow nearly parallel trends. Obviously, if the linear correlation holds for reactants and products in the gas phase, then $\Delta H^o(5)$ is constant and so is $[\overline{D}(M0L) - D(L{-}H)]$.

The method described above has been tested for a variety of organic, inorganic, and organometallic thermochemical data[20-25] and there is no apparent reason why it should not work—even approximately—for the lithium compounds. An example involving the lithium halides is shown in Figure 6.1. The fit is given by Eq. 7.

$$\Delta H^o_f(LiX,c) = (1.157 \pm 0.020)\,\Delta H^o_f(HX,g) - (303.6 \pm 2.9) \quad (7)$$

The derived enthalpies of formation of the gas phase alkyl lithiums versus the corresponding alkanes are plotted in Figure 6.2. As stressed elsewhere,[20-21] the observation of a linear correlation between $\Delta H^o_f(ML_n)$ and $\Delta H^o_f(LH)$ means that the energetics of a series of molecules ML_n reflect the energetics of LH, so that a variation of the enthalpy of formation of L is proportional to a change in the enthalpy of formation of ML_n. Accordingly, we can consider the *n*-alkyl lithiums as constituting one ''family,'' but the energetics of this series is clearly different from the other classes of alkyl lithiums which do not lie on the *n*-alkyl line. It is worth noting that methyl lithium is stabilized relative to the *n*-alkyl family, whereas the secondary and tertiary lithium alkyls are destabilized.

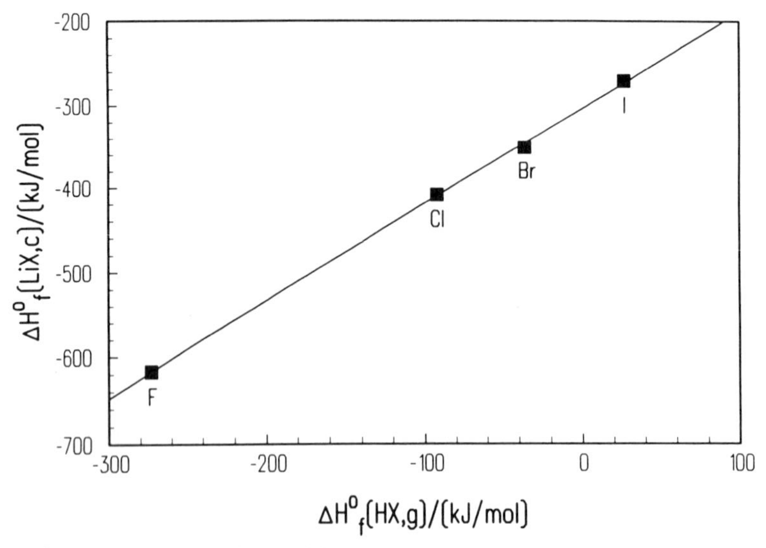

Figure 6.1 Standard enthalpies of formation of crystalline lithium halides, LiX, versus the standard enthalpies of formation of gaseous HX.

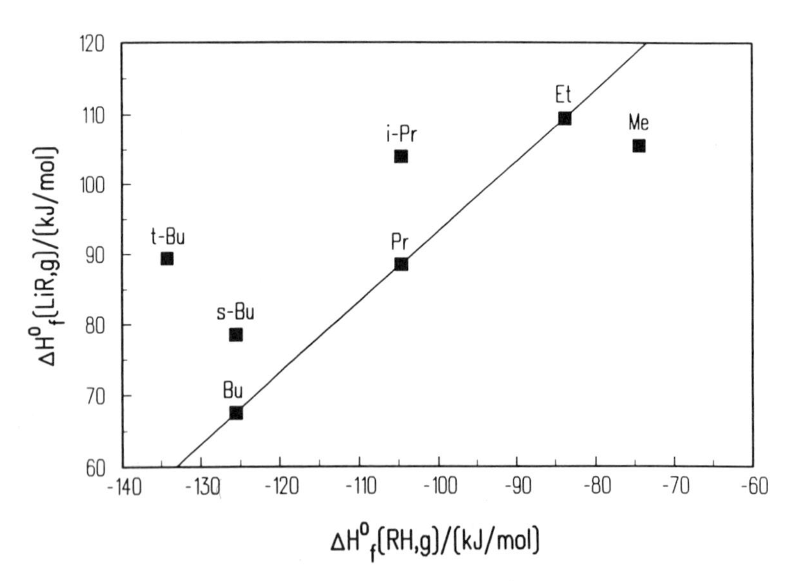

Figure 6.2 Quantum chemically derived standard enthalpies of formation of gaseous lithium alkyls versus the standard enthalpies of formation of gaseous alkanes.

An obvious difficulty is what defines a family. Experience has shown that halogen and oxygen-, sulfur-, and carbon-bonded ligands all define separate families.[20–22] However, it is also clear from the Periodic Table that oxygen is to sulfur as fluorine is to chlorine. It is just that we have many oxygen and sulfur-bonded ligands, but only one type of fluorine- and chlorine-bonded ligand, namely F and Cl themselves. In the case of lithium compounds containing carbon-bonded ligands there is a paucity of data, and the available data is complicated by questions of accuracy and aggregation. Combining all of the organolithiums into one family shows considerable scatter. However, considering the primary organolithiums as the family, and as a benchmark, shows methyl lithium to be surprisingly stable and secondary and tertiary species to be unstable with respect to their primary isomers. It is also to be acknowledged that in the large carbon limit the slope relating an organolithium and its corresponding hydrocarbon must be unity. Acknowledging that there is little interest and even less information about tetradecyl and pentadecyl lithium means that the difference of their gas phase enthalpies of formation (per monomer), which must be ca. 20.6 kJ/mol,[5,17] is really quite irrelevant.

The fact that the primary *n*-butyl lithium is more stable than either its secondary or tertiary isomer, in contrast to what is observed for the alkanes, may be related to the relative electropositivity of lithium and to the rather ionic character of the lithium–carbon bond. In other words, perhaps the lithium alkyls are better described by Li^+R^-, in which case one would expect that a series of these molecules would reflect the variation in the energetics of the anions R^-. Indeed, a plot of ΔH_f^o(LiR,g) against $\Delta H_f^o(R^-,g)$, shown in Figure 6.3 and fitted by Eq. 8, supports this idea (auxiliary data[27–30] in Table 6.3). However, since other slopes are so close to integers (unity for Li and $n = 1$), this slope appears anomalous.

$$\Delta H_f^o(\mathrm{LiR,g}) = (0.514 \pm 0.009)\, \Delta H_f^o(\mathrm{R^-,g}) + (34.9 \pm 1.0) \qquad (8)$$

The values for *tert*-butyl and isopropyl were not included in the correlation. The former lies clearly above the line, which may indicate that the structures of the *tert*-butyl moiety in the lithium-containing molecule and in the ground state of the anion are substantially different. This can also occur in the case of isopropyl, but it is noted that the point for the likewise secondary anion, 2-butyl, is fit by Eq. 8.

The correlations in Figures 6.2 and 6.3 seem to support the reliability of the enthalpies of formation of the lithium alkyls derived by Leroy and coworkers[15] as well as the derived Li—R bond dissociation enthalpies displayed in Table 6.2. The value of D(Li—Me) = 201 kJ/mol looks sensible when compared with D(Li—H) = 237 ± 1 kJ/mol, calculated from the enthalpy of formation of gaseous lithium hydride, 140.62 ± 0.04 kJ/mol.[10] The relatively small difference D(Li—H) − D(Li—Me), 36 kJ/mol, is in the range expected for electropositive elements (e.g., see ref 24 and references cited therein). By

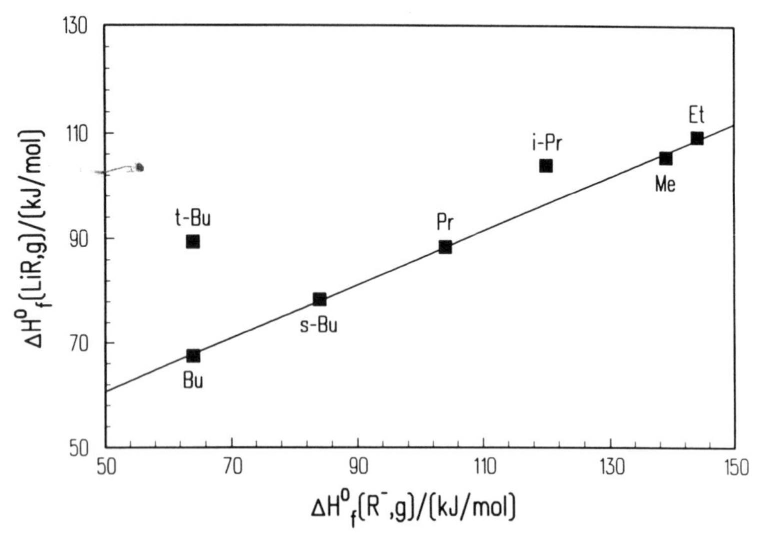

Figure 6.3 Quantum chemically derived standard enthalpies of formation of gaseous lithium alkyls versus the standard enthalpies of formation of gaseous alkyl anions.

contrast, a value of D(Li—Me) recommended by Smith and Patrick,[31] 276 kJ/mol, seems too high. That result was estimated by adding 21 kJ/mol [the approximate difference between D(C—H) in methane and butane] per bond to the value of D(Li—Bu) obtained from the enthalpy of formation of gaseous LiBu. The data in Table 6.1 leads to D(Li—Bu) = 235 ± 9 kJ/mol—which

Table 6.3 Selected Enthalpies of Formation of Alkyl Radicals and Alkyl Anions (kJ/mol)

R	$\Delta H_f^0(R, g)^a$	$\Delta H_f^0(R^-, g)^b$
Me	146.9 ± 0.6	139 ± 4
Et	119 ± 4	144 ± 6
Pr	96 ± 4	104 ± 6
i-Pr	89 ± 3[c]	120 ± 5
Bu	(74 ± 8)	(64)
s-Bu	71 ± 2[c]	84 ± 5
t-Bu	49 ± 2[c]	64 ± 5

[a]Data from reference 27, unless otherwise indicated. The value in parentheses is an estimate.
[b]Data calculated from $\Delta H_f^0(R^-, g) = \Delta H_{acid}(RH, g) - \Delta H_f^0(H^+, g) + \Delta H_f^0(RH, g)$. The acidities and the enthalpies of formation of the alkanes were taken from references 29 and 30, respectively. The value for $\Delta H_f^0(Bu^-, g)$ relies on an estimate of the acidity of butane (1720 kJ/mol).
[c]Value from reference 28.

is also high when compared to $D(\text{Li—H})$ and to the theoretical value in Table 6.2. Assuming that the enthalpy of formation of the liquid LiBu compound is correct (-132 kJ/mol), this indicates some degree of association in the gas phase, so that the derived ΔH_f^o(LiBu,g) is a lower limit of the true value and the derived $D(\text{Li—Bu}) = 235$ kJ/mol is an upper limit. The same reasoning may be applied to the enthalpy of formation of gaseous LiEt shown in Table 6.1, which yields $D(\text{Li—Et}) = 215$ kJ/mol.

The enthalpies of formation of dilithium cyclooctatetraene ($Li_2C_8H_8$, Li_2 cot), lithium carbide (dilithium acetylide, Li_2 C_2), and lithium 7,7,8,8-tetracyano-*p*-quinodimethane [Li *p*-$(NC)_2CC_6H_4C(CN)_2$, Li tcnq] are also shown in Table 6.1. The value for dilithium cyclooctatetraene was derived from reaction–solution calorimetry studies by Stevenson and Valentin,[7] which led to the enthalpy of reaction shown in Eq. 9, $\Delta H^o(9) = -395.4 \pm 8.4$ kJ/mol, and thus to $\Delta H_f^o(Li_2\ \text{cot, c}) = -140.9 \pm 8.5$ kJ/mol.

$$2\text{Li(c)} + \text{cot(l)} \rightarrow \text{Li}_2\ \text{cot(c)} \tag{9}$$

The value for lithium carbide, $\Delta H_f^o(Li_2C_2,c) = -67.7 \pm 4.3$ kJ/mol, was determined by Kudo[8] by high-temperature equilibration of lithium, graphite, and the compound of interest, and is in good agreement with an early measurement by Katskov and his co-workers, -65.2 ± 7.5 kJ/mol.[32] Both are slightly more negative than the value recommended in NBS tables (-59.4 kJ/mol).[33] Aronson and Mittleman[9] reported the ΔG^0 of reaction (10) as -324.3 kJ/mol.

$$\text{Li(c)} + \text{tcnq(c)} \rightarrow \text{Li tcnq(c)} \tag{10}$$

They argue that ΔS^0 should be small for this solid state reaction and deduce $\Delta H^o(10) = -324.3 \pm 8.4$ kJ/mol. From $\Delta H_f^o(\text{tcnq,c}) = 665.1 \pm 2.6$ kJ/mol,[30] we conclude that $\Delta H_f^o(\text{Li tcnq,c}) = 340.8 \pm 8.7$ kJ/mol.

As we observed previously, those alkyl lithiums that are destabilized relative to the corresponding alkane appear above the line defined by the *n*-alkyl groups in the plot of RLI(g) versus RH(g); methyl lithium, thought to be stabilized relative to methane, appears below that line. We ask if this is a general phenomenon in which all stabilized species lie below that line. Accordingly, Figure 6.4 shows the plot of ΔH_f^o (RLi,c/l) versus ΔH_f^o (RH,g).[30] Again, we see that the *sec*- and *tert*-butyl lithiums are to the left of the EtLi/*n*-BuLi line, while MeLi is to the right, as are all the remaining RLi for which there are data. Each of these compounds may also be stabilized by bonding to electropositive lithium, relative to bonding to hydrogen, which parallels the well-established acidity of hydrocarbons which produce *sp*- or sp^2-hybridized carbanions or resonance-stabilized carbanions.

Applequist[34] performed the low-temperature ($-70°$C) equilibration reaction.[11]

$$\text{RLi(soln)} + \text{PhI(soln)} \longrightarrow \text{RI(soln)} + \text{PhLi(soln)} \tag{11}$$

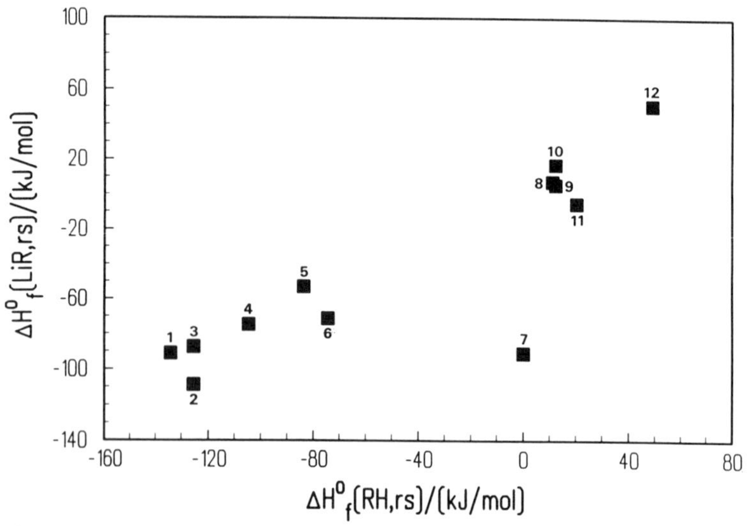

Figure 6.4 Standard enthalpies of formation of lithium hydrocarbyls in their reference states versus the standard enthalpies of formation of corresponding hydrocarbons. 1 = *t*-Bu; 2 = Bu; 3 = *s*-Bu; 4 = *i*-Pr; 5 = Et; 6 = Me; 7 = H; 8 = *p*-Cl-C_6H_4; 9 = *p*-Me-C_6H_4; 10 = $PhCH_2$; 11 = CH_2CHCH_2; 12 = Ph. Note that we have ignored the data for Li_2cot, Li_2C_2, and Li tcnq because they do not belong to the same family.

Although we are given the equilibrium constants and the Gibbs energies for numerous reactions, we cannot immediately derive the enthalpies of reaction because of unknown solvation effects (e.g., enthalpy and entropy of aggregation of the RLi) and lack of enthalpies of formation for most of the relevant iodides.

The data for the lithium alkoxides, obtained through reaction–solution calorimetry measurements of their hydrolysis reactions,[12] are assessed in Figure 6.5. Equation 12 shows the least-squares fitting of the three points for the *n*-alkoxides.[35]

$$\Delta H_f^o(\text{LiOR,c}) = (0.894 \pm 0.071)\ \Delta H_f^o(\text{ROH,l}) - (221.3 \pm 20.0) \qquad (12)$$

A value for the enthalpy of formation of LiOEt, −456.5 kJ/mol, quoted from NBS tables,[33] is about 17 kJ/mol higher than that shown in Table 6.1. This NBS value probably relies on the work by Blanchard and his co-workers, who measured the enthalpy of hydrolysis of lithium ethoxide in a 0.05 M aqueous sulfuric acid solution[36] and determined that ΔH_f^o(LiOEt,c) = −532.2 ± 4.2 kJ/mol. The large discrepancy between this original result and the one given by NBS is probably due to a calculation error in ref. 36 and also to the fact that the Blanchard group did not account for the dilution of the sulfuric acid solution in their calculation. Unfortunately, this correction cannot be evaluated exactly because of the lack of experimental information in the original

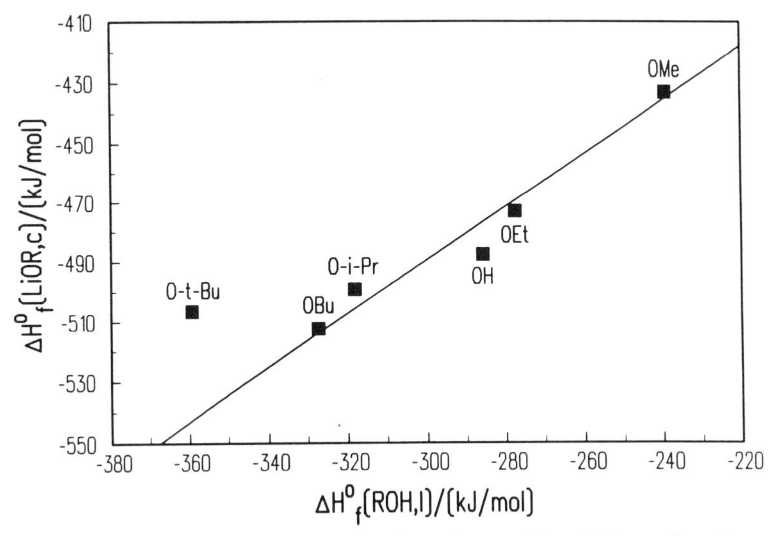

Figure 6.5 Standard enthalpies of formation of crystalline lithium alkoxides versus the standard enthalpies of formation of liquid alcohols.

paper. Tel'noi and Rabinovich also corrected Blanchard's value in their review and state that ΔH_f^o(LiOEt,c) = −456.1 ± 4.2 kJ/mol,[37] close to that of NBS.

The importance of the energetics of lithium alkoxides in organic synthesis is stressed in several publications by Arnett and co-workers. A brief account of some of his most recent work[38–40] is given below. The enthalpies of deprotonation of several 2-substituted-l-phenyl-l-propanols by lithium bis-(trimethylsilyl)amide, yielding the respective lithium alkoxides, were measured by titration calorimetry in THF, benzene, and dioxane.[38] Although the enthalpies of formation cannot be derived, the reaction enthalpies enabled a comparison between the energetics of chelating and nonchelating alkoxides. In benzene, for example, the enthalpies of deprotonation of alcohols that can form chelating alkoxides are 26–46 kJ/mol more exothermic that those measured for alcohols that form alkoxides that are incapable of chelation. The second paper, by Arnett and Moe,[39] gives data for the enthalpies of deprotonation of a series of carboxylic acids, alcohols, amides, and so on by lithium bis-(trimethylsilyl)amide in THF. Vapor pressure osmometry experiments provided information on the aggregation numbers of those compounds and their lithium salts. A fairly linear correlation between the enthalpies of deprotonation and pK_a values was found, suggesting, as pointed out by the authors, that these enthalpies are not strongly dependent on the aggregation numbers (which vary from one to seven in THF). The last paper, also by Arnett and Moe,[40] reports the enthalpies of deprotonation of isopropanol by a variety of organolithium bases in a hexane–diethyl ether solution. The most exothermic reactions were observed for lithium alkyls [e.g., −235.1 ± 7.5 kJ/mol for Li(t-Bu)], followed by secondary and tertiary lithium amides and lithium alkoxides. As could be expected, the reaction with LiO-*t*-Bu was almost thermoneutral (-8.8 ± 2.5 kJ/mol).

In close relation to Arnett's first paper[38] are the results reported by Klumpp.[41] The differences between the enthalpies of protonation in benzene of organolithium compounds containing methoxy groups and reference compounds without these moities were measured by reaction–solution calorimetry and the results are given in Figure 6.6. It is observed that the protonation of the chelated molecules is less exothermic than that of their references, the negative of the difference being attributed to enthalpy of the Li—O interaction. A detailed discussion of those number has been made in terms of stereochemical features of the molecules, as well as their degrees of association in solution.[41] Further energetic data for Li—O interaction are available in Klumpp's review and other publications. For instance, the calorimetric work by Beak and Siegel[42] led to the enthalpy of the formal isomerization reaction 13 in dibutyl ether, $+34.7 \pm 2.3$ kJ/mol, which reflects the Li—O intramolecular bond enthalpy. In con-

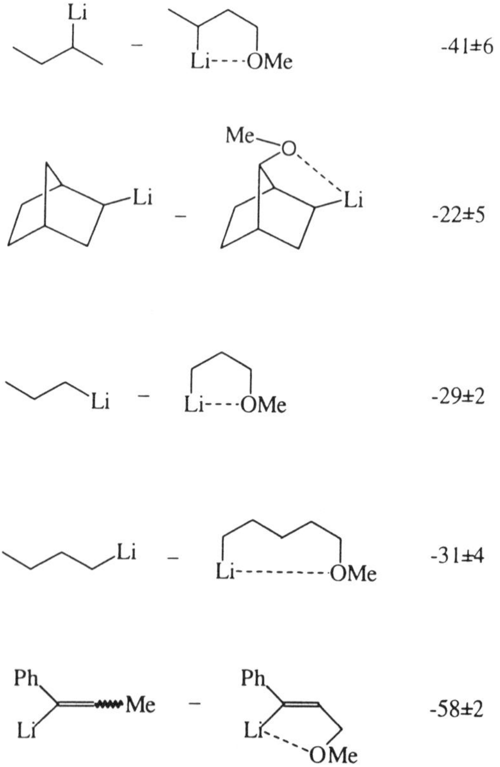

Figure 6.6 Differences between enthalpies of protonation of organolithium compounds, reflecting the Li–O bond enthalpy in each case. Data from reference 41.

trast, the isomerization enthalpy of *m*- to *p*-lithioanisole is almost identically zero.

$$[\text{Li}\cdots\text{OMe-}C_6H_4]_2 \cdot n(Bu_2O) \longrightarrow [p\text{-MeO-}C_6H_4\text{-Li}]_2 \cdot n(Bu_2O) \quad (13)$$

There are few gas phase thermochemical studies involving organolithium compounds. In one of these studies, Staley and Beauchamp used ion–cyclotron resonance mass spectrometry to probe the binding energies of Li^+ to a variety of π-, n- and σ-donor bases (B).[43] The bond dissociation enthalpies $D(Li^+{-}B)$ (lithium cation affinities) were anchored on $D(Li^+{-}H_2O)$ = 142 kJ/mol and include bases such as benzene (155 kJ/mol), cyclohexane (100 kJ/mol), alkyl halides, olefins, and amines. Many of the $Li^+{-}B$ bond dissociation enthalpies correlate linearly with the proton affinities of the bases. Deviations from this correlation were observed for those bases that may interact covalently with H^+ but whose bonding to Li^+ is mainly ionic.

A related study[44] deals solely with the lithium cation affinities of alcohols and ethers. It was shown that these enthalpies generally correlate strongly with the substituent polarizability parameter, r_α, for the ROR' species. Interestingly, the simplest alcohol/ether, H_2O, follows this pattern. Also, the gas phase basicity (the ΔG^0 counterpart of proton affinities) and the corresponding ΔG^0_{Li+} generally correspond, although in this case alcohols and ethers form two parallel lines and water is somewhat deviant. We say "generally" because the two exceptions noted were for CF_3CH_2OH and *n*-BuOH. The former is sensible because trifluoroethanol has multiple basic sites—the oxygen and the carbon-bound fluorines are both relatively negatively charged and so have attractive interactions with protons and lithium ions alike. A cyclic chelate-like structure for the lithium complex is entirely reasonable. However unusual such a structure may be for lithiated butanol, the suggested formation of a 6-membered ring containing four carbons, the oxygen, and the lithium is most consistent with the available experimental data. Indeed, this terminal "bridging" $CH_3{-}Li$ interaction is reminiscent of that invoked in oligomeric hydrocarbyl lithiums[45,46] and so gains additional plausibility.

Another study by Piedade, Beauchamp. and co-workers addresses the implications of the thermochemistry of lithium compounds on the fragmentation of lithium—peptide complexes in the gas phase.[47] The authors have used literature data to estimate proton affinities of several lithium molecules LiL [L = NH_2, MeNH, Me_2 N, OH, MeO, HCOO, MeCOO, and MeC(O)NH], as in reaction 14.

$$LHLi^+(g) \longrightarrow LiL(g) + H^+(g) \quad (14)$$

The enthalpies of formation of $LHLi^+$(g) were derived from the Li cation affinities of LH reported by Taft and co-workers[48] and the enthalpies of formation of LiL(g) were quoted from Leroy's paper[15] or obtained from new theoretical calculations.

We now consider the thermochemistry of inorganic lithium compounds. One of the major archives for inorganic thermochemistry[33] devotes nine pages of tables to these species. Admittedly, extolling this length is somewhat disingenous because lithium appears as element 98 in the "standard order of arrangement of the elements and compounds based on the periodic classification of the elements." As such, the chemical thermodynamic networks for compounds containing "only" the other 97 elements had to come first, and only compounds of Li with the other alkali metals do not appear under lithium.[49]

We feel it benefits neither the reader nor the authors to review all of the data in this source. However, for an archetypal (solid) Li salt, LiX, one can hope to interrelate its enthalpy with that of other salts where in either or both ions are changed. While it is rare to find a set of cations for which there are constant differences between the enthalpies of formation for salts of two chosen anions,[50] general success is achieved when comparing the enthalpies of salts of three anions.[51] For example, lithium shows no anomalies in the use of the general relationship

$$1/\mathrm{a}\Delta H_\mathrm{f}^o(\mathrm{M_aX_b,c}) = \mathrm{m}\Delta H_\mathrm{f}^o(\mathrm{MCl,c}) + 1/2(1 - m)\Delta H_\mathrm{f}^o(\mathrm{M_2O}),\mathrm{c}) + \mathrm{C} \quad (15)$$

for all of the X studied by these authors.[52] Likewise, it is rare to find a set of cations for which there are constant differences between the enthalpies of formation for salts of two chosen cations[53]; general success is found when comparing the enthalpies of three cations. For example, the thermochemistries of lithium, sodium, and potassium are interrelatable by the highly accurate equation

$$\Delta H_\mathrm{f}^o(\mathrm{Na_nX,c}) = 0.639\Delta H_\mathrm{f}^o(\mathrm{K_nX,c}) + 0.361\ \Delta H_\mathrm{f}^o(\mathrm{Li_nX,c}) + 5.1\mathrm{n} \quad (16)$$

It is not obvious why these equations work as well as they do. One can invoke an ionic model and correlate the ionic lattice energy to the reciprocal of the interion distance. It is found that all of the alkali halides fall on one line and the corresponding hydrides fall on a line nearby to the right. Also nearby—but now to the left—is yet another straight line for the elements themselves where they are described as MM or M^+M^- ion pairs. No explanation is apparent.

It is not unreasonable to suspect that many of these regularities arise because of the high symmetry of the inorganic solid state. Whether one invokes a periodic lattice and therefore bond additivity (intra- and intercomponent) or composition-independent Madelung constants, solids are easier to understand than small "chunks" of matter, that is, $(MX)_n$ oligomers. These have numerous structural possibilities which can correspond to regular polygons or polyhedra. For example, the tetramer can be T_d or D_{4h} while the hexamer can be D_{3d} or

D_{6h}. Other possibilities are less regular. For example, both the tetramer and the hexamer can be linear (i.e., $C_{\infty v}$), corresponding to a perfectly aligned arrangement of the monomeric dipoles. To thoroughly evaluate all of these possibilities would take us into details of calculational theory, molten salts, and molecular beams. It should suffice to discuss here the tetramers and hexamers of LiF, LiOH, and $LiNH_2$. No experimental data seems to be available for these species. However, calculational theory[54] at the HF/6-31G + *sp* + *d* level[55] on the tetramers show the T_d to be more stable than the D_{4h} for LiF and LiOH (by 40.0 and 67.8 kJ/mol), but less stable for $LiNH_2$ (37.2 kJ/mol). These results follow from the observation that both F^- and OH^- have (at least three) suitable lone pairs to complex with three lithiums in the T_d structure, while NH_2^- has only two lone pairs and so ''prefers'' the D_{4h} form. However, such an explanation does not seem particularly applicable to the hexamers since corresponding calculations[56] for all three lithium species show the D_{3d} structure to be more stable—the energy differences are 134.3, 184.1, and 84.5 kJ/mol respectively. Of course, one can invoke strain energy considerations, but that adds a major component of conceptual, if not calculational, complexity. How then does one calibrate theory and experiment? In principle one can compare the calculated bonding energies of nLiX molecules and one $(LiX)_n$ with the experimental enthalpy of oligomerization. One can also compare one molecule of LiX with $1/n$ of $(LiX)_n$. This is somewhat more conceptually useful because these ''normalized'' results correct for differing degrees of oligomerization. In fact, there are but few experimental data to compare with theory, namely, those for $(LiF)_2$, $(LiF)_3$, and $(LiOH)_2$. Table 6.4 presents the derived ''normalized'' enthalpies from the three major thermochemical data compendia, the

Table 6.4 Normalized Enthalpy of Oligomerization to Form $(LiX)_n$(X=F, OH) (kJ/mol of monomer, LiX)

	298 K			0 K			
n, X	JANAF[a]	NBS[b]	TPIS[c]	JANAF[a]	NBS[b]	TPIS[c]	QC[d]
2, F	−130.6	−128.1	−122.3	−128.5	−128.1	−124.8	−136.0
3, F	−164.9	−164.9	−167.3	−162.9	−164.9	−165.3	−175.7
4, F							−196.4
6, F							−218.0
∞, F	−276.1	−276.2	−277.4	−273.8	−273.8	−275.0	
2, OH	−121.4		−116.2	−117.7		−114.5	−140.5
3, OH							−173.6
4, OH							−198.7
6, OH							−214.2
∞, OH	−250.6	−246.8	−239.3	−247.1	−243.3	−236.1	

[a]See reference 10.
[b]See reference 33.
[c]See reference 58.
[d]See references 54 and 56.

JANAF,[10,57] NBS,[33] and TPIS[58] tables, where all values are given for both 298 and 0 K. The quantum chemical calculated results are also included (and labelled QC) and, as a prod to the experimentalist, we include the calculated normalized oligomerization energies for $(LiF)_n$ and $(LiOH)_n$ for $n = 2, 3, 4$, and 6. We also include the experimental value for $n = \infty$, which is to be identified as the enthalpy of sublimation of LiX at the two temperatures of interest—it should also correspond to the limiting energy for ''large'' enough cluster size.

Despite the thermochemical standard temperature of 298 K, it is most meaningful to consider the experimental results at 0 K when comparing theory and experiment. At that temperature calculational quantum chemical theory provides good agreement for LiF and its oligomers but much poorer agreement for LiOH and its dimer. However, zero-point energy corrections should also be included. The necessary—albeit generally estimated—vibrational frequencies for the lower lithium fluoride and hydroxide oligomers can be obtained from the JANAF tables. For LiF, $(LiF)_2$, and $(LiF)_3$, the normalized zero-point energies (i.e., per monomer) increase in the order 5.5, 9.5, and 10.2 kJ/mol (cf. Eq. 17).

$$\mathrm{ZPE}_{\mathrm{norm}} = \sum_i \tfrac{1}{2} h\nu_i / n \tag{17}$$

For LiOH and $(LiOH)_2$, the normalized zero-point energies are 30.0 and 44.8 kJ/mol. The calculated normalized binding energy for $(LiF)_2$ is too high by 9.5–5.5 = 4.0 kJ/mol, whereas for $(LiF)_3$ and $(LiOH)_2$ the calculated binding energies are too high by 4.7 and 14.8 kJ/mol respectively. Agreement between theory and experiment, already good, is made better by the inclusion of zero-point energies. This agreement is fortunate since the necessary experiments are difficult and interpretations are often obscure, while quantum chemical calculations (even for the zero-point energies) are more accessible and directly interpretable. Regrettably, there are no experimental measurements for the enthalpies of formation to complement, accompany, and therefore test the predictions of calculational theory for the various oligomers of $LiNH_2$ or for the structurally glorious organolithium species shown in references 45 and 46 and elsewhere in the current volume. Such studies are eagerly awaited.

A possible conclusion to this necessarily brief review on organolithium thermochemistry could be a simple recognition that the energetics of these substances is of relatively low importance. The scarcity of thermochemical data is, however, more likely owing to several experimental difficulties, including the characterization, purification, and handling of lithium compounds, which are encountered if enthalpies of formation are to be determined. These problems are in part overcome by measuring enthalpies of reactions in solution, without attempting to derive those quantities. Theoretical methods have, on the other hand, been improving the reliability of their quantitative predictions, so that they may be rightly considered the best way to acquire information on

Table 6.5 Calculated Enthalpies of Sublimation or Vaporization of Some Lithium Compounds (kJ/mol)

LiL	ΔH_s^0 or ΔH_v^0
LiMe,c	177[a]
LiEt,c	163[a]
Li(*i*-Pr),l	179[a]
LiBu,l	176[a]
	200[b]
Li(*s*-Bu),l	166[a]
Li(*t*-Bu),l	181[a]
LiOMe,c	218[c]
LiNHMe,c	153[c]

[a]Calculated from ΔH_f^0(c/l) in Table 6.1 (ref. 2) and ΔH_f^0(g) in Table 6.2.
[b]Calculated from ΔH_f^0(c/l) in Table 6.1 (refs. 5 and 6) and ΔH_f^0(g) in Table 6.2.
[c]Calculated from ΔH_f^0(c/l) in Table 6.1 and ΔH_f^0(g) in reference 15.

the energetics of lithium molecules in the gas phase. Theoretically derived enthalpies of formation, taken together with calorimetric data for the enthalpies of formation of crystalline or liquid substances, show that the (few) experimental values of enthalpies of vaporization or sublimation should be used with great caution. This is evidenced in Table 6.5, which collects those phase-change enthalpies that are calculated from ΔH_f^o(l/c) in Table 6.1, ΔH_f^o(g) in Table 6.2, and reference 15. The values for ΔH_s^o(LiEt) and ΔH_v^o(LiBu) in Table 6.5 are significantly large than those in Table 6.1, suggesting strong association in the vapor phase. Some of the vaporization enthalpies for the lithium alkyls in Table 6.5 seem, however, too high, which, according to the analysis given above, is likely to result from errors in the experimental enthalpies of formation of the crystalline compounds. It should also be noted that the enthalpies of formation calculated for gas phase ethyl lithium and butyl lithium from their respective enthalpies of sublimation and vaporization differ greatly from those derived in Table 6.2.

Acknowledgments

JFL thanks the Chemical Science and Technology Laboratory, National Institute of Standards and Technology, for partial support of his research. JAMS thanks Junta Nacional de Investigação Científica e Tecnológica, Portugal (Project PMCT/C/CEN/42/90) for financial support. A travel grant from the Luso-American Foundation for Development, Portugal, is also acknowledged.

References

1. Holm, T. *J. Organomet. Chem.* **1974**, *77*, 27.
2. Pedley, J. B.; Rylance, J. *Sussex N.P.L. Computer Analysed Thermochemical Data: Organic and Organometallic Compounds*; University of Sussex: Brighton, 1977. We cite here the older edition of the compendium by the first-named author because values for organometallic compounds do not appear in the 1986 volume.
3. Lebedev, Yu. A.; Miroschnichenko, E. A.; Chaikin, A. M. *Dokl. Akad. Nauk SSSR* **1962**, *145*, 1288.
4. Pilcher, G.; Skinner, H. A. In *The Chemistry of the Metal–Carbon Bond*, Hartley, F. R.; Patai, S., Eds.; Wiley: New York, 1982.
5. Cox, J. D.; Pilcher, G. *Thermochemistry of Organic and Organometallic Compounds*; Academic Press: London, 1970.
6. Fowell, P. A.; Mortimer, C. T; *J. Chem. Soc.* **1961**, 3793.
7. Stevenson, G. R.; Valentin, J. *J. Phys. Chem.* **1978**, *82*, 498.
8. Kudo, H. *Chem. Lett.* **1989**, *1611*.
9. Aronson, S.; Mittleman, J. S. *J. Solid State Chem.* **1981**, *36*, 221.
10. Chase, Jr., M. W.; Davies, C. A.; Downey, Jr., J. R.; Frurip, D. J.; McDonald, R. A.; Syverud, A. N.; JANAF Thermochemical Tables, *J. Phys. Chem. Ref. Data* **1985**, *14*, Supplement No. 1.
11. Konings, R. J. M.; Cordfunke, E. H. P.; Ouweltjes, W. *J. Chem. Thermodyn.* **1989**, *21*, 415.
12. Leal, J. P.; Martinho Simões, J. A. unpublished results.
13. Juza, R; Hillenbrand, E. *Z. Anorg. Chem.* **1953**, *273*, 297.
14. Brubaker, G. R.; Beak, P. *J. Organomet. Chem.* **1977**, *136*, 147, reacted hydrocarbyl lithiums with ethanol. From this they derived the following enthalpies of formation (kJ/mol) of RLi(ether): R = -29.8 ± 5.4; R = allyl, 18.0 ± 5.4; R = (E)-l-propenyl, 22.6 ± 2.1; R = 2-propenyl, 84.1 ± 11.7; R = phenyl, 80.3 ± 3.4. These values are about 30 kJ/mol higher than those in Table 6.1, which suggests some discrepancy in reference states, very possibly the enthalpy of formation of lithium ethoxide in ether.
15. Sana, M.; Leroy, G.; Wilante, C. *Organometallics* **1991**, *10*, 264.
16. Pilcher, G. In Energetics of Organometallic Species; Martinho Simões, J. A. Ed.; NATO ASI Series, Kluwer: Dordrecht, 1992.
17. Montgomery, R. L.; Rossini, F. D. *J. Chem. Thermodyn.* **1978**, *10*, 471.
18. Direct application of the Montgomery and Rossini analysis results in the choice of electronegativity for lithium of either 1.7 or 5.3. The latter is clearly wrong—Li is not more electronegative than F. The former is also seemingly wrong. If the electronegativity of H is approximately 2.1, Li is too different from H in its observed chemistry for there to be only an electronegativity difference of 0.4; after all, that is the approximate difference between C and H. Admittedly, Montgomery and Rossini lacked the necessary experimental data to refine their equation to include electropositive groups.
19. Luo, Y.-R.; Benson, S. W. *Acc. Chem. Res.* **1992**, *25*, 375; *J. Phys. Chem.* **1988**, *92*, 5255.

20. Dias, A. R.; Martinho Simões, J. A.; Teixeira, C.; Airoldi, C.; Chagas, A. P. *J. Organometal. Chem.* **1987**, *335*, 71.
21. Dias, A. R.; Martinho Simões, J. A.; Teixeira, C.; Airoldi, C.; Chagas, A. P. *J. Organometal Chem.* **1989**, *361*, 319.
22. Dias, A. R.; Martinho Simões, J. A.; Teixeira, C.; Airoldi, C.; Chagas, A. P. *Polyhedron* **1991**, *10*, 1433.
23. Griller, D.; Martinho Simões, J. A. Wayner, D. D. M. In *Sulfur-Centered Reactive Intermediates in Chemistry and Biology*, Chatgilialoglu, C.; Asmus, K.-D. Eds.; Plenum: New York, 1991.
24. Martinho Simões, J. A. In *Energetics of Organometallic Species*, Martinho Simões, J. A., Ed; NATO ASI Series, Kluwer: Dordrecht, 1992.
25. Leal, J. P.; Pires de Matos, A.; Martinho Simões, J. A. *J. Organometal. Chem.* **1991**, *403*, 1.
26. The analysis in Chickos, J. S.; Hesse, D. G.; Liebman, J. F.; Panshin, S. Y. *J. Org. Chem.* **1988**, *53*, 3424 suggests that $[\Delta H^o_{V1} - n\Delta H^o_{V4}]$ should be a constant related to the b value for the element M. For numerous monosubstituted hydrocarbon derivatives they showed that the enthalpy of vaporization (kJ/mol) equals $4.7\tilde{n}_c + 1.3n_Q + 3.0 + b(X)$, where $\tilde{n}_c$ is the number of nonquaternary carbons, n_Q is the number of quaternary carbons, and $b(X)$ is a substituent-dependent parameter. For liquid ML_n where L is a hydrocarbyl ligand, the enthalpy of vaporization of ML_n is $n(4.7)\tilde{n}_c + n(1.3n_Q) + 3.0 + b(M)$ and the enthalpy of vaporization of n molecules of LH is $n(4.7)\tilde{n} + n(1.3n_Q) + 3.0n$. The difference between ML_n and nLH is $3.0(1 - n) + b(M)$, a number independent of the precise member of the family of ligands LH.
27. Martinho Simões, J. A.; Beauchamp, J. L.; *Chem. Rev.* **1990**, *90*, 629.
28. Seetula, J. A.; Russell, J. J.; Gutman, D. *J. Am. Chem. Soc.* **1990**, *112*, 1347.
29. (a) DePuy, C. H.; Gronert, S.; Barlow, S. E.; Bierbaum, V. M.; Damrauer, R. *J. Am. Chem. Soc.* **1989**, *111*, 1968. (b) Damrauer, R. In *Selective Hydrocarbon Activation: Principles and Progress*; Davies, J. A.; Watson, P. L.; Greenberg, A.; Liebman, J. F. Eds.; VCH: New York, 1990.
30. Pedley, J. B.; Naylor, R. D.; Kirby, S. P. *Thermochemical Data of Organic Compounds*, Chapman and Hall: London, 1986.
31. Smith, G. P.; Patrick, R. *Int. J. Chem. Kinet.* **1983**, *15*, 167.
32. Katskov, D. A.; L'vov, B. V.; Danilkin, V. I. *Russ. J. Appl. Spectr.* **1977**, *27*, 1313.
33. Wagman, D. D.; Evans, W. H.; Parker, V. B.; Schumm, R. H.; Halow, I.; Bailey, S. M.; Churney, K. L; Nuttall, R. L. The NBS Tables of Chemical Thermodynamic Properties: Selected Values for Inorganic and C_1 and C_2 Organic Substances in SI Units, *J. Phys. Chem. Ref. Data* **1982**, *11*, Supplement No. 2.
34. Applequist, D. E. *J. Am. Chem. Soc.* **1963**, *85*, 743.
35. The large error of the intercept is caused by the long extrapolation of the very negative enthalpies of formation of the alcohols used in the correlation.
36. Blanchard, J. M.; Joly, R. D.; Lettoffe, J. M.; Perachon, G.; Thourey, J. *J. Chim. Phys. et Phys.-Chim. Biol.* **1974**, *71*, 472.
37. Tel'noi, V. I.; Rabinovich, I. B. *Russ. Chem. Rev.* **1980**, *49*, 603.

38. Nichols, M. A.; McPhail, A. T.; Arnett, E. M. *J. Am. Chem. Soc.* **1991**, *113*, 6222.
39. Arnett, E. M.; Moe, K. D. *J. Am. Chem. Soc.* **1991**, *113*, 7288.
40. Arnett, E. M.; Moe, K. D. *J. Am. Chem. Soc.* **1991**, *113*, 7068.
41. Klumpp, G. W. *Rec. Trav. Chim. Pays-Bas* **1986**, *105*, 1.
42. Beak, P.; Siegel, B. *J. Am. Chem. Soc.* **1974**, *96*, 6803.
43. Staley, R. H.; Beauchamp, J. L. *J. Am. Chem. Soc.* **1975**, *97*, 5920.
44. Abboud, J.-L. M.; Yañez, M; Elguero, J.; Liotard, D.; Essefar, M.; el Mouhtadi, M.; Taft, R. W. *New J. Chem.*, **1992**, *16*, 739.
45. Schleyer, P. v. R. *Pure Appl. Chem.* **1983**, *55*, 355; **1984**, *56*, 151.
46. Setzer, W.; Schleyer, P. v. R. *Adv. Organomet. Chem.* **1985**, *24*, 353.
47. Minas da Piedade, M. E.; Irikura, K. K.; Walsh, S.; Beauchamp, J. L.; Goddard, W. A. III, unpublished results.
48. Taft, R. W.; Anvia, F.; Gal, J.-F.; Walsh, S.; Capon, M.; Homes, M. C.; Hosn, K.; Oloumi, G.; Vasanwala, R.; Yazdani, S. *Pure & Appl. Chem.* **1990**, *62*, 17.
49. For example, oxygen is element 1 and so the first data page in reference 33 deals with the neutral and some of the ionic forms of O, O_2, and O_3. H appears second and so the various states of atomic and molecular hydrogen appear on the page for hydrogen as well as neutral and ionic forms of hydrogen combined with oxygen, for example, water (liquid, gas, and 0 K solid), hydrogen peroxide (liquid, gas, and 12 stages of dilution in water including ''infinite dilution'') and both gaseous and solvated H^+ and OH^-. Chlorine appears in table 10 and the entries for HCl and the various chlorine oxyacids and their anions appear here. However, should the reader want thermochemical data on LiCl or $LiClO_4$, then he or she must wait until table 98 for compounds containing lithium.
50. A particularly evocative example is that of Cl^- and CN^-, for which the difference of the enthalpies of formation of their derivatives is nearly constant for ionic species, but changes dramatically with increasing covalency, cf. Meot-Ner (Mautner), M.; Cybulski, S. M.; Scheiner, S.; Liebman, J. F. *J. Phys. Chem.* **1988**, *92*, 2738.
51. Hisham, M. W. M.; Benson, S. W. In *From Atoms to Polymers: Isoelectronic Analogies*; Liebman, J. F.; Greenberg, A. Eds.; VCH Publishers:New York, 1989.
52. We have recast the enthalpy of formation expression from reference 51 so as to use enthalpies of formation of the ''real compound,'' rather than ''normalized,'' that is, per lithium atom.
53. An example is that of K^+ and NH_4^+, cf. Liebman, J. F.; Romm, M. J.; Meot-Ner (Mautner), M.; Cybulski, S. M.; Scheiner, S. *J. Phys. Chem.* **1991**, *95*, 1112.
54. Sapse, A.-M.; Raghavachari, K.; Schleyer, P. v. R.; Kaufmann, E. *J. Am. Chem. Soc.* **1985**, *107*, 6483.
55. The authors of reference 54 showed that MP2 and MP3 correlation corrections are small (under 10 kJ/mol) for the normalized oligomerization energies at the 6-31G level. These corrections may be expected to be smaller at the 6-31G + *sp* + *d* level and so it is neither surprising, disappointing, nor all that necessary that MPn calculations be performed.

56. Raghavachari, K.; Sapse, A.-M.; Jain, D. C. *Inorg. Chem.* **1989**, *26*, 2585.

57. The error bars for the *unnormalized* ΔH_f values, which are the same at both 0 and 298 K, are LiF(c), ± 0.8; LiF(g), $+8.4$; $(LiF)_2$, ± 16.7; $(LiF)_3$, ± 25.1; LiOH(c), ± 0.4; LiOH(g), ± 6.3; $(LiOH)_2$, ± 33.5 kJ/mol.

58. Gurvich, L. V.; Veits, I. V.; Medvedev, V. A.; Khachkuruzov, G. A.; Yungman, V. S.; Bergman, G. A. *Termodinamicheskie Svoistva individual'nykh Vesgchestv (Thermodynamic Properties of Individual Substances)*; Glushko, V. P., Gen. Ed.; Izdatel'stvo, ''Nauka'': Moscow, Vols. 1–4 (in 8 parts) (1978–1982).

7

FROM "CARBANIONS" TO CARBENOIDS: THE STRUCTURE OF LITHIATED AMINES AND LITHIATED ETHERS

GERNOT BOCHE, JOHN C. W. LOHRENZ, AND ACHIM OPEL

Fachbereich Chemie
Philipps-Universität Marburg
Marburg, Germany

Lithium Chemistry: A Theoretical and Experimental Overview, Edited by Anne-Marie Sapse and Paul von Ragué Schleyer.
ISBN 0-471-54930-4

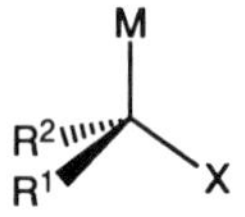

M

R^2 R^1 X

1. Typical acceptor substituents X:
   X = C(O)R, SO_2R, S(O)(NR)R, S(O)R, SR, CN, NO_2, C(N-NR)R, NC, $P(O)(OR)_2$, etc.
2. M / halide carbenoids
   X = F, Cl, Br, I
3. X = NR_2 and OR ($OSiMe_3$)

Scheme 1 Acceptor-substituted organolithium compounds; M = metal; X = acceptor substituent.

1 INTRODUCTION

A body of knowledge about the structures and reactivities of organolithium compounds has accumulated within the last 15 years. Of special interest synthetically are compounds with acceptor substituents X at the anionic carbon atom (see Scheme 1).

Seebach[1] and Williard[2] summarized the results on enolates, and a review of lithium sulfones, sulfoximides, sulfoxides, thioethers, dithianes, nitriles, nitro compounds, and hydrazones was published by Boche.[3] Undoubtedly, the best studied class of compounds to date is the one in which the anionic carbon bears a "typical" acceptor substituent X. It is well known that these substituents stabilize a negative charge very well.[4] One important common feature of the more recent structural investigations of these "carbanions" is the shortened C—X bond compared to the C—X bond in the corresponding nonlithiated species.[1-3] An example is given in Figure 7.1.

When X is a halide, one is dealing with carbenoids,[5] a class of lithium (metal) organic compounds with a name of their own. The chemical behavior of carbenoids justifies their special denotation: these compounds, for example, react not only with electrophiles E (as all "carbanions" do), but also with nucleophiles like RLi whereby X is substituted for R (see Scheme 2).

The ambident nature of the known Li/halide carbenoids led to many inves-

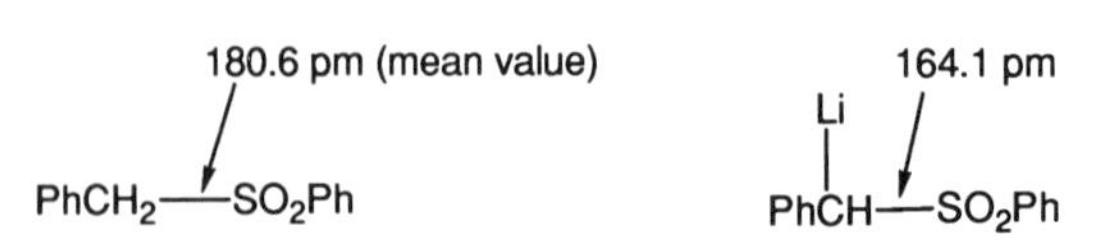

Figure 7.1 Shortening of the C—X bond in organolithium compounds with "normal" acceptor substituents.[3]

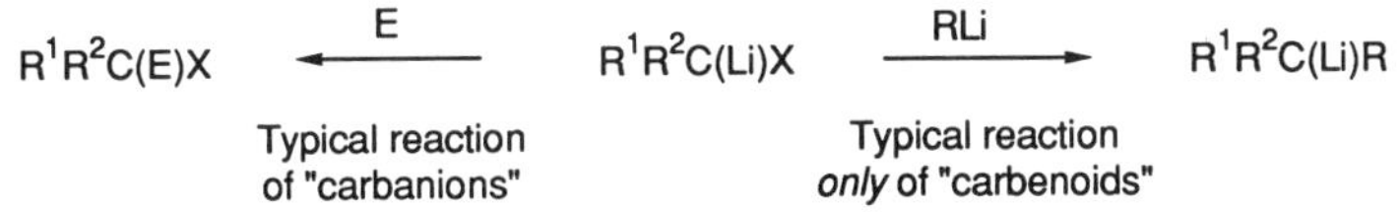

$R^1R^2C(E)X \xleftarrow{E} R^1R^2C(Li)X \xrightarrow{RLi} R^1R^2C(Li)R$

Typical reaction of "carbanions" — Typical reaction *only* of "carbenoids"

Scheme 2 "Carbanionic" and "carbenoid" reactivities of carbenoids.

tigations of their structure. Significant insights came from ^{13}C NMR investigations by Seebach et al.[6] These authors discovered that the ^{13}C signal of the anionic carbon of carbenoids is strongly shifted to the lower field in comparison to the C atom of the nonlithiated compounds. From this result Seebach et al. concluded that the C—Hal bond in these carbenoids should be *elongated* (see Fig. 7.2), just opposite to the situation in carbanions with typical acceptor substitutents, as mentioned above.

Theoretical calculations by Schleyer et al. as well as by others on Li–halide carbenoids are in agreement with this conclusion.[7,8] In all of the structural isomers of, for example, $LiCH_2F$, the C—F bond is elongated as compared to CH_3F. Table 7.1 contains our MP2/6-311++G(*d,p*)//MP2/6-311++G(*d,p*) + ZPE values together with the earlier data[7a,8] in parentheses.

Unfortunately, the notorious instability of Li—Hal carbenoids[5] so far prevented the experimental verification of the conclusions drawn from the NMR studies[6] and the calculations.[7,8] The closest structure to that of a Li—Hal carbenoid to date is that of the dimeric "Wittig–Furukawa" reagent **1** published by Denmark[9] (see Fig. 7.3). In this Zn/I complex, however, the C—I bonds are not elongated at all: the four C—I distances (213, 215, 216, and 221 pm) are clearly within the range of normal C—I bond lengths (214–221 pm).[9]

In the following sections we report on structural investigations and reactions of acceptor-substituted organolithium compounds in which X corresponds to NR_2 and OR, respectively. We studied these compounds first because not much was known about their structures in spite of many important applications in the stereoselective synthesis of organic compounds, and second because we supposed that especially RO-substituted organolithium compounds might be closely related to Li/F (Cl, Br, I) carbenoids. This turned out to be the case, as shown below.

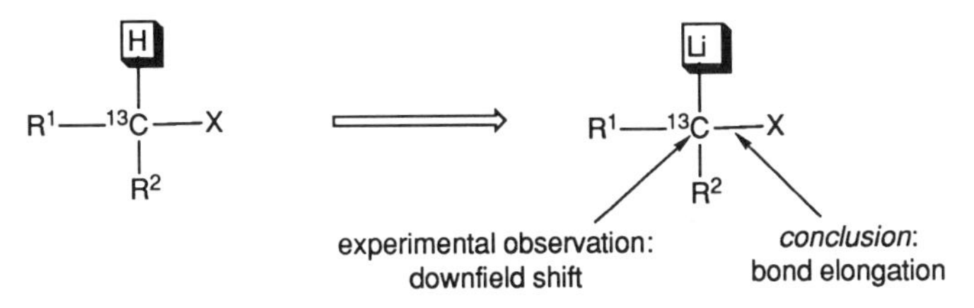

Figure 7.2 ^{13}C NMR investigations of Li/Hal carbenoids.[6]

TABLE 7.1 MP2/6-311++G(*d,p*)//MP2/6-311++G(*d,p*)+ZPE Values of CH_3F and $LiCH_2F$ with Earlier[7a] MP2/4-31G//4-31G Values in Parentheses

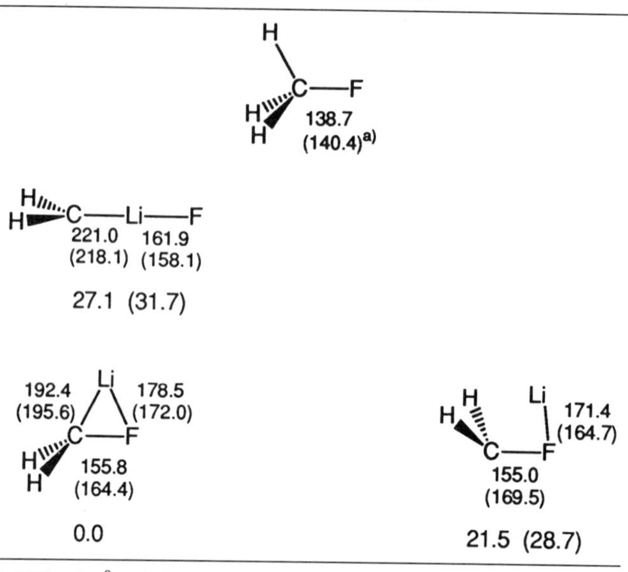

[a]3-21G value[8]; relative energies in kcal/mol; bond lengths in pm.

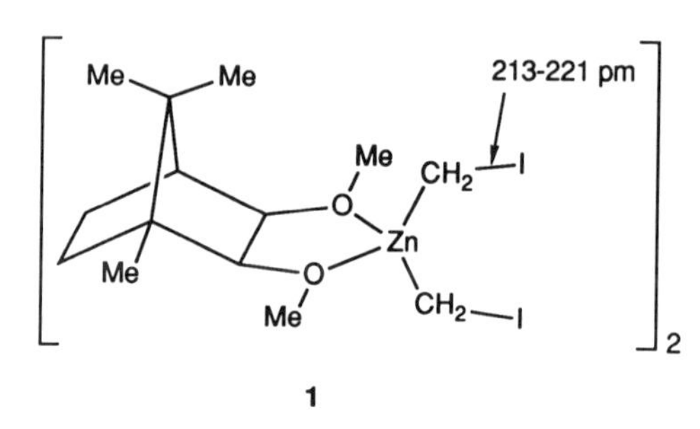

Figure 7.3 Crystal structure of the "Wittig–Furukawa" reagent **1**.[9]

2 α-LITHIATED AMINES

There are many reports on α-nitrogen substituted organolithium compounds in the literature, mainly because of the preparative significance of these intermediates. In this chapter we restrict ourselves to amine-substituted representatives.[10] These compounds are strongly related to what were originally called "dipole stabilized" carbanions.[11,12] Other work in this field concentrates on amine-substituted allyl-[13] and benzyllithium species,[14] lithium compounds with an amine and another acceptor substituent at the same carbon atom,[15] and BF_3-

activated, amine-substituted organolithium reagents.[16] There are also reports on amine-substituted alkyllithium species,[17] α-lithiated pyrroles, and indoles.[18]

Structural investigations on amine-substituted organo*lithium* compounds, especially under the conditions outlined in the introduction, do not exist so far. What structural types do we expect? The first model calculations of the structures of $LiCH_2NH_2$ were published by Clark et al. 11 years ago.[19] In Table 7.2 we summarize our MP2/6-311+ +G(*d*,*p*)//MP2/6-311+ +G(*d*,*p*) + ZPE calculations together with the results of the earlier calculations (in parentheses),[8,19] which were performed with the smaller basis of MP2/6-31G(*d*)// 3-21G.

In the most stable isomer, **2A**, lithium bridges the anionic carbon and the nitrogen atom. The C—N bond is *elongated* to 154.5 pm. The calculated value for $H_3C—NH_2$ amounts to 146.3 pm, while the experimentally determined C—N distance is reported to be 146.5 pm.[20] In the next stable isomer, **2B**, the Li—N bond is broken, which raises the energy of **2B** by 13.7 kcal/mol. If lithium is only bonded to nitrogen as in **2C**, the energy is even higher: +15.0 kcal/mol. In both **2B** and **2C** the C—N bonds are also elongated (to 150.8 and 152.7 pm, respectively).

TABLE 7.2 MP2/6-311+ +G(*d*,*p*)//MP2/6-311+ +G(*d*,*p*)+ZPE Values of CH_3NH_2 and $LiCH_2NH_2$ Isomers with Earlier[19] MP2/6-31G(*d*)//3-21G Values in Parentheses

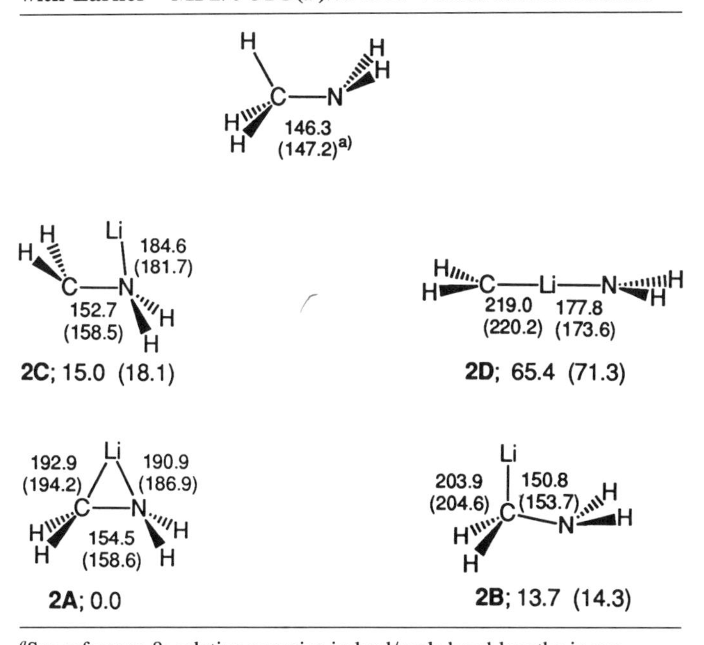

[a]See reference 8; relative energies in kcal/mol; bond lengths in pm.

When we calculated [MP2/6-311G(*d*,*p*)//3-21G] the dimerization energy of **2A** it became evident that the formation of the dimer **(2A)**$_2$ is very favorable (by 58.0 kcal/mol): Dimerization occurs along the C—Li bonds. The C—N bonds are also elongated (155.9 pm), and the C—Li bonds of the bridging Li-atoms are longer (CA—LiA, 227.2 pm) than those of the nonbridging lithiums (CB—LiA, 215.3 pm).

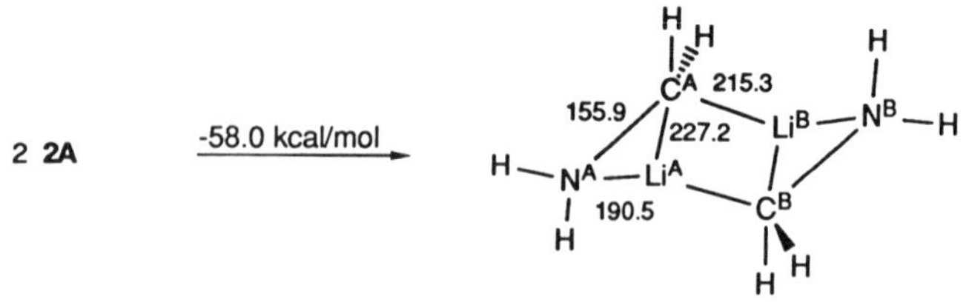

[2A]$_2$

The structure of a "normal" lithiated amine thus should correspond to the theoretical model **(2A)**$_2$. Only if the more unfavorable situation, as in **2B** or **2C,** is compensated for by other effects, should the experimentally detected structures be related to these model structures. That this indeed is observed is shown by the following crystal structure determinations.

2.1 Crystal Structure of [α-(Dimethylamino)benzyllithium · Diethylether]$_2$ [3 · OEt$_2$]$_2$

The crystal structure of [**3**·OEt$_2$]$_2$[21a], in which each lithium is coordinated to one molecule of diethyl ether, is shown in Figure 7.4. As one can see from this figure, C1 and N1 are bridged by Li1, and a dimer is formed along the carbon–lithium bonds. The distance of C1 to the bridging Li1 [247.5(7)pm] is longer than C1—Li1A [223.0(7) pm]. This corresponds exactly to the situation in the calculated structure of **(2A)**$_2$.

It is also of interest that a carbon–lithium bond to one of the phenyl carbon atoms, as is normally observed in benzyllithium compounds,[22] does not exist [e.g., Li1—C2A amounts to 272.1(7) pm]; apparently the lithium cation prefers in the solid state complexation with nitrogen to η^2- or η^3-coordination, including the phenyl ring.[21b] The benzylic carbon is strongly pyramidalized: the sum of the angles at C1 amounts to 341° (sp^3: 328.5°; sp^2: 360°). The C1—N1 bond is 146.9(4) pm long, which corresponds exactly to the mean value (146.8 pm) of C—N bonds in sp^3(!)-hybridized PhCH(R)—NMe$_2$ compounds.[23] The mean value for C_{sp^2}—NMe$_2$ bonds, as in enamines, is shorter (140.0 pm).[24] A rough estimate considering the hybridization of C1 in [**3**·OEt$_2$]$_2$, and the different bond lengths mentioned above, leads to the conclusion that C1—N1 in [**3**·OEt$_2$]$_2$ should be slightly elongated (~3 pm) compared to a C—N bond in a neutral compound with a similar hybridization of the carbon atom. This result is also in agreement with the calculations, al-

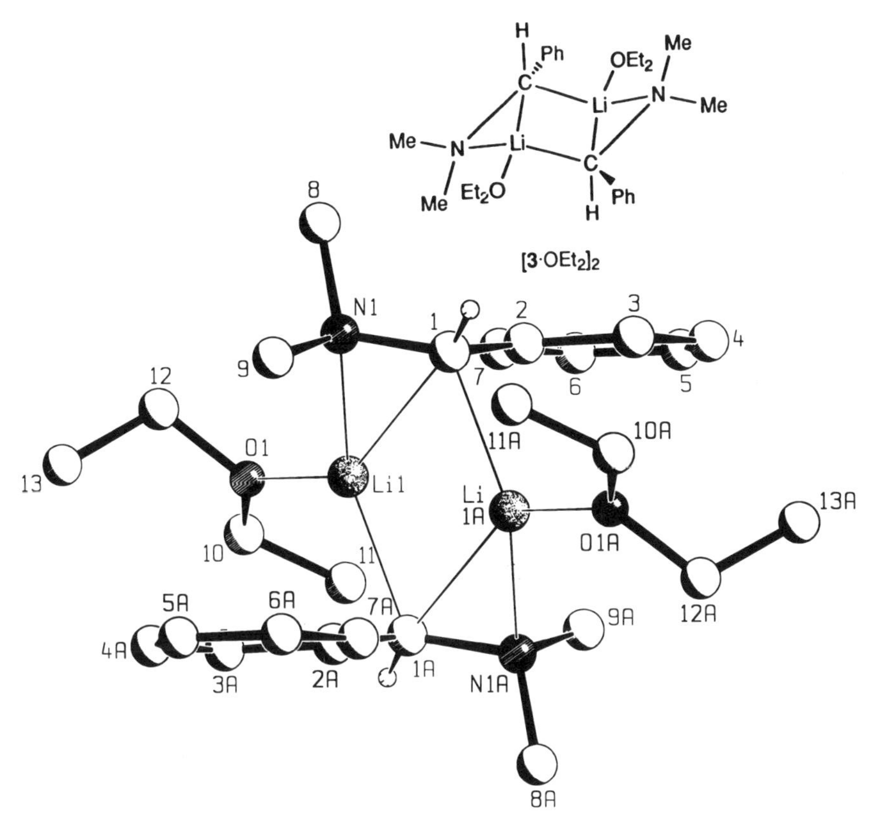

Figure 7.4 Crystal structure of [α-(dimethylamino)benzyllithium·diethyl ether]$_2$ $[\mathbf{3}\cdot OEt_2]_2$.[21a]

though the experimentally found bond elongation is smaller than the one in the calculated models **2** [see Table 7.2 and $(\mathbf{2A})_2$].

2.2 Crystal Structure of (1*S*)-1-(*N*-Pivaloyl-*N*-methylamino)benzyllithium·(−)-Sparteine 4·sparteine

What happens when one of the two methyl groups in α-(dimethylamino)benzyllithium **3** is replaced by a pivaloyl group is shown in the crystal structure of (1*S*)-1-(*N*-pivaloyl-*N*-methylamino)benzyllithium·(−)-sparteine **4**·sparteine[25] (see Fig. 7.5). In **4**·sparteine Li1 is attached to the benzylic carbon C1, to oxygen O1 of the carbonyl group, and to N2 and N3 of (−)-sparteine. The chiral amine induces the *S* configuration at C1. Most significantly, lithium Li1 does *not bridge* C1—N1 as in $[\mathbf{3}\cdot OEt_2]_2$. Rather a more favorable five-membered ring chelate is formed. The crystal structure of **4**·sparteine thus contains for the first time in an organo*lithium* compound the structural element invoked for "dipole-stabilized" lithiated amines by Beak in

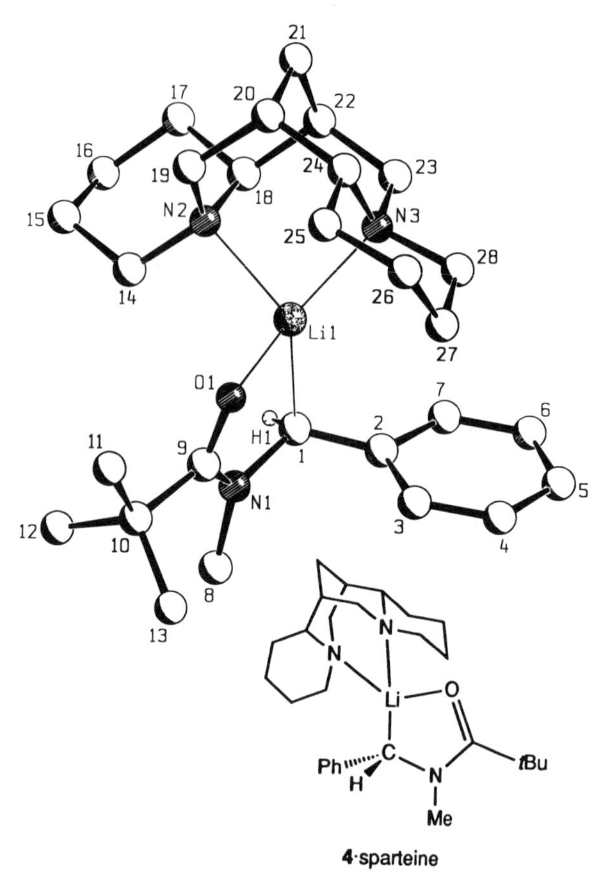

Figure 7.5 Crystal structure of (1*S*)-1-(*N*-pivaloyl-*N*-methylamino)benzyllithium·(−)-sparteine **4**·sparteine.[25]

1978.[11] Is there any "dipole stabilization," which means a shortening of N1—C9 and a lengthening of C9—O1 in the lithiated species? Both structural changes, however small, indeed are observed: the mean value of C—N in nonlithiated amides corresponding to **4** is 135.3 pm[23] while C9—N1 in **4**·sparteine is shortened to 133.7(4) pm; the mean value of C—O in the neutral species is 123.4 pm[23] whereas C9—O1 is lengthened to 125.0(4) pm. According to the model calculations shown below (4-31G),[26] however, the major part of the stabilization should be due to the chelation effect resulting from the formation of the C1—N1—C9—O1—Li1 five-membered ring. It is interesting to note that the calculated energy gained by chelation (−13.2 kcal/mol) corresponds nicely to the energy required to go from the most stable model structure **2A** to **2B** (+13.7 kcal/mol) (see Table 7.2).

The sum of the bond angles at C1 in **4**·sparteine adds up to 341°, a value also found in [**3**·OEt_2]$_2$. Again, C1 is strongly pyramidalized. In comparison to a neutral species with the same hybridization at C1, the C1—N1 bond

-13.2 kcal/mol

[146.2(4) pm] in the "anion" **4**·sparteine should therefore be slightly lengthened, as also observed in [**3**·OEt_2]$_2$.

2.3 Crystal Structure of [3-Iodo-2-lithio-*N*-methylindole·2Tetrahydrofuran]$_2$ [5·2THF]$_2$

The crystal structures of the lithiated iodo-indole dimer [3-iodo-2-lithio-*N*-methylindole·2 tetrahydrofuran]$_2$ [**5**·2THF]$_2$[27] is shown in Figure 7.6. Since [**5**·2THF]$_2$ does not contain "special" substituents as does the pivaloyl group in **4**·sparteine, one might have expected a structure similar to that of α-(dimethylamino)benzyllithium [**3**·OEt_2]$_2$ (see Fig. 7.4). Instead, the Li atoms Li1 and Li1A are only attached to the anionic carbons C1 and C1A. A coordination of Li1 and Li1A to N1 and/or N1A is not observed: Li1—N1, for example,

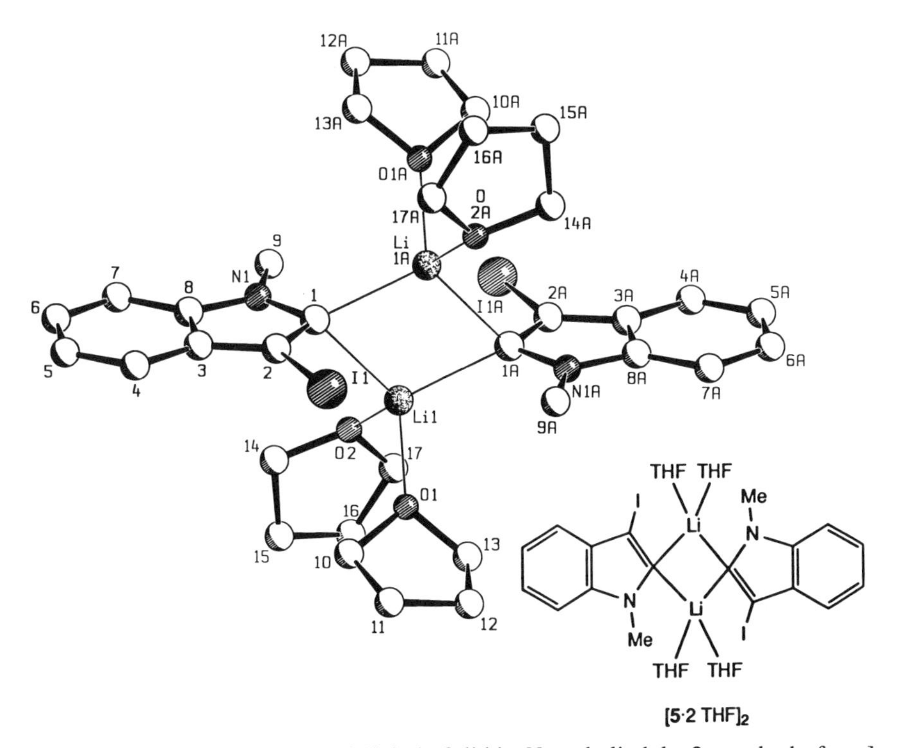

Figure 7.6 Crystal structure of [3-iodo-2-lithio-*N*-methylindole·2 tetrahydrofuran]$_2$ [**5**·2THF]$_2$.[27]

amounts to 310.5(7) pm. The other two coordination sites at Li are occupied by two THF molecules. The reason for the difference to the situation in $[\mathbf{3}\cdot OEt_2]_2$ is evident: the lone-pair on nitrogen in $[\mathbf{3}\cdot OEt_2]_2$ has much better donor qualities than the electron pair at N1 (N1A) in $[\mathbf{5}\cdot 2THF]_2$, which is involved in the formation of a five-membered aromatic ring system. We shall come back to this situation when we discuss the crystal structures of the lithiated benzofurans **11** and **12** in Section 3.1.

The C1—N1 bond length in $[\mathbf{5}\cdot 2THF]_2$ is in agreement with what is already known from the C—N bond lengths in $[\mathbf{3}\cdot OEt_2]_2$ and **4** sparteine. A length of 141.1(5) pm means a longer bond than the mean value of "C1—N1" bonds in indoles (139.0 pm). The C8—N1 bond in $[\mathbf{5}\cdot 2THF]_2$ is also shorter [138.5(5) pm].

It is also worth mentioning that $[\mathbf{5}\cdot 2THF]_2$ contains an iodine atom at C2 which does not undergo a facile β-elimination with a lithium at C1 to give the corresponding hetarine. Accordingly, the C2—I1 bond [209.0(4) pm] does not show any lengthening: the mean value of C_{sp^2}—I bonds is 209.5 pm.[23] As we discuss in Section 3.1 for $[\mathbf{11}\cdot O\textit{i}Pr_2]_2$, this situation is common to five-membered ring systems. The formation of a "triple bond" would introduce too much strain.

2.4 Structures of Other α-Lithiated Amines

Since in the structures of the three α-lithiated amines, $[\mathbf{3}\cdot OEt_2]_2$, $\mathbf{4}\cdot$sparteine, and $[\mathbf{5}\cdot 2THF]_2$, a slight lengthening of the C—N bonds is observed, one has to raise the question whether this is a general situation in α-lithiated(metallated) amines. There are four more crystal structure investigations of such compounds in the literature (see Scheme 3).

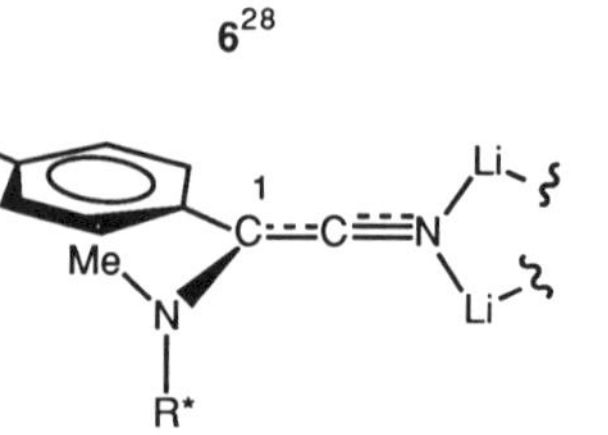

(THF)$_4$ Mg Br ... **6**[28]

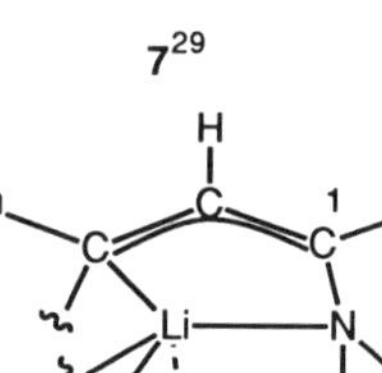

TMEDA ... Ph ... **7**[29]

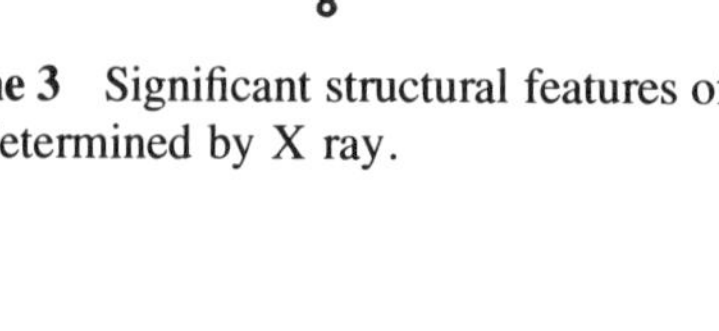

MeO ... Me ... R* ... **8**[30]

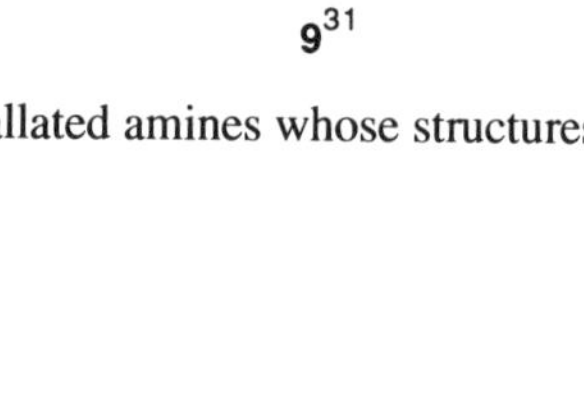

Ph ... Ph ... OMe ... **9**[31]

Scheme 3 Significant structural features of α-metallated amines whose structures have been determined by X ray.

The first crystal structure of an α-metalated amine, the Grignard species **6**, was reported by Seebach.[28] C1—N in **6** amounts to 148.2(13) pm; in a compound with an sp^3-hybridized "C1" atom, used for comparison, "C1"—N is 147.2(4) pm long. As far as the marginal lengthening of C1—N in **6** is concerned, one should consider that the position of the hydrogen atom at C1 in **6** has not been determined. Consequently, the hybridization at this carbon atom, which is important for a comparison of the bond lengths, is unknown. Any hybridization of C1 in **6** with a higher *s*-character than sp^3 corresponds to a more pronounced lengthening of the C1—N bond. The chelation of magnesium agrees well with the chelation of lithium in **4**·sparteine.

The crystal structure of the dimeric [2-lithio-*N*-phenylpyrrole·TMEDA]$_2$ **7**[29] has the advantage of an intramolecular comparison of C—N bond lengths: C2—N (141.2 pm) is longer than C5—N (139.5 pm). Since Li in **7** is also not coordinated to the pyrrole-*N*-atoms, this structure corroborates nicely the structural features of **5**·2THF.

A lengthened C1—N bond is also observed in **8** and **9** if the bond lengths are compared to the mean value of C—N bond lengths in enamines (140.0 pm):[24] in **8** 145.2(5) pm were determined for C1—N30;[30] C1 is planar and thus sp^2-hybridized. In **9**, C1—N [146(1) pm][31] is also longer than in enamines.

2.5 Summary

In the crystal structures of the α-lithiated (α-"magnesiated") amines known to date [**3**·OEt_2]$_2$, **4**·sparteine, [**5**·2THF]$_2$, **6**, **7**, **8**, and **9**, the bond lengths from the anionic carbon atom C1 to the adjacent N atom are generally slightly (3–5 pm) lengthened if compared to the bond lengths in the nonlithiated species. This corresponds to model calculations (Table 7.2). Furthermore, the calculated structures of $LiCH_2NH_2$ (Table 7.2) are in agreement with the essential features of the experimentally determined structures of [**3**·OEt_2]$_2$, **4**·sparteine, [**5**·2THF]$_2$, **6**, and **7**. In the lithiated amino nitrile **8** the position of lithium is determined by the nitrile functionality,[3] and in the amine-substituted allyl "anion" **9**, Li bridges four rather than only two atoms, a situation similar to that in **4**·sparteine.

The (slight) bond lengthening of the C—N bonds as found experimentally should be one of the characteristics of carbenoids (see the introduction). However, since we are not aware of any reaction in which an α-lithiated amine reacts as an electrophile with a nucleophile (RLi) under substitution of (the very bad leaving group) $LiNR_2$, the analogous reaction is characteristic of Li/

$$R^1R^2C(Li)NR_2 \xrightarrow[\quad]{RLi} \!\!\!\!\!\!\!/\;\; R^1R^2C(Li)R + LiNR_2$$

Hal carbenoids[5]—*α-lithiated* amines are *not carbenoids*. Rather they behave like normal "carbanions."

3 α-LITHIATED ETHERS

What are the changes on going from α-lithiated amines to α-lithiated ethers? The calculations of the $LiCH_2OH$ structures **10**, first performed by Schleyer and Houk et al.[19] and repeated by us on a higher level, show a similar overall situation for **10A-D** as for **2A-D** (see Table 7.3). The C—O bridged isomer **10A** is the most stable one with a C—O bond elongated to 152.2 pm—the CH_3OH value amounts to 142.1 pm; the experimental CH_3OH value[32] is 142.5 pm.[33] Structure **10B** with Li only attached to carbon is 13.6 kcal/mol higher in energy, and the C—O bond is 146.7 pm long. Structure **10C**, which has only an O—Li bond, is 16.4 kcal/mol above **10A**, and the C—O bond is elon-

TABLE 7.3 MP2/6-311++G(*d,p*)//MP2/6-311++G(*d,p*)+ZPE[34] Values of CH_3OH and $LiCH_2OH$ (10A-D) with Earlier[19] MP2/6-31G(*d*)//3-21G Values in Parentheses, Relative Energies in kcal/mol, and Bond Lengths in pm

H
C—O
H H
H 142.1 (144.1)[a)] H

H H Li 171.7 (165.1)
C—O
151.4 (156.7) H

10C; 16.4 (19.6)

H C—Li—O—H
H 221.0 162.5
(220.5) (155.5)

10D; 44.4 (50.0)

194.1 (196.4) Li 177.5 (169.6)
C—O
H H 152.2 (155.6) H

10A; 0.0

Li
201.8 (204.1) (149.1) 146.7
C—O—H
H H

10B; 13.6 (13.8)

2 **10A** —-56.6 kcal/mol→

H H
C[A] 215.2 Li[B]
H 153.9 O[A] 234.8 O[B] H
172.2 Li[A] C[B]
H H

$[\mathbf{10A}]_2$

gated to 151.4 pm. Structure **10D**, a "metallocarbenium" isomer, is 44.4 kcal/mol higher in energy than **10A**. This is clearly less than the 65.4 kcal/mol of the corresponding $LiCH_2NH_2$ derivative **2D**, which could indicate a more pronounced electrophilic character of the carbon atom in lithiated ethers than in lithiated amines. The dimerization of two **10A** isomers via the C—Li bonds again is very favorable: according to MP2/6-31/G(*d*)//3-21G calculations the "C—Li dimer" is 56.6 kcal/mol more stable than two monomers **10A;** the C—Li bond to the bridging Li is longer (e.g., CA—LiA = 234.8 pm) than the other C—Li bond (e.g., CA—LiB = 215.2 pm).

As shown below, the results of the model calculations agree nicely with the experimental results of the crystal structure investigations of α-lithiated ethers.

3.1 Crystal Structures of α-Lithiated Ethers

To date the crystal structures of four α-lithiated ethers have been determined: (1) [2-lithio-3-bromo-benzofuran·diisopropylether]$_2$ [**11**·OiPr_2]$_2$,[36] (see Fig. 7.7); (2) [2-benzofuryl-lithium·TMEDA]$_2$ [**12**·TMEDA]$_2$,[37] (see Fig. 7.8); (3) η^1-(1*S*,2*E*)-1-(*N*,*N*-diisopropylcarbamoyloxy)-3-trimethylsilylallyllithium·(−)-sparteine **13**·sparteine,[38] (see Fig. 7.9); (4) diphenyl-(trimethylsilyloxy)methyllithium·3 THF **14**·3 THF,[35b] (see Fig. 7.10).

The most stable calculated structure **10A** (see Table 7.3), and especially its favorable dimer [**10A**]$_2$, agree well with the crystal structures of [**11**·O*i*Pr_2]$_2$ (Fig. 7.7) and [**12**·TMEDA]$_2$ (Fig. 7.8). In [**11**·O*i*Pr_2]$_2$ dimerization occurs along the C—Li bonds, and lithium bridges C and O. The distance to the anionic C of the bridging Li (250.2 pm) is longer than to the other C—Li bond (212.2 pm). The situation of [**11**·O*i*Pr_2]$_2$ differs significantly from that of the related [3-iodo-2-lithio-*N*-methyl*indole*·2THF]$_2$ (Fig. 7.7) in which a bridging of the C—N bond by Li is not observed: the additional pair of electrons at the benzofuran oxygen in [**11**·O*i*Pr_2]$_2$ makes the difference. The fourth coordination site at Li in [**11**·O*i*Pr_2]$_2$ takes the oxygen atom of diisopropyl ether.[39] Most significantly, the C(anionic)—O bond is elongated to 147.0 pm: for comparison the other C—O bond is 137.5 pm long, and the mean value for

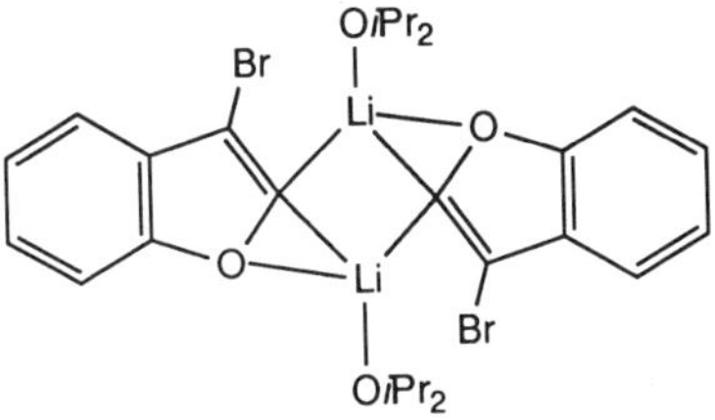

[11·O*i*Pr_2]$_2$

Figure 7.7 Crystal structure of [**11**·O*i*Pr_2]$_2$.[36]

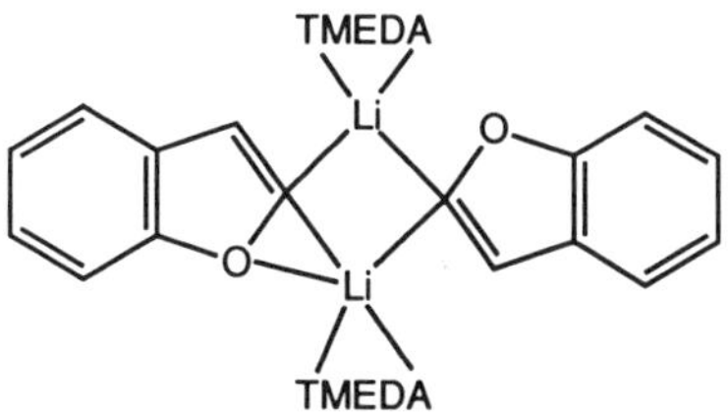

[12·TMEDA]$_2$

Figure 7.8 Crystal structure of [**12**·TMEDA]$_2$.[37]

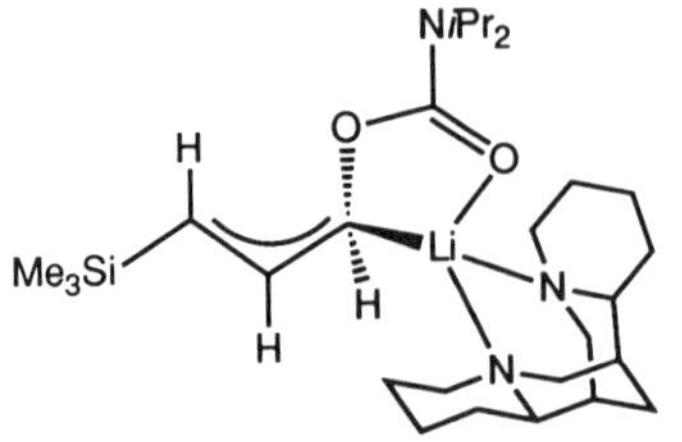

Figure 7.9 Crystal structure of **13**·sparteine.[38]

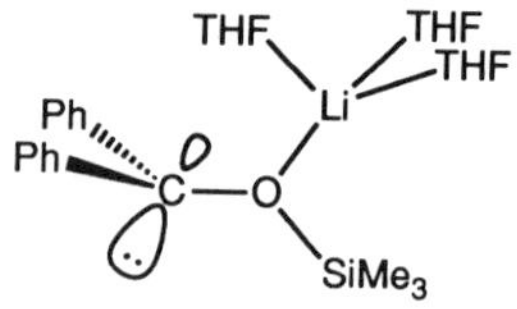

Figure 7.10 Crystal structure of **14**·3THF.[35b]

"C—O" bonds in 27 benzofurans amounts to 138.5 pm.[23] Although in the case of [**12**·TMEDA] (Fig. 7.8) only one bridging Li is observed, this structure still supports the importance of Li-bridging in lithiated ethers: since each Li is coordinated to one TMEDA molecule and to two carbon atoms this Li accepts even penta-coordination in order to bridge C and O; normally Li prefers to be tetra-coordinated.[1-3] C(anionic)—O again is elongated to 145.3 pm.

The crystal structure of the sparteine complexed carbamoyloxy-3-trimethylsilylallyllithium **13**·sparteine (Fig. 7.9) resembles closely the structure of the (*N*-pivaloyl-*N*-methyl-amino)-benzyllithium **4**·sparteine (Fig. 7.5): Li in **13**·sparteine (Fig. 7.9) is not bridging Li-O because a more favorable dipole-stabilized five-membered ring chelate is formed which overcomes the energy difference between the most stable calculated isomer **10A** and **10B** (+13.6 kcal/mol; Li only bonded to carbon, see Table 7.3). Once again C—O is elongated (147.6 pm); for comparison, the mean value of 11 C_{sp^3} (six C_{sp^2})—OCb bonds is 143.6 (139.9) pm.[23,38]

Diphenyl(trimethylsilyloxy)methyllithium·3THF **14**·3THF (Fig. 7.10) fulfills the requirements for a structural type corresponding to the theoretical model **10C** (Table 7.3). The stabilization of the negative charge at the anionic C by delocalization into the two phenyl rings is the prerequisite for the removal of Li from this carbon atom (Li—C 280.7 pm). It is noteworthy that the two phenyl groups at C1 are slightly bent *toward* Li1 (and not away from it!), exactly as predicted by the calculated model **10C**. The C—O bond is strongly elongated to 148.8 pm. The mean (maximum) value for the C—O bond in 135 C_{sp^3}–$OSiMe_3$ groups amounts to 140.2 (146.1 pm); in 24 C_{sp^2}–$OSiMe_3$ cases, the values are 137.5(142.8) pm. Since the hybridization at the anionic C is much closer to sp^2 than to sp^3, the elongation in the "anion" **14**·3THF roughly amounts to 10 pm.[35b]

In conclusion, the crystal structures of the α-lithiated ethers [**11**·$OiPr_2$]$_2$, [**12**·TMEDA]$_2$, **13**·sparteine, and **14**·3THF correspond nicely to the theoretical models of $LiCH_2OH$ **10A–C.** The structures show a significant (~7–10 pm) lengthening of the C—O bonds, as expected for Li/oxygen (and Li/halogen) carbenoids (see Section 1). That α-lithiated ethers indeed are Li/oxygen carbenoids is shown in the following section.[40]

3.2 Reactions of α-Lithiated Ethers with Nucleophiles RLi, Olefines, and C—H Bonds

One of the characteristic differences between a "carbanion" and a carbenoid is the reaction of the latter as an *electrophile* with *nucleophiles* as, for example, RLi (see Scheme 2). As shown below, this reaction is also observed with α-lithiated ethers and thus characterizes them indeed as Li/oxygen carbenoids.

The first reaction of this type was described by Lüttringhaus in 1938 in what is probably the first paper on α-lithiated ethers.[41a] In the reaction of phenyllithium **15** with benzylphenylether **16**, first the α-lithiated ether **17** is formed by deprotonation of **16** by **15**, see reaction (1) in Scheme 4. In a subsequent reaction (2) the α-lithiated ether **17** reacts with PhLi **15**, which now acts as a nucleophile to give diphenylmethyllithium **18** and lithium phenoxide.

In a third reaction (3) the lithiated ether **17** reacts with the newly formed "anion" **18** to give **19** and lithium phenoxide. On protonation **18** led to diphenylmethane and **19** to 1,1,2-triphenylethane. The lithiated ether **17** thus behaves quite clearly as a *Li/oxygen carbenoid*! Wittig has confirmed these results in his paper describing the discovery of the Wittig rearrangement.[42] He also discussed for the first time the increased properties as a leaving group of lithium phenoxide in **17** ("die gesteigerte Beweglichkeit der Phenolatgruppe").[42] Later Schöllkopf[43] investigated the Li/oxygen carbenoid **17** in reactions with olefins to give the cyclopropanes **20** and **21** (see Scheme 5).

The formation of **20** and **21** is another proof of the carbenoid nature of **17**. It is not clear whether the carbenoid or the carbene (as originally suggested[43]) undergoes the cyclopropane formation. Other products isolated by Schöllkopf resulting from reactions of **17** with nucleophiles RLi once again support the

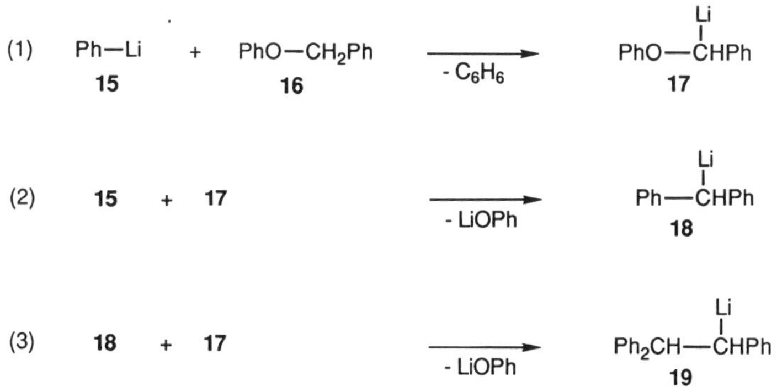

(1) Ph—Li (**15**) + PhO—CH_2Ph (**16**) $\xrightarrow[-\ C_6H_6]{}$ PhO—CH(Li)Ph (**17**)

(2) **15** + **17** $\xrightarrow[-\ LiOPh]{}$ Ph—CH(Li)Ph (**18**)

(3) **18** + **17** $\xrightarrow[-\ LiOPh]{}$ Ph_2CH—CH(Li)Ph (**19**)

Scheme 4 Reaction sequence in the reaction of phenyllithium **15** with benzyl phenyl ether **16** to give diphenylmethane (from **18**) and 1,1,2-triphenylethane (from **19**) after acidic workup.

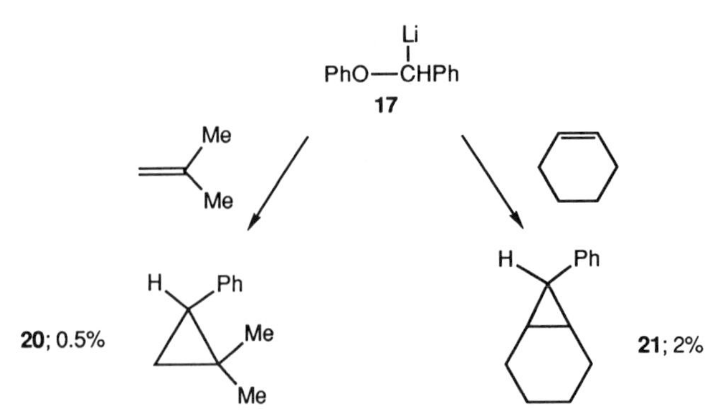

Scheme 5 Reactions of the Li/oxygen carbenoid **17** with olefins to give the cyclopropanes **20** and **21**.[43]

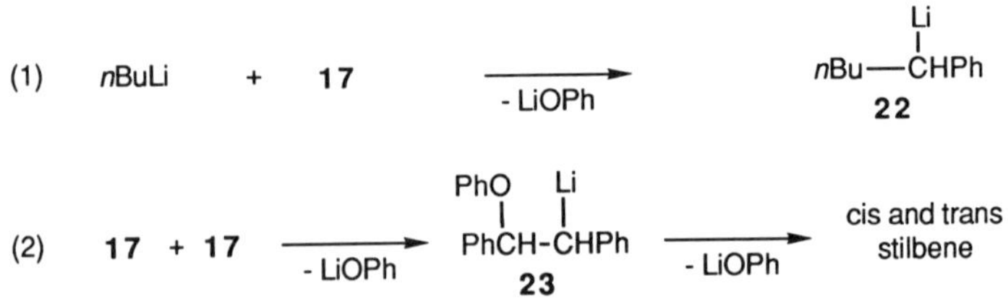

(1) *n*BuLi + **17** $\xrightarrow[\text{- LiOPh}]{}$ *n*Bu—CH(Li)Ph **22**

(2) **17** + **17** $\xrightarrow[\text{- LiOPh}]{}$ PhCH(OPh)-CH(Li)Ph **23** $\xrightarrow[\text{- LiOPh}]{}$ cis and trans stilbene

Scheme 6 Further reaction of the carbenoid **17**.[43]

carbenoid nature of this species (see Scheme 6).[43] If **17** is prepared from benzyl phenyl ether **16** with *nBuLi*, the latter functions also as a nucleophile and reacts with **17** to provide **22** [reaction (1)] which has also been trapped with D_2O (!). In reaction (2) **17** reacts both as an *electrophile* and a *nucleophile* to give **23**, which eliminates lithium phenoxide to give *cis*- and *trans*-stilbene.

At this point one could argue that the carbenoid nature of the α-lithiated benzyl phenyl ether **17** might be exceptional for two reasons: (1) nucleophilic substitution reactions of RLi take place at a benzylic carbon atom which is generally very favorable, and (2) lithium phenoxide is a much better leaving group than lithium alkoxide. However, dimethyl ether also is cleaved by organolithium reagents RLi via the same α-lithiation–substitution reaction sequence outlined above. Ziegler[44] showed this for the first time in 1950 (see Scheme 7).

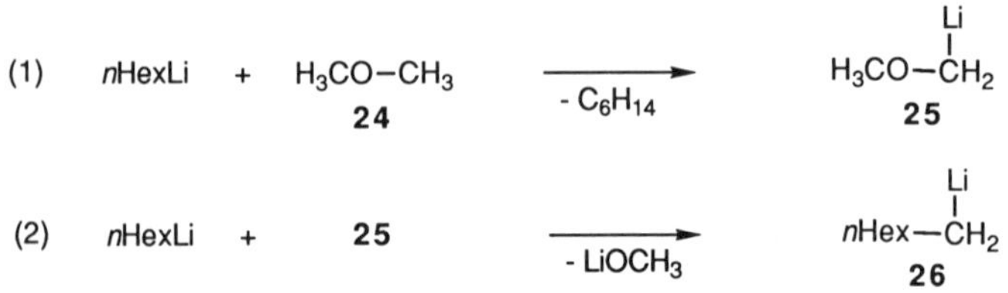

(1) *n*HexLi + $H_3CO{-}CH_3$ **24** $\xrightarrow[\text{- } C_6H_{14}]{}$ $H_3CO{-}CH_2Li$ **25**

(2) *n*HexLi + **25** $\xrightarrow[\text{- } LiOCH_3]{}$ *n*Hex—CH_2Li **26**

Scheme 7 Reaction of *n*BuLi with dimethylether.[44]

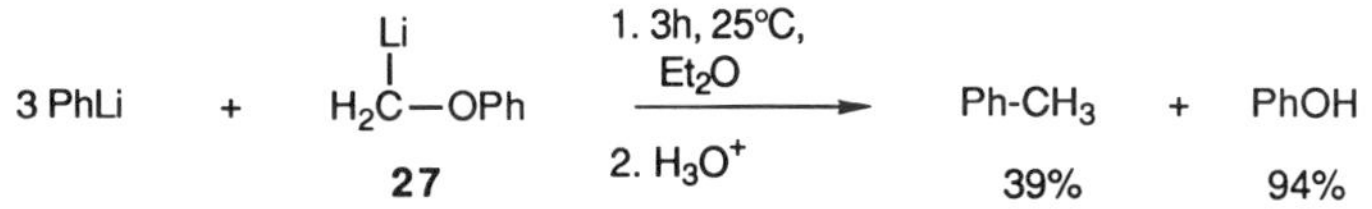

$$3\ PhLi + \underset{\textbf{27}}{H_2\overset{Li}{\overset{|}{C}}-OPh} \xrightarrow[2.\ H_3O^+]{1.\ 3h,\ 25°C,\ Et_2O} \underset{39\%}{Ph\text{-}CH_3} + \underset{94\%}{PhOH}$$

Scheme 8 Reaction of the Li/oxygen carbenoid **27** with PhLi.[47]

Ziegler reacted *n*-hexyllithium with dimethyl ether **24,** which resulted in the formation of the carbenoid **25** [reaction (1)]; in the subsequent step the carbenoid **25** reacted with *n*-hexyllithium to give *n*-heptyllithium **26** [reaction (2)], which in turn reacted with **25** to give *n*-octyllithium, and so on. Hoberg trapped such a methylene carbenoid with cyclohexen to give norcarane, albeit in low yield.[45] Systematic investigations of dialkyl ether cleavage reactions were subsequently carried out by Maercker.[46] Very recently, we reacted the lithiated phenyl methyl ether **27** with 3 mol equiv. phenyllithium in diethyl ether for 3 h at 25°C. On protonation 39% toluene and 94% phenole were obtained,[47] (see Scheme 8).

Carbenoid reactivity is also shown by α-lithiated *cyclic unsaturated ethers*,[48] although the first example seems to be that of a cyclic unsaturated MgBr/oxygen carbenoid.[49] Kocienski[50] summarized the reactions of lithiated dihydropyran **28** and especially dihydrofuran **30**, which react with RLi to give stereospecifically the substitution products **29** and **32**, both of which are utilized in natural products synthesis (see Scheme 9). One should mention that the CuCN-catalyzed reaction is synthetically more useful because it leads to higher yields of the ring-opened products **29** and **32**.

Kocienski invokes a mechanism in which the incoming RLi first forms a metallate (e.g., with **30**) of the type **31** in which an intramolecular 1,2-migration of R and a substitution of alcoholate take place with inversion of configuration at the carbenoid carbon atom.[50] Some years earlier Duraisamy and

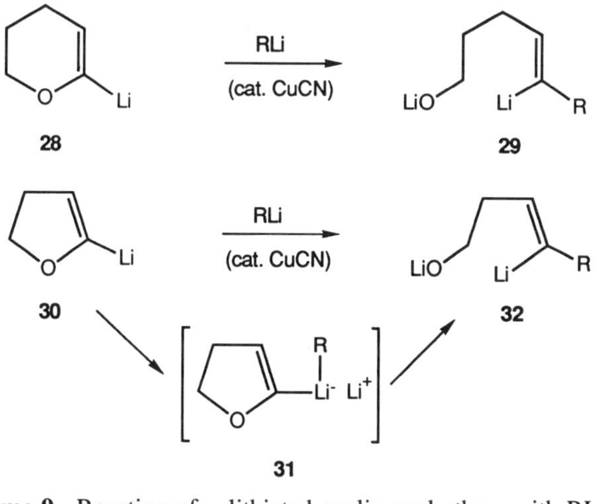

RLi (cat. CuCN)
28
29
LiO Li R
RLi (cat. CuCN)
30
32
LiO Li R
R Li⁻ Li⁺
31

Scheme 9 Reaction of α-lithiated cyclic enol ethers with RLi.[50]

Walborsky found a similar substitution reaction at a noncyclic vinylic Li/Cl carbenoid which also occurs with inversion of configuration.[51] These authors proposed a "normal" substitution reaction (which so far has not been observed at a "natural," nonlithiated sp^2-hybridized carbon atom!), facilitated in that case by "metal-assisted ionization." A similar explanation for the facile substitution of X (halide) in carbenoids had been given by Köbrich.[52] The "metal assistance," of course, applies also for Li/oxygen carbenoids.

Of special interest to us was whether one of the α-lithiated ethers whose crystal structure had been determined ([**11**'O*i*Pr_2]$_2$, [**12**·TMEDA]$_2$, **13**·sparteine, and **14**·3 THF; see Fig. 7.7–10) also reacted as an electrophile with nucleophiles RLi. We selected α-lithiated benzofuran **12** for this study.[53] The results are listed in Scheme 10.

In our hands reactions occurred with CH_3Li, nBuLi, and PhLi; independently of our studies Nguyen and Negishi investigated the same reaction.[54] The yields obtained by these authors are listed in parentheses (Scheme 10).[55] The strongly lengthened C(anionic)—O bonds observed in [**12**·TMEDA]$_2$, as well as in the related [**11**·OiPr_2]$_2$, are obviously in good agreement with these substitution reactions to occur so easily in the Li/oxygen carbenoid **12**. If the bridging of C—O by Li as observed in the crystal structures of [**11**·OiPr_2]$_2$ and [**12**·TMEDA]$_2$ also exists in the THF solution of **12** (the calculations also predicted the bridged **10A** to be the most stable isomer; see Table 7.3), then there is an experimental hint for the "metal assisted ionization" to occur in the substitution reaction.[51] One should, however, not forget that in the case of the reactions of lithiated vinyl ethers with RLi one has to invoke a mechanistic alternative which has not been excluded rigorously: an addition–elimination sequence.

The Li/oxygen carbenoid nature of the intermediate is also quite clear in a reaction investigated by Cope et al. in 1960.[56] These authors showed by means

12 + R-Li **33** ⟶ **34**

R	yields [%] of **34**
CH_3	45 (not reported)
*n*Bu	64 (90)
*s*Bu	- (78)
*t*Bu	- (92)
Ph	42 (almost 100)

Scheme 10 Reaction of lithiated benzofuran **12** with RLi **33** to give the substitution products **34** (isolated after protonation, deuteration, or carboxylation); the results of Nguyen and Negishi are given in parentheses.[54]

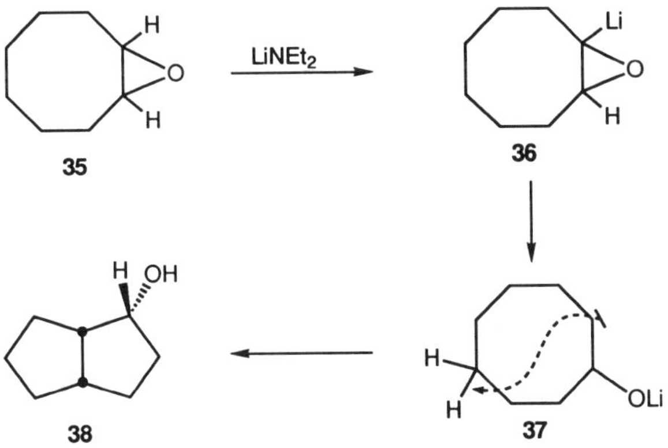

Scheme 11 A C–H insertion reaction across the ring involving the Li/O carbenoid **36**[56]

of deuterium labelling studies that the reaction of the epoxide **35** with bases like $LiNEt_2$ first leads to the α-lithiated epoxide **36** which then reacts with a C—H bond across the ring to give the bicyclic alcohol **38** (see Scheme 11).

Cope et al. assumed that the carbene **37** formed from the Li/O carbenoid **36** should be the intermediate to undergo the C—H insertion reaction. The high reactivity of such epoxide-derived Li/oxygen carbenoids is not surprising: the formation of a carbene such as **37** (as well as the substitution of alkoxide by RLi) profits from the loss of strain in the three-membered ring. Consequently, α-lithiated expoxides have been widely used in synthesis.[56–62] The "reductive alkylation" of these species, which finally leads to substituted alkenes, was discovered by Crandall,[57a] who also wrote a detailed review article on reactions of epoxides with bases[57b] (see Scheme 12).

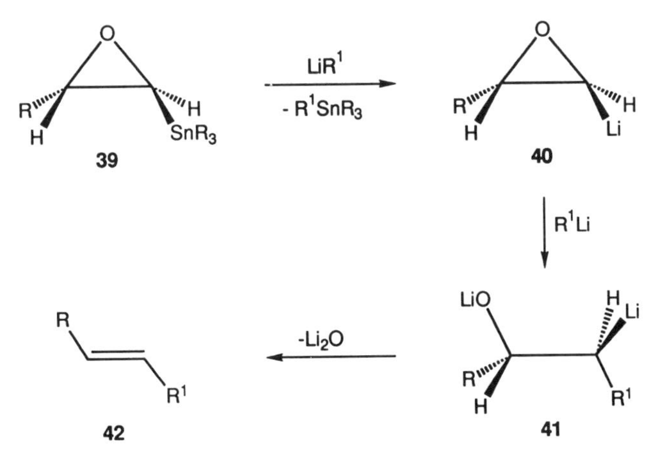

Scheme 12 "Reductive alkylation" of the lithiated epoxide **40** (prepared from **39**) to give **42** via **41**.

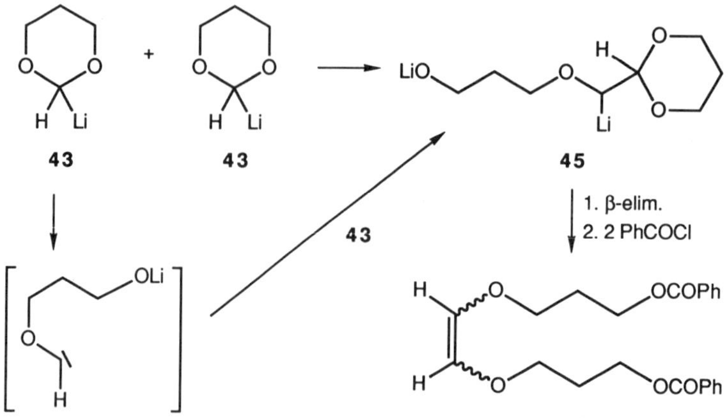

Scheme 13 ''Thermal decomposition'' of the dialkoxy carbenoid **43** to give **46** (after reaction with PhCOCl).[63]

The question of whether LiR^1 attacks the carbenoid **40** directly or whether **40** first transforms into a carbene which then adds to R^1Li, is still not settled (as in the case of **36** and **37**). A similar situation holds for a reaction discovered by Shiner et al.,[63] namely, the thermal decomposition of the *di*alkoxymethyllithum species **43** (see Scheme 13).

Warming a solution of **43** in THF to room temperature followed by reaction with benzoyl chloride (PhCOCl) yielded the (*E*) and (*Z*) dibenzoates **46** in equal amounts. Compound **45** (which after β-elimination and reaction with two PhCOCl provides **46**) could have been formed either by attack of one **43** at another **43**, or by intermediate formation of the carbene **44** which then is trapped by **43**.[64–66]

In summary, α-lithiated ethers show typical carbenoid behavior in their reactivity: One observes reactions with RLi which substitutes LiOR, additions to C=C double bonds to give cyclopropanes, insertion into C—H bonds, and—last but not least—the Fritsch–Buttenberg–Wiechell rearrangement **50** → **51**[50a] (see Scheme 14).

All of these reactions characterize α-lithiated ethers indeed as Li/oxygen carbenoids. The comparatively facile breaking of the C—O bond is in accord

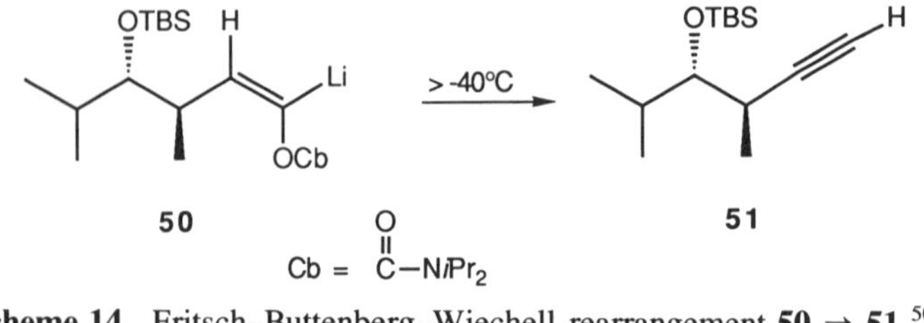

Cb = C(=O)—N*i*Pr₂

Scheme 14 Fritsch–Buttenberg–Wiechell rearrangement **50** → **51**.[50a]

with the elongation of this bond as found in the solid state structures outlined in Section 3.1. In the following section we show how Li/oxygen carbenoids also fulfill the NMR criteria so far elaborated for Li/Hal carbenoids.[6]

3.3 ^{13}C NMR Investigations and IGLO Calculations of α-Lithiated Ethers

Seebach et al. were the first to study Li/Hal carbenoids by means of NMR spectroscopy. A summary of their results is given in reference 6d; a brief description is shown in Scheme 15. Some of their results are listed with our data in Table 7.4.

The downfield shift $\Delta\delta$ of the carbenoid ^{13}C atom in $LiCCl_3$ **52** (65.9 ppm) and $LiCHCl_2$ **53** (50.0 ppm) depends on the number of Cl atoms in the carbenoid; $LiCH_2Cl$ **54** (32.3 ppm) fits into this pattern. Most interestingly, the Li/oxygen carbenoids also show the downfield shift $\Delta\delta$ as observed in the Li/Cl carbenoids **52–54:** with two oxygen substituents at the carbenoid carbon **(55)** $\Delta\delta$ is more pronounced (40.0 ppm) than with one oxygen—**56** ($\Delta\delta$ = 26.2 ppm), **57** ($\Delta\delta$ = 26.4 ppm), **58** ($\Delta\delta$ = 27.8 ppm), and **59** ($\Delta\delta$ = 28.0 ppm). The smaller $\Delta\delta$ values in the Li/oxygen carbenoids **55–59** than in the Li/Cl carbenoids **52–54** obviously parallel the observation of alkoxide being a far worse leaving group than chloride. Compound **60,** measured by Seebach et al,[67] is in accord with **56–59**. It is interesting to point to the carbamoyloxy derivative **61,** which also in solution should exist as a five-membered ring chelate, at least according to the crystal structure determinations of **4**·sparteine (Fig. 7.5) and **13**·sparteine (Fig. 7.9). Its $\Delta\delta$ value (7.0 ppm) is much smaller than the downfield shifts of **56–60.** In the cases of the vinylic carbenoids **11** and **12**, the $\Delta\delta$ values are much larger (73.3 and 70.8 ppm, respectively) than in the sp^3 examples discussed above. Such large shifts are generally observed in sp^2-hybridized systems like phenyllithium or vinyllithium.[6d] Thus, in the case of the former, $\Delta\delta$ amounts in the monomer to 68.2[68] and in the dimer to 60.3 ppm;[6d] also for vinyllithium the $\Delta\delta$ value depends on the aggregation state: dimer, 63.3 ppm; tetramer, 54.4 ppm.[69] A comparison of **11** and **12** with phenyl- and vinyllithium shows that the sp^2 carbenoids **11** and **12** have slightly larger $\Delta\delta$ values, as expected from the results of **55–60.**

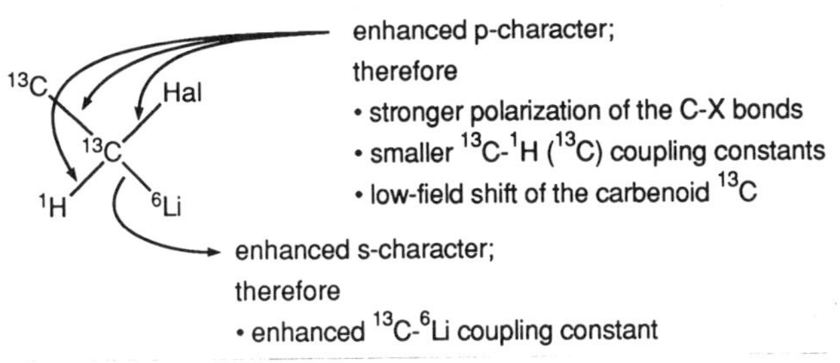

Scheme 15 Brief summary of the NMR investigations of Li/Hal carbenoids.[6]

TABLE 7.4 δ[ppm] Values of the Carbenoids Listed; Measurements in [D_8]THF and/or 2-Methyltetrahydrofuran at −80 to −120°C. Δδ[ppm] Refers to the Corresponding Nonlithiated Compounds

	carbenoid	δ [ppm]	Δδ [ppm]	$^1J(^{13}C,^6Li)$ [Hz]
52[a)]	$LiCCl_3$	145.9	65.9	17.0 (t)
53[a)]	$LiCHCl_2$	105.5	50.0	16.3 (t)
54	$LiCH_2Cl$	57.9	32.3	(sb)
55	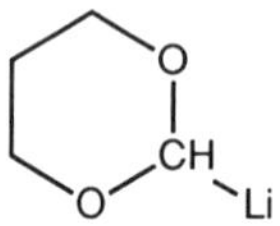 O, CH, O, Li	134.8	40.0	(sb)
56	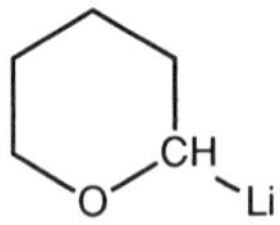 CH, O, Li	95.7	26.2	(sb)
57	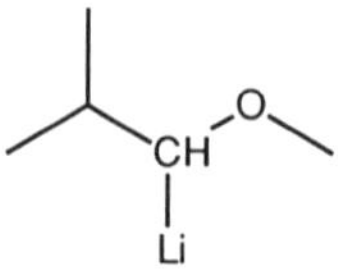 CH, O, Li	106.4	26.4	6.8 (quin)
58[b)]	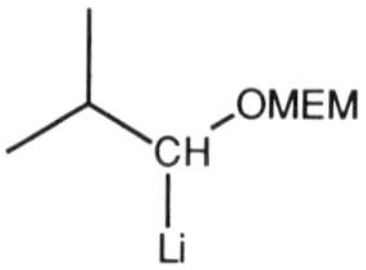 CH, OMEM, Li	102.4	27.8	(sb)
59[c)]	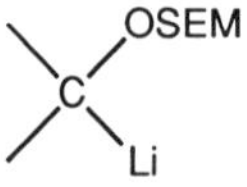 C, OSEM, Li	96.7	28.0	(svb)
60[d,e)]	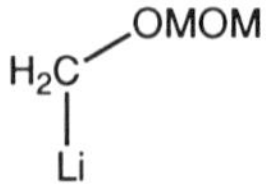 H_2C, OMOM, Li	74.0	19.0	(sb)
61[f)]	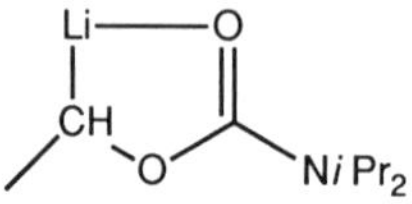 Li—O, CH, O, N*i*Pr_2	67.3	7.0	14.4 (t)

TABLE 7.4 (*Continued*)

	carbenoid	δ [ppm]	Δδ [ppm]	$^1J(^{13}C,^6Li)$ [Hz]
11	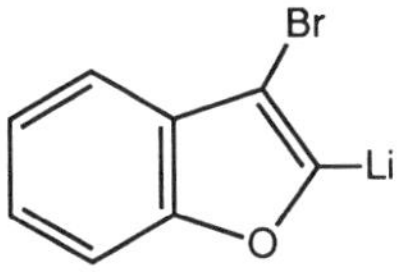 3-bromo-2-lithiobenzofuran	218.4	73.3	13.9 (t)
12[g,h]	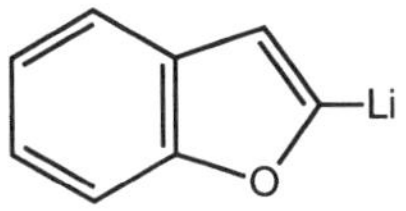 2-lithiobenzofuran	215.6	70.8	7.7 (quin)

[a]Determined by Seebach et al.[6]
[b]MEM = 2-methoxyethoxymethyl.
[c]SEM = 2-(trimethylsilyl)ethoxymethyl.
[d]See reference 67.
[e]MOM = methoxymethyl.
[f]In $[D_{10}]$ diethyl ether.
[g]Determined by Harder et al.[37]
[h]In toluene; t = triplet; quin = quintuplet; sb = broad singulet; svb = very broad singulet

In the case of **57** we have been able to observe the ^{13}C signal in THF solution as a *quintuplet* which results from the coupling of *two* ^6Li atoms with this carbon atom. As we have pointed out earlier, a very stable structure of α-lithiated ethers should be the dimer of the calculated **10A**, that is, $[\mathbf{10A}]_2$,

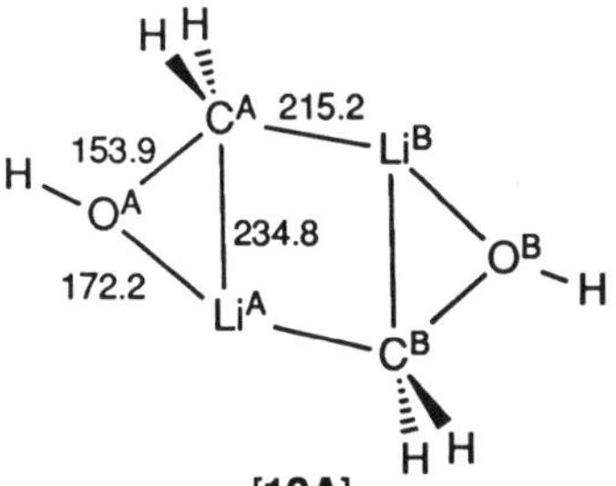

$[\mathbf{10A}]_2$

which thus would agree with the NMR observation.[70] Such a structural type was also found in the crystal structure of $[\mathbf{11}\cdot O\mathit{i}Pr_2]_2$ (Fig. 7.7). On the other hand, the ^{13}C *triplet*, which we observed in the case of the carbamoyloxy derivative **61,** is in agreement with the *monomers* found in the solid state structures of the related **4**·sparteine (Fig. 7.5) and **13**·sparteine (Fig. 7.9).

^{13}Carbon *triplet* signals have also been observed by Seebach et al.[6d] in the spectra of $^6LiCCl_3$ **(52)** and $^6LiCHCl_2$ **(53)** which, however, should *not be due to a monomeric structure similar to that of* **61.**[71]

To check whether the experimentally determined ^{13}C chemical shifts of the carbenoids listed in Table 7.4 agree with theoretically determined values we used the IGLO (Individual Gauge for Localized Orbitals) method developed by Kutzelnigg and Schindler.[72] IGLO turned out to be very successful for determining the structures of carbocations.[73] Recently it was applied to organolithium compounds.[74] The results of the first calculations of carbenoids are listed in Table 7.5.

TABLE 7.5 IGLO Calculated ^{13}C Chemical Shift Differences $\Delta\delta$[ppm] Between the Compounds Listed and the Corresponding Nonlithiated Species

carbenoid	$\Delta\delta$ [ppm][a)]	opt. geometry
10A	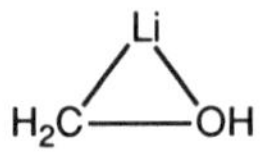 29.7	MP2/6-311++g(d,p)
$[\mathbf{10A}]_2$	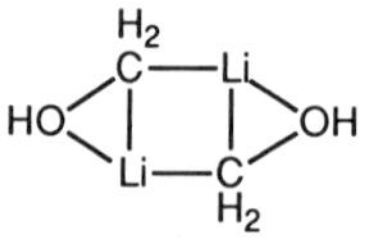 34.4	3-21G
$[\mathbf{10A}]_3$	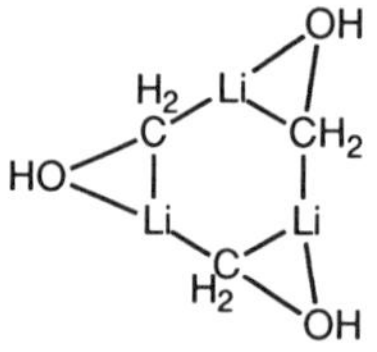 27.6	3-21G
10B	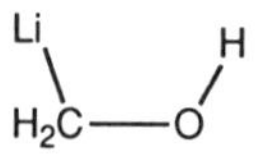 6.3	MP2/6-311++G(d,p)
62	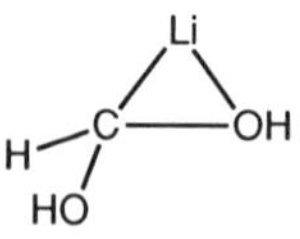 43.3	6-311++g(d,p)
63	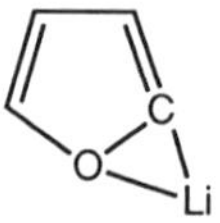 76.8	6-31G(d)

[a]Basis II+*sp* except for $[\mathbf{10A}]_2$ and $[\mathbf{10A}]_3$ for which DZ+*sp* was used; *s* and *p* coefficients and exponents were taken from the 6-31+G(d) basis sets.

As one can see from Table 7.5, the IGLO values calculated for the bridged **10A** (29.7 ppm), its bridged dimer $[\mathbf{10A}]_2$ (34.4 ppm), and the bridged trimer $[\mathbf{10A}]_3$ (27.6 ppm) agree well with the experimental data of **56–60**. For the nonbridged isomer **10B** with lithium only attached to carbon the IGLO value (6.3 ppm) agrees well with the one measured for **61** (7.0 ppm), which should have the chelated five-membered ring structure also in solution and not a lithium bridged C—O bond. This suggests that—except for special conditions as in the case of **61**—Li/oxygen carbenoids should have a *bridged structure*, probably one similar to that of the bridged dimer $[\mathbf{10A}]_2$, in agreement with the other results outlined in more detail above. The agreement between the values of the calculated dihydroxy compound **62** ($\Delta\delta$ = 43.3 ppm) and the lithiated acetal **55** ($\Delta\delta$ = 40.0 ppm) is also very satisfactory.

As far as the ''vinylic'' carbenoids **11** ($\Delta\delta$ = 73.3 ppm) and **12** ($\Delta\delta$ = 70.8 ppm) are concerned (see Table 7.4) the calculated value for the bridged model structure **63** ($\Delta\delta$ = 76.8 ppm) also agrees nicely with the experimental data of **11** and **12** in solution. They are definitively bridged in the solid state (see Figs. 7.7 and 7.8).

4 SUMMARY

Structural investigations of α-lithiated amines and α-lithiated ethers lead to the following conclusions:

1. The solid state structures of the α-lithiated amines $[\mathbf{3}\cdot OEt_2]_2$, **4**· sparteine, and $[\mathbf{5}\cdot 2THF]_2$ agree nicely with model calculations of $LiCH_2NH_2$ **2.** In all cases the bond of the anionic carbon to nitrogen is slightly (2–5 pm) lengthened. Other crystal structures of α-lithiated(metallated) amines are in agreement with these findings (see **6–9**). A lengthening of the C—X bond is expected for $LiCH_2X$ carbenoids (X = Hal). However, since α-lithiated amines do not react with nucleophiles RLi under substitution of LiX (X = NR_2) (a reaction which similarly occurs very easily in the case of Li/Hal carbenoids), α-lithiated amines should not be regarded as Li/N carbenoids.

2. In the solid state structures of the α-lithiated ethers $[\mathbf{11}\cdot O\textit{i}Pr_2]_2$, $[\mathbf{12}\cdot TMEDA]_2$, **13**·sparteine, and **14**·3THF the C—O bonds are remarkably elongated (7–10 pm). A literature review shows that even in the first preparation of a lithiated ether in 1938 the typical reaction of Li/Hal carbenoids, namely the substitution reaction with RLi, also occurs with the α-lithiated ether.

Many more examples indicate the generality of this displacement reaction. Furthermore, other typical carbenoid reactions have been observed: C—H insertion, cyclopropane formation with olefins, and Fritsch–Buttenberg–Wiechell rearrangement in the case of a lithiated vinylic ether. This shows quite clearly that α-lithiated ethers are carbenoids: LiOR, is a better leaving group than $LiNR_2$; it is not as good as LiHal.

^{13}C NMR investigations corroborate the findings given above. As in Li/Hal carbenoids, the ^{13}C signal of the carbenoid C atom in Li/oxygen carbenoids is shifted to lower field compared to the nonlithiated compound. The quintuplet signal of the carbenoid ^{13}C observed in the case of **57**, resulting from two ^{6}Li nuclei bonded to this carbon, is in agreement with a dimer structure as calculated to be very stable for $LiCH_2OH$ ([**10A**]$_2$) and as found in the solid state structure of the dimer [**11**·O*i*Pr_2]$_2$.

Finally, for the first time IGLO calculations for carbenoids are presented. They agree nicely with the experimentally determined ^{13}C chemical shifts thus supporting the structural proposal for Li/oxygen carbenoids deduced from solid state structures, reactivities, and ^{13}C NMR investigations. α-Lithiated ethers—and this was somewhat surprising to us at the outset of our investigations of α-lithiated ethers—thus turn out to be *Li/oxygen carbenoids*, the first class of carbenoids which have been fully structurally characterized.

REFERENCES

1. Seebach, D.; *Angew. Chem. Int. Ed. Engl.* **1988**, *27*, 1624.
2. Williard, P. G. In *Comprehensive Organic Synthesis*; Trost, B. M., Ed.; Pergamon: Oxford, 1991; Vol. 1, Part 1.
3. Boche, G. *Angew. Chem. Int. Ed. Engl.* **1989**, *28*, 277.
4. Cram, D. J. *Fundamentals of Carbanion Chemistry*; Academic Press: New York, 1985.
5. For reviews see: (a) Kirmse, W. *Angew. Chem. Int. Ed. Engl.* **1965**, *4*, 1. (b) Köbrich, G. *Angew. Chem. Int. Ed. Engl.* **1967**, *6*, 41. (c) Köbrich, G. *Angew. Chem. Int. Ed. Engl.* **1972**, *11*, 473. (d) Siegel, H. *Top. Curr. Chem.* **1982**, *106*, 55.
6. (a) Siegel, H.; Hiltbrunner, K.; Seebach, D. *Angew. Chem. Int. Ed. Engl.* **1979**, *18*, 785. (b) Seebach, D.; Siegel, H.; Müllen, K.; Hiltbrunner, K. *Angew. Chem. Int. Ed. Engl.* **1979**, *18*, 784. (c) Seebach, D.; Siegel, H.; Gabriel, J.; Hässig, R. *Helv. Chim. Acta* **1980**, *63*, 2046. (d) Seebach, D.; Hässig, R.; Gabriel, J. *Helv. Chim. Acta* **1983**, *66*, 308.
7. (a) Clark, T.; Schleyer, P. V. R. *J. Chem. Soc. Chem.* **1979**, 883. (b) Clark, T.; Schleyer, P. V. R. *Tetrahedron Lett.* **1979**, 4963. (c) Clark, T.; Schleyer, P. v. R. *J. Am. Chem. Soc.* **1979**, *101*, 7747. (d) Rohde, C.; Clark, T.; Kaufmann, E.; Schleyer, P. v. R. *J. Chem. Soc. Chem. Commmun.* **1982**, 882. (e) Vincent, M. A.; Schaefer III, H. F. *J. Chem. Phys.* **1982**, *77*, 6103. (f) Luke, B. T.; Pople, J. A.; Schleyer, P. v. R.; Clark, T. *Chem. Phys. Lett.* **1983**, *102*, 148. (g) Mareda, J.; Rondan, N. G.; Houk, K. N.; Clark, T.; Schleyer, P. v. R. *J. Am. Chem. Soc.* **1983**, *105*, 6997. (h) Wang, B.; Deng, C. *Chem. Phys. Lett.* **1988**, *147*, 99. (i) Wang, B.; Deng, C.; Xu, L.; Tao, F. *Chem. Phys. Lett.* **1989**, *161*, 388.
8. Schleyer, P. v. R.; Clark, T.; Kos, A. J.; Spitznagel, G. W.; Rohde, C.; Arad, D.; Houk, K. N.; Rondan, N. G. *J. Am. Chem. Soc.* **1984**, *106*, 6467, cit. lit.
9. Denmark, S. E.; Edwards, J. P.; Wilson, S. R. *J. Am. Chem. Soc.* **1991**, *113*, 723.

10. We do not discuss here (a) α-lithiated isocyanides: for a short review and crystal structure of lithiodiphenylmethylisocyanide·(-)-sparteine·2THF see Ledig, B.; Marsch, M.; Harms, K.; Boche, G. *Angew. Chem. Int. Ed. Engl.* **1992**, *31*, 79. (b) α-lithiated nitro compounds: for a short review and crystal structure of [α-nitrobenzyllithium·ethanol]$_\infty$ see Klebe, G.; Böhn, K. H.; Marsch, M.; Boche, G. *Angew. Chem. Int. Ed. Engl.* **1987**, *26*, 78; see also reference 3.

11. Review articles on dipole–chelate stabilization: (a) Beak, P.; Reitz, D. B. *Chem. Rev.* **1978**, *78*, 275. (b) Beak, P.; Zajdel, W. J.; Reitz, D. B. *ibid.* **1984**, *84*, 471.

12. A selection of more recent publications is listed below. Dipole–chelate stabilization by *amides*: (a) Huber, I. M. P.; Seebach, D. *Helv. Chim. Acta* **1987**, *70*, 1944. (b) Bartolotti, L. J.; Gawley, R. E. *J. Org. Chem.* **1989**, *54*, 2980. *Carbamates*: (c) Beak, P.; Lee, W.-K. *Tetrahedron Lett.* **1989**, *30*, 1197. (d) Gawley, R. E.; Rein, K.; Chemburkar, S. *J. Org. Chem.* **1989**, *54*, 3002. (e) Comins, D. L.; LaMunyon, D. H. *Tetrahedron Lett.* **1989**, *30*, 5053. (f) Pearson, W. H.; Lindbeck, A. C. *J. Org. Chem.* **1989**, *54*, 5651. (g) Beak, P.; Lee, W. K. *ibid.* **1990**, *55*, 2578. (h) Beak, P.; Kerrick, S. T. *J. Am. Chem. Soc.* **1991**, *113*, 9708. Chong, J. M.; Park, S. B. *J. Org. Chem.* **1992**, *57*, 2220. *Derivatives of urea*: (i) Pearson, W. H.; Lindbeck, A. C. *J. Am. Chem. Soc.* **1991**, *113*, 8546. *Formamidines*: (j) Meyers, A. I.; Dickman, D. A.; Bailey, T. R. *ibid.* **1985**, *107*, 7974. (k) Gawley, R. E.; Hart, G. C.; Goicoechea-Pappas, M.; Smith, A. L. *J. Org. Chem.* **1986**, *51*, 3076. (l) Rein, K.; Goicoechea-Pappas, M.; Anklekar, T. V.; Hart, G. C.; Smith, G. A.; Gawley, R. E. *J. Am. Chem. Soc.* **1989**, *111*, 2211. (m) Gawley, R. E.; Hart, G. C.; Bartolotti, L. J. *J. Org. Chem.* **1989**, *54*, 175. (n) Meyers, A. I.; Shawe, T. T. *ibid.* **1991**, *56*, 2751. (o) Meyers, A. I.; Gonzalez, M. A.; Struzka, V.; Akahane, A.; Guiles, J.; Warmus, J. S. *Tetrahedron Lett.* **1991**, *32*, 5501. (p) Meyers, A. I.; Guiles, J.; Warmus, J. S.; Gonzales, M. A. *ibid.* **1991**, *32*, 5505. (q) Meyers, A. I.; Warmus, J. S.; Gonzales, M. A.; Guiles, J.; Akahane, A. *ibid.* **1991**, *32*, 5509. *N-nitrosoamines*: (r) Seebach, D.; Renger, B.; Hügel, H.; Wykypiel, W. *Chem. Ber.* **1978**, *111*, 2630, cit. lit.

13. (a) Ahlbrecht, H.; Eichler, J. *Synthesis* **1974**, 672. (b) Martin, S. F.; DuPriest, M. T. *Tetrahedron Lett.* **1977**, 3925. (c) Evans, D. A.; Mitch, C. H.; Thomas, R. C.; Zimmerman, D. M, Robey, R. L. *J. Am. Chem. Soc.* **1980**, *102*, 5955.

14. (a) Oakes, F. T.; Sebastian, J. F. *J. Organomet. Chem.* **1978**, *159*, 363. (b) Enders, D.; Gerdes, P.; Kipphardt, H. *Angew. Chem. Int. Ed. Engl.* **1990**, *29*, 179. (c) Ahlbrecht, H.; Harbach, J.; Hauck, T.; Kalinowski, H.-O. *Chem. Ber.* **1992**, *125*, 1753.

15. (a) Smith, R. E.; Morris, G. F.; Hauser, C. R. *J. Org. Chem.* **1968**, *33*, 2562. (b) Touzin, A. M. *Tetrahedron Lett.* **1975**, 1477. (c) Craig, J. C.; Ekwuribe, N. N. *ibid.* **1980,** 2587. (d) Häner, R.; Olano, B.; Seebach, D. *Helv. Chim. Acta* **1987**, *70*, 1676.

16. Kessar, S. V.; Singh, P.; Singh, K. N.; Dutt, M. *J. Chem. Soc., Chem. Commun.* **1991**, 570, cit. lit.

17. (a) Peterson, D. J.; Ward, J. F. *J. Organomet. Chem.* **1974**, *66*, 209. (b) Ahlbrecht, H.; Dollinger, H. *Tetrahedron Lett.* **1984**, *25*, 1353. (c) Elissondo, B.; Verlhac, J.-B.; Quintard, J.-P.; Pereyre, M. *J. Organomet. Chem.* **1988**, *339*, 267. (d) Murakami, M.; Hayashi, M.; Ito, Y. *J. Org. Chem.* **1992**, *57*, 793 (a samarium compounds).

18. (a) Shirley, D. A.; Roussel, P. A. *J. Am. Chem. Soc.* **1953**, *75*, 375. (b) Müller, H. Dissertation, University of Heidelberg, 1964, citated in Hoffmann, R. W. *Dehydrobenzene and Cycloalkynes*; Verlag Chemie: Weinheim, Academic Press: New York, 1967. (c) Gjoes, N.; Gronowitz, S. *Acta Chem. Skand.* **1971**, *25*, 2596. (d) Chadwick, D. J.; Willbe, C. *J. Chem. Soc. Perkin Trans. 1*, **1977**, 887. (e) Levy, A. B. *Tetrahedron Lett.* **1979**, 4021, cit. lit. (f) Caixach, J.; Capell, R.; Galvez, C.; Gonzalez, A.; Roca, N. *J. Heterocycl. Chem.* **1979**, *16*, 1631. (g) Minato, A.; Tamao, K.; Hayashi, T.; Suzuki, K.; Kumada, M. *Tetrahedron Lett.* **1981**, *22*, 5319. (h) Brittain, J. M.; Jones, R. A.; Arques, J. S.; Saliente, T. A. *Synth. Commun.* **1982**, *12*, 231. (i) Fraser, R. R.; Mansour, T. S.; Savard, S. *Can. J. Chem.* **1985**, *63*, 3505.
19. Clark, T.; Schleyer, P. v. R.; Houk, K. N.; Rondan, N. G. *J. Chem. Soc. Chem. Commun.* **1981**, 579; see also reference 8.
20. Sette, F.; Stör, J.; Hitchcock, A. P. *J. Chem. Phys.* **1984**, *81*, 4906.
21. (a) Boche, G.; Marsch, M.; Ahlbrecht, H.; Harbach, J. unpublished results. (b) NMR investigations of **3** in solution led to the detection of an equilibrium between an η^1- and an η^3-coordinated species, see reference 14c; the applied method did not allow the detection of a Li—N coordination.
22. Summary: Zarges, W.; Marsch, M.; Harms, K.; Koch, W.; Frenking, G.; Boche, G. *Chem. Ber.* **1991**, *124*, 543.
23. Cambridge Structural Data Base.
24. Brown, K. L.; Damm, L.; Dunitz, J. D.; Eschenmoser, A.; Hobi, R.; Kratky, C. *Helv. Chim. Acta* **1978**, *61*, 3108.
25. Boche, G., Monde, M.; Harbach, J.; Harms, K.; Ledig, B.; Scheibert, F.; Lohrenz, J. C. W.; Alhbrecht, H. Chem. Bcr. **1993**, *126*, 1887–1894.
26. Rondan, N. G.; Houk, K. N.; Beak, P.; Zajdel, W. J.; Chandrasekhar, J.; Schleyer, P. v. R. *J. Org. Chem.* **1981**, *46*, 4108.
27. See. Ref. 25.
28. Seebach, D.; Hansen, J.; Seiler, P.; Gromek, J. M. *J. Organomet. Chem.* **1985**, *285*, 1.
29. Bauer, W.; Müller, G.; Pi, R.; Schleyer, P. v. R. *Angew. Chem. Int. Ed. Engl.* **1986**, *25*, 1103.
30. Boche, G.; Marsch, M.; Enders, D. unpublished results; see also reference. 3.
31. Ahlbrecht, H.; Boche, G.; Harms, K.; Marsch, M.; Sommer, H. *Chem. Ber.* **1990**, *123*, 1853.
32. See reference 20.
33. Other [8,34] and our[35] calculations of the anion $^-CH_2OH$ also predict a longer C—O bond (150.5 and 149.4 pm[35]).
34. (a) Spitznagel, G. W.; Clark, T.; Chandrasekhar, J.; Schleyer, P. v. R. *J. Comput. Chem.* **1982**, *3*, 363. (b) Bernardi, F.; Bottoni, A.; Venturini, A.; Mangini, A. *J. Am. Chem. Soc.* **1986**, *108*, 8171.
35. (a) Boche, G.; Haller, F.; Harms, K.; Hoppe, D.; Koch, W.; Lohrenz, J.; Marsch, M.; Opel, A.; Thümmler, C.; Zschage, O. In *Proceedings of the Fifth International Kyoto Conference on New Aspects of Organic Chemistry, Nov. 11–15, 1991*; Yoshida, Z.; Ohshiro, Y. Eds.; VCH Verlagsgesellschaft: Weinheim; Kodanha Ltd.: Tokyo, 1992, 159–179. (b) Boche, G.; Opel, A.; Marsch, M.; Harms, K.;

Haller, F.; Lohrenz, J.; Thümmler, C.; Koch, W.; Boche, G. *Chem. Ber.* **1992**, *125*, 2265–2273. With regard to the crystal structure of **14**·3THF described also in this publication, it is interesting to add that $F_2C{-}Cl^-$ has been investigated recently in the gas phase by Paulino, J. A.; Squires, R. R. *J. Am. Chem. Soc.* **1991**, *113*, 1845. The C—Cl bond is very weak and according to calculations very long (309 pm). Thus this compound is called a carbene anion complex. Quite similarly, **14**·3THF can be understood as a carbene $-OSiMe_3$ (Li^+·3THF) complex; **14**·3THF is also an "oxygen ylid."

36. Boche, G.; Bosold, F.; Zulauf, P.; Marsch, M.; Harms, K.; Lohrenz, J. *Angew. Chem. Int. Ed. Engl.* **1991**, *30*, 1455. It is interesting to mention that in the structure of the (benzyloxymethyl)zirconocene chloride shown below, the metal also bridges the C—O bond, which is 145.5(8) pm long; Buchwald, S. L.; Nielsen, R. B.; Dewar, J. C. *Organometallics* **1989**, *8*, 1593.

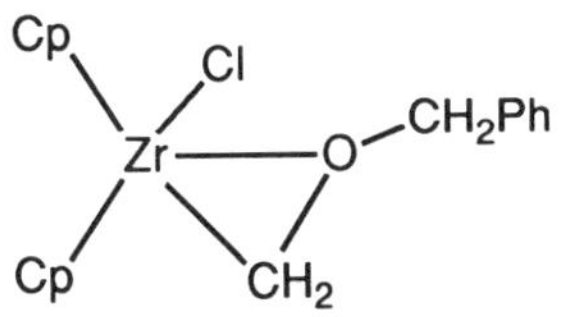

37. Harder, S.; Boersma, J.; Brandsma, L.; Kanters, J. A.; Bauer, W.; Pi, R.; Schleyer, P. v. R.; Schöllhorn, H.; Thewalt, U. *Organometallics* **1989**, *8*, 1688; see also Harder, S. Dissertation, University of Utrecht, 1990.

38. Marsch, M.; Harms, K.; Zschage, O.; Hoppe, D.; Boche, G. *Angew. Chem. Int. Ed. Engl.* **1991**, *30*, 321.

39. As in the case of the C—I bond in the lithiated iodoindole [**5**·2THF]$_2$ (Fig. 7.6) in the lithiated bromobenzofuran [**11**·O*i*Pr$_2$]$_2$ (Fig. 7.7) a lengthening of the C2—Br1 bond (189.3 pm) and thus a facile elimination of LiBr is not observed.[36]

40. At the end of this chapter it might be interesting to mention that we also calculated the structure of the methanol dianion $^-CH_2O^-$ with Mg^{2+} as the gegenion (3-21G*). In the dimer Mg bridges C and O. As in the case of **10A** dimerization is very favorable. However, in contrast to **10A**—but again not unexpectedly—dimerization occurs along the Mg—O bonds. It was very satisfying to see that the recently published ytterbium(II)benzophenone–dianion complex is exactly of this structural type; see Hou, Z.; Yamazaki, H.; Kobayashi, K.; Fujiwara, Y.; Taniguchi, H. *J. Chem. Soc., Chem. Commun.* **1992**, 722.

41. (a) Lüttringhaus, A.; Sääf, G. v. *Angew. Chem.* **1938**, *51*, 915; see also (b) Lüttringhaus, A.; Wagner-v. Sääf, G.; Sucker, E.; Borth, G. *Liebigs Ann. Chem.* **1947**, *557*, 46.

42. Wittig, G.; Löhmann, L. *Liebigs. Ann. Chem.* **1942**, *550*, 260.

43. Schöllkopf, U.; Eisert, M. *Liebigs Ann. Chem.* **1963**, *664*, 76. The low yields of **20** and **21** are due to competing reactions, as are those outlined in Scheme 4 and the Wittig rearrangement of **17.**

44. Ziegler, K.; Gellert, H.-G. *Liebigs Ann. Chem.* **1950**, *567*, 185.

45. Hoberg, H. *Liebigs Ann. Chem.* **1962**, *656*, 1.

46. (a) Review: A. Maercker *Angew. Chem. Int. Ed. Engl.* **1987**, *26*, 972. (b) Maercker, A. *Liebigs Ann. Chem.* **1969**, *730*, 91. (c) Maercker, A.; Demuth, W.

Angew. Chem. Int. Ed. Engl. **1973,** *12*, 75. (d) Maercker, A.; Demuth, W. *Liebigs Ann. Chem.* **1977**, 1909.

47. Boche, G.; Bosold, F.; Lohrenz, J. C. W.; Opel, A.; Zulauf, P. *Chem. Ber.* **1993**, *126*, 1873–1885.
48. (a) Pattison, F. L. M.; Dear, R. E. A. *Can. J. Chem.* **1963**, *41*, 2600. (b) Stähle, M.; Hartmann, J.; Schlosser, M. *Helv. Chim. Acta* **1977**, *60*, 1730.
49. Hill, C. M.; Senter, G. W.; Haynes, L.; Hill, M. E. *J. Am. Chem. Soc.* **1954**, *76*, 4538.
50. (a) Review: Kocienski, P.; Barber, C. *Pure & Appl. Chem.* **1990**, *62*, 1933. (b) Kocienski, P.; Wadman, S. *J. Am. Chem. Soc.* **1989**, *111*, 2363.
51. Duraisamy, M.; Walborsky, H. M. *J. Am. Chem. Soc.* **1984**, *106*, 5035.
52. Köbrich wrote[5b]: "Wegen seines Kationen- und damit Lewis-Säure-Charakters tritt es (Li^+) mit den ungebundenen Elektronen des α-Halogens in Wechselwirkung und zieht so das Halogen vom Carbenoid-Kohlenstoff ab."
53. See ref. 47.
54. Nguyen, T.; Negishi, E. *Tetrahedron Lett.* **1991**, *32*, 5903.
55. A possible reason for the different results could be TMEDA which was present in our experiments. TMEDA could favor deprotonation reaction of *s*BuLi and *t*BuLi.
56. Cope, A. C.; Berchtold, G. A.; Peterson, P. E.; Sharman, S. H. *J. Am. Chem. Soc.* **1960**, *82*, 6370; see also Cope, A. C.; Brown, M.; Lee, H.-H. *ibid.* **1958**, *80*, 2855. Cope, A. C.; Lee, H.-H.; Petree, H. E. *ibid.* **1958**, *80*, 2849.
57. (a) Crandall, J. K.; Lin, L.-H. *J. Am. Chem. Soc.* **1967**, *67*, 4526. (b) Detailed review: Crandall, J. K.; Apparu, M. *Org. React.* **1983**, *29*, 345.
58. (a) Eisch, J. J.; Galle, J. E. *J. Am. Chem. Soc.* **1976**, *98*, 4646. (b) Eisch, J. J.; Galle, J. E. *ibid.* **1988**, *341*, 293. (c) Eisch, J. J.; Galle, J. E. *J. Org. Chem.* **1990**, *55*, 4835.
59. Molander, G. A.; Mautner, K. *J. Org. Chem.* **1989**, *54*, 4042.
60. Lohse, P.; Loner, H.; Acklin, P.; Sternfeld, F.; Pfaltz, A. *Tetrahedron Lett.* **1991**, *32*, 615.
61. Taniguchi, M.; Oshima, K.; Utimoto, K. *Tetrahedron Lett.* **1991**, *32*, 2783.
62. Soderquist, J. A.; Lopez, C. *Tetrahedron Lett.* **1991**, *32*, 6305.
63. Shiner, C. S.; Tsunoda, T.; Goodman, B. A.; Ingham, S.; Lee, S.; Vorndam, P. E. *J. Am. Chem. Soc.* **1989**, *111*, 1381.
64. Although to our knowledge[3] carbenoid reactivity of α-lithiated *thio*ethers has never been observed, it should be mentioned that Cohen et al.[65] reported such a reaction in the case of lithiated *dithio*acetals. Thus **47** leads to **48,** a reaction which is the *intra*molecular analogue of the transformation **43** → **46**. In a similar fashion, tri(phenylthio)methyllithium **49** behaves also as a carbenoid at ambient temperatures.[66]

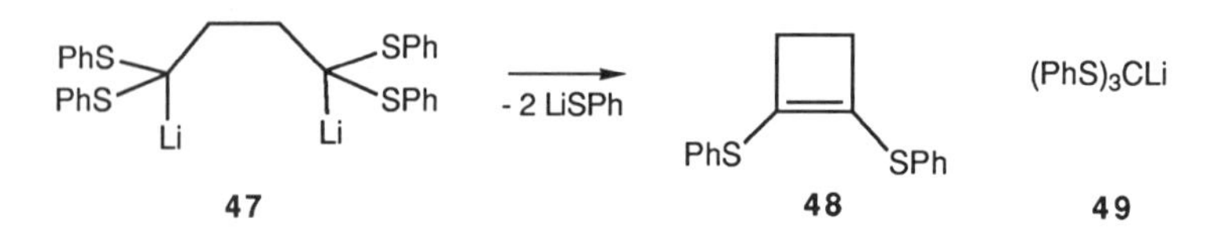

65. (a) Cohen, T.; Ouellette, D.; Pushpanandu, K.; Senaratne, A.; Yu, L.-C. *Tetrahedron Lett.* **1981**, *22*, 3377. (b) See also Ritter, R. H.; Cohen, T. *J. Am. Chem. Soc.* **1986**, *108*, 3718.

66. (a) Nitsche, M.; Seebach, D.; Beck, A. K. *Chem. Ber.* **1978**, *111*, 3644. (b) Wildschut, G. A.; Bos, H. J. T.; Brandsma, L.; Arens, J. F. *Monatsh. Chem.* **1967**, *98*, 1043. (c) Beak, P.; Worley, J. W.; *J. Am. Chem. Soc.* **1972**, *94*, 597. (d) Seebach, D. *Chem. Ber.* **1972**, *105*, 487.

67. Nayera, C.; Yus, M.; Hässig, R.; Seebach, D. *Helv. Chim. Acta* **1984**, *67*, 1100.

68. Bauer, W.; Winchester, W. R.; Schleyer, P. v. R. *Organometallics* **1987**, *6*, 2371.

69. Bauer, W.; Hampel, F. *J. Chem. Soc. Chem. Commun.* **1992**, 903.

70. Preliminary cryoscopic measurements of the aggregation of **57** in THF at −108°C lead to $n = 2.2–2.9$, which clearly tells that **57** is not a monomer. At the moment we have however no explanation for the experimental result, which could mean that higher aggregates than dimers are also present.

71. When we calculated (3-21G) a dimer structure of the Li/Hal model carbenoid $LiCH_2F$ we found a stable dimer structurally related to that of $[\mathbf{10A}]_2$, that is, one in which dimerization occurred along the C—Li bonds. There is, however, one important difference to $[\mathbf{10A}]_2$: the C—Li bond length of the bridging Li atom is rather long (253.7 pm)—the other C—Li bond amounts to 212.1 pm. This could mean that in Li/Hal carbenoids also dimer structures like the one of $[\mathbf{10A}]_2$ exist; however, one of the C—Li bonds is too long to lead to $^{13}C^{6}Li$ coupling. On solvation of each Li with one H_2O this C—Li bond becomes even longer (285.9 pm). In earlier theoretical investigations[7d] a $LiCH_2F$ dimer was found in which the dimerization occurred along Li—F.

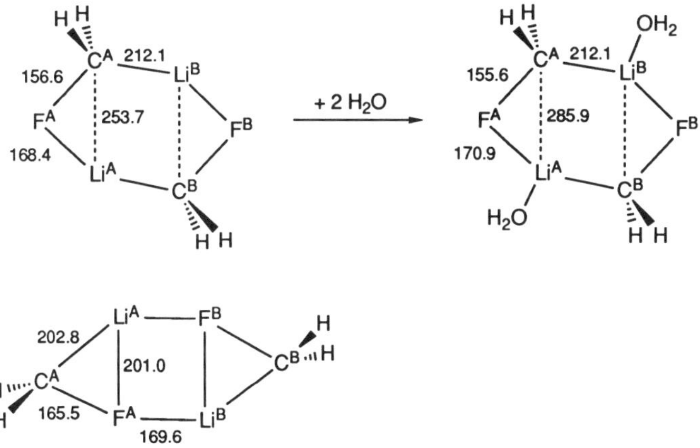

72. (a) Kutzelnigg, W. *Isr. J. Chem.* **1980**, *19*, 193. (b) Schindler, M.; Kutzelnigg, W. *J. Chem. Phys.* **1982**, *76*, 1919. (c) Kutzelnigg, W. *J. Mol. Struct.* **1989**, *202*, 11. (d) Review: Kutzelnigg, W.; Fleischer, U.; Schindler, M. *NMR, Basic Principles and Progress*; Springer: New York, 1990, Vol. 23, p. 165.

73. (a) Schindler, M. *J. Am. Chem. Soc.* **1987**, *109*, 1020. (b) See also Wiberg, K. B.; Hadad, C. M.; Sieber, S.; Schleyer, P. v. R. *J. Am. Chem. Soc.* **1992**, *114*, 5820, cit. lit.

74. (as) Bühl, M.; van Eikema Hommes, N. J. R.; Schleyer, P. v. R.; Fleischer, U.; Kutzelnigg, W. *J. Am. Chem. Soc.* **1991**, *113*, 2459. (b) Paquette, L. A.; Bauer, W.; Sirik, M. R.; Bühl, M.; Feigel, M.; Schleyer, P. v. R. *J. Am. Chem. Soc.* **1990**, *112*, 8776.

Note Added in Proof

Since the delivery of the manuscript we have been able to characterize by X-ray crystallography two "classical" carbenoids: 1-Chloro-2,2-bis(4-chlorophenyl)-1-lithioethene·TMEDA·2 THF, Boche, G.; Marsch, M.; Müller, A.; Harms, K. *Angew. Chem. Int. Ed. Engl.* **1993**, *32*, 1032, and 9-Bromo-9[(bromomagnesium)methylene]fluorene·4 THF, Boche, G.; Harms, K.; Marsch, M.; Müller, A. *J. Chem. Soc., Chem. Commun.* **1994**, 1393.

In the Li/Cl carbenoid the C—Cl bond is elongated by 13 pm, while in the MgBr/Br carbenoid the C—Br bond is 10 pm longer than in the comparable nonmetallated compound. These results agree nicely with those of the Li/oxygen carbenoids discussed above.

It is also worth to mention that in a Li/oxygen *nitrenoid* analogous structural changes have been found: Boche, G.; Boie, C.; Bosold, F.; Harms, K.; Marsch, M. *Angew. Chem. Int. Ed. Engl.* **1994**, *33*, 115.

Furthermore, the study of *carbenoids* and *nitrenoids* have led to the conclusion that compounds of the general type O(M)X (e.g., M = Li and X = OR) should be *oxenoids*. Quantum chemical calculations and reactions with RLi to give ROLi—the analogous reactions are typical for carbenoids and nitrenoids—clearly demonstrate the oxenoid nature of such compounds. The calculations also show nicely why carbenoids, nitrenoids, and oxenoids, although being "anions," are more *electrophilic*(!) than their nonlithiated analogs; Boche, G.; Bosold, F.; Lohrenz, J. C. W., *Angew. Chem. Int. Ed. Engl.* **1994**, *33*, 1161.

8

COMPLEXES OF INORGANIC LITHIUM SALTS

RONALD SNAITH AND DOMINIC S. WRIGHT

University Chemical Laboratory
Cambridge, United Kingdom

ABBREVIATIONS

THF	Tetrahydrofuran	C_4H_8O
HMPA	Hexamethylphosphoramide	$(Me_2N)_3P{=}O$
Pyr	Pyridine	C_5H_5N
TMEDA	Tetramethylethylenediamine	$Me_2N(CH_2)_2NMe_2$
TMPDA	Tetramethylpropylenediamine	$Me_2N(CH_2)_3NMe_2$
PMDETA	Pentamethyldiethylenetriamine	$(Me_2NCH_2CH_2)_2NMe$

Lithium Chemistry: A Theoretical and Experimental Overview, Edited by Anne-Marie Sapse and Paul von Ragué Schleyer.
ISBN 0-471-54930-4

1 INTRODUCTION: SCOPE AND LIMITS OF THE COVERAGE

The coordination chemistry of lithium and the other alkali metals was viewed until the 1960s as a rather barren area, made so since the singly charged and relatively large M^+ ions were held to have rather poor coordinating ability. The majority of known metal salt complexes were hydrates, although a few chelate complexes (e.g., with β-diketonate and salicylaldehyde ligands) were prepared by Sidgwick as long ago as 1925. Two developments have transformed this research area over the past 25 or so years. The first began with the Nobel prize-winning discovery by Pedersen of the macrocyclic polyether or "crown" ether ligands,[1,2] and continued with a second generation of related ligands such as the cryptands[3,4] and the lariat ethers.[5] Such ligands wrap around the metal ion, encapsulating it to greater or lesser degrees, and they are highly selective regarding which metal cation can be accommodated. They deaggregate the $(M^+X^-)_\infty$ lattice and often ultimately so, such that a single M^+ ion is found within the ligand and there remains only slight (if any) interaction with a single anion, X^-. Such features have led to important uses in cation separation, ion-specific measuring devices,[6] and phase-transfer catalysis,[7] and as mimics of naturally occurring macrocyclic antibiotics.[8] The second major development was brought about by extensive investigation of nonaqueous alkali metal chemistry, prompted largely by the burgeoning importance of lithium reagents in organic synthesis.[9-12] Excluding water, and providing instead acyclic Lewis base ligands [either neat (when liquid) or in limited and specified amounts within a noninteracting hydrocarbon solvent (e.g., hexane, toluene)] has revealed a rich structural chemistry of alkali metal complexes of the general type $(M^+X^- \cdot xL)_n$, where L is the neutral ligand and X^- is an organic (e.g., alkyl, aryl, imide, amide, alkoxide, enolate) or inorganic anion.[13-19]

This chapter concerns itself mainly with complexes of *lithium* salts having *inorganic* anions [e.g., halide (Cl^-, Br^-, I^-), thiocyanate (NCS^-), borohydride and related (BH_4^-, BF_4^-), nitrate (NO_3^-), chlorate (ClO_4^-), hexafluorophosphate (PF_6^-)] and *acyclic* neutral Lewis base ligands [e.g., monodentate Et_2O, THF, HMPA; bidentate TMEDA,; tridentate PMDETA (see Abbreviations)]. Other chapters deal with the syntheses, structures, and bonding of lithium species containing organic anions (e.g., Chapters 1, 5, 7, and 9) and one (Chapter 10) covers specifically complexes with multidentate cyclized ligands. Such complexes are mentioned here only when a significant lithium–anion interaction is retained. Similarly, complexes containing organic anions (amide especially) appear only when inorganic anions are present also ("mixed-anion" species). Finally, reference is made to sodium and potassium inorganic salt complexes only when there are interesting contrasts to be made with the lithium congeners.

The chapter is organized as follows. Section 2 presents general structural and bonding features pertaining to lithium salts with inorganic anions and to the complexes of such salts. Such features lead to several uses (potential or realized) of these complexes, and are discussed herein. The syntheses of complexes by direct or so-called in situ routes are described in Section 3, which includes discussion of the advantages and disadvantages of both types of route and possible mechanisms for the in situ route. Most of the chapter concerns the structures of complexes and the bonding within them, and consequently Section 4 is subdivided according to the type of anion present. Within each subsection results of solid-state (largely X-ray diffraction), calculational (largely ab initio MO methods) and solution (largely IR and NMR spectroscopy, and colligative measurements) studies are presented.

2 GENERAL STRUCTURAL AND BONDING FEATURES

Although still the subject of some controversy, it is now generally agreed that alkali metal–X bonds (X = halogens, C, N, O, etc.) are essentially ionic.[20–22] This being so, a useful concept envisages a gradual buildup of an infinite six-coordinate $(M^+X^-)_\infty$ lattice (as found, for example, for all the alkali metal halides except CsCl, CsBr, CsI) from an initial ion pair, M^+X^- ($M^+{=}Li^+$; Fig. 8.1*a*). Electrostatically, it is inevitable that these ion pairs will associate, so dissipating the electronic charges on each center. The most efficient way of doing this is by ring formation, each center then becoming two-coordinate. Dimers (especially) and trimers are favored (Fig. 8.1*b* and *c*, respectively). Angles at X^- are more acute and so larger rings, which would reduce repulsions between M^+ ions and between X^- ions, are not in general found owing to untenably large angles at M^+. What then happens to such rings depends on the precise nature of X^-. If the groups constituting X^- are relatively flat and for the most part coplanar with the ring, then ring stacking[18,19,23] can occur, for example, two dimers to give a cubane tetramer (Fig. 8.1*d*), two trimers to give a pseudo-octahedral hexamer (Fig. 8.1*e*), numerous rings to give a polymer (Fig. 8.1*f*). Anions X^- such as aryl, alkynyl ($RC{\equiv}C^-$), imide ($R_2C{=}N^-$), and enolate [$RC({=}CH_2)O^-$] are of this type. Amide anions (R_2N^-) cause very different behavior since their groups (R) project above and below $(R_2N^-M^+)_n$ rings, precluding stacking. Instead, depending largely on the precise stereochemical constraints posed by the R groups, the rings themselves are isolated or association occurs laterally, giving ladders (Fig. 8.1*g*).[18,19,24] Specific to this chapter, halide anions—being point charge centers and ones of high electron density—can both stack *and* ladder. The result (Fig. 8.1*h*) is the six-coordinate, NaCl-type lattice typical of most alkali metal halides. Many other inorganic anions are of course group anions (e.g., NCS^-, BH_4^-) and so the dimensionality of metal salts containing them is usually less than that of $(M^+Hal^-)_\infty$ lattices. Nonetheless, buildup of aggregated polymers from ion pairs can be envisaged in much the same way.

There are essentially just two methods available for the partial or full deaggregation of $(M^+X^-)_\infty$ lattices. The first involves energy input to break some or all of the $M^+ - X^-$ bonds and hence to produce, in the melt or vapor, fragments of the erstwhile three-dimensional infinite lattice. For example, ap-

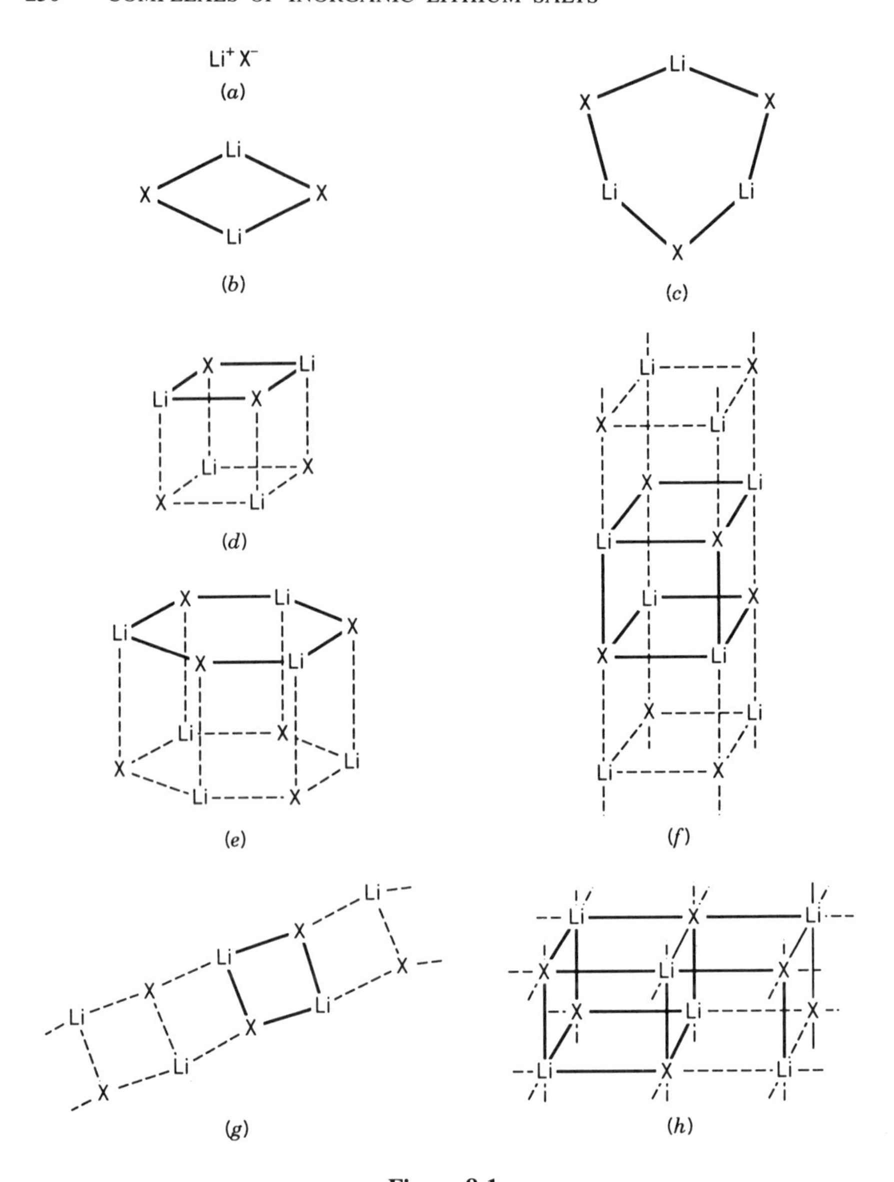

Figure 8.1

preciable quantities of $(M^+Hal^-)_2$ dimers (Fig. 8.1*b*) occur in the vapors of alkali metal halides.[25,26] In addition, larger clusters can be produced using pulsed lasers and then detected by mass spectrometry, for example, species such as $Na_{14}Cl_{13}^+$, $Cs_{14}I_{13}^+$, and $Na_{38}F_{37}^+$. Such clusters are sufficiently large that they are recognizable as having in essence the $(M^+Hal^-)_\infty$ crystalline structure (Fig. 8.1*h*), and hence they are termed nanometer-scale crystals or "nanocrystals."[27] The second method of producing fragments of $(M^+X^-)_\infty$ lattices—

and the method of most concern here—is by the introduction of Lewis base ligands (L). These add further steric constraints to the aggregation processes illustrated in Figure 8.1 so that extensive stacking and/or laddering is prevented. Complexed structures of limited aggregation [e.g., complexed cubanes $(M^+X^-.xL)_4$, complexed dimeric rings $(M^+X^-.xL)_2$, and complexed monomers $M^+X^-.xL$] or lower dimensionality polymers are then found (see Section 4).

The properties of these alkali metal inorganic salt complexes are often those associated with covalently bonded and discrete molecular species, for example, relatively low melting points and reasonable to good solubility in weakly polar organic solvents such as hexane, benzene, and toluene. This has led to statements (still quite widespread) that the alkali metals, and lithium especially, can show appreciable or even dominant covalent character. However, such statements are a misconception. These properties do not reflect bonding type, but rather the small and molecular nature of many $(M^+X^-.xL)_n$ complexes (e.g., n = 4, 2, 1) and the organic nature of their molecular peripheries. For example, the cubane complex $(LiCl.HMPA)_4$ (Fig. 8.2; see also Section 4.2) melts at around 140°C [cf., 614°C for $(LiCl)_\infty$] and it is fairly soluble in noninteracting organic solvents.[28] And yet the center of this structure must be highly ionic/polar, with each lithium being bonded to three chlorines and one oxygen. The *central* position of these ionic bonds is an important point since, taken as a whole, this structure is that of a small *molecule*, with weak intermolecular forces (hence the low melting point) and with an organic periphery provided largely by methyl groups of the HMPA ligands (hence the solubility in organic solvents). Such conceptually interesting ionic yet molecular complexes have been termed "supramolecules."[17]

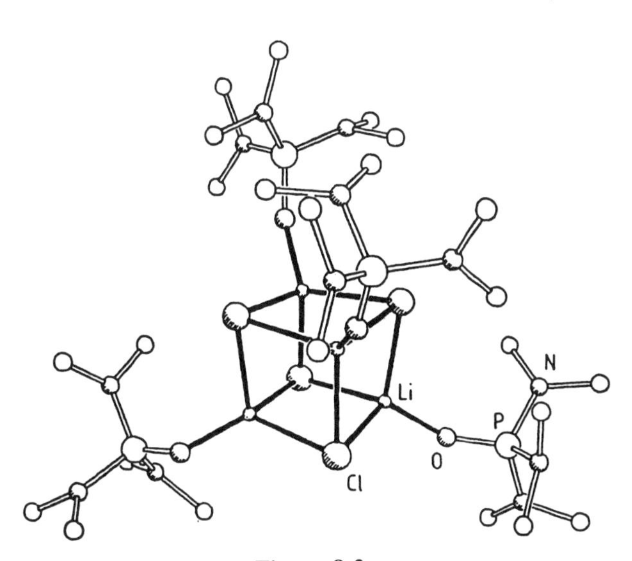

Figure 8.2

The properties discussed above have led to several realized or projected uses of these complexes:

1. *As Halogenating Agents.* Organically soluble halogenating agents would be of great value for converting hydrocarbons into halogenated products, themselves required for the manufacture of thermally stable polymers. Currently, suspensions of alkali metal halides [e.g., $(Cs^+F^-)_\infty$ in sulfolane or glymes[29-31]] are often used for this purpose, but there is little stoichiometric control of the reaction and a relatively low proportion of the halide is available for use, whether at the solid surface or by slight dissolution. For example, $(Li^+F^-)_\infty$ dissolves in neat HMPA at ~200°C (the lattice energy of this salt is ~1030 kcal mol^{-1}!), but the slightest cooling precipitates $(Li^+F^-)_\infty$ alone, and not a complex of it.[32] Preisolated solid complexes, weighed out for desired stoichiometric reactions and soluble in the organic reaction media employed (so providing potentially 100% of the halide ion taken) could have considerable advantages.

2. *As Low-Energy Electrolytic Sources of Metals.* Currently lithium metal, used to manufacture light yet tough alloys for the defense and aerospace industries, as a source of tritium for nuclear fusion processes, and as a nuclear reactor coolant, is obtained by electrolysis of molten salts, for example, eutectics such as LiCl + KCl at 450°C.[12] Complexes $(Li^+X^-.xL)_n$, with much lower melting points and isolated as highly pure crystalline materials, could offer economic advantages providing they have reasonable melt conductivities. Measurements on the series $(Li^+X^-.4HMPA)_n$ [$X^-=Br^-, I^-, BF_4^-$] as melts at ~140°C give specific conductivities of the order of $10^{-2}\ \Omega^{-1}\ cm^{-1}$, about one-tenth those of uncomplexed salts at ~400°C.[33]

3. *As Battery Materials.* There is currently intense interest in the application of lithium and sodium systems as components of batteries, particularly for off-peak electricity storage and in electric cars—the latter use will be prompted by environmental legislation. For example, in California alone by the year 2003 10% of all new cars (over 200,000) must be emission free.[34] Development of suitable batteries will be a major challenge. Those produced to date give limited range (having to be recharged frequently) and poor acceleration, and they are expensive because they require elaborate installation engineering. For example, Li/FeS_x batteries employ Li and FeS_x electrodes with a molten electrolyte of LiCl/KCl at 400°C.[12,35,36] It would be better to avoid molten systems which require elevated temperatures and pose severe containment problems. Hence the search is on for solid electrolytes with good ion conductivity. Materials of promise include metal salt complexes, the best so far being $(MeOH)_4Li^+.I^-$ which has a relatively high specific conductivity of $2.7 \times 10^{-3}\ \Omega^{-1}\ cm^{-1}$ due to highly mobile Li $(MeOH)_4^+$ complexed cations.[37]

4. *As Specific Reagents in Organic Syntheses.* Many alkali metal salts are employed as solution reagents in organic syntheses and are commonly used in polar solvents, especially ethers. Thus it is likely that salt *complexes* are the actual reagents which operate. An important point is that the mechanisms of

reactions and the selectivities (both regio- and stereo-) of reagents will only be understood when the precise identities and structures of these complexes $(M^+X^-.xL)_n$ are known. Such knowledge should in turn permit tailoring of reagent complexes to give different reactivities and selectivities. Complexed reagents with organic anions (alkyl, amide, enolate, etc.) are covered elsewhere (see comments in Section 1). Those with inorganic anions include, for example, $LiBH_4$ complexes whose structures (and hence presumably reactivities and selectivities as reducing agents) vary according to which Lewis base ligand is present (see Section 4.4). "Mixed-anion" (organic and inorganic) reagents are also relevant here since the presence of, for example, alkali metal halide units can affect reactivity, as demonstrated in lithium amide chemistry. The stereoselectivity of enolization is known to be affected dramatically by added inorganic salts,[38] and the production of mixed aggregates, such as those between LDA ($^iPr_2NLi$) and LiCl,[39] may be responsible (see Section 4.5).

3 SYNTHETIC METHODS

3.1 Introduction

As noted in Section 2, the introduction of a Lewis base ligand (L) will often inhibit the usual aggregation of Li^+X^- ion pairs to $(Li^+X^-)_\infty$ polymeric lattices. Conceptually, and indeed practically, there are two general ways of effecting such an introduction:

1. The **direct** method, whereby extant solid $(Li^+X^-)_\infty$ salts are treated with the Lewis base, either neat (if liquid) and so in vast excess, or in solution in an organic solvent. The strategy is to attempt to break down highly aggregated species such as those in Figure 8.1*f–h* (see Section 2).
2. The **in situ** method or methods, whereby a reaction is performed (usually at an initially low temperature) to generate Li^+X^-, presumably first as this simple ion pair. However, the reaction mix contains stoichiometric amounts of the ligand L, the strategy being to intercept usual lattice growth and "capture" the ion pair or a limited aggregate (e.g., Fig. 8.1*a–c*).

This section gives examples of complexes prepared by both methods (the structures of complexes are given in Section 4, and there each complex is given a number). The merits and drawbacks of each of the two general methods are discussed. It can be said that the in situ routes (and, especially, the so-called **ammonium salt route**) are superior on several counts. However, neither route has succeeded in providing a lithium fluoride complex, $(Li^+F^-.xL)_n$, and the thermodynamic reasons for this failure are illustrated with the aid of ab initio MO calculational results. Finally, other calculations and some ESR spectroscopic findings are presented to illustrate why and how the ammonium salt route is usually so effective.

3.2 The Direct Route

The direct method involves the seemingly simple dissolution, often with the application of heat, of a salt $(M^+X^-)_\infty$ in a donor (L) or donor–solvent mixture (Eq. 1).

$$^1/_\infty(M^+X^-)_\infty + xL \rightarrow {}^1/_n(M^+X^-.xL)_n \tag{1}$$

Its most widespread application to alkali metal coordination chemistry has been reported in a series of papers entitled "Lewis-Base Adducts of Main Group 1 Metal Compounds."[40–53] For the most part, lithium halides $(Li^+X^-)_\infty$, with $X^- = Cl^-$, Br^-, and I^-, have been treated with N-donor ligands such as pyridine (Pyr)[52] and substituted pyridines (2-Me;[42] 4-Bu^t;[43] 2,6-[45,46] and 3,5-Me$_2$;[42] 2,4,6-Me$_3$[52]), quinoline,[43,44,52] 2,2′-bipyridine,[51] acetonitrile,[49] and TMEDA.[48] Variations have included changes of the metal to sodium [as in the direct synthesis of $(NaI.PMDETA)_2$[50]], of the anion to nitrate {as in the monomer $LiNO_3.(3,5\text{-}Me_2Pyr)_3$ and the polymer $[LiNO_3.(2\text{-}MePyr)_2]_\infty$[47]}, and of the Lewis base to an O-centered one {as in $[LiCl.(sulfolane)]_\infty$[53]}.

Although crystalline complexes can be isolated and interesting structures solved (see especially Section 4.2 for lithium halide complexes), the direct route does have several drawbacks and inadequacies. The first is the failure of many salts to dissolve fully or partially even in neat liquid donor ligands. In part this is a thermodynamic problem, the lattice energy of the preformed salt presumably exceeding any attainable complexation energy. However, there are also kinetic difficulties associated with a two-phase reaction system of an infinitely latticed solid suspended in a liquid (neat L or L plus solvent). Most of the studies noted above[40–53] employ neat liquid ligands in vast molar excess over solid salts; even then, heating (often to reflux) is required to cause even partial dissolution. The second drawback is the difficulty of excluding water. Even if inert-atmosphere (dry N_2 or Ar) Schlenk techniques are followed during preparation of complexes, H_2O can be present in the system either in the solvent and/or liquid ligand L, or in the supposedly anhydrous $(M^+X^-)_\infty$ salt. Attempts can be made to circumvent this problem by rigorous drying of liquids and by dehydration of salts by heating in vacuo, but these are not always successful; witness the isolation of $[(Quin)_2Li(H_2O)_2]^+.Br^-.2Quin$[44] from heating $(Li^+Br^-)_\infty$ in quinoline, and of $[(H_2O)_2Li.(H_2O)_2.Li(H_2O)_2]^{2+}.2Cl^-.(2,6\text{-}Me_2Pyr)_6$[45] by similar treatment of $(Li^+Cl^-)_\infty$ with the pyridine. An especially worrying feature of H_2O inclusion in the isolated complex is that of inconsistency: the stoichiometric amount of H_2O present (from L and/or the solvent and/or the salt) may well vary over successive syntheses. Illustrating this is the case of the LiCl/HMPA system. The anhydrous cubane $(LiCl.HMPA)_4$ [Section 2, Fig. 8.2; complex **(13)** in Section 4.2] can be prepared easily by in situ methods, but extensive refluxing of "anhydrous" $(Li^+Cl^-)_\infty$ in HMPA/toluene mixtures causes only slight dissolution.[28] Cooling of solutions obtained affords a series of aqua complexes having variable amounts of H_2O, the only well-defined complex being $[Li(H_2O)_2(HMPA)_2]^{2+}.2Cl^-$.[54]

One final problem with the direct method concerns the possible number of products and their yields. Conceptually, a $(M^+X^-)_\infty$ lattice might be fragmented in many ways and treatment with a vast excess of ligand L seems particularly likely to encourage a whole range of such fragmentation processes. Thus it seems probable that solutions obtained may contain a series of complexes, that isolated being simply the most thermodynamically stable one as a solid under the crystallization conditions employed. This is always the case, of course, but it is an especially worrying feature of direct routes since yields of isolated solid complexes are usually low.

3.3 Early In Situ Routes

Early in situ routes were not recognized as such, at least not in terms of the concept of generating M^+X^- ion pairs in the presence of a Lewis base ligand which might prevent their usual lattice growth. Indeed, the first report (in 1983) of a reaction of this type, between Bu^nBr and Bu^nLi in hexane containing PMDETA and giving $(LiBr.PMDETA)_2$ (Eq. 2),[40] noted that the main advantage was the increased reactivity of Bu^nLi in the presence of PMDETA.

$$^{n}BuLi + {}^{n}BuBr \xrightarrow{PMDETA} {}^{1}/_{2}(LiBr.PMDETA)_2 + Bu{-}Bu \qquad (2)$$

The in situ label for a reaction of this type was first coined in 1984 with a report of the ''utterly fortuitous in situ preparation'' of $(LiCl.HMPA)_4$ [Section 2, Fig. 8.2; complex **(13)** in Section 4.2] from a 3 : 1 reaction of the iminolithium $Bu^t_2C{=}NLi$ and HMPA in toluene with $AlCl_3$ in Et_2O.[28] The intention had been to prepare the tris(imino)alane, $(Bu^t_2C{=}N)_3Al$—in retrospect, this is probably a liquid— but the only isolable crystalline product turned out to be the lithium chloride complex. This finding prompted the thoughts, outlined at the beginning of this section and in Section 2, that in situ routes might be valuable for syntheses of metal salt complexes, $(M^+X^-.xL)_n$. In general, until the development of the ammonium salt route (discussed below), the metal (M^+) has been lithium provided by use of Bu^nLi solution and the anion (X^-) has been halide (Cl^-, Br^-, I^-), provided by an organic or silyl halide. For example, complexes LiI.PMDETA and $(LiCl)_4$.3PMDETA were prepared from Bu^nLi/PMDETA with MeI and Me_3SiCl respectively,[41] and $(LiX.TMEDA)_2$, X = Br and I, and $(LiCl)_6$.4TMEDA were prepared from Bu^nLi.TMEDA with Bu^nBr, MeI, and Bu^tCl respectively.[48] Attempts to synthesize a lithium fluoride complex by reacting Bu^nLi with $BF_3.Et_2O$ in HMPA failed (Eq. 3); it seems that a Li^+F^- unit or a small oligomer $(Li^+F^-)_n$ is indeed generated, but it adds BF_3 as well as HMPA to give the final lithium tetrafluoroborate complex.[32]

$$Bu^nLi + 2BF_3.Et_2O \xrightarrow{4HMPA} LiBF_4.4HMPA + Bu^nBF_2.Et_2O \qquad (3)$$

These in situ routes employing organolithiums as metal sources and organic or main group metal halides as anion sources do have several advantages over direct methods, but they retain disadvantages also. One advantageous feature is that they may avoid the thermodynamic and kinetic problems associated with attempted breakdown of an extant $(M^+X^-)_\infty$ lattice. At the onset of these reactions, ion pairs M^+X^- must be formed even if only transiently, and they are likely to be soluble, that is, a one-phase system is in operation. Of course, if aggregation/lattice energies far exceed complexation energies, the end result will be precipitation of $(M^+X^-)_\infty$ or, at best, a lattice containing included ligand molecules. A second favorable feature is that in situ routes may permit better control of product stoichiometries and of the number of products. As noted above, direct syntheses usually employ a vast excess of the neat donor L and several products may well be present in solutions obtained since $(M^+X^-)_\infty$ lattices have numerous conceivable fragmentation pathways; it is noticeable that solid complexes isolated are usually low in yield. In contrast, these in situ routes operate in solvents containing specific and stoichiometric amounts (which can be varied) of L. Under these conditions, buildup of M^+X^- ion pairs may have a limited number of preferred pathways, leading to fewer final solution products and higher yields of specific solid complexes. The final advantage of in situ methods is that exclusion of H_2O is rather easier. In direct methods any H_2O in the final complex probably originates mostly from partial hydration of $(M^+X^-)_\infty$ precursors. Here, small M^+X^- systems are *assembled*, and so the problem is circumvented. However, careful exclusion of moisture is still necessary: the metal sources (usually organolithium reagents) are very air sensitive, organic halides especially are hygroscopic as are most solvents and liquid donors. The major real and residual disadvantages of these early in situ reactions concern cost, efficiency, and applicability. Their hygroscopic natures apart, organic halides, main group element halides, and so on are quite expensive. Organolithium sources are also inefficient since they are produced using two equivalents of the metal (e.g., $Bu^nBr + 2Li \rightarrow Bu^nLi + LiBr$), that is, at best, any metal salt complex synthesis has an overall metal-based yield of 50%. Finally, halides (and perhaps BF_4^- via BF_3, Eq. 3) are the only obvious anion precursors available. These drawbacks, particularly inauspicious if potential uses of alkali metal inorganic salt complexes are to be realized (see Section 2), prompted development of the in situ ammonium salt route.

3.4 The Ammonium Salt Route

It was discovered in 1986 and reported in 1987[55] that ammonium salts $NH_4^+X^-$ suspended in toluene containing stoichiometric amounts of a donor ligand L will react with Bu^nLi (hexane solution) (Eq. 4).

$$Bu^nLi + NH_4^+X^-(s) \xrightarrow[0\text{–}60°C]{xL} {}^1/_n(Li^+X^-.xL)_n + NH_3 + BuH \qquad (4)$$

Initially, the route was applied only to lithium halide[55,56] complexes ($X^- = Cl^-, Br^-, I^-$) with HMPA (L) ligands, but it was shown subsequently that a range of $NH_4^+X^-$, $(NH_4^+)_2.X^{2-}$ salts (e.g., $X^- = NCS^-$, BF_4^-, PF_6^-, NO_3^-; $X^{2-} = CO_3^{2-}$) are applicable also, as are a variety of O- and N- Lewis bases (e.g., THF, Dioxan, Diglyme; TMEDA, TMPDA; PMDETA).[33,57–60] The variety of anions available overcomes one disadvantage of earlier in situ methods noted above. Ammonium salts are also cheap (cf. organic and main group element halides) and they can be used as supplied since most such salts are not hygroscopic; in any case, rapid purification can be achieved by sublimation under vacuum. Finally, early exploration of the ammonium salt route revealed that alternative complexes are available simply by variation of the amount of ligand L available (e.g., Eq. 5). Such alternatives are not possible using direct routes (e.g., 6).

$$Bu^nLi + NH_4^+I^-(s) \xrightarrow[\text{toluene}]{\text{xHMPA}} {}^1/_n(Li^+I^-.xHMPA)_n \qquad (n = 2 \text{ or } 4) \qquad (5)$$

$$Li^+I^-(s) + 4HMPA \xrightarrow{\text{toluene}} {}^1/_n(Li^+I^-.4HMPA)_n, \text{ only} \qquad (6)$$

Why and how these ammonium salt routes work so well are discussed in detail later in this section, but several experimental observations can be noted now. First, the best solvent system appears to be toluene and the presence of hexane (added with the commercially supplied Bu^nLi) slows the reaction, particularly when halides are involved. For this reason, syntheses of lithium halide complexes were effected best by preparing a chilled solution of Bu^nLi/L in toluene and then adding solid $NH_4^+X^-$ ($X^- = Cl^-$, Br^-, I^-). Under such conditions, brief warming of the mixture led to vigorous reaction and evolution of ammonia and butane. During this process the ammonium salt "dissolves" gradually to leave a (usually) colorless solution, the cooling of which (in some cases with prior addition of hexane) affords crystals of the desired complex, usually in good yield (60–80% is typical). In contrast, thiocyanate complexes could readily be prepared in hexane/toluene media by adding the Bu^nLi solution to a suspension of the ammonium salt in toluene and the ligand L. A final experimental point is that these reactions proceed very frequently via elaborate color changes. For example, the NH_4Br/Bu^nLi/HMPA mix in toluene changes from red → violet → sepia → and finally to colorless; the colorless complex $(LiBr)_2.3HMPA$ is isolated in 60% yield.[56] Similarly, NH_4NCS/Bu^nLi/TMEDA in hexane/toluene becomes bright yellow at room temperature as the salt reacts, a colorless solution being obtained at the end of the reaction; the colorless complex $(LiNCS.TMEDA)_\infty$ is isolated in 85% yield.[58] Taken together, these combined observations involving Bu^nLi as the metal source shed rather little light on the mechanisms of these reactions. The polarity of the medium may be important. Reactions may occur by a purely surface process which might involve radicals of some sort, or they may be propagated by small amounts of the ammonium salts being continuously dissolved into solution as the reactions proceed. However, prior to the onset of the reactions (by addition of Bu^nLi and/or by allowing the chilled mixture to warm), there is no apparent dissolution of the NH_4X solid, that is, at least initially, the NH_4X lattice does not appear to give a "molecular" and soluble $NH_4X.yL$ complex of some sort which then reacts *in solution* with the organometallic.

Despite the success of the ammonium salt route with Bu^nLi as the metal source, the route still has disadvantages. Foremost is that it represents an inefficient use of the metal since each equivalent of Bu^nLi requires two equivalents of Li for its preparation. In addition, extension of the route to other metals would require in-house syntheses of their organometallics and many of these (e.g., Bu^nNa, Bu^nK[9,11,61]) are exceedingly air-sensitive species. For these reasons the route was developed to allow the use of first the solid metal hydrides (which are manufactured quantitatively by the reaction $2M + H_2 \rightarrow 2MH$) and then, ultimately, the solid metals themselves (Eqs. 7 and 8, respectively).

$$MH_{(s)} + NH_4^+X^-_{(s)} \xrightarrow[\text{toluene}]{\text{xL}} {}^1/_n(M^+X^-.xL)_n + NH_3 + H_2 \quad (7)$$

$$M_{(s)} + NH_4^+X^-_{(s)} \xrightarrow[\text{toluene}]{\text{xL}} {}^1/_n(M^+X^-.xL)_n + NH_3 + {}^1/_2H_2 \quad (8)$$

Use of solid lithium hydride as opposed to Bu^nLi solution slows reactions considerably, in part because of the noted low reactivity of commercially supplied LiH.[62] For example, $(LiCl.HMPA)_4$ could be obtained in only 50% yield after 4 days of refluxing LiH and NH_4Cl in toluene/HMPA; nonetheless, this represents an improved yield in terms of lithium consumption over the 83% obtained (albeit after only a few minutes reflux) using Bu^nLi. More significantly, the use of heavier alkali metal hydrides which are more reactive (e.g., NaH, KH) allowed efficient and facile syntheses of numerous complexes.[33,63,64] For example, addition of HMPA (2 equivalents) to a mixture of NH_4NCS and NaH solids (1 equivalent of each) in toluene gave an immediate blue color followed by gas evolution and color changes through turquoise to green and finally to pale yellow. At this stage no solids remained and refrigeration of the solution gave $[(NaNCS.HMPA)_2]_\infty$ and $(NaNCS.2HMPA)_2$ in yields of 30 and 57% respectively.

The use of metals themselves has proved even more valuable, especially in allowing syntheses of inorganic salt complexes of other than the alkali metals, for example, the alkaline earths[57,65] and transition metals and lanthanides.[57,66] In typical syntheses, the metals (e.g., Li, Na, K; Ca, Sr, Ba; La, Y, Eu; Mn) are taken with the appropriate number of equivalents of solid NH_4X salts (e.g., 1, 2, 3, 2–4, respectively) in toluene and the Lewis base (2–4 equivalents) is then added. The reaction can then be initiated and brought to completion by heating to reflux. Reaction times vary; for example, the K/NH_4NCS/HMPA system requires only a few minutes heating to give $(KNCS)_3.5HMPA$[64] in 90% yield, whereas a Sr/NH_4I/HMPA mixture affords $SrI_2.4HMPA$[65] in 85% yield after one hour's reflux, and the La/NH_4NCS/HMPA[66] reaction needs prior treatment with ultrasound and then refluxing for one day to give $La(NCS)_3.4HMPA$ in 50% yield. Notwithstanding such conditions, these are remarkable reactions, not the least in being between two solids in a liquid medium. They may proceed via electron reductions of NH_4^+ ions by M since colors are often observed (the alkali and some alkaline earth metals are well known to dissolve in *neat* ligands such as HMPA, affording blue solutions due to the presence of solvated electrons[67,68]). Whatever the mechanism, the advantages of this ammonium salt route are illustrated even more clearly for these heavier-metal systems than they were for alkali metal ones. Thus the smaller and more charged M^{2+} and M^{3+} ions impart high lattice energies, making direct dissolution of inorganic salts difficult if not impossible, and they also have high hydration energies, which makes the preparation of fully anhydrous salts an arduous business. In situ assembly of anhydrous salt units MX_2, MX_3 and then their capture by complexation avoids such difficulties. Overall, it can be stated with confidence that the ammonium salt route is the best available method for the synthesis of inorganic metal salt complexes.

Notwithstanding the comments made above, even the ammonium salt route has failed to generate a lithium fluoride complex (see also comments in Sections 2 and 4.2).[32] Addition of solid NH_4F to a red solution of Bu^nLi/HMPA in toluene followed by heating to reflux causes the gradual dissolution of NH_4F to give a pale yellow solution. On cooling, a microcrystalline solid is deposited, but this is essentially $(Li^+F^-)_\infty$ with some HMPA incorporated in the lattice. Clearly, even with such a good ligand as HMPA, the thermodynamics of lattice formation outweigh those of complexation of monomeric or even small aggregated $(Li^+F^-)_n$ units. To explore this point further, ab initio MO calculations (6-31G and 6-31G** basis sets[69,70]) have been carried out on LiF monomers and $(LiF)_2$ dimers and on some of their complexes, namely, $LiF.H_2O$, $(LiF.H_2O)_2$, $LiF.O{=}PH_3$, and $LiF.O{=}P(NH_2)_3$.[71] Favored optimized structures are illustrated in Figures 8.3 and 8.4. For the LiF monomer (Fig. 8.3*a*), with a Li—F distance of 1.573 Å (1.550 Å at 6-31G**) the charges show that the "molecule" is highly polarized; emphasizing the near-100% ionic nature of the bond, optimization with only a 1*s* shell on Li (so effectively Li^+) gives a near-identical distance of 1.572 Å. Dimerization is favored greatly (Fig. 8.3*b*), lowering the energy by 75.1 kcal mol^{-1} (73.3 kcal mol^{-1} at 6-31G**). The charges on Li and on F change little from those found in the monomer, and the dimer remains highly ionic. Not surprisingly, therefore, further association is undergone eagerly. Earlier work had shown that the energies of tetramerization and hexamerization of LiF to tetrahedral $(Li^+F^-)_4$ and pseudo-octahedral $(Li^+F^-)_6$ are about 190 kcal mol^{-1},[72] and 310 kcal mol^{-1},[73] respectively. The association energies *per monomer* thus *increase*, from approximately 37, to 47, to 52 kcal mol^{-1} for dimer, tetramer, and hexamer respectively. The key question remaining is whether complexation is a viable alternative way for each Li center to increase its coordination number, for example, LiF to $(LiF)_2$ or to $LiF.H_2O$, both with two-coordinate Li^+; LiF to $(LiF)_4$, $(LiF)_6$, or $(LiF.H_2O)_2$, all with three-coordinate Li^+. In fact, monohydration as seen in the optimized structure of $LiF.H_2O$ (Fig. 8.3*c*) is favorable only to the extent of 28.8 kcal mol^{-1}. Figure 8.3*d* shows the best

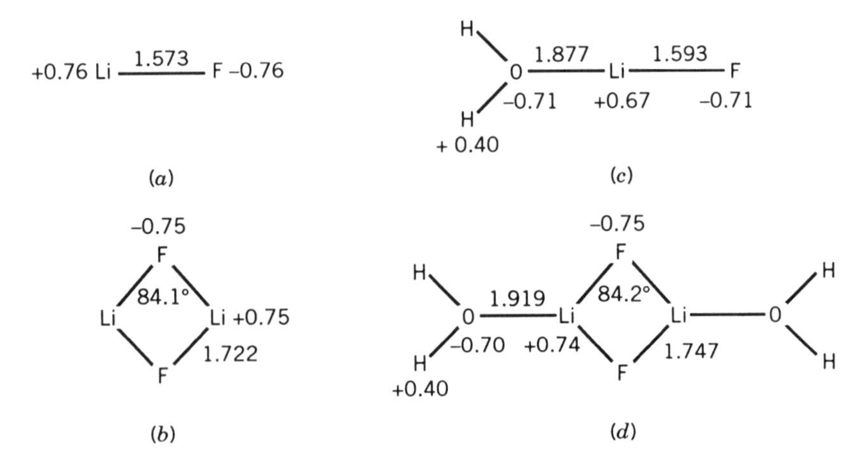

Figure 8.3

structure of $(LiF.H_2O)_2$, this all-planar form being 3.7 kcal mol^{-1} lower in energy than one in which the H_2O ligands lie perpendicular to the $(LiF)_2$ ring plane. The energy gain by *solvation* of $(LiF)_2$ with $2H_2O$ is 48.8 kcal mol^{-1} while that by *association* of $LiF.H_2O$ to $(LiF.H_2O)_2$ is 66.3 kcal mol^{-1}, and both values can be contrasted with the aforementioned tetramerization and hexamerization energies. It is clear that aqua complexation is not a viable competitor to association for this highly ionic Li^+F^- system.

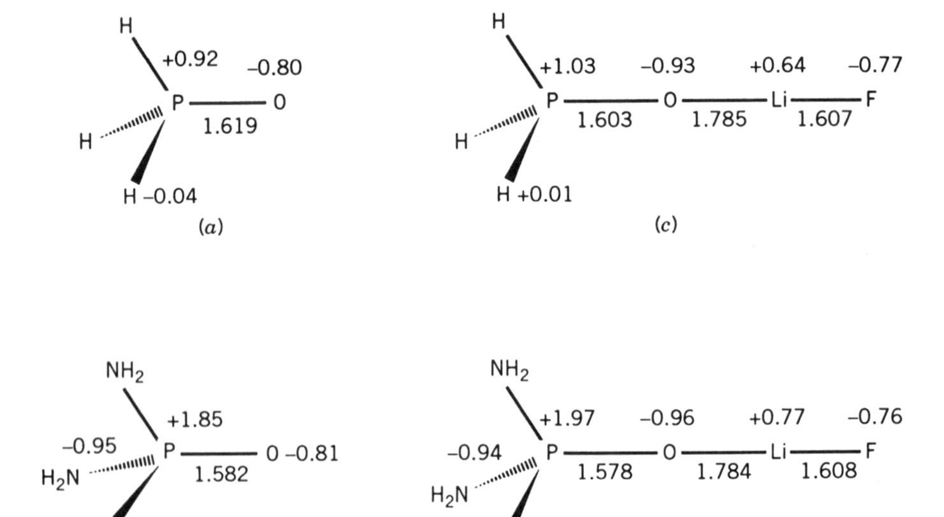

Figure 8.4

It remained to be seen if H_2O was a reasonable mimic for HMPA. The optimized structures of $H_3P{=}O$ and $(H_2N)_3P{=}O$ (Fig. 8.4*a* and *b*, respectively) reveal very similar charges on their O atoms, and both ligands should reflect the electronic behavior of HMPA quite closely. Both solvate Li^+ more strongly than does H_2O: for the optimized structures of $LiF.O{=}PH_3$ and $LiF.O{=}P(NH_2)_3$ (Fig. 8.4*c* and *d*, respectively) solvation energies are similar at 38.6 and 37.0 kcal mol^{-1} respectively (cf., 28.8 kcal mol^{-1} for H_2O solvation). These ligands are thus better than H_2O for solvation of Li^+ but they still do not come even close to matching the alternative for Li^+F^-, self-association to oligomers and thence to $(Li^+F^-)_\infty$. Failure to isolate a complex of lithium fluoride is perhaps not surprising.

Notwithstanding this failure to provide a complex of lithium fluoride, the ammonium salt route does work exceptionally well for a broad range of metals, most other anions, and Lewis base ligands. Other ab initio MO calculations (6-31G basis set[69,70]) have been used to probe the likely thermodynamics of these reactions.[71] As a model, the interactions of LiH and MeLi were examined separately with both NH_4^+ and NCS^- ions, before finally investigating the reactions of each reagent with $NH_4^+NCS^-$. Initially, the reaction of LiH with NH_4^+ was scrutinized by optimizing 11 potential intermediates involving contacts of the Li and/or H ends of LiH with N or H of NH_4^+. Only two viable structures resulted, and both involve contact of the H atom of LiH with NH_4^+; species of the type $(H{-}Li\cdot\cdot\cdot NH_4)^+$ exhibit high repulsion and energies greater than those of the separate components. Figure 8.5*a* shows a structure in which the hydridic LiH center is linked very loosely to the N atom by a long N—H "bond;" its length and bond index (in parentheses) contrast with values of 1.64 Å and 0.90 in NH_4^+ itself. The Li—H bond is only slightly elongated and weakened (cf. bond length 1.64 Å, bond index 0.96 in LiH itself). Although this potential intermediate is lower in energy by 31.0 kcal mol^{-1} than the separate components, it is less stable (by 8.6 kcal mol^{-1}) than the structure shown in Figure 8.5*b*. Here the NH_4^+ ion attacks the H end of LiH forming a linear $Li{-}H\cdot\cdot\cdot H{-}N$ backbone. Significantly, the possible elimination of H_2 is being set up here. The Li—H bond and the central N—H bond are longer and weaker than in separate LiH and NH_4^+, and an H—H bond is forming (length, 0.81 Å; cf. 0.74 Å in molecular H_2). At the extremities of this intermediate, the N—H bonds are almost single ones and the metal is essentially Li^+. Hence it is primed to give H_2 and NH_3, leaving Li^+, which would condense with the anion and the ligand present in the system, so affording an $Li^+X^-.xL$ product. For completeness, the reaction of LiH with the other component of the mix, NCS^-, was also examined. Of seven models (including attachment at the Li^+ end by N, by S, and by both, and including formation of $H\cdot\cdot\cdot S$ and $H\cdot\cdot\cdot N$ contacts), a linear form involving a Li—N interaction, $[S{\overset{...}{-}}C{\overset{...}{=}}N\cdot\cdot\cdot Li{-}H]^-$ was found to be by far the most stable (52.2 kcal mol^{-1} lower in energy than the separate entities). In essence, NCS^- solvates LiH and thereby weakens slightly the Li—H bond. This may assist in the processes implicated above for the $(Li\cdot\cdot\cdot H{-}H\cdot\cdot\cdot NH_3)^+$ intermediate

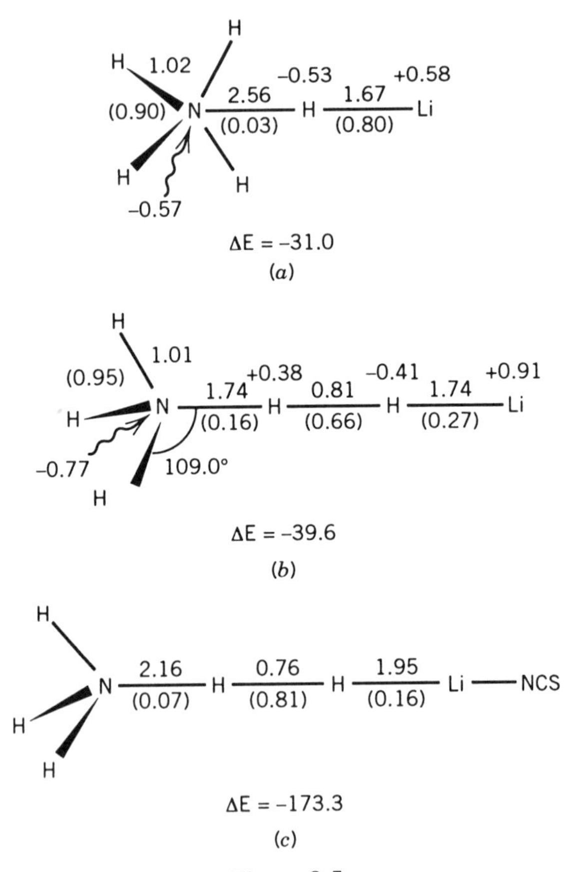

Figure 8.5

and so a species representing simultaneous attack on LiH by NH_4^+ *and* NCS^- was optimized (Fig. 8.5*c*). Compared to the species in Figure 8.5*b*, NCS^- coordination has weakened yet further the Li· · ·H and (central) N· · ·H contacts, but strengthened the H—H one. Indeed this latter system is a massive 173.3 kcal mol^{-1} lower in energy than separated LiH + NH_4^+ + NCS^- reactants.

A model structure for the reaction of NH_4^+ with MeLi was also optimized (Fig. 8.6*a*). It is analogous to that shown for LiH·NH_4^+ in Figure 8.5*b*, and has a linear N—H· · ·C—Li backbone. Association of MeLi with NH_4^+ causes an energy reduction of 58.5 kcal mol^{-1} (cf. 39.6 kcal mol^{-1} for LiH.NH_4^+). The intermediate is primed to eliminate CH_4, with both the involved N—H bond and the C—Li bond being weakened severely (cf. C—Li lengths of 2.01 Å and indices of 0.82 in MeLi) and with the central C—H bond length (1.10 Å) being very similar to that found in CH_4 (1.09 Å). The CH_3 group has also been inverted by electrophilic attack of NH_4^+ on the opposite side of the C—Li bond; thus the reaction is the electrophilic analogue

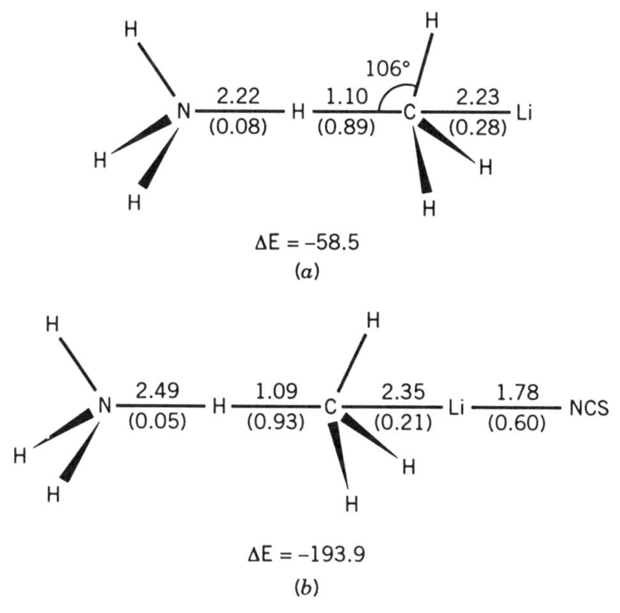

Figure 8.6

of S_N2-type reactions. The possibility of anion involvement was considered also. The most stable structure of NCS^- with MeLi is linear, ($S{=}C{\equiv}N\cdot\cdot\cdot Li{-}CH_3$), and is 49.4 kcal mol^{-1} lower in energy than the separated components. Such solvation of Li^+ by NCS^- may again assist the overall reaction by weakening Li—C bonds (e.g., the C—Li bond length is 2.09 Å in this structure; cf. 2.01 Å in MeLi). Finally, a composite structure for MeLi + NH_4NCS was calculated (Fig. 8.6*b*). As in the system LiH + NH_4NCS, the backbone is linear and it shows the essential features needed for deprotonation of NH_4^+, coproduction of CH_4 and NH_3, and assembly of LiNCS. Thus, N—Li and C—H bonds are forming, C—Li and N—H bonds are breaking. Indeed, judging by the bond lengths and indices shown, these processes are nearly complete, a fact reflected in the huge energy gain (−193.9 kcal mol^{-1}) when this system forms from separate MeLi and NH_4NCS components.

The MO calculational results given above illustrate some of the thermodynamic reasons concerning *why* the ammonium salt route works so well. In essence, there are huge enthalpy gains to be made by inserting LiH or MeLi into $NH_4^+NCS^-$ bonds, by formation of H_2 or MeH, and by complexation of Li^+ by NCS^- (and, of course, in the experimental systems also by the Lewis base L present). Precisely *how* the route operates is a much more difficult question. Relevant experimental observations were noted earlier, among them that unusual colors are seen during many of these reactions. In this regard, three particular reaction systems, each involving solid NH_4NCS and HMPA (1:2 equivalents) in toluene, were examined in some detail, namely, those with Bu^nLi solution,[59] with solid NaH,[63] and with K.[64] The $NH_4NCS/Bu^nLi/2HMPA$/toluene system changes from red-orange at −196°C, through violet, then to colorless toward room temperature. Binary mixtures of NH_4NCS + HMPA in toluene and of NH_4NCS + Bu^nLi in toluene are colorless, and no reactions occur. The final possible binary mixture, Bu^nLi + HMPA in toluene, is deep red (probably owing to some $PhCH_2Li$ formation), but this color is stable from −186 to 25°C. These observations implied that the color changes ensue from reactions giving the final LiNCS complex and that HMPA must be

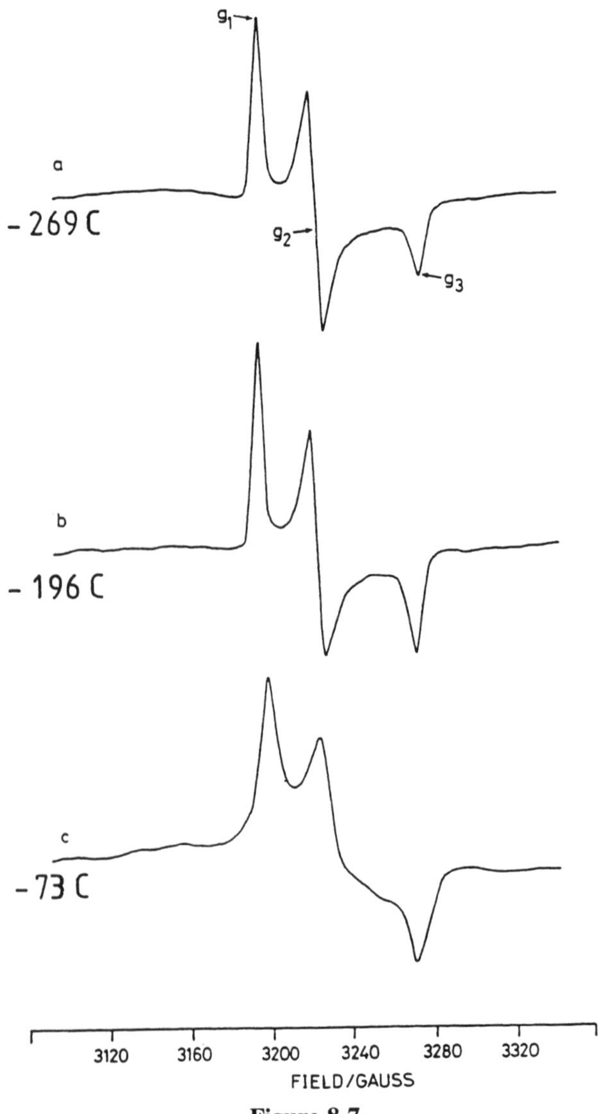

Figure 8.7

present for such reactions to occur. To explore this further, all four components were loaded into an ESR tube at $-196°C$. A blue color is seen near the bottom of the tube, just above the solid NH_4NCS, and ESR spectra could be recorded on this section of the mixture over the range -269 to $-73°C$ (Fig. 8.7). The g value is found to be anisotropic, with features present for the three principal values of g. No hyperfine coupling is observed between the unpaired spin and the spin of any nuclei present. At $-269°C$ (Fig. 8.7*a*), three g tensors are detected: $g_1 = 2.006$, $g_2 = 2.009$, and $g_3 = 1.980$, giving $g_{av} = 2.005$. Such spectra prove that radical species are produced during this unusual reaction; microwave power saturation measurements showed in fact that there is only one type of radical present.

Similar experiments were carried out on the NH_4NCS/NaH/2HMPA/toluene system. This mixture gives an immediate blue color at 25°C and its UV visible spectrum contains an intense band at 618 nm (cf. $\lambda_{max} = 745$ nm for the solvated electron[67,68]). The blue mixture also affords ESR spectra at -269 to $-100°C$, the signals having g tensors at precisely the same field as in the lithium system. The same radical was pinpointed again in the K/NH_4NCS/2HMPA/toluene reaction system, which also gives a blue color.

These findings show that all these NH_4NCS reaction systems involve the intermediacy of a common radical species, possibly a thiocyanate-based one such as $SCN^{\dot{-}}$ or $(SCN)_2^{\dot{-}}$ formed by HMPA-assisted single-electron transfer from the metal source to the NCS^- anion. Other than this, there is no real insight into the mechanism(s) of the ammonium salt route in general.

4 STRUCTURAL AND BONDING STUDIES

4.1 Introduction

As a broad class, complexes containing inorganic lithium salts Li^+X^- [X^- = halide (Cl^-, Br^-, I^-), pseudo-halide (OCN^-, SCN^-), and group anions (e.g., BH_4^-, ClO_4^-)] form an extremely varied and structurally diverse family of compounds. Perhaps the most intriguing and conceptually interesting feature of all these complexes is the realization of a class of compounds which is archetypically ionic and, at the same time, frequently molecular. In respect to the latter, many of the structures of the salt complexes discussed here can be regarded as extruded molecular fragments of the parent salt lattices $(Li^+X^-)_\infty$ in which normal lattice growth has been prevented (see Section 2). There are two major ways by which this "capturing" of a salt fragment can be achieved, and familiarity with them is valuable when interpretating the solid-state structures discussed in this chapter. First, the presence of a Lewis base donor ligand (L) can interrupt lattice growth by surrounding the fragment as an oligomeric $(LiX.xL)_n$ complex in which the lithium cations are solvated. Such complexes have been studied fairly extensively in the past 20 years and, in many cases, their structures in the solid state and in solution are related to those of lithium organometallic derivatives. Thus they can be regarded as, or are related to,

Lewis base-solvated molecular rings, stacks, and ladders which (in some cases) can further associate into continuous polymeric arrays by Lewis base donor bridging and/or by anion bridging. A second method by which the salt fragments can be captured is by assembly into a heterometallic or mixed organometallic/inorganic complex in the presence of an organometallic species. Such cocomplexes have frequently been prepared fortuitously in metathesis reactions of transition metal, lanthanide, early main group metal, and p-block metal halide salts with lithium organometallics. An interesting feature of these complexes is that "anion solvation" by donation from X^- to these metal centers, as well as Li^+ Lewis base solvation, commonly occur.

This section aims to provide a general, though not comprehensive, structural view of this broad class of complexes in the solid state and in solution. Additionally, the theoretical interpretations of their structurally preferred options are discussed from the point of view of model MO calculations. The complexes are ordered and discussed in terms of the anions present and their structural types (i.e., Lewis base donor complexes in Sections 4.2–4.4 and cocomplexes in Section 4.5). Ion-contacted species only are considered, in which there are clear Li—X contacts in the solid state (<3 Å), and complexes with macrocyclic and large polydentate ligands are only discussed in relation to similar complexes containing small acyclic ligands.

4.2 Lithium Halide Complexes

Without any doubt, lithium halide-containing complexes are by far the most studied class of inorganic salt complexes. There has been a large range of compounds characterized in the solid state containing all the halide salts LiX ($X = Cl^-$, Br^-, and I^-) except for LiF. These structural studies in particular have provided much of our knowledge concerning the factors governing the structures and stabilities of inorganic salt complexes in general.

The lack of structural and solution work on LiF complexes is unsurprising bearing in mind the extremely large lattice energy of the salt $(LiF)_\infty$ which is bound to outweigh thermodynamically any solvation energy by Lewis base donor ligands (see Section 3). Indeed, structurally characterized complexes of any of the alkali metal fluorides are rare, exceptions being the highly stable (O—H· · ·F· · ·H—O) hydrogen-bonded KF complexes with acetic and other organic acids which have been characterized in the solid state.[74,75] Some solution studies have addressed the nature of CsF suspended in sulfolane, which is an efficient fluorinating agent and in which a donor complex may be present.[29-31] However, despite this lack of structural information, a range of organic fluoro-silylamide and -phosphide complexes which contain Li· · ·F contacts have been characterized in the solid state.[76-83] Two such species, the dimeric $[Bu^t_2SiP(Ph)(F)Li.2THF]_2$ **(1)**[76] containing an eight-membered heteroatomic ring and $\{[Bu^t_2Si(F)]_2N\}Li.2THF$ **(2)**[77] which has a heteroatomic ladder core, are shown in Figure 8.8*a* and *b* respectively. The observation of ^7Li–^{19}F NMR coupling in these complexes is an intriguing feature as regards the nature

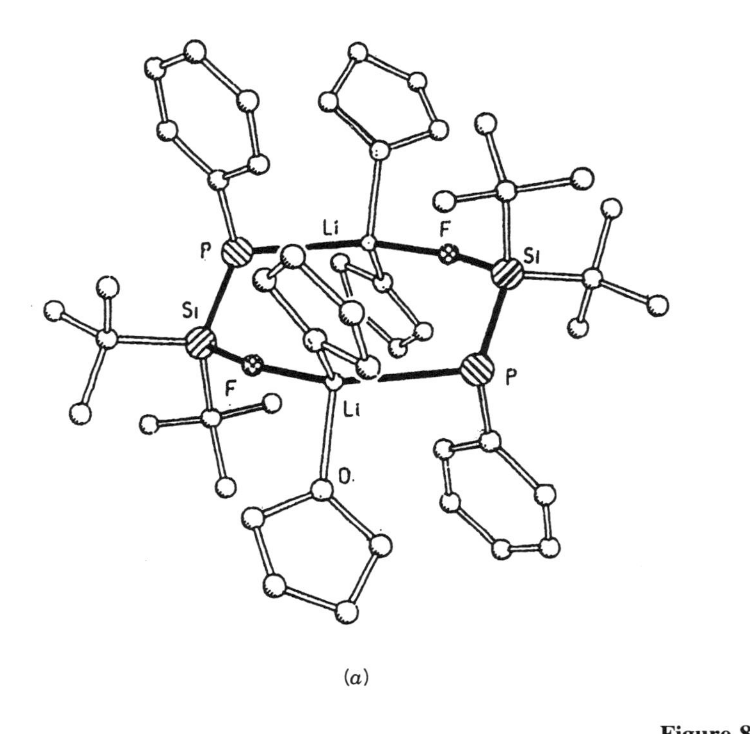

(a)

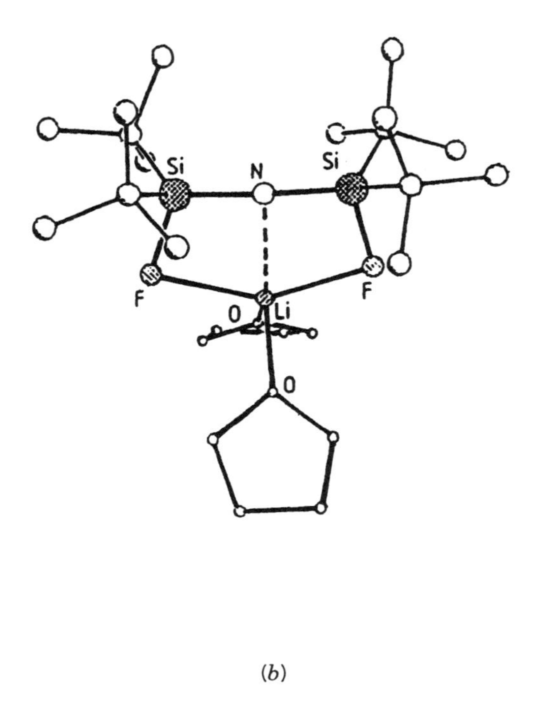

(b)

Figure 8.8

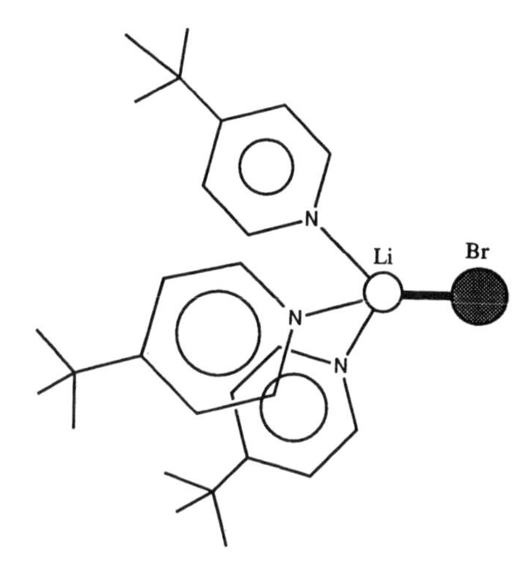

Figure 8.9

of the Li—F bonding involved and of the charge distributions and nature of this fragment in these species.

In contrast to lithium fluoride complexes, other lithium halide–Lewis base donor complexes are numerous. Monomeric ion-contacted complexes involving monodentate O and N ligands and in which Li is pseudo-tetrahedral are a common structural type for all the halides. They represent a situation in which the basic monomeric units of $(LiX)_\infty$ have been "stripped" from the infinite lattice.[41–43,51,52] Pyridine and substituted pyridine ligands in which the alkyl substituents are remote from the N donor site (e.g., 3,5-dimethyl[42] and 4-tert-butyl-pyridine,[43] respectively) produce monomeric halide complexes (LiX.3PyrN), **(3)** and **(4)** (Fig. 8.9) respectively.

In a few rare examples, where bidentate and sterically undemanding donor ligands are present, the Li centers can assume a five-coordinate distorted geometry, for example, in $[(MeOCH_2)_2]_2LiBr$ **(5)**.[84] The same coordination number, although with a more regular square-based pyramidal geometry, is achieved in the structure of (18-crown-4)LiCl **(6)** (Fig. 8.10).[85]

The aggregation of monomeric LiX units into higher aggregates is highly dependent on the effectiveness of Lewis base donation, mainly influenced by the steric bulk and the particular heteroatom (O, N, etc.) present, as well as on the effectiveness of halide (μ_2- and μ_3-) bridging. The extent of association can therefore be rationalized, although not predictively, in terms of the subtle balance between these competing effects. A good example of the influence of the steric bulk of the Lewis base donor is that of the 2-methyl-pyridine lithium halide complexes $[(2\text{-MePyr})LiX]_2$ **(7)** ($X = Cl^-, Br^-, I^-$) (Fig. 8.11) which, in contrast to the aforementioned monomeric 3,5-dimethyl **(3)**[42] and 4-tert-butyl-pyridine complexes **(4)**,[43] are dimeric in the solid state.[42] Presumably the slightly increased steric bulk of the pyridine ligands makes disolvation of Li^+ followed by dimerization more favorable than trisolvation of Li^+ and the formation of monomeric complexes.

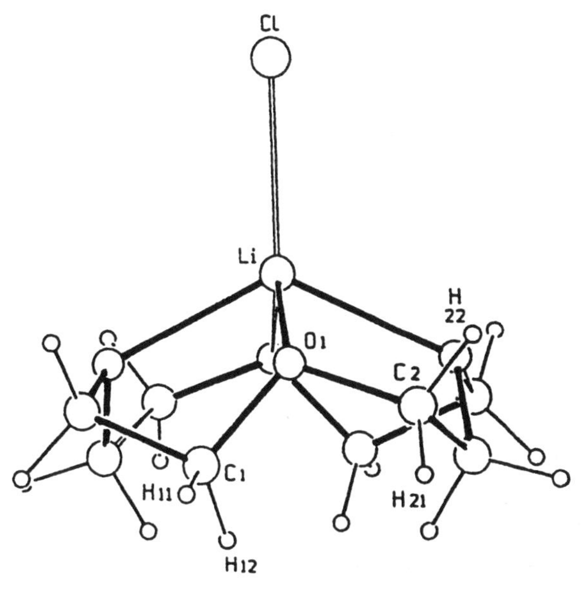

Figure 8.10

The previous trend is continued in the structures of the tetrameric $[LiBr]_4.6(2,6\text{-}Me_2Pyr)$ **(8)** ($2,6\text{-}Me_2Pyr$ = 2,6-dimethylpyridine)[46] (Fig. 8.12) and in $[LiI]_4.6(2,4,6\text{-}Me_3Pyr)$ **(9)** ($2,4,6\text{-}Me_3Pyr$ = 2,4,6-trimethylpyridine),[52] which have similar staggered ladder structures reminiscent of those found in certain lithium organometallic complexes.[18,19,24,86] Lateral association into ladders in these complexes, as opposed to vertical stacking into cubane structures, is evidently influenced by the Lewis base : LiX ratio and by the steric bulk of the ligands involved. It is interesting to note that a 2 : 1 complex of $2,6\text{-}Me_2Pyr$: LiI has also been characterized in the solid state, having a dimeric ring structure $[LiI.2(2,6\text{-}Me_2Pyr)]_2$ **(10)**.[46]

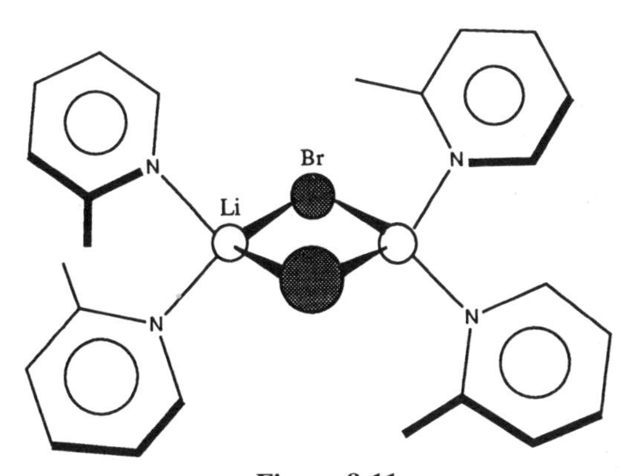

Figure 8.11

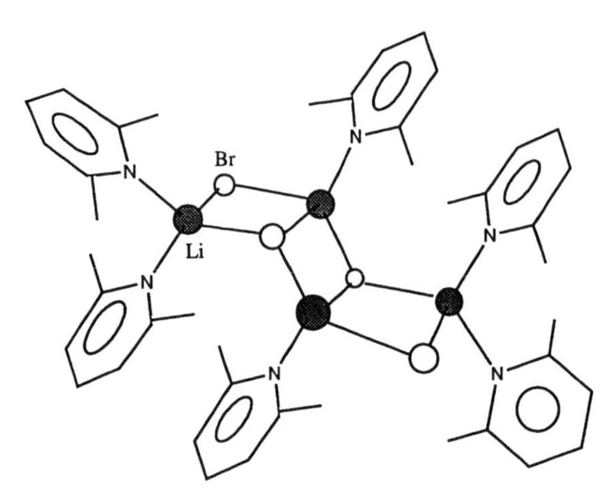

Figure 8.12

The steric bulk of the coordinating ligands is evidently not the sole influence dictating these structures, as can be concluded from the observation of dimeric structures of $(LiCl.2THF)_2$ **(11)**[87,88] (Fig. 8.13) and $[2Me_2C{=}O.Li(\mu{-}Br)]_2$ **(12)**[89] where the steric requirements of the ligands incorporated are very different.

The relative strengths of Li–donor versus Li–halide interactions and the donicity of the ligands are also of importance. These additional influences are highlighted in a number of examples, such as in a comparison of the structures of $[LiX.xHMPA]_n$ [X = Cl^-, Br^-, I^-; HMPA = $O{=}P(NMe_2)_3$] in which, as the Li—halide interactions become weaker ($Cl^- > Br^- > I^-$), the structure "disintegrates" from the cubane structure of $[LiCl.HMPA]_4$ **(13)**[28] to the dimeric structure of $[LiBr]_2.3HMPA$ **(14)**,[56] and then to the ion-separated species $Li(HMPA)_4^+.I^-$ **(15)**.[33]

The $[LiBr]_2{\cdot}3HMPA$ **(14)** complex illustrates an additional feature, that of ligand bridging competing with halide bridging. The structure (Fig. 8.14*a*)

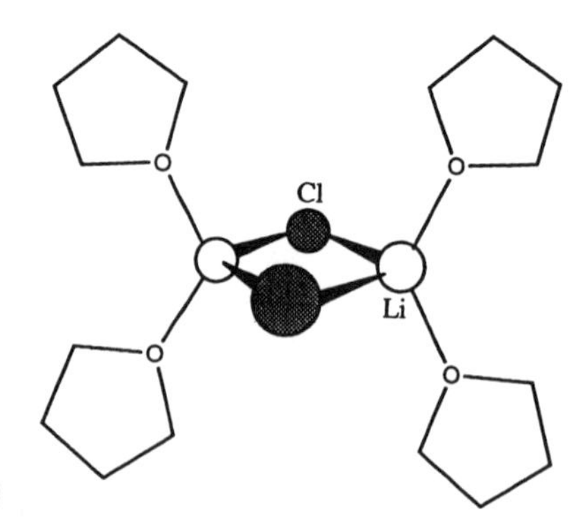

Figure 8.13

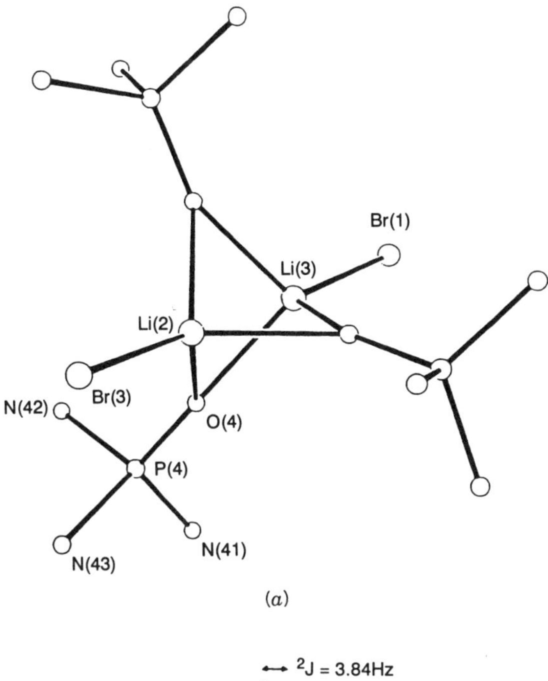

(*a*)

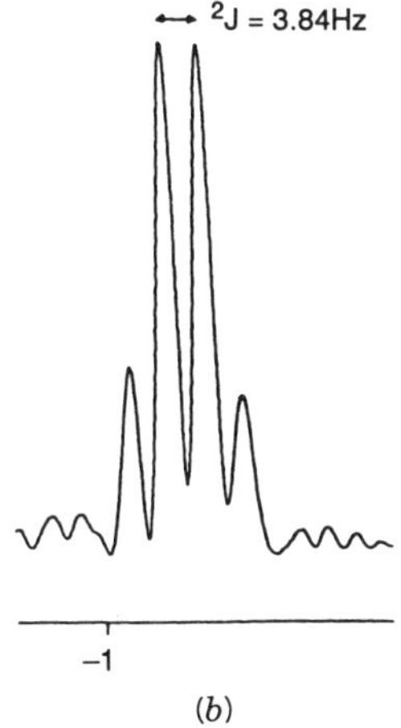

(*b*)

Figure 8.14

is that of a dimer of two separate monomeric LiBr units, Br—Li.[μ—O=P(NMe$_2$)$_3$]$_3$.Li—Br.[56] The strong preference for HMPA bridging, which is also seen in the ion-separated structure of the aqua complex [(H$_2$O)$_2$Li{μ—O=P(NMe$_2$)$_3$}]$_2^{2+}$.2Cl$^-$ **(16)** (Fig. 8.15),[54] is explained by ab initio calculations which show that the ligand resembles the ylid form [$^-$O—P$^+$(NMe$_2$)$_3$] rather than the phosphine oxide form [O=P(NMe$_2$)$_3$][71] (see Figs. 4*a* and *b*). The strength of the Li—O interactions is borne out in the first

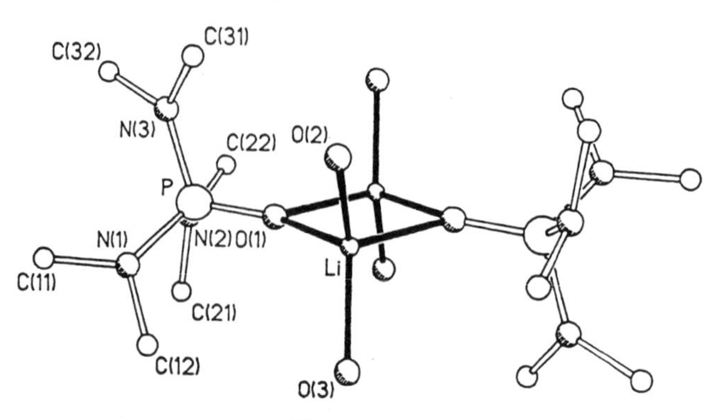

Figure 8.15

observation of $^2J(^7Li-^{31}P)$ coupling in the ^7Li NMR spectrum of **(14)** at low temperature (Fig. 8.14*b*).[56]

The tetrameric cubane structure of **(13)** (see Fig. 8.2 in Section 2), which can be regarded as a stack of two solvated $[LiCl \cdot HMPA]_2$ rings, is as yet the only example of such a lithium halide cubane characterized in the solid state.[28] The structure of **(13)** is possibly the most obvious example illustrating the concept of an extruded fragment of a salt lattice. The $(LiCl)_4$ units can be compared to those present in the parent lattice of $(LiCl)_\infty$. It is interesting to note that this structure is highly stable to additional solvation by HMPA. This contrasts with the behavior of **(14)** in which complete solvation of Li^+ occurs to give $Li(HMPA)_4^+.Br^-$ on addition of excess ligand.[33] This behavior again testifies to the competition between Li–halide interactions versus Li–ligand interactions as an influence on the structures and stabilities of these complexes.

With the latter principles in mind, it is to be expected and it is generally found that the largest molecular or polymeric $(LiX)_n$ aggregates will occur where weak ligation of the Li^+ cations occurs and/or where Li–halide interactions are maximized ($Cl^- > Br^- > I^-$). However, bearing in mind particularly the complicated coordinative alternatives available to multidentate ligands, it is unlikely that any set of rules would be sufficient in predicting absolutely the structures of unknown complexes in general. A particular case in point is that of the LiX ($X = Cl^-, Br^-, I^-$) complexes with bidentate TMEDA $[(Me_2NCH_2)_2]$[48] and with tridentate PMDETA $[(Me_2NCH_2CH_2)_2NMe]$.[40,41] Whereas the $[Li(\mu-X).TMEDA]_2$ **(17)** complexes ($X = Br^-, I^-$) are simple dimers in the solid state, the LiCl complex $(LiCl)_6 \cdot (TMEDA)_2$ **(18)**, which is isolated in low yield, has a complicated polymeric structure based on a "double-cubane" $(LiCl)_6$ core (Fig. 8.16).[48]

The solid-state structure of $[Li(\mu-Br).PMDETA]_2$ **(19)** is a surprising dimer in which an unusual five-coordinate geometry is achieved for the Li^+ cations (Fig. 8.17).[40] However, it is interesting to note that this species is involved in a monomer/dimer equilibrium in benzene solutions and that the equilibrium constant for the dissociation into monomers (K_S), in which Li^+ is

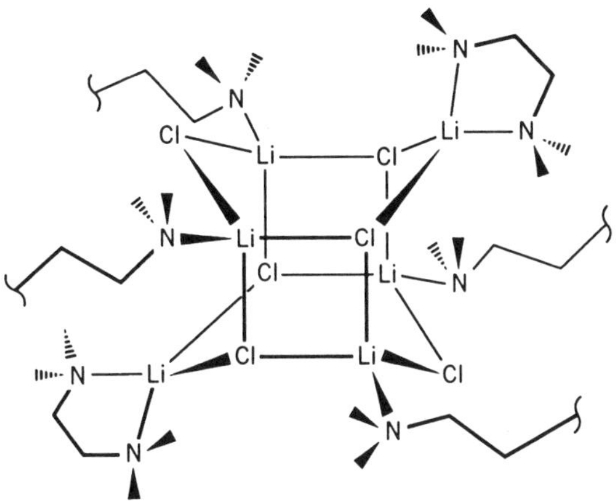

Figure 8.16

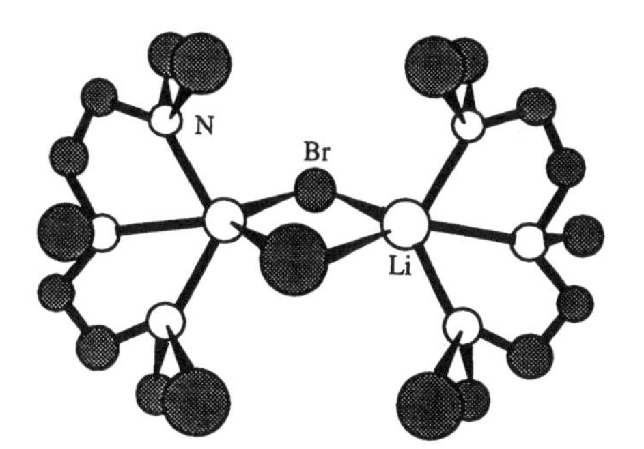

Figure 8.17

four-coordinate, is as expected favorable as compared to dimers in which Li^+ is four-coordinate.[90] This finding is of importance in that solution studies on Lewis base donor alkali metal halide complexes, on which there are few experimental handles, are extremely rare and most studies have concerned the solid state. The study also exemplifies a new cryoscopic method for determining the equilibrium constants and related thermodynamic data in which measurements of the association number (n), for example, for $[LiBr.PMDETA]_n$, at various concentrations (C_S) are fitted to a theoretical equation for complex dissociation (Fig. 8.18).

The LiCl complex $(LiCl)_4$.3PMDETA **(20)**, in contrast to the dimeric LiBr complex **(19)**, has a highly unusual structure which can be regarded as a ''split cubane,'' in which the two contacted $(LiCl)_2$ fragments are bridged together by a PMDETA ligand (Fig. 8.19).[41] The role of this bridging PMDETA is similar to that of Diglyme $[(MeOCH_2CH_2)_2O]$ in the solid-state structure of $[NaI.Diglyme]_\infty$ **(21)**, where a zig-zag $(NaI)_\infty$ chain is further associated by terminal O—Na linkages and by an unusual μ_2—O bridging mode (Fig. 8.20).[91]

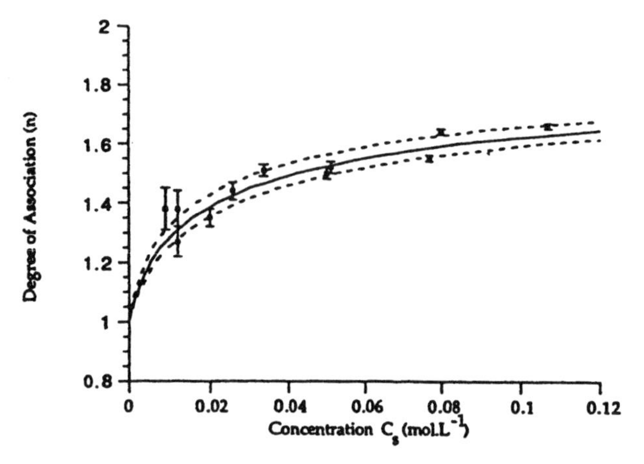

Figure 8.18

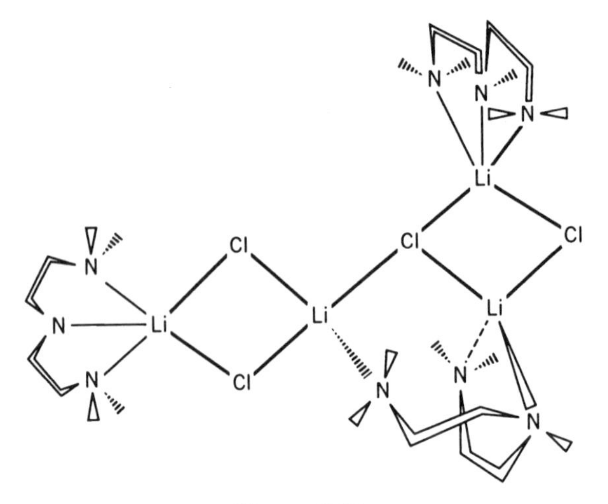

Figure 8.19

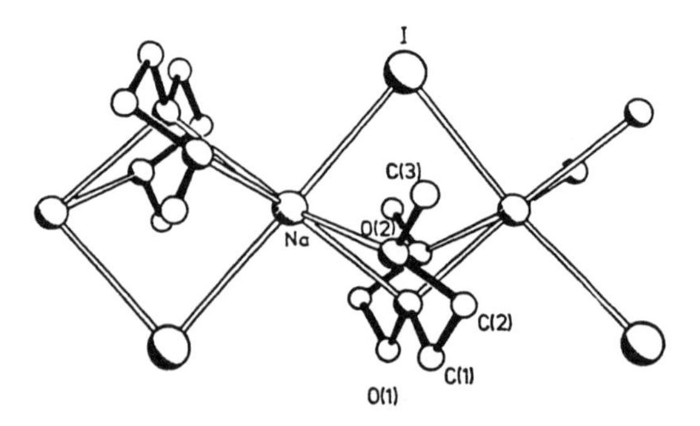

Figure 8.20

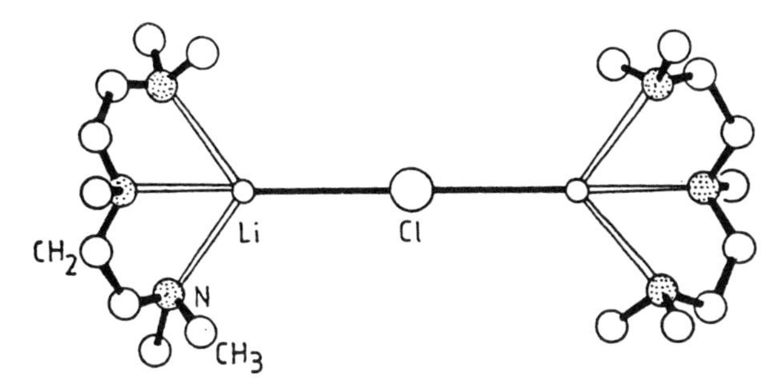

Figure 8.21

A further interesting class of halide complexes are the cationic derivatives. These are found within the structures of compounds containing large organometallic anions. The simplest example is that of the PMDETA.-Li(μ—Cl)Li.PMDETA$^+$ cation **(22)** (Fig. 8.21), unique in containing a linear Li—Cl—Li bridge.[92]

Two other examples are seen in the structures of $[(Et_2O)_{10}.Li_4Cl_2]^{2+}.2[Li_2Cu_3Ph_6]^-$ **(23)**[93] and $[Li_6Br_4.(Et_2O)_{10}]^{2+}.2[Ag_3Li_2Ph_6]^-$ **(24)**.[94] The tetrametallic dication of **(23)** (Fig. 8.22) contains a central Et_2O-solvated $(LiCl)_2$ core in which μ_3—Cl bridging anions are attached to terminal Et_2O-solvated Li cations. As such, the dication resembles a "broken" cube $(LiCl)_4$ in which two Cl^- anions have been lost from one $(LiCl)_2$ face.[93] The remarkable hexametallic dication of **(24)** contains in Li_6Br_4 fragment which can be regarded as two three-rung ladders sharing their terminal Br anions (Fig.

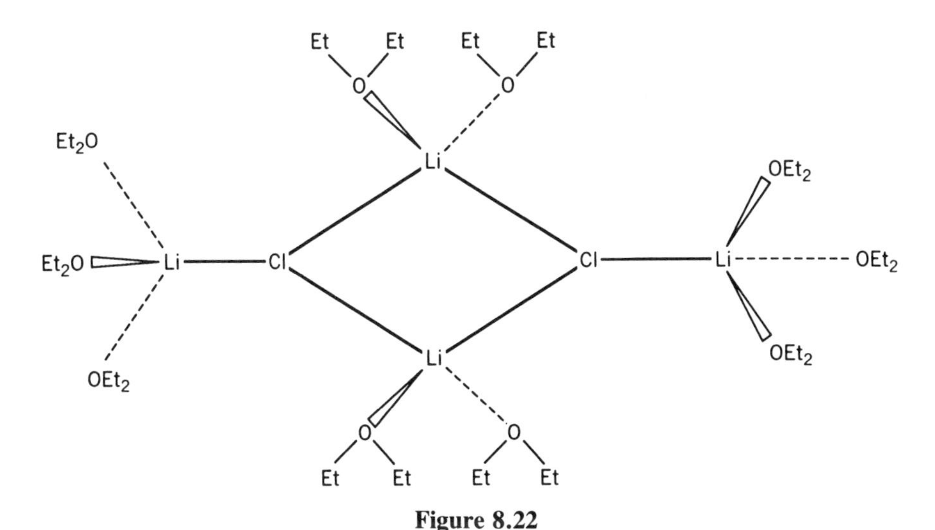

Figure 8.22

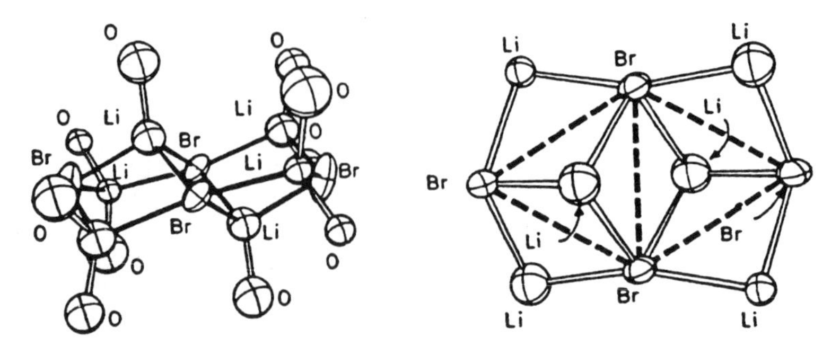

Figure 8.23

8.23). This is the largest molecular lithium halide fragment (or complex) so far structurally characterized.[94]

Surprisingly, few polymeric lithium halide complexes have been characterized in the solid state in which Li—X contacts are maintained. Such species are far more common for lithium salts with multifunctional anions (e.g., NCS^-), where multidentate ligands are present, and for the heavier alkali metals. This situation occurs mainly because Li^+ is generally coordinatively satisfied by a pseudo-tetrahedral geometry and further association, by ligand or halide bridging to increase the coordination number, is rarely favorable. However, the recent X-ray structure of $[LiBr.THF]_\infty$ **(25)**[95] sheds some additional light on the factors governing whether such lithium halide complexes adopt oligomeric cubane or polymeric ladder structures. The structure (Fig. 24*a* and *b*) is that of a unique "corrugated" ladder in which $(LiBr)_2$ units are linked together by μ_3—Br bridges. Oligomeric (generally tetrameric) ladder structures are now fairly common for the organometallic complexes of Li, where they adopt "staggered" or slightly "folded"[18,19,24,86] conformations. Polymeric ladder structures have also been observed for a variety of other alkali metal organometallic derivatives.[15] However, the continuous ladder structure of **(25)** is the only example of such a polymer for any alkali metal halide.

An important comparison can be made between the polymeric structure of **(25)** and that of the tetrameric cubane structure of $[LiCl.HMPA]_4$ **(13)** (see Fig. 8.2, Section 2).[28] The same 1:1 ligand to LiX ratio is present in both complexes and, at first sight, on the basis of ring stacking and ring laddering principles, it is difficult to understand why **(25)** does not have a cubane structure (especially bearing in mind that there is probably a similar balance between the ligand solvation and Li—X association in both complexes). However, the structure of **(25)** can be regarded as (and resembles) a polymer of $(LiBr.THF)_4$ cubes. The formation of a polymer for **(25)** can be seen to relieve $Br^- \cdots Br^-$ repulsion in the cubane while still maintaining a similar four-coordinate ligand sphere for Li^+. In **(13)**, the sterically demanding HMPA ligands probably preclude such ladder association despite $Cl^- \cdots Cl^-$ repulsion, and so a cubane structure is maintained.

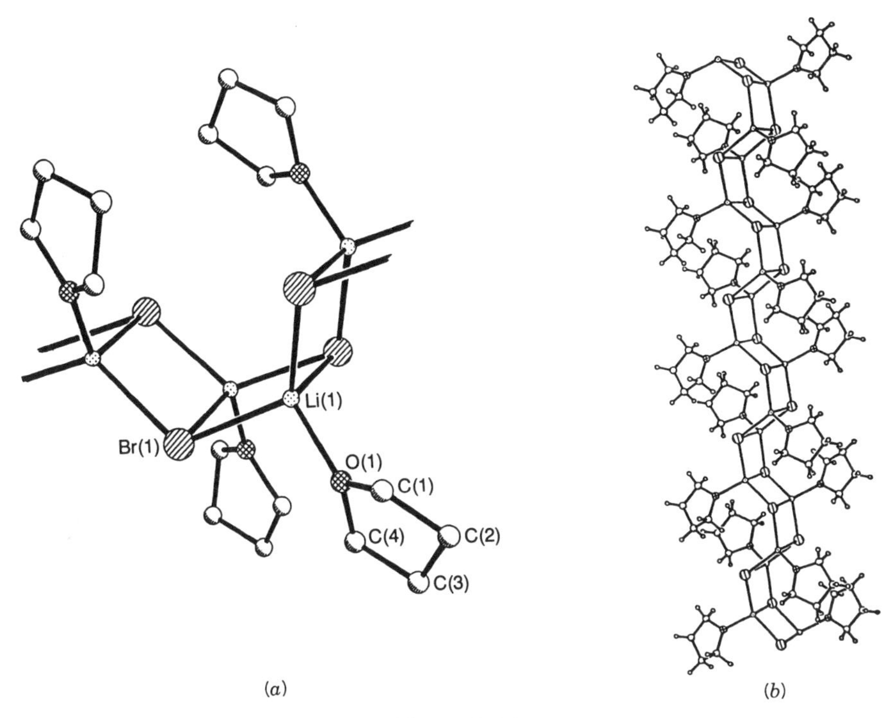

Figure 8.24

4.3 Lithium Thiocyanate Complexes

As mentioned in Section 4.2, where there is a potential for multifunctionality in the anion (e.g., NCS^-) the structures of salt complexes are likely to become even more complicated and less predictable owing to the different coordinative alternatives available for anion complexation to Li^+ and to the possibility of anion bridging. This feature is discussed in Sections 4.3 and 4.4.

A good example illustrating the increased likelihood of anion bridging is seen in the structure of $[LiNCS.TMEDA]_\infty$ **(26)**.[58] The individual (TMEDA)LiNCS monomers of **(26)** have near-linear backbones and are linked together by Li· · ·S coordinative bonds (Fig. 8.25). In this way, Li^+ ions achieve a distorted tetrahedral geometry and zig-zagged one-dimensional strands are formed. Such a bifunctional mode for the NCS^- ion is still comparatively rare for main group metal species although it does occur in certain macrocyclic complexes of thiocyanate salts and is seen in the structure of $K[Me_3Al\text{-}NCS\text{-}AlMe_3]$[96] and in $[Pt(SCN)_2Cl_2(PPr_3)_2]$.[97] The distances within the thiocyanate anion in **(26)** prompt a formulation as $[S \overset{..}{-} C \overset{..}{=} N]^-$.

Although there are no contacts between the individual (· · ·Li—NCS· · ·Li· · ·) strands of **(26)**, they run alternately in opposite directions forming, pictorially, layers separated by about 3.3 Å (Fig. 8.26).

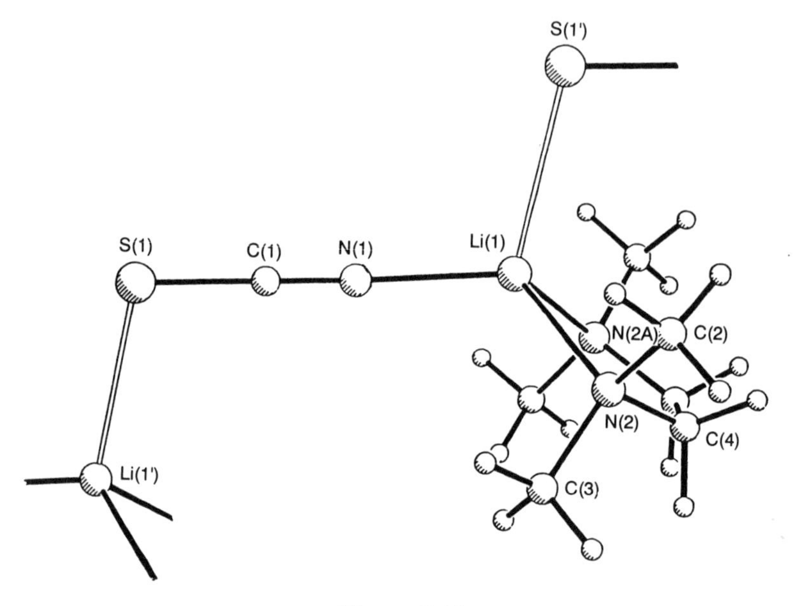

Figure 8.25

Specific heat capacity measurements on this complex (Fig. 8.27) indicate that a number of phase changes and/or conformational rearrangements in the polymer structure occur as the temperature is reduced to approximately 4 K.[98]

Solution and solid-state infrared measurements have proved a particularly useful tool in the investigation of the structures of thiocyanate complexes. Cha-

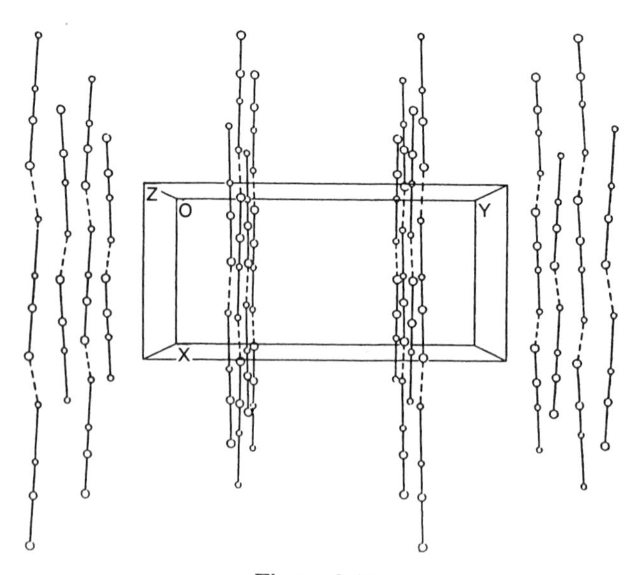

Figure 8.26

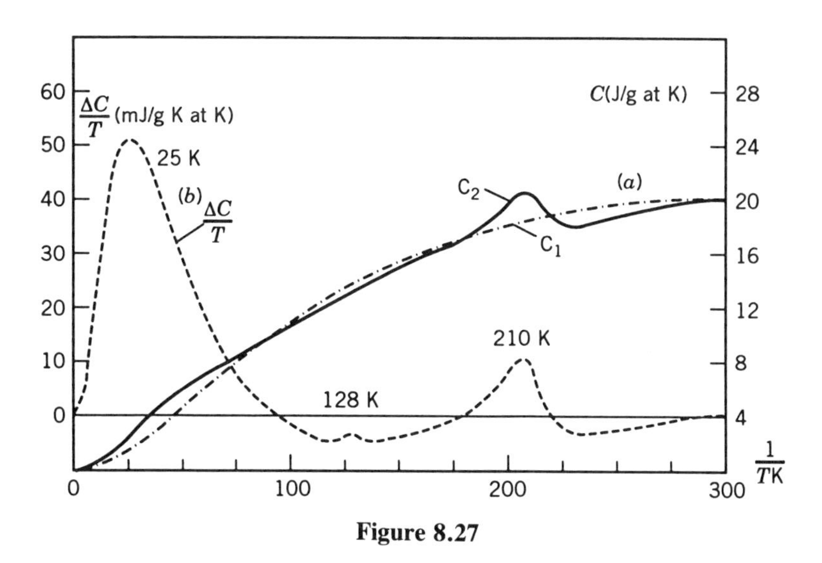

Figure 8.27

banel has shown by both infrared and Raman studies of LiNCS in asymmetric hindered ether solvents that a terminal linkage Li—NCS is characterized by a band at 2065 cm^{-1} and that a μ_2—NCS bridging mode is found at 2032 cm^{-1}.[99,100] Additionally, a μ_3—NCS bridging mode, which has not yet been realized in the solid state, has been shown to arise at about 1993 cm^{-1} and cubane structures containing such μ_3-anions have been detected by a combination of infrared, Raman, and vapor-phase osmometry measurements.[101] Interestingly, **(26)** forms [TMEDA.Li(μ_2—NCS)]$_2$ dimers when dissolved in aromatic solvents, as is shown by a combination of cryoscopic molecular mass measurements, ^{7}Li NMR spectroscopy, and infrared spectroscopy (Fig. 8.28). Thus there is a solid-state/solution tautomerism occurring.[58]

Changing the Lewis base donor ligand from bidentate TMEDA to tridentate PMDETA produces a complex [LiNCS.PMDETA]$_n$ **(27)**. Although the crystal structure of **(27)** has not been determined, cryoscopic measurements and solution and solid-state infrared spectroscopy show that this complex is a dimer [PMDETA.Li(μ_2—NCS)]$_2$ in the solid [similar to [Li(μ—Br).PMDETA]$_2$ **(19)** in which Li is also five-coordinate[40]] which is involved in a monomer–dimer equilibrium in benzene. Thus the association number increases from 1.41 $\pm$ 0.06 (0.026 mol.1^{-1}) to 2.09 $\pm$ 0.07 (0.073 mol.1^{-1}) with increased concentration, and both terminal-NCS (2059 cm^{-1}) and μ_2—NCS (2038 cm^{-1}) absorptions are observed in such solutions.[33]

A dramatic structural change from that of the polymeric array of [LiNCS.TMEDA]$_\infty$ **(26)** occurs also when the bidentate TMEDA ligand is replaced by the bidentate TMPDA [$Me_2N(CH_2)_3NMe_2$], which has a larger ligand ''bite'' and is more sterically demanding. The resulting complex is dimeric [TMPDA.Li(μ_2—NCS)$_2$]$_2$ **(28)** in the solid state (Fig. 8.29).[60] Surprisingly, the μ_2—NCS bridge is found in an unprecedented bent mode, which contrasts with the symmetrical μ_2—NCS bridging observed in alkali metal

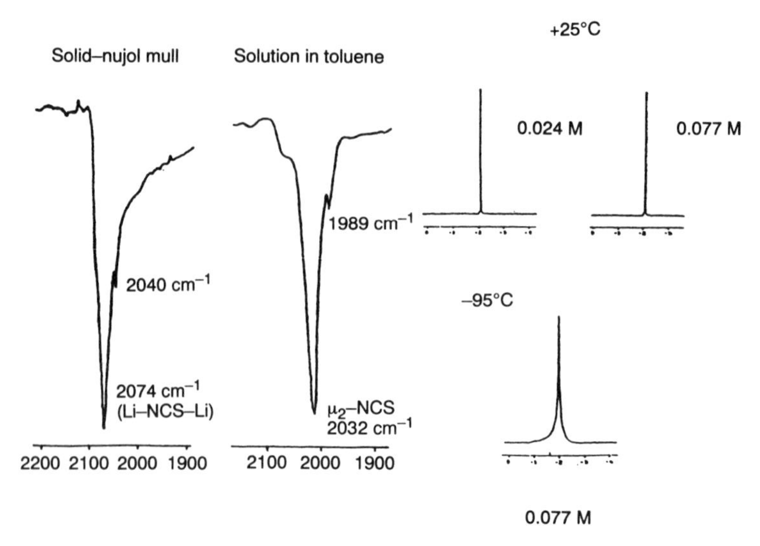

Figure 8.28

polyether complexes and in macrocyclic polyether complexes.[102] The central Li_2N_2 core of **(28)** is planar, but highly asymmetric. The structure can be described as a dimer of two loosely linked monomers and is significantly dissociated into monomers in benzene solutions. Cryoscopic and infrared studies on these solutions[60] suggest that the assignments of infrared stretching frequencies to the various thiocyanate bonding modes in LiNCS complexes may not be fully correct.[99–101]

The dimeric $\{(HMPA)_2.Li[\mu_2{-}O{=}P(NMe_2)_3]_2.Li(NCS)_2\}$ **(29)** complex (Fig. 8.30) has a highly unusual asymmetric structure in the solid state in

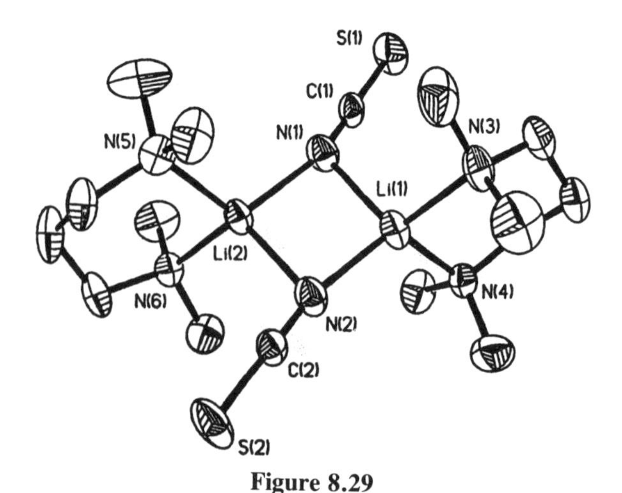

Figure 8.29

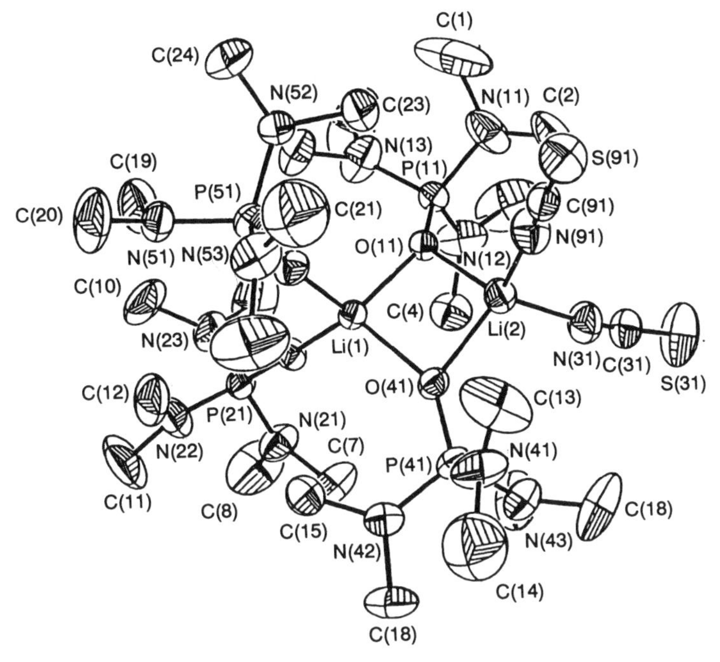

Figure 8.30

which one of the pseudo-tetrahedral Li^+ cations is coordinated by two terminal HMPA ligands and by two μ_2—O attached HMPA ligands. The latter in turn are linked to the other Li center to which is attached two terminal NCS^- anions.[59] This molecular arrangement, rather than a symmetrical one expected on ionic principles, together with ESR spectroscopic investigations of the ammonium salt reaction of NH_4NCS, Bu^nLi, and HMPA (1 : 1 : 2) producing **(29)** (see Section 3.4), suggests that the product may be kinetically controlled. Thus, in the initial step of the reaction an intermediate complex $[Li(HMPA)_2(NCS)_2]^-.NH_4^+$ **(30)** may be formed which allows solubilization of the ammonium salt. Complex **(30)** then could react further with Bu^nLi to produce the asymmetric product **(29)**. The isolation and characterization of **(30)** by changing the stoichiometry of the reaction appears to confirm this hypothesis, as does the X-ray characterization of $(NH_4^+)_2.[Ba(NCS)_4.2HMPA]^{2-}$ **(31)** (Fig. 8.31) from a similar reaction of BaH_2 with NH_4NCS.[57]

An interesting comparison can be made between the asymmetric HMPA—LiNCS complex **(29)** and the HMPA adducts of NaNCS and of KNCS; these are the polymer $[\{(SCN)Na[\mu{-}O{=}P(NMe_2)_3]\}_2]_\infty$ **(32)** (Fig. 8.32),[63] the symmetrical dimer $\{(HMPA)(SCN)Na[\mu{-}O{=}P(NMe_2)_3]\}_2$ **(33)** (Fig. 8.33),[63] and the trimeric cluster $\{KNCS.[\mu_2{-}O{=}P(NMe_2)_3]\}_3.[\mu_3{-}O{=}P(NMe_2)_3]_2$ **(34)** (Fig. 8.34).[64] In contrast to **(29)**, both the NaNCS complexes [**(32)** and **(33)**] adopt expected symmetrical *trans* arrangements in which μ_2—O bridging of the Na^+ cations is present.[63] As the size of the alkali

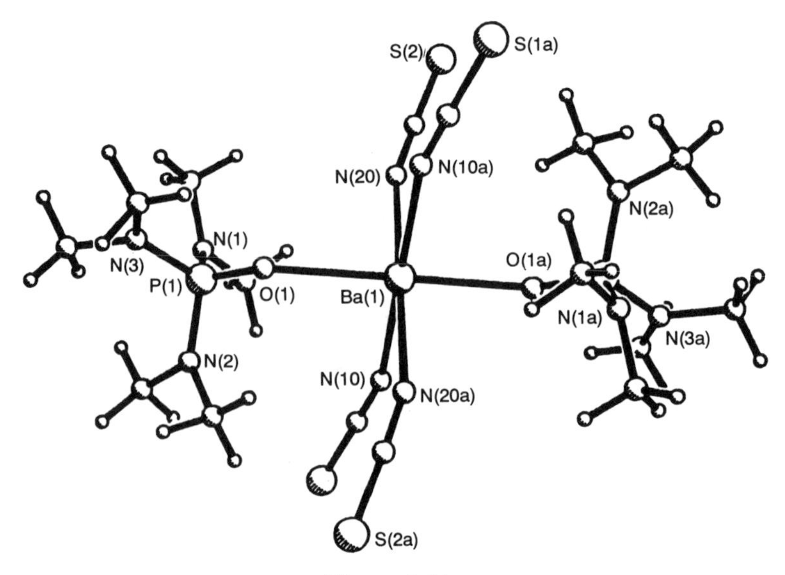

Figure 8.31

metal cation increases from Li to K there is an increase in aggregation. This increase presumably reflects the resulting increase in strain in the $\{M[\mu_2{-}O{=}P(NMe_2)_3]\}_x$ rings which grow from $x = 2$ (dimer cores) for **(29)**, **(32)**, and **(33)** to $x = 3$ (a trimer core) for **(34)**. Complex **(34)** was the first compound in which $\mu_3{-}O$ HMPA ligands had been observed in the solid state and is as yet the only trimeric inorganic alkali metal salt complex.[64]

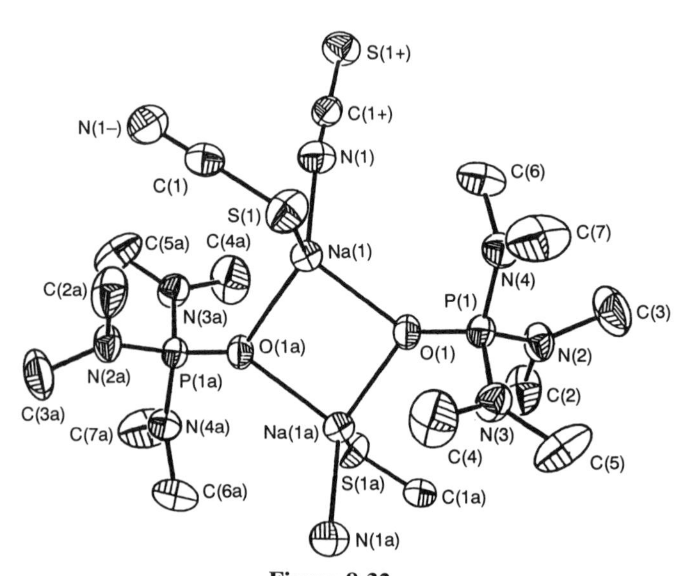

Figure 8.32

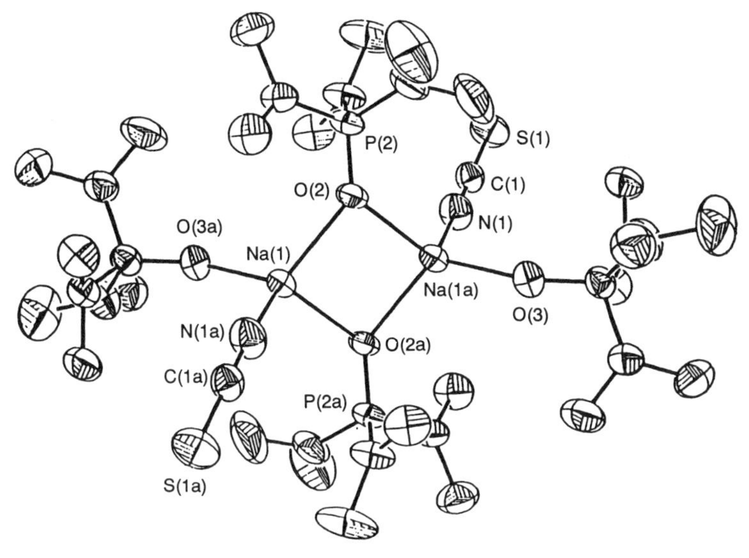

Figure 8.33

Concerning the various bonding modes available to the thiocyanate ligand observed in all the previous complexes, ab initio MO calculations have been used to probe their relative favorabilities. The key calculational results (6-31G with *d* orbitals on S) relating to these species are illustrated in Figure 8.35.[60]

For an uncomplexed monomer, the most stable optimized structure (relative

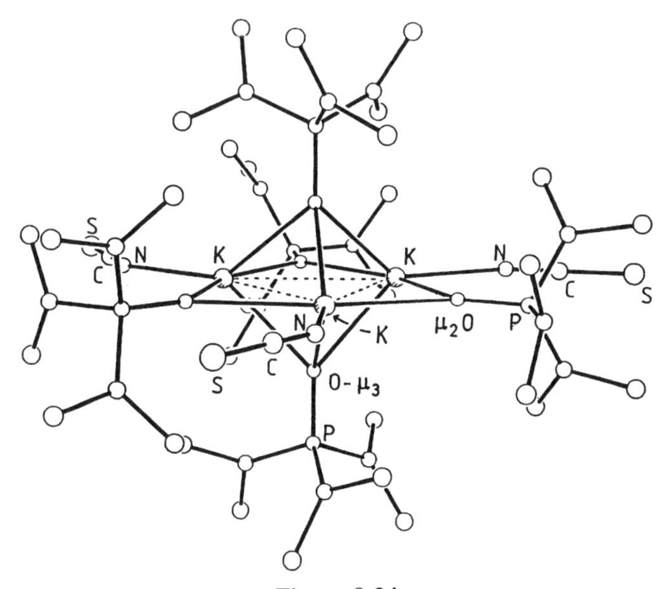

Figure 8.34

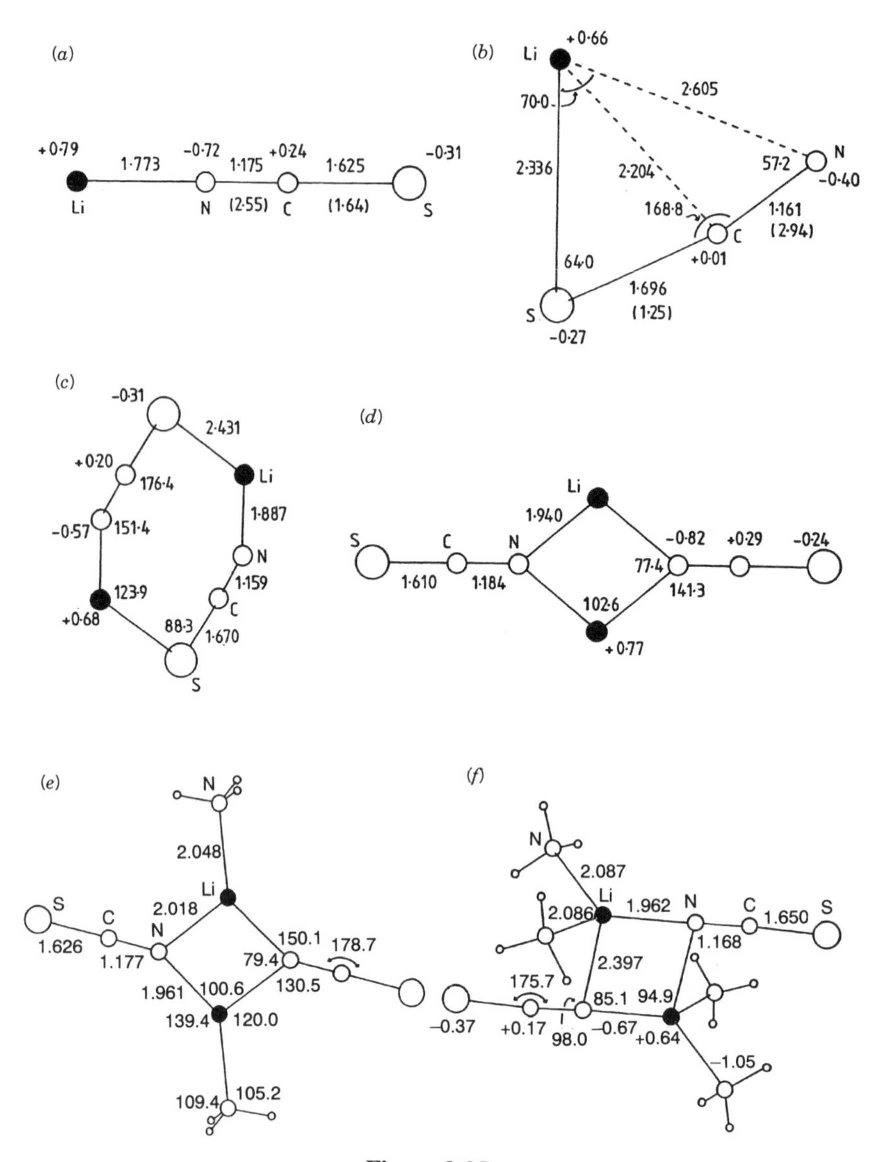

Figure 8.35

energy 0.0 kcal mol^{-1}) has a fully linear Li—NCS arrangement (Fig. 8.35*a*). A linear Li—SCN structure, which is not in fact a local minimum, is less stable by 29.9 kcal mol^{-1}. Of intermediate energy (13.6 kcal mol^{-1}) is the bent arrangement shown in Figure 8.35*b*. Complexation of the two most stable forms by two NH_3 ligands [a model for the TMEDA and TMPDA ligands of **(26)** and **(28)** respectively] narrows the gap between these species, the linear solvated structure now being only 8.0 kcal mol^{-1} more stable than the bent one.

Calculations on the uncomplexed dimer structures show that the $[Li—NCS\cdots Li]_2$ eight-membered ring (Fig. 8.35*c*) is 4.4 kcal mol^{-1} more stable than the μ_2—N bridged Li_2N_2 ring dimer (Fig. 8.35*d*). This result relates to the observation of the difunctional NCS bridging mode in **(26)**. However, further calculations on the solvated dimers illustrate that the competition between these two bridging modes is finely balanced. Monosolvation of these dimer structures by NH_3 narrows the gap between the two bridging thiocyanate modes. Now the solvated μ_2—N bridged Li_2N_2 ring dimer structure (Fig. 8.35*e*) is slightly more stable than the solvated $[Li—NCS\cdots Li]_2$ eight-membered ring by 0.23 kcal mol^{-1}. Additionally, the most favorable linkage is now bent, although it is favorable only by 0.06 kcal.mol^{-1} over a C_{2v} dimer in which a symmetrical μ_2—N bridge is present. Solvation of each Li with two NH_3 gives the most stable asymmetric bridge structure (shown in Fig. 35*f*) which is 2.0 kcal mol^{-1} more stable than the solvated $[Li—NCS\cdots Li]_2$ eight-membered ring dimer. This predicted bonding pattern is realized in the structure of the asymmetric dimer **(28)** and closer inspection of these calculations validates the formulation of this complex as a loose dimer.

It is interesting to note with regard to the calculations above that the most favorable bridging mode appears to be highly dependent on whether N or O donor molecules solvate Li. Thus, 6-31G calculations on the $[Li—NCS\cdots Li]_2$ eight-membered ring dimer and the μ_2—N bridged dimer solvated with one H_2O on each Li give, in contrast to the analogous NH_3 solvated species, the $[Li—NCS\cdots Li]_2$ ring as the most stable form by 2.1 kcal mol^{-1} over the solvated μ_2—N bridged dimer (Fig. 8.36).[71]

In summary, the bonding mode adopted by the thiocyanate anion and the eventual structures of the Lewis base adducts formed are subject to a number of complicated factors. Ab initio calculations and the solid-state structures suggest that the various structural options available are finely balanced and energetically close.

4.4 Group Anion Complexes

Structural work on Lewis base donor complexes of lithium salts containing group anions such as BH_4^-, AlH_4^-, BF_4^-, ClO_4^-, NO_3^-, and CO_3^{2-} is particularly sparse, and few studies have addressed the way such structures vary depending on the donor employed for a specific salt. As such, a discussion of the comparatively few complexes which have been structurally characterized must inevitably concern, except in a few isolated cases, the various ways in which such anions can interact with Li^+ cations rather than to attempt to find any underlying trends.

Hydride contacts to Li^+ have been observed in a variety of borohydride, aluminum hydride, and transition metal hydride complexes. A particular interest in isolable borohydride and aluminum hydride complexes stems from their potential use as stoichiometrically controlled reagents in organic synthesis[9] and in the theoretical natures of $Li\cdots H$ bonding in these species, particularly

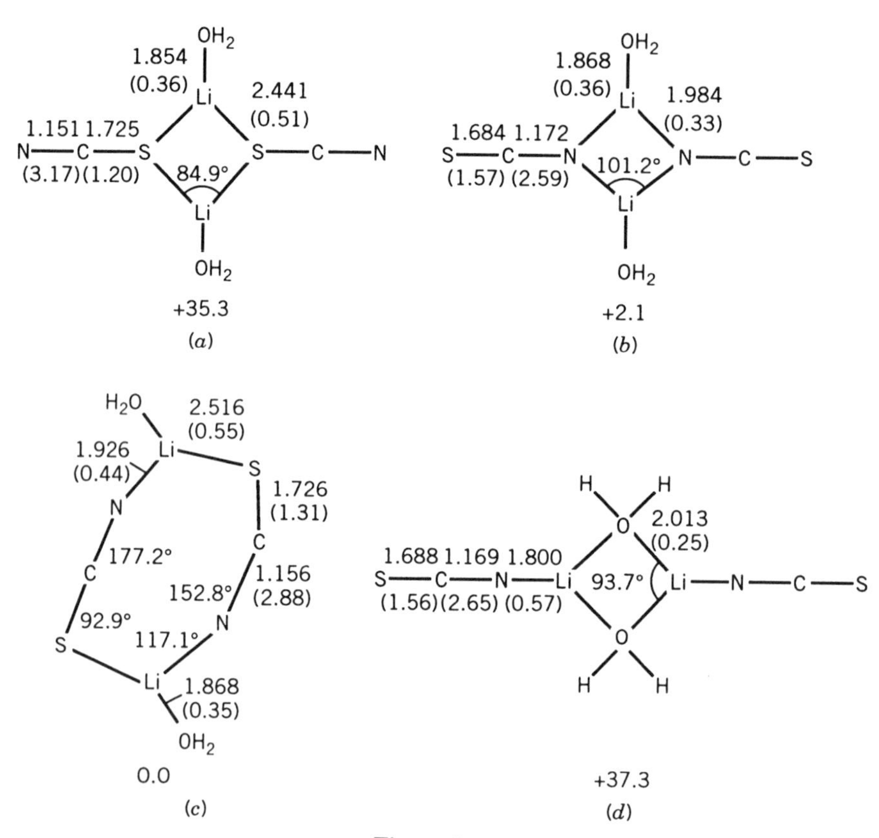

Figure 8.36

in relation to similar interactions in the solid-state structures of MeLi[103] and Me_4BLi.[104]

For the borohydrides only a few Lewis base donor complexes have been characterized in the solid state and the difficulty in accurately positioning the hydridic H atoms in such species has hampered the rationalization of Li—H—B bridging. In the monomeric complexes (Glyme)Li(μ_2—H)$_2$B(2,4,6-$Me_3C_6H_2$)$_2$ **(35)**[105] and $(THF)_3Li(\mu_2-H)_3BC(SiMe_2Ph)_3$ **(36)**[106] (Fig. 8.37*a* and *b* respectively), in which bulky organic groups are attached to B, contacts between Li^+ and two and three μ_2-bridging hydrides are observed. Such is borne out by an examination of the Li· · ·B contact distance in each of these complexes (the H atoms having not been located in both structures), this being about 0.3 Å shorter in **(36)** than in **(35)**. Comparison of the $Li(\mu_2-H)_3B$ attachment in **(36)** [Li· · ·B, 2.19(4) Å] with the $Li(\mu_2-H)_3C$ bonding in MeLi[103] and the $Li(\mu_2-H)_3B$ bonding in Me_4BLi[104] (in both, Li· · ·C, 2.36 Å) suggests, though not conclusively, that these interactions are similar in origin.

A dimeric structure is observed in the solid state for $[LiBH_4.TMEDA]_2$ **(37)** (Fig. 8.38), the only simple borohydride complex so far structurally charac-

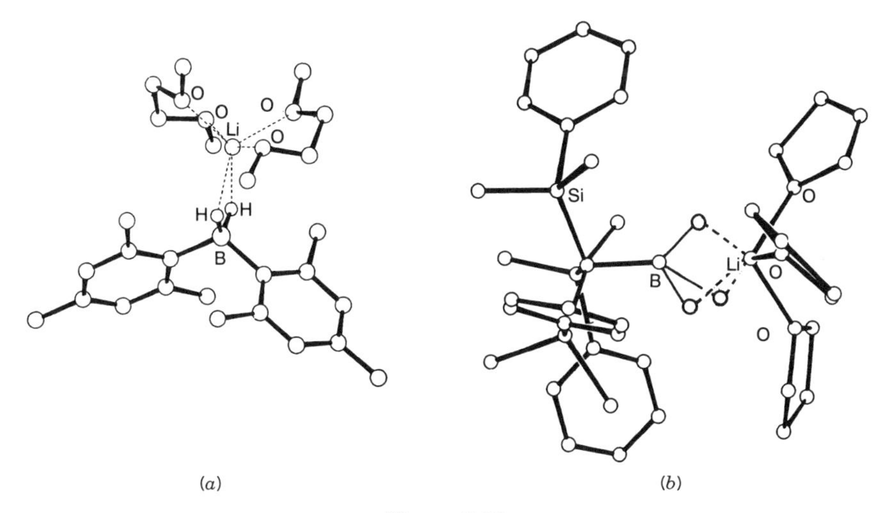

Figure 8.37

terized.[107] A combination of Li(μ_2—H)Li and Li—H—B(H_2)—H—Li bridging links the two Li centers together.

Owing to extremely low-energy fluxional processes occurring in complexes **(35)**–**(37)**, even low-temperature NMR experiments have failed to confirm conclusively the natures of these species in solution, though they suggest that all resemble the solid-state structures in a range of organic solvents. Such conclusions and the observation of the markedly different modes of association of Li^+ and borohydride anions in the latter structures, go part way to explaining the well-known solvent-dependent behavior of $LiBH_4$ as a reducing agent in organic synthesis.[9]

To date, no borohydride complexes have been structurally characterized containing μ_3—H bridges to Li^+. However, this bridging mode has been observed in the elegant tetrameric cubane structure of $[Na(\mu_3{-}H)BMe_3]_4.Et_2O$ **(38)**, which contains a Na_4H_4 core (Fig. 8.39).[108]

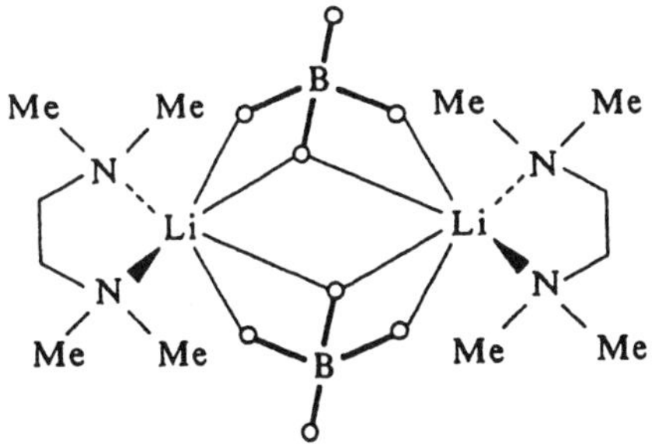

○ = H

Figure 8.38

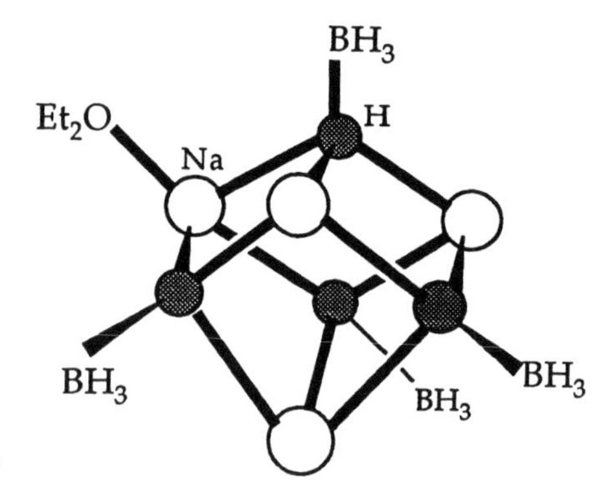

Figure 8.39

Comparatively little work has been done on the aluminum hydride complexes of Li. In the solid-state structures of dimeric $[(Bu^t)_3Al(\mu_2{-}H)Li]_2$ **(39)**[109] and $[(1,4\text{-}Me_2Morph)Li(\mu_2{-}H)Al[CH(SiMe_3)_2]_2(Bu^t)]$ **(40)**[110] (Fig. 8.40*a* and *b* respectively), both μ_2- and terminal Li/Al bridging are observed. Evidently, as in the borohydride complexes [**(35)** and **(36)**], the choice of bridging mode adopted is a function of the steric bulk of the organic groups attached to Al (or B) and of the Lewis base donors employed. The dimeric $(LiH)_2$ core of **(39)** can be viewed as a fragment of lithium hydride, and indeed the Li—H distances in this complex (1.68–1.92 Å)[109] are on average shorter than those in both the $(LiH)_\infty$ lattice (av. 1.78 Å)[111] and in $LiAlH_4$ itself (1.88–2.16 Å).[112]

Recently two new structural options have been realized for lithium aluminum hydride complexes. In a remarkable study of the metallation of $(Me_3Si)_2NH$ with $LiAlH_4$, Stalke[113] showed that the thermodynamic product formed after prolonged reaction is the monomeric complex $[(Me_3Si)_2N]_2Al(\mu_2{-}H)_2Li.2OEt_2$ **(41)** (Fig. 8.41*a*) in which two $\mu_2{-}H$ bridges link the Al and Li centers together. The kinetically unstable monosubstituted intermediate $[(Me_3Si)_2AlH_3Li.2OEt_2]_2$ **(42)** (Fig. 8.41*b*) was characterized by low-temperature X-ray crystallography and shown to have a dimeric structure in which Li—H—Al(H)—H—Li bridging occurs, giving an eight-membered ring structure. As in $[(Bu^t)_3Al(\mu_2{-}H)Li]_2$ **(39)**, the Li—H distances are short (average 1.78 Å) compared to those in $(LiH)_\infty$ and $LiAlH_4$.

The observation of a triplet at −50°C in the ^7Li NMR spectrum of **(41)** in THF solution due to $^2J(^7Li{-}^1H)$ coupling (10.5 Hz) confirms that the structure is largely intact in such solutions. However, the narrow temperature band for this observation suggests that monomer–dimer equilibrium and/or an intramolecular fluxional process involving the bridging mode of the AlH_4^- anion may also be occurring.

A rare example of a hydride complex containing a Group 3 element other than B or Al has been structurally characterized for an indium species. Reduction of $(THF)_3Li(\mu_2{-}Cl)In[C(SiMe_3)_3]Cl_2$ with $LiAlH_4$ in THF gives the un-

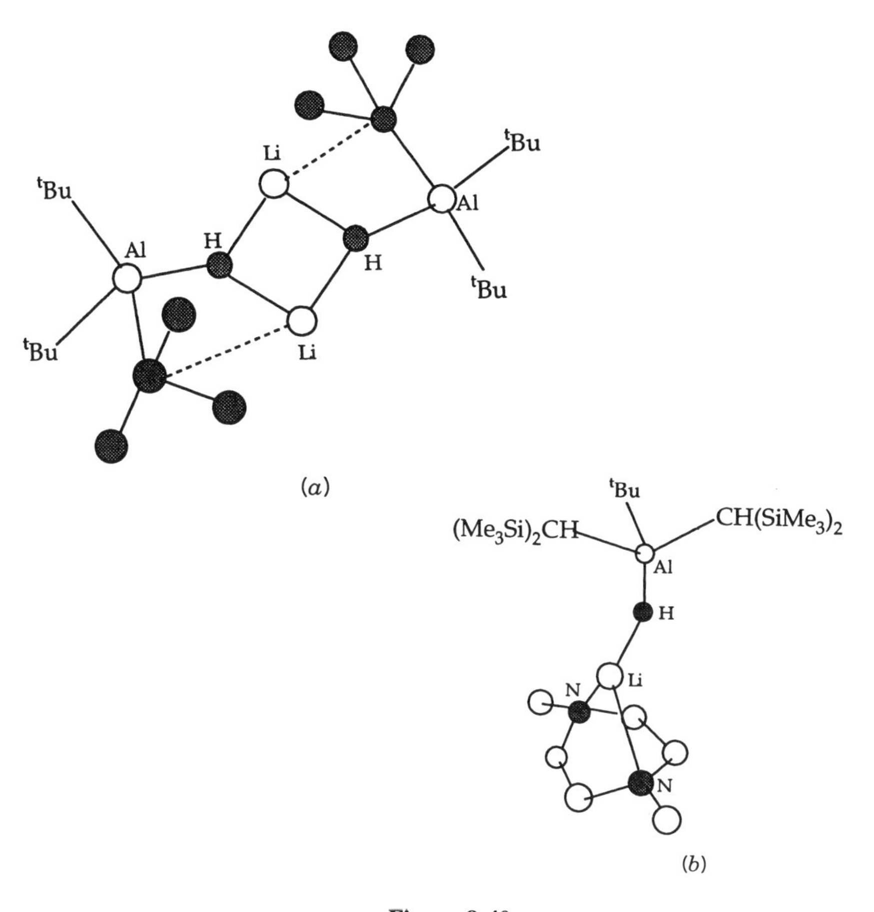

Figure 8.40

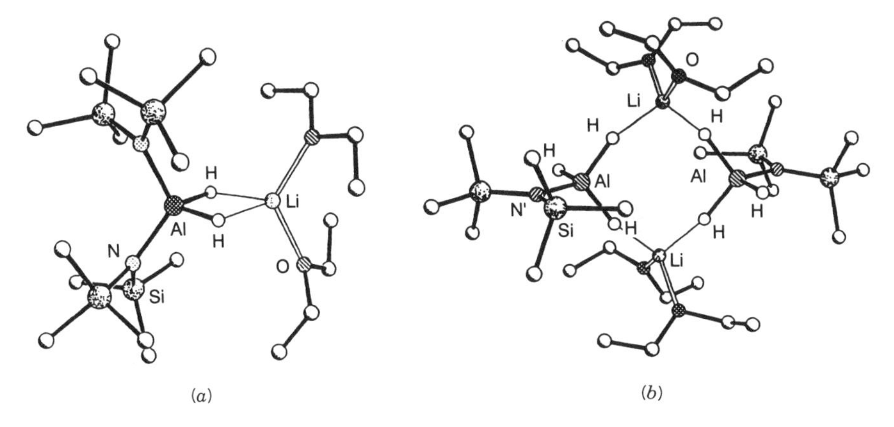

Figure 8.41

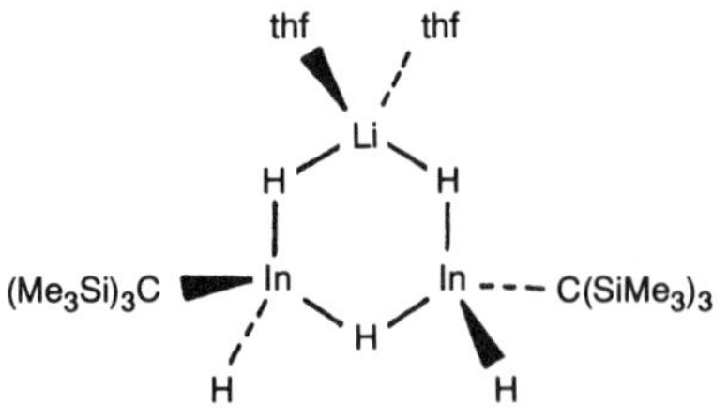

Figure 8.42

usual $[(Me_3Si)_3CIn]_2H_3(\mu_2{-}H)_2Li.2THF$ **(43)** complex (Fig. 8.42), which contains an In_2Li core in which the three metal centers are held together by $\mu_2{-}H$ bridges in a six-membered ring.[114]

Finally, probably the most spectacular example of a Li· · ·H contacted complex is that of the tetrameric tungstate complex $[\{W(PMe_3)_3H_5Li\}_4]$ **(44)** (Fig. 8.43).[115] The complex has a macrocyclic ring structure formed from an intricate network of $W(\mu_2{-}H)\cdot\cdot\cdot Li$ contacts. The central core of **(44)** is an eight-membered $[Li(\mu_2{-}H)]_4$ ring in which three additional Li· · ·H contacts are made to each Li, thus giving them a distorted five-coordinate geometry.

The coordinations of tetrafluoroborate ions in the few structurally characterized ion-contacted complexes with small Lewis base donor ligands are similar to those of the borohydride ions. Thus complexes in which one and two $Li(\mu_2{-}F)B$ contacts are present have been characterized in the solid state. In monomeric $[PMDETA.LiBF_4]$ **(45)** (Fig. 8.44), the BF_4^- anion is disordered over two orientations in the crystal. Surprisingly, both these arrangements have bent $Li{-}F{-}BF_3$ bridges (Li—F, ca. 1.91 Å; Li—F—B, ca. 148°).[116] The reasons for this bending are not clear, but semiempirical MNDO calculations on the full monomer indicate that the bent Li—F—B contact only differs by approximately 1 kcal mol^{-1} from the linear arrangement. Bending of the bridge may therefore be a consequence simply of crystal packing effects.[73] $^7Li{-}^{19}F$ NMR coupling is not observed in variable-temperature solution studies on **(45)**. However, measurement of the T_1 values for ^{11}B and ^7Li at various temperatures indicate that both nuclei are tumbling at the same rate (E_{act} = 3.2 ± 0.1 kcal

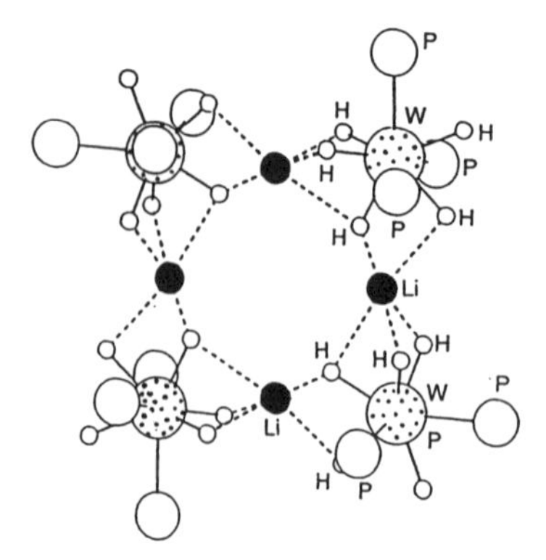

Figure 8.43

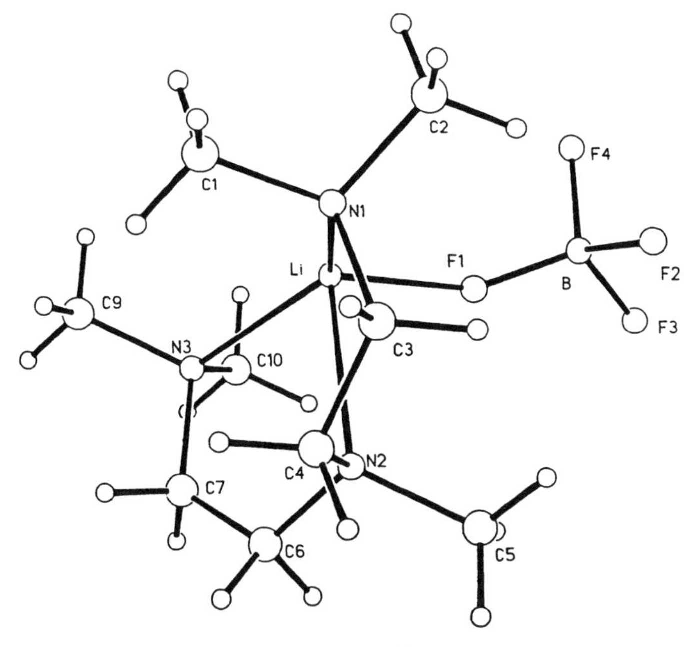

Figure 8.44

mol^{-1}) and thus that a close ion pair is sustained in solution between Li^+ and BF_4^-.

Surprisingly, variable-concentration cryoscopic molecular mass measurements on benzene solutions of $[LiBF_4.HMPA]_n$ **(46)**, which has an ion-separated structure in the solid state, that is $[Li(HMPA)_4^+.BF_4^-]$, indicate that even here $Li—F—BF_3$ association, giving a five-coordinate Li^+ cation, occurs.[32] Such is seen in the increase in the association number (n) above n = 0.5 (ion separation) with increased concentration for this complex in benzene in which it is extremely soluble.

A dibridge between Li^+ and BF_4^- is observed in the solid-state structure of $[(Ph_3P)_2Pt(\mu_2—O)]_2Li(\mu_2—F)_2BF_2$ **(47)** (Fig. 8.45) which is composed of a μ_2—O bridged Pt_2O_2 dimer core straddled by a Li^+ cation, which is in turn coordinated by two F atoms of a BF_4^- anion.[117]

Ab initio MO calculations (6-31G)[71] on models of $LiBF_4$ (Fig. 8.46*a–d*) indicate that the degree of μ_2—F bridging between the anion and Li^+ is dictated by the dual effects of the desire to maximize Li· · ·F interaction, so as to reduce the net charge on Li, and the need to minimize the strain of the BF_4^- anion. Thus, in the dibridged model *b* these two effects are optimized and this is the most stable unsolvated geometry. In contrast, although maximizing cation–anion interaction, the tetra-bridged structure *d* leads to unacceptably large strain in the BF_4^- anion and this model is least stable.

The most stable optimized geometries of the solvated species $(H_2O)_xLiBF_4$

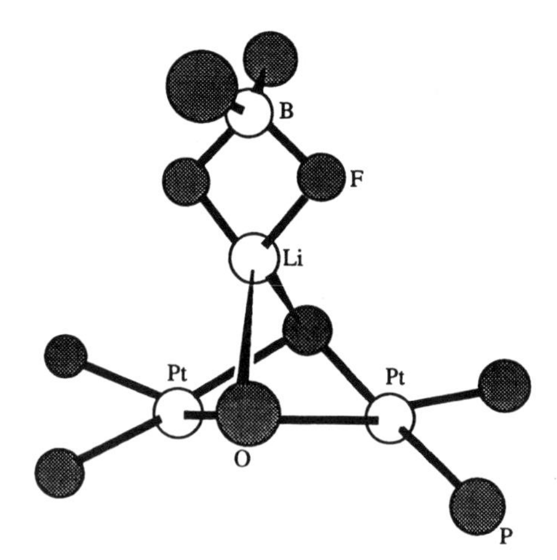

Figure 8.45

(x = 1–4) are shown in Figure 8.47. The main conclusion which can be drawn from these calculations is that association between Li^+ and BF_4^- ions is highly favored and thus is an expected structural feature in such complexes. This is exemplified particularly in the tetra-solvated model *d*, which adopts a surprising ion-paired structure in which Li^+ is five coordinate rather than having the expected ion-separated structure $(H_2O)_4Li^+.BF_4^-$. The total complexation energy of $(H_2O)_4Li^+$ and BF_4^- in *d* is 100.6 kcal mol^{-1}. This value is pertinent particularly to the observed tendency for ion pairing between $(HMPA)_4Li^+$ and BF_4^- ions in complex **(46)** in concentrated benzene solutions in which a species similar to model *d* must be present.[32] The calculational predictions that the most stable disolvated model will have a dibridge (model *b*) and that the most stable trisolvated model will possess a monobridge (model *c*) are in broad agreement with the observation of a dibridge in $[(Ph_3P)_2Pt(\mu_2{-}O)]_2Li(\mu_2{-}F)_2BF_2$ **(47)**[117] and of a monobridge in $[PMDETA.LiBF_4]$ **(45)**.[116]

Few inorganic salt complexes containing oxyanions (e.g., NO_3^-, CO_3^{2-}, ClO_4^-) in which there are interactions with Li^+ cations have been investigated in the solid state. The most well-characterized complexes are of lithium nitrate and perchlorate.

The solid-state structures of the monomeric $[(3,5\text{-}Me_2Pyr)_3Li(\mu_2{-}ONO_2)]$ **(48)** (Fig. 8.48) and the polymeric $[(2\text{-}MePyr)_2Li(\mu_2{-}O)NO_2]_\infty$ **(49)** (Fig. 8.49) nitrate complexes have been obtained.[47] Presumably, the reason for the different stoichiometries in these complexes is due to the different steric bulks, and therefore donicities, of the methylated pyridine ligands involved. In **(48)** one O of each NO_3^- is attached to Li^+, thus giving a typical four-coordinate geometry. In contrast, the NO_3^- anion does not chelate Li^+ in **(49)**, which would give a four-coordinate monomer. Rather, the anions link the monomeric units into a polymeric strand with a Li—O—N(O)—O—Li backbone. There

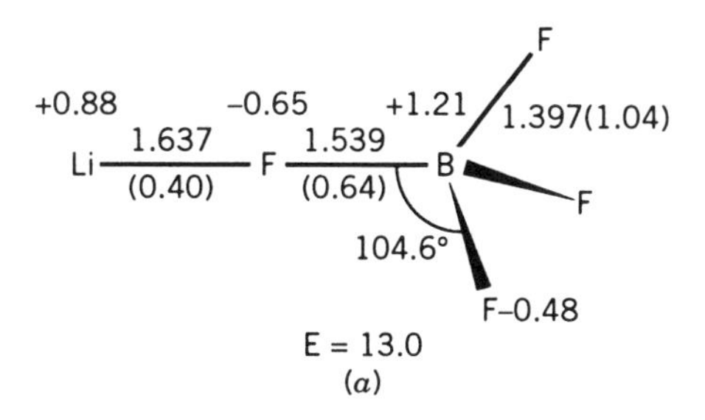

+0.88
-0.65
+1.21
F
1.397(1.04)
Li
1.637
(0.40)
F
1.539
(0.64)
B
F
104.6°
F-0.48
E = 13.0
(a)

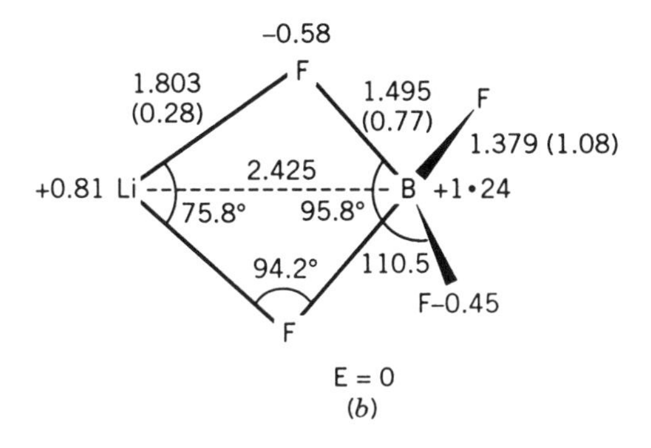

-0.58
F
1.803
(0.28)
1.495
(0.77)
F
1.379 (1.08)
+0.81 Li
2.425
B +1•24
75.8°
95.8°
94.2°
110.5
F-0.45
F
E = 0
(b)

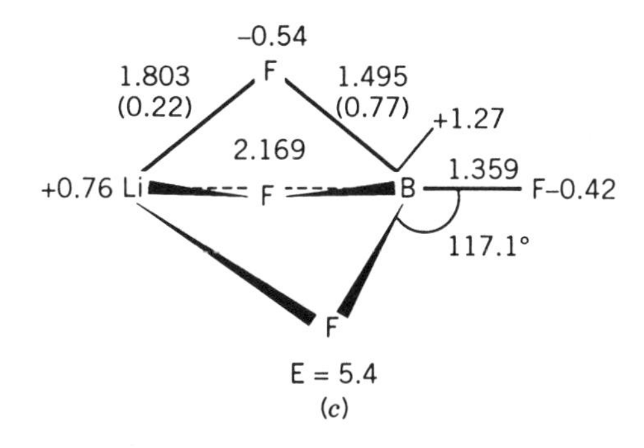

-0.54
F
1.803
(0.22)
1.495
(0.77)
+1.27
2.169
+0.76 Li
F
B
1.359
F-0.42
117.1°
F
E = 5.4
(c)

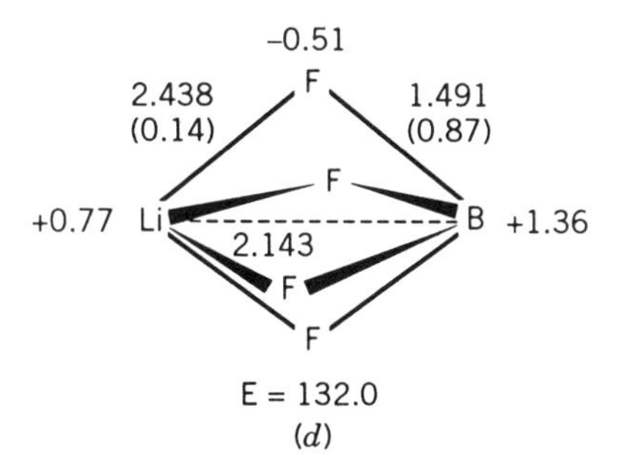

-0.51
F
2.438
(0.14)
1.491
(0.87)
F
+0.77 Li
B +1.36
2.143
F
F
E = 132.0
(d)

Figure 8.46

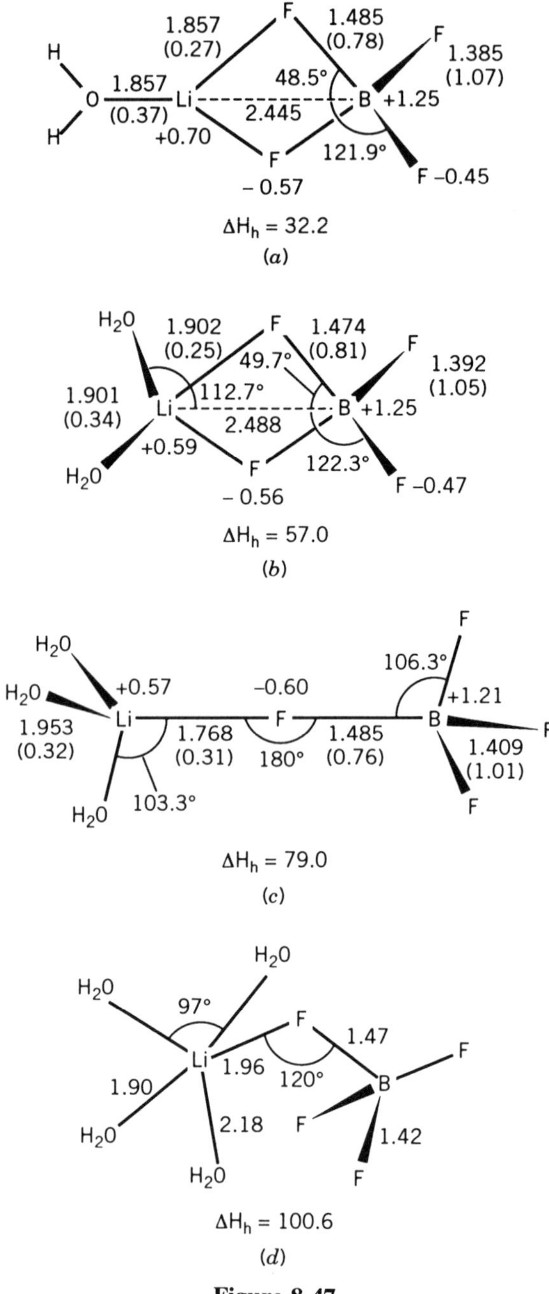
H
H
O
1.857
(0.37)
Li
+0.70
1.857
(0.27)
F
1.485
(0.78)
48.5°
2.445
B
+1.25
F
1.385
(1.07)
121.9°
F
- 0.57
F -0.45
ΔH_h = 32.2
(a)
H_2O
1.902
(0.25)
F
1.474
(0.81)
49.7°
F
1.392
(1.05)
1.901
(0.34)
Li
112.7°
2.488
B
+1.25
+0.59
H_2O
F
- 0.56
122.3°
F -0.47
ΔH_h = 57.0
(b)
H_2O
H_2O
+0.57
1.953
(0.32)
Li
1.768
(0.31)
-0.60
F
180°
1.485
(0.76)
106.3°
+1.21
B
F
F
1.409
(1.01)
F
H_2O
103.3°
ΔH_h = 79.0
(c)
H_2O
H_2O
97°
F
1.47
F
Li
1.96
120°
B
1.90
H_2O
2.18
F
1.42
H_2O
F
ΔH_h = 100.6
(d)

Figure 8.47

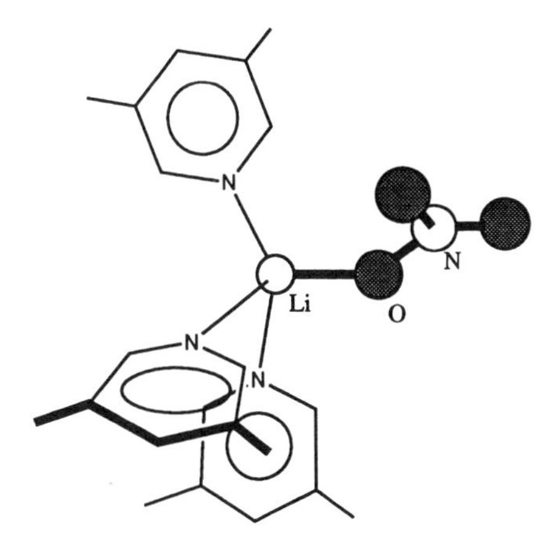

Figure 8.48

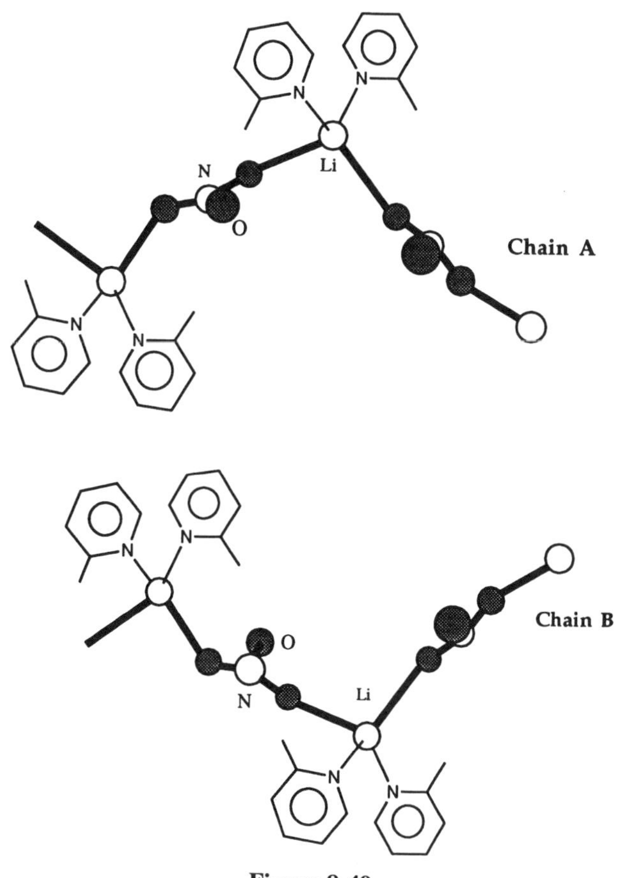

Figure 8.49

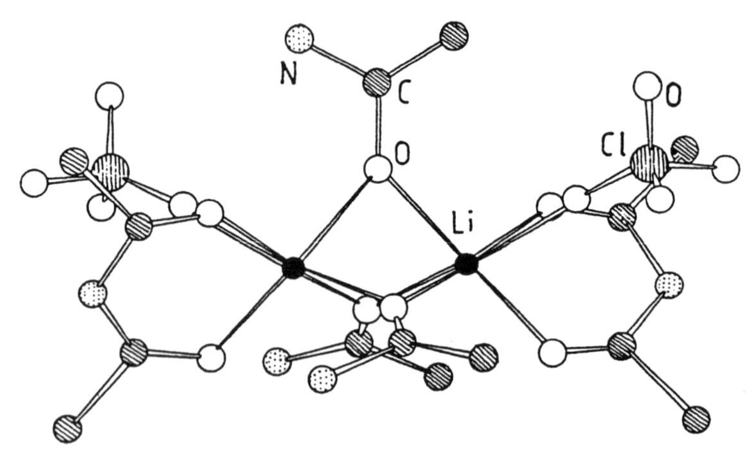

Figure 8.50

are two crystallographically independent polymeric units in **(49)** (A and B) which differ essentially in the orientation of their backbones.

In the polymeric perchlorate complex [$LiClO_4$. (1.5acetamide).(diacetamide)]$_2$ **(50)**, which is prepared by heating $LiClO_4$ in acetamide, the basic unit is that of a dimer in which three μ_2—O attached acetamide ligands bridge the two Li^+ cations together (Fig. 8.50).[118] One terminal bidentate diacetamide ligand, which arises through coupling of acetamide during the preparation of **(50)**, chelates each Li center of the dimeric unit and one terminal singly attached ClO_4^- anion completes the five-coordinate geometry around Li^+. The ClO_4^- anions, in a similar way to the NO_3^- anions in **(49)**, link the dimer units of **(50)** together by Li—O—Cl(O)$_2$—O—Li bridges.

4.5 Mixed-Anion and Heterometallic Complexes

As mentioned in Section 4.1, mixed-anion and heterometallic complexes form an extremely diverse category of compounds in which there is the common feature that lithium salt fragments (very commonly halide fragments) are trapped by their incorporation into another, frequently organometallic, species. The role of the "host" complex in fragmenting the parent lattices of the lithium halide salts can be seen as similar to that of Lewis base donor molecules (Sections 4.2–4.4). An additional important feature of such "molecular sponges" is that commonly anion solvation as well as cation solvation of the lithium salt fragment can occur.

Both methyllithium (MeLi) and phenyllithium (PhLi) are commercially supplied as "halide-rich" ether solutions. The halide inclusion is an inevitable consequence of the metal–halogen exchange reactions of alkyl or aryl halides with lithium metal used in their commercial production, although it should be pointed out here that halide-free generation of the lithium organometallics is

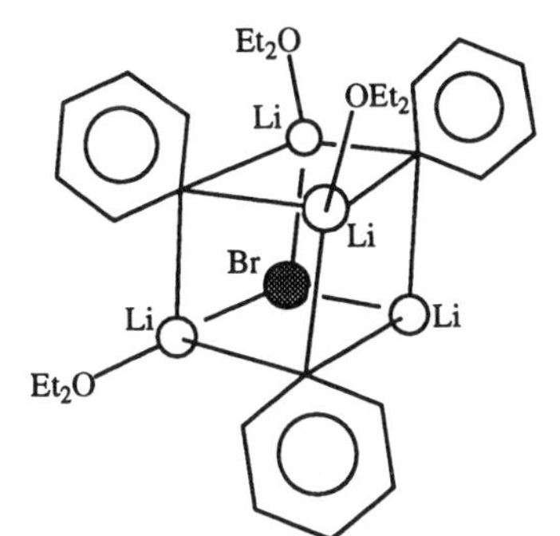

Figure 8.51

possible.[119] As early as 1961, with the characterization of $[MeLi.LiBr.2Et_2O]_n$, it had been appreciated that such solutions contain stoichiometric mixed-anion cocomplexes of the organometallics and the by-product halide salts.[120, 121] One of the first such structurally characterized complexes of this type was that of $[(PhLi.Et_2O)_3.LiBr]$ **(51)** (Fig. 8.51), isolated from the reaction of excess Li with PhBr, and in which a cubane-like Li_4C_3Br core is present.[122] This core motif is the same as that found in $[PhLi.Et_2O]_4$ **(52)** in which the Li—C and Li—O distances are similar to those in **(51).**[122] Thus it can be seen that the host cubane of **(52)** acts as a template in the formation of **(51).** A further feature is that LiBr monomer inclusion in **(51)** has been facilitated by both conventional Li^+ solvation by Et_2O and by Br^- solvation by Li^+ ions in the parent structure of **(52)**.

A $(LiBr)_2$ dimer unit is incorporated into the structure of cyclopropyllithium in the etherate complex $[(cyclopropylLi.Et_2O)_2\ (LiBr.Et_2O)_2]$ **(53)** (Fig. 8.52), prepared in moderate yield by the reaction of excess Li with cyclopropylbromide. Again the same concept of host cation–anion solvation can be applied to this species in which a now more distorted cubane $Li_4C_2Br_2$ core than that in **(51)** is present.[123]

A lithium thiocyanate monomer is incorporated into the cubane framework of $[PhOLi.HMPA]_4$ **(54)**[124] (Fig. 8.53*a*) by the ammonium salt reaction of phenol and NH_4NCS with Bu^nLi (3 : 1 : 4 equivalents respectively) in toluene and HMPA (4 equivalents). The remarkable $[(PhOLi.HMPA)_3$.-

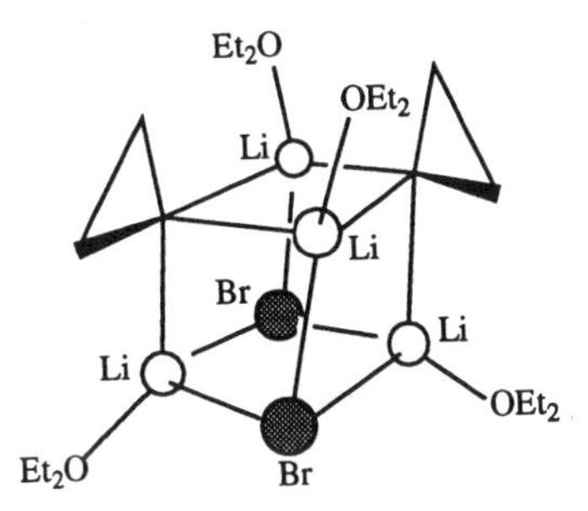

Figure 8.52

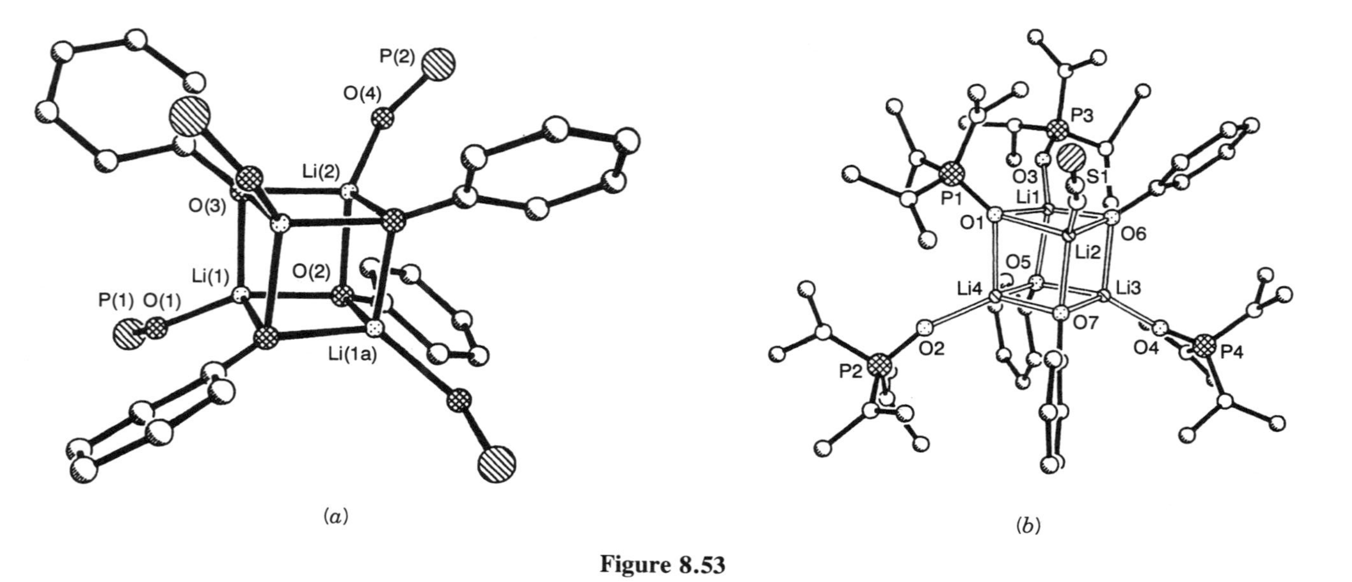
P(2)
O(4)
Li(2)
O(3)
Li(1)
O(2)
P(1) O(1)
Li(1a)
(a)
P3
O3
S1
P1
Li1
O1
O6
O5
Li2
Li4
Li3
O7
O2
O4
P2
P4
(b)

Figure 8.53

(LiNCS).μ_3—O=P(NMe_2)$_3$] complex **(55)** (Fig. 8.53*b*) has a distorted cubane Li_4O_4 core akin to **(54)** but in which three μ_3—O phenoxide anions and one μ_3—O HMPA ligand hold the four Li^+ cations together.[125] The only other observation of the μ_3—O HMPA bridging mode is in the inorganic complex {KNCS.[μ_2—O=P(NMe_2)$_3$]}$_3$.[μ_3—O=P(NMe_2)$_3$]$_2$ **(34)** (Fig. 8.34, Section 4.3).[64] The occurrence of this bridging mode in **(54)** testifies to the electronic similarity of the phenoxide anion and HMPA, as alluded to earlier in connection with lithium halide donor complexes (Section 4.2). Evidently such a bridge is more favorable than the incorporation of a μ_3—NCS apex in the structure of **(54)**, and the NCS^- anion is found here in a terminal bonding mode. However, despite the presence of a terminal, rather than μ_3—NCS, anion in the solid-state structure of **(55)**, ^{31}P and ^7Li NMR studies show that a fluxional process involving all the statistically possible terminal-NCS and μ_3—NCS isomers is occurring in aromatic solvents.[124, 125] This study serves to illustrate the usefulness of the ^{31}P NMR chemical shift of the HMPA ligand in the assignment of this ligand's bonding. Where the HMPA ligand is found in a terminal mode (e.g., in $Li(HMPA)_4^+.X^-$, $X^- = Br^-$ and I^-[33]) the ^{31}P chemical shift is at about δ24–25 and for a μ_2—O HMPA bridging mode [e.g., in Br—Li.[μ—O=P(NMe_2)$_3$]$_3$.Li—Br **(14)**[56]] the HMPA resonance is found at approximately δ28 (both relative to 10% H_3PO_4 in water). On the basis of the previous chemical shifts, a μ_3—O HMPA ligand would be expected to come at about δ31. In the low-temperature ^{31}P NMR spectra of complex **(55)**, resonances centered at about δ24 and 30 are observed. Although complicated, the detection of two resonances at about δ30–31 shows the presence both of the "1,2 isomer," the same as that in the solid-state structure of **(55)** in which the terminal NCS^- anion is attached to a Li^+ cation next to a μ_3—O HMPA ligand, and of the "1,4 isomer," in which the terminal NCS^- anion is attached to a Li^+ cation diagonally opposite to a μ_3—O HMPA ligand (Fig. 8.54).

The direct absorption of solid LiI by a solution of the trimeric lithium phen-

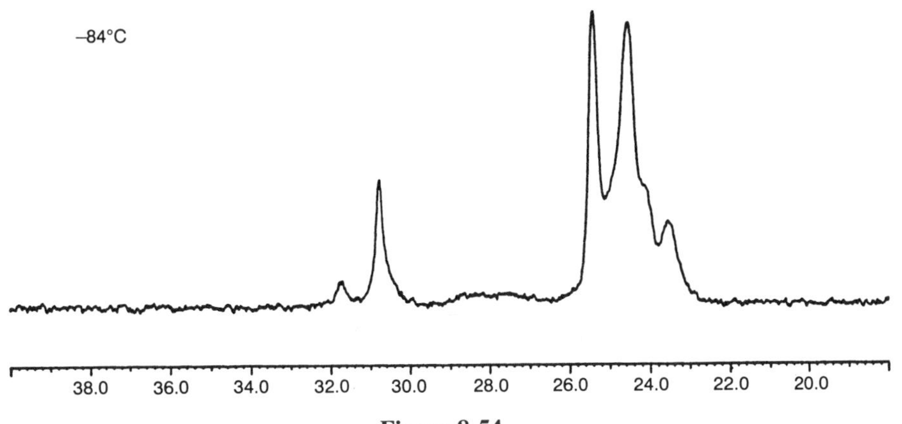

Figure 8.54

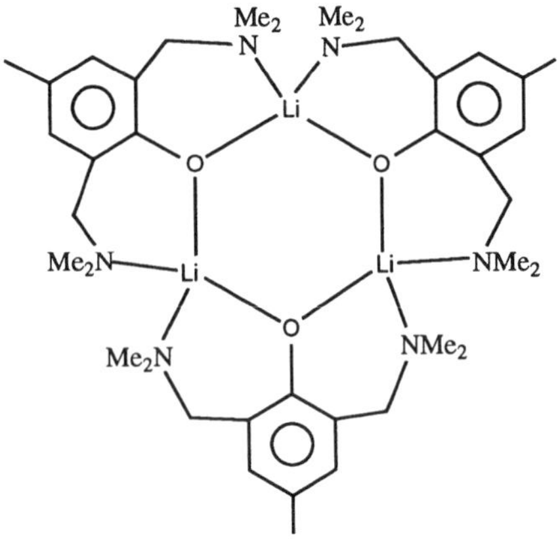

Figure 8.55

oxide **(56)** (Fig. 8.55) results in the mixed phenoxide/LiI complex **(57)** (Fig. 8.56) in which both the anion and the cation of the two Li—I monomers have been incorporated [in contrast to the situation occurring in **(55)**], into the framework of the cubane.[126]

This tendency for phenoxides to form cubanes, as seen in complexes **(54)**, **(55)**, and **(57)**, is echoed in the adoption of a cubane structure for the unusual heterometallic halide–phenoxide complex $[(Me_3Si)_3C]Cd(LiBr)_3(OSiMe_3).2THF$ **(57)**,[127] which contains a tetrametallic $CdLi_3Br_3O$ cubane-like core (Fig. 8.57) in which one μ_3—O phenoxide and three μ_3—Br anions are present.

Mixed lithium organometallic–inorganic complexes have recently become a particularly active area of study, especially concerning cocomplexes with lithium amide derivatives. This renewed interest stems from the realization (from synthetic, mechanistic, and NMR spectroscopic studies[38, 128]) that lithium amide–lithium halide aggregates have superior stereoselectivity over the parent amide bases in enolization reactions. Such interest has no doubt been spurred by the recent structural characterizations of two lithium diisopropylamide (LDA) complexes[129, 130] and by the likelihood that increased knowledge concerning related mixed halide complexes may give rise to enhanced reagent selectivity in organic synthesis.[17]

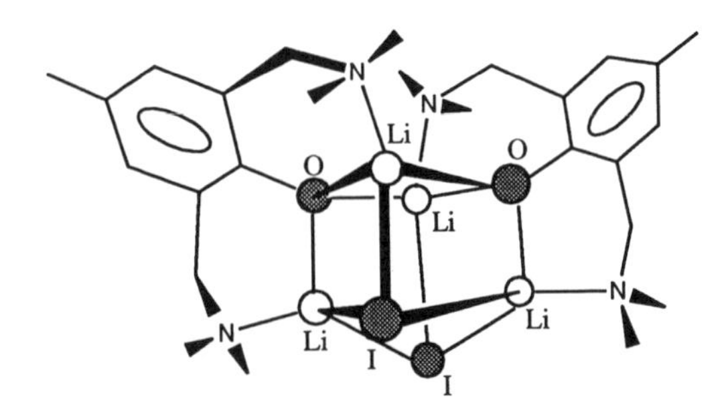

Figure 8.56

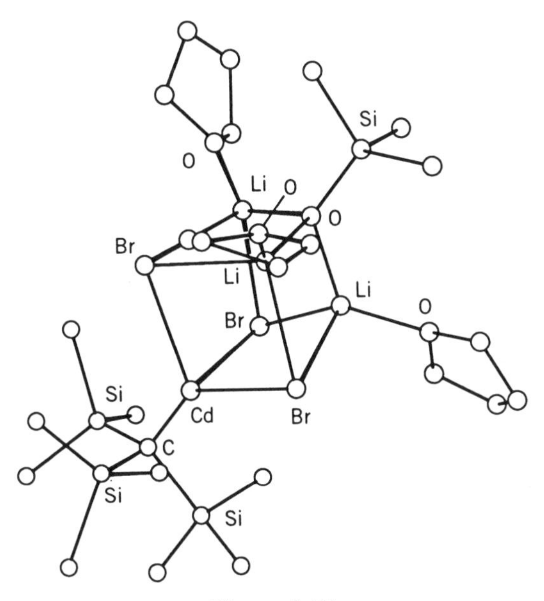

Figure 8.57

The first structurally characterized complex of this type was the simple dinuclear complex $[(C_6H_4)_2(NH)NLi.LiCl.4THF]$ **(59)** (Fig. 8.58) in which one Li—Cl monomer has been trapped.[131]

Recently, a modified version of the ammonium salt route (see Section 3) involving low-temperature deprotonation of hydrohalide salts ($R_2NH_2^+.X^-$; $X^- = Cl^-$ and Br^-) has been employed in the synthesis of mixed halide–amide aggregates[38] and the solid-state structure of the trinuclear LiCl

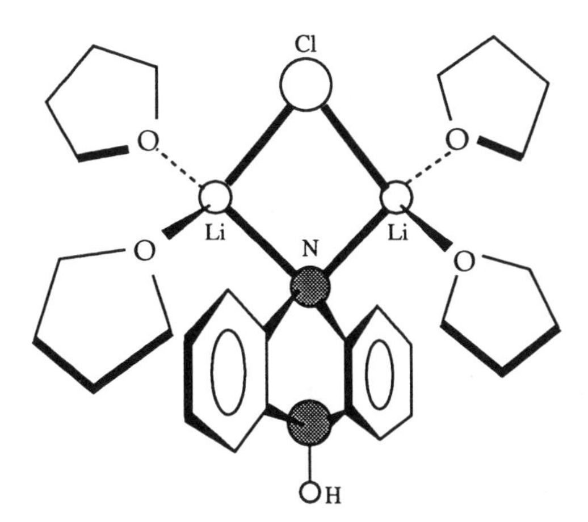

Figure 8.58

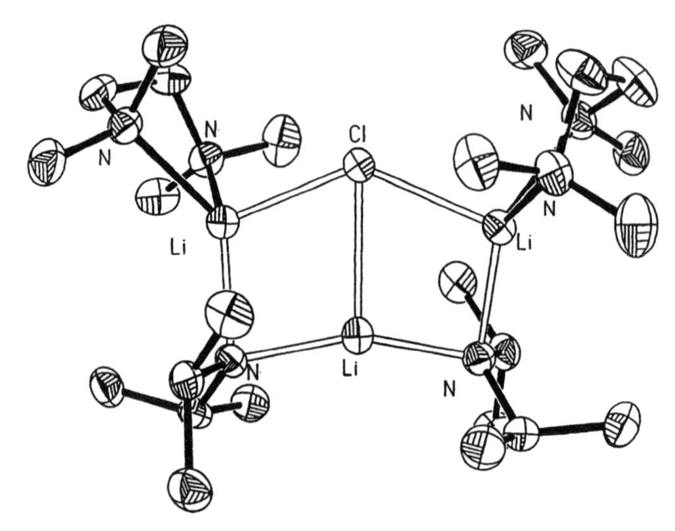

Figure 8.59

ladder complex of LDA, $[(Me_2CH)_2NLi]_2.LiCl(TMEDA)_2$ **(60)**, prepared in this way has been reported.[39] The structure can be regarded as a Li—Cl monomer trapped between two LDA monomers (Fig. 8.59). The three-coordinate central Li adopts an unusual planar geometry in this species and short agostic C—H· · ·Li interactions from the Me groups stabilize this geometry.

Of the multitude of the second class of cocomplexes, heterometallic complexes, which have been investigated and characterized mainly in the solid state, there are no really apparent structural trends. However, all these species contain early main group (other than Li) metal, p block metal, transition metal, or lanthanide or actinide metal (generally organometallic) complexes which, in a similar way to the behavior of the lithium organometallics in mixed-anion complexes, function as hosts in trapping lithium salt fragments. A common prerequisite for such host species to cocomplex lithium fragments, especially where large lanthanide, actinide, or transition metals are involved, is that the host metal (M) is not coordinately saturated. The anion of the inorganic lithium salt fragment (X^-) therefore plays the role of a donor to these metal centers.

A representative example of this class is the bimetallic Cp'_2-U.$(\mu_2$—$Cl)_2$Li.PMDETA [Cp′ = 1,3-$(Me_3Si)_2C_5H_3$] **(61)** in which the Li^+ cation is linked through two μ_2—Cl bridges to U (Fig. 8.60).[132] The unusual five-coordinate geometry at Li in **(61)** is similar to that in the inorganic halide complex $[Li(\mu_2$—$Br).PMDETA]_2$ **(19)** (Section 4.2).[40] The function of the neutral organometallic U(III) fragment in **(61)** Cp'_2UCl is clearly related, as has been noted above, to the function of the metalloorganic fragments in mixed-anion complexes in trapping the LiCl fragment. However, the concept of the host complexes "capturing" the lithium salt fragments by a combination of

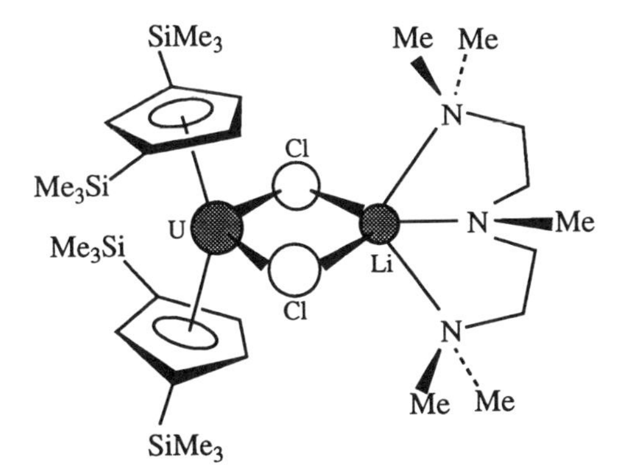

Figure 8.60

cation and anion solvation is perhaps most clearly illustrated by a few structurally characterized heterometallic species in which pendant heteroatom groups act as intramolecular donors to Li. Good examples of this type are $\{(Me)_2Al[(Me_2Si)_2C.PMe_2](\mu_2{-}Cl)Li.TMEDA\}$ **(62)** (Fig. 8.61*a*)[133] and $CpTi(\mu_2{-}NSiMe_3)_2Si(Ph)N(SiMe_3)(\mu_2{-}Cl)Li(OEt_2)$ **(63)** (Fig. 8.61*b*).[134]

Similar structural $M(\mu_2{-}X)_2Li$ ($X^- = Cl^-, Br^-$) motifs are observed in a large variety of bimetallic complexes which criss-cross the Periodic Table. Such are seen in the structures of the bimetallic complexes $[(THF)(Me_3Si)_3CMg(\mu_2{-}Br)_2Li.2(THF)]$ **(64)**,[127] $(THF)_2(Cl)_2Y(\mu_2{-}Cl)_2Li.2THF]$ **(65)**,[135] $[(Me_3Si)_3C.Ga(Cl)(\mu_2{-}Cl)_2Li.2THF]$ **(66)**,[136] and $(Bu^t_3CO)_2Zr(Cl)(\mu_2{-}Cl)_2Li.2Et_2O$ **(67)**,[137] in which electron-deficient, coordinately unsaturated metal (acceptor) centers are present.

A mono $(\mu_2{-}Cl)$ bridge is observed in the structure of $(PMDETA)Li.(\mu_2{-}Cl)U[CH(SiMe_3)_2]_3$ **(68)** (Fig. 8.62*a*)[138] which can be compared to that present in the $PMDETA.Li(\mu_2{-}Cl)Li.PMDETA^+$ cation **(22)** (Section 4.2).[92] Evidently, the presence of sterically bulky ligands restricts the extent of halide bridging even to the extent that the lanthanide center assumes a merely four-coordinate geometry in **(68)**.

A similar bent $(\mu_2{-}Cl)$ bridge is also observed in $(THF)_3Li(\mu_2{-}Cl)Lu(Cp^*)(CH_2SiMe_3)[CH(SiMe_3)_2]$ **(69)** ($Cp^* = C_5Me_5$),[139] (Fig. 8.62*b*). The influence of steric factors on the nature of halide bridging is exemplified in the solid-state structure of the cocomplex $Cp^*_2Y(\mu_2{-}Cl)_2Li.2THF].[(Cp^*)_2Y(Cl)(\mu_2{-}Cl)Li.3THF]$ **(70)**, which contains both the di- and monobridged forms (Fig. 8.63).[140] Monobridging of a Cl^- anion also occurs, although less commonly than for the lanthanides and actinides, in certain p block complexes, such as in $[(Me_3Si)_3Si]_2Sn(\mu_2{-}Cl)Li.3THF$ **(71)**[141] and in $[(Me_3Si)_3C](Cl)_2In(\mu_2{-}Cl)Li.3THF$ **(72)**.[136]

Chains and clustered polymetallic species, containing combinations of Li and other metallic centers linked through a variety of halide bridges, are also

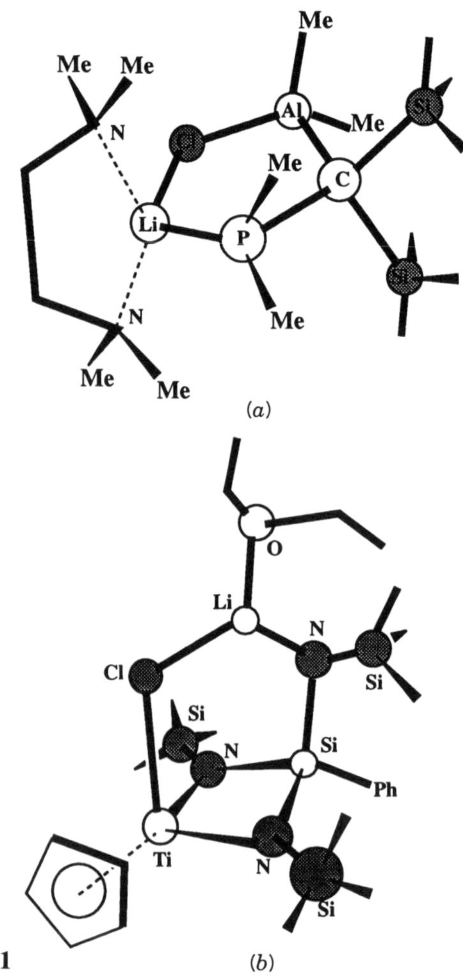

Figure 8.61

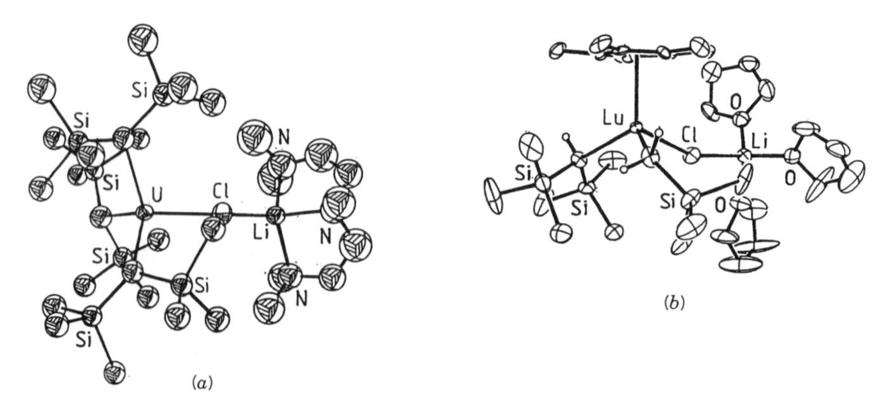

Figure 8.62

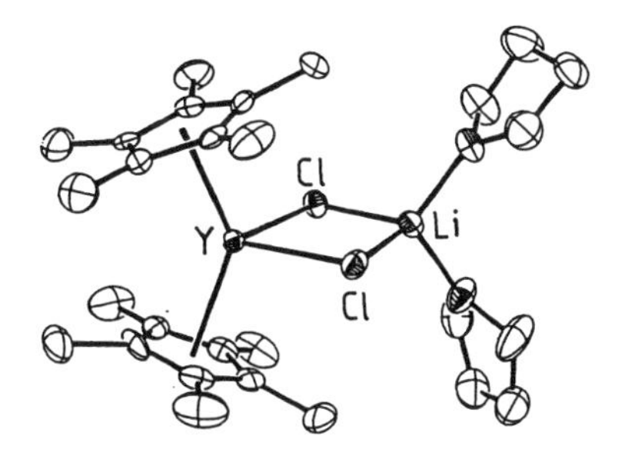

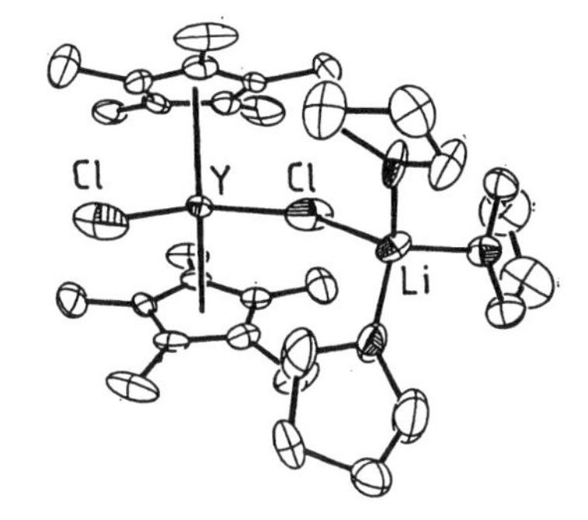

Figure 8.63

common and form an extremely structurally diverse class of complexes. For the purposes of this discussion, it will suffice to mention those complexes containing particularly unusual and structurally interesting features.

In the asymmetric molecular complex $Cl_3(Glyme)La(\mu_2-Cl)_2$.-$(\mu_3-Cl-Li.2THF)La(Glyme)(\mu_2-Cl)_2Li.2THF$ **(73)** (Glyme = Me-OCH_2CH_2OMe) two LiCl fragments are incorporated in two very distinct ways (Fig. 8.64).[142] The Cl^- anion of one LiCl unit functions as a μ_3-Cl bridge in holding the two La centers and Li^+ together, and the other adopts a more conventional terminal dibridged $(\mu_2-Cl)_2$ contact to one La. A similar μ_3-Cl attachment to that present in **(73)** has also been seen in the dication complex $[(Et_2O)_{10}.Li_4Cl_2]^{2+}$ of **(23)** (Section 4.2).[93]

Chain arrangements related to that of **(73)**, involving μ_2-Cl bridging of the acceptor metals to terminal Lewis base complexed Li^+ cations, are fairly

Figure 8.64

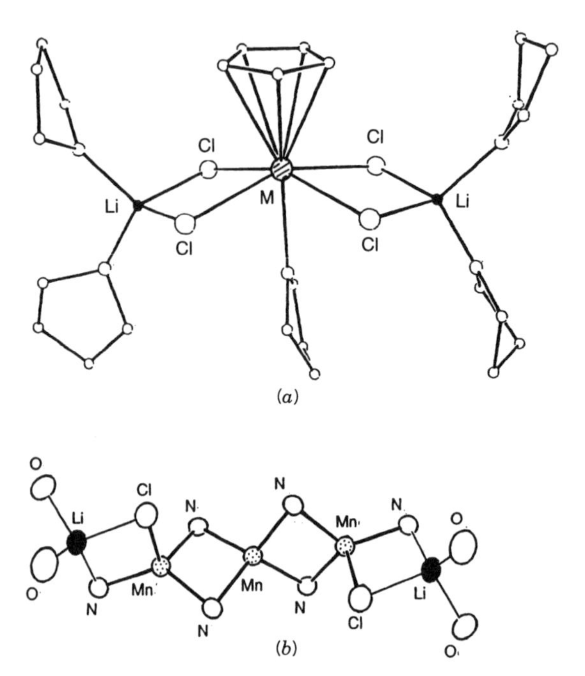

Figure 8.65

common. Simple examples are those of the trimetallic complexes $[(THF)CpM\{(\mu_2{-}Cl)_2Li.2THF\}_2]$ (M = Nd, La) **(74)** (Fig. 8.65*a*).[143] A spectacular pentametallic chain arrangement is also present in $\{Mn(NEt_2)_2\}_3(LiCl.2THF)_2$ **(75)** (Fig. 8.65*b*).[144]

Both $\mu_2{-}Cl$ and $\mu_3{-}Cl$ bridges between the host metal and Li are present in the remarkable trimetallic cluster of $[(C_5H_4CH_2C_5H_4)U]_2$-$(\mu_2{-}Cl)_3.(\mu_3{-}Cl)_2Li.2THF]$ **(76)** in which one LiCl fragment is present (Fig. 8.66).[145] The structure of the complex is based on a trigonal bipyramidal $U_2(\mu_3{-}Cl)_2Li$ core in which the metal centers are further associated by equatorial $(\mu_2{-}Cl)$ bridges.

A similar trimetallic system to that occurring in **(76)** is observed in the trimetallic complex anion of $\{[(Me_3Si)_3CCd(\mu_2{-}Cl)]_2(\mu_2{-}Cl)_2$-$Li.2THF\}^-.Li(THF)_4^+$ **(77)** (Fig. 8.67).[146] Interestingly, and in contrast to the structure of **(76)**, the Cd-associated $\mu_2{-}Cl$ bridges of **(77)** do not become further involved (as axial $\mu_3{-}Cl$ bridges) with the Li^+ center.

Perhaps the most dramatic cluster cocomplex yet characterized is that of the Y alkoxide/Li alkoxide/LiCl adduct $[Y_4(\mu_3{-}OCMe_3)_3(\mu_2{-}OCMe_3)_4$-$(OCMe_3)_4(\mu_4{-}O)]_2.[(LiOCMe_3)_2.(LiCl)]_2$ **(78)** (Fig. 8.68), prepared by the transmetallation of YCl_3 with $LiOCMe_3$ in THF, in which the two neutral tetranuclear Y_4 oxyalkoxide fragments sandwich a "slipped-hexameric" Li-alkoxide/LiCl core.[147]

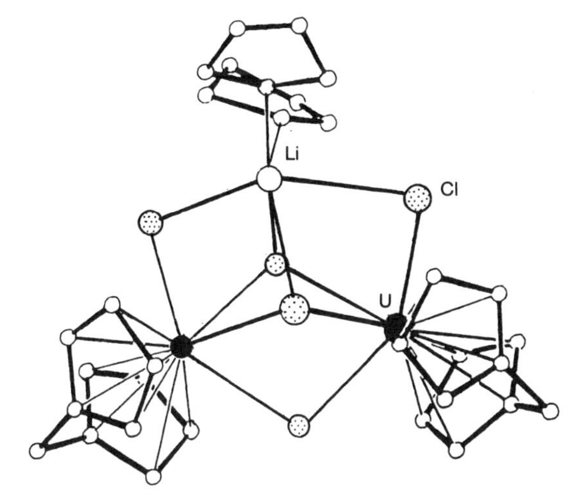

Figure 8.66

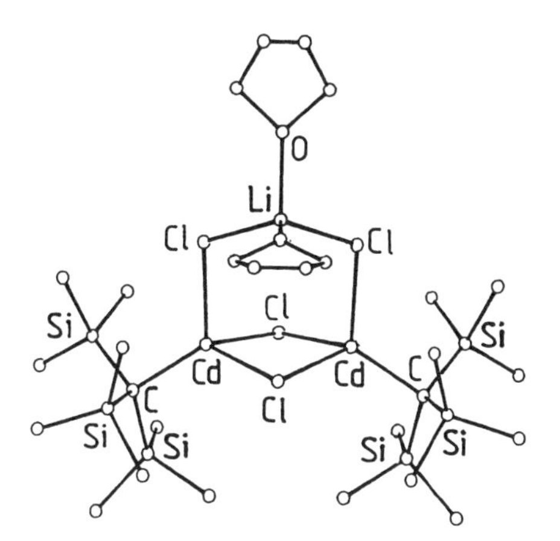

Figure 8.67

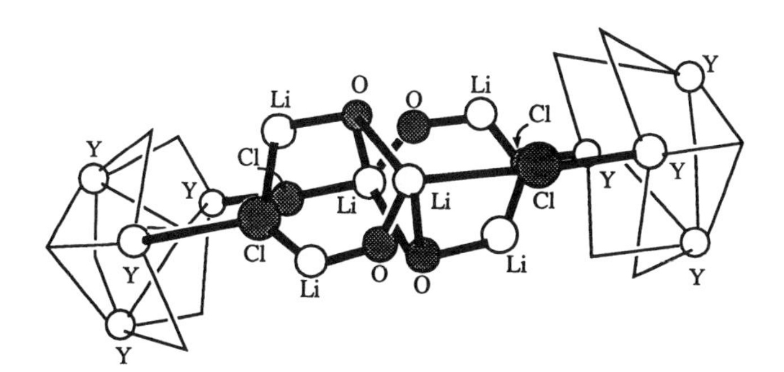

Figure 8.68

REFERENCES

1. Pedersen, C. J. *J. Am. Chem. Soc.* **1967**, *89*, 2495, 7017.
2. Pedersen, C. J. *Angew. Chem., Int. Ed. Engl.* **1988**, *27*, 1021.
3. Dietrich, B.; Lehn, J.-M.; Sauvage, J.-P. *Tetrahedron Lett.* **1969**, 2885, 2889.
4. Lehn, J.-M. *Angew. Chem., Int. Ed. Engl.* **1988**, *27*, 89.
5. Gokel, G. W. *Chem. Soc. Rev.* **1992**, 39.
6. Cooper, S. R. *Crown Compounds—Towards Future Applications*; VCH: Cambridge, 1992.
7. Dehmlow, E. V.; Dehmlow, S. S. *Phase Transfer Catalysis*, 2nd ed.; VCH: London, 1983.
8. Cram, D. J. *Angew. Chem., Int. Ed. Engl.* **1988**, *27*, 1009.
9. Fieser, M. *Reagents for Organic Synthesis*; Wiley-Interscience: New York, 1990; Vol. 15. This and earlier volumes give specific uses of alkali metal reagents arranged according to individual metal compounds.
10. Wakefield, B. J. *Organolithium Methods*, Academic Press: New York, 1988.
11. Brandsma, L.; Verkruijsse, H. D. *Preparative Polar Organometallic Chemistry*, Springer: Berlin, 1987; Vol. 1.
12. *Lithium: Current Applications in Science, Medicine, and Technology*; Bach, R. O., Ed., Wiley-Interscience: New York, 1985.
13. Poonia, N. S.; Bajaj, A. V. *Chem. Rev.* **1979**, *79*, 389.
14. Setzer, W.; Schleyer, P. v. R. *Adv. Organomet. Chem.* **1985**, *24*, 353.
15. Schade, C.; Schleyer, P. v. R. *Adv. Organomet. Chem.* **1987**, *27*, 169.
16. Wardell, J. L. In *The Chemistry of the Metal–Carbon Bond*; Hartley, F. R., Ed.; Wiley: Chichester, 1987; Vol. 4.
17. Seebach, D. *Angew. Chem., Int. Ed. Engl.* **1988**, *27*, 1624.
18. Gregory, K.; Schleyer, P. v. R.; Snaith, R. *Adv. Inorg. Chem.* **1991**, *37*, 47.
19. Mulvey, R. E. *Chem. Soc. Rev.* **1991**, *20*, 167.
20. Schleyer, P. v. R. *Pure Appl. Chem.* **1983**, *55*, 355; **1984**, *56*, 151.
21. Streitwieser, A. *Acc. Chem. Res.* **1984**, *17*, 353.
22. Reed, A. E.; Weinstock, R. B.; Weinhold, F. *J. Chem. Phys.* **1985**, *83*, 735.
23. Armstrong, D. R.; Barr, D.; Snaith, R.; Clegg, W.; Mulvey, R. E.; Wade, K.; Reed, D. *J. Chem. Soc., Dalton Trans.* **1987**, 1071.
24. Armstrong, D. R.; Barr, D.; Clegg, W.; Hodgson, S. M.; Mulvey, R. E.; Reed, D.; Snaith, R.; Wright, D. S. *J. Am. Chem. Soc.* **1989**, *111*, 4719.
25. Sannigrahi, A. B.; Kar, T.; Niyogi, B. G.; Hobza, P.; Schleyer, P. v. R. *Chem. Rev.* **1990**, *90*, 1061.
26. *Alkali Halide Vapors*, Davidovits, P.; McFadden, D. L., Eds.; Academic Press: New York, 1979.
27. Whetten, R. L. *Acc. Chem. Res.* **1993**, *26*, 49.
28. Barr, D.; Clegg, W.; Mulvey, R. E.; Snaith, R. *J. Chem. Soc., Chem. Commun.* **1984**, 79.
29. Bayliff, A. E.; Bryce, M. R.; Chambers, R. D.; Matthews, R. S. *J. Chem. Soc., Chem. Commun.* **1985**, 1018.

30. Ichikawa, J.; Sujimoto, K.-I.; Sonoda, T.; Kobayashi, H. *Chem. Lett.* **1987**, 1985.
31. Clark, J. H.; Hyde, A. J.; Smith, D. K. *J. Chem. Soc., Chem. Commun.* **1986**, 791.
32. Barr, D.; Hutton, K. B.; Morris, J. H.; Mulvey, R. E.; Reed, D.; Snaith, R. *J. Chem. Soc., Chem. Commun.* **1986**, 127.
33. Snaith, R.; Wright, D. S. unpublished observations.
34. "The Electric Car's Achilles' Axle," in *The Economist*, September 19, 1992, p. 133.
35. *Lithium Batteries*; Gabano, J. P., Ed.; Academic Press: New York, 1983.
36. *Lithium Battery Technology*; Venkatasetty, H. V., Ed.; Wiley: New York, 1984.
37. Weppner, W.; Welzel, W.; Kniep, R.; Rabenau, A. *Angew. Chem., Int. Ed. Engl.* **1986**, *25*, 1087.
38. Hall, P. L.; Gilchrist, J. H.; Collum, D. B. *J. Am. Chem. Soc.* **1991**, *113*, 9571.
39. Mair, F. S.; Clegg, W.; O'Neil, P. A. *J. Am. Chem. Soc.* **1993**, *115*, 3388.
40. Hall, S. R.; Raston, C. L.; Skelton, B. W.; White, A. H. *Inorg. Chem.* **1983**, *22*, 4070.
41. Raston, C. L.; Skelton, B. W.; Whitaker, C. R.; White, A. H. *J. Chem. Soc., Dalton Trans.* **1988**, 987.
42. Raston, C. L.; Whitaker, C. R.; White, A. H. *J. Chem. Soc., Dalton Trans.* **1988**, 991.
43. Raston, C. L.; Skelton, B. W.; Whitaker, C. R.; White, A. H. *Aust. J. Chem.* **1988**, *41*, 341.
44. Raston, C. L.; Skelton, B. W.; Whitaker, C. R.; White, A. H. *Aust. J. Chem.* **1988**, *41*, 409.
45. Raston, C. L.; Whitaker, C. R.; White, A. H. *Aust. J. Chem.* **1988**, *41*, 413.
46. Raston, C. L.; Whitaker, C. R.; White, A. H. *Inorg. Chem.* **1989**, *28*, 163.
47. Raston, C. L.; Whitaker, C. R.; White, A. H. *Aust. J. Chem.* **1988**, *41*, 1917.
48. Raston, C. L.; Skelton, B. W.; Whitaker, C. R.; White, A. H. *Aust. J. Chem.* **1988**, *41*, 1925.
49. Raston, C. L.; Whitaker, C. R.; White, A. H. *Aust. J. Chem.* **1989**, *42*, 201.
50. Raston, C. L.; Whitaker, C. R.; White, A. H. *Aust. J. Chem.* **1989**, *42*, 1393.
51. Skelton, B. W.; Whitaker, C. R.; White, A. H. *Aust. J. Chem.* **1990**, *43*, 755.
52. Raston, C. L.; Robinson, W. T.; Skelton, B. W.; Whitaker, C. R.; White, A. H. *Aust. J. Chem.* **1990**, *43*, 1163.
53. Harvey, S.; Skelton, B. W.; White, A. H. *Aust. J. Chem.* **1991**, *44*, 309.
54. Barr, D.; Clegg, W.; Mulvey, R. E.; Snaith, R. *J. Chem. Soc., Chem. Commun.* **1984**, 974.
55. Barr, D.; Snaith, R.; Wright, D. S.; Mulvey, R. E.; Wade, K. *J. Am. Chem. Soc.* **1987**, *109*, 7891. The route has been patented: Associated Octel Co. Ltd., European Patent Nos. 88309913.7, 1987 and 8915531.1, 1989.
56. Barr, D.; Doyle, M. J.; Mulvey, R. E.; Raithby, P. R.; Reed, D.; Snaith, R.; Wright, D. S. *J. Chem. Soc., Chem. Commun.* **1989**, 318.
57. Brooker, A. T.; Snaith, R.; Wright, D. S. unpublished observations.

58. Barr, D.; Doyle, M. J.; Mulvey, R. E.; Raithby, P. R.; Snaith, R.; Wright, D. S. *J. Chem. Soc., Chem. Commun.* **1988**, 145.

59. Barr, D.; Doyle, M. J.; Drake, S. R.; Raithby, P. R.; Snaith, R.; Wright, D. S. *J. Chem. Soc., Chem. Commun.* **1988**, 1415.

60. Armstrong, D. R.; Khandelwal, A. H.; Raithby, P. R.; Snaith, R.; Stalke, D.; Wright, D. S. *Inorg. Chem.* **1993**, *32*, 2132.

61. Schade, C.; Bauer, W.; Schleyer, P. v. R. *J. Organomet. Chem.* **1985**, *295*, C25.

62. Klusener, P. A. A.; Brandsma, L.; Verkruijsse, H. D.; Schleyer, P. v. R.; Friedl, T.; Pi, R. *Angew. Chem., Int. Ed. Engl.* **1986**, *25*, 465.

63. Barr, D.; Doyle, M. J.; Drake, S. R.; Raithby, P. R.; Snaith, R.; Wright, D. S. *Polyhedron* **1989**, *8*, 215.

64. Barr, D.; Doyle, M. J.; Drake, S. R.; Raithby, P. R.; Snaith, R.; Wright, D. S. *Inorg. Chem.* **1989**, *28*, 1768.

65. Barr, D.; Brooker, A. T.; Doyle, M. J.; Drake, S. R.; Raithby, P. R.; Snaith, R.; Wright, D. S. *J. Chem. Soc., Chem. Commun.* **1989**, 893.

66. Barr, D.; Brooker, A. T.; Doyle, M. J.; Drake, S. R.; Raithby, P. R.; Snaith, R.; Wright, D. S. *Angew. Chem., Int. Ed. Engl.* **1990**, *29*, 285.

67. Edwards, P. P. *Adv. Inorg. Chem. Radiochem.* **1982**, *25*, 135.

68. Dye, J. L. *Prog. Inorg. Chem.* **1984**, *32*, 327.

69. Hehre, W. J.; Ditchfield, R.; Pople, J. A. *J. Chem. Phys.* **1972**, *56*, 2257; Hariharan, P. C.; Pople, J. A. *Theor. Chim. Acta* **1973**, *28*, 213; Dill, J. D.; Pople, J. A. *J. Chem. Phys.* **1975**, *62*, 2921.

70. Dupuis, M.; Spangler, D.; Wendoloski, J. J. *GAMESS*, N.R.C.C. Software Catalog, Vol. 1, Program No. 2 GO1, 1980; Guest, M. F.; Kendrick, J.; Pope, S. A. *GAMESS* Documentation, Daresbury Laboratory, 1983.

71. Armstrong, D. R.; Snaith, R.; Wright, D. S. unpublished observations.

72. Sapse, A.-M.; Raghavachari, K.; Schleyer, P. v. R.; Kaufmann, E. *J. Am. Chem. Soc.* **1985**, *107*, 6483.

73. Raghavachari, K.; Sapse, A.-M.; Jain, D. C. *Inorg. Chem.* **1987**, *26*, 2585.

74. Emsley, J.; Jones, D. J. *J. Chem. Soc., Chem. Commun.* **1980**, 703.

75. Emsley, J.; Jones, D. J.; Kuroda, R. *J. Chem. Soc., Dalton Trans.* **1982**, 1179.

76. Stalke, D.; Meyer, M.; Adrianarson, M.; Klingebiel, U.; Sheldrick, G. M. *J. Organomet. Chem.* **1989**, *366*, C15.

77. Pieper, U.; Walter, S.; Klingebiel, U.; Stalke, D. *Angew. Chem., Int. Ed. Engl.* **1990**, *29*, 209.

78. Dippel, K.; Klingebiel, U.; Sheldrick, G. M.; Stalke, D. *Chem. Ber.* **1987**, *120*, 611.

79. Stalke, D.; Keweloh, N.; Klingebiel, U.; Noltemeyer, M.; Sheldrick, G. M. *Z. Naturforsch B* **1987**, *42*, 1237.

80. Stalke, D.; Klingebiel, U.; Sheldrick, G. M. *J. Organomet. Chem.* **1988**, *344*, 37.

81. Adrianarson, M.; Stalke, D.; Klingebiel, U. *J. Organomet. Chem.* **1990**, *381*, C38.

82. Stalke, D.; Whitmire, K. H. *J. Chem. Soc., Chem. Commun.* **1990**, 833.

83. Klingebiel, M.; Meyer, M.; Pieper, U.; Stalke, D. *J. Organomet. Chem.* **1991**, *408*, 19.
84. Rodgers, R. D.; Bynum, R. V.; Atwood, J. L. *J. Cryst. Spect. Res.* **1984**, *14*, 29.
85. Borgholte, H.; Dehnicke, K. *Z. Anorg. Allg. Chem.* **1991**, *606*, 91.
86. Banister, A. J.; Clegg, W.; Gill, W. R. *J. Chem. Soc., Chem. Commun.* **1987**, 850; Hey-Hawkins, E.; Sattler, E. *J. Chem. Soc., Chem. Commun.* **1992**, 775; Boche, G.; Langlotz, I.; Marsch, M.; Harms, K. *Angew. Chem., Int. Ed. Engl.* **1992**, *31*, 1205.
87. De Angelis, S.; Solari, E.; Gallo, E.; Floriani, C.; Chiesi-Villa, A.; Rizzdi, C. *Inorg. Chem.* **1992**, *31*, 2520.
88. Schmuck, A.; Leopold, D.; Wallenhauer, S.; Seppelt, K. *Chem. Ber.* **1990**, *123*, 761.
89. Amstutz, R.; Dunitz, J. D.; Laube, T.; Schweizer, W. B.; Seebach, D. *Chem. Ber.* **1986**, *119*, 434.
90. Davidson, M. G.; Snaith, R.; Stalke, D.; Wright, D. S. *J. Org. Chem.* **1993**, *58*, 2810.
91. Mulvey, R. E.; Clegg, W.; Barr, D.; Snaith, R. *Polyhedron* **1986**, *5*, 2109.
92. Buttrus, N. H.; Eaborn, C.; Hitchcock, P. B.; Smith, J. D.; Stamper, J. G.; Sullivan, A. C. *J. Chem. Soc., Chem. Commun.* **1986**, 969.
93. Hope, H.; Oram, D.; Power, P. P. *J. Am. Chem. Soc.* **1984**, *106*, 1149.
94. Chiang, M. Y.; Böhlem, E.; Bau, R. *J. Am. Chem. Soc.* **1985**, *107*, 1679.
95. Edwards, A. J.; Paver, M. A.; Raithby, P. R.; Russell, C. A.; Wright, D. S. unpublished observations.
96. Shakir, R.; Zaworotko, M. J.; Atwood, J. L. *J. Organomet. Chem.* **1979**, *171*, 9.
97. Owston, P. G.; Rowe, J. M. *Acta Crystallogr.* **1960**, *13*, 253.
98. Edwards, P. P.; Loram, J.; Snaith, R.; Wright, D. S. unpublished observations.
99. Paoli, D.; Luçon, M.; Chabanel, M. *Spectrochim. Acta* **1978**, *34*, 1087.
100. Paoli, D.; Luçon, M.; Chabanel, M. *Spectrochim. Acta* **1979**, *35*, 593.
101. Chabanel, M.; Luçon, M.; Paoli, D. *J. Phys. Chem.* **1981**, *85*, 1058; Goralski, P.; Chabanel, M. *Inorg. Chem.* **1987**, *26*, 2169.
102. Izatt, R. M.; Pawlak, K.; Bradshaw, J. S.; Bruening, R. L. *Chem. Rev.* **1991**, *91*, 1721.
103. Weiss, E.; Lucken, E. A. C. *J. Organomet. Chem.* **1964**, *2*, 197; Weiss, E.; Hencken, G. *J. Organomet. Chem.* **1970**, *21*, 265.
104. Rhine, W.; Stucky, G. D.; Patterson, S. W. *J. Am. Chem. Soc.* **1975**, *97*, 6401.
105. Hooz, J.; Akiyama, S.; Cedar, F. J.; Bennet, M. J.; Tuggle, R. M. *J. Am. Chem. Soc.* **1974**, *96*, 274.
106. Eaborn, C.; El-Kheli, M. N. A.; Hitchcock, P. B.; Smith, J. D. *J. Chem. Soc., Chem. Commun.* **1984**, 1673.
107. Armstrong, D. R.; Clegg, W.; Colquhoun, H. M.; Daniels, J. A.; Mulvey, R. E.; Stephenson, I. R.; Wade, K. *J. Chem. Soc., Chem. Commun.* **1987**, 630.
108. Bell, N. A.; Shearer, H. M. M.; Spencer, C. B. *J. Chem. Soc., Chem. Commun.* **1980**, 711.

109. Uhl, W. *Z. Anorg. Allg. Chem.* **1989**, *570*, 37.
110. Uhl, W.; Schnepf, J. E. O. *Z. Anorg. Allg. Chem.* **1991**, *595*, 225.
111. Zintl, E.; Harder, A. *Z. Phys. Chem., Sect. B.* **1935**, *28*, 478.
112. Sklar, N.; Post, B. *Inorg. Chem.* **1967**, *6*, 669.
113. Heine, A.; Stalke, D. *Angew. Chem., Int. Ed. Engl.* **1992**, *31*, 854.
114. Arent, A. G.; Eaborn, C.; Hitchcock, P. B.; Smith, J. D.; Sullivan, A. C. *J. Chem. Soc., Chem. Commun.* **1986**, 988.
115. Berry, A.; Green, M. L. H.; Bandy, J. A.; Prout, K. *J. Chem. Soc., Dalton Trans.* **1991**, 2185.
116. Doyle, M. J.; Mann, B. E.; Raithby, P. R.; Snaith, R.; Wright, D. S. unpublished observations.
117. Li, W.; Barnes, C. L.; Sharp, P. R. *J. Chem. Soc., Chem. Commun.* **1990**, 1634.
118. Gentile, P. S.; White, J. G.; Cavalluzzo, D. D. *Inorg. Chim. Acta* **1976**, *20*, 37.
119. Schlosser, M.; Ladenberger, V. *J. Organomet. Chem.* **1967**, *8*, 193.
120. Talaleeva, T. V.; Rodionov, A. N.; Kocheshkov, K. A. *Dokl. Chem.* (Engl. Trans.) **1961**, *140*, 985.
121. Batalov, A. P. *J. Gen. Chem. USSR* (Engl. Trans.) **1971**, *41*, 154.
122. Hope, H.; Power, P. P. *J. Am. Chem. Soc.* **1983**, *105*, 5320.
123. Schmidbaur, H.; Schier, A.; Schubert, U. *Chem. Ber.* **1983**, *116*, 1938.
124. Khandelwal, A. H.; Snaith, R.; Stalke, D.; Wright, D. S. unpublished observations.
125. Raithby, P. R.; Reed, D.; Snaith, R.; Wright, D. S. *Angew. Chem., Int. Ed. Engl.* **1991**, *30*, 1011.
126. van der Schaaf, P. A.; Hogerheide, M. P.; Grove, D. M.; Spek, A. L.; van Koten, G. *J. Chem. Soc., Chem. Commun.* **1992**, 1703.
127. Buttrus, N. H.; Eaborn, C.; El-Kheli, M. N. A.; Hitchcock, P. B.; Smith, J. D.; Sullivan, A. C.; Tavakkoli, K. *J. Chem. Soc., Dalton Trans.* **1988**, 381.
128. Galiano-Roth, A. S.; Kim, Y.-J.; Gilchrist, J. H.; Harrison, A. T.; Fuller, D. J.; Collum, D. B. *J. Am. Chem. Soc.* **1991**, *113*, 5053.
129. Barnett, N. D. R.; Mulvey, R. E.; Clegg, W.; O'Neil, P. A. *J. Am. Chem. Soc.* **1991**, *113*, 8187.
130. Bernstein, M. P.; Romesberg, F. E.; Fuller, D. J.; Harrison, A. T.; Collum, D. B.; Liu, Q.-Y.; Williard, P. G. *J. Am. Chem. Soc.* **1992**, *114*, 5100.
131. Engelhardt, L. M.; Jacobsen, G. E.; White, A. H.; Raston, C. L. *Inorg. Chem.* **1991**, *30*, 3978.
132. Blake, P. C.; Lappert, M. F.; Atwood, J. L.; Zhang, H. *J. Chem. Soc., Chem. Commun.* **1988**, 1436.
133. Karsch, H. H.; Zellner, K.; Müller, G. *J. Chem. Soc., Chem. Commun.* **1991**, 466.
134. Bauer, D. J.; Bürger, H.; Liewald, G. R. *J. Organomet. Chem.* **1986**, *307*, 177.
135. Guan, J.; Jin, S.; Shen, Q. *Organometallics* **1992**, *11*, 2483.
136. Atwood, J. L.; Bott, S. G.; Hitchcock, P. B.; Eaborn, C.; Shariffudin, R. S.; Smith, J. D.; Sullivan, A. C. *J. Chem. Soc., Dalton Trans.* **1987**, 747.

137. Lubben, V.; Wolczanski, P. T.; van Duyne, G. B. *Organometallics* **1984**, *3*, 977.

138. Atwood, J. L.; Lappert, M. F.; Smith, R. G.; Zhang, H. *J. Chem. Soc., Chem. Commun.* **1988**, 1308.

139. van der Heijden, A.; Pasman, P.; de Boer, E. J. M.; Schaverien, C. J. *Organometallics* **1989**, *8*, 1459.

140. Evans, W. J.; Boyle, T. J.; Ziller, J. W. *Inorg. Chem.* **1992**, *31*, 1120.

141. Arif, A. M.; Cowley, A. H.; Elkins, T. M. *J. Organomet. Chem.* **1987**, *325*, C11.

142. Xiang-Gao, L.; Jing-Zhi, L.; Ning-Hai, H.; Zhang-Sheng, J.; Guo-Zhi, L. *Acta Chim. Sinica* **1989**, *47*, 1191.

143. Zhongsheng, J.; Ninghai, H.; Yi, L.; Xiaolong, X.; Guozhi, L. *Inorg. Chim. Acta* **1988**, *142*, 333.

144. Belforte, A.; Calderazza, F.; Englert, U.; Strähle, J.; Wurst, K. *J. Chem. Soc., Dalton Trans.* **1991**, 2419.

145. Secauer, C. A.; Day, V. W.; Ernst, R. D.; Kennelly, W. J.; Marks, T. J. *J. Am. Chem. Soc.* **1976**, *98*, 3713.

146. Al-Juaid, S. S.; Buttrus, N. H.; Eaborn, C.; Hitchcock, P. B.; Smith, J. D.; Tavakkoli, K. *J. Chem. Soc., Chem. Commun.* **1988**, 1389.

147. Evans, W. J.; Sollberger, M. S.; Hanusa, T. P. *J. Am. Chem. Soc.* **1988**, *110*, 1841.

9

STRUCTURES OF LITHIUM SALTS OF HETEROATOM COMPOUNDS

FRANK PAUER AND PHILIP P. POWER

Department of Chemistry
University of California
Davis, California

Lithium Chemistry: A Theoretical and Experimental Overview, Edited by Anne-Marie Sapse and Paul von Ragué Schleyer.
ISBN 0-471-54930-4

ABBREVIATIONS

1-Ad	1-adamantanyl
am.	amido
BHT	butylated hydroxytoluene, 2,6-*t*-Bu_2-4-MeC_6H_2
bipy	2,2′-Bipyridine
br	bridging
c	cyclo
Cp	cyclopentadienyl
Cp*	pentamethylcyclopentadienyl
Dipp	2,6-diisopropylphenyl
DME	dimethoxy ethane
E	element
Et	ethyl
HMPA	hexamethylphosphorictriamide
HMCTS	hexamethylcyclotrisilazane
im.	imido
i-Pr	isopropyl
LDA	lithiumdiisopropylamide
M	metal
Me	methyl
Mes	mesityl, 2,4,6-trimethylphenyl
Mes*	supermesityl, 2,4,6-tri-tert-butylphenyl
MO	molecular orbital
n-Bu	*n*-butyl
NMR	nuclear magnetic resonance
OEP	octaethylporphyrinate
OMCTS	octamethylcyclotetrasilazane
PE	photoelectron spectroscopy
Ph	phenyl
PMDETA	pentamethyldiethylenetriamine
py	pyridine
R	substituent
t	terminal
t-Bu	tertiary butyl
THF	tetrahydrofuran
TMEDA	*N,N*′-tetramethylethylenediamine
TMP	2,2,6,6-tetramethylpiperidinate
TriMEDA	trimethylethylenediamine
Triph	2,4,6-triphenylphenyl
Trip	2,4,6-triisopropylphenyl
UV-Vis	electronic absorption spectroscopy
VT	variable temperature

1 INTRODUCTION

The importance of lithium derivatives of organic and related ligands as versatile reagents for the synthesis of novel inorganic, organometallic, organic, and bioorganic compounds continues to grow. The widespread use of these compounds has led to considerable interest in their structures. The most recent comprehensive review of the subject by W.N Setzer and P.v.R. Schleyer[1] (with coverage to 1983, and a partial listing for 1984) cited over 200 references. In the ensuing 10-year period the growth in the number of structural determinations may be described as explosive. It is not surprising that further reviews on lithium compounds have been published since, dealing with different aspects of the subject, for example, derivatives of amides,[2] enolates,[3] and sulfones[4].

The dramatic increase in the structural knowledge of the title compounds may readily be appreciated if it is borne in mind that the earlier[1] review also included coverage of Li—C compounds and that the Li—E compounds (E = heteroatom, i.e., N, O, P, S, Se, etc.) comprised only about a quarter (ca. 50) of these references. Even though the latter species do not generally have Li—C bonds, they are often considered in the same context as organolithium species. There are good reasons, however, to review the structures of lithium heteroatomic species separately from their organolithium analogs. The major one is that, in many instances, different structural arrangements are observed owing to the presence of one or more lone pairs of electrons, and the different steric requirements of the groups, at the heteroatom. Thus the electron-deficient aggregate structures usually observed in the Li—C species are generally not seen with the related Li—N or Li—O compounds.

The coverage of this review is limited to lithium heteroatom compounds that are electronically (if not structurally) related to organolithium species. As a result, the structures of many lithium compounds which might have been included in the title are not discussed. These include lithium hydrides, halides, carboxylates or related ligands, and also compounds in which a lithium ion is solvated only by amines, phosphines, ethers, or thioethers, that is, where there is no other type of lithium–element bond, other than a dative one, present.

The rapid expansion in the number of available crystal structures over the past 10 years has been in part due to developments in the area of low-temperature X-ray data collection and crystal mounting techniques. These have greatly facilitated the handling of air-sensitive crystals and increased the rate at which data sets can be collected.[5] We have attempted to survey crystal structures of lithium compounds available up the end of 1991. The major source of information is the Cambridge Structural Database.[6] Recent data from the primary literature (for 1992), from our laboratory and those of a number of colleagues, who generously provided data prior to publication, are also included.

One of the most important features of lithium compounds is the nature of the bond to lithium. According to MO calculations, Li—C and Li—N bonds are mostly of an ionic nature, with ionicities of 89 and 90%, respectively.[2] For elements with lower electronegativity, the degree of ionicity may be somewhat lower, but not considerably. Thus the interaction that occurs in lithium species discussed here is primarily an electrostatic one. The simultaneous interaction of the metal center with more than one anion leads to aggregation. Because of the noncovalent nature of these entities, they have been termed "supramolecules."[2,3]

The most general case of a cation–anion interaction is in a salt where there are no organic groups attached to the anion, thus both ions are roughly spherical, for example, as in LiF. The result is a polymeric array with the ions in a symmetrical environment. This can be envisioned as infinite stacking in three dimensions. The more numerous and bulky the substituents at the anionic center become, the higher the steric crowding and the lower the extent of the aggregation. The stacking becomes finite in or more directions, thus affording two-dimensional networks, one-dimensional chains, or discrete molecules. If stacking occurs in molecules, a two-layer arrangement is the most common result. A hexamer can be visualized as a stacking of a trimer, a cubic tetramer the result of the interaction of two dimers. If the size of the substituents at the central atom in the anion prevents stacking or if additional donor solvents saturate the coordination sphere of the metal ion, trimeric, dimeric, or eventually, monomeric structures are observed. In this review the compounds are arranged according to the degree of their aggregation.[7] Different aggregation states can be classified in terms of general structural types they often assume. These are presented in Figure 9.1.

In general, monomers (type 1) are observed only if strong chelating bases and bulky substituents at the anion are employed. Exceptions to this rule may occur with delocalized anions or when the solvating donor binds more strongly to the Li^+ ion than the anionic center. The latter phenomenon is often seen in the case of heavier element anions such as PR_2^-, AsR_2^-, SR^-, SeR^-, or TeR^-.

Generally, dimerization is the result of the interaction of a metal center in one unit with the anion of another in an (ideally) symmetrical fashion, giving rise to a four-membered Li—X—Li—X ring of type 2. Owing to the chelating properties of some anionic ligands, a variety of subtype structures can be adopted. In 2a, the outer rings afforded by the anions chelating the lithium are fused to adjacent sides of the central $(Li—X)_2$ ring, resulting in a tricyclic structure, with the rings twisted with respect to each other. This arrangement produces metal ions in chemically different environments. In 2b, the peripheral rings are fused to opposite sides of the central ring, while in 2c the dimerization is a result of the Li^+ ions interacting with the different coordination sites of separate anions. In 2b and 2c, the metals ideally adopt identical environments.

Cyclic trimers (type 3) and tetramers (type 4a) are the result of sterically less demanding substituents on the anionic centers. Type 4b tetramer is the simplest example of a stacked compound, which is composed of four-membered rings that are essentially type 2 dimers. The same is true for hexamers (type 5), with two layers of type 3 trimers.

In less-hindered anions an infinite one-dimensional chain structure is often observed (type 6a). If the one-dimensional polymerization occurs between type

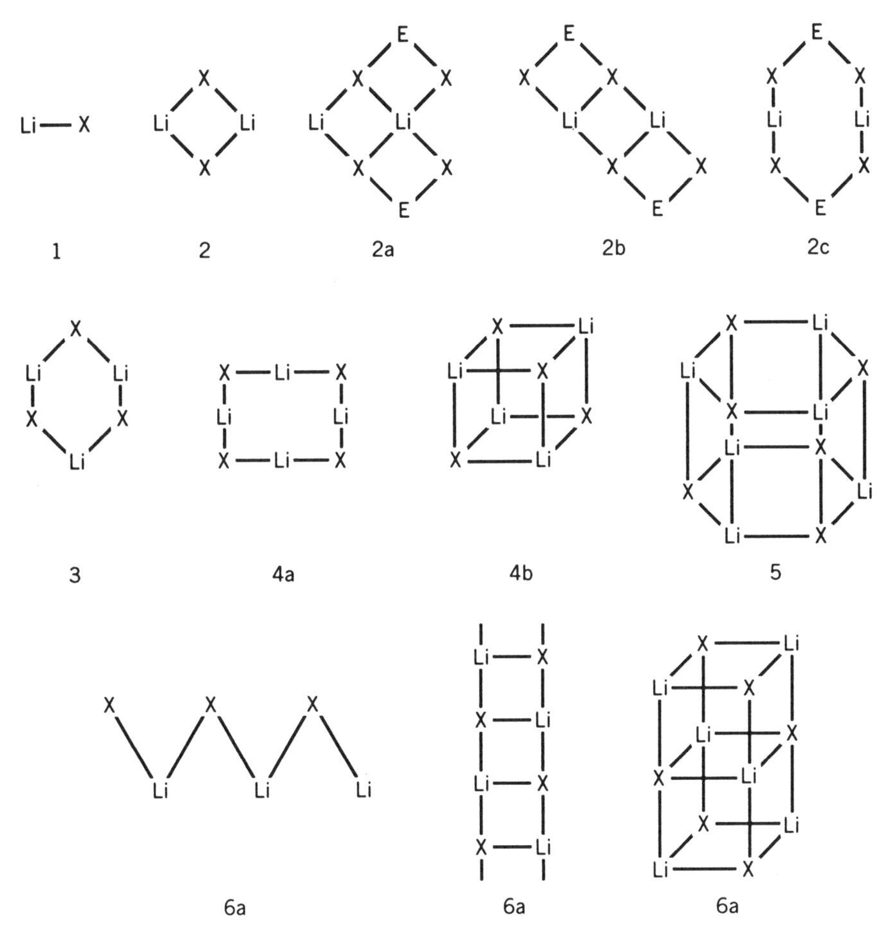

Figure 9.1 Schematic illustration of different types of aggregation in lithium salts of heteroatom compounds.[2a]

2 dimeric frameworks, the result is a double chain or "ladder" (type 6b); the tetrameric case is depicted in type 6c. Often, significant deviations from these ideal cases are observed, and many of these are discussed in the main body of the chapter.

2 LITHIUM COMPOUNDS OF THE MAIN GROUP 6 ELEMENTS

2.1 Lithium Alkoxides, Aryloxides, Silanolates, and Stannanolates

2.1.1 Simple LiOR Species. Anionic oxygen donors display a strong attraction for Li^+ ions. Generally speaking, such interactions are much stronger than those observed with common donor solvents, which often have difficulty solvating the Li^+ ion at the expense of a reduced number of $Li^+\cdots OR^-$ interactions. Thus, in contrast to alkyls, aryls, amides, phosphides, arsenides, or thiolates, only a few monomeric lithium alkoxides, aryloxides, or related derivatives are known. Reduced steric crowding must also play a role in encouraging strong $Li^+—OR^-$ bonds, since the oxygen atom carries only one organic group. Thus, LiOR species are less bulky than the corresponding $LiNR_2$ or $LiCR_3$ compounds in which the nitrogen or carbon atom have two or three substituents. Important structural parameters for simple LiOR species are provided in Table 9.1.

2.1.2 Monomeric Structures. The use of tetradentate donor molecule 16-crown-4 results in a monomeric Li(16-crown-4)O{2,4,6-$(NO_2)_3C_6H_2$}[8] structure **1**. The mononuclear nature of this species is presumably the result of the macrocyclic effect of the crown ether ligand. Owing to the high coordination number of the Li^+ ion, the Li—OR bond is fairly long (1.88 Å). The Li—O bonds to the crown ether are, on average, 2.09 Å in length. A mononuclear structure may also be obtained through the use of a combination of a bidentate donor molecule and bulky substituents, as exemplified by the monomeric species Li(TMEDA)OB{CH$(SiMe_3)_2$}$_2$[9] **2**. In this case, the Li—O bond is extremely short (1.67 Å) owing to the low coordination at Li. The B—O bond length of 1.31 Å is also short. This is probably a consequence of the increased effective negative charge at oxygen which enhances the ionic contribution to the B—O bond.

The species [Li(H_2O)O{2,4,6-$(NO_2)_3C_6H_2$}]$_2$-dibenzo-36-crown-12[10a] **3a** is dimeric only in the formal sense the Li^+ ions are not bound to common O centers. The Li^+ ion coordinates two neighboring oxygen atoms on opposite sides of the crown ether (Li—O 1.98, 2.23 Å). In addition to the bond to the anionic O atom (1.94 Å), Li^+ is coordinated by an O atom of a nitro group in ortho position (2.06 Å) and a donating water molecule (1.96 Å). When the Li^+ ion is coordinated by bipy, a dimeric structure of formula [Li(bipy){2,4,6-$(NO_2)_3C_6H_2$}]$_2$,[10b] **3b**, is observed. Oddly, the Li^+ ion appears to be more strongly bound (1.95 Å) to an oxygen from an ortho-NO_2 group than to the formally anionic oxygen (2.16 Å).

2.1.3 Dimeric Structures. Dimeric species constitute the largest class of LiOR compounds. The dimeric motif is observed even in the presence of large substituents at oxygen such as -C(*t*-Bu)$_3$. Both [Li(THF)OC(*t*-Bu)$_3$]$_2$[11] **4** and [LiOC(*t*-Bu)$_3$]$_2$[9] **5** are type 2 dimers with very short Li—O bonds (1.67, 1.83 Å) in **5**. In **4** the Li—O distance is 1.84 Å. Few differences are observed between the structures of the dimers [Li(Et_2O)O(4-Me-2,6-*t*-$Bu_2C_6H_2$)]$_2$[12] **6** and [Li(THF)O(2,6-*t*-Bu_2-C_6H_3)]$_2$[13] **7**. The Li—O bonds have average values of 1.86 and 1.85 Å, and the Li—O_{donor} interactions are 1.96 and 1.90 Å, the latter value owing, presumably, to the better donor characteristics of THF in

Table 9.1 Simple LiOR Species

	Aggr.	CN_{Li}	Li—O (Å)	L	Li—L (Å)	Add. Coord.	Li—E (Å)	Compound/ Ref.
Li(16-crown-4)O{2,4,6-$(NO_2)_3C_6H_2$}	1	5	1.88	16-crown-4	2.09[a]			**1**/8
Li(TMEDA)OB{$CH(SiMe_3)_2$}$_2$	1	3	1.67	TMEDA	2.11[a]			**2**/9
{Li(H_2O)O{2,4,6-$(NO_2)_3$}-C_6H_2(dibenzo-36-crown-12)	1	5	1.94	L[b]	1.98, 2.23	o-NO_2	2.06	**3a**/10a
				H_2O	1.96			
[Li(bipy){2,4,6-$(NO_2)_3C_6H_2$}]$_2$	2	5	2.16	bipy	1.95 (Li—N)	o-NO_2	1.95	**3b**/10b
[Li(THF)OC(t-Bu)$_3$]$_2$	2	3	1.84	THF				**4**/11
[LiOC(t-Bu)$_3$]$_2$	2	2	1.67, 1.83					**5**/9
[Li(Et_2O)O{4-Me-2,6-(t-Bu)$_2C_6H_2$}]$_2$	2	3	1.86[a]	Et_2O	1.96			**6**/12
[Li(THF)O{2,6-(t-Bu)$_2$-C_6H_2}]$_2$	2	3	1.85	THF	1.90			**7**/13
[(LiOPh)$_2$(15-crown-5)]$_2$	2	4	1.94, 1.87, 1.90	15-crown-5	2.14			**8**/14
		6	1.84	15-crown-5	2.33[a]			
[(LiOPh)$_2$(18-crown-6)]	2	5	1.88[a]	18-crown-6	2.15[a]			**9**/15
[Li(THF)OSi(Me)$_2$C($SiMe_3$)$_3$]$_2$	2	3	1.76[a]	THF	2.14			**10**/16
		3	1.87[a]	THF	1.97			
[Li(THF)OSi(t-Bu)$_2$OSi(t-Bu)$_2$F]$_2$	2	3	1.86	THF	1.97			**11**/17
[Li(THF)$_2$OSi(t-Bu)$_2$Cl]$_2$	2	4	1.93[a]	THF	2.05[a]			**12**/18
[LiOB{CH($SiMe_3$)$_2$}$_2$]$_2$	2	2	1.78					**13**/9
[Li(THF)OBMes_2]$_2$	2	3	1.85	THF	1.92			**14**/19
[LiOSi(t-Bu)$_2$OSiMe_2OSi(t-Bu)$_2$F]$_2$	2	3	1.82			F	1.88	**15**/17
[Li(py)OSiPh_2OSiPh_2OLi(py)]$_2$	2	3	1.80, 1.87	py	2.05			**16**/20
		4	1.94, 1.99, 2.01	py	2.06			

Table 9.1 (*Continued*)

	Aggr.	CN_{Li}	Li—O (Å)	L	Li—L (Å)	Add. Coord.	Li—E (Å)	Compound/ Ref.
$[LiOC(t\text{-}Bu)_2CH_2P(Me)_2]_2$	2	3	1.79			P	2.50	**17**/21
$[LiOC(t\text{-}Bu)_2CH_2P(Ph)_2]_2$	2	3	1.79			P	2.65	**18**/21
$[LiOC(t\text{-}Bu)_2CH_2P(Ph)_2]_2(tBu_2CO)$	2	3	1.80, 1.87	L[c]	1.97			**19**/21
		4	1.82[a]			P	2.61	
$[Li(THF)OC(Me)CH_2P(Ph)_2\text{-}CH_2C(Me)O]_2$	2	4	1.94, 2.05	THF	2.05	C=O	1.76	**20**/22
$[LiO\{4\text{-}Me\text{-}2,6\text{-}CH{=}N(i\text{-}Pr)_2\text{-}C_6H_2\}]_3$	3	4	1.87[a]			L[d]	2.12	**21**/23a
$[LiOSi(t\text{-}Bu)_2OSi(t\text{-}Bu)_2F]_3$	3	3	1.80[a], 1.88[a]			F	1.99[a]	**22a**/18
$[LiO(2,6\text{-}Me_2NCH_2\text{-}4\text{-}MeC_6H_2)]_3$	3	4	1.86[a]	*o*-CH_2NMe_2	2.34			**22b**/23b
$[LiOSi(t\text{-}Bu)_2NH_2]_4$	4	4	1.91–2.11			H_2N	2.20–2.30	**23**/24
$[LiOSi(t\text{-}Bu)_2OH]_4$	4	4	1.89–2.12			HO	2.02–2.09	**24**/17
$[LiOSi(t\text{-}Bu)_2OC(O)t\text{-}Bu]_4$	4	4	1.99[a]			C=O	1.92	**25**/25
$[LiOCH(t\text{-}Bu)CH_2C(O)t\text{-}Bu]_4$	4	4	1.90–1.95			C=O	1.96[a]	**26**/26
$[LiOSi(Me)_2naphthyl]_6$	6	3	1.88–1.98					**27a**/27
$[LOC(Me)_2Ph]_6$	6	3	1.92 (av.)					**27b**/28a
$[LiOSi(Me)_2PEt_2]_6$	6	3	1.93					**27c**/28b

[a]Denotes average values.
[b]Dibenzo-36-crown-12.
[c]$t\text{-}Bu_2C{=}O$.
[d]Ortho—C=N-*i*-Pr.

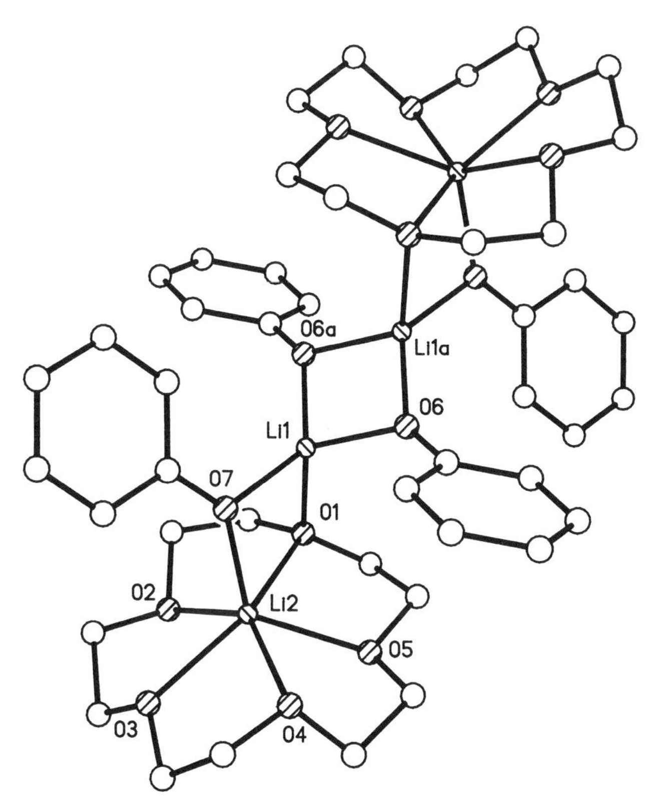

Figure 9.2 Structure of $[Li(OPh)_2(15\text{-crown-}6)]_2$, **8**.[14] H atoms are not shown for clarity.

comparison to Et_2O. In $[(LiOPh)_2(15\text{-crown-}5]_2$[14] **8** (Fig. 9.2) and $[(LiOPh)_2(18\text{-crown-}6)]$[15] **9** (Fig. 9.3), there are central type 2 dimeric $(Li—O)_2$ units. Owing to its larger size, 18-crown-6 can coordinate both Li^+ ions, each through three oxygen atoms (Li—O 1.89–2.42 Å; 2.15 Å, av.). These Li^+ ions are bridged by phenoxide groups (Li—O 1.88 Å). The structure of **8** has a very similar $(LiOPh)_2$ type 2 core (Li—O, 1.87 and 1.90 Å). Each Li^+ ion, however, is coordinated to another -OPh group as well as one ether oxygen from a 15-crown-5 donor. This ether oxygen atom also bridges to a further Li^+ ion whose coordination is completed by the remaining four oxygen donors of the 15-crown-5 ligands, as illustrated in Figure 9.2. Four different Li—O distances are observed in the type 2 dimer $[Li(THF)OSiMe_2\{C(SiMe_3)_3\}]_2$[16] **10**, which has no crystallographically imposed symmetry. One Li^+ ion has shorter Li—O bonds within the four-membered ring (1.72 and 1.79 Å) and has a long Li—THF interaction of 2.14 Å. The situation is reversed for the other Li^+ ion, where the endocyclic Li—O bonds are longer (1.85, 1.88 Å), and the THF contact is shorter (1.97 Å).

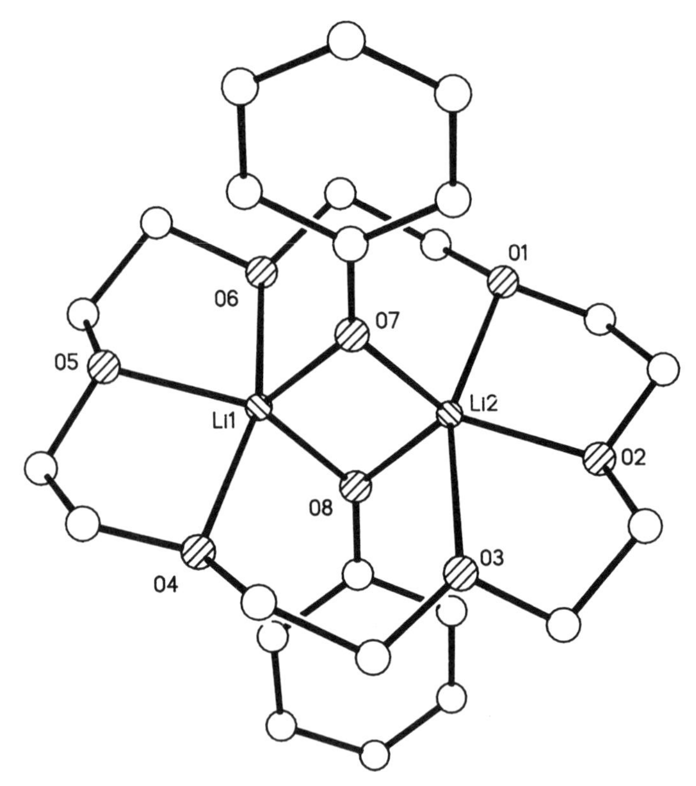

Figure 9.3 Structure of $[(LiOPh)_2(18\text{-crown-}6)]$, **9**.[15] H atoms are not shown for clarity.

The introduction of a fluorine atom in ligands offers an additional coordination site for Li^+. However, in $[Li(THF)OSi(t\text{-}Bu)_2OSi(t\text{-}Bu)_2F]_2$[17] **11**, no significant Li—F interactions are present, in contrast to what is observed in **22** (vide infra). The three-coordinate Li^+ ions display endocyclic Li—O contacts of 1.86 Å and longer Li—THF distances of 1.97 Å (av.). In the type 2 dimer $[Li(THF)_2OSi(t\text{-}Bu)_2Cl]_2$[18] **12**, two THF molecules per Li atom are present, with rather long ($Li—O_{anion}$ 1.93 Å, av., $Li—O_{THF}$ 2.05 Å) Li—O distances. This may be attributable in part to the presence of an electron-withdrawing atom on silicon.

The solvent-free species $[LiOB\{CH(SiMe_3)_2\}_2]_2$[9] **13** has the centrosymmetric type 2 dimeric structure. In spite of the two-coordination at Li, there are longer Li—O contacts (1.78 Å) than those seen in **2**, where Li is three-coordinate. This may reflect the decreased electron density on the now three-coordinate O center in **13**. Also, the B—O bonds in **13** are 0.04 Å longer than those in **2**. The transannular Li· · ·Li distance is only 2.21 Å. When an additional donor molecule is bound to Li^+ as in $[Li(THF)OBMes_2]_2$[19] **14,** the endocyclic Li—O bond lengths increase to 1.85 Å. The B—O bond, however, is shorter than that in **13** (1.32 Å).

The lithium salt of a trisiloxy species $[LiOSi(t\text{-}Bu)_2OSiMe_2OSi(t\text{-}Bu)_2F]_2$[17] **15** has an unusual eight-membered ring structure that includes a Li—F interaction of 1.88 Å. Dimerization occurs through a bridging Li—O unit, resulting in two eight-membered rings fused to opposite sides of a central type 2 four-membered ring, which has almost equal Li—O sides (1.82 Å in the monomeric eight-membered ring, 1.81 Å in the bridging unit). The species $[Li(py)OSiPh_2OSiPh_2OLi(py)]_2$[20] **16** (Fig. 9.4) is essentially a type 2 dimer, with a central four-membered $(Li—O)_2$ ring with Li—O bonds of 2.00 Å (av., four-coordinate Li). The O—Si—O—Si—O backbone acts as a dianionic chelate (the other Li—O contact to the chelating anion is 1.94 Å). Thus, two six-membered rings are fused to opposite sides of the central ring, forming an overall puckered ladder or stair-shaped structure. The second type of Li^+ is three-coordinate and behaves as a bridge between the nonbridging O atom (Li—O, 1.80 Å) of one monomeric unit and the bridging one of the other (Li—O, 1.87 Å). Both Li^+ ions are also coordinated by pyridine.

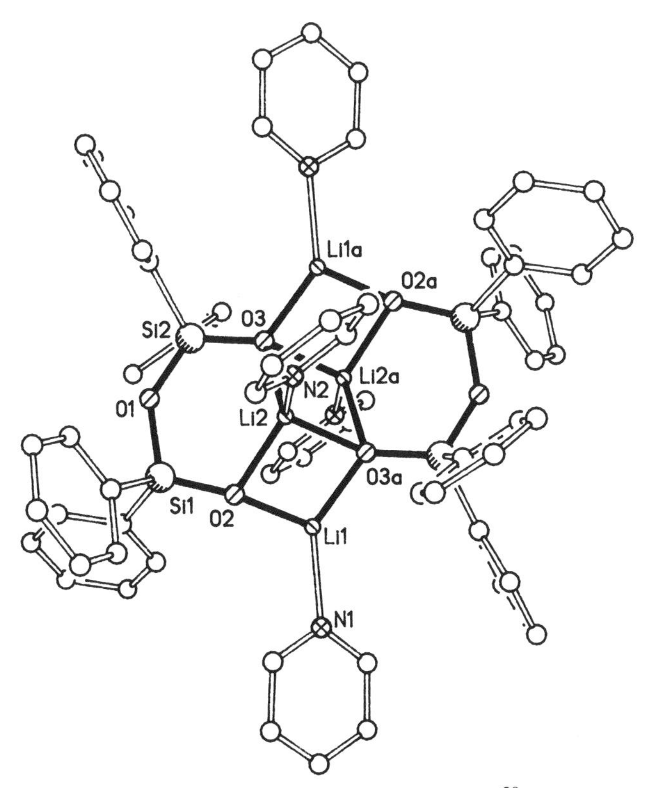

Figure 9.4 Structure of $[Li(py)OSPh_2OSiPh_2OLi(py)]_2$, **16**.[20] H atoms are not shown for clarity.

Phosphino groups act as chelating donors in $[LiOC(t\text{-}Bu)_2CH_2PR_2]_2$;[21] R = Me: **17**; R = Ph: **18**; $[LiOC(t\text{-}Bu)_2CH_2PPh_2]_2(t\text{-}Bu_2C{=}O)$ **19** to afford dimeric structures. While the donor-free structures **17** and **18** exhibit a stair-shaped type 2b dimer motif with a central $(Li{-}O)_2$ ring (Li—O is 1.79 Å in both cases), the presence of tert-butyl ketone affords an asymmetric dimer of type 2a, with five-membered rings fused to adjacent sides of the central Li—O unit, rendering one Li atom central (Li—O, 1.82 Å av.; Li—P, 2.61 Å av.), the other peripheral (Li—O slightly asymmetric, with 1.80 and 1.87 Å). This Li^+ ion is also coordinated (Li—O 1.97 Å) to a ketone donor.

A stair-shaped structure is observed in $[Li(THF)OCMeCH_2PPh_2CH_2C(Me)O]_2$[22] **20**, with two eight-membered rings fused to a central four-membered ring, being the result of an anionic, chelating O—C=C—P=C—C=O backbone and dimerization (Li—O, 2.05 Å) along the Li—O bond (1.94 Å). The Li—O(=C) distance is 1.76 Å. Coordination of THF gives rise to four-coordination of the Li atom (2.05 Å).

2.1.4 Trimeric Structures. Trimeric structures have rarely been observed for LiOR species. Only three have been reported to date. One example is [LiO{4-

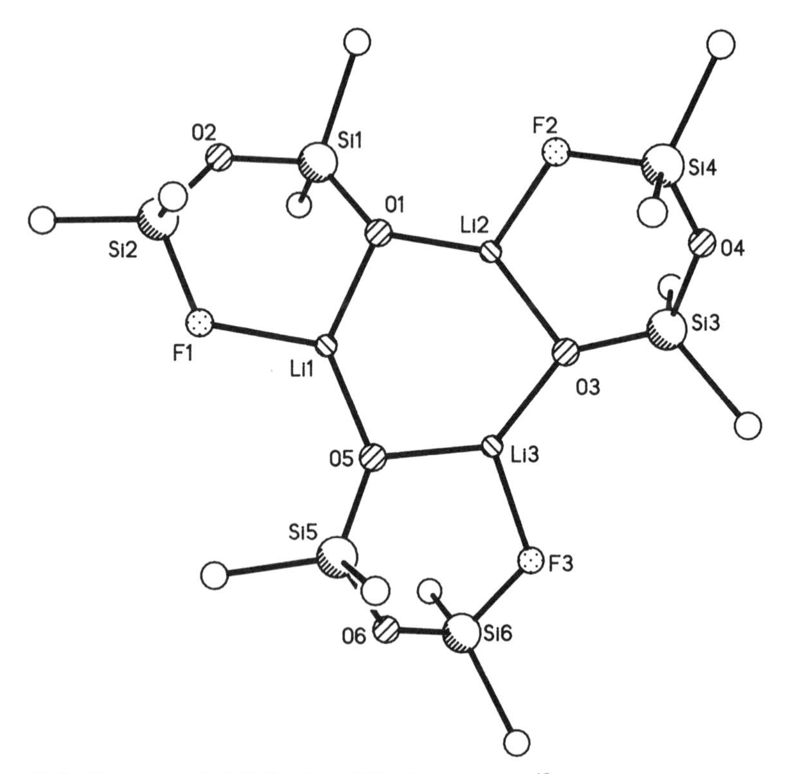

Figure 9.5 Structure of $[LiOSi(t\text{-}Bu)_2OSi(t\text{-}Bu)_2F]_3$, **22**.[18] H atoms are not shown for clarity.

Me-2,6-(CH=N-*i*-Pr)$_2$C$_6$H$_2$}]$_3$[23a] **21**. In this type 3 trimer, which has Li—O distances of 1.87 Å, the ortho imino groups coordinate to the Li^+ ions to which the O atom of the same anion forms a bridge (Li—N = 2.12 Å). The central six-membered ring is planar with a mean deviation of 0.02 Å from the plane.

A type 3 trimeric structure is also observed for [LiOSi(*t*-Bu)$_2$OSi(*t*-Bu)$_2$F]$_3$,[18] **22a** (Fig. 9.5), which in contrast to the dimeric THF-coordinated **11** has short Li—F contacts. Three O—Si—O—Si—F—Li six-membered rings are fused along alternate Li—O edges to the central (Li—O)$_3$ ring. In the central ring, the Li—O bonds are alternately 1.80 and 1.88 Å long. The Li—F contacts average 1.99 Å. In [LiO(2,6-Me$_2$NCH$_2$-4-MeC$_6$H$_2$]$_3$[23b] **22b**, there is a planar Li_3O_3 core with an average Li—O distance of 1.86 Å^2. The Li^+ ions are also coordinated by nitrogens from the 1—CH$_2$NMe$_2$ groups.

2.1.5 Tetrameric Structures. Cubane-type tetramers are more readily obtained for alkoxides than for amides (vide infra). This may be a result of the lower steric requirements of alkoxides and the availability of three electron pairs on oxygen. In [LiOSi(*t*-Bu)$_2$NH$_2$]$_4$[24] **23** (Fig. 9.6), the Li—O distances in the distorted cube have lengths in the range 1.91–2.11 Å. The presence of -NH$_2$ groups on the oxygen-bound silicon atoms gives rise to a chelating anion

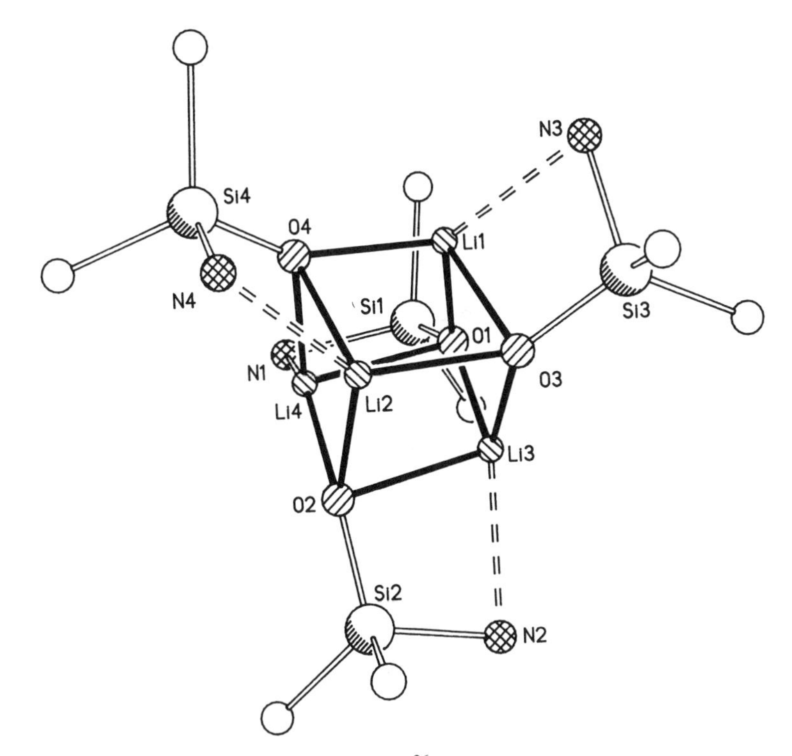

Figure 9.6 Structure of [LiOSi(*t*-Bu)$_2$NH]$_4$[26], **23**. H atoms are not shown for clarity.

(Li—N, 2.20–2.30 Å), which results in the formation of two four-membered rings. These are arranged in two sets, each being part of a tricyclic system with two opposite faces of the cubane forming the central ring. The two tricyclic rings are roughly perpendicular to one another.

$[LiOSi(t\text{-}Bu)_2OH]_4$[17] **24** is isoelectronic to **23**, and has Li—O distances within the range 1.89–2.12 Å in the cube and 2.02–2.09 Å in the four-membered rings. The same arrangement is present in $[LiOSi(t\text{-}Bu)_2OC(O)t\text{-}Bu]_4$[25] **25**, with six-membered Li—O—Si—O—C=O rings fused to four edges of the cube. The Li—O distances in the cube average 1.99 Å, the additional intramolecular ketone coordination affords a very short Li—O contact of 1.92 Å, perhaps because of the two-coordination of this O center. A similar structure was observed for $[LiOCH(t\text{-}Bu)CH_2C(O)t\text{-}Bu]_4$[26] **26**. In this case, Li—O bonds within the cubes have values of 1.90–1.95 Å, whereas those in the rings are 1.96 Å (av.) long.

2.1.6 Hexameric Structures. $[LiOSiMe_2(naphthyl)]_6$[27] **27a** (Fig. 9.7), $[LiOCMe_2Ph]_6$[28a] **27b** and $[LiOSiMe_2PEr_2]_6$[28b], **27c** are, to our knowledge, the

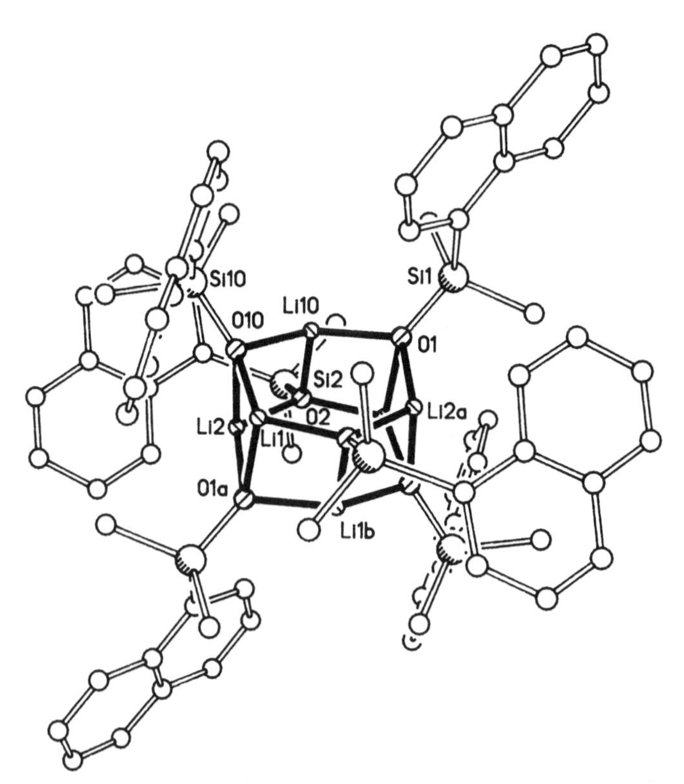

Figure 9.7 Structure of $[LiOSiMe_2(naphthyl)]_6$, **27**.[27a] H atoms are not shown for clarity.

only examples of hexameric LiOR structures. They consist of a type 5 Li—O framework, with Li—O bonds ranging from 1.88 to 1.98 Å (1.92 Å, av.), with three-coordinate Li and four-coordinate O centers. In **27c** there is a distorted Li_6O_6 core with an average Li—O distance of 1.92 Å. There are no close interactions between the Li^+ ions and the phosphorus donors.

2.2 Lithium Enolates

Examination of the structure of enolates as reaction intermediates may help elucidate their reactivity and mechanism of product formation. Unlike the notation often used to describe the reactions of enolates, which depicts the negative charge on the carbon atom, the deprotonation of the enol results in a negatively charged oxygen center and an essentially double C—C bond. A comprehensive review on enolates has been published recently.[3] For this reason, only a tabular list of lithium enolates (Table 9.2) is presented here. Most compounds adopt typical structures of their aggregation state, as illustrated by the structures depicted in Figures 9.8–9.10.

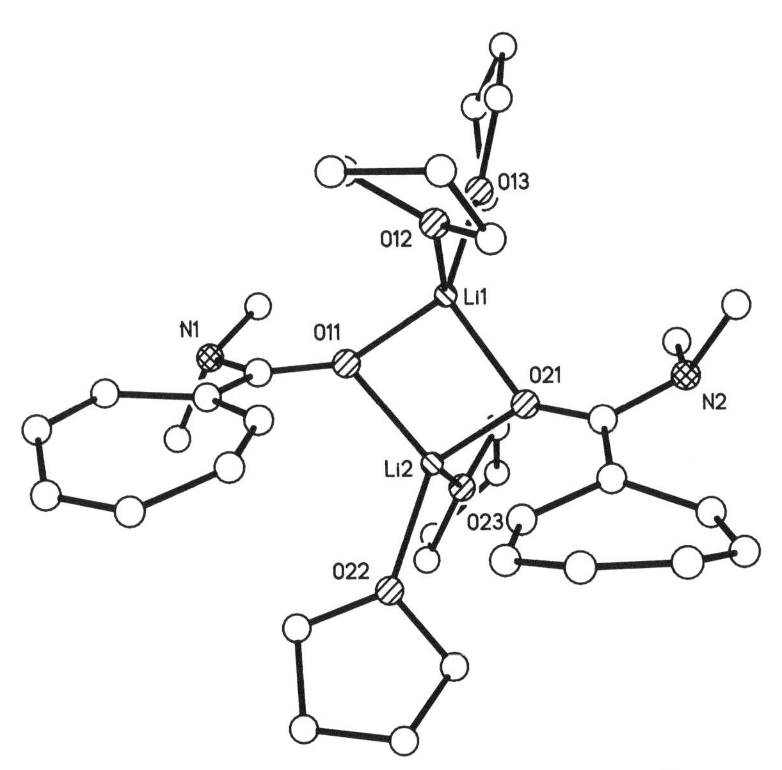

Figure 9.8 Structure of $[Li(THF)_2OC(NMe_2)(c\text{-}2,4,6\text{-}C_7H_6)]_2$, **28**.[29] H atoms are omitted for clarity.

TABLE 9.2 Lithium Enolates

	Aggr.	CN_{Li}	Li—O (Å)	Donor	Li—X (X) (Å)	Compound/ Ref.
$[Li(THF)_2OC(NMe_2)(c\text{-}2,4,6\text{-}C_7H_6)]_2$	2	4	1.88, 1.92	THF	1.99 (O)	**28**/29
$[Li(O2,6\text{-}Me_2\text{-}c\text{-}\{CN\text{-}C_3H_6\text{-}N\})\text{-}OCH(t\text{-}Bu)CHC(CH_2)t\text{-}Bu]_2$	2	3	1.85–1.93	HMPU	1.86 (O)	**29**/30
$[Li(TMEDA)OC(CHMe)O(t\text{-}Bu)]_2$	2	4	1.95, 1.90	TMEDA	2.18 (N)	**30**/31
$[Li(TMEDA)OC(CMe_2)O(t\text{-}Bu)]_2$	2	4	1.90, 1.91	TMEDA	2.15, 2.31 (N)	**31**/31
$[Li(TriMEDA)OC(CHMe)NMe_2]_2$	2	4	1.84, 1.90	TriMEDA	2.14 (N)	**32**/32
$[Li(TriMEDA)OC(CH_2)t\text{-}Bu]_2$	2	4	1.87, 1.90	TriMEDA	2.10 (N)	**33**/32
$[Li(THF)OC(OMe)CHC(CHMe)\text{-}c\text{-}2,5\text{-}Me_2N\text{-}C_4H_6]_2$	2	4	1.87, 1.89	L[b] THF	2.17 (N) 1.95 (O)	**34**/33
$[Li(THF)OC\{CH(t\text{-}Bu)\}OMe]_4$	4	4	1.96[a]	THF	1.93 (O)	**35**/31
$[LiOC(CH_2C_6H_4CH_2NMe_2]_4$	4	4	1.97[a]	(NMe_2)	2.12 (N)	**36**/34
$[LiOC(OMe)CHC(CHMe)\text{-}c\text{-}N\text{-}C_4H_8]_4$	4	4	1.94[a]	L[c]	2.06 (N)	**37**/33
$[LiOCCH_2(t\text{-}Bu)]_6$	6	3	1.84, 1.96[d] 1.95[e]			**38**/35
$[LiOC(OEt)CHNEt_2]_6$	6	4	1.97	NEt_2	2.12 (N)	**39**/36

[a]Denotes average values; ( ) denotes an intramolecular donor site.
[b]$c\text{-}NCH(Me)(CH_2)_2CH(Me)$.
[c]$c\text{-}N(CH_2)_4$.
[d]Alternating Li—O distances (av.) in the six-membered ring of one stack.
[e]Li—O distance (av.) in the rungs.

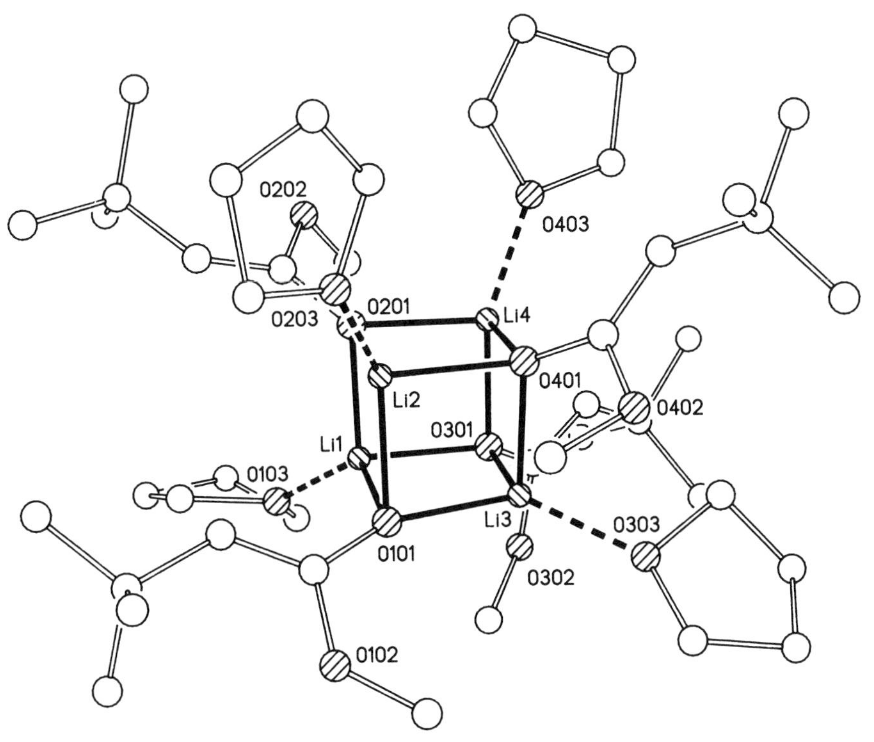

Figure 9.9 Structure of [Li(THF)OC{CH(*t*-Bu)}OMe], **35**.[31] H atoms are omitted for clarity.

2.3 Miscellaneous Alkoxides and Related Species

A large number of compounds are known that possess Li—O interactions but cannot be classified as simple LiOR species. Many of these involve metals other than Li. Prior to describing these, a few compounds (Table 9.3) that cannot be conveniently classified as mixed metal alkoxides are mentioned briefly.

A spirocyclic anionic system with a central Li^+ ion is found in $[NEt_4][Li\{HOC_6H_4C(O)CHC(O)C_6H_4OH\}_2]_2$[37] **40**. The ammonium cation is well separated from the anion, which has Li—O bonds in the range 1.85 to 1.91 Å and C—O (1.29 Å, av.) and C—C (1.40 Å, av.) bond lengths that suggest a delocalized structure. In the dimer $[Li_2N(i\text{-Pr})_2OC(CH_2)CMe_2CH_2CH_2OSiMe_2(t\text{-Bu})]_2$[38] **41**, there is no crystallographically imposed symmetry. The two Li^+ ions display two different environments that are reminiscent of those in **16**. One [Li(1)] is part of a seven-membered ring arising from the chelating property of a monoanionic oxygen donor (Li—O, 1.92 Å). The Li(2) center is involved in a type 2 dimerization with the anionic oxygen atom. The diisopropylamide anions coordinate to both Li^+ ions in the monomeric unit [Li(1)—N, 1.96 Å; Li(2)—N, 2.01 Å]. The fused rings adopt

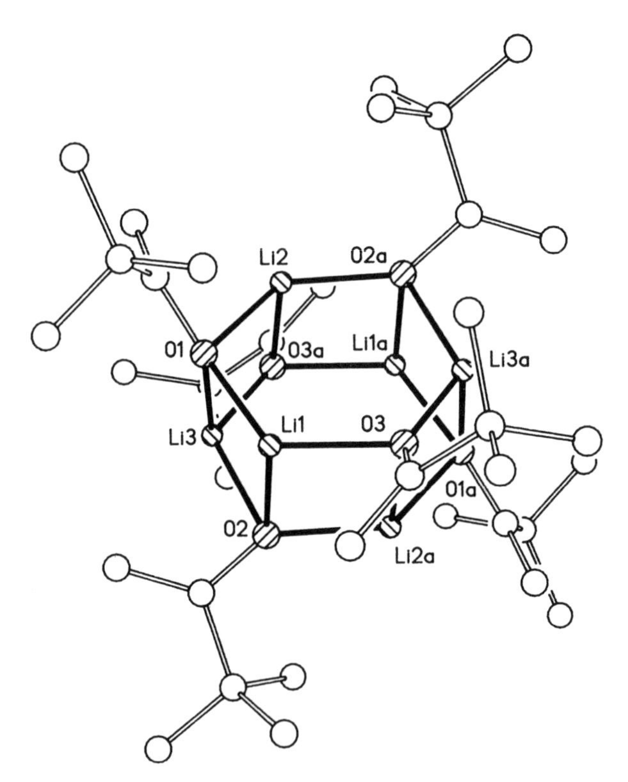

Figure 9.10 Structure of $[LiOC(CH_2)t\text{-}Bu]_6$, **38**.[35] H atoms are omitted for clarity.

a boatlike conformation. All values are averaged between the corresponding atoms in the crystallographically independent monomers.

The mixed-anion compound $[LiO(t\text{-}Bu)]_4\{Li(n\text{-}Bu)\}_4$[39] **42** displays the stacked structure depicted schematically in Figure 9.11. The four central Li^+ ions are coordinated by an α-C atom of n-Bu (Li—C, 2.22 Å) and three oxygens of the tert-butoxy anions (Li—O, 1.89 Å). The four outer lithiums are coordinated by one oxygen (Li—O, 1.84 Å), two α-C atoms (Li—C, 2.19 Å), and a β-CH_2 group of one of the two n-butyl anions (Li—C, 2.35 Å; Li—H, 2.34 Å).

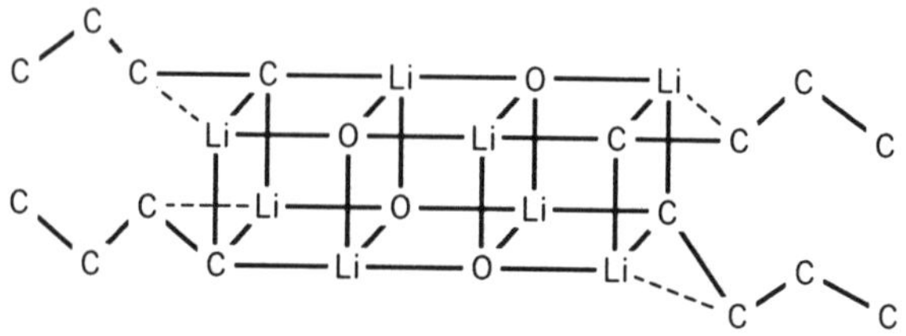

Figure 9.11 Schematic drawing of the framework atoms of $\{LiO(t\text{-}Bu)\}_4\{Li(n\text{-}Bu)\}_4$, **42**.[39]

Table 9.3 Miscellaneous LiOR Species

	Aggr.	CN_{Li}	Li—O (Å)	L	Li—L (Å)	Add. Coord.	Li—E (Å)	Compound/ Ref.
$[NEt_4][Li\{HOC_6H_4C(O)CH\text{-}C(O)C_6H_4OH\}_2]$	1	4	1.85–1.91					**40**/37
$[Li_2N(i\text{-}Pr)_2OC(CH_2)CMe_2\text{-}(CH_2)_2OSiMe_2t\text{-}Bu]_2$	2	3	1.92[a]			$\text{-}OSiMe_2t\text{-}Bu$	1.98	**41**/38
			1.94[a]			$i\text{-}Pr_2N$	1.96[a]	
		3	1.96[a]			$i\text{-}Pr_2N$	2.01[a]	
$[LiO(2,6\text{-}Me_2NCH_2\text{-}4\text{-}MeC_6H_2)\cdot LiI]_2$	2	4	1.92	$o\text{-}CH_2NMe_2$	1.99	I^-	2.78	**41b**/23b
$\{LiO(t\text{-}Bu)_4Li(n\text{-}Bu)\}_4$	4	3	1.89[a]			α-C	2.22[a]	**42**/39
	3	3	1.84[a]			α-C	2.19[a]	
						β-C	2.35[a]	
$[LiOS(Ph)(CSiMe_3)NSiMe_3]_4$	4	5	2.41, 1.93, 2.10			Li-N	2.11	**43**/40
						Li-C	2.43	

[a]Denotes average values.

$[Li\{Me_3SiCH(Ph)S(O)NSiMe_3\}]_4$,[40] **43**, is the result of the deprotonation of a CH_2 group in a position α to the sulfur. This results in a delocalized system involving the sulfur center (S—N, 1.51; S—C, 1.64; S—O, 1.50 Å) and a type 4b tetrameric Li—O cubane with Li—O interactions of, 1.93, 2.10, and 2.41 Å. Lithium is additionally coordinated by the methylene C and the imido nitrogen of separate anions, displaying contacts of 2.43 (Li—C) and 2.11 Å (Li—N).

2.4 Mixed-Metal Alkoxides and Related Species

In these compounds an -OR group bridges Li^+ ions and various other metals. The most important structural data are listed in Table 9.4. It should be noted, however, that some mixed-metal species are not included here. For example, where there is a possibility of a direct interaction between the lithium and the heteroatom, as in $LiSn\{O(2,6\text{-}Ph_2C_6H_3)\}_3$ (**276** vide infra), such compounds are discussed in the section appropriate for that interaction. Generally speaking, in mixed-metal alkoxides and aryloxides, the bonds to bridging ligands are longer than those to terminal ligands of the same kind. An example of this behavior is provided by $\{(Et_2O)Li\}\{AlMe(BHT)_2\{OC(Me)Ph_2\}$[41a], **44a** where the Al—O bonds to the bridging and terminal BHT groups are 1.81 and 1.74 Å. Oddly, the Li—O bonds to the bridging -OR groups (1.91 and 1.92 Å) are only marginally shorter than the $Li—OEt_2$ distance, 1.93 Å. In $Li(THF)_2Cl(Er_3CO)Al(OCEr_3)_2$[41b], **44b**, the Li^+ ion is bound to two THF's and two bridging $—OCEr_3$ oxygens.

In $LiMeU\{OCH(t\text{-}Bu)_2\}_4$[42] **45** the uranium is coordinated by two bridging and two terminal $\text{-}OCH(t\text{-}Bu)_2$ ligands as well as a methyl group. The Li^+ ion does not carry a solvent donor and is coordinated by only two oxygen atoms from a bridging $\text{-}OCH(t\text{-}Bu)_2$ group. In $Li\{OC(t\text{-}Bu)_3\}_2MnN(SiMe_3)_2$[43] **46**, the Li^+ ion has similar coordination and is bound to two tri(tert-butyl)methoxy ligands that bridge to manganese, which is also bound to a bis(trimethylsilyl)amido ligand. The lithium center is very crowded and there are short Li—C (2.52 Å) and Li—H (2.1–2.2 Å) contacts to the methyl groups of the $\text{-}OC(t\text{-}Bu)_3$ unit. As a result of this steric hindrance the Li^+ ions are incapable of bonding to further ligands. One of the Li^+ ions in $Li\{OC(t\text{-}Bu)_3\}_2MnBr_2Li(THF)_2$[43] **47** has a very similar environment to that just described, whereas the other is bridged to the metal center by two Br^- ions and further coordinated by two THF donors. In $Li(THF)\{(t\text{-}Bu)_2HCO\}_2Cr\{OCH(t\text{-}Bu)_2\}_2$[44] **48** and $Li\{(t\text{-}Bu)_2CHOH\}Fe\{OCH(t\text{-}Bu)_2\}_4$[44] **49**, and Cr and Fe centers are four-coordinate. Two of the $\text{-}OCH(t\text{-}Bu)_2$ groups bridge to the Li^+ ion which is coordinated by THF in the chromium species **48**, or the alcohol $HOCH(t\text{-}Bu)_2$ in the case of the iron compound **49**. In $\{Li(THF)\}_2\text{-}(ODipp)_2V(ODipp)_2$[45] **50**, where V is in the +2 oxidation state, there is no terminal anionic ligand present. Two -ODipp ligands bridge two Li^+ ions, both of which are further coordinated by THF.

In the V(III) derivative $Li(THF)(ODipp)_2V(ODipp)_2$[46] **51**, there is only one

Table 9.4 Heterometallic LiOR Derivatives

	CN_{Li}	Li—O (Å)	M	M—O_t (Å)	M—O_{br} (Å)	L	Li—L (Å)	Compound/ Ref.
Li(Et$_2$O)[Al(Me){O(2,6-t-Bu$_2$-4-Me-C$_6$H$_2$)}]O(MePh$_2$)O(2,6-t-Bu$_2$-4-Me-C$_6$H$_2$)	3	1.92	Al	1.74	1.81	THF	1.93	**44a**/41a
L(THF)$_2$(Cl)(Et$_3$Co)Al(OCEr$_3$)$_2$	4	1.97	Al	1.69	1.73	THF	1.99	**44b**/41b
LiMeU{OCH(t-Bu$_2$)}$_4$	2	1.83	U	2.10	2.27			**45**/42
Li{OC(t-Bu)$_3$}$_2$MnN(SiMe$_3$)$_2$	2	1.81	Mn		2.00			**46**/43
Li{OC(t-Bu$_3$)$_2$}MnBr$_2$Li(THF)$_2$	2,4	1.85	Mn		2.02	(THF)		**47**/43
Li(THF){OCH(t-Bu)$_2$}$_2$Cr{OCH(t-Bu)$_2$}$_2$	3	1.84	Cr	1.83	1.94	THF	1.94	**48**/44
Li(t-Bu$_2$CHOH)Fe{OCH(t-Bu)$_2$}$_4$	3	1.87	Fe	1.84	1.93	HOCH(t-Bu)$_2$	1.95	**49**/44
{Li(THF)}$_2$V(ODipp)$_4$	3	1.83	V		2.02	THF	1.89	**50**/45
Li(THF)(ODipp)$_2$V(ODipp)$_2$	3	1.84	V	1.83	1.95	THF	1.82	**51**/46
[Li(12-crown-4)$_2$]$^+$[V(ODipp)$_4$]$^-$	8[c]		V	1.87		12-crown-4	2	**52**/46
{Li(DME)}$_2$V(OPh)$_6$	4	1.89	V		2.02	DME	2.05	**53**/46
Li(THF)$_2$(PhO)$_2$W(OPh)$_4$	4	1.94	W	1.92	2.01	THF	1.89, 1.98	**54**/47
Li(THF)$_2${OC(t-Bu)$_3$}Cr(Cl)OC(t-Bu$_3$)	4	1.92	Cr	1.88	1.99	THF	1.96	**55**/48
{Li(MeOH)$_2$}$_2$Cu{(2-OC$_6$H$_4$)$_2$}$_2$	4	1.92, 1.97	Cu		1.93 2.22	MeOH	1.89–2.05	**56**/49
[Li(EtO$_4$)Nb(OEt)$_2$]$_n$	4	1.94	Nb	1.88[a]	1.98[a]			**57**/50
{Li(TMEDA)}$_2$Co{(OSiPh$_2$OSiPh$_2$O)$_2$}	4	1.94	Co		1.97	TMEDA	2.15	**58**/51
Li(py)$_2$Co(py){OSiPh$_2$OSiPh$_2$O}$_2$-Co(py)Cl	4	1.86, 1.93	Co		1.98–2.20	py	2.08, 2.13	**59**/51
Li(py)Li(py)$_2$Zr{OSiPh$_2$OSiPh$_2$O}$_3$	4	2.00 2.07	Zr	2.01	2.10	2py; 1py	2.09, 2.00	**60**/52
{Li(THF)]$_3$(μ_3-Br)$_3$(μ_3-OSiMe$_3$)Cd-C(SiMe$_3$)$_3$	4	1.81–1.88	Cd[d]			THF	1.97[a]	**61**/53
Li(THF)$_2$Cl$_2$Co(OBMes$_2$)$_2$Li(THF)$_2$	4, 4	1.96	Co		1.95	THF	2.00 (1.94)	**62**/19

[a]Denotes average values.
[b]The Li cation is separated from the anion by two 12-crown-4 molecules.
[c]Disordered crown ether molecule.
[d]The Cd atom has no oxygen contact; three Br atoms bridge between Li and Cd.
[e]Coordination of the chlorine-bridged Li atom.

Li^+ ion, and this has a very similar environment to that seen in **50**. Addition of 12-crown-4, however, results in complete Li^+ ion separation to give [Li(12-crown-4)$_2$] [V(ODipp)$_4$][46] **52**. In contrast to the examples given above, V is coordinated by three PhO—Li—OPh units in [{Li(DME)}$_3$V(OPh)$_6$],[46] **53**. This is a rare example of a homoleptic six-coordinate phenolate complex and the environment of V may be described as a distorted octahedron. The Li—O and V—O bonds in **53** are longer than in those of **50–52** owing to the higher coordination at V. In {Li(THF)$_2$}W(OPh)$_6$,[47] **54**, a Li^+ ion is associated with a $[W(OPh)_6]^-$ moiety through bridging to two phenoxides. Two THF molecules also coordinate the Li^+ ion.

A T-shaped geometry is observed at Cr in Li(THF)$_2${(*t*-Bu)$_3$CO}Cr(Cl)OC(*t*-Bu)$_3$[48] **55**, wherein the Li^+ ion is bridged to Cr by a Cl^- and a OC(*t*-Bu)$_3^-$ group. Two methanol molecules serve as donor solvents in {Li(MeOH)$_2$}$_2$Cu{(2-OC$_6$H$_4$)$_2$S}$_2$,[50] **56**. The central Cu atom is bonded by the central sulfur atoms of two thiodiphenolato groups and the phenoxy moieties (Fig. 9.12).

In [LiNb(OEt)$_6$]$_n$[50] **57**, Nb is bonded to two nonbridging, axial ethoxy groups, whereas the equatorial -OEt group bridges to two Li^+ ions in a *trans*

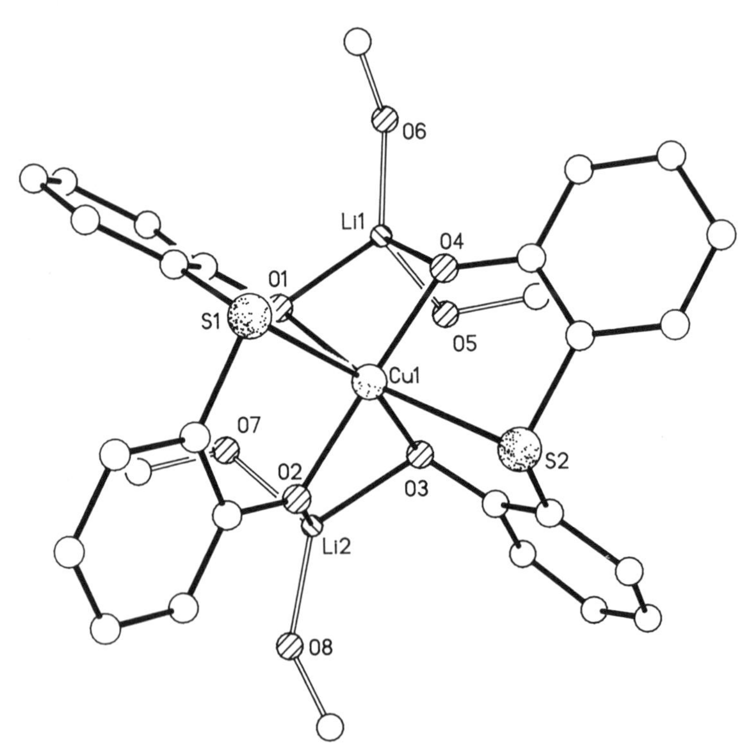

Figure 9.12 Structure of {Li(MeOH)$_2$}$_2$Cu{(2-OC$_6$H$_4$)$_2$S}$_2$, **56**.[50] H atoms are not shown for clarity.

configuration. This results in a polymeric chain structure with a Li-$(OEt)_2$NbLi$(OEt)_2$Nb backbone. The cobalt center in {Li(TMEDA)}$_2$-Co$(OSiPh_2OSiPh_2O)_2$[51] **58** is chelated by two O—Si—O—Si—O di-anionic units. Two sets of O atoms bridge to the TMEDA-coordinated Li^+ ions. In Li$(py)_2$Co(py)$(OSiPh_2OSiPh_2O)_2$Co(py)Cl[51] **59**, one of the two Li^+ ions in **58** is formally replaced by a terminal CoCl unit. Pyridine (two molecules) coordinates the Li^+ ion and both the central and peripheral Co atoms. In Li(py)Li$(py)_2$Zr$(OSiPh_2OSiPh_2O)_3$[52] **60** (Fig. 9.13), the two Li atoms adopt different environments. One is coordinated by three O atoms of three separate disiloxandiol dianions, and is bound to only one pyridine molecule. The other Li^+ ion, which is bound to two pyridine molecules, is coordinated by only two of the three remaining O atoms of the anions, leaving one O atom two-coordinate and, thus, in a ''terminal'' position.

The structure of {Li(THF)}$_3$(μ_3-Br)$_3$(μ_3-$OSiMe_3$)Cd{C$(SiMe_3)_3$}[53] **61** has a distorted cubane core composed of the atoms $CdBr_3Li_3O$. The Br^- ions bear no terminal ligands, whereas Cd and O are bound to -C$(SiMe_3)_3$ and -$OSiMe_3$ groups, respectively, and each Li^+ ion is bound to a THF. The situation in

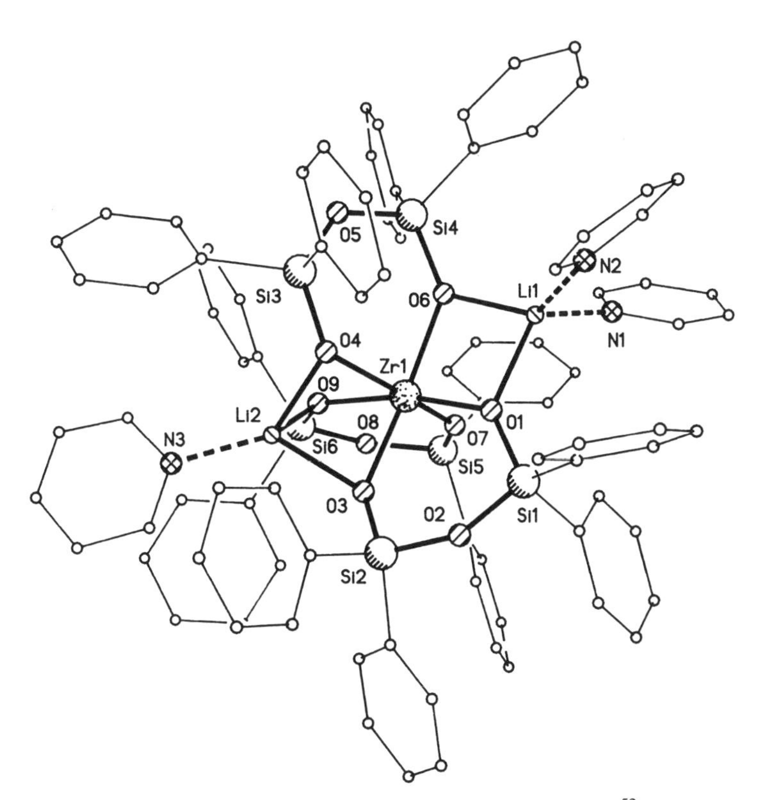

Figure 9.13 Structure of Li(py)Li$(py)_2$Zr$(OSiPh_2OSiPh_2O)_3$, **60**.[52] H atoms are not shown for clarity.

$Li(THF)_2Cl_2Co(OBMes_2)_2Li(THF)_2$[19] **62** is reminiscent of that in **14**. Formally, one of the two Li^+ ions in the dimer is replaced by a Co—Cl—Li—Cl four-membered ring, with Co adopting a position between the two bridging boryloxy groups. Both Li^+ ions are now coordinated by two THF donors and the Li—O bond lengths elongate considerably (0.11 Å for the anions; 0.08 Å for the THF contacts).

2.5 Lithium Thiolates

Lithium thiolates have significantly weaker metal–chalcogenide interactions than their oxygen analogues. One consequence is that monomeric structures are more readily obtained for thiolates and their heavier congeners. In essence, the formation of extra Li—S interactions through further bridging is not as energetically attractive as the formation of additional Li^+–solvate ligand bonds, which usually involve interactions with O or N donor atoms. An example of this behavior is provided by the monomeric species $Li(THF)_3SMes^*$[54] **63**, which has an Li—S distance of 2.45 Å. This distance is marginally longer than the sum of covalent radii (2.41 Å). The structure may be contrasted with that of $[Li(THF)O\{2,6\text{-}t\text{-}Bu_2C_6H_3\}]_2$ **7**, which is a strongly associated dimer.

A similar structure to **63** is observed in $Li(THF)_3STriph$[55] **64** (Fig. 9.14) which has an Li—S bond length of 2.43 Å. With $\text{-}CH(SiMe_3)_2$ and $\text{-}C(SiMe_3)_3$ groups attached to sulfur, the dimeric structures $[Li(THF)_2SCH(SiMe_3)_2]_2$,[56] **65**, and $Li_2(THF)_{3.5}\{SC(SiMe_3)_3\}_2$,[56] **66** are observed. These species were the

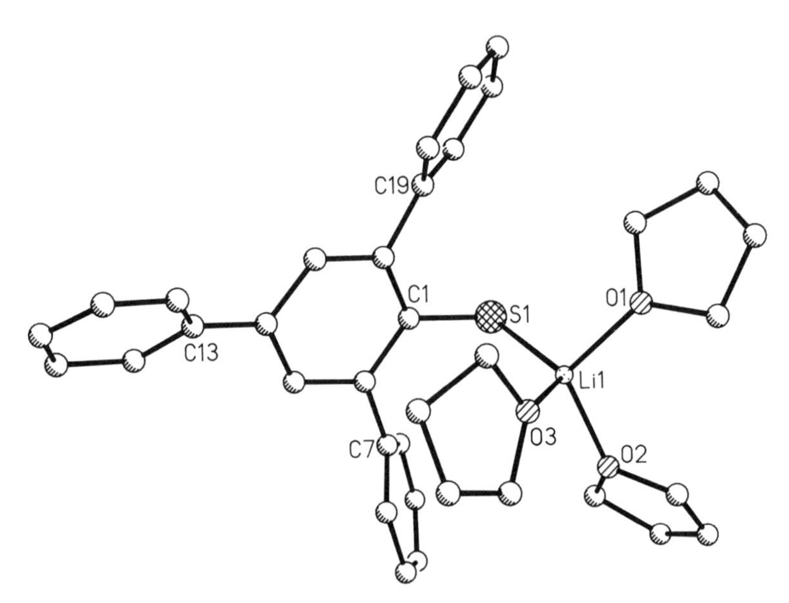

Figure 9.14 Compound $Li(THF)_2STriph$, **64**.[55] H atoms are not shown for clarity.

first X-ray crystal structural reports for lithium thiolates. In **66** the two Li^+ ions have different environments, one is coordinated by two THF molecules, whereas the second Li^+ ion is coordinated by one THF molecule with 100% and another with 50% occupancy. In both **65** and **66**, THF is prone to disorder. In **65** there is a positional disorder of one of the donor molecules and different Li—O distances (2.12 and 1.94 Å) are seen. In **66** the THF molecule with half occupancy has an extremely long interaction (2.46 Å) with the Li^+ ion, which has an effective coordination number of 3.5. This situation implies considerable weakness of the THF coordination to the sterically shielded Li^+ ions. The 3.5-coordinate Li^+ ion displays extremely short Li—S distances of 2.34 Å on average. The dimeric lithium tritylthiolate $[Li(THF)_2SCPh_3]_2$[57] **67** has a similar structure to **65** and displays Li—S distances of 2.49 and 2.46 Å.

In addition to the size of substituents, their arrangement relative to the central atoms can have an impact on the structure. This is exemplified by the series of $[Li(py)_2SPh]_n$,[58] **68**, $[Li(py)SCH_2Ph]_n$,[58] **69**, and $Li(py)_3S$-2-Me-C_6H_4,[58] **70**. The phenyl thiolato species, **68** has a polymeric chain structure with Li—S bonds of 2.46 and 2.51 Å, whereas the benzyl derivative has a polymeric double chain or ladder structure. It has Li—S contacts of 2.48 Å (av.) in the chain and 2.51 Å (av.) in the rungs. The isomeric *o*-tolyl derivative, however, is monomeric, with a shorter Li—S bond of 2.41 Å, which is close to the calculated value (vide supra). In all these compounds, a corresponding number of pyridine molecules are coordinated to preserve a coordination number of four at Li. A dianionic S_6 chain acts as a chelate in $Li_2(TMEDA)_2S_6$,[59,60] **71**, and is symmetrically bound to two Li^+ ions (Li—S, 2.50 Å). TMEDA also coordinates Li^+, resulting in a total coordination number of four for the metal.

Several mixed-metal complexes featuring interactions between lithium and thiolate ligands are known. As in many other mixed-metal cases, Li^+ in $Li(Et_2O)_2(TripS)_2Ga(STrip)_2$,[61] **72**, and the isostructural $Li(Et_2O)_2$-$(TripS)_2In(STrip)_2$,[61] **73**, is coordinated by bridging anions, while the central metal atom also binds terminal ligands of the same kind. The Li—S bonds are particularly short in the indium species (2.39 Å). Two tert-butyl thiolato groups serve as bridges between the moieties $(THF)_2Li^+$ and $LuCp_2^{*+}$ in $Li(THF)_2$(*t*-$BuS)_2LuCp_2^*$,[62,63] **74**. The coordination is asymmetric, as can be seen in the different Li—S distances (2.41 and 2.49 Å) and the Lu—S contacts (2.81 and 2.72 Å). The monomeric unit in $[Li_2(THF)_2Cp^*TaS_3]_2$,[64] **75**, consists of two four-membered Ta—S—Li—S rings fused along one Ta—S bond. Association through further Li—S interactions gives rise to dimerization, affording a hexagonal-prismatic arrangement, reminiscent of a type 5 hexameric structure (Fig. 9.15).

No Li—S contacts are present in $[Li(Et_2O)_3][Co_2(STrip)_5]$,[65] **76**. The coordination sphere of Li consists of only three Et_2O molecules in a slightly irregular trigonal planar arrangement (sum of angles is 360°). The coordination of the Co atoms consists of one terminal and three bridging STrip groups. A listing of structural parameters for metal thiolates is provided in Table 9.5.

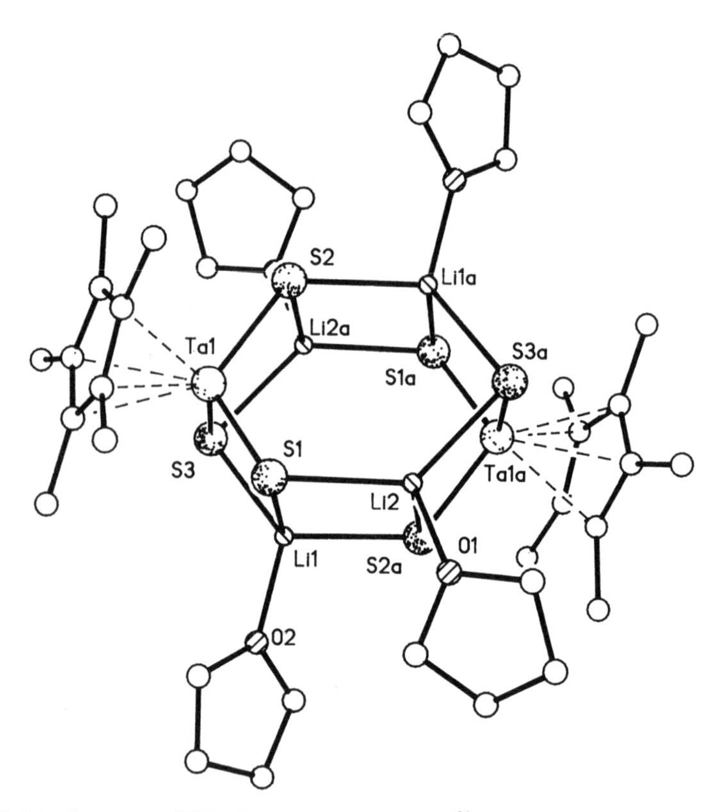

Figure 9.15 Structure of $[Li_2(THF)_2Cp^*TaS_3]_2$, **75**.[64] H atoms are omitted for clarity. Selected bond distances (Å) are Li1-S2 2.54; Li1-S3 2.44; Li1-S1a 2.45; Li2-S1 2.50; Li2-S3 2.44; Li2-S2a 2.49; Ta-S1 2.27; Ta-S2 2.27; Ta-S3 2.30.

2.6 Lithium Selenolates and Tellurolates

The X-ray crystal structure of $Li(THF)_3SeMes^*$,[66,67] **77** (Fig. 9.16) was reported simultaneously in two publications. It is isostructural to its sulfur analogue **63** and has an Li—Se bond length of 2.57 Å, which is significantly longer than the sum of the covalent radii (2.51 Å). A dimeric structure is adopted in $[Li(DME)SeSi(SiMe_3)_3]_2$,[68] **78**, even though the substituent on Se is a bulky $-Si(SiMe_3)_3$ group, and Li^+ is bound to a chelating donor (DME). One of the Li—Se bonds is as long as the one observed in **77** (2.57 Å), the other is slightly longer (2.61 Å). The Li—O interactions (2.02 Å, av.) are in the normal range. The species $Li(THF)_2(PhSe)_2LuCp_2$,[63] **79**, is the selenium analogue of **74** except that tert-butyl groups are replaced by phenyls. There is also asymmetry in the Li-chalcogenide distances (2.60 and 2.70 Å), whereas the Lu atom is symmetrically bound to the selenolate groups.

Similar trends are observed in the structures of lithium tellurolates. A monomeric $Li(THF)_3TeMes^*$, **80**, was reported in preliminary form,[69a] with a Li—Te distances of 2.82 Å (cf. 2.79 Å for the sum of the covalent radii). The Li—THF contacts average 1.95 Å. The species $Li(DME)Te(C_6H_4$-2-

Table 9.5 Lithium Thiolates and Related Species

	Aggr.	CN_{Li}	Li—S (Å)	L	Li—L (Å)	Compound/ Ref.
$Li(THF)_3SMes^*$	1	4	2.45	THF	1.97	**63**/54
$Li(THF)_3STriph$	1	4	2.43	THF	1.95	**64**/55
$[Li(THF)_2SCH(SiMe_3)_2]_2$	2	4	2.45	THF	1.95	**65**/56
$Li_2(THF)_{3.5}\{SC(SiMe_3)_3\}_2$	2	3.5[b]	2.34[a]	THF	1.92[b]	**66**/56
		4	2.49	THF	1.96	
$[Li(THF_2SCPh_3]_2$	2	4	2.84	THF	1.95	**67**/57
$[Li(py)_2SPh]_n$	n[c]	4	2.49[b]	py	2.07	**68**/58
$[Li(py)SCH_2Ph]_n$	n[d]	4	2.48[a]	py	2.06[a]	**69**/58
$Li(py)_3S(2\text{-}MeC_6H_4)$	1	4	2.41[a]	py	2.08[a]	**70**/58
$\{Li(TMEDA)\}_2S_6$	1	4	2.50[e]	TMEDA	2.13[c]	**71**/59,60
$Li(Et_2O)_2Ga(STrip)_4$	1	4	2.42	Et_2O	2.01	**72**/61
$Li(Et_2O)_2In(STrip)_4$	1	4	2.39	Et_2O	1.92	**73**/61
$Li(THF)_2\{S(t\text{-}Bu)\}_2LuCp_2^*$	1	4	2.44[a]	THF	1.95	**74**/62,63
$[\{Li(THF)\}_2(Cp^*)TaS_3]_2$	2	4	2.48[a]	THF	1.93[a]	**75**/65
$[Li(Et_2O)_3][Co_2(STrip)_5]$	3			Et_2O	1.91	**76**/65

[a]Denotes average values.
[b]Li atom displays contact to THF molecule with half occupancy (2.46 Å).
[c]Polymeric chain.
[d]Polymeric ladder.
[e]Average values over the two independent molecules.

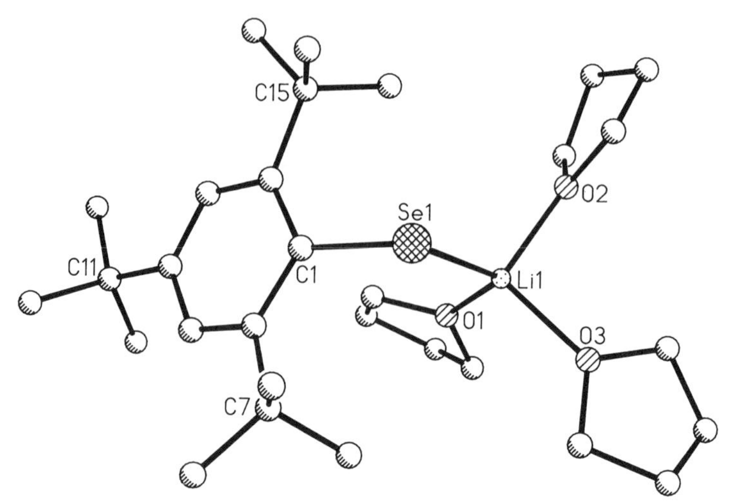

Figure 9.16 Structure of $Li(THF)_3SeMes^*$, **77**.[66,67] H atoms are not shown for clarity.

$CH_2NMe_2)$[70] **81** is also monomeric. The amino group in the ortho position of the phenyl ring coordinates Li^+, which results in a chelate structure with a short Li—Te bond of 2.72 Å and a Li—N bond of 2.04 Å. The Li—O distances average 1.95 Å. Bridging Li—Te bonds were structurally characterized in the dimeric $[Li(THF)_2TeSi(SiMe_3)_3]_2$,[69a] **82a** (Fig. 9.17) and in $[Li(DME)\text{-}TeSi(SiMe_3)_3]$[69b], **82b**. The solvent separated ion pair $[Li(12\text{-crown-}4)_2]\text{-}[TeSi(SiMe_3)_3]$[69a], **83**, was observed when **82a** was treated with 12-crown-4. Structural data for lithium selenolates and tellurolates are given in Table 9.6.

Figure 9.17 Structure of $[Li(THF)_2\text{-}TeSi(SiMe_3)_3]_2$, **82a**.[69] H atoms are not shown for clarity. Slightly different Li—Te bond lengths of 2.82 and 2.88 Å were observed. A similar structure was seen for the compound $[Li(DME)TeSi(SiMe_3)]_2$,[69b] **82b**, which has Li—Te bonds that average 2.83 Å. Ion pair separation occurred upon treatment of **82** with 12-crown-4 to give $[Li(12\text{-crown-}4)_2][TeSi(SiMe_3)_3]$ **83**.[69a] There is no significant difference in the Si—Te bond lengths in each structure, which supports the idea that the Li—Te bond consists of a rather weak ionic interaction.

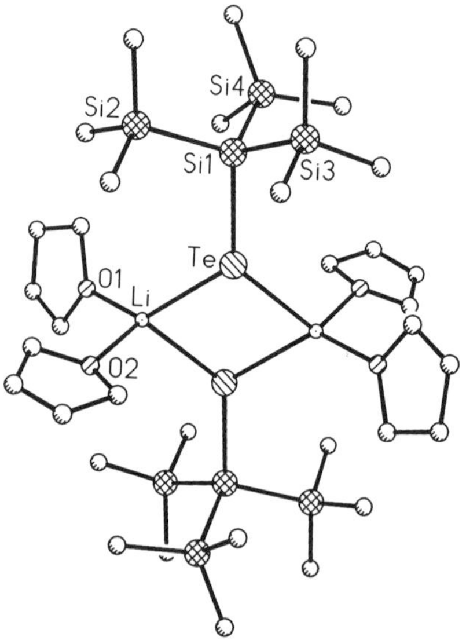

Table 9.6 Lithium Selenolates and Tellurolates

	Aggr.	CN_{Li}	E	Li—E (Å)	L	Li—L (Å)	Compound/ Ref.
$Li(THF)_3SeMes^*$	1	4	Se	2.57[b]	THF	1.97[b]	**77**/66,67
$[Li(DME)SeSi(SiMe_3)_3]_2$	2	4	Se	2.59[a]	DME	2.02[a]	**78**/68
$Li(THF)_2(SePh)_2LuCp_2$	1	4	Se	2.65[a]	THF	1.90[a]	**79**/63
$Li(THF)_3TeMes^*$	1	4	Te	2.82	THF	1.95	**80**/70
$Li(DME)Te(2\text{-}CH_2NMe_2C_6H_4)$	1	4	Te	2.72	DME	1.95[c]	**81**/70
$[Li(THF)_2TeSi(SiMe_3)_3]_2$	2	4	Te	2.85[a]	THF	1.95[a]	**82a**/69a
$[Li(DME)TeSi(SiMe_3)_3]_2$	2	4	Te	2.83[a]	DME	2.0	**82b**/69b
$[Li(12\text{-crown-4})_2]$-$[TeSi(SiMe_3)_3]$	1	8[d]	Te		12-crown-4	2.33[a]	**83**/69a

[a]Denotes average values.
[b]Average values over the two independnet molecules.
[c]Average over both Li—O contacts; displays an additional, intramolecular Li—N contact of 2.04 Å.
[d]12-crown-4 coordination.

3. LITHIUM COMPOUNDS OF THE MAIN GROUP 5 ELEMENTS

3.1 Lithium Amides

Amido groups have been widely used throughout the periodic table to stabilize unusual types of bonding, coordination numbers, and oxidation states.[71] Lithium amides, which are generally prepared by the reaction of *n*-BuLi with the amine, have played a key role in the synthesis of these compounds. In spite of their importance, structural studies of lithium amides had, until recently at least, progressed much more slowly than those on Li—C species. A comprehensive study of metal amides[71] published in 1979 listed only one lithium amide structure, the trimeric $[LiN(SiMe_3)_2]_3$.[72,73] By 1983 the structures of approximately 10 lithium amide species were known. At present, about 100 crystal structures have been determined. Thus the growth in our knowledge of the structures of these compounds over the past 10 years has been very rapid.

Lithium amides have been the subject of recent reviews, with coverage to mid-1990,[2] in which the principles underlying their structures were thoroughly discussed. Nonetheless, many lithium amide structures have been published in the intervening period. In addition, there have been no recent reviews of lithium amides that have covered functional groups on nitrogen. Bearing these comments in mind, many of the compounds dealt with in previous reviews are not discussed in detail but are listed in the tables. In the discussion, emphasis is placed primarily on major developments and more recent results.

The structures of amides are characterized by a strong tendency to associate. As mentioned above, the manner of their association was discussed very lucidly in a recent review[2a] and needs no further elaboration here. Monomeric structures are observed only in the presence of very bulky groups at nitrogen and/or donor molecules that coordinate strongly to lithium. In most cases, strong interactions between lithium and nitrogen are observed, although in a very few instances, such as $[Li(12\text{-crown-}4)_2][N(SiPh_3)_2]$,[74] complete separation of the Li^+ ion has been achieved (vide infra).

*3.1.1 **Structures with Terminal Amido Ligands.*** The most numerous lithium amides involve simple monodentate ligands of the type -NHR, $-NR_2$, or -NRR′. The simplest compounds are derivatives of the primary amine RNH_2. When the substituent group is Mes* or Dipp, a dimeric structure is observed when the Li^+ is crystallized from ether. If, however, TMEDA is added to solutions of LiNHMes*, a monomeric structure of formula Li(TMEDA)NHMes* **84** is obtained.[75] A very short N—Li bond (1.90 Å) is observed. The Li^+ ion is three-coordinate. Monomeric structures have also been observed in the secondary amide derivatives Li(TMEDA)NPh(naphthyl)[76] **85** and Li-(PMDETA)NPh(naphthyl)[76] **86**, which have Li—N bond lengths of 1.97 and 2.00 Å, respectively. The longer bond in the latter species results from the higher coordination number of Li^+, which is coordinated by PMDETA. When there are smaller substituents on N, more highly aggregated species are observed even in the presence of chelating donor ligands like TMEDA. An example is the dimeric species $[Li(TMEDA)NMePh]_2$,[76] **87**.

A further example of the use of a multidentate ligand to obtain monomeric structures is the reaction of 12-crown-4 with $LiN(SiMe_3)_2$ or $LiNPh_2$ to afford Li(12-crown-4)$N(SiMe_3)_2$,[77] **88**, or Li(12-crown-4)NPh_2,[78] **89**. In these compounds the Li^+ ion is five-coordinate with approximately square pyramidal coordination by one nitrogen and four oxygen donors. This results in slightly longer Li—N distances of 1.97 and 2.03 Å, respectively. The Li—O distances are near 2.18 Å for **89** and 2.23 Å for **88**. A similar structure is observed for Li(12-crown-4)$N(SiMePh_2)_2$,[79] **90**. It is clear that in the structures **84–86** and **88–90**, mononuclearity has been induced by the strong coordination of the multidentate donor ligands with some assistance from bulky groups. It is also possible to induce monomeric structures solely by the use of large substituents. This behavior is exemplified by the structures of $Li(THF)_2N(SiMePh_2)_2$,[79] **91**, and $Li(THF)_2N(SiPh_3)_2$,[79] **92**, both of which possess trigonally coordinated Li and N centers with Li—N bonds of 1.95 and 1.99 Å, respectively. Their monomeric structures may be contrasted with those of $[Li(Et_2O)N(SiMe_3)_2]_2$,[80,81] **93**, and $[Li(THF)N(SiMe_3)_2]_2$,[82] **94**. The weaker donor ligand Et_2O in **93** results in longer Li—N bonds (2.06 Å) than in the THF solvated species **94** (2.02 Å). In addition, the Li—O interactions are much stronger in the latter (1.88 vs. 1.96 Å in **93**). The Si—N—Si angles are the same within the standard deviations (av. 122.1°). The dimeric structures observed for both of these are presumably a consequence of the smaller size of the $-SiMe_3$ substituents.

The large size of the $-SiPh_3$ group is underlined by the result of the addition of 12-crown-4 to THF solutions of $LiN(SiPh_3)_2$. In this case, complete ion pair separation is effected to afford the anion $[N(SiPh_3)_2]^-$ and cation $[Li(12\text{-crown-}4)_2]^+$, **95** (Fig. 9.18) (cf. structure of **88** and **90**). The anion, which is isoelectronic to $[N(PPh_3)_2]^+$, has an extremely wide SiNSi angle of 154.9° and a short Si—N distance of 1.64 Å.

Similarly, the reaction of $LiN\{Si(t\text{-Bu})_2F\}_2$ and 12-crown-4 gives the ion pair $[Li(12\text{-crown-}4)_2][N\{Si(t\text{-Bu})_2F\}_2]$,[83a] **96a**. In this instance, the steric and electronic effects of the $-Si(t\text{-Bu})_2F$ substituent result in an even wider angle of 162.6° at nitrogen and shorter Si—N distances that are near 1.62 Å. In the absence of the crown ether induced ion pair separation, the Li^+ ion interacts with the fluorine atoms (Li—F, 2.12 and 2.21 Å) in $Li(THF)_2N\{Si(t\text{-Bu})_2F\}_2$,[83] **97** (Fig. 9.19), and an almost linear Si—N—Si backbone (176.7°) is observed. The Li—N bond (2.11 Å) is weaker (cf. **93** and **94**), perhaps as a result of the Li—F interactions. The Li—F interactions also induce the lengthening of the Si—F bonds to about 1.66 Å. The solvated Li* cation and the carbazole anion are linked by H-bonding in the compound $[Li(THF)_2(t\text{-BuOH}][NC_{12}Hg]$[83b], **96b**.

Several other mononuclear structures involving halogenated silyl substit-

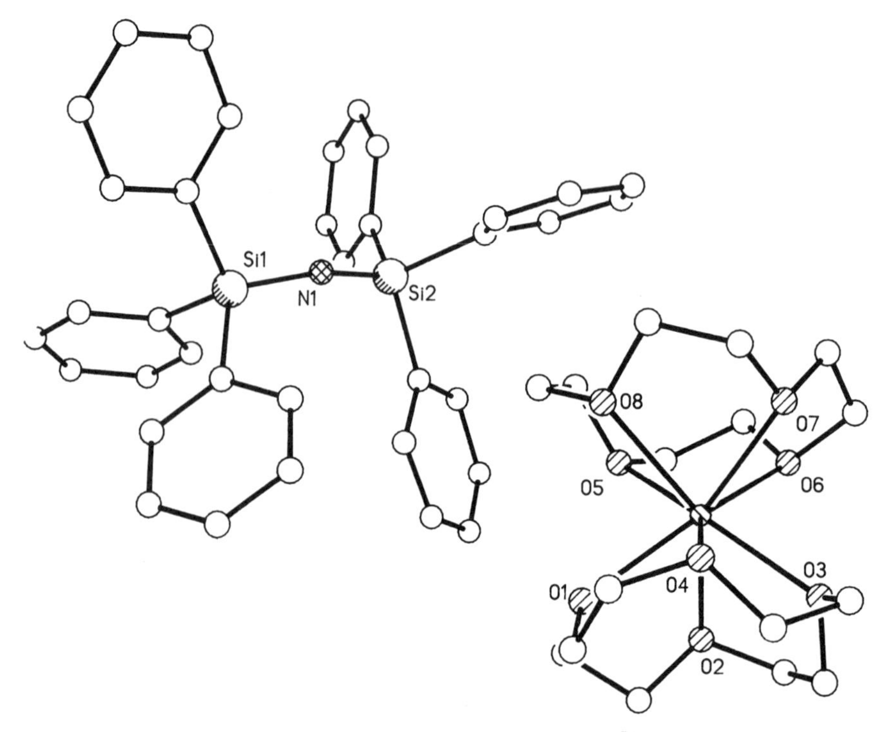

Figure 9.18 Structure of $[Li(12\text{-crown-}4)_2][N(SiPh_3)_2]$, **95**.[74] H atoms are not shown for clarity.

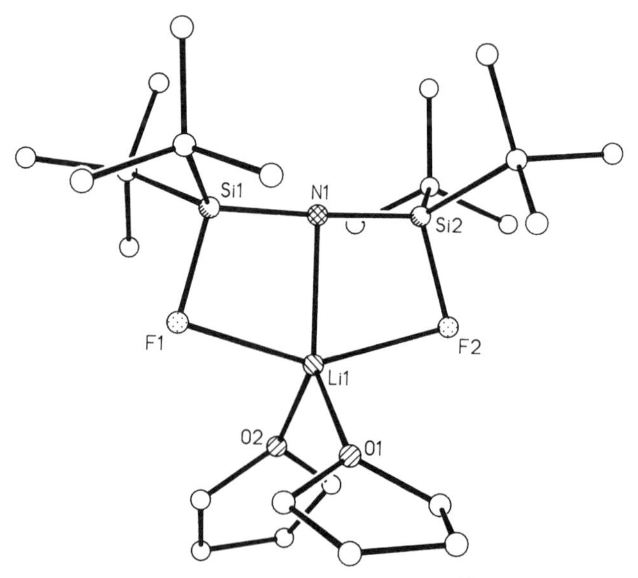

Figure 9.19 Structure of $Li(THF)_2N\{Si(t\text{-Bu})_2F\}_2$, **97**.[83a] H atoms are not shown for clarity.

uents, in which ion pair separation could probably be induced by the addition of crown ether, have been reported. These include Li(TMEDA)-[N{Si(*t*-Bu)$_2$F}(4-FC$_6$H$_4$)],[84] **98**, Li(THF)$_2$N(C$_6$F$_5$){Si(*t*-Bu)$_2$F},[85] **99**, Li(THF)$_3$NMes*{Si(*i*-Pr)$_2$F},[86] **100**, and Li(THF)$_2$NMes*{Si(*i*-Pr)$_2$Cl},[86] **101**.

In the species **100**, where Mes* is the substituent at nitrogen, there is no Li—N interaction. Instead, lithium is bound to the F atom of the anion and additional THF molecules. The Si—F bond, however, remains strong enough (although elongated to 1.69 Å) to avoid cleavage and concomitant Li—F elimination. The strong Li—F interaction is inferred from the bond length of 1.82 Å, while the short Si—N distance is in the range of a Si—N double bond (1.62 Å). There is also an almost linear (172.1°) nitrogen coordination. Unlike the latter species, there is no halogen–lithium contact in **101**, owing to the weaker donor properties of chlorine towards lithium. Thus the Li^+ ion is bound to nitrogen in the usual fashion, and has a bond length of 1.99 Å. It is notable that the Si—N distance is only marginally longer than in **100** (1.64 Å), whereas the Si—N—C angle is much narrower (138.7°).

The mercaptobenzoxazolyl-lithium Li(TMEDA)(H$_2$O){*c*-NC(S)OC$_6$H$_4$},[87] **102**, displays H_2O as an unusual donor molecule besides TMEDA, rendering Li^+ four-coordinate and affording a monomeric structure. The anionic Li—N bond is rather long (2.05 Å); the Li—O contact is in the normal range of Li—O interactions for donor molecules such as Et_2O and THF. Mononuclear structures may also be observed when the amide ligand is a chelating species as in Li(THF)$_2$Me$_3$SiNSi(F)$_2$NSiMe$_2$-{*c*-(NSiMe$_2$)$_2$}SiMe$_3$,[88] **103**, Li-(THF)$_2$Me$_3$SiNP(Ph)$_2$NSiMe$_3$,[89] **104**, Li(TMEDA)-2-Me-6-(Me$_3$Si)N-pyridine,[90] **105**, Li(TMEDA)quinoline-8-N(SiMe$_3$),[90] **106**, Li(THF)$_2$N{C(CH$_2$)$_2$-naphthyl}{*c*-N(CH$_2$)$_4$}CH$_2$OMe, **107**,[91] Li-phthalocyanine, **108**,[92] and [Li-(THF)$_4$][LiOEP], **109**[93] (Fig. 9.20).

Treatment of the amino-iminosilane *t*-BuSi(NMes*)(NHMes*) with *n*-BuLi and 12-crown-4 results in [Li(THF)12-crown-4][Si(*t*-Bu){N(Mes*)}$_2$],[94a] **110a**. The Li^+ ion is coordinated by a crown ether and a THF molecule. The Si—N bonds are quite short (1.59 and 1.63 Å), which suggests considerable delocalization in the NSiN skeleton. The average Li—O bond distance in the crown ether adduct is 2.04 Å. A related, monomeric Li^+ salt of an amino-imino phosphine may also be obtained. The compound Li(THF)-N{C(Ph)$_3$}PNMes*[94b], **110b** features a three-coordinate Li^+ ion bound to a THF and two nitrogens with almost equal Li—N distances of 1.97 and 1.98 Å.

The structure of Li{C$_6$H$_4$N-2-N(H)Ph}NPh(2-py),[95] **111**, consists of an {NPh(2-py)}-anion, which coordinates to Li^+ with both of its N donors (Li—N = 2.09 Å to the pyridinyl moiety and 2.11 Å to the ortho-substituted Ph—N group). A protonated, neutral C$_6$H$_4$N-2-N(H)Ph molecule coordinates to its nitrogen atom (Li—N = 2.08 Å). There is a short Li—O contact (1.82 Å) to the HMPA donor molecule. Structural data for mononuclear lithium amides are given in Table 9.7.

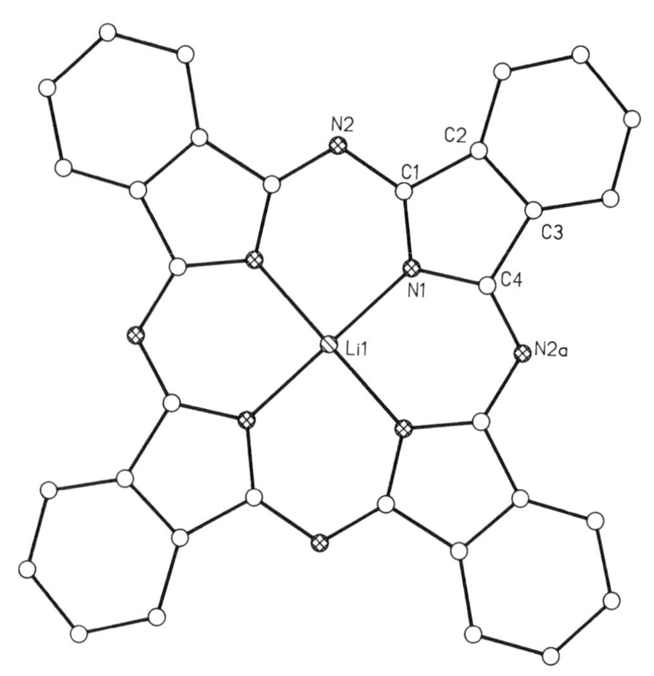

Figure 9.20 Structure of the anion $[LiOEP]^-$ in **109**.[93] H atoms are not shown for clarity.

3.1.2 Dimeric Amides. The simplest dimeric amides are derived from primary amines. The type 2 structure is observed in $[Li(Et_2O)NHR]_2$ (R = Mes*,[96] **112**; Dipp,[97] **113**). The former species has Li—N bonds of 1.99 and 2.04 Å, a planar $(LiN)_2$ framework, and a narrow Li—N—Li angle of 77.8°. The low degree of aggregation is, presumably, a consequence of the large size of Mes*. For R = Dipp, **113**,[97] the corresponding structural parameters are 1.99 and 2.01 Å and 75.9°. A type 2 dimeric structure without a solvating molecule was observed in $[LiN(Dipp)SiMe_3]_2$,[98] **114**. The Li—N distances are fairly long for a species involving two-coordinate Li (1.96 Å). This is presumably a consequence of the bulk of the substituents. The Li—N—Li angle (73.7°) is narrower than those in $[Li(Et_2O)N(H)R]_2$, **112** and **113**.

The reaction of 7,8-benzoquinoline with *n*-BuLi results in the addition of an *n*-butyl group to the α-C atom, while Li coordinates to N in $[Li(Et_2O)$-$c\{NCH(n\text{-Bu})CHCH\text{-}2,3\text{-}C_{10}H_6\}]_2$,[99] **115**. A dimeric type 2 structure with Li—N bonds of 2.04 and 2.10 Å is observed. The aromaticity of the ring is disrupted, as indicated by the puckering of the ring plane. Lithium dibenzylamide also adopts a type 2 structure in the presence of monodentate donors like Et_2O and HMPA, as seen in the compounds $[Li(Et_2O)N(CH_2Ph)_2]_2$,[100a,b] **116**, and $[Li(HMPA)N(CH_2Ph)_2]_2$,[100a,b] **117**. The Li—N distances are 1.98 and 1.99 Å in the former and 2.00 and 2.01 Å in the latter. Presumably, the small difference is due to the different bonding strength of the donor (Li—OEt_2,

Table 9.7 Mononuclear Lithium Amides

	CN_{Li}	Li—N (Å)	L	Li—L (Å)	Add. Coord.	Li—X (Å)	Compound/ Ref.
Li(TMEDA)NHMes*	3	1.90	TMEDA	2.15			**84**/75
Li(TMEDA)N(Ph)naphthyl	3	1.97	TMEDA	2.13[a]			**85**/76
Li(PMDETA)N(Ph)naphthyl	4	2.00	PMDETA	2.20[a]			**86**/76
Li(12-crown-4)NPh_2	5	1.97	12-crown-4	2.23[a]			**88**/77
Li(12-crown-4)$N(SiMe_3)_2$	5	2.03	12-crown-4′	2.18[a]			**89**/78
Li(12-crown-4)$N(SiMePh_2)_2$	5	2.06	12-crown-4	2.20[a]			**90**/79
$Li(THF)_2N(SiMePh_2)_2$	3	1.95	THF	1.91			**91**/79
$Li(THF)_2N(SiPh_3)_2$	3	1.99	THF	1.95			**92**/79
$[Li(12\text{-crown-}4)_2][N(SiPh_3)_2]$	8		12-crown-4	2.36[a]			**95**/74
$[Li(12\text{-crown-}4)_2][N\{Si(t\text{-Bu})_2F\}_2]$	8		12-crown-4	2.37[a]			**95**/83a
$[Li(THF)_2\{OH(t\text{-Bu})\}_2][NC_{12}M_9]$	4		THF	1.93	t-BuOH	1.91	**96**/83b
$Li(THF)_2N\{Si(t\text{-Bu})_2F\}_2$	5	2.11	THF	1.94	F	2.12, 2.21	**97**/83a
$Li(TMEDA)[N\{Si(t\text{-Bu})_2F\}4\text{-F-}C_6H_4$	4	1.98	TMEDA	2.09	F	2.04	**98**/84
$Li(THF)_2N(C_6F_5)\{Si(t\text{-Bu})_2F\}$	5	1.90	THF	1.93	O—C_6F_5	2.27	**99**/85
					SiF	2.39	
$Li(THF)_3NMes^*\{Si(i\text{-Pr})_2F\}$	4		THF	1.96	SiF	1.82	**100**/86
$Li(THF)_3NMes^*\{Si(i\text{-Pr})_2Cl\}$	3	1.99	THF	1.95			**101**/86
$Li(TMEDA)(H_2O)\text{-}c\text{-}\{NC(S)OC_6H_4\}$	4	2.05	TMEDA	2.13			**102**/87
			H_2O	1.95			

Table 9.7 ***(Continued)***

	CN_{Li}	Li—N (Å)	L	Li—L (Å)	Add. Coord.	Li—X (Å)	Compound/ Ref.
$Li(THF)_2Me_3SiNSiF_2NSiMe_2$-*c*-$(NSiMe_2)_2SiMe_3$	4	1.94	THF	1.88	SiF	2.48	**103**/88
$Li(THF)_2Me_3SiNP(Ph)_2NSiMe_3$	4	2.06	THF	1.93			**104**/89
Li(TMEDA)-2-Me-6-(Me_3Si)N-pyridine	4	2.02[a]	TMEDA	2.08[a]	N	2.07[a]	**105**/90
Li(TMEDA)quinoline-8-N($SiMe_3$)	4	2.03	TMEDA	2.15[a]	pyr	2.01	**106**/90
$Li(THF)_2N\{C(CH_2)_2$-naphthyl$\}N(CH_2)_4$-CH_2OMe	4	1.96	THF	1.99	O	2.00	**107**/91
Li-phthalocyanine	4	1.94					**108**/92
$[Li(THF)_4][LiOEP]$	4	2.04					**109**/93
			THF	1.88			
[Li(THF)(12-crown-4)]-$[Si(tBu)(NMes^*)_2]$	5		12-crown-4	2.04[a]			**110a**/94a
			THF	1.93			
$L(Et_2O)N\{c(Ph)_3\}PNMes^*$	3	1.97, 1.98	Et_2O	1.95			**110b**/94b
Li(HMPA)(2-N(Ph)H-pyr)-2-N(Ph)pyridine	4	2.09	HMPA	1.82	N	2.11	**111**/95
			L[b]	2.08			

[a]Denotes average values.
[b]*o*-NH(Ph)-pyridine (protonated anion acts as additional donor molecule).

2.01 Å vs. Li{HMPA}, 1.85 Å). $[Li(Et_2O)NPhCH_2(t\text{-}Bu)_2]_2$,[101] **118** (type 2 structure), has Li—N bond lengths of 2.08 and 2.00 Å. Weak coordination of the Li^+ ion by the phenyl ring is also observed with Li—C distances of 2.44 and 2.66 Å involving the ipso- and ortho-C atoms, respectively. There are not interactions apparent between the Li^+ ion and the C—C double bond. The π-bond seems to be localized in the C—C (1.33 Å) rather than the N—C (1.44 Å) bond. The introduction of a phosphido group as a nitrogen substituent provides an additional coordination site for the Li^+ ion. In $[LiN(Ph)PPh_2]_2$,[102] **119**, it can be argued that the lithium center bonds side-on to the phosphorus-nitrogen bond. The dimerization occurs along a central $(LiN)_2$ ring with Li—N bonds of 2.02 Å, which is in the normal range for lithium amides. The shortest Li—P distance is 2.68 Å, which is significantly longer than those seen in lithium phosphides (vide infra).

A typical lithium fluorosilylamide structure is represented by $[LiN(t\text{-}Bu)Si(t\text{-}Bu)_2F]_2$,[103a] **120** (Fig. 9.21). It has a type 2b structure with a $(LiN)_2$ core and slightly asymmetric Li—N bonds of 2.09 and 1.99 Å. In addition, there are Li—F contacts of 1.90 Å, which are significantly shorter than those observed in **97**. Unlike other dimers with chelating anions (vide infra), **120a** has a boat-like conformation, with the four-membered Si—N—Li—F rings fused to opposite sides of the central ring. In the type 2 dimeric structure of $[Li(TMEDA)N(H)Si(t\text{-}Bu)_2F]_2$, **120b**,[103b] no Li—F contacts are observed. It has C_2 symmetry and displays Li—N bond lengths of 2.06 Å (av.) in the four-membered ring. The Li—N bonds to the solvating TMEDA molecule are rather long (Li—N = 2.27 Å, av.). In $[LiFSi(i\text{-}Pr)_2NSi(t\text{-}Bu)_2OSi(Me)\{N(SiMe_3)_2\}F]_2$,[104] **121**, an Li—N—Si—O—Si—F framework forms a six-

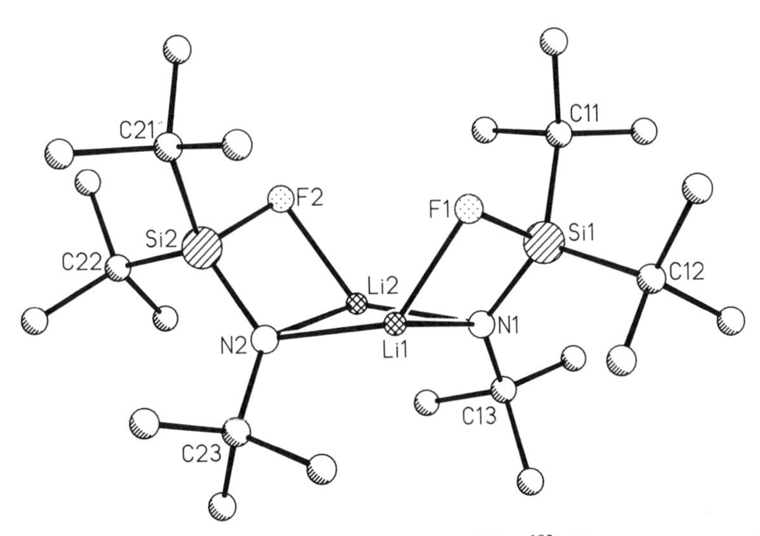

Figure 9.21 Structure of $[LiN(t\text{-}Bu)Si(t\text{-}Bu)_2F]_2$, **120a**.[103a] H atoms are not shown for clarity.

membered ring. The F atom of a fluorosilyl group bound to the anionic N atom serves as a bridge to the Li^+ ion of the other monomer, thus affording dimeric structures with a central $(Li—N—Si—F)_2$ eight-membered ring. The Li—N distance is in the normal range for three-coordinate Li atoms (1.98 Å); the Li—F interaction in the six-membered ring is longer (1.97 Å) than the bridging one (1.82 Å). $[Me_3SiOSi(i\text{-}Pr)_2NSi(t\text{-}Bu)_2FLi]_2$,[105] **122**, essentially consists of a chelating N—Si—O unit that coordinates to the Li^+ ion of the same monomeric unit (Li—N, 1.95; Li—O 2.00 Å). Dimerization is effected by Li—F bridges, resulting in a $(Li—N—Si—F)_2$ eight-membered ring with unusually strong Li—F interactions (1.80 Å). NMR investigations also show a strong lithium–fluorine coupling in solution ($^1J_{LiF}$ = 95 Hz). In **123** $[Li\{HN(i\text{-}Pr)_2\}N\text{-}1\text{-}c\text{-}C_6H_9\text{-}Ph]_2$,[106] there are short Li—C distances of 2.57 and 2.99 Å involving the α-C atoms of the 1-cyclohexenyl group. The Li—N distances are 2.02 and 2.08 Å, respectively.

Cyclosilazanes $(RR'SiNH)_n$ are formed by the reaction of NH_3 with dichlorosilanes. Their rings have several NH moieties available for deprotonation. Monolithiation of hexamethylcyclotrisilazane (HMCTS) or its -$SiMe_3$ derivative results in dimeric products $[Li(THF)HMCTS]_2$,[107] **124** and **125**[108] (Fig. 9.22).

In the latter species, one of the ring nitrogens is substituted by an $SiMe_3$ group. The central four-membered $(LiN)_2$ ring is planar and the six-membered cyclosilazane moiety has a distorted boat confirmation. The Li—N bonds are fairly long for three-coordinate lithium (2.02–2.06 Å; 2.04 Å, av.). The coordination sphere of Li is saturated by one THF molecule per monomer. The

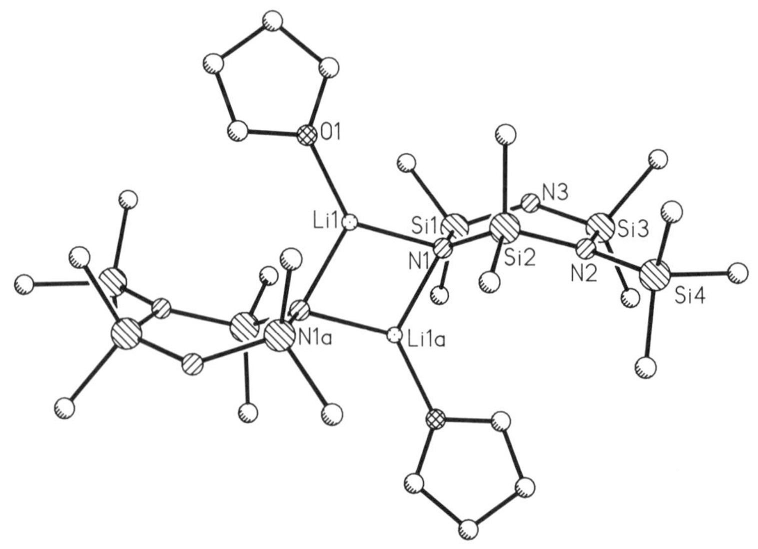

Figure 9.22 Structure of $[Li(THF)HMCTS]_2$, **125**.[108] H atoms are not shown for clarity.

compound $[Li(TMEDA)\text{-}indolyl]_2$,[109a] **126b**, displays a dimeric structure with C_2 symmetry. Although one Li—N bond has a normal (2.00 Å) length, the other is rather long (2.23 Å). The participation of one of the nitrogen lone pairs in the aromatic indolyl system may account for the longer Li—N distance. The structure of the fluorenyl analogue, **127**,[110] with nitrogen in the 1-position was reported in 1987; it dimerizes through lithium bridges. The Li^+ ions are each coordinated by two THF molecules. As in **126**, one of the Li—N bonds is considerable shorter (2.03 Å) than the other (2.16 Å). The C—N bond lengths are consistent with a delocalized system (1.39 Å). TMEDA-coordinated fluorenyllithium[111] is also a dimer. In contrast to **127**, the lithium–element bonds are equally long (2.29 Å). The essentially tridentate ligand in $[LiN\{CH(Ph)CH_2OMe\}_2N]_2$,[112] **128** (Fig. 9.23) acts as a doubly bidentate anion in a chelating fashion.

The dimerization is via a central type 2 $(LiN)_2$ ring, while the outer O atoms of both anions each coordinate to one Li^+ ion. It is noteworthy that the bonding strength of the anions is slightly different; the Li—N contacts to one anion are shorter (2.02 Å) than those of the other (2.06 Å), not as great is the difference in the Li—O contacts (1.98 and 1.99 Å vs. 2.02 and 1.98 Å).

In contrast to the TMEDA adducts **105** and **106**, $[Li\text{-}8\text{-}Me_3SiN\text{-}quinoline]_2$,[90] **129**, $[Li(Et_2O)2\text{-}Me_3SiN\text{-}py]_2$,[90] **130**, and $[Et_2O\text{-}2\text{-}Me\text{-}6(Me_3Si)N\text{-}py]_2$,[90] **131** are dimeric in the absence of donor solvents. $[Li(\mu^1\text{-}HMPA)\text{-}2\text{-}PhN\text{-}py]_2[Li(\mu^2\text{-}HMPA)\text{-}2\text{-}PhN\text{-}py]_2$,[113] **132**, has two different arrangements of the same cation–anion pair; one with a stair-shaped tricyclic dimer with a type 2b central $(LiN)_2$ unit and a terminal HMPA donor, the other with bridging HMPA oxygen atoms and chelating anions in spiral tricyclic fashion. In $[Li(THF)S(Ph)(NSiMe_3)_2]_2$,[114] **133**, the use of THF results in a type 2b stair-shaped tricyclic structure. One THF molecule coordinates each Li^+ ion.

The only chemical difference between $[LiS(t\text{-}Bu)(NSiMe_3)_2]_2$,[115] **134**, and $[LiS(t\text{-}Bu)N\text{-}t\text{-}Bu)(NSiMe_3)]_2$,[115] **135a**, is the substitution of one imido-$SiMe_3$ in **134** with a *t*-Bu group in **135a**. This minor change, however, results in different structures. The employment of the tert-butyl group instead of -$SiMe_3$ on N results in a dimerization forming a type 2c eight-membered ring rather

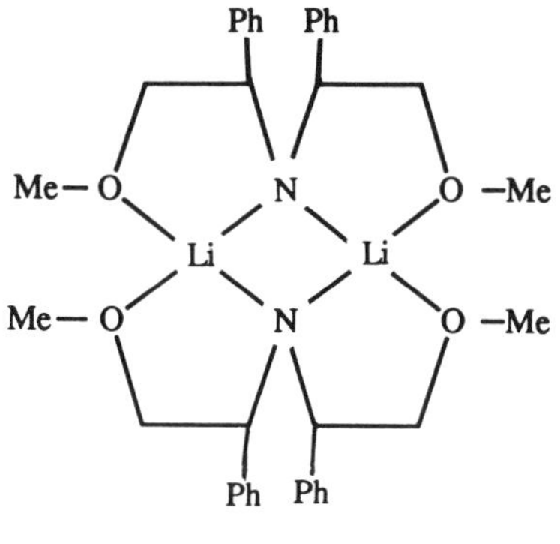

Figure 9.23 Schematic illustration of the structure $[LiN\{CH(Ph)CH_2OMe\}_2N]_2$, **128**.[112]

than a tricyclic structure. The transannular Li—N distances (2.80 Å) in **135** are too long to involve significant bonding. The Li and N centers are thus two- and three-coordinate, respectively. This view is in harmony with the short Li—N bonds of 1.91 Å. Similar, type 2C, structures are observed for the dimers $\{Li(t\text{-}Bu)NPNMes^*\}_2$[96b], **135b**, and $\{Li(1\text{-}Ad)NPNMes^*\}_2$[94b], **135c**, slightly asymmetric Li—N distances, 1.85 and 1.90 Å in **135b** and 1.87 and 1.94 Å in **135c** were observed for the two coordinate Li^+ ions which have NLiN angles near 168°. In **134**, the transannular Li—N distances are shorter (2.37 Å) and close to the value observed in **133**. The structure may be viewed as a stair-shaped tricycle. Since the Li^+ ion is three-coordinate and one of the N atoms is four-coordinate, the Li—N distances in the outer eight-membered ring unit of the tricycle are elongated (1.92 and 1.97 Å) with respect to **135**, but, because of the absence of a donor molecule at lithium, remain shorter than in **133**. The addition of 12-crown-4 to $[LiS(Ph)(NSiMe_3)_2]_2(Et_2O)$ does not result in a conventional solvent-separated ion pair in the crystal structure. Instead, the species $[Li(12\text{-crown-}4)_2][Li\{S(Ph)(NSiMe_3)_2\}_2]$,[116a] **136a**, is obtained. One Li^+ ion atom is coordinated between two crown ether molecules, the other is coordinated by the two chelating anions in a spirocyclic fashion. Since there is increased interelectronic repulsion in the anionic unit, the Li—N distances are significantly longer (2.09–2.17 Å). In ^{7}Li-MAS-NMR solid state experiments, the coexistence of two species with three different Li environments was detected. These were assigned to the Li^+ ion environment in the crystal structure and the presence of the naked iminosulfinamide anion and a crown ether coordinated Li^+ ion. The structure of $[Li(THF)(NSiMe_3)_2C\text{-}4\text{-}MeC_6H_4]_2$, **136b**,[116b] has slightly different lithium environments. It deviates slightly from a type 2b dimeric structure, and has Li—N bond lengths of 2.12 Å (av.).

Monodeprotonation of the 1,7-diaza-12-crown-4 molecule results in a dimeric structure, **137**, in which the Li^+ ion is centrally bonded to one of the two crown ether molecules ($Li—N_{prot.}$, 2.19 Å; $Li—N_{deprot.}$, 2.00 Å), and laterally to a deprotonated N atom of the other crown ether in the dimer $[1\text{-Li-}1,7\text{-N-}12\text{-crown-}4]_2$.[117] This leads to the adoption of a variation of a type 2 structure. The Li—O contacts (2.26 and 2.23 Å) are in the range of those observed in other crown ether–lithium compounds (cf. **88–90**, **95**, **96**).

In $Li[t\text{-}BuNSiMe_2O(t\text{-}Bu)]_2Li(THF)$,[118] **138** (Fig. 9.24) the dimerization occurs along a central four-membered $(LiN)_2$ ring, to which two (Li—N—Si—O) rings are fused to afford a type 2a structure. The peripheral, three-coordinate Li atom is also bound to a THF donor molecule, whereas the central Li is four-coordinate. Note that in **133**, the use of THF affords a type 2b dimer. The lithium–donor bonds are fairly long ($Li_{centr.}—N$, 2.11 Å; $Li_{centr.}—O$, 2.01 Å; $Li_{periph.}—N$, 2.04 Å; $Li_{periph.}—O_{THF}$, 2.01 Å).

The species $Li(THF)_2Li[OC(OMe)CC\{c\text{-}(C_4H_8)\text{-}N(NMe_2)\}]_2$,[106] **139**, is another instance of a type 2a dimer. One Li atom is chelated by the O and N atoms of the N—C—C—C—O backbone of both anions, resulting in two spirocyclic six-membered rings with the Li^+ ion as the spirocyclic center, which

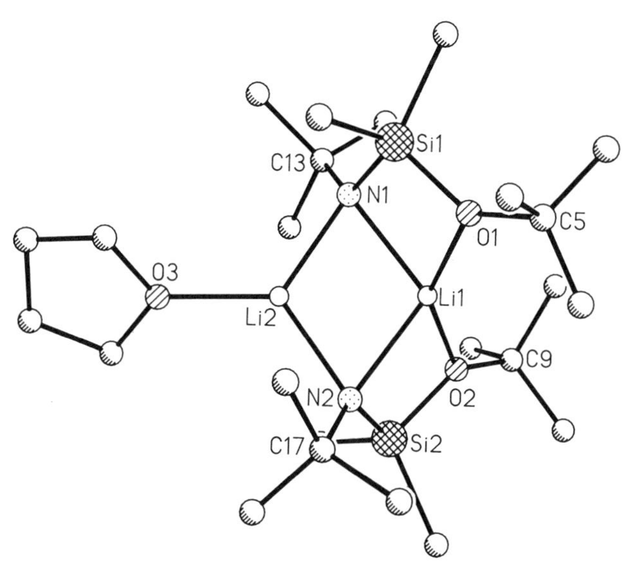

Figure 9.24 Structure of Li[*t*-BuNSiMe$_2$O(*t*-Bu)]$_2$Li(THF), **138**.[118] H atoms are not shown for clarity.

displays Li—N bond lengths of 2.02 Å and fairly short Li—O distances (1.92 Å). The other Li atom is coordinated by these two O atoms and two THF molecules. The Li—O distances to the peripheral Li^+ ion are fairly short (1.93 Å), the same value is present in the Li–THF contacts.

The crystallization of **120** from THF solutions affords a dimer with two different Li environments, Li(THF)Li[N(*t*-Bu)Si(*t*-Bu)$_2$F]$_2$,[119] **140** (Fig. 9.25). In this structure, one Li^+ ion is coordinated by two amide nitrogens with Li—N bond lengths of 1.95 and 2.03 Å. There is also a weak Li—F interaction of 2.24 Å. The other Li^+ ion interacts with the two F atoms (Li—F, 1.85 and 2.07 Å), a THF molecule (Li—O, 1.87 Å), and the nitrogen atom of one of the anions (Li—N, 2.10 Å). VT-NMR investigations have shown that the Li—N bonding alternates between the two anions, while retaining two different Li environments. The three-coordinate Li^+ center also shows two different Li—F contacts. One has an Li—F distance of 2.09 Å, whereas the other Li—F distance is short (1.85 Å). There is also a weak transannular Li—N contact of 2.10 Å.

The structures of Li(Et$_2$O)Li[S(Ph)(NSiMe$_3$)$_2$]$_2$,[114] **141**, and Li(Et$_2$O)-Li[S(Ph)N(*t*-Bu)(NSiMe$_3$)]$_2$,[114] **142**, differ from that of [Li(THF)S(Ph)-(NSiMe$_3$)$_2$]$_2$, **133**, owing to the different donor capacities of Et$_2$O and THF. The use of Et$_2$O affords type 2a twist-tricycle with Et$_2$O bonded to the outer Li^+ ion only. As in **133**, the four-coordinate N atoms are associated with the longer Li—N bonds. The use of tert-butyl instead of -SiMe$_3$ results in no marked difference bewteen the structures of **141** and **142**.

The presence of an ethoxy group in β-position to the anionic N atom in

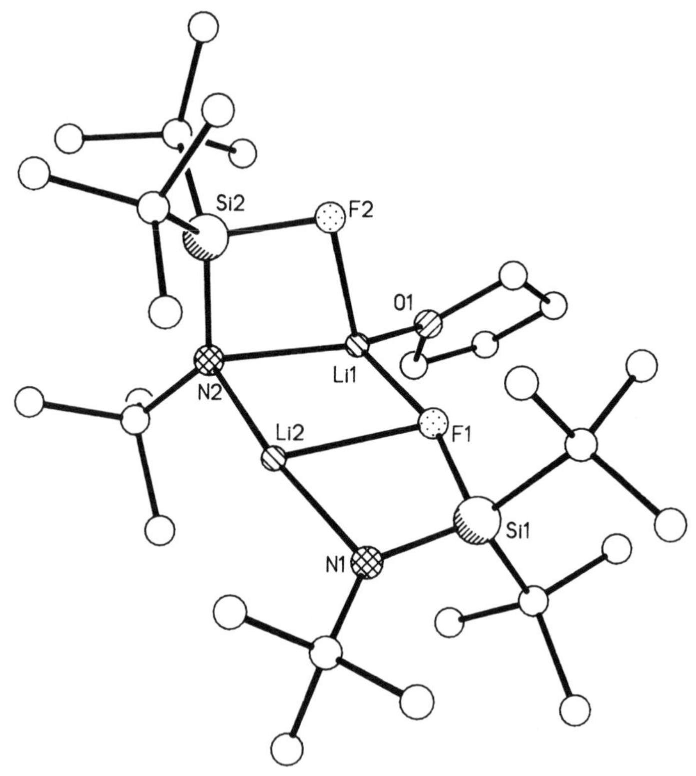

Figure 9.25 Structure of Li(THF)Li[H(*t*-Bu)Si(*t*-Bu)$_2$F]$_2$, **140**.[119] H atoms are not shown for clarity.

Li(THF)$_2$Li(THF)[*c*-NCH(Me)C(OEt)NC(Me)C(OEt)]$_2$,[120] **143**, introduces a new structural type. The two N-containing six-membered rings are twisted with respect to each other in such a fashion that both ethoxy groups coordinate to the same Li^+ ion in the central $(LiN)_2$ ring, forming rather weak Li—O contacts (2.24 and 2.25 Å). This Li^+ ion is coordinated by one THF molecule, whereas the other is bound to two THF molecules. The Li—N distances are equivalent (2.05–2.06 Å). Structural data for dimeric lithium amides are provided in Tables 9.8 and 9.9.

3.1.3 Higher Aggregates. The structures of lithium amides with aggregation number three is at present limited to three compounds. The first such species to be described was [LiN(SiMe$_3$)$_2$]$_3$,[72,73] **144**, which was also the first structural determination of a substituted lithium amide. It has a planar ring arrangement of three Li and three N centers. The Li—N bonds are 2.00 Å long with internal angles of 148° at Li and 92° at N (average values). An almost identical structure is observed for the germanium analogue [LiN(GeMe$_3$)$_2$]$_3$,[121] **145**. In this case, however, the Li—N bonds are shorter (ca. 1.96 Å), perhaps because of the less crowding -$GeMe_3$ substituents. A similar bond length, 1.95 Å, is observed in the structure of $[LiN(CH_2Ph)_2]_3$,[100a,b] **146**, which also has a planar Li_3N_3 core arrangement with internal ring angles of about 144° at Li and 95° at N.

The structures of very few higher aggregates of $LiNR_2$ compounds have been published. Presently known examples are confined to the tetramers, **147** $(LiTMP)_4$[80] (Fig. 9.26) and **148**, $Li_4(TMEDA)_2\{c\text{-}N(CH_2)_4\}_4$[122] (Fig. 9.27), and the hexamers, **149** $[Li\{c\text{-}N(CH_2)_6)\}]_6$[123] (Fig. 9.28) and **150**, $Li_6(PMDETA)_2\{c\text{-}N(CH_2)_4\}_6$[122,124] (Fig. 9.29). The type 4a tetramer **147** features a planar Li_4N_4 array with a wide, almost linear angle (168.5°) at Li and an internal angle of 101.5° at N. The average Li—N distance of 2.00 Å is almost identical to that in $[LiN(SiMe_3)_2]_3$. The structure of **148** is an example of a limited ladder structure in which two Li_2N_2 rings associate laterally with further association blocked by TMEDA coordination. The structure of the tetrameric $[Li\{HNc\text{-}(CH)_5\}\{Nc\text{-}(CH)_5\}]_4$[122b], **148b** also possesses a type Gb ladder structure but has an overall concave shape. The terminal four-coordinate Li^+ ions are complexed by the piperidines. In the case of the hexamer **150** (Fig. 9.29) the use of PMDETA in the appropriate ratio allows a ladder array of three Li_2N_2 rings to be isolated. In marked contrast, lithiation of the cyclic aliphatic amine H-c-$N(CH_2)_6$ in the absence of donor ligands leads to the iso-

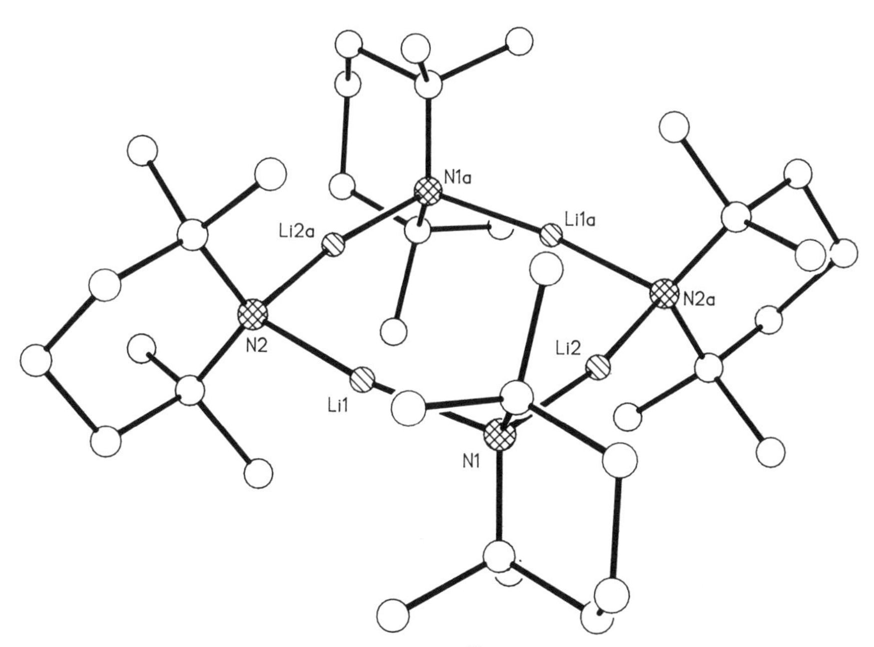

Figure 9.26 Structure of $(LiTMP)_4$, **147**.[80] H atoms are not shown for clarity.

Table 9.8 Symmetric Dimeric Lithium Amides

	CN_{Li}	Li—N (Å)	L	Li—L (Å)	Add. Coord.	Li—X (Å)	Compound/ Ref.
$[Li(Et_2O)NHMes^*]_2$	3	2.02[a]	Et_2O	1.91			**112**/96
$[Li(Et_2O)NHDipp]_2$	3	2.00[a]	Et_2O	1.92			**113**/97
$[Li(TMEDA)NMePh]_2$	4	2.08	TMEDA	2.30			**86**/76
$[LiN(Dipp)SiMe_3]_2$	2	1.96[a]					**114**/98
$[Li(Et_2O)\text{-}c\{NCH(n\text{-}Bu)CHCH\text{-}2,3\text{-}C_{10}H_6\}]_2$	3	2.07[a]	Et_2O	1.95			**115**/99
$[Li(Et_2O)N(CH_2C_6H_5)_2]_2$	3	1.99	Et_2O	2.01			**116**/100
$[Li(HMPA)N(CH_2C_6H_5)_2]_2$	3	2.01	HMPA	1.85			**117**/100
$[Li(Et_2O)N(SiMe_3)_2]_2$	3	2.06	Et_2O	2.01			**93**/80,81
$[Li(THF)N(SiMe_3)_2]_2$	3	2.02	THF	1.88			**94**/82
$[Li(Et_2O)NPhCCH_2(t\text{-}Bu)]_2$	3	2.08, 2.00	Et_2O	1.94	Li—C	2.44	**118**/101
$[LiNPhPPh_2]_2$	4	2.02	Et_2O	1.90	PPh_2	2.68	**119**/102
$[LiN(t\text{-}Bu)Si(t\text{-}Bu)_2F]_2$	3	2.09, 1.99					**120a**/103a
$[Li(TMEDA)N(H)Si(t\text{-}Bu)_2F]_2$	4	2.06[a]	TMEDA	2.27			**120b**/103b
$[LiFSi(i\text{-}Pr)_2NSi(t\text{-}Bu)_2OSi\text{-}\{MeN(SiMe_3)_2F]_2$	3	1.98			F	1.82, 1.07	**121**/104
$[LiFSi(t\text{-}Bu)_2NSi(i\text{-}Pr)_2OSiMe_3]_2$	3	1.95			O F	2.00 1.80	**122**/105
$[Li\{HN(i\text{-}Pr)_2\}N\text{-}1\text{-}c\text{-}C_6H_9Ph]_2$	3	2.05	$i\text{-}Pr_2NH$	2.11			**123**/106
$[Li(THF)HMCTS]_2$	3	2.04	THF	1.93			**124**/107
$[Li(THF)NSiMe_3HMCTS]_2$	3	2.04	THF	1.91			**125**/108
$[Li(TMEDA)indolyl]_2$	4	2.00, 2.23	TMEDA	2.16, 2.23			**126**/109
$[Li(THF)_2\text{-}fluorolyl]_2$	4	2.03, 2.16	THF	1.96			**127**/110
$[Li\{MeOCH_2CH(Ph)\}_2N]_2$	4	2.02			O	1.99	**128**/112

	4	2.06			O	2.00	
	4	2.06			O	2.00	
[LiN($SiMe_3$)(quinoline-8)]$_2$	3	2.06			N	2.01	**129**/90
[Li(Et_2O)N($SiMe_3$)(quinoline-8)]$_2$	4	2.07, 2.21	Et_2O	1.94	N	2.03	**130**/90
[Li(Et_2O)-2-Me-6-(Me_3Si)N-py]$_2$	4	2.08	Et_2O	1.98	N	2.08	**131**/90
[Li(μ^1-HMPA)-2-PhN-py]$_2$-[Li(μ^2-HMPA)-2-PhN-py]$_2$	4	2.14	HMPA	1.87			**132**/113
	4	1.98	HMPA	1.98			
[Li(THF)Me_3SiNS(Ph)NSiMe_3]$_2$	4	2.02, 2.05, 2.39	THF	1.97			**133**/114
[Li(Me_3S)NS(*t*-Bu)NSiMe_3]$_2$	3	1.92, 1.97, 2.37					**134**/115
[Li(t-Bu)NS(*t*-Bu)NSiMe_3]$_2$	2	1.91[a]					**135**/115
[Li(12-crown-4)$_2$][Li{Me_3SiNS(Ph)-NSiMe_3}$_2$]	4	2.13[a]					**136a**/116a
	8		12-crown-4	2.36[a]			
[Li(THF)Me_3SiNC(4-Me-C_6H_4)-NSiMe_3]$_2$	4	2.12[a]	THF				**136b**/116b
[Li-1,7-diaza-12-crown-4]$_2$	5	2.11			N	2.19	**137a**/117a
		2.00			O	2.29[a]	
[LiNCH_2(t-Bu)CHPhCH$_2$N*c*-(CH)$_5$]$_2$	3						**137b**/117b
[Li(TMEDA)N*c*-(CH)$_4$cc(R)Me]$_2$	4	2.10	TMEDA	2.14			**137c**/117c
R = M							
R − Me	4	2.09	TMEDA	2.13			**137d**/117c

[a]Denotes average values.

Table 9.9 Asymmetric Dimeric Lithium Amides

	CN_{Li}	Li—N (Å)	L	Li—N (Å)	Add. Coord.	Li—X (Å)	Compound/ Ref.
Li(THF)Li[N(*t*-Bu)$SiMe_2$O(*t*-Bu)$]_2$	3	2.04	THF	2.01			**138**/118
	4	2.11			O	2.01	
$Li(THF)_2$Li[OC(OMe)CC{*c*-C_4H_8N(NMe_2}$]_2$	4	2.02			O	1.92	**139**/106
	4		THF	1.93	O	1.93	**140**/119
Li(THF)Li[N(*t*-Bu)Si(*t*-Bu$)_2$F$]_2$	3	1.95, 2.03			F	2.24	**140**/119
		2.10	THF	1.87	F	1.85, 2.07	
$Li(Et_2O)Li[Me_3SiNS(Ph)NSiMe_3]_2$	3	2.01	Et_2O	1.92			**141**/114
	4	1.98					
$Li(Et_2O)$Li[*t*-BuNS(Ph)$NSiMe_3]_2$	3	2.03	Et_2O	1.90			**142**/114
	4	1.99, 2.25					
Li(THF)Li(THF$)_2$–	4	2.05	THF	1.98			**143**/120
[*c*-NCHMeC(OEt)N-CMeC(OEt)$]_2$	5	2.05	THF	1.92	O	2.25[a]	

[a]Denotes average values.

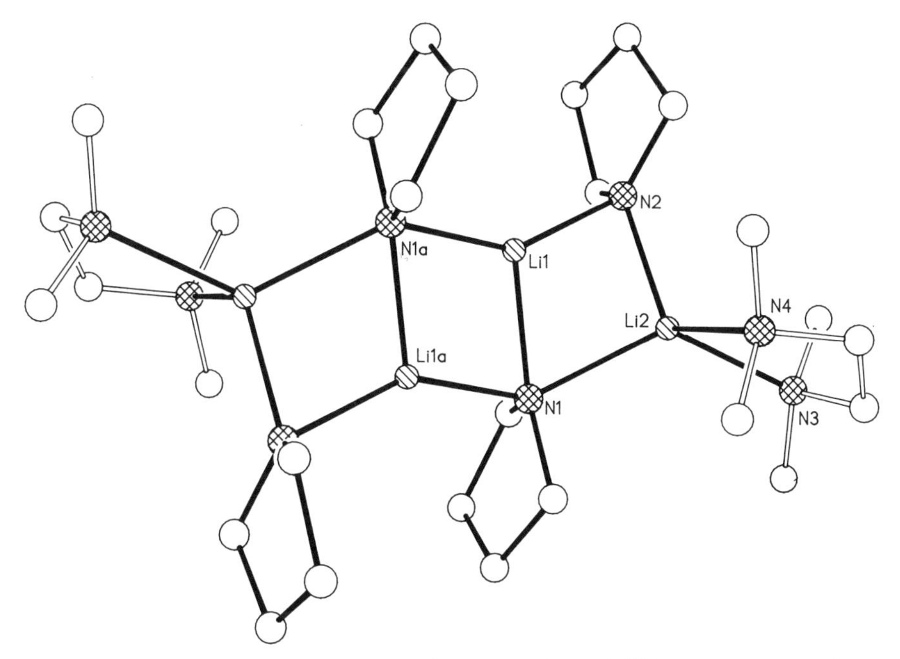

Figure 9.27 Structure of $Li_4(TMEDA)_2\{c\text{-}N(CH_2)_4\}_4$, **148**.[122a] H atoms are not shown for clarity.

lation of the hexameric **149** (Fig. 9.28), which has a stacked structure with type 5 aggregation. This involves the association of two $[Li\text{-}c\text{-}N(CH_2)_6)]_3$ units to form a hexagonal-prismatic Li_6N_6 array in which all N atoms have the relatively rare coordination number five. Lithiation of a mixture (5:2) of 6-methyl-2-trimethylsilylaminopyridine (Hmtmsap) and 2-amino-6-methyl-pyridine (Hamp) gives the unique heptametallic species **151** $[Li_7(mtmsap)_5(amp)_2]$.[90] Structural data for the higher lithium amide aggregates are given in Table 9.10.

3.2 Molecules with Multiple Lithium Amide Functions

In this section we deal with the structural properties of lithium amides that contain two or more lithium amide functions per molecule. In some instances, it is apparently not possible to lithiate all amine sites. The compounds where only one amide site is metallated are discussed with lithium amides bearing only one amide function. Structural data are provided in Table 9.11.

Lithiation of octamethylcyclotetrasilazane (OMCTS) affords a dilithiated species as the main product. The structural parameters of their THF and TMEDA solvates $[Li(THF)_2Li(OMCTS)]_2$,[125] **152**, and $[Li(TMEDA)Li(OMCTS)]_2$,[126] **153**, are very similar. Both feature a terminal Li^+ ion bound to one anionic nitrogen atom, and a Li^+ ion bridging both anionic nitrogen

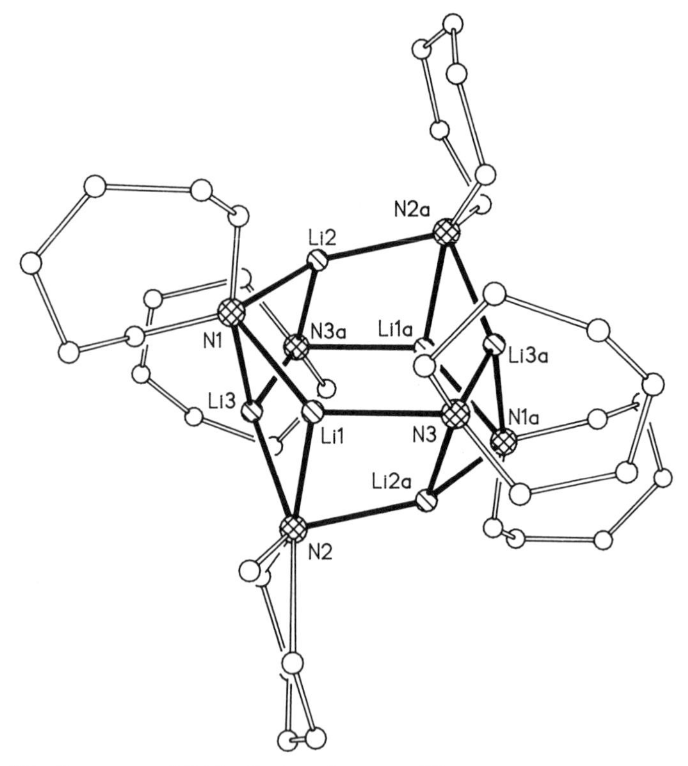

Figure 9.28 Structure of $[Li\{c\text{-}N(CH_2)_6\}]_6$, **149**.[123] H atoms are not shown for clarity.

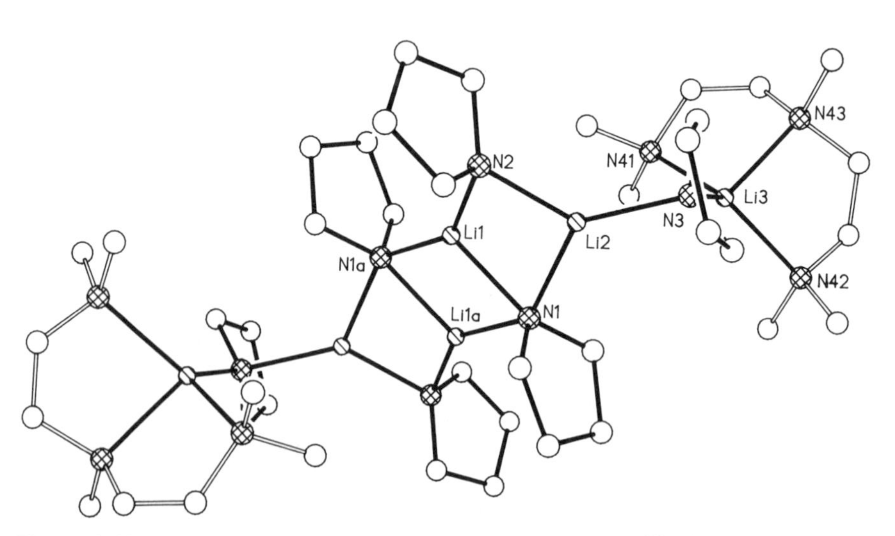

Figure 9.29 Structure of $Li_6(PMDETA)_2\{c\text{-}N(CH_2)_4\}_6$, **150**.[122a,124] H atoms are not shown for clarity.

Table 9.10 Higher Lithium Amide Aggregates

	Aggr.	CN_{Li}	Li—N (Å)	L	Li—L (Å)	Compound/ Ref.
$[LiN(SiMe_3)_2]_3$	3	2	2.00[a]			**144**/72, 73
$[LiN(GeMe_3)_2]_3$	3	2	1.96[a]			**145**/121
$[LiN(CH_2Ph)_2]_3$	3	2	1.95[a]			**146**/100
$(LiTMP)_4$	4	2	2.00			**147**/80
$Li_4(TMEDA)_2\{c\text{-}N(CH_2)_4\}_4$	4	3	2.01[a]			**148a**/122a
			2.11[a]	TMEDA	2.20[a]	
$[Li\{HNc\text{-}(CH_2)_3\}\{N\text{-}c(CH_2)_5\}_4$	3, 4		1.95–2.08	$HNc\text{-}(CH_2)_5$	2.14–2.21	**148b**/122b
$[Li\text{-}c\text{-}N(CH_2)_6]_6$	6	3	2.05[a,b]			**149**/123
			2.07[a,c]			
$Li_6(PMDETA)_2\{c\text{-}N(CH_2)_4\}_6$	6	3	1.93–2.12[d]			**150a**/124, 122
		4	1.97	PMDETA	2.19[a]	
$[Li_7(mtmsap)_5(amp)_2]$[e,f]	7	3	2.04,[a]			**150b**/90a
			2.04[a,g]			
		4	2.13,[a]			
			2.17[a,g]			
$[\{LiN(i\text{-}Pr)_2\}_2\text{-}(Me_2NCH_2CH_2NMe_2)]_\infty$		3	2.02	TMEDA	2.16	**150c**/90b

[a]Denotes average values.
[b]Average distance in the six-membered rings.
[c]Average distance in the rungs.
[d]All values in caption of Figure 9.
[e]mtmsap = 6-methyl-2-trimethylsilylaminopyridinate.
[f]amp = 2-amino-6-methyl-pyridinate.
[d]Distance between Li and the pyridyl nitrogen.

Table 9.11 Molecules with Multiple Lithium Amide Functions

	CN_{Li}	Li—N (Å)	L	Li—L (Å)	Add. Coord.	Li—X (Å)	Compound/ Ref.
$[Li(THF)_2LiOMCTS]_2$	3	2.15[a]					**152**/125
	3	1.95	THF	1.93[a]			
$[Li(TMEDA)LiOMCTS]_2$	3	2.15[a]					**153**/126
	3	2.15[a]	TMEDA	2.14[a]			
$[\{Li(t\text{-}Bu)N\}_2SiMe_2]_2$	3	2.02, 2.10					**154**/127
$[\{LiN(Mes)\}_2SiMe_2]_2$	3	2.01[a]					**155**/128
	9[b]	1.96			Mes-C	η_6: 2.39[a] η_2: 2.39[a]	
$[LiDippN(CH_2)_2NDippLi]_2$	3	2.02, 2.15, 1.95					
	2	1.94, 2.04					
$[Li(12\text{-crown-}4)_2][Li(THF)Si(t\text{-}Bu)_2$	3	1.98	THF	1.98			**157**/129
$\{NSi(t\text{-}Bu)_2Me\}_2]$	8		12-crown-4	2.24–2.43			
$Li_2(t\text{-}Bu)N\text{-}c\text{-}\{Si(Me)N(t\text{-}Bu)\}_2N(t\text{-}Bu)$	3	2.09, 2.16					**158**/130
$\{Li(Et_2O)\}_2(t\text{-}Bu)N\text{-}c\text{-}\{Si(Me)N(t\text{-}Bu)\}_2\text{-}N(t\text{-}Bu)$	3	1.89, 2.18	Et_2O	2.00			**159**/130
$Li_2(HMPA)_3PhN(CH_2)_2NPh$	4	2.07, 2.14	$HMPA_{br}$	2.03			**160**/131
			$HMPA_t$	1.88			
$[\{Li(Me_3Si)N\}_3SiMe]_2$	3	2.03, 2.06					**161**/132
$[\{Li(Me_3Si)N\}_3Si(t\text{-}Bu)]_2$	3	1.99					**162**/132
		2.12					
$[\{Li(Me_3Si)N\}_3SiPh]_2$	3	2.03					**163**/132
		2.11					
$[\{Li(t\text{-}Bu)N\}_3SiPh]_2$	3	2.00					**164**/127
		2.12					

[a]Denotes average values.
[b]On the assumption of an η^2 and an η^6 coordination to two different Mes groups.

atoms in 1,5-positions of the eight-membered ring. Donor ligands, either two THF's or one TMEDA, occupy two coordination sites of the terminal Li^+ ions. Thus all Li^+ ions in the dimer are three-coordinate, with the shortest Li—N bonds associated with terminal ions (1.95 Å). This N center also has a long contact to the bridging Li (2.17 Å). The other contacts are 2.16 and 2.10 Å (the latter effecting the dimerization). Because of electrostatic attractions, the protonated N atoms are oriented toward the bridging Li^+ ion. This is revealed by an Li—N distance of 2.41 Å to one of the N atoms in **152**.

Dilithiation of a diaminosilane results in a dimeric structure in the case of $[\{LiN(t\text{-}Bu)\}_2SiMe_2]_2$,[127] **154**. The Li^+ ions are on opposite sides of the planes defined by the N—Si—N units and are essentially symmetrically coordinated by the nitrogen atoms in a chelating fashion with Li—N bonding distances in the range of 2.09–2.13 Å. Bridging takes place in such a manner that the $\{LiN(t\text{-}Bu)\}_2SiMe_2$ units are perpendicular to each other and bridged by four Li—N contacts. This affords a dimer with D_{2d} symmetry. The bridging Li—N bonds are slightly shorter (2.00 and 2.04 Å). The N atoms are five-coordinate and, thus, formally electron deficient. This is reflected in the rather long Si—N bonds for a negatively charged system (1.73 Å, cf. **88**).

In a related compound, the substitution of *t*-Bu for Mes produces a totally different structural type. In the center of $[\{LiN(t\text{-}Bu)\}_2SiMe_2]_2$,[128] **155** (Fig. 9.30) is composed of a stair-shaped $(NSiNLi)_2$ framework with a central $(LiN)_2$ ring of type 2 dimerization. The π-system of the N-bonded Mes groups behaves as a donor such that the environment of the outer Li^+ ions consists of the peripheral N atom (1.96 Å), the ipso (2.28 Å) and ortho (2.50 Å) carbon atoms of one Mes ring, and the entire π-system of a second Mes ring (Li—C = 2.35–2.43 Å).

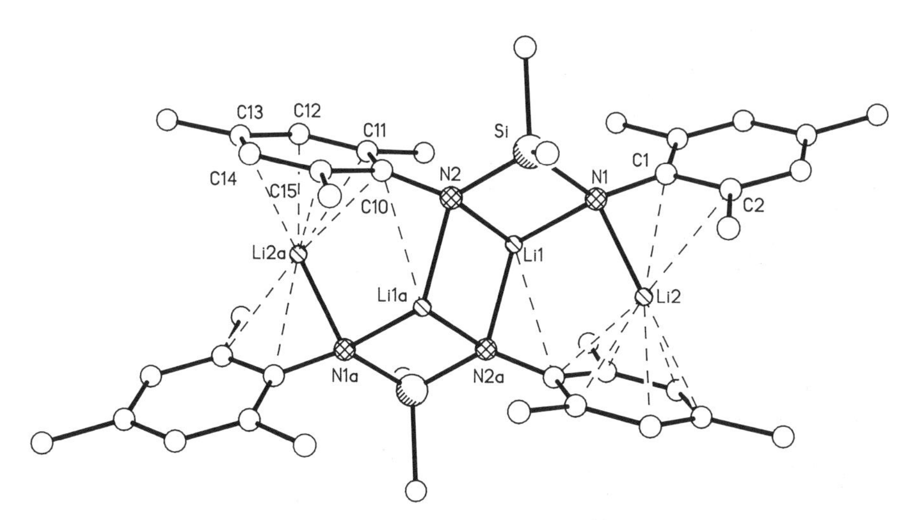

Figure 9.30 Structure of $[\{LiN(t\text{-}Bu)\}_2SiMe_2]_2$, **155**.[128] H atoms are not shown for clarity.

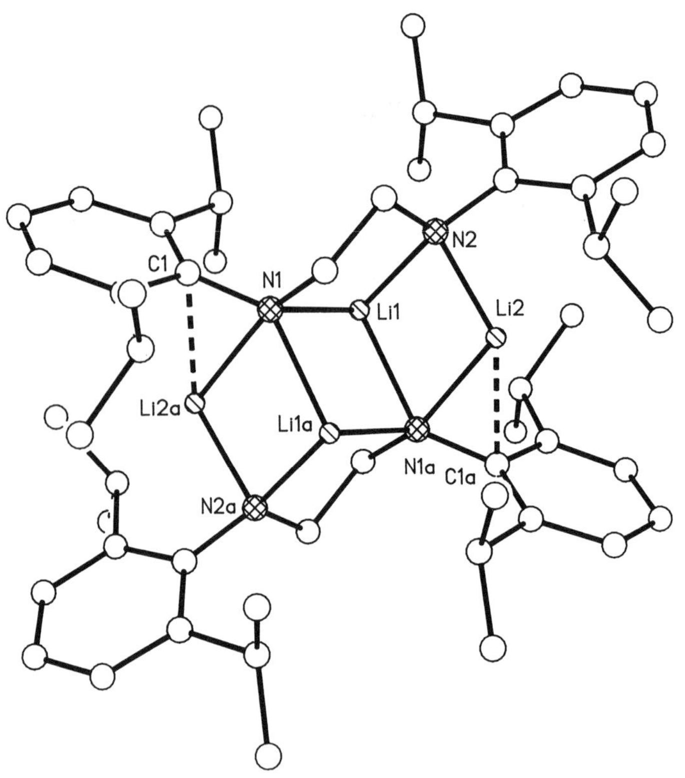

Figure 9.31 Structure of $[LiDippNCH_2CH_2NDippLi]_2$, **156.**[128] H atoms are not shown for clarity.

The structure of $[LiDippNCH_2CH_2NDippLi]_2$,[128] **156** (Fig. 9.31) differs from **155** in that the Dipp groups do not act as η^6 donors to Li, presumably because of the steric crowding resulting from the *i*-Pr substituents. The outermost Li^+ ions interact with an N atom of one ligand and an N atom of the central type 2 $(LiN)_2$ ring. The latter nitrogen is five-coordinate, which is reflected in a long N—Li contact (2.15 Å). Short Li· · ·H distances of 2.24 Å to the *i*-Pr protons in the Dipp groups add to the coordination sphere of the outer Li^+ ions.

If 12-crown-4 is added to a dilithiated $[R'NSiR_2NR']^{2-}$ species (R and R′ = large groups) in THF (R = *t*-Bu; R′ = *t*-Bu_2MeSi), only one of the Li^+ ions is removed from the dianion to afford [Li(12-crown-4)$_2$][Si(*t*-Bu)$_2${NSi(*t*-Bu)$_2$Me}$_2$Li(THF),[129] **157**. The anion-bound Li^+ ion is chelated by the two N donors and a THF. It has fairly short Li—N bonds of 1.98 Å (av.) and the Si—N distances are near 1.69 Å (av.). The presence of a $(SiN)_2$ four-membered ring, wherein the silicon atoms bear two types of substituent, gives rise to *cis–trans* isomers. Dilithiation of the *cis* isomer results in an elongated cubane-type structure *t*-BuN(Li){c-[SiMeN(*t*-Bu)]$_2$}N(Li)*t*-Bu,[130] **158**, formed from $(SiN)_2$ and $(NLi)_2$ subunits. The Li^+ ions are three- and the nitrogen atoms four-coordinated. The $(Li—N)_2$ unit has the stronger Li—N bond of 2.09 Å, whereas the Li—N distances in the Li—N—Si—N moiety are longer at about 2.16 Å. Intermolecular CH· · ·Li interactions between 2.00 and 2.24 Å result in a weak association of two cubane units.

The *trans* isomer $\{Li(Et_2O)\}_2$*t*-BuN{*c*-[SiMeN(*t*-Bu)]$_2$}N(*t*-Bu),[130] **159**, crystallizes with an Et_2O donor molecule attached to each Li^+ ion. These are strongly bound to the anionic exocyclic N atoms (Li—N, 1.88 Å) and more loosely to the endocyclic N atoms (2.18 Å) to give a chelate structure. Four-membered Li—N—Si—N rings are fused to the opposite sides of the central $(SiN)_2$ ring to afford a type 2 stair-shaped tricyclic framework. In $Li_2(HMPA)_3PhN(CH_2)_2NPh$,[131] **160**, the Li^+ ions are asymmetrically located above and below the N—C—C—N backbone; the shorter set of Li—N bonds displays values of 2.06–2.09 Å, while the longer contacts are as long as 2.14–2.15 Å. A terminally bonded HMPA molecule coordinates each Li^+ ion (Li—O, 1.88, av.) and another HMPA bridges two Li^+ ions (Li—O, 1.99 and 2.07 Å).

The basic framework in the structural type in $[\{LiN(SiMe_3)\}_3SiR]_2$[132] (**161**: R = Me; **162**: R = *t*-Bu; **163**: R = Ph) and [{LiN(*t*-Bu)}$_3$SiPh)$_2$,[127] **164**, is a result of the dimerization of a trisamidosilane (see Fig. 9.32). It exhibits molecular D_{3d} symmetry. Each of the three Li^+ ions in a monomeric unit is coordinated by two nitrogen atoms in a chelating fashion, while the bridging occurs through single Li—N contacts. This causes all N atoms to be five-coordinate, which leads to a weakening in the Li—N and Si—N bonds (Si—N = 1.72–1.75 Å).

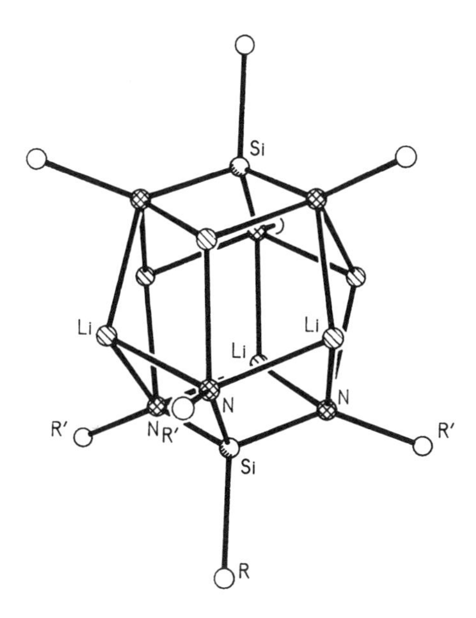

Figure 9.32 Drawing of the core atoms in $\{Li(R')N\}_3Si(R)]_2$, **161–164**.[127,132] R′ = $SiMe_3$; **161**[132]: R = Me; **162**[132]: R = *t*-Bu; **163**[132]: R = Ph; **164**[127]: R′ = *t*-Bu, R = Ph. H atoms and organic substituents are not shown for clarity.

3.3 Lithium Borylamides

Borylamides are species in which one or both of the organic substituents in conventional amides are replaced by the boryl group(s) -BR_2. There is generally a fairly strong interaction between the boron *p*-orbital and the nitrogen lone pair, which results in a shorter B—N bond. Another consequence of the B—N π-bonding is that the basicity of the nitrogen center is lowered and this leads to a reduced tendency toward aggregation in lithium salts of borylamides.

The compound $[Li(Et_2O)N(H)BMes_2]_2$,[133] **165**, is a type 2 dimer, and is currently the sole structural representation of a dimeric lithium borylamide. The B—N bond is quite short (1.39 Å) in spite of the four-coordination at nitrogen. The small C—B—C angle of 117.4° is in agreement with B—N double bonding character. The Li—N bonds of 2.01 and 2.02 Å are similar to those in many other dimeric amides.

Replacing the hydrogen at nitrogen by an organic group such as phenyl affords the monomeric species $Li(Et_2O)_2NPhBMes_2$,[134] **166**. With larger substituents this structure is, not surprisingly, retained in $Li(Et_2O)_2NMesBMes_2$,[135] **167** and Li(TMEDA)N(*t*-Bu)B(*t*-Bu)$_2$,[136] **168**. The Li—N bond lengths are about 1.95 Å. This is in the normal range for three-coordinate lithium bonded to three-coordinate nitrogen (cf. **85**; Li—N, 1.90 Å). The decreased basicity at N may account for the slightly longer values seen here. The steric crowding in **168** is manifested in a large angle between the planes at boron and nitrogen (see Fig. 9.33).

The crystal structure of only one lithium salt of a diborylamide

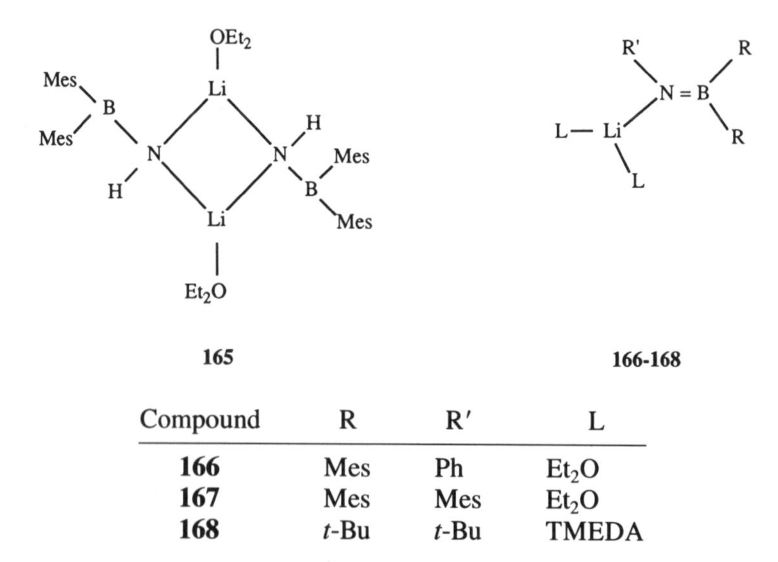

Compound	R	R′	L
166	Mes	Ph	Et_2O
167	Mes	Mes	Et_2O
168	*t*-Bu	*t*-Bu	TMEDA

Figure 9.33 Schematic drawings of compounds **165–168**.[133–135]

$[Li(Et_2O)_3][Mes_2BNBMes_2]$,[133] **169**, has been published. In this species cation–anion separation is readily achieved without the use of ligands that strongly bind Li^+ ions. The B—N—B array has an average B—N distance of 1.35 Å and a virtually linear B—N—B angle (176.2°). The angle between the two CBC planes is 88°. The anion may thus be viewed as an analogue of allene, with which it is isoelectronic. The Li^+ ion is coordinated to three Et_2O molecules with an average Li—O distance of 1.90 Å. Structural data for the borylamides are given in Table 9.12.

3.4 Lithium Imides

Lithium imides show a greater tendency toward higher aggregation numbers than amides. The observed structures generally involve three-dimensional frameworks of Li and N atoms, which can be formally regarded as electron-deficient oligomers with multicenter bonding. This situation is reminiscent of the less bulky organolithium compounds.[1] In many respects, the structures of lithium imides may be regarded as being intermediate between the electron-deficient Li—C species and electron-precise amides.

The principles underlying the bonding in lithium imides has been discussed in the literature.[137–140] Thus the information in this section is mainly tabular (see Tables 9.13 and 9.14) with only brief notes on the chief structural types and more recent developments.

As in the amide case, the use of donor molecules effects lower degrees of aggregation, although no monomeric imides are yet known. The addition of py or HMPA to $[LiN{=}CPh_2]_n$ results in an Li_4N_4 cubane tetramer $[Li(py)N{=}CPh_2]_4$,[141,142] **174** (as seen in $(LiMe)_4$[1,143,144] or $[Li(THF)OC_5H_7]_4$[1,145]), which is an electron-deficient compound, as exemplified by the long Li—N contacts (2.02–2.17 Å). The use of HMPA as donor molecule affords a different structure from **174**. $[Li(HMPA)N{=}C(t\text{-}Bu)_2]_2$,[146,142] **175**, is an electron-precise dimeric structure of type 2 and has short Li—N bonds of 1.92 and 1.95 Å. It is also noteworthy that the C=N bond is shorter than in the higher aggregates (1.23 Å). Strong bonding of HMPA to the Li^+ ions (Li—O, 1.86 Å) is also evident (cf. compounds, **146**, etc.).

The lower steric crowding at nitrogen in imides does not give a monomeric species when treated with a bidentate ligand. This is exemplied by the structure of $[Li(TMEDA)N{=}C(H)CH_2Ph]_2$,[147] **176** (Fig. 9.34). The Li—N contacts are rather long (2.05 Å) for a four-coordinate Li^+ ion. The extremely short C—N

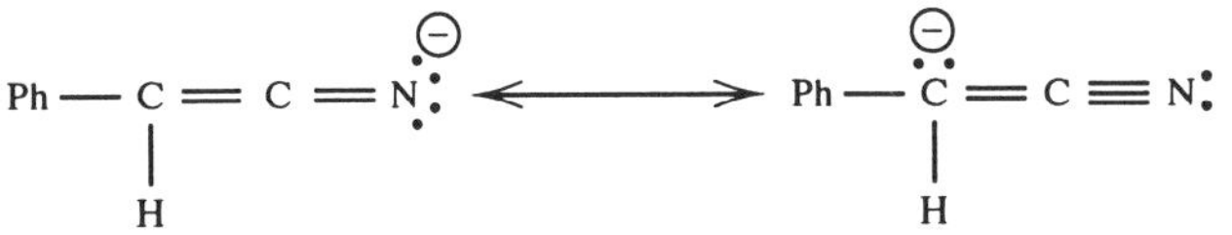

Ph—C(H)=C=N:⁻ ⟷ Ph—C⁻(H)=C≡N:

Figure 9.34 Mesomeric structures of **176**.[147]

Table 9.12 Lithium Borylamides

	CN_{Li}	Li—N (Å)	B—N (Å)	CBC°	CBC°	Add. Donor	Li—L (Å)	Compound/ Ref.
$[Li(Et_2O)N(H)BMes_2]_2$	3	2.02[a]	1.39	117.4		Et_2O	1.96	**165**/133
$Li(Et_2O)_2NPhBMes_2$	3	1.94	1.39	122.0	8.8	Et_2O	1.93	**166**/134
$Li(Et_2O)_2NMesBMes_2$	3	1.94	1.36	121.1	10.9	Et_2O	1.95	**167**/135
Li(TMEDA)N(*t*-Bu)B(*t*-Bu)$_2$	3	1.97	1.38	116.9	30.5	TMEDA	2.20[a]	**168**/136
$[Li(Et_2O)_3][Mes_2BNBMes_2]$	3		1.35	117.4		Et_2O	1.90[a]	**169**/133

[a]Denotes average values.

Table 9.13 Uncomplexed Lithium Imides

	R	R′	CN_{Li}	Li—N (Å)	C=N (Å)	Compound/ Ref.
$[LiN{=}C(t\text{-}Bu)_2]_6$	*t*-Bu	*t*-Bu	3	2.06[a]	1.30[a]	**170**/137
$[LiN{=}C(NMe_2)_2]_6$	NMe_2	NMe_2	3	2.00[a]	1.24[a]	**171**/138
$[LiN{=}CPh(t\text{-}Bu)]_6$	Ph	*t*-Bu	3	2.02[a]	1.25[a]	**172**139, 140
$[LiN{=}CPhNMe_2]_6$	Ph	NMe_2	3	2.01[a]	1.26[a]	**173**/139, 140

[a] Denotes average values.

bond reveals that the deprotonation has occurred on the C atom in β-position with only one free electron pair on nitrogen, as illustrated by the resonance formulas.

A combination of HMPA as donor solvent and phenyl groups attached to carbon result in an unusual structure wherein one Li^+ ion is removed from a hexameric framework and is coordinated by four HMPA molecules. The anion has an asymmetric structure with five Li^+ ions in three different environments in a distorted trigonal pyramidal structure. The overall formula is $[Li(HMPA)_4][Li_5\{N{=}CPh_2\}_6(HMPA)]$,[148] **177**. In the anion an HMPA-coordinated Li(1) caps three μ^3-N atoms, three Li^+ ions each bridge two μ^3-nitrogen atoms and one from the LiN_3 base unit. The three basal N atoms are only bound to two Li^+ and this is reflected in the Li—N distances of 2.05 and 2.02 Å. $Li\{Na(HMPA)\}_3\{N{=}C(NMe_2)_2\}_4$,[149] **178**, is a mixed-metal compound. A distorted $LiNa_3N_4$ cubane structure is observed. Although the structure is similar to that of **174**, the Li—N bonds are relatively short. This may be attributed to a different polarization of Li—N and Na—N bonds. In contrast to **174**, the Li^+ ion is not coordinated to an additional donor molecule. This may be a result of crowding by the methyl groups of the dimethylamino substituents. The C=N distance is in the expected range (1.24 Å). $Li_4Na_2\{N{=}CPh(t\text{-}Bu)\}_6$[150] **179** may be regarded as a stack of three four-membered $(MN)_2$ rings, with the $(NaN)_2$ ring in the center. A Li—Na disorder results in large standard deviations in the positions of the metal atoms.

3.5 Azaallylic Lithium Compounds

Strictly speaking, azaallylic compounds are organolithium derivatives, since it is a carbon center that is deprotonated. Owing to delocalization, however, a partially negatively charged N atom is often involved in the coordination to Li.

The type 2 $(LiN)_2$ framework in **180** $[Li(Et_2O)(2\text{-}CH_2\text{-}3,3\text{-}Me_2\text{-}indolenide)]_2$[151] displays Li—N bond lengths in the typical range (2.00 and 2.09 Å). In addition to coordinating Et_2O, the Li^+ ions interact with the double bond in the α-position to N in the indolenide system. This results in fairly short Li—C contacts of 2.37 Å to the vicinal C atom to N and of 2.51 Å to

Table 9.14 Complexed Lithium Imides

	Type	CN_{Li}	Li—N (Å)	C=N (Å)	Li—L (Å)	Compound/ Ref.
$[Li(py)N{=}CPh_2]_4$	4b	4	2.09[a]	1.26[a]	2.09[a] (py)	**174**/141, 142
$[Li(HMPA)N{=}C(t\text{-}Bu)_2]_2$	2	3	1.94[a]	1.23	1.86 (HMPA)	**175**/146, 142
$[Li(TMEDA)N{=}C(H)CH_2Ph]_2$	2	4	2.04[a]	1.15[a]	2.05[a] (TMEDA)	**176**/147
$[Li(HMPA)_4][Li_5\{N{=}CPh_2\}$-(HMPA)]		3	2.02[a]	1.26[a]	[b]	**177**/148
		4	2.10[a]		1.83 (HMPA)	
		4[c]			1.84[a] (HMPA)	
$Li\{Na(HMPA)_3\}\{N{=}C(NMe_2)_2\}_4$		4b[c]	[d]	1.98[a]	1.24[a]	**178**/149
$Li_4Na_2\{N{=}C(Ph)(t\text{-}Bu)\}_6$			[d]	2.12[a]	1.26[a]	**179**/150

[a] Denotes average values.
[b] Li in the cationic species.
[c] Type 4b structure with Na occupying three corners of the cubane.
[d] Stack of three dimers.

the CH_2 group. Thus the negative charge is somewhat delocalized in the η^3-azaallyl unit, which displays an N—C length of 1.39 Å and a C—C distance of 1.34 Å. The species $[Li(Et_2O)\{c\text{-N-CH}(CHSiMe_3)C_4H_4\}]_2$,[152,153] **181**, is a type 2 dimer with Et_2O coordinated to lithium. In $Li\{c\text{-NCH}(C[SiMe_3]_2)C_4H_4\}\{c\text{-NC}(C[SiMe_3]_2)C_4H_4\}$,[152,153] **182**, the Li^+ ion is coordinated to the pyridyl moiety of a nondeprotonated molecule, which affords a monomeric structure for this molecule.

The large bidentate donor molecule in $Li(sparteine)\{c\text{-NCH}(CHSiMe_3)C_4H_4\}$,[153] **183**, also affords a monomeric structure, whereas the use of TMEDA, as a donor ligand, leads to an asymmetric dimerization in the species $[Li(TMEDA)\{c\text{-NCH}(CHSiMe_3)C_4H_4\}]_2$,[153] **184**, which has C_2 symmetry with two different sets of Li—N contacts. The Li—C distances are rather long; there is no η^3-allylic coordination to the Li^+ ion present. This is also reflected in the short C—C bond. The species $\{Li(TMEDA)\}_2\{c\text{-NCH}(CHSiMe_3)(CH)_3CH(CHSiMe_3)\}$,[154] **185**, is unique in this series of structures, it contains deprotonated benzylic groups in both ortho positions in the pyridine subunit, resulting in a dianion. The use of a TMEDA donor is sufficient to obtain a monomeric structure, with the Li^+ ions on opposite sides of the ring plane (see Fig. 9.35 and Table 9.15).

3.6 Miscellaneous Amides and Imides Involving Li—N Bonds

The structures of three lithium derivatives that display a tautomeric -N(H)CS unit in ring systems are known. In $[Li\{c\text{-C(H)}(CH_2)_2NC(S)N\}]_n$,[155] **186**, there is a polymeric arrangement involving the coordination of the second N atom in the ring, and there are Li—N, Li—S, and Li—N′ distances of 2.09, 2.80, and 2.06 Å, respectively. The compound $[Li(TMEDA)\{c\text{-}(CH_2)_2SC(S)N\}]_2$,[155] **187**, is a dimer in which the anion is not chelating, owing to competitive coordination of the Li^+ ion by a TMEDA molecule. Bridging occurs along a Li—S contact (2.49 Å). The Li—N distance is 2.04 Å. In $[C_6H_4\{c\text{-NLi}(HMPA)_2C(S)NLi(HMPA)\}]_2$,[155] **188**, a doubly chelating dianion is present, in which one Li^+ ion is coordinated by two terminal HMPA ligands; the other by two HMPA molecules, bridging to another subunit, afford an overall dimeric structure. The terminal Li^+ ions have bonds of 2.01 Å to N and 2.84 Å to S, the bridging Li^+ ion has distances of 2.01 and 2.52 Å to N and S.

Three water-complexed Li^+ ions are found in lithiated 2-mercaptobenzoxazolyl species. In the TMEDA complexed compound $Li(TMEDA)(H_2O)\{c\text{-}C_6H_4OC(S)N\}$, **102**, mentioned earlier, no Li—S interactions are observed. In $[Li(HMPA)(H_2O)\{c\text{-}C_6H_4OC(=S)N\}]_2$,[156] **189**, two monomeric units are bridged by two HMPA molecules to afford a dimeric structure. The Li—N distance is lengthened (2.07 Å), and the Li—OH_2 distance is shorter (1.91 Å) than that in **102**. The absence of one H_2O molecule in the dimer of **189** results in a Li—S interaction of 2.76 Å in $[Li(HMPA)\{c\text{-}C_6H_4OC(S)N\}]_2(H_2O)$[156] **190**. The Li—N distance in the H_2O complexed (Li—O 1.93 Å) moiety is 2.02 Å long.

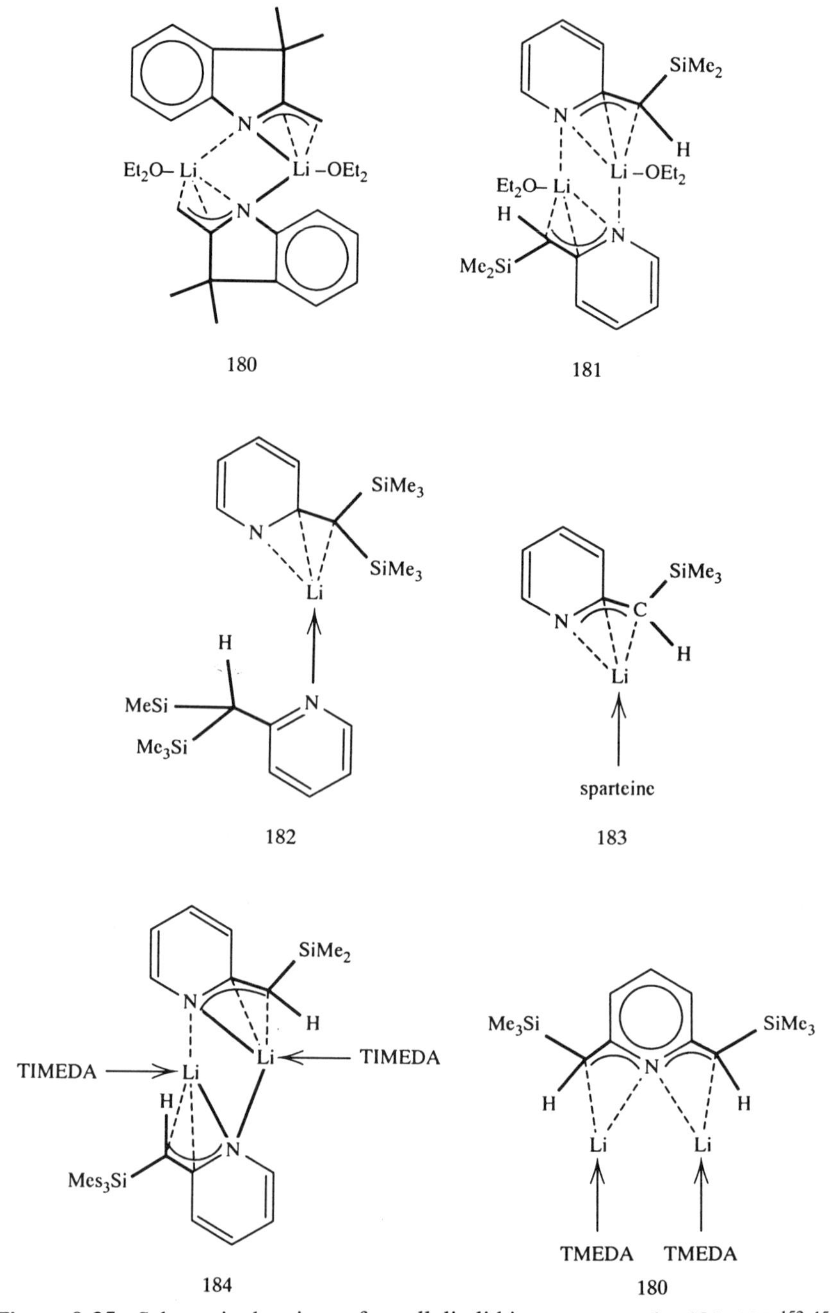

Figure 9.35 Schematic drawings of azaallylic lithium compounds, **181–185**.[152–154]

Table 9.15 Azaallylic Lithium Compounds[a]

	CN_{Li}	Li—N (Å)	Li—C_{ring} (Å)	Li—C_{benz} (Å)	Li—L (Å)	C_{ring}—C_{benz} (Å)	Compound/ Ref.
[Li(Et_2O)-2-CH_2-3,3-Me_2-indolenide	5	2.09 2.11	2.37	2.51	1.94 (Et_2O)	1.34	**180**/151
[Li(Et_2O){*c*-N-CH($CHSiMe_3$)-C_4H_4}]$_2$	2	2.19[b], 2.04	2.34	2.35	1.91 (Et_2O)	1.40	**181**/152, 153
Li[*c*-NCH{C($SiMe_3$)$_2$}C_4H_4]-[*c*-N-C-{C($SiMe_3$)$_2$}C_4H_4]	1	2.00	2.32	2.22	2.01	1.43	**182**/152, 153
Li(sparteine){*c*-NCH($CHSiMe_3$)-C_4H_4}	1	1.98	2.35	2.32	2.04[c]	1.40	**183**/153
[Li(TMEDA){*c*-NCH($CHSiMe_3$)-C_4H_4]$_2$	2	2.08, 2.25	2.82, 2.95	2.91, 3.23	2.19[a]	1.37	**184**/153
{Li(TMEDA)}$_2${2,6-($CHSiMe_3$)$_2$py}	1	2.01	2.26	2.27	2.09[a]	1.41	**185**/154

[a] cBond lengths in Angstroms; average values for the two subunits in FERKOC.
[b] Denotes average values.
[c] Sparteine (average Li—N distance).
[d] TMEDA (average Li—N distance).

In $[\{Li(HMPA)\}_2SC(NPh)(NPh)]_2$,[157] **191**, the central carbon atom is surrounded by an anionic amido group (C—N, 1.43 Å), an imino group (C—N, 1.29 Å), and an anionic sulfido center (C—S, 1.72 Å). The overall structure is a dianionic dimer, with one Li^+ ion chelated by the different N atoms in the monomer ($Li—N_{am}$, 2.24; $Li—N_{im.}$, 1.98 Å), and the other coordinated to the amido nitrogen (Li—N, 2.05 Å) and the bridging sulfur of the other monomeric moiety (Li—S, 2.43 Å). This results in a central eight-membered ring. Both Li^+ ions are coordinated by HMPA donor molecules.

$\{Li(TMEDA)\}_2N(i\text{-}Pr)_2N{=}C{=}CHPh$,[158] **192**, can be considered the LDA adduct of a lithiated phenylacetonitrile with the anions bridged by the two Li atoms. Although the phenylacetonitrile anion is deprotonated at the C atom, the coordination site is at nitrogen; the negative charge appears to be delocalized, as indicated by the short (Ph)C—C distance (1.38 Å). The C—N bond length remains consistent with that of a triple bond (1.17 Å). This is in agreement with the longer bonds to the Li^+ ions (2.08 and 2.11 Å; cf. 2.05 Å in **176**), while the Li—N bonds in the diisopropylamide unit are slightly shorter (2.04 Å). Remarkably long Li-TMEDA bonds (2.20–2.35 Å) are observed. The most prominent structural feature in $Li(NH_2CH_3)Li[c\text{-}\{N(Me)(CH_2)_2\}_3][c\text{-}\{N(Me)(CH_2)_2\}_2N(CH_2)_2]^+Na^-$,[159] **193**, is the anionic nature of the uncoordinated Na. The two Li^+ ions have different environments: one is coordinated to three sites of a 1,4,7-trimethyl-1,4,7-triazacyclononane (2.05–2.15 Å) and the deprotonated N atom of a 1,4-dimethyl-1,4,7-triazacyclononane (1.89 Å); the second Li^+ ion with bonds to the latter N center (1.97 to $N_{deprot.}$, 2.06 and 2.07 Å) and to a methylamine molecule (2.09 Å) is also four-coordinate. Another intriguing structure arises from the reaction of *p*-tert-butylcalix[4]areneP(H)N(H)Me_2[154b] with butyllithium. In this case the $NHMe_2$ bound to phosphorus is deprotonated to afford $\{Li(THF)_2\}\{p\text{-tert-butylcalix[4]areneP(H)NMe}_2\}]$, **193b**. The Li^+ ion bridges the -NMe_2 group Li—N (2.05 Å) and two oxygens Li—O (2.25 Å) from the calixarene and is also bound to two THF donors. Structural data for the compounds discussed in this section are given in Table 9.16.

3.7 Heterometallic Lithium Amides and Imides

As in the case of the alkoxide and aryloxide ligands, several heterometallic derivatives are known. Four amido groups are bound to aluminum in $Li(THF)_2Al(NMe_2)_4$,[160] **194**, two of which serve as bridges to the Li^+ ion, while the other two are in terminal positions. The terminal Al—N bonds are shorter (1.85 Å) than the bridging ones (1.90 Å). The Li—N distances are rather long (2.13 Å) for a four-coordinate Li^+ ion. In $[LiAl(H)(NEt_2)_3]_n$,[161] **195**, an amido ligand is replaced by hydride, and there are no solvating molecules. The observed structure is polymeric. The Li—N contacts are remarkably long (2.17 and 2.24 Å on the all-amido side and 2.29 Å on the side where the hydride is present). $LiAl\{N(H)CPh_3\}_4$,[162] **196**, has long Li—N (2.03 Å, av.) and Al—N distances (1.94 Å av.). Lithium also coordinates to the ipso- and ortho-C donors of two phenyl rings bound to the two N donors (Li—C, 2.24–2.57 Å).

In the imido species, $Li(t\text{-}Bu_2C{=}N)_2Al\{N{=}C(t\text{-}Bu)_2\}_2$,[163] **197**, two bridging and two terminal imido groups are observed. In addition to the coordination by two nitrogen ligands the Li^+ ion interacts with one methyl group of the tert-butyl substituent. These have calculated Li—H distances of 1.98 Å (assuming ideal geometry at the C atom and a C—H bond length of 1.09 Å). This results in a ring-like arrangement of the bridging $t\text{-}Bu_2C{=}N$ groups around the Li^+ ion. The Li—N distances are 1.95 Å, Al—N = 1.87 (br.) and 1.78 Å (t.).

Treatment of **161** with $CpTiCl_3$ affords $Li[Si(Ph)(NSiMe_3)_3Ti(Cl)Cp]$,[164] **198**, with the tridentate silicontriamide acting as a chelating donor to Ti with two coordination sites, while the third nitrogen atom coordinates to the Li^+ ion, which in turn forms a bridging unit through a chlorine atom to titanium. Coordination of an Et_2O molecule renders the Li^+ ion three-coordinate. It is noteworthy that the Si—N bonds for the Ti-coordinating N atoms are longer than in neutral aminosilanes (1.79 Å), while the Li-bound N donor displays an Si—N bond length of 1.64 Å, which is short for an anionic silylamido group. The Li—N distance is in the range of a three-coordinate Li^+ ion bound to a three-coordinate N donor (1.95 Å). In monomeric $LiVMe_2\{NSi(t\text{-}Bu)_3\}_2$,[165] **199**, Li is two-coordinate in a four-membered Li—N—V—N ring, because of the steric requirements of the -$Si(t\text{-}Bu)_3$ substituents on the, formally dianionic, nitrogen donor. The Li—N distances are long for bonding between two-coordinate Li and three-coordinate N (1.99 Å), and it appears that imido groups remain very strongly bound to V (V—N, ~1.71 Å). Two different Li environments and differently coordinated NMes groups are present in $\{Li(THF)_2\}_2Nb(n\text{-}Bu)(NMes)_3$,[166] **200** (Fig. 9.36).

In this structure two mesitylimido groups bridge to one Li^+ ion affording four-coordinate lithium with long Li—N contacts (2.08 Å). Only one NMes group bridges to the second Li^+ ion, which is three-coordinate, and has a considerably shorter Li—N bond (1.90 Å). This difference is not strongly reflected in the Nb—N bonds, the shortest of which involves the three-coordinate N atom (1.82–1.98 Å).

Excess $LiN(SiMe_3)_2$ affords a three-coordinate homoleptic manganese center in $Li(THF)Mn\{N(SiMe_3)_2\}_3$,[167] **201**. Two of the amido groups bridge to the Li^+ ion, while the third amide ligand bonds solely to Mn^{2+}. The Li^+ ion is further coordinated by one THF molecule. The Li—N distances of 2.10 Å are long for three-coordinate Li^+. The $Mn{-}N_{term}$ distance is 2.02 and the $Mn{-}N_{br}$ length is 2.14 Å. The species $LiMn\{MesNSiMe_2NMes\}_2$-$MnN(SiMe_3)_2$,[128] **202**, has a unique structure. As in $[\{LiN(Mes)\}_2$-$SiMe_2]_2$, **155**, the Li^+ ion is coordinated by two Mes groups in a sandwich-like η^6 fashion with Li—C distances in the range 2.25–2.57 Å. One Mn^{2+} ion is bound to an -$N(SiMe_3)_2$ group and to an N donor from each $\{MesNSiMe_2NMes\}^{2-}$ ligand. The other Mn center is coordinated to the remaining N sites of these anions. In $\{Li(THF)_2Cl\}_2\{Mn(NEr_2)_2\}_3$,[168] **203**, three Mn atoms are bridged by two sets of two diethyl amido groups. At both ends of this array,

Table 9.16 Miscellaneous Amides and Imides Involving Li—N Bonds

	CN_{Li}	Li—N (Å)	Li—L (Å)	M	$M—N_t$ (Å)	$M—N_{br}$ (Å)	Add.	Coord.	Compound
$[c\text{-}C(H)(CH_2)_2NC(S)NLi]_n$	4	2.09, 2.06	1.87 (HMPA)				Li—S	2.80	**186**/155
$[c\text{-}(CH_2)_2SC(S)NLi(TMEDA)]_2$	4	2.04	2.18^{a} (TMEDA)				LiS	2.49	**187**/155
$[C_6H_4\{NLi(HMPA)_2C(S)NLi$-	4	2.01	1.88^{a} ($HMPA_t$)				Li—S	2.84	**188**/155
$(HMPA)\}]_2$	4	2.01	1.98^{a} ($HMPA_{br}$)				Li—S	2.52	
$[Li(HMPA)(H_2O)\{c\text{-}C_6H_4OC(S)N\}]_2$	4	2.07	1.97^{a} ($HMPA_{br}$)						**189**/156
			1.91 (H_2O)						
$[Li(HMPA)\{c\text{-}C_4H_4OC(S)N\}]_2(H_2O)$	4	2.02	1.94^{a} ($HMPA_{br}$)						**190**/156
			1.93 (H_2O)						
	4	2.06	1.94^{a} ($HMPA_{br}$)						
$[Li_2(HMPA)_2PhNC(S)NPh]_2$	4	2.24	1.95 ($HMPA_{br}$)				$Li—Ni_m$	1.98	**191**/157
	4	2.05	1.97 ($HMPA_{br}$)				Li—S	2.43	
			1.84 ($HMPA_t$)						
$\{Li(TMEDA)\}_2N(i\text{-}Pr)_2N{=}C{=}CHPh$	4	2.10^{a}	2.20–2.35						**192**/158
			2.04 (TMEDA)						
$Li(NH_2CH_3)Li[c\text{-}\{N(Me)(CH_2)_2\}]_3$-	4	1.89	$2.05\text{–}2.15^{b}$	Na					**193**/159
$[c\text{-}\{N(Me)(CH_2)_2]_2\text{-}N(CH_2)_2]^+Na^-$	4	1.97	2.09 ($MeNH_2$)						
$[Li(THF)_2Li\{p$-tert-butylcalix[4]arene-									
$P(H)NMe_2\}]$	4	2.05	2.26 (calix)				THF		**193b**/15
$Li(THF)_2Al(NMe_2)_4$	4	2.13	2.00 (THF)	Al	1.85	1.90			**194**/160
$LiAl(H)(NEt_2)_3$	4^{d}	2.21^{a}		Al		1.87^{a}			**195**/161
		2.29^{e}							
$LiAl\{N(H)C(Ph)_3\}_2$	2^{f}	2.03		Al		1.94			**196**/162
$Li(t\text{-}Bu_2C{=}N)_2Al\{N{=}C(t\text{-}Bu)_2\}_2$	2^{g}	1.97		Al	1.78	1.87			**197**/163
$Li\{PhSi(NSiMe_3)_3Ti(Cl)Cp\}$	3	1.95	1.95 (Et_2O)	Ti	1.86		LiCl	2.46	**198**/164

$LiVMe_2\{NSi(t\text{-}Bu)_3\}_2$	2	1.99		V		1.71			**199**/165
$[Li(THF)_2]_2Nb(n\text{-}Bu)(NMes)_3$	3	1.90	THF[h]Nb	1.89	1.85[a]				**200**/166
	4	2.08							
$Li(THF)Mn\{N(SiMe_3)_2\}_3$	3	2.10	THF	Mn	2.02	2.14			**201**/167
$LiMn\{MesNSiMe_2NMes\}_2MnN(SiMe_3)_2$	[i]			Mn	2.05[a]		[i]	[i]	**202**/128
$\{Li(THF)_2Cl\}_2\{Mn(NEt_2)_2\}_3$	4[j]	2.07	1.99[a] (THF)	Mn		2.13[a]	LiCl	2.38	**203**/168
$\{Li(THF)\}_2Mo(NMe_2)_6$	4	2.08[a]	1.90 (THF)	Mo		2.12[a]			**204**/169
$\{Li(THF)\}_2Zr(NMe_2)_6$	4	2.16	1.91 (THF)	Zr		2.22[a]			**205**/170
$Li_4W_2\{N(t\text{-}Bu)\}_8$	3	1.89		W	1.75[a]	1.88[a]			**206**/171
		2.23							
$Li(TMEDA)Re\{N(t\text{-}Bu)\}_4$	4	1.94, 2.14	2.23[a] TMEDA	Re	1.78[a] 1.80	1.87,			**207**/171
$Li(THF)_3\text{-}c\text{-}\{NP(F)(Cp){=}NP(F)(Cp\cdots Fe){=}P(H)(BEt_3)\}$	4	2.06	1.98[a] (THF)	Fe					**208**/172

[a]Denotes average values.
[b]1,4,7-Trimethyl-1,4,7-triazacyclononane.
[c]1,4-Dimethyl-1,4,7-triazacyclononane anion.
[d]Li—H coordination (1.95 Å).
[e]Li—N distance on hydrido side.
[f]Additional Li—C coordination.
[g]Additional Li—H coordination.
[h]Li—O distance not available.
[i]Li—C contacts present (2.25–2.57 Å).
[j]Additional Li—Cl coordination.

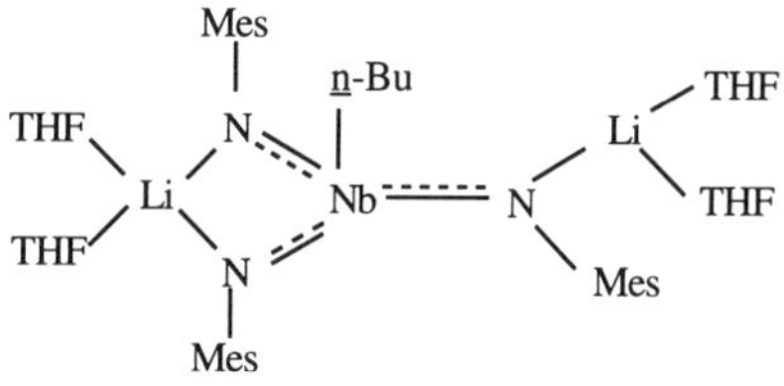

Figure 9.36 Schematic illustration of the bonding of $\{Li(THF)_2\}_2Nb(n\text{-}Bu)(NMes)_3$, **200**.[166]

Mn and Li are bridged by another amido group (Li—N, 2.07 Å) and a Cl^- ion. The Li^+ ion is also coordinated by two THF molecules.

In $\{Li(THF)\}_2Mo(NMe_2)_6$,[169] **204** and $\{Li(THF)\}_2Zr(NMe_2)_6\}$,[170] **205** (Fig. 9.37), there are two Li^+ ions bound to two MN_3 faces of the MN_6 octahedron (M = Mo and Zr). One THF donor molecule completes the coordination at each Li^+ ion. In the Mo compound, the Li—N distances are rather long (2.08 Å, av.), and even longer in the Zr derivative (2.12, 2.15, and 2.20 Å). Two tungsten(VI) atoms in dimeric $Li_4W_2\{N(t\text{-}Bu)\}_8$,[171] **206**, are apically coordinated by a terminal $t\text{-}BuN^{2-}$ group. Three of these coordinate to four Li^+ ions in a central rectangular array with short Li—Li distances of 2.10–2.20 Å; these serve as bridges between the monomeric units. All bridging N atoms are four-, and all Li^+ ions are three-coordinate. The Li—N bonds are in the range 1.89–2.23 Å.

The Re(VII) center in $Li(TMEDA)Re\{N(t\text{-}Bu)\}_4$,[171] **207**, is surrounded by four formally dianionic N donors, leaving a negative charge to be balanced by a Li^+ ion. This is coordinated between two of the imide ligands and assumes four-coordination by also bonding to TMEDA. The terminal Re—N bonds are very short (1.77 and 1.78 Å), suggesting triple bond character. One of the $Re—N_{br}$ bonds is also quite short (1.80 Å) and is associated with a long Li—N bond of 2.14 Å. The remaining bridging imide has the longest Re—N bond (1.87 Å) in the molecule and a much shorter Li—N contact (1.92 Å), which is short for a four-coordinate Li^+ ion.

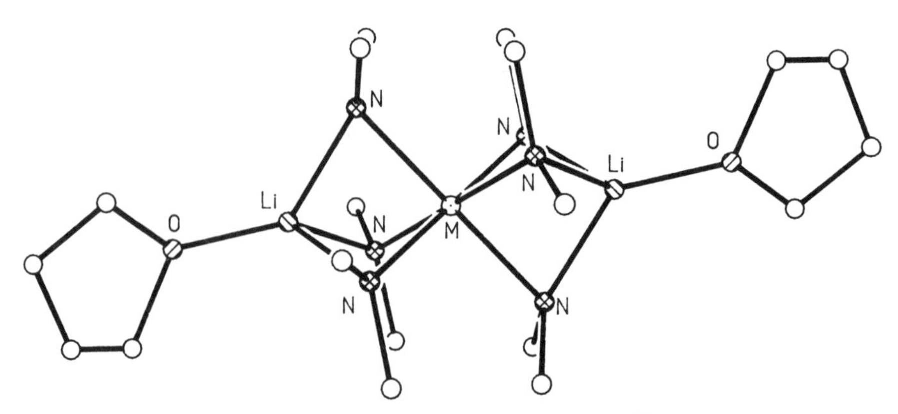

Figure 9.37 Illustration of the structures of **204**[169] and **205**.[170]

Lithium is bonded to an anionic nitrogen in a six-membered $(PN)_3$ ring unit in $Li(THF)_3\{c\text{-}NP(F)(Cp)NP(F)(\eta^5\text{-}C_5H_5)NP(H)(BEt_3)\}$,[172] **208**, in which two P atoms are substituted with Cp groups, which in turn sandwich an iron atom. The third P atom forms an adduct to a BEt_3 unit. The Li—N bond is fairly long (2.06 Å), which somewhat reflects the situation in solution, where it fluctuates between two nitrogen atoms.

3.8 Lithium Phosphides

Like their nitrogen counterparts, lithium phosphides are useful reagents for the introduction of phosphido groups into a large variety of systems. A major difference between lithium amides and phosphides is the relative weakness of the Li—P interaction. Thus, Li^+ ions can be more readily separated from the phosphide anion (cf. reaction of 12-crown-4 with $LiNPh_2$ and $LiPPh_2$ to give (12-crown-4)$LiNPh_2$ **89** and [Li(12-crown-4)$_2$][PPh_2], vide infra). Also, unlike their amide counterparts, the coordination geometry of phosphorus in terminal lithium phosphides is a pyramidal one, whereas planar coordination is normally observed for the nitrogen center in amides. This is presumably due to the much greater inversion barrier at the heavier pnictide center. Structural data for simple lithium phosphides are given in Table 9.17.

Table 9.17 Simple Lithium Phosphides

	Aggr.	CN_{Li}	Li—P (Å)	12-crown-4	Li—L (Å)	Compound/ Ref.
$[Li(DME)PH_2]_n$	n^b	4	2.58^a	DME	2.05^a	**209**/173, 174
$[Li(Et_2O)PPh_2]_n$	n^b	3	2.48	Et_2O	1.94^a	**210**/175
$[Li(THF)_2PPh_2]_n$	n^b	4	2.63	THF	1.96^a	**211**/175
$[Li(THF)P(c\text{-}C_6H_{11})_2]_n$	n^b	3	2.50	THF	1.94	**212**/175
$[Li(12\text{-crown-}4)][PPh]_2$	1	8^c	–	12-crown-4	2.37^a	**213**/176
$[Li(TMEDA)PPh_2]_2$	2	4	2.61	TMEDA	2.14^a	**214**/177
$Li(PMDETA)PPh_2$	1	4	2.57	PMDETA	2.12^a	**215**/177
$Li(THF)_3P(H)Mes$	1	4	2.53	THF	1.94^a	**216**/178
$[Li(THF)_2P(H)Mes]_n$	n^b	4	2.65	THF	1.98^a	**217**/179
$[Li(Et_2O)PMes_2]_2$	2	3	2.50	Et_2O	1.94^a	**218**/178
$[LiP(t\text{-}Bu)_2]_4(THF)_2$	4	3	2.56^a	THF	1.92^a	**219**/180
$[LiP\{CH(SiMe_3)_2\}_2]_2$	2	2	2.47			**220**/181
$Li(THF)_3P(PPh)_2C_6H_4$	1	4	2.58	THF	1.93^a	**221**/182
$[Li(THF)_2P(SiMe_3)_2]_2$	2	4	2.62	THF	1.98^a	**222**/183
$[Li(DME)P(SiMe_3)_2]_2$	2	4	2.56	DME	2.02^a	**223**/174
$[LiP(SiMe_3)_2]_4(THF)_2$	4	3	2.53^a	THF	1.89^a	**224**/183
$[LiP(SiMe_3)_2]_6$	6					**225**/184
$[Li(TMEDA)PPh(SiMe_3)]_2$	2	4	2.62^a	TMEDA	2.16^a	**226**/185

[a]Denotes average values.
[b]Polymeric chains.
[c]Separate ion pair.

3.8.1 Simple Lithium Phosphides. The simplest lithium phosphide is $LiPH_2$, and this has been crystallized as its DME adduct. The lack of significant steric bulk around the P atom affords a polymeric chain structure $[Li(DME)PH_2]_n$,[173,174] **209**, with a backbone of alternating Li and P centers. The Li—P distances are 2.56 and 2.60 Å, the Li—P—Li angle is almost linear (177°). Similar Li—P chains are present in $[Li(Et_2O)PPh_2]_n$,[175] **210** (Fig. 9.38), $[Li(THF)_2PPh_2]_n$,[175] **211**, and $[Li(THF)P(c\text{-}C_6H_{11})_2]_n$,[175] **212**.

Such chains may be disrupted or terminated in several ways. For example, the addition of 12-crown-4 affords ion pair separation in the case of $[Li(12\text{-crown-}4)_2][PPh_2]$,[176] **213**. The addition of the multidentate ligands TMEDA or PMDETA also reduces the degree of aggregation to dimeric or monomeric species as exemplified by the formulas $[Li(TMEDA)PPh_2]_2$,[177] **214**, and $Li(PMDETA)PPh_2$,[177] **215**. Even monodentate donors such as THF are capable of producing a monomeric structure, as in the case of $Li(THF)_3P(H)Mes$,[178] **216**. However, a polymeric structure $[Li(THF)_2P(H)Mes]_n$,[179] **217**, is observed if the THF is partly removed under reduced pressure. It is noteworthy that greater bulk at phosphorus in combination with the weaker donor results in the dimer $[Li(Et_2O)PMes_2]_2$,[178] **218**. However, the bulky $LiP(t\text{-}Bu)_2$ crystallizes as a tetrameric aggregate of formula $[LiP(t\text{-}Bu)_2]_4(THF)_2$,[180] **219**, which has a ladder structure. This compound represented (along with **213**) the first structural determinations for lithium diorganophosphides.

Donor-free $[LiP\{CH(SiMe_3)_2\}_2]_2$,[181] **220**, assumes a dimeric structure. The two-coordinate nature of lithium affords short Li—P bonds of 2.47 Å. In this compound, the Li—P—Li angle is small (72.2°), perhaps owing to the large

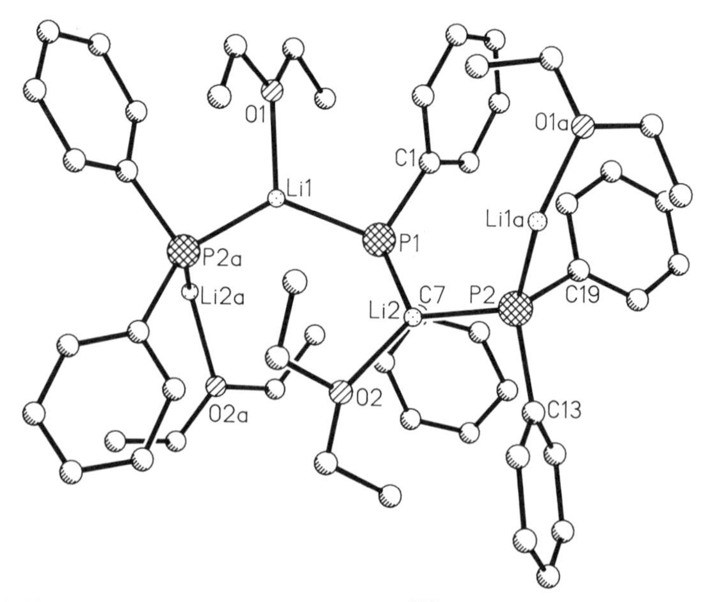

Figure 9.38 Structure of $[Li(Et_2O)PPh_2]_n$, **210**.[175] H atoms are not shown for clarity.

substituents on P. A P_3 unit, the central P atom of which formally bears the negative charge, is fused to a benzene ring in $Li(THF)_3P(PPh)_2C_6H_4$,[182] **221**. Compared to **216**, the Li—P distance (2.58 Å) is slightly longer, but close to that found in PMDETA-coordinated **215**. The P—P distances in the anion (2.16 Å) are slightly shorter than P—P single bond (2.22 Å). The Li^+ ions in $[Li(THF)_2P(SiMe_3)_2]_2$,[183] **222** are four-coordinate because of bonds to two THF molecules (Li—P, 2.62 Å). The DME analogue $[Li(DME)P(SiMe_3)_2]_2$,[174] **223**, has significantly shorter Li—P contacts (2.56 Å), while the Li—O contacts are somewhat longer (2.02 vs. 1.98 Å, av.). The partial removal of THF from the phosphide $[Li(THF)_2P(SiMe_3)_2]_2$[183] by evacuating for 20 h at 10^{-3} torr results in a structure of formula $[LiP(SiMe_3)_2]_4(THF)_2$,[183] **224** (Fig. 9.39), that is very similar to that observed for the *t*-Bu analogue **219**. The formation of $[LiP(SiMe_3)_2]_6$,[184] **225**, results in a hexamer in a stair- or ladderlike arrangement. Whereas **222** can be envisioned as a ladder structure with two rungs and **224** with four rungs, there are six rungs in the structure of **225**. This occurs in accord with the principle that the lower the number of donor molecules, the higher the aggregation state becomes. In this instance, it leads to a one-dimensional extension of the number of fused four-membered

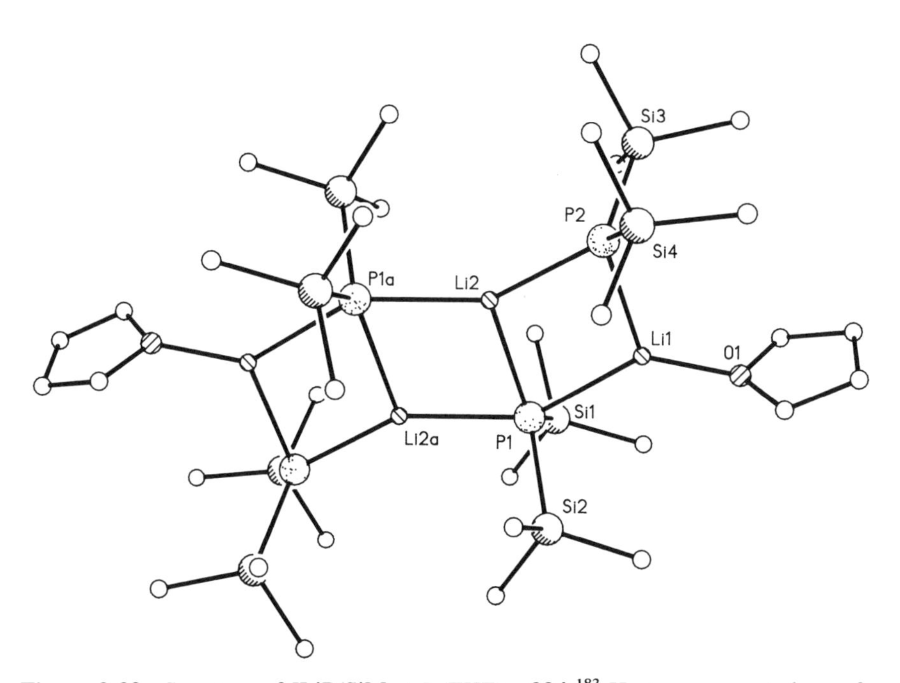

Figure 9.39 Structure of $[LiP(SiMe_3)_2]_4(THF)_2$, **224**.[183] H atoms are not shown for clarity. Its structural similarity to **219** may be gauged from the following data: Li1-P1, 2.55 [2.60]; Li1-P2, 2.48 [2.48]; Li1-O_{THF}, 1.89 [1.92]; Li2-P1, 2.64 [2.67]; Li2-P2, 2.44 [2.50]; Li2-P1′, 2.54 [2.59] Å. The corresponding values for the previously mentioned $[LiP(t\text{-}Bu)_2]_4(THF)_2$, **219**,[180] are given in brackets.

rings. Two different phosphorus substituents in $[Li(TMEDA)PPhSiMe_3]_2$,[185] **226**, results in *cis–trans* isomers in a type 2 dimeric structure. The Li—P distances in the central four-membered ring are essentially the same, with values of 2.62 Å.

*3.8.2 **Lithium Fluorosilylphosphides.*** As in the case of lithium amides, the introduction of a fluorine atom in the silyl groups has not only steric, but also electronic, consequences. The interplay between bulkiness of the Si and P substituents is exemplified in monomeric $Li(THF)_3P(t\text{-}Bu)Si(t\text{-}Bu)_2F$,[186] **227**, versus dimeric $[Li(THF)_2PPhSi(t\text{-}Bu)_2F]_2$,[187] **228,** which has a central eight-membered $(Li—P—Si—F)_2$ ring. Although there is no L—F contact in **227**, the Li—F interaction in **228** apparently weakens the Li—P and Si—F bonds. With a larger mesityl substituent on P and the use of polydentate ligands, aggregation in $Li(PMDETA)PMesSi(t\text{-}Bu)_2F$,[188] **229**, and $[Li(12\text{-}crown\text{-}4)_2][P(Mes)Si(t\text{-}Bu)_2F]$,[188] **230**, is precluded. The use of the crown ether results in separate ion pairs, as seen in **230** (cf. **213**).

In the P analogue of the amide **97** a different structure is seen. The species $Li(THF)_2P\{Si(t\text{-}Bu)_2F\}_2$,[189] **231**, has no Li—P interaction in the crystal structure. The Si—P—Si angle is 103.5°, in marked contrast to the linear Si—N—Si framework in **97**. The Li^+ ion is coordinated by the F atoms (Li—F, 1.86 Å) and two THF molecules, to form a six-membered ring. These Li—F interactions are considerably stronger than in **97** (2.12 and 2.21 Å). Structural data for compounds **227–231** are given in Table 9.18.

*3.8.3 **Boryl- and Gallylphosphides.*** Like borylamides, borylphosphides are interesting species, not least because there is a possibility of *p-p* π-overlap involving the phosphorus lone pair and the empty *p*-orbital on boron. This interaction may be reflected in structural parameters such as the B—P distance, the angle between the planes at boron and phosphorus, the RBR bond angle, and rotational barriers around the B—P bond. The use of a lithium substituent on phosphorus in borylphosphides enhances the electron density on P, reduces it inversion barrier, and improves π-overlap. The B—P distances in boryl phosphides are in the range 1.81–1.85 Å (cf. sum of radii of B and P = 1.96 Å).

The use of different substituents on P or B as in **232–238**[190-193] has little influence on the B—P distance. NMR studies show that there are rotation barriers as high as 23 kcal mol^{-1} around the B—P bond. In **237** the use of THF as donor solvent leads to a Li—P elongation of ca. 0.01 Å with respect to the average values of the Et_2O-containing structures, owing to the higher coordination number of the Li^+ ion. The shortest B—P distance was observed when the other B substituent was a phosphino group.[192a]

A dimeric structure is observed for the species $[Li(DME)PHB\{N(i\text{-}Pr)_2\}_2]_2$,[192b] **237b**. The Li—P distance, 2.54 Å, is longer than most of those observed in the other B—P compounds but is consistent with the higher coordination number of the Li^+ ion. The B—P distance, 1.92 Å, is consistent

Table 9.18 Lithium Fluorosilylphosphides

	Aggr.	CN_{Li}	Li—P (Å)	L	Li—L (Å)	Li—F	Compound/ Ref.
$Li(THF)_3P(t\text{-}Bu)Si(t\text{-}Bu)_2F$	1	4	2.55	THF	1.96[a]		**227**/186
$[Li(THF)_2PPhSi(t\text{-}Bu)_2F]_2$	2	4	2.62	THF	1.98[a]	1.90	**228**/187
$Li(PMDETA)PMesSi(t\text{-}Bu)_2F$	1	4	2.55	PMDETA	2.13[a]		**229**/188
$[Li(12\text{-}crown\text{-}4)][PMesSi(t\text{-}Bu)_2F]$	1	8[b]		12-crown-4	2.36[a]		**230**/188
$Li(THF)_2P\{Si(t\text{-}Bu)_2F\}_2$	1	4		THF	1.90	1.86	**231**/189

[a]Denotes average values.
[b]Separate ion pair.

with single bonding. Presumably, the -N(i-Pr)$_2$ substituents on B contribute to weakening the B—P π-bonding and thus the P lone-pair is more available for bridging.

The reaction of t-Bu$_2$GaCl with Li$_2$P(t-Bu) results in different products. Only those having lithium are discussed here. A monodeprotonated, planar, four-membered (GaP)$_2$ ring is present in [Li(Et$_2$O)$_4$][H{Ga(t-Bu)$_2$P(t-Bu)}$_2$],[194] **239**, which has equal Ga—P bonds of 2.42 Å. There are no Li—P contacts in the solid state structure, owing to an ion pair separation. The corresponding doubly deprotonated [(t-Bu)$_2$GaP(t-Bu)$_2$]$_2^{2-}$ dianion in [Li(12-crown-4)$_2$]$_2$[{Ga(t-Bu)$_2$P(t-Bu)}$_2$],[194] **240**, has two 12-crown-4 sandwiched Li$^+$ ions as counter-cations. The Ga—P bonds in the ring are slightly shorter (Ga—P, 2.40 Å) than those in **239**.

A different structure is observed in Li(Et$_2$O)$_2${P(t-Bu)}$_2$Ga(t-Bu)Ga(t-Bu)$_2$,[194] **241** (Fig. 9.40), which has a puckered Ga$_2$P$_2$ four-membered ring. One Ga atom is four- and the other is three-coordinate. The former bonds are considerably shorter (2.29 Å) than the latter (2.45 Å). The Li—P bonds (2.66 Å) are rather long, perhaps because of the four-coordinate nature of the Li and P centers. Structural data the boryl and gallyl-phosphides are given in Table 9.19.

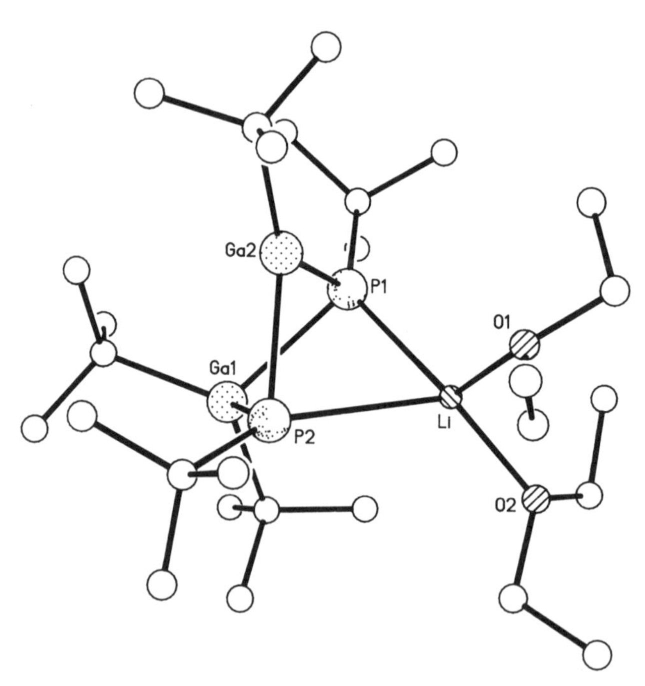

Figure 9.40 Structure of Li(Et$_2$O){P(t-Bu)}$_2$Ga(t-Bu)Ga(t-Bu)$_2$, **241**.[194] H atoms are not shown for clarity.

Table 9.19 Lithium Boryl- and Gallylphosphides

	CN_{Li}	L	Li—P (Å)	B=P (Å)	B—P (Å)	Twist Angle[b]	Compound/ Ref.
$Li(Et_2O)_2P(c\text{-}C_6H_{11})BMes_2$	3	Et_2O	2.45	1.83		18.3	**232**/190, 191
$Li(Et_2O)_2PMesBMes_2$	3	Et_2O	2.45	1.82		2.4	**233**/190, 191
$[Li(12\text{-crown-}4)_2]$ $[PMesBMes_2]$	8[d]	12-crown-4		1.84		(0.4[c])	**234**/190, 191
$Li(Et_2O)_2P(t\text{-Bu})BTrip_2$	3	Et_2O	2.46	1.84		36.0	**235**/192a
$Li(Et_2O)_2P(Mes)B\{PHMes\}Trip$	3	Et_2O	2.48	1.83	1.92	14.7	**236**/192a
$Li(THF)_3P(SiMe_3)B(Mes)_2$	4	THF	2.54	1.81		23.5	**237a**/192a
$[Li(OME)PHB\{N(c\text{-}Pr)_2\}_2]_2$	4	DME	2.54		1.92		**237b**/192b
$Li(Et_2O)_2P(1\text{-Ad})BMes_2$	3	Et_2O	2.44	1.82		13.8	**238**/193
$[Li(Et_2O)_4][H\{Ga(t\text{-Bu})_2\text{-}P(t\text{-Bu})\}_2]$	4[e]	Et_2O					**239**/194
$[Li(12\text{-crown-}4)_2]$ $[\{Ga(t\text{-Bu})_2P(t\text{-Bu})\}_2]$	8[d]	12-crown-4					**240**/194
$Li(Et_2O)_2\{P(t\text{-Bu})\}_2Ga(t\text{-Bu})\text{-}Ga(t\text{-Bu})_2$	4	Et_2O	2.66[a]				**241**/194

[a]Denotes average value.
[b]"Twist" denotes the angle between the RBR and the RPLi planes.
[c]Deviation of the C-P vector from the R-B-R plane.
[d]12-crown-4 coordination of the Li^+ ion.
[e]Coordination by four Et_2O molecules.

3.8.4 Multiphosphido Ligands. Unlike its monomeric HMPA complexed N analogue **160**, $[\{Li(THF)_2\}_2PPh(CH_2)_2PPh]_2$,[195,196] **242**, is a dimer with a Li^+ ion in two different environments. One Li^+ ion is bound to two P atoms in a chelating fashion (Li—P, 2.53 and 2.56 Å), the other serves as a bridge between the ligands and has Li—P contacts of 2.57 Å. This results in an eight-membered $(LCCP)_4$ ring, with two sets of ethylene groups bridging two P atoms. Each Li^+ ion is also coordinated by one THF molecule. The use of TMEDA in $\{Li(TMEDA)\}_2\{1,2\text{-}(PPh)_2C_6H_4\}$,[197] **243**, precludes a bridging arrangement as in **242**. Instead, the same structural type as in **160** is observed, with the Li^+ ion above and below the plane defined by the rings. The Li—P distances are in the range 2.55–2.59 Å. Remarkably short Si—P bonds (2.11 Å) are present in $Li(Et_2O)_2Si(t\text{-}Bu)(PMes^*)_2$,[198] **244**, suggesting a delocalized π-bond in the P—Si—P framework. Note that there is only one substituent on each phosphorus atom, so that **244** can be considered a lithiated derivative of Mes*PSi(*t*-Bu)PMes*. The Li—P distances are at the longer end of the range for four-coordinated Li^+ ions (2.64 and 2.67 Å). Structural data for these compounds are given in Table 9.20.

3.8.5 Lithium Salts of Cage Structures Containing Phosphorus. In $\{Li(TMEDA)\}_3P_7$,[199] **245**, the anion consists of a seven-membered phosphorus cage, which has a monocapped trigonal prismatic structure. The Li^+ ions coordinate to the capped P_3 triad and have Li—P bonds of 2.55 Å. In $\{[Li(12\text{-crown-}4)_2]\}_3P_{21}$,[200] **246**, addition of crown ether results in ion pair separation. The trianionic P_{21}^{3-} cage structure has three two-coordinate P centers. The P—P bonds involving these are shortest (2.14 Å), the others display values of about 2.26 Å.

3.8.6 Other Lithium Phosphides. The anion $PC_4Me_4^-$ can be considered an analogue of $C_5Me_5^-$, in which phosphorus replaces a CMe moiety. Its TMEDA-solvated lithium salt $Li(TMEDA)PC_4Me_4$,[201] **247**, features a virtually planar anionic ring with C—C distances in the aromatic range (1.40–1.42 Å). The P—C bonds are short (ca. 1.85 Å), emphasizing the aromatic properties of the anionic ring. The Li—P distance is fairly short (2.53 Å) compared to the values of the simple lithium phosphides (vide supra). The Li—C distances are in the range of 2.37–2.43 Å (2.39 Å av.), so that coordination of Li^+ to the C_4P ring is essentially an η^5 one. TMEDA additionally coordinates to Li (Li—N, 2.14 Å, av.).

Strong Li—N interactions, in the range of 2.02–2.05 Å, are observed in $[Li(THF)P\text{-}c\text{-}\{N(CH_2)_2\}_4)_n$,[202] **248**. The negative charge formally resides on the central P atom, which displays long bonds to Li (2.67–3.29 Å; 2.87 Å av.). The influence of the Li-coordination of N is reflected in different P—N bond lengths, 1.74 Å (av.) for the noncoordinating and 1.98 Å (av.) for the Li-coordinating N centers.

Table 9.20 Miscellaneous Phosphides

	Aggr.	CN_{Li}	Li—P (Å)	L (Å)	Li—L (Å)	Add. Coord.	Li—L′	Compound
$[\{Li(THF)_2\}_2PPh(CH_2)_2PPh]_2$	2	4	2.58[a]	THF	1.93[a]			**242**/195
$\{Li(TMEDA)\}_2$-1,2-$(PPh_2)_2C_6H_4$	1	4	2.58[a]	TMEDA	2.07[a]			**243**/197
$Li(Et_2O)_2(t\text{-}Bu)Si(PMes^*)_2$	1	4	2.66[a]	Et_2O	1.94			**244**/198
$\{Li(TMEDA)\}_3P_7$	1	4	2.55[a]	TMEDA	2.07[a]			**245**/199
$[\{Li(12\text{-crown-}4)_2\}]_3\ P_{21}$	1	8[b]		12-crown-4	2.39			**246**/200
Li(TMEDA){c-P(CMe)$_4$}	1	7[c]	2.53	TMEDA	2.14[a]	4C	2.29[a]	**247**/201
[Li(THF)P(c-{N(CH$_2$)$_2$}]$_n$	n	4	2.89[a]	THF	1.93[a]	N	2.87	**297a**/202a
$[LiP(H)Mes^*LiMes^*]_2$	2	2.7	(η'-P), (η_6-Mes*)		2.47	η,η^6-Mes*	2.09, 2.31	**248b**/202b

[a]Denotes average values.
[b]12-crown-4 coordination.
[c]η^5 coordination to aromatic ring + additional TMEDA coordination.

3.8.7 Lithium Derivatives of Diphosphino Methanide Ligands. In diphosphino methanides, the central carbon atom formally carries the negative charge. However, some delocalization is suggested by the shorter P—C distances (vide infra) that in a single bond (cf. sum of P and C sp^2 radii 1.95 Å). In most instances, the coordination of this ligand is through the phosphorus rather than the carbon centers. Structural data are given in Table 9.21.

In dimeric $[Li(THF)C(PMe_2)_3]_2$,[203] **249**, the Li^+ ion is chelated by the two P centers and a THF. Dimerization is the result of the η^3-interaction to the delocalized P—C—P moiety. The same type of Li^+ environment is adopted in $[Li(THF)(PMe_2)_2CSiMe_3]_2$,[204] **250**, and $[Li(THF)(PMe_2)_2CH]_2$,[204] **251**, with Me_3Si or H being the carbon substituents. A chelating donor precludes aggregation in $Li(TMEDA)(PMe_2)_2CSiMe_3$,[204] **252**, and $Li(TMEDA)(PMe_2)_2CH$,[205] **253**, by saturating the coordination sphere of the lithium ion. A short Li—P bond is observed in **252**, which may be attributed to its monomeric formulation. In the dimers, the Li—C contacts are remarkably short. In all cases there is no significant influence on the P—C—P angle, which displays values from 102.7 to 109.9°.

3.8.8 Heterometallic Lithium Phosphides. The species $Li(THF)M\{P(t\text{-}Bu)_2\}_3$; M = Sn: **254**[206] and M = Pb: **255**[206] are tin and lead analogues of diphosphinomethanides with a phosphide group as the third ligand on the main group 4 metal. The Li—P bonds are fairly short, 2.48 and 2.46 Å respectively, owing perhaps to the electropositive nature of the central atoms. However, the Pb compound displays a slightly asymmetric bonding situation with one Li—P contact being considerably longer (2.57 Å). Unlike the C analogues, the compounds are monomeric even in THF, reflecting the weak donor capacity of the metal centers. $Li(TMEDA)(PPh_2)_2LuCp_2$,[207] **256**, was one of the first structures where phosphido groups serve as bridges between a rare metal and Li. The chelating nature of TMEDA precludes oligomerization, resulting in a monomeric structure. The Li—P contacts (2.61 and 2.69 Å) are rather long, presumably owing to the strength of the Lu—P bonds. A $\text{-}P(t\text{-}Bu)_2$ group bridges Cu and Li (Li—P, 2.54 Å) in $Li(THF)_3(t\text{-}Bu)_2PCuMe$,[208] **257**. The Li^+ ion interacts with the phosphorus center and three THF molecules. An infinite chain arrangement is the result of a bridging $\text{-}P(t\text{-}Bu)_2$ group between Cu and Li in $[Li(THF)_2Cu\{P(t\text{-}Bu)_2\}_2]_n$,[209] **258**. The Li—P distances in

Table 9.21 Lithium Diphosphinomethanides and Related Species

	Aggr.	CN_{Li} (Å)	Li—P (Å)	L (Å)	Li—C (Å)	Compound/ Ref.
$[Li(THF)C(PMe_2)_3]_2$	2	6	2.81[a]	THF	2.29	**249**/203
$[Li(THF)(Me_3Si)C(PMe_2)_2]_2$	2	6	2.82[a]	THF	2.30	**250**/204
$[Li(THF)HC(PPh_2)_2]_2$	2	6	2.84[a]	THF	2.24	**251**/204
$Li(TMEDA)(Me_3Si)C(PMe_2)_2$	1	4	2.52	TMEDA		**252**/204
$Li(TMEDA)HC(PPh_2)_2$	1	4	2.58	TMEDA		**253**/205

[a]Denotes average values.

this structure of 2.83 Å (av.) are considerably longer than in **257**. A Hf—P—Li—P four-membered ring is observed in $Li(DME)Hf(PPh_2)_5$,[210] **259**. Three phosphido groups are terminally bonded to Hf, and two bridge to Li. The bridging is slightly asymmetric, with Li—P contacts of 2.65 and 2.58 Å, Hf—P bonds of 2.64 and 2.71 Å to the bridging phosphido groups, and 2.50 Å (av.) to the terminal ones. Structural data are given in Table 9.22.

3.9 Lithium Arsenides, Stibinides, and Bismuthides

In comparison to their phosphorus analogues, relatively few lithium arsenide, stibinide, and bismuthide structures have been reported. The species $[Li(DME)As(SiMe_3)_2]_2$,[211a] **260**, has a type 2 dimeric structure (Li—As, 2.59 Å), whereas the higher congeners $[Li(DME)E(SiMe_3)_2]_n$ (E = Sb;[211b] **216b** Bi[211b] **261**, display a polymeric structure in a zig-zag chain arrangement (Li—Sb, 2.93 Å; Li—Bi, 2.93 Å) similar to that illustrated in Figure 9.38. In both cases, Li is chelated by one DME molecule, rendering it four-coordinate.

Although $Li(dioxane)_3AsPh_2$,[78] **262**, is a monomer (Li—As, 2.66 Å), a type 2 dimer is observed for $[Li(Et_2O)_2AsPh_2]_2$,[78] **263**, with Li—As bond lengths of 2.71 and 2.76 Å. The ion pairs $[Li(12\text{-crown-}4)_2][AsPh_2]$,[176] **264**, and $[Li(12\text{-crown-}4)_2][SbPh_2]$,[78] **265**, are similar to the phosphorus analogue **214**. Ion pair separation is also seen in $[Li(12\text{-crown-}4)_2][Sb_3Ph_4]$,[78] **266**. The anion is the result of a disproportionation of $SbPh_3$ upon the addition of Li metal. In dimeric $[Li(THF)As(t\text{-Bu})As(t\text{-Bu})_2]_2$,[212] **267**, the Li—As distance (2.58 Å) is remarkably short, owing to the three-coordination (cf. four-coordinate Li in **262** and **263**) of the Li^+ ion. The compound $Li(THF)_2As\{Si(SiMe_3)\}\text{-}Si(F)Trip_2$,[212b] **267b**, features three coordinate Li^+ with a short Li—As distance of 2.46 Å. $Li(TMEDA)(AsPh_2)_2LuCp_2$,[213] **268**, is isomorphous to its P analogue **256**, with Li—As bonds of 2.65 and 2.73 Å.

The species $Li(THF)_3AsPhBMes_2$,[214] **269a**, is an arsenide analogue of the borylphosphides (vide supra). It has an Li—As bond of 2.67 Å and a B—As bond of 1.93 Å. This distance (cf. sum of B and As radii = 2.06 Å) and the angle between the CBC and the Li—As—C plane of 1.1° indicate strong *p-p* π-overlap. ^1H VT-NMR studies indicate that the barrier of rotation is ca. 21 kcal mol^{-1}. The sum of angles at As is 341.4°, the Li—As vector deviates from the plane defined by C_2BAsC. The use of TMEDA leads to ion pair separation in the structure of $[Li(TMEDA)_2][AsPhBMes_2]$,[214] **269b**, which has a B—As bond of 1.94 Å. Structural data for the above compounds are given in Table 9.23.

4 LITHIUM COMPOUNDS OF THE MAIN GROUP 4 ELEMENTS

Unlike lithium derivatives of various carbon ligands, there are few structurally characterized species that have Li—Si, Li—Ge, Li—Sn, or Li—Pb bonds.

Table 9.22 Lithium Phosphides Containing Other Metals[a]

	CN_{Li}	Li—P (Å)	L	Li—L (Å)	M	M—P_t	M—P_{br}	Compound/ Ref.
$Li(THF)Sn\{P(t\text{-}Bu)_2\}_3$	4	2.48	THF	1.91	Sn	2.68	2.69[b]	**254**/206
$Li(THF)Pb\{P(t\text{-}Bu)_2\}_3$	4	2.52[b]	THF	1.87	Pb	2.77	2.79[b]	**255**/206
$Li(TMEDA)(Ph_2P)_2LuCp_2$	4	2.65[b]	TMEDA	2.11[b]	Lu	–	2.80[b]	**256**/207
$Li(THF)_3(t\text{-}Bu)_2PCuMe$	4	2.54	THF	NA	Cu		2.22	**257**/208
$[Li(THF)_2Cu\{P(t\text{-}Bu)_2\}_2]_n$	4	2.83	THF	NA	Cu		2.26[b]	**258**/209
$Li(DME)Hf(PPh_2)_5$	4	2.62[b]	DME	2.05	Hf	2.50[b]	2.67[b]	**259**/210

[a]Bond lengths in Angstroms.
[b]Denotes average values.

Table 9.23 Lithium Arsenides, Stibinides, and Bismuthides

	Aggr.	CN_{Li}	Li—E (Å)[b]	L	Li—L (Å)	Compound/ Ref.
$[Li(DME)As(SiMe_3)_2]_2$	2	4	2.59	DME	2.10[a]	**260**/211a
$[Li(DME)Sb(SiMe_3)_2]_n$	n	4	2.93	DME	2.01	**261a**/211b
$[Li(DME)Bi(SiMe_3)_2]_n$	n	4	2.92	DME	2.10	**261b**/211b
$Li(dioxane)_3AsPh_2$	1	4	2.66	dioxane	1.94[a]	**262**/78
$[Li(Et_2O)_2AsPh_2]_2$	2	4	2.74[a]	Et_2O	1.99[a]	**263**/78
$[Li(12\text{-crown-}4)_2][AsPh_2]$	1	8[c]		12-crown-4	2.37[a]	**264**/176
$[Li(12\text{-crown-}4)_2][SbPh_2]$	1	8[c]		12-crown-4	2.41[a]	**265**/78
$[Li(12\text{-crown-}4)_2][Sb_3Ph_4]$	1	8[c]		12-crown-4	2.37[a]	**266**/78
$[Li(THF)\{As(t\text{-Bu})As(t\text{-Bu})_2\}]_2$	2	3	2.58	THF	1.89	**267a**/212a
$Li(THF)_2As\{Si(SiMe_3)_3\}Si(F)Trip_2$	1	3	2.46	TMF	1.92	**267b**/212b
$Li(TMEDA)(Ph_2As)_2LuCp_2$	1	4	2.69[a]	TMEDA	2.09	**268**/213
$Li(THF)_3AsPhBMes_2$	1	4	2.67	THF	1.94[a]	**269a**/214
$[Li(TMEDA)_2][AsPhBMes_2]$	1	4		TMEDA	2.10[a]	**269b**/214

[a]Denotes average values.
[b]E = P, As, Sb, or Bi.
[c]12-crown-4 coordination.

The solvent-free, hexameric species $(LiSiMe_3)_6$,[215,216] **270**, has been known for a number of years. It has bridging Li—Si bond lengths averaging 2.69 Å, which are considerably longer than the sum of the covalent radii (2.51 Å) of Li and Si. Structural data are given in Table 9.24.

4.1 Lithium Silicides

The compound $Li(THF)_3SiPh_3$,[217] **271**, displays a terminal Li—Si bond that has a length of 2.67 Å, which is nearly as long as that seen in **271**. There are no interactions between lithium and the phenyl rings, as there are in trityllithium.[218] The C—Si—C bond angles are considerably lower (101.3° av.) than the ideal tetrahedral value, while the Li—Si—C angles are correspondingly larger (116.8° av.). This indicates location of the lone pair mainly on silicon and increased *s*-character of the lone pair orbital. In $[\{Li(THF)_3\}_2Si_4Ph_8]$,[219] **272**, the Li^+ ions are bonded to both ends of the Si_4 backbone. The Li—Si distances are marginally longer (2.72 Å) than in **271**, the Si—Si bonds are also elongated (2.42 Å av.) with respect to a single bond of 2.34 Å. $Li(THF)_3Si(SiMe_3)_3$,[220] **273** (Fig. 9.41) displays a Li—Si distance of 2.67 Å, which is exactly the same value as in **271**. The Li—O contacts are 1.94 Å long.

Whereas aromatic and quasi-aromatic rings systems with main groups 3, 5, and 6 elements as ring members are known, no stable benzene analogues of the heavier main group 4 congeners have been reported. The unstable species SiC_5H_6 has been shown to exhibit aromatic properties by means of PE spectroscopy and UV-Vis data,[221,222] while greater stability has been achieved by the use of bulkier groups[223] on Si. However, no X-ray structure analysis has

Table 9.24 Lithium Compounds of Main Group 4 Elements

	Aggr.	CN_{Li}	Li—E (Å)	L	Li—L (Å)	Add. Coord.	Compound/ Ref.
$(LiSiMe_3)_6$	6						**270**/1, 215, 216
$Li(THF)_3SiPh_3$	1	4	2.67	THF	1.98[a]		**271**/217
$[Li(THF)_3]_2Si_4Ph_8$	1	4	2.72	THF	1.96[a]		**272**/219
$Li(THF)_3Si(SiMe_3)_3$	1	4	2.67	THF	1.94		**273**/220
$[Li(12\text{-crown-}4)_2][SiMe_2C_5H_5]$	1	8[c]		12-crown-4	2.37[a]		**274**/224
$Li(PMDETA)SnPh_3$	1	4	2.87	PMDETA			**275**/225
$LiSn(O\text{-}2,6\text{-}Ph_2\text{-}C_6H_3)_3$	1	3[d]	2.78			O (1.99)	**276**/226
$[LiSn\{O(t\text{-}Bu)_3\}]_2$	2	3	3.09			O (2.02)	**277**/227
$[Li(dioxane)_4][Li\{Sn(furyl)_3\}_2]$	1	4		Dioxane	1.92[a]		**278**/228
		6	3.81[d]			2.10	
						2.29	
$Li(PMDETA)PbPh_3$	1	4	2.86	PMDETA			**279a**/229a
$[Li(12\text{-crown-}4)_2][Pb_2Ph_5]$	1	8[c]		12-crown-4	2.32[a]		**279b**/229b

[a]Denotes average values.
[b]E = Si, Ge, Sn, or Pb.
[c]12-crown-4 coordination.
[d]Assuming three-coordination to the O atoms, but the Li—Sn distance rather nonbonding (see text).
[e]Nonbonding distance.

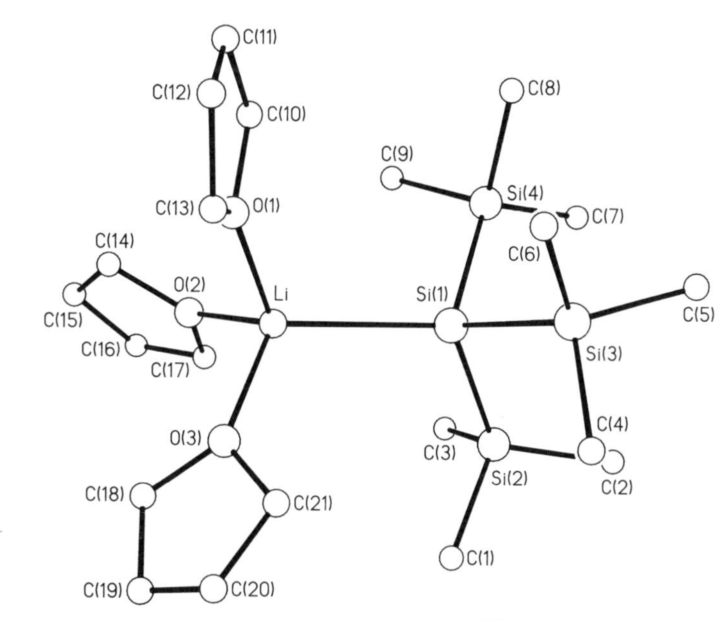

Figure 9.41 Structure of $Li(THF)_3Si(SiMe_3)_3$, **273**.[220] H atoms are not shown for clarity.

been performed to date. A formal addition of organolithium derivatives affords anionic species isoelectronic to λ^5-phosphabenzenes. The quasi-aromatic character of these silabenzene anions has been postulated in the ion pair $[Li(12\text{-crown-}4)_2][c\text{-}SiMe_2C_5H_5]$,[224] **274**. This compound has a planar SiC_5 framework, short endocyclic (1.83 Å av.) and long exocyclic (1.89 Å av.) Si—C bonding distances. NMR data support some degree of aromatic character.

4.2 Lithium Stannides and Plumbides

The species $Li(PMDETA)SnPh_3$, **275**, is the tin analogue of **271** and the corresponding trityllithium species.[218] It features a terminal Li—Sn bond of 2.87 Å and very narrow C—M—C angles (96.1°). Thus the lone pair orbital probably has very high *s*-character. Support for this view comes from $^7Li^{117}Sn$ and ^{119}Sn couplings in solution, in which the Li—Sn bond is maintained intact. Monomeric $LiSn\{O(2,6\text{-}Ph_2C_6H_3)\}_3$,[226] **276**, displays Li—O interactions of 1.99 Å. The Li—Sn distance is 2.78 Å, which is shorter than in **275**. However, it is arguable whether or not this distance constitutes an Li—Sn bond or that the approach of Li and Sn is the result of Li^+ ion coordination by the oxygen centers. The latter point of view receives support from the fact that there is apparently no Li—Sn bond in dimeric $[LiSn(O\text{-}t\text{-}Bu)_3]_2$,[227] **277** (with average Li—Sn distances of 3.09 Å). The tridentate $Sn(O\text{-}t\text{-}Bu)_3$ moieties form two OSnOLi four-membered rings in a chelating fashion, thus causing one O atom to be three- and other two two-coordinate. The Li^+ ions bridge the two anions. Li—O contacts range from 1.93 to 2.13 Å (2.02 Å av.). Two different Li environments are observed in $[Li(dioxane)_4][Li\{Sn(furyl)_3\}_2]$,[228] **278**. The cationic Li—O bonds are fairly short (1.86–1.94 Å) for four-coordinate Li. In the anion, two different sets of Li—O bonds can be distinguished and Li is coordinated in a distorted octahedral fashion, with shorter equatorial (2.10 Å) and longer axial (2.29 Å) distances. The Li—Sn distances of 3.81 Å (av.) are outside the range of bonding interactions.

The species $Li(PMDETA)PbPh_3$,[229a] **279a**, has a similar structure to its analogue, **275**. The Li—Pb distance is 2.86 Å. In $[Li(12\text{-crown-}4)_2][PbPh_2PbPh_3]$,[229b] **279b**, there is no Li—Pb contact owing to ion pair separation. It displays a Pb—Pb bond length of 3.01 Å and the angles at the anionic Pb atom average 96.2°.

5 LITHIUM COMPOUNDS OF MAIN GROUP 3 ELEMENTS

5.1 Lithium–Boron Compounds

Several species are now known that have Li—B distances that are close to the sum of the covalent radii (2.19 Å) of these elements. No terminal Li—B bonds are known at present. In all the compounds discussed here, the boron center generally has some formal negative charge and the Li^+ ion has a bridging function. These considerations must be taken into account in comparisons involving the strength of the Li—B interaction. Among these is the species $\{Li(OEt_2)\}_2Mes_2BBPhMes$,[230] **280**, which is a doubly reduced boron analogue of ethylene that contains a B—B double bond of length 1.64 Å. It was the first example of a structurally characterized boron species that corresponded to the general formula $(LiBR_2)_n$. Owing to the electronic characteristics of boron, association (in the dimer at least) takes place through strong B—B bonding rather than bridging through lithium. The closest Li—B contacts in this molecule are 2.34 and 2.36 Å. The shortest Li—C contact is 2.32 Å. Bearing in mind that the radius of B is greater than that of C (0.85 vs. 0.77 Å), the Li—B bond could be regarded as having comparable strength. A similar arrangement is present in $\{Li(OEt_2)\}_2Me_2NBPhBPhNMe_2$,[231] **281**. Short Li—B contacts (2.28 and 2.34 Å) are present and Li^+ is coordinated to the N atom (1.98 Å), to the ipso-C (2.25 Å), and to one of the ortho C atoms (2.45 Å) of the phenyl group. While the B—B bonds have approximately the same length (1.63 Å) as in the tetraaryl species, the B—N π-bonding has been destroyed and single B—N bond lengths are observed (1.56 Å). Lithium is symmetrically coordinated to both B atoms in $Li(OEt_2)[C(BMes)B\{CH(SiMe_3)_2\}Mes]$,[232] **282** (Fig. 9.42). The Li—B distances (2.49 Å) are rather long, whereas the distance to the central C atom is considerably shorter (2.09 Å). Thus, the Li^+ ion interacts primarily with the carbon center. The nearly linear BCB angle (176.5°) suggests *sp* hybridization of the central C atom, with one B—C bond display-

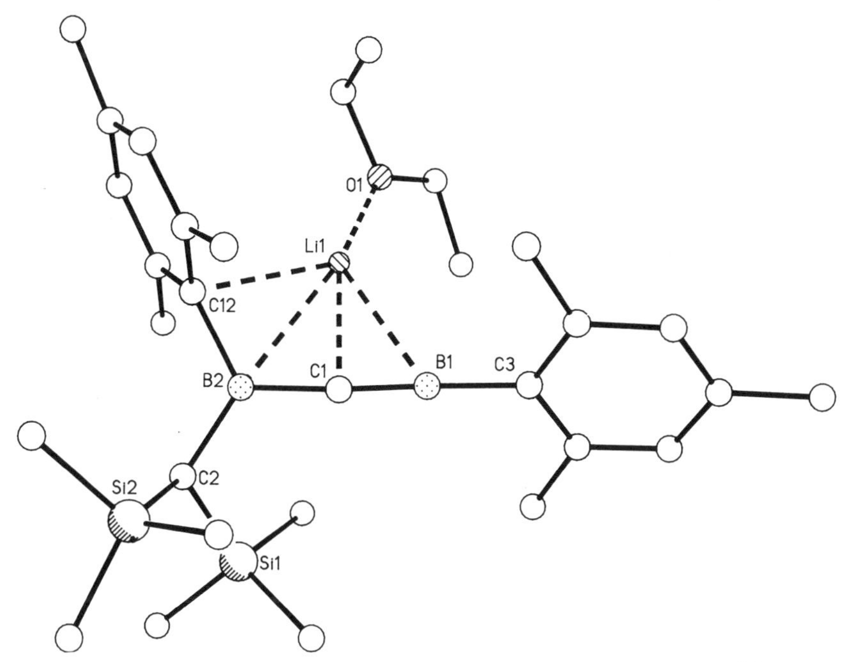

Figure 9.42 Structure of $Li(OEt_2)_2[C(BMes)B\{CH(SiMes)_2\}Mes]$, **282**.[232] H atoms are not shown for clarity.

ing more multiple character (1.34 Å) than the other (1.49 Å); however, the latter value is still shorter than a B—C (sp^2) single bond (1.59 Å).

The anion in $\{Li(Et_2O)\}_2C\{B(t\text{-Bu})Mes\}_2$,[233] **283**, is doubly negatively charged, but the B=C double bonds are rather long (1.45 Å) in comparison to that in **282**, presumably owing to electrostatic repulsion. The Li^+ ions are asymmetrically coordinated, and have Li—B distances of 2.39 and 2.72 Å. The distance between Li and the central C atom (2.04 Å) is very short. In addition to its Et_2O coordination, Li displays short distances to the ipso- and ortho C atoms of the B bonded Mes groups, with 2.34 and 2.67 Å, respectively. The BCB angle is not quite linear (168.4°).

The species $Li[(Me_3Si)_2CBC\{B(2,3,5,6\text{-}Me_4\text{-}C_6H)_2\}\{B(2,3,5,6\text{-}Me_4\text{-}C_6H)\text{-}t\text{-}Bu\}]$,[234] **284**, contains a nearly linear (175.7°) C—B—C unit. The C—B bond lengths (1.43 and 1.45 Å) are comparable to those in **283**, whereas the B—C bonds to the boryl groups are longer (1.53 and 1.56 Å), close to the values of B—C single bonds. Li^+ displays short contacts to the Me_3Si-substituted C atom (2.17 Å), to the π-system of one of the 2,3,5,6-tetramethylphenyl groups (2.33–2.40 Å) and to the B atom (2.56 Å).

Lithium has two different environments in $Li(C_7H_8)LiB(2,3,5,6\text{-}Me_4C_6H)\{C(SiMe_3)_2\text{-}2,3,5,6\text{-}Me_4\text{-}5\text{-}(SiMe_3)_2C{=}B(\text{cyclohexa-2,5-dienyl})$,[234] **285** (Fig. 9.43). One Li^+ ion coordinates to the central cyclohexa-1,4-diene

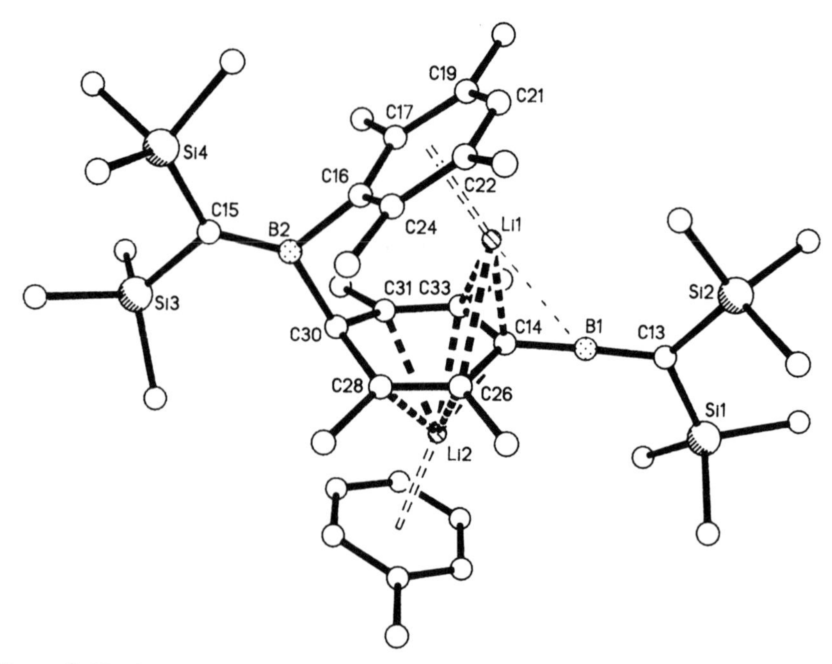

Figure 9.43 Structure of $Li(C_7H_8)LiB(2,3,5,6\text{-}Me_4C_6H)\{C(SiMe_3)_2(2,3,5,6\text{-}Me_4\text{-}5\text{-}SiMe_3)_2$ (C = B cyclohexa-2,5-dienyl), **285**.[231] H atoms omitted for clarity.

moiety, with Li—C contacts of 2.26–2.44 Å to the double-bonded C atoms. The coordination, in a η^6 fashion, to a toluene molecule is unusual with Li—C distances of 2.47–2.53 Å. The second Li^+ ion is situated between the cyclohexa-1,4-diene ring, with a short contact to the single-bonded C atom (2.11 Å) and the 2,3,5,6-tetramethylphenyl substituent as part of the boryl substituent on the other side of the central ring (Li—C, 2.32–2.49 Å). Essentially, there are no Li—B contacts present; the shortest Li—B distance being 2.77 Å. Structure **285** displays short B—C bonds (1.40, 1.43, and 1.39 Å), the latter two as part of the >C=B=C< double bond system.

A butadiene-like arrangement is present in $Li_2\{BMesC(SiMe_3)_2\}_2$,[234] **286**. The B—B distance in the center is 1.86 Å, and the CBBC torsion angle is 45.3°. Thus, no delocalization over the C=B—B=C framework is apparent in **286**. The B=C distances are relatively long for double bonds, displaying values of 1.48 and 1.51 Å. The two Li^+ ions coordinate in different ways to the dianion; one shows short contacts to the carbon–boron backbone with Li—C bonds of 2.13 and Li—B distances of 2.17 Å. The other Li^+ ions is situated between the boron-substituted Mes groups, with short Li—C contacts between 2.24 and 2.59 Å. A B—C double and single bond are present in $\{Li(Et_2O)\}_2(2,3,5,6\text{-}Me_4C_6H)B{=}CB(2,3,5,6\text{-}Me_4C_6H)\{C(SiMe_3)_2\}$,[235] **287** (Fig. 9.44) with B—C bond lengths of 1.32, 1.54, and 1.51 Å. The Li^+ ions coordinate to the central C atom and to different B atoms, with carbon and

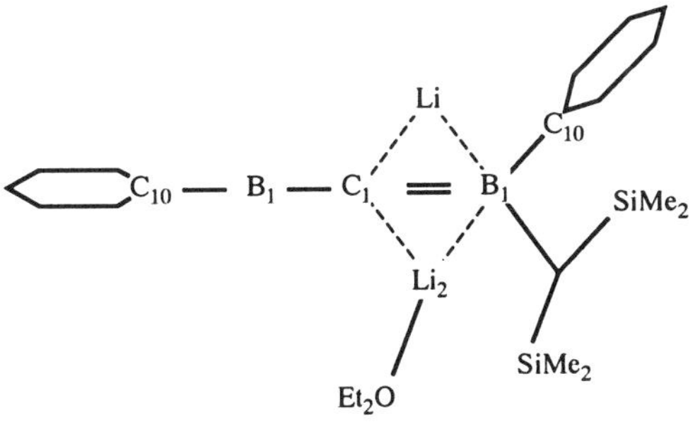

287

B(1)-C(2)	1.32 Å	B(1)-C(10)	1.56 Å	Li(2)-C(2)	2.15 Å	C(10)-B(11)-C(2)	176.3°
C(2)-B(3)	1.55 Å	B(1)-Li(1)	2.45 Å	Li(2)-B(3)	2.28 Å	B(1)-C(2)-B(3)	177.2°
B(3)-C(4)	1.51 Å	C(2)-Li(1)	2.10 Å	Li(2)-C(4)			
		Li(1)-C(30)	2.28 Å				

Figure 9.44 Compound $[\{Li(Et_2O)\}_2(2,3,5,6\text{-}Me_4C_6H)B{=}CB(2,3,5,6\text{-}Me_4C_6H)\text{-}\{C(SiMe_3)_2\}$, **287**.[235]

boron contacts of Li1-C2 (2.10 Å) and Li1-B1 (2.45 Å), Li2-C2 (2.15 Å) and Li2-B3 (2.28 Å). Li1 exhibits a short contact to the ipso-C atom of the 2,3,5,6-tetramethylphenyl group (2.28 Å), Li2 binds to C4 in the BCBC backbone (2.26 Å). In addition to this, each Li^+ ions coordinated by an Et_2O molecule.

Partial B—N multiple bonding is observed in Li(TMEDA){*c*-N(*t*-Bu)BMeN(*t*-Bu)BMe_2},[136] **288**. A TMEDA-coordinated Li^+ ion is bonded to one of the nitrogen atoms (2.00 Å) and, more loosely, to the boron atom bearing only one methyl group (2.51 Å). The B—N distances between this boron and both nitrogen atoms are relatively short (1.46 and 1.41 Å). Not surprisingly, the B—N bonds to the other boron atom, being substituted by two methyl groups and thus four-coordinate, are rather long (1.57 and 1.63 Å). In both cases, the shorter B—N bonds are those to the N atoms not coordinating the Li^+ ion. Thus the boron atom bearing only one methyl group participates in a delocalized multiple bonding system.

An example of a boron-centered radical anion is observed in [Li(12-crown-

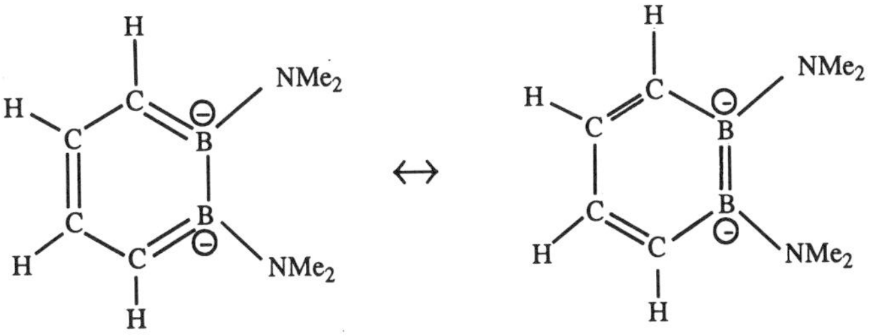

Figure 9.45 Some resonance formulas of **290**.[237]

Table 9.25 Lithium Compounds of Main Group 3 Elements

	Aggr.	CN_{Li}	Li—B (Å)	L	Li—L (Å)	Add. Coord.	Compound/ Ref.
$\{Li(Et_2O)\}_2Mes_2BBPhMes$	1	4[b]	2.35[a]	Et_2O	1.93[a]	C(2.32)	**280**/230
$\{Li(Et_2O)\}_2Me_2NPhBBPhNMe_2$	1	5[c]	2.31[a]	Et_2O	1.92	C(2.25, 2.45) N(1.98)	**281**/231
$Li(Et_2O)[C(BMes)B\{CH(SiMe_3)_2\}$-Mes]	1	4	2.49	Et_2O	1.87	C(2.09)	**282**/232
$\{Li(Et_2O)\}_2C\{B(t\text{-}Bu)Mes\}_2$	1	5[d]	2.39 2.72	Et_2O	1.96	C(2.04) C_{ipso}(2.34) C_{ortho}(2.72)	**284**/233
$Li[(Me_3Si)_2CBC\{B(2,3,5,6\text{-}Me_4\text{-}C_6H)_2\}$-$\{B(2,3,5,6\text{-}Me_4\text{-}C_6H)t\text{-}Bu\}$	1	8[e]	2.56			C(2.17) C(η^6)(2.33–2.40)	**284**/234
$Li(C_7H_8)LiB(2,3,5,6\text{-}Me_4\text{-}C_6H)$-$\{C(SiMe_3)_2(2,3,5,6\text{-}Me_4\text{-}5\text{-}(SiMe_3)_2C{=}B$-(cyclohexa-2,5-dienyl)	1	7[f]	2.77	–		C(2.11) C(η^6)(2.32–2.49)	**285**/234
		7[g]				C^6	
$Li_2\{BMesC(SiMe_3)_2\}_2$	1	6[h]	2.17			C(2.13) C_{SiMe3}(2.34)	**286**/234
		6[i]				C_{Mes}(2.24–2.59)	
$[\{Li(Et_2O)\}_2(2,3,5,6\text{-}Me_4\text{-}C_6H)B{=}C$ $B(2,3,5,6\text{-}Me_4\text{-}C_6H)\{C(SiMe_3)_2\}$							**287**/235
Li(TMEDA)-*c*-$\{BMeN(t\text{-}Bu)BMe_2N$-$(t\text{-}Bu)\}$	1	4	2.51	TMEDA	2.17*	N(2.00)	**288**/136

°

[Li(12-crown-4)$_2$][BMes$_3$]	1	8[j]		12-crown-4	2.37[a]		**289**/236
{Li(TMEDA)}$_2$[*c*-{B(NMe$_2$)}$_2$C$_4$H$_4$]	1	8[k]	2.47	TMEDA	2.16	C(2.36[a])	**290**/237
			2.55				
Li(TMEDA)BNMe$_2$C$_9$M$_7$	1	8[k]	2.59	TMEDA	2.08	C(2.42)	**291**/238
[Li(TMEDA){B(NEt$_2$)C$_4$H$_4$}Li]$_2$	2	7[l]	2.34	TMEDA	2.16[a]	C(2.22[a])	**292**/239
		7[m]	2.27				C(2.16[a])
			2.58				N(2.03)
Li(TMEDA){*c*-B(Me)N(SiMe$_3$)-(CH)$_2$C(SiMe$_3$)}	1	7[l]	2.38	TMEDA	2.14	C(2.31[a])	**293**/240
						N(2.26)	
[Li$_2${*c*-B(*t*-Bu)C(SiMe$_3$)B(*t*-Bu)C-(SiMe$_3$)}]$_2$	2	8[n]	2.35[a]			C(2.15[a])	**294**/241
		4[o]	2.55[a]			C(2.25[a])	
		2[p]				C(2.31[a])	

[a]Denotes average values.
[b]Coordination to both B atoms, the ipso C atom of one Mes group, and Et$_2$O.
[c]Coordination to both B atoms, one N atom, the ipso C atom of one Mes group, and Et$_2$O.
[d]Coordination to one B atom, the central C atom, the ipso and ortho C atoms of one Mes group, and Et$_2$O.
[e]Coordination to the central B atom, to the (SiMe$_3$)$_2$-substituted C atom bonded to this, and to the aryl substituent in an η^6 fashion.
[f]Coordination to the boron-bonded C atom of the hexadienyl system and to one C atom bonded to this; in addition η^6 coordination to an aryl group.
[g]η^3 toluene 2.47–2.53; η^4 cyclohexadienyl 2.26–2.44.
[h]η^4 coordination to the 2,3-dibora-butadienyl backbone and to two Me groups of separate SiMe$_3$ substituents.
[i]η^3 coordination to two separate Mes groups.
[j]Coordination to 12-crown-4.
[k]η^6 coordination to the six-membered ring and TMEDA coordination.
[l]η^5 coordination to the five-membered ring and TMEDA coordination.
[m]η^5 coordination to the ring and contacts to the B atom of the second monomer and the N atom of the amino group on it.
[n]Coordination to all atoms in both rings.
[o]Coordination to one B and one C atom in each ring.
[p]Two Li atoms display coordination to two separate sets of carbon atoms, each in a bridging manner between the two ring systems.

4)$_2$[BMes$_3$],[236] **289**. Coordination with 12-crown-4 affords ion pair separation. The B coordination is planar and the B—C bonds are in the range of 1.59–1.61 Å. The presence of an unpaired electron seems to have no considerable impact on the geometry of the anion in comparison to the neutral species (B—C, 1.58 Å). A dianionic dilithio-1,2-diborabenzene {Li(TMEDA)}$_2$[*c*-{B(NMe$_2$)}$_2$C$_4$H$_4$],[237] **290**, represented schematically in Figure 9.45, was the first lithium heteroaromatic system containing two adjacent boron atoms. Two Li$^+$ ions coordinate in an η^6 fashion from both sides of the ring, with Li—B distances of 2.47 and 2.55 Å. The Li atoms are additionally coordinated by TMEDA molecules. The B—B distances are normal (1.71 Å) for a B—B single bond.

The species Li(TMEDA)B(NMe$_2$)$_2$C$_9$H$_7$,[238] **291** can be considered an anionic naphthalene analogue. The Li$^+$ ion coordinates the boron-containing moiety of the ring system and is additionally bonded to a TMEDA molecule. The Li—B interaction is significantly longer (2.59 Å) than the Li—C distances (2.36–2.52 Å; 2.35 Å av.). The bond lengths in the ring indicate some delocalization involving boron and the exocyclic B—N-bond is lengthened (1.45 Å) in agreement with this view.

A considerable B—C bond shortening occurs in the case of a borole anion [Li(TMEDA){B(NEt$_2$)C$_4$H$_4$}Li]$_2$,[239] **292**, which is a boron-containing analogue of the Cp$^-$ ion. The structure of **292** reveals η^5-coordinated Li$^+$ ions, one of which carries TMEDA. This is more weakly coordinated to the ring system [Li(1)-B(1), 2.34 Å; Li(1)-C(av.), 2.22 Å] than the other Li$^+$ ion which has [Li(2)-B(1) and Li(2)-C(av.)] distances of 2.27 and 2.16 Å. The B—C distances are 1.52 Å, indicating some multiple bonding character and the B—N interaction (1.51 Å) is considerably weakened.

The structure of Li(TMEDA){*c*-B(Me)N(SiMe$_3$)(CH)$_2$C(SiMe$_3$)},[240] **293** is monomeric. Lithium has interactions with boron (2.38 Å), nitrogen (2.26 Å), and three C atoms in the five-membered ring (2.31 Å av.). The Li$^+$ ion is also coordinated by TMEDA (Li—N, 2.14 Å). The B—N (1.49 Å) and the B—C bonds (1.60 Å) are fairly long, whereas the C—C bond (1.34 Å) opposite the boron is essentially a double one, indicating that delocalization is not extensive.

A four-membered, dianionic ring system as in Li$_4$[*c*-{B(*t*-Bu)C-(SiMe$_3$)}$_2$]$_2$,[241] **294**, can also be envisioned as a potentially aromatic system. Two B—C—B—C rings are asymmetrically bridged by four Li$^+$ ions. The bonding in this compound suggests a delocalized system, but the ring is bent along the B—B transannular axis with an angle of 36.7° between the normals of the B—C—B planes.

Structural data for the above compounds are given in Table 9.25.

ACKNOWLEDGMENTS

We thank J. Arnold, Berkeley and D. Stalke, Göttingen for providing structural data prior to publication. FP would like to thaank the Deutscher Akademischer Austauschdienst for a NATO fellowship, in the course of which this chapter was prepared.

REFERENCES

1. Setzer, W. N.; Schleyer, P. v. R. *Adv. Organomet. Chem.* **1985**, *24*, 353.
2. (a) Gregory, K.; Schleyer, P. v. R.; Snaith, R. *Adv. Inorg. Chem.* **1991**, *37*, 47; (b) Mulvey, R. E. *Chem. Soc. Rev.* **1991,** *20*, 167.
3. Seebach, D. *Angew. Chem., Int. Ed. Engl.* **1988**, *27*, 1624; *Angew. Chem.* **1988**, *100*, 1685.
4. Boche, G. *Angew. Chem., Int. Ed. Engl.* **1989**, *28*, 277; *Angew. Chem.* **1989**, *101*, 286.
5. Hope, H. *Acta Crystallogr.* **1988**, *B44*, 22.
6. Allen, F. H.; Bellard, S.; Brice, M. D.; Cartwright, B. A.; Doubleday, A.; Higgs, H.; Hummelink, T.; Hummelink-Peters, B. G.; Kennard, O.; Motherwell, W. D. S. The Cambridge Crystallographic Center, Cambridge Structural Database; Version 4.5, July 1991.
7. This is the way in which most previous review have classified Li compounds. Tables 9.1–9.25 are also classified in this manner. These provide a brief summary of some structural details. Distances to lithium are given to the nearest hundredth of an Ångstrom.
8. van Beylen, M.; Roland, B.; King, G. S. D.; Aerts, J. *J. Chem. Res.* **1985**, *388*, 4201.
9. Becker, G.; Hitchchock, P. B.; Lappert, M. F.; MacKinnon, I. A. *J. Chem. Soc., Chem. Commun.* **1989**, 1312.
10. (a) Doughty, S. M.; Stoddart, J. F.; Colquhoun, H. M.; Slawin, A. M. Z.; Williams, D. J. *Polyhedron 4* **1985**, 567; (b) Hundal, M. S.; Sood, G.; Kapoor, P.; Poonia, N. S. *J. Cryst. Spectr. Res.* **1991**, *21*, 201.
11. Hvoslef, J.; Hope, H.; Murray, B. D.; Power, P. P. *J. Chem. Soc., Chem. Commun.* **1983**, 1438.
12. Cetinkaya, B.; Gümrükcü, I.; Lappert, M. F.; Atwood, J. L.; Shakir, R. *J. Am. Chem. Soc.* **1980**, *102*, 2086.
13. Huffman, J. C.; Geerts, R. L.; Caulton, K. G. *J. Crystallogr. Spectrosc.* **1984**, *14*, 541.
14. Murchie, M. P.; Bovenkamp, J. W.; Rodrigue, A.; Watson, K. A.; Fortier, S. *Can. J. Chem.* **1988**, *66*, 2515.
15. Watson, K. A.; Fortier, S.; Murchie, M. P.; Bovenkamp, J. W.; Rodrigue, A.; Buchanan, G. W.; Ratcliffe, C. J. *Can. J. Chem.* **1990**, *68*, 1201.
16. Hitchcock, P. B.; Buttrus, N. H.; Sullivan, A. C. *J. Organomet. Chem.* **1986**, *303*, 321.
17. Schmidt-Bäse, D.; Klingebiel, U. *Chem. Ber.* **1989**, *122*, 815.
18. Schmidt-Bäse, D.; Klingebiel, U. *Chem. Ber.* **1990**, *123*, 449.
19. Weese, K. J.; Bartlett, R. A.; Murray, B. D.; Olmstead, M. M.; Power, P. P. *Inorg. Chem.* **1987**, *26*, 2409.
20. Hursthouse, M. B.; Hossain, M. A.; Motevalli, M.; Sanganee, M. *J. Organomet. Chem.* **1990**, *381*, 293.

21. Engelhardt, L. M.; MacB. Harrowfield, J.; Lappert, M. F.; MacKinnon, I. A.; Newton, B. H.; Raston, C. L.; Skelton, B. W.; White, A. H. *J. Chem. Soc., Chem. Commun.* **1986**, 846.
22. Angelova, O.; Kirilov, E. M. G.; Kirilov, M.; Petrov, G.; Kaneti, J.; Macicek, J. *J. Chem. Soc., Perkin Trans.* **1989**, *2*, 1405.
23. (a) Korobov, M. S.; Minkin, V. I.; Nivorozhkin, L. E.; Kompan, O. E.; Struchkov, Y. T. *Zh. Obshch. Khim.* **1989**, *59*, 429; (b) van den Schaaf, P. A.; Hogenheide, M. P.; Grove, D.; Spek, A. L.; van Koten, G. *J. Chem. Soc. Chem. Commun.* **1992**, 1703.
24. Graalmann, O.; Klingebiel, U.; Clegg, W.; Haase, M.; Sheldrick, G. M. *Angew. Chem., Int. Ed. Engl.* **1984**, *23*, 891.
25. Dippel, K.; Keweloh, N. K.; Jones, P. G.; Klingebiel, U.; Schmidt, D. *Z. Naturforsch.* **1987**, *42b*, 1253.
26. Williard, P. G.; Sabrino, J. M. *Tetrahedron Lett.* **1985**, *26*, 3931.
27. (a) Bazhenova, T. A.; Lobhovskaya, R. M.; Shobaeva, R. P.; Shilov, A. E.; Shilova, A. K. *J. Organomet. Chem.* **1987**, *330*, 9; (b) Bazhenova, T. A.; Lobkovskaya, R. M.; Shobaeva, R. P.; Shilov, A. E.; Shilova, A. K. *Koord. Khim.* **1988**, *14*, 1542.
28. (a) Chisolm, M. H.; Drake, S. R.; Naini, A. A.; Sheib, W. *Polyhedron* **1991**, *10*, 805; (b) Jones, R. A.; Koschmeider, S. U.; Atwood, J. L.; Bott, S. G. *J. Chem. Soc. Chem. Commun.* **1992**, 726.
29. Bauer, W.; Laube, T.; Seebach, D. *Chem. Ber.* **1985**, *118*, 764.
30. Amstutz, R.; Dunitz, J. D.; Laube, T.; Schweizer, W. B.; Seebach, D. *Chem. Ber.* **1986**, *119*, 434.
31. Seebach, D.; Amstutz, R.; Laube, T.; Schweizer, W. B.; Dunitz, J. D. *J. Am. Chem. Soc.* **1985**, *107*, 5403.
32. Laube, T.; Dunitz, J. D.; Seebach, D. *Helv. Chim. Acta* **1985**, *68*, 1373.
33. Williard, P. G.; Tata, J. R.; Schlessinger, R. H.; Adams, A. D.; Iwanowicz, E. J. *J. Am. Chem. Soc.* **1988**, *110*, 7901.
34. Jastrzebski, J. T. B. H.; van Koten, G.; Christophersen, M. J. N.; Stam, C. H. *J. Organomet. Chem.* **1985**, *292*, 319.
35. Williard, P. G.; Carpenter, G. B. *J. Am. Chem. Soc.* **1985**, *107*, 3345.
36. Jastrzebski, J. T. B. H.; van Koten, G.; van de Mierop, W. F. *Inorg. Chim. Acta* **1988**, *142*, 169.
37. Teixidor, F.; Llobet, A.; Casabo, J.; Solans, X.; Font-Altaba, M.; Aguilo, M. *Inorg. Chem.* **1985**, *24*, 2315.
38. Williard, P. G.; Hintze, M. J. *J. Am. Chem. Soc.* **1987**, *109*, 5539.
39. Marsch, M.; Harms, K.; Lochmann, L.; Boche, G. *Angew. Chem., Int. Ed. Engl.* **1990**, *29*, 308.
40. Gais, H. J.; Dingerdissen, U.; Krüger, C.; Angermund, K. *J. Am. Chem. Soc.* **1987**, *109*, 3775.
41. (a) Power, M. B.; Bott, S. G.; Atwood, J. L.; Barron, A. R. *J. Am. Chem. Soc.* **1990**, *112*, 3446; (b) Evans, W. J.; Boyle, T. J.; Ziller, J. W. *Polyhedron* **1992**, *11*, 1093.
42. Stewart, J. L.; Andersen, R. A. *J. Chem. Soc., Chem. Commun.* **1987**, 1846.
43. Murray, B. D.; Power, P. P. *J. Am. Chem. Soc.* **1984**, *106*, 7011.

44. Bochmann, M.; Wilkinson, G.; Young, G. B.; Hursthouse, M. B.; Malik, K. M. A. *J. Chem. Soc., Dalton Trans.* **1980**, 1863.
45. Scott, M. J.; Wilisch, W. C. A.; Armstrong, W. H. *J. Am. Chem. Soc.* **1990**, *112*, 2429.
46. Wilisch, W. C. A.; Scott, M. J.; Armstrong, W. H. *Inorg. Chem.* **1988**, *27*, 4333.
47. Davies, J. J.; Gibson, J. F.; Skapski, A. C. *Polyhedron* **1982**, *1*, 641.
48. Hvoslef, J.; Hope, H.; Murray, B. D.; Powers, P. P. *J. Chem. Soc., Chem. Commun.* **1983**, 1438.
49. Holzmann, A.; Berges, P.; Hinrichs, W.; Klar, G. *J. Chem. Res.* **1987**, *42*, 328.
50. Eichorst, D. J.; Payne, D. A.; Wilson, S. R.; Howard, K. E. *Inorg. Chem.* **1990**, *29*, 1458.
51. Hursthouse, M. B.; Mazid, M. A.; Motevalli, M.; Sanganee, M.; Sullivan, A. C. *J. Organomet. Chem.* **1990**, *381*, C43.
52. Hossain, M. A.; Hursthouse, M. B.; Ibrahim, A.; Mazid, M.; Sullivan, A. C. *J. Chem. Soc., Dalton Trans.* **1989**, 2347.
53. Buttrus, N. H.; Eaborn, C.; El-Kheli, M. N. A.; Hitchcock, P. B.; Smith, J. D.; Sullivan, A. C.; Tavakkoli, K. *J. Chem. Soc., Dalton Trans.* **1988**, 381.
54. Sigel, G. A.; Power, P. P. *Inorg. Chem.* **1987**, *26*, 2819.
55. Ruhlandt-Senge, K.; Power, P. P. *Bull. Soc. Chim. France* **1992**, *129*, 594.
56. Aslam, M.; Bartlett, R. A.; Block, E.; Olmstead, M. M.; Power, P. P.; Sigel, G. E. *J. Chem. Soc., Chem. Commun.* **1985**, 1674.
57. Shoner, S.; Power, P. P., unpublished results, 1990.
58. Banister, A. J.; Clegg, W.; Gill, W. R. *J. Chem. Soc., Chem. Commun.* **1987**, 850.
59. Banister, A. J.; Barr, D.; Brooker, A. T.; Clegg, W.; Cunnington, M. J.; Doyle, M. J.; Drake, S. R.; Gill, W. R.; Manning, K.; Raithby, P. R.; Snaith, R.; Wade, K.; Wright, D. S. *J. Chem. Soc., Chem. Commun.* **1990**, 105.
60. Tatsumi, K.; Inoue, Y.; Nakamura, A.; Cramer, R. E.; VanDoorne, W.; Gilje, J. W. *Angew. Chem., Int. Ed. Engl.* **1990**, *29*, 422.
61. Ruhlandt-Senge, K.; Power, P. P., unpublished results.
62. Schumann, H.; Albrecht, I.; Hahn, E.; *Angew. Chem., Int. Ed. Engl.* **1985**, *24*, 985.
63. Schumann, H.; Albrecht, I.; Gallagher, M.; Hahn, E.; Muchmore, C.; Pickardt, J. *J. Organomet. Chem.* **1988**, *349*, 103.
64. Tatsumi, K.; Inoue, Y.; Nakamura, A. *J. Am. Chem. Soc.* **1989**, *111*, 782.
65. Ruhlandt-Senge, K.; Power, P. P. *J. Chem. Soc. Dalton. Trans.*, in press.
66. Ruhlandt-Senge, K.; Power, P. P. *Inorg. Chem.* **1991**, *30*, 3683.
67. duMont, W. W.; Kubiniok, S.; Lange, L.; Pohl, S.; Saak, W.; Wagner, I. *Chem. Ber.* **1991**, *124*, 1315.
68. Gindelberger, D. E., unpublished results.
69. (a) Bonasia, P. J.; Gindelberger, D. E.; Dabbousi, B. O.; Arnold, J. *J. Am. Chem. Soc.* **1992**, *114*, 5209; (b) Becker, G.; Klinkhammer, K. W.; Lartiges, S.; Böttcher, P.; Pohl, W. *Z. Anorg. Allg. Chem.* **1992**, *613*, 7.

70. Arnold, J. and co-workers, work submitted.

71. Lappert, M. F.; Power, P. P.; Sanger, A. R.; Srivastava, R. C. *Metal and Metalloid Amides*; Ellis-Horwood: Chichester, England, 1980.

72. Mootz, D.; Zinnius, A.; Böttcher, B. *Angew. Chem.* **1969**, *81*, 398.

73. Rogers, R. D.; Atwood, J. L.; Grüning, R. *J. Organomet. Chem.* **1978**, *157*, 229.

74. Bartlett, R. A.; Power, P. P. *J. Am. Chem. Soc.* **1987**, *109*, 6509.

75. Fjeldberg, T.; Hitchcock, P. B.; Lappert, M. F.; Thorne, A. J. *J. Chem. Soc., Chem. Commun.* **1984**, 822.

76. Barr, D.; Clegg, W.; Mulvey, R. E.; Snaith, R.; Wright, D. S. *J. Chem. Soc., Chem. Commun.* **1987**, 716.

77. Power, P. P.; Xiaojie, X. *J. Chem. Soc., Chem. Commun.* **1984**, 358.

78. Bartlett, R. A.; Dias, H. V. R.; Hope, H.; Murray, B. D.; Olmstead, M. M.; Power, P. P. *J. Am. Chem. Soc.* **1986**, *108*, 6921.

79. Chen, H.; Bartlett, R. A.; Dias, H. V. R.; Olmstead, M. M.; Power, P. P. *J. Am. Chem. Soc.* **1989**, *111*, 4338.

80. Lappert, M. F.; Slade, M. J.; Singh, A.; Atwood, J. L.; Rogers, R. D.; Shakir, R. *J. Am. Chem. Soc.* **1983**, *105*, 302.

81. Engelhardt, L. M.; May, A. S.; Raston, C. L.; White, A. H. *J. Chem. Soc., Dalton Trans.* **1983**, 1671.

82. Engelhardt, L. M.; Jolly, B. S.; Punk, P. C.; Raston, C. L.; Skelton, B. W.; White, A. H. *Austr. J. Chem.* **1986**, *39*, 1337.

83. (a) Pieper, U.; Walter, S.; Klingebiel, U.; Stalke, D. *Angew. Chem., Int. Ed. Engl.* **1990**, *29*, 209; (b) Lambert, C.; Hampel, F.; Schleyer, P. V. R. *Angew. Chem. Int. Ed. Engl.* **1992**, *31*, 1209.

84. Klingebiel, U.; Stalke, D.; Vollbrecht, S. *Z. Naturforsch.* **1992**, *47b*, 27.

85. Stalke, D.; Klingebiel, U.; Sheldrick, G. M. *Chem. Ber.* **1988**, *121*, 1457.

86. Boese, R.; Klingebiel, U. *J. Organomet. Chem.* **1986**, *315*, C27.

87. Barr, D.; Raithby, P. R.; Schleyer, P. v. R.; Snaith, R.; Wright, D. S. *J. Chem. Soc., Chem. Commun.* **1990**, 643.

88. Werner, E.; Klingebiel, U.; Pauer, F.; Stalke, D.; Riedel, R.; Schaible, S. *Z. Anorg. Allg. Chem.* **1991**, *596*, 35.

89. Recknagel, A.; Steiner, A.; Noltemeyer, M.; Brooker, S.; Stalke, D.; Edelmann, F. T. *J. Organomet. Chem.* **1991**, *414*, 327.

90. Engelhardt, L. M.; Jacobson, G. E.; Junk, P. C.; Raston, C. L.; Skelton, B. W. *J. Chem. Soc., Dalton Trans.* **1988**, 1011.

91. (a) Enders, D.; Bachstätter, G.; Kremer, K. A. M.; Marsch, M.; Harms, K.; Boche, G. *Angew. Chem., Int. Ed. Engl.* **1988**, *27*, 1522; (b) Bernstein, M. P.; Romesberg, F. E.; Fuller, D. J.; Harrison, A. T.; Collum, D. P.; Liu, Q-Y.; Williard, P. G. *J. Am. Chem. Soc.* **1992**, *114*, 5100.

92. Sugimoto, H.; Mori, M.; Masudo, H.; Taga, T. *J. Chem. Soc., Chem. Commun.* **1986**, 962.

93. Arnold, J. *J. Chem. Soc., Chem. Commun.* **1990**, 976.

94. (a) Undiner, G. E.; Tan, R. P.; Powell, D. R.; West, R. *J. Am. Chem. Soc.* **1991**, *113*, 8437; (b) Detsch, R.; Niecke, E.; Nieger, E.; Schoeller, W. W.; *Chem. Ber.* **1992**, *125*, 1119.

95. Barr, D.; Clegg, W.; Mulvey, R. E.; Snaith, R. *J. Chem. Soc., Chem. Commun.* **1984**, 469.
96. Cetinkaya, B.; Hitchcock, P. B.; Lappert, M. F.; Misra, M. C.; Thorne, A. J. *J. Chem. Soc., Chem. Commun.* **1984**, 148.
97. Bartlett, R. A.; Power, P. P. unpublished results.
98. Brooker, S.; Kennepohl, D.; Roesky, H. W.; Sheldrick, G. M. *Chem. Ber.* **1991**, *124*, 2223.
99. Setzer, W. N.; Schleyer, P. v. R.; Mahdi, W.; Dietrich, H. *Tetrahedron* **1988**, *44*, 3339.
100. (a) Barr, D.; Clegg, W.; Mulvey, R. E.; Snaith, R. *J. Chem. Soc., Chem. Commun.* **1984**, 285. (b) Armstrong, D. R.; Mulvey, R. E.; Walker, G. T.; Barr, D.; Snaith, R.; Clegg, W.; Reed, D. *J. Chem. Soc., Dalton Trans.* **1988**, 617.
101. Dietrich, H.; Mahdi, W.; Knorr, R. *J. Am. Chem. Soc.* **1986**, *108*, 2462.
102. Ashby, M. T.; Li, Z. *Inorg. Chem.* **1992**, *31*, 1321.
103. (a) Stalke, D.; Keweloh, N.; Klingebiel, U.; Noltemeyer, M.; Sheldrick, G. M. *Z. Naturforsch.* **1987**, *42b*, 1237. (b) Kottke, T.; Klingebiel, U.; Noltemeyer, M.; Pieper, U.; Walter, S.; Stalke, D. *Chem. Ber.* **1991**, *124*, 1941.
104. Dippel, K.; Klingebiel, U.; Sheldrick, G. M.; Stalke, D. *Chem. Ber.* **1987**, *120*, 611.
105. Walter, S.; Klingebiel, U.; Noltemeyer, M. *Chem. Ber.* **1992**, *125*, 783.
106. Wanat, R. A.; Collum, D. B.; van Duyne, G.; Clardy, J.; De Pue, R. T. *J. Am. Chem. Soc.* **1986**, *108*, 3415.
107. Haase, M.; Sheldrick, G. M. *Acta Crystallogr.* **1986**, *C42*, 1009.
108. Egert, E.; Kliebisch, U.; Klingebiel, U.; Schmidt, D. *Z. Anorg. Allg. Chem.* **1987**, *548*, 89.
109. (a) Gregory, K.; Bremer, M.; Bauer, W.; Schleyer, P. v. R.; Lorenzen, N. P.; Kopf, J.; Weiss, E. *Organometallics* **1990**, *9*, 1485. (b) Rhine, W. E.; Stucky, G. D. *J. Am. Chem. Soc.* **1972**, *94*, 7346.
110. Hacker, R.; Kaufmann, E.; Schleyer, P. v. R.; Mahdi, W.; Dietrich, H. *Chem. Ber.* **1987**, *120*, 1533.
111. Streitwieser, Jr., A. *Acc. Chem. Res.* **1984**, *17*, 353.
112. Barr, D.; Berrisford, D. J.; Jones, R. V. H.; Slawin, A. M. Z.; Snaith, R.; Stoddart, J. F.; Williams, D. J. *Angew. Chem., Int. Ed. Engl.* **1989**, *28*, 1044.
113. Barr, D.; Clegg, W.; Mulvey, R. E.; Snaith, R. *J. Chem. Soc., Chem. Commun.* **1984**, 700.
114. Knösel, F.; Edelmann, F. T.; Pauer, F.; Stalke, D.; Bauer, W. *J. Organomet. Chem.*, submitted.
115. Pauer, F.; Stalke, D. *J. Organomet. Chem.* **1991**, *431*, C1.
116. (a) Pauer, F.; Rocha, J.; Stalke, D. *J. Chem. Soc., Chem. Commun.* **1991**, 1477. (b) Wedler, M.; Edelmann, F. T.; Stalke, D. *J. Organomet. Chem.* **1992**, *431*, C1.
117. Barr, D.; Berrisford, D. J.; Mendez, L.; Slawin, A. M. Z.; Snaith, R.; Stoddart, J.; Williams, D. J.; Wright, D. S. *Angew. Chem., Int. Ed. Engl.* **1991**, *30*, 82.
118. Veith, M.; Bönlein, J.; Huch, V. *Chem. Ber.* **1989**, *122*, 841.

119. Stalke, D.; Klingebiel, U.; Sheldrick, G. M. *J. Organomet. Chem.* **1988**, *344*, 37.

120. Seebach, D.; Bauer, W.; Laube, T.; Schweizer, W. B.; Dunitz, J. D. *J. Chem. Soc., Chem. Commun.* **1984**, 853.

121. Rannenberg, M.; Hausen, H. D.; Weidlein, J. *J. Organomet. Chem.* **1989**, *376*, C27.

122. (a) Armstrong, D. R.; Barr, D.; Clegg, W.; Hodgson, S. M.; Mulvey, R. E.; Reed, D.; Snaith, R.; Wright, D. S. *J. Am. Chem. Soc.* **1989**, *111*, 4719; (b) Boche, G.; Langlotz, I.; Mansch, M.; Harms, K.; Nudelman, N. S. *Angew. Chem. Int. Ed. Engl.* **1992**, *31*, 1205.

123. Barr, D.; Clegg, W.; Hodgson, S. M.; Lamming, G. R.; Mulvey, R. E.; Scott, A. J.; Snaith, R.; Wright, D. S. *Angew. Chem., Int. Ed. Engl.* **1989**, *28*, 1241.

124. Armstrong, D. R.; Barr, D.; Clegg, W.; Mulvey, R. E.; Reed, D.; Snaith, R.; Wade, K. *J. Chem. Soc., Chem. Commun.* **1986**, 869.

125. Dippel, K.; Klingebiel, U.; Noltemeyer, M.; Pauer, F.; Sheldrick, G. M. *Angew. Chem., Int. Ed. Engl.* **1988**, *27*, 1074.

126. Dippel, K.; Klingebiel, U.; Kottke, T.; Pauer, F.; Sheldrick, G. M.; Stalke, D. *Chem. Ber.* **1990**, *123*, 237.

127. Brauer, D. J.; Bürger, H.; Lienwald, G. R. *J. Organomet. Chem.* **1986**, *308*, 119.

128. Chen, H.; Bartlett, R. A.; Dias, H. V. R.; Olmstead, M. M.; Power, P. P. *Inorg. Chem.* **1991**, *30*, 2487.

129. Tecklenburg, B.; Klingebiel, U.; Schmidt-Bäse, D. *J. Organomet. Chem.* **1992**, *426*, 287.

130. Veith, M.; Goffing, F.; Huch, V. *Chem. Ber.* **1988**, *121*, 943.

131. Armstrong, D. R.; Barr, D.; Brooker, A. T.; Clegg, W.; Gregory, K.; Hodgson, S. M.; Snaith, R.; Wright, D. S. *Angew. Chem., Int. Ed. Engl.* **1990**, *29*, 410.

132. Brauer, D. J.; Bürger, H.; Liewald, G. R.; Wilke, J. *J. Organomet. Chem.* **1985**, *287*, 305.

133. Bartlett, R. A.; Chen, H.; Dias, H. V. R.; Olmstead, M. M.; Power, P. P. *J. Am. Chem. Soc.* **1988**, 110.

134. Bartlett, R. A.; Feng, X.; Olmstead, M. M.; Power, P. P.; Weese, K. J. *J. Am. Chem. Soc.* **1988**, *109*, 4851.

135. Chen, H.; Bartlett, R. A.; Olmstead, M. M.; Power, P. P.; Shoner, S. C. *J. Am. Chem. Soc.* **1990**, *112*, 1048.

136. Paetzold, P.; Pelzer, C.; Boese, R. *Chem. Ber.* **1988**, *121*, 51.

137. (a) Shearer, H. M. M.; Wade, K.; Whitehead, G. *J. Chem. Soc., Chem. Commun.* **1979**, 943; (b) Sato, D.; Kawasaki, H.; Shimada, I.; Arata, Y.; Okamura, K.; Date, J.; Koga, K. *J. Am. Chem. Soc.* **1992**, *114*, 761; (c) Anders, E.; Optiz, A.; Boese, R. *Chem. Ber.* **1992**, *125*, 1267.

138. Clegg, W.; Snaith, R.; Shearer, H. M. M.; Wade, K.; Whitehead, G. *J. Chem. Soc., Chem. Commun.* **1983**, 1309.

139. Barr, D.; Clegg, W.; Mulvey, R. E.; Snaith, R.; Wade, K. *J. Chem. Soc., Chem. Commun.* **1986**, 295.

140. Armstrong, D. R.; Barr, D.; Snaith, R.; Clegg, W.; Mulvey, R. E.; Wade, K.; Reed, D. *J. Chem. Soc., Dalton Trans.* **1987**, 1071.

141. Barr, D.; Clegg, W.; Mulvey, R. E.; Snaith, R. *J. Chem. Soc., Chem. Commun.* **1984**, 79.
142. Barr, D.; Snaith, R.; Clegg, W.; Mulvey, R. E.; Wade, K. *J. Chem. Soc., Dalton Trans.* **1987**, 2141.
143. Weiss, E.; Lucken, E. A. C. *J. Organomet. Chem.* **1964**, *2*, 197.
144. Weiss, E.; Hencken, G. *J. Organomet. Chem.* **1970**, *21*, 265.
145. Amstutz, R.; Schweizer, W. B.; Seebach, D.; Dunitz, J. D. *Helv. Chem. Acta* **1981**, *64*, 2617.
146. Barr, D.; Clegg, W.; Mulvey, R. E.; Reed, D.; Snaith, R. *Angew. Chem., Int. Ed. Engl.* **1985**, *24*, 328.
147. Boche, G.; Marsch, M.; Harms, K. *Angew. Chem., Int. Ed. Engl.* **1986**, *25*, 373.
148. Barr, D.; Clegg, W.; Mulvey, R. E.; Snaith, R. *J. Chem. Soc., Chem. Commun.* **1984**, 226.
149. Clegg, W.; Mulvey, R. E.; Snaith, R.; Toogood, G. E.; Wade, K. *J. Chem. Soc., Chem. Commun.* **1986**, 1740.
150. Barr, D.; Clegg, W.; Mulvey, R. E.; Snaith, R. *J. Chem. Soc., Chem. Commun.* **1989**, 57.
151. Jackman, L. M.; Scarmoutzos, L. M.; Smith, B. D.; Williard, P. G. *J. Am. Chem. Soc.* **1988**, *110*, 6058.
152. Colgen, D.; Papasergio, R. I.; Raston, C. L.; White, A. H. *J. Chem. Soc., Chem. Commun.* **1984**, 1708.
153. Papasergio, R. I.; Skelton, B. W.; Twiss, P.; White, A. H.; Raston, C. L. *J. Chem. Soc., Dalton Trans.* **1990**, 1161.
154. Hacker, R.; Schleyer, P. v. R.; Reber, G.; Müller, G.; Brandsma, L. *J. Organomet. Chem.* **1986**, *316*, C4.
155. Armstrong, D. R.; Mulvey, R. E.; Barr, D.; Porter, R. W.; Raithby, P. R.; Simpson, T. R. E.; Snaith, R.; Wright, D. S.; Gregory, K.; Mikulcik, P. *J. Chem. Soc., Dalton Trans.* **1991**, 765.
156. Armstrong, D. R.; Barr, D.; Raithby, P. R.; Snaith, R.; Wright, D. S. *Inorg. Chim. Acta* **1991**, *185*, 163.
157. Armstrong, D. R.; Mulvey, R. E.; Barr, D.; Snaith, R.; Wright, D. S.; Clegg, W.; Hodgson, S. M. *J. Organomet. Chem.* **1989**, *362*, C1.
158. Zarges, W.; Marsch, M.; Harms, K.; Boche, G. *Angew. Chem., Int. Ed. Engl.* **1989**, *28*, 1392.
159. Huang, R. H.; Ward, D. L.; Dye, J. L. *Acta Crystallogr.* **1990**, *C46*, 1835.
160. Böck, S.; Nöth, H.; Rahm, P. *Z. Naturforsch.* **1988**, *43b*, 53.
161. Linti, G.; Nöth, H.; Rahm, P. *Z. Naturforsch.* **1988**, *43b*, 1101.
162. Petrie, M.; Ruhlandt-Senge, K.; Power, P. P. unpublished results.
163. Rhine, W. E.; Stucky, G.; Peterson, S. W. *J. Am. Chem. Soc.* **1975**, *97*, 6401.
164. Brauer, D. J.; Bürger, H.; Lienwald, G. R. *Organomet. Chem.* **1986**, *307*, 177.
165. de With, J.; de Horton, A.; Orpen, A. G. *Organometallics* **1990**, *9*, 2207.
166. Smith, D. P.; Allen, K. D.; Carducci, M. D.; Wigley, D. E. *Inorg. Chem.* **1992**, *31*, 1319.
167. Murray, B. D.; Power, P. P. *Inorg. Chem.* **1984**, *23*, 4584.

168. Belforte, A.; Calderazzo, F.; Englert, U.; Strähle, J. *J. Chem. Soc., Chem. Commun.* **1989**, 801.

169. Chisholm, M. H.; Hammond, C. H.; Huffman, J. C. *Polyhedron* **1988**, *7*, 399.

170. Chisholm, M. H.; Hammond, C. H.; Huffman, J. C. *Polyhedron* **1988**, *7*, 2515.

171. Danopoulos, A. A.; Wilkinson, G.; Hussain, B.; Hursthouse, M. B. *J. Chem. Soc., Chem. Commun.* **1989**, 896.

172. Manners, I.; Coggio, W. D.; Mang, M. N.; Parvez, M.; Allcock, H. R. *J. Am. Chem. Soc.* **1989**, *111*, 3481.

173. Jones, R. A.; Koschmieder, S. U.; Nunn, C. M. *Inorg. Chem.* **1987**, *26*, 3610.

174. Becker, G.; Hartmann, H. M.; Schwarz, W. *Z. Anorg. Allg. Chem.* **1989**, *577*, 9.

175. Bartlett, R. A.; Olmstead, M. M.; Power, P. P. *Inorg. Chem.* **1986**, *25*, 1243.

176. Hope, H.; Olmstead, M. M.; Power, P. P.; Xu, X. *J. Am. Chem. Soc.* **1984**, *106*, 819.

177. Mulvey, R. E.; Wade, K.; Armstrong, D. R.; Walker, G. T.; Snaith, R.; Clegg, W.; Reed, D. *Polyhedron* **1987**, *6*, 987.

178. Bartlett, R. A.; Olmstead, M. M.; Power, P. P.; Sigel, G. A. *Inorg. Chem.* **1987**, *26*, 1941.

179. Hey, E.; Weller, F. *J. Chem. Soc., Chem. Commun.* **1988**, 782.

180. Jones, R. A.; Stuart, A. L.; Wright, T. C. *J. Am. Chem. Soc.* **1983**, *105*, 7459.

181. Hitchcock, P. B.; Lappert, M. F.; Power, P. P.; Smith, S. J. *J. Chem. Soc., Chem. Commun.* **1984**, 1669.

182. Schmidpeter, A.; Burget, G.; Sheldrick, W. S. *Chem. Ber.* **1985**, *118*, 3849.

183. Hey, E.; Hitchcock, P. B.; Lappert, M. F.; Rai, A. K. *J. Organomet. Chem.* **1987**, *325*, 1.

184. Hey-Hawkins, E.; Sattler, E. *J. Chem. Soc., Chem. Commun.* **1992**, 775.

185. Hey, E.; Raston, C. L.; Skelton, B. W.; White, A. H. *J. Organomet. Chem.* **1989**, *362*, 1.

186. Boese, R.; Bläser, D.; Klingebiel, U.; Andrianarison, M. *Z. Naturforsch.* **1989**, *44b*, 265.

187. Stalke, D.; Meyer, M.; Andrianarison, M.; Klingebiel, U.; Sheldrick, G. M. *J. Organomet. Chem.* **1989**, *366*, C15.

188. Andrianarison, M.; Stalke, D.; Klingebiel, U. *J. Organomet. Chem.* **1990**, *381*, C38.

189. Klingebiel, U.; Meyer, M.; Pieper, U.; Stalke, D. *J. Organomet. Chem.* **1991**, *408*, 19.

190. Bartlett, R. A.; Feng, X.; Power, P. P. *J. Am. Chem. Soc.* **1986**, *108*, 6817.

191. Power, P. P. *Angew. Chem., Int. Ed. Engl.* **1990**, *29*, 449.

192. (a) Bartlett, R. A.; Dias, H. V. R.; Feng, X.; Power, P. P. *J. Am. Chem. Soc.* **1989**, *111*, 1306; (b) Dow, D.; Wood, G. L.; Duesler, E. N.; Paine, R. T.; Nöth, H. *Inorg. Chem.* **1992**, *31*, 1695.

193. Pestana, D. C.; Power, P. P. *J. Am. Chem. Soc.* **1991**, *113*, 8426.

194. Petrie, M.; Power, P. P. unpublished results.

195. Brooks, P.; Craig, D. C.; Gallagher, M. J.; Rae, A. D.; Sarroff, A. *J. Organomet. Chem.* **1987**, *323*, C1.

196. Anderson, D. M.; Hitchcock, P. B.; Lappert, M. F.; Moss, I. *Inorg. Chim. Acta* **1988**, *141*, 157.
197. Anderson, D. M.; Hitchcock, P. B.; Lappert, M. F.; Leung, W. P.; Zora, J. A. *J. Organomet. Chem.* **1987**, *333*, C13.
198. Niecke, E.; Klein, E.; Nieger, M. *Angew. Chem., Int. Ed. Engl.* **1989**, *28*, 751.
199. Hönle, W.; von Schnering, H. G.; Schmidtpeter, A.; Burget, G. *Angew. Chem., Int. Ed. Engl.* **1984**, *23*, 817.
200. Fritz, G.; Schneider, M. W.; Hönle, W.; von Schnering, H. G. *Z. Naturforsch.* **1988**, *43b*, 561.
201. Douglas, T.; Theopold, K. H. *Angew. Chem., Int. Ed. Engl.* **1989**, *28*, 1367.
202. (a) Lattman, M.; Olmstead, M. M.; Power, P. P.; Rankin, D. W. H.; Robertson, H. E. *Inorg. Chem.* **1988**, *27*, 3012. (b) Kunz, S.; Hey-Hawkins, E. *Organometallics* **1992**, *11*, 2729.
203. Kasch, H. H.; Müller, G. *J. Chem. Soc., Chem. Commun.* **1984**, 569; Karsch, H. H.; Weber, L.; Wewers, D.; Boese, R.; Müller, G. *Z. Naturforsch.* **1984**, *39b*, 1518.
204. Karsch, H. H.; Deybelly, B.; Müller, G. *J. Organomet. Chem.* **1988**, *352*, 47.
205. Brauer, D. J.; Hietkamp, S.; Stelzer, O. *J. Organomet. Chem.* **1986**, *299*, 137.
206. Arif, A. M.; Cowley, A. H.; Jones, R. A.; Power, J. M. *J. Chem. Soc., Chem. Commun.* **1986**, 1446.
207. Schumann, H.; Palamidis, E.; Schmid, G.; Boese, R. *Angew. Chem., Int. Ed. Engl.* **1986**, *25*, 718.
208. Martin, S. F.; Fishpaugh, J. R.; Power, J. M.; Giolando, D. M.; Jones, R. A.; Nunn, C. M.; Cowley, A. H. *J. Am. Chem. Soc.* **1988**, *110*, 7226.
209. Cowley, A. H.; Giolando, D. M.; Jones, R. A.; Nunn, C. M.; Power, J. M. *J. Chem. Soc., Chem. Commun.* **1988**, 206.
210. Baker, R. T.; Krusic, P. J.; Tulip, T. H.; Calabrese, J. C.; Wreford, S. S. *J. Am. Chem. Soc.* **1983**, *105*, 6763.
211. (a) Becker, G.; Witthauer, C. *Z. Anorg. Allg. Chem.* **1982**, *492*, 28. (b) Mundt, O.; Becker, G.; Rossler, M.; Witthauer, C. *Z. Anorg. Allg. Chem.* **1983**, *506*, 42.
212. (a) Arif, A. M.; Jones, R. A.; Kidd, K. B. *J. Chem. Soc., Chem. Commun.* **1986**, 1440; (b) Driess, M.; Pritzkow, H. *Angew. Chem. Int. Ed. Engl.* **1992**, *31*, 316.
213. Schumann, H.; Palamidis, E.; Loebel, J.; Pickardt, J. *Organometallics* **1988**, *7*, 1008.
214. Petrie, M. A.; Shoner, S. C.; Dias, H. V. R.; Power, P. P. *Angew. Chem., Int. Ed. Engl.* **1990**, *29*, 1033.
215. Schaaf, T. R.; Butler, W.; Glick, M. D.; Oliver, J. P. *J. Am. Chem. Soc.* **1974**, *96*, 7593.
216. Ilsley, W. H.; Schaaf, T. F.; Glick, M. D.; Oliver, J. P. *J. Am. Chem. Soc.* **1980**, *102*, 3769.
217. Olmstead, M. M.; Power, P. P. unpublished results.
218. Brooks, J. J.; Stucky, G. D. *J. Am. Chem. Soc.* **1972**, *94*, 7333.
219. Becker, G.; Hartmann, H. M.; Hengge, E.; Schrank, F. *Z. Anorg. Allg. Chem.* **1989**, *572*, 63.

220. Dias, H. V. R.; Power, P. P. and Heine, A.; Stalke, D. unpublished results.
221. Solouki, B.; Rosmus, P.; Bock, H.; Maier, G. *Angew. Chem., Int. Ed. Engl.* **1980**, *19*, 51.
222. Maier, G.; Mihm, G.; Reisenauer, H. P. *Angew. Chem., Int. Ed. Engl.* **1980**, *19*, 52.
223. Jutzi, P.; Meyer, M.; Reisenauer, H. P.; Maier, G. *Chem. Ber.* **1989**, *122*, 1227.
224. Jutzi, P.; Meyer, M.; Dias, H. V. R.; Power, P. P. *J. Am. Chem. Soc.* **1990**, *112*, 4841.
225. Reed, D.; Stalke, D.; Wright, D. S. *Angew. Chem., Int. Ed. Engl.* **1991**, *30*, 1459.
226. Smith, G. D.; Fanwick, P. E.; Rothwell, I. P. *Inorg. Chem.* **1989**, *28*, 618.
227. Veith, M.; Rösler, R. *Z. Naturforsch.* **1986**, *41b*, 1071.
228. Veith, M.; Ruloff, C.; Huch, V.; Töllner, F. *Angew. Chem., Int. Ed. Engl.* **1988**, *27*, 1381.
229. (a) Armstrong, D. R.; Davidson, M. G.; Moncrieff, D.; Stalke, D.; Wright, D. S. *J. Chem. Soc. Chem. Commun.* **1992**, 1413. (b) Bartlett, R. A.; Power, P. P. unpublished results.
230. Moezzi, A.; Olmstead, M. M.; Power, P. P. *J. Am. Chem. Soc.* **1992**, *114*, 2715.
231. Moezzi, A.; Power, P. P. *Angew Chem.* **1992**, *31*, 1082.
232. Hunold, R.; Allwohn, J.; Baum, G.; Massa, W.; Berndt, A. *Angew. Chem., Int. Ed. Engl.* **1988**, *27*, 961.
233. Pilz, M.; Allwohn, J.; Hunold, R.; Massa, W.; Berndt, A. *Angew. Chem., Int. Ed. Engl.* **1988**, *27*, 1370.
234. Pilz, M.; Allwohn, J.; Willershausen, P.; Massa, W.; Berndt, A. *Angew. Chem., Int. Ed. Engl.* **1990**, *29*, 1030.
235. Allwohn, J.; Pilz, M.; Hunold, R.; Massa, W.; Berndt, A. *Angew. Chem., Int. Ed. Engl.* **1990**, *29*, 1032.
236. Olmstead, M. M.; Power, P. P. *J. Am. Chem. Soc.* **1986**, *108*, 4235.
237. Herberich, G. E.; Hessner, B.; Hostalek, M. *Angew. Chem., Int. Ed. Engl.* **1986**, *25*, 642.
238. Paetzold, P.; Finke, N.; Wennek, P.; Schmid, G.; Boese, R. *Z. Naturforsch.* **1986**, *41b*, 167.
239. Herberich, G. E.; Hostalek, M.; Laven, R.; Boese, R. *Angew. Chem., Int. Ed. Engl.* **1990**, *29*, 317.
240. Schmid, G.; Zaika, D.; Lehr, J.; Augart, N.; Boese, R. *Chem. Ber.* **1988**, *121*, 1873.
241. Schmidt, G.; Baum, G.; Massa, W.; Berndt, A. *Angew. Chem., Int. Ed. Engl.* **1986**, *25*, 1111.

10

SYNTHETIC IONOPHORES FOR LITHIUM IONS

RICHARD A. BARTSCH AND VISVANATHAN RAMESH
Department of Chemistry and Biochemistry
Texas Tech University
Lubbock, Texas

RICARDO O. BACH
Lake Wylie, South Carolina

TOSHIYUKE SHONO
Department of Applied Chemistry
Faculty of Engineering
Osaka University
Osaka, Japan

KEIICHI KIMURA
Chemical Process Engineering
Faculty of Engineering
Osaka University
Osaka, Japan

Lithium Chemistry: A Theoretical and Experimental Overview, Edited by Anne-Marie Sapse and Paul von Ragué Schleyer.
ISBN 0-471-54930-4

1 INTRODUCTION

Expanding applications in science, medicine, and technology[1] have led to increased interest in lithium and its compounds. Lithium salts have been used in treatment of manic depression and other neurological and psychiatric disorders.[1-4] Antiviral activity against DNA-type viruses is exhibited by Li^+.[5] Energy generation through primary and secondary batteries, as well as in fuel cells, depends in many cases on the presence of lithium.[1] Also, organolithium reagents are anionic polymerization catalysts and play a very important role in modern organic synthesis.[1,6]

In light of such applications, methodology for the separation and determination of Li^+ is receiving considerable attention. For many of these efforts, a common theme is the use of a synthetic multidentate ligand to complex Li^+. Through the efforts of several research groups, insight into the design and preparation of synthetic ionophores with Li^+ selectivity has been obtained. Ionophores have been developed which exhibit selectivity for Li^+ in separations of alkali metal cations and other metal ion species by solvent extraction and liquid membrane transport processes. Other multidentate ligands have been prepared which show marked Li^+ selectivity in metal ion determination by ion-selective electrode and spectrophotometric methods.

It is the purpose of this chapter to provide the reader with a summary of the available information regarding synthetic lithium ion-selective ionophores and their applications. The coverage is illustrative rather than exhaustive and surveys the literature through the end of 1992 with some mention of publications which appeared in the first half of 1993.

Following this introductory section, attention will be focused initially upon ionophores for complexation of Li^+ in homogeneous media. Subsequently solid state structures of ionophore-complexed Li^+ will be examined. Information gained in these two sections will then be applied in the following section to the determination of Li^+ by ion-selective electrodes and by spectrophotometry with chromo- and fluoroionophores. In the next section, the separation of Li^+ from other metal ion species in nonhomogeneous systems by solvent extraction, liquid membrane transport, and sorption processes will be examined. A portion of this section will summarize results reported for the separation of lithium isotopes by ionophore-bases processes.

Synthetic multidentate ligands may be separated into several classes, as shown in Figure 10.1. According to current nomenclature,[7a] acyclic multidentate ligands are termed podands. Common examples are the glymes and polyethylene glycols. Monocyclic multidentate ligands are coronands, of which crown ethers are familiar examples. Crown ethers that have a side arm with one or more coordination site attached to the macrocyclic ring are podando-coronands, commonly called lariat ethers. Cryptands are multicyclic multidentate ligands and spherands are macrocyclic ligands with a set of preorganized, inward-facing methoxy groups.

Several reviews and monographs which contain important information about the structures and applications of lithium compounds and complexes are available. These include:

1. Christensen, J. J.; Eatough, D. J.; Izatt, R. N., "The Synthesis and Ion Binding of Synthetic Multidentate Macrocyclic Compounds;" *Chem. Rev.* **1974**, *74*, 351.[8] This review of thermodynamic data for interactions of metal

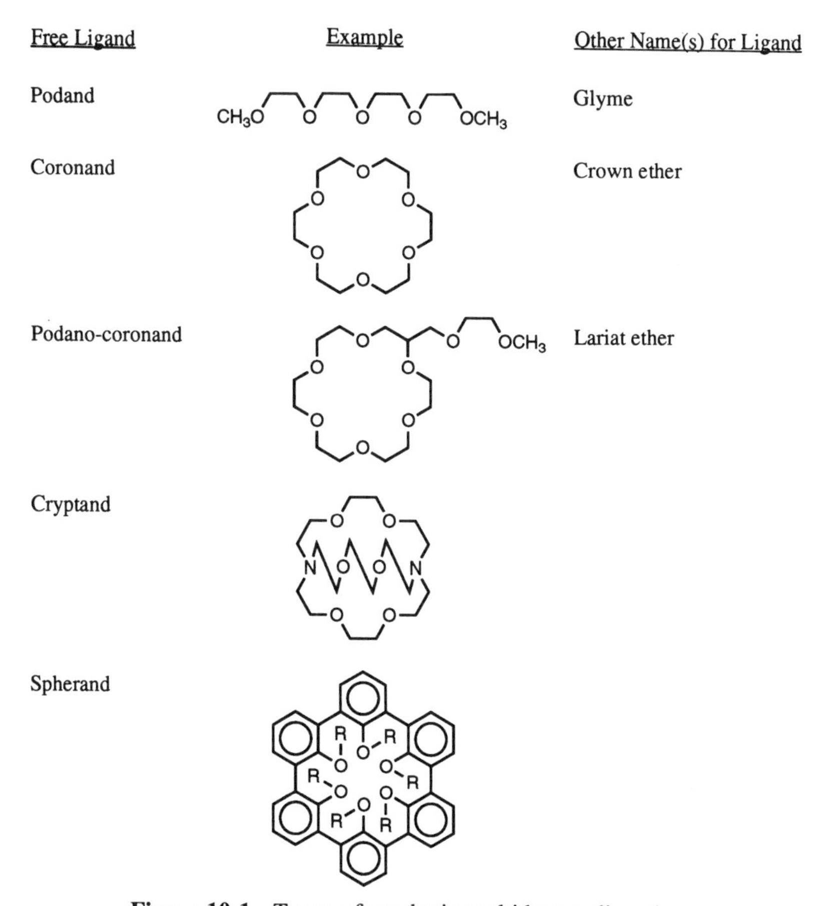

Figure 10.1 Types of synthetic multidentate ligands.

ions with synthetic multidentate compounds covers the literature through 1972 and contains occasional listings for Li^+ complexation.

2. Hubberstey, P., "Compounds of the Alkali Metals Containing Organic Molecules or Complex Ions;" *Coord. Chem. Rev.* **1985**, *66*, 1.[9] The review deals with complexes of acyclic lipophilic ionophores, crown ethers, salts of carboxylic acids, heterobimetallic complexes containing alkali metals, and organolithium compounds.

3. Izatt, R. M.; Bradshaw, J. S.; Nielson, S. A.; Lamb, J. D.; Christensen, J. J., "Thermodynamic and Kinetic Data for Cation Macrocyclic Interaction;" *Chem. Rev.* **1985**, *85*, 271.[10] This review includes compilations of the reported thermodynamic and kinetic data regarding interactions of metal ions with natural ionophores and synthetic multidentate ligands. This large review supplements the data contained in the 1974 review.

4. Bach, R. O., Ed. *Lithium—Current Applications in Science, Medicine, and Technology*; Wiley: New York, 1985.[1] This monograph contains an excellent coverage of industrial and medical applications of lithium and its compounds.

5. Hubberstey, P., "Compounds of the Alkali Metals Containing Organic Molecules or Complex Ions;" *Coord. Chem. Rev.* **1988**, *85*, 1.[11] Crown ether, cryptate, and spherand Li^+ complexes, salts of carboxylic acids, and organolithium compounds are reviewed with a major emphasis on organolithium compounds.

6. Bajaj, A. V.; Poonia, N. S., "Comprehensive Coordination Chemistry of Alkali and Alkaline Earth Coordination with Macrocyclic Multidentates: Latest Position;" *Coord. Chem. Rev.* **1988**, *87*, 55.[12] This large review deals with the interactions of crown ether and crown ether-related macrocycles with alkali and alkaline earth metal cations in the solid state and in solution.

7. Olsher, U.; Izatt, R. M.; Bradshaw, J. S.; Dalley, N. K., "Coordination Chemistry of Lithium Ion: A Crystal and Molecular Structure Review;" *Chem. Rev.* **1991**, *91*, 137.[13] Solid-state structures of Li^+ complexed by acyclic multidentate ligands, crown ethers, lariat ethers, cryptands, and spherands, as well as other ligands, are surveyed in this comprehensive review.

8. Izatt, R. M.; Pawlak, K.; Bradshaw, J. S.; Bruening, R. L., "Thermodynamic and Kinetic Data for Macrocyclic Interaction with Cations and Anions;" *Chem. Rev.* **1991**, *91*, 1721.[14] This large review includes thermodynamic and kinetic data for the interactions of metal ions with natural ionophores and synthetic multidentate ligands. Literature coverage is for articles which were not included in the 1974 or 1985 reviews.

2 LITHIUM ION COMPLEXATION BY IONOPHORES IN HOMOGENEOUS SOLUTION

The complexation reactions of an ionophore (L) and an inorganic salt are the equilibrium reactions shown in Eqs. 1 and 2 for the 1:1 complex and 2:1

complex (two ionophores and one metal ion in a ''sandwich'' complex), respectively. In the generalized equation (Eq. 3), S is a solvent molecule and k_c and k_d are the rate constants for complexation and decomplexation, respectively.

$$L + M^+X^- \underset{}{\overset{K_1}{\rightleftharpoons}} (LM)^+ + X^- \tag{1}$$

$$L + (LM)^+ \overset{K_2}{\rightleftharpoons} (L_2M)^+ \tag{2}$$

$$(L)_{solv} + (M^{n+},mS) \underset{k_d}{\overset{k_c}{\rightleftharpoons}} (L,M^{n+})_{solv} + mS \tag{3}$$

The stability constant (also known as the complexation constant or complex formation constant) indicates the degree of complexation and the stability of the resulting complex in the solution.

The stability constant, K_{th}, in the thermodynamic sense is that given in Eq. 4. However the stability constants which have usually been

$$K_{th} = \frac{a_C[L,M^{n+}]}{a_L[L]a_M[M^{n+}]} \tag{4}$$

reported are those obtained with Eq. 5, using the concentrations of the

$$K = K_{th}\frac{a_La_M}{a_C} = \frac{[L,M^{n+}]}{[L][M^{n+}]} \tag{5}$$

components, since the activity coefficients of the complex (a_C), crown ether (a_L), and cation (a_M) are usually unknown values.

Methods for measuring the stability constants of metal ion–ionophore complexes include calorimetry, potentiometry with ion-selective electrodes, spectrophotometry, conductivity, nuclear magnetic resonance spectroscopy, cyclic voltammetry, and polarography.[7b] By comparing stability constants for different combinations of metal ion, ionophore, solvent, temperature, and so on, the influence of these variables upon the complexation phenomena may be assessed.

2.1 Podands

Since Li^+ is defined as a hard acid, it will prefer hard donor atoms in Li^+ complexes.[13] Therefore, the stability of Li^+ complexes is expected to decrease as the ligand donor atoms are changed as follows: 0 > N >> S.

A comparison of the stability constants for association of alkali metal perchlorates with tetraethylene glycol dimethyl ether (tetraglyme, **1**) and its cyclic analogue 15-crown-5 **(2)** in methanol and with tetraglyme in acetonitrile is presented in Table 10.1.[8,15,16] Stability constants for the alkali metal complexes with tetraglyme are uniformly low and that for Li^+ is the lowest. For

Table 10.1 Stability Constants of Alkali Metal Complexes with Tetraglyme (1) and 15-Crown-5 (2) in Methanol and Acetonitrile at 25°C[8,15,16]

		log *K*				
Solvent	Ionophore	Li^+	Na^+	K^+	Rb^+	Cs^+
Methanol	1	0.89	1.11	1.68	1.52	1.45
	2	1.23	3.30	3.35		2.62
Acetonitrile	1	2.17	2.40	2.02	1.92	1.53

the change from tetraglyme to 15-crown-5 as the ionophore, the association constants increase for all of the alkali metal cations. However, the complexation of Li^+ is still very weak. For complexation by tetraglyme in acetonitrile, the stability constants decrease in the order $Na^+ > Li^+ > K^+ > Rb^+ > Cs^+$. The data presented in Table 10.1 indicate that selectivity for Li^+ complexation will not be achieved with podands based only upon ethyleneoxy units.

Incorporation of phosphoryl groups at the ends of the polyether chain is reported to enhance Li^+ complexation.[17,18] Olsher, Elgavish, and Jagur-Grodzinski[19] reported strong complexation of Li^+ by the lipophilic diether diamide **3** in acetonitrile and nitromethane. Also, Hiratani found that the addition of lithium perchlorate to methanol or acetonitrile solutions of the acyclic polyether quinolyl compounds **4** and **5** produced a marked enhancement in fluorescence intensity.[20,21] The enhancement was much less when sodium perchlorate was added and potassium perchlorate produced no increase in fluorescence intensity. These observations suggest that selective Li^+ complexation may be obtained with suitably designed podands. Similar, specific fluorescence increases were reported by Hiratani and co-workers[22] when lithium perchlorate was added to solutions of 2,9-dibutyl-1,10-phenanthroline **(6)** in acetonitrile.

Beer and co-workers[23] have observed that the addition of lithium tetrafluoroborate to an acetonitrile solution of the ferrocene bis-tertiary amide **7** produced considerable shifts (up to 2 ppm) of the carbonyl carbon signal in the carbon-13 NMR spectrum. Furthermore, addition of the lithium salt to a solution of **7** in acetonitrile gave a shift of the ferrocene oxidation wave to a more positive potential and the appearance of a new redox couple associated with the Li^+ complex. Echegoyen, Gokel, and co-workers[24] have noted that the redox behavior of podands **8** in acetonitrile responds differently to the presence of lithium, sodium, and potassium perchlorates.

Electronic and fluorescence spectra of the alkali metal salts of podand **9**, which has a quinolyl group at one end of the polyether chain and a benzoic acid unit at the other, have been determined by Hiratani.[25] In the electronic spectra, the absorption maximum at 303 nm for **9** is shifted to 280 nm for the alkali metal salts. In dichloromethane, the absorption intensity for the alkali metal salts decreases in the order $Li^+ > Na^+ > K^+ > Rb^+ > Cs^+$. The variation in absorption intensity for the five alkali metal cations was markedly

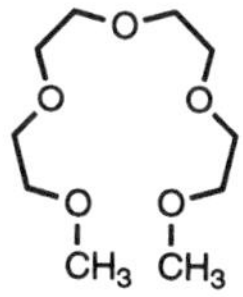

1 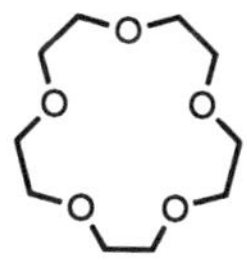 **2** 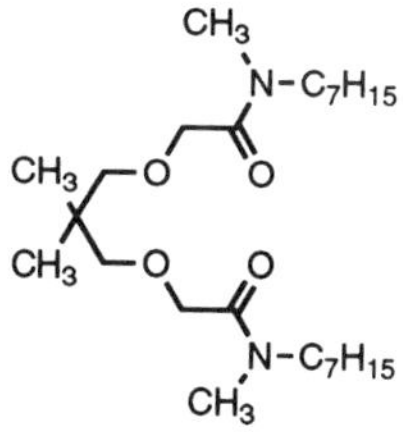 **3**

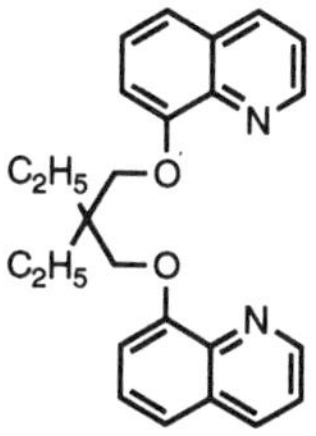

4 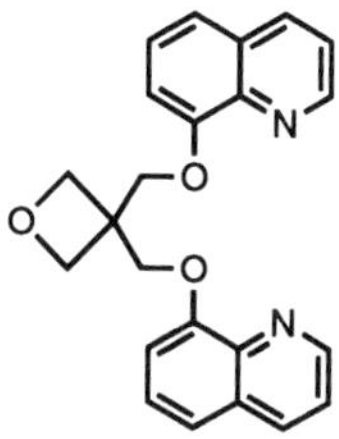 **5** 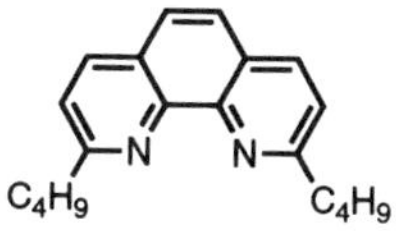 **6**

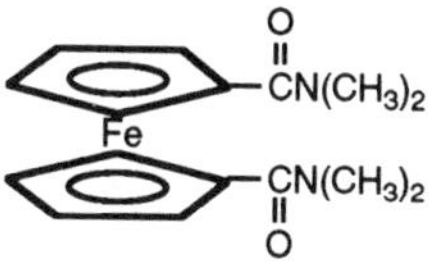

7 **8** **9**

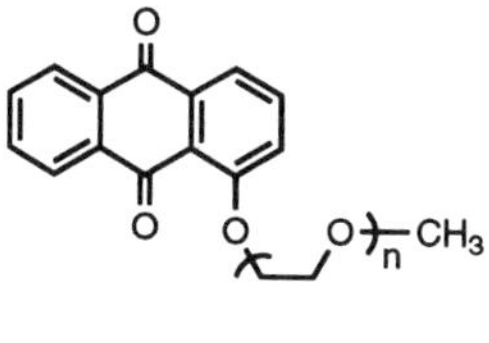

(n = 0 - 4)

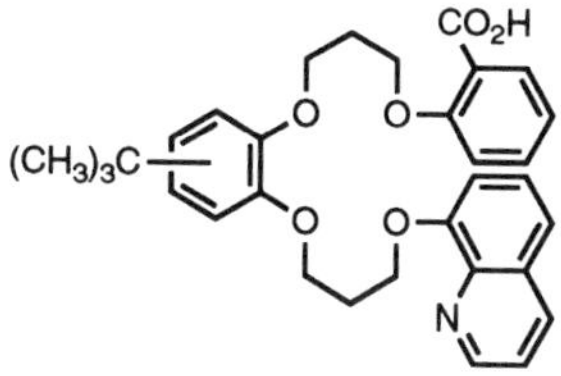

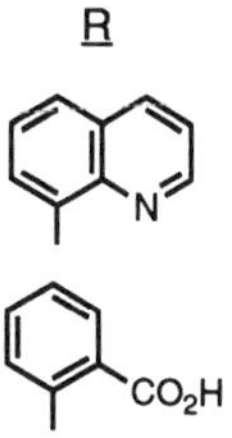

10 **11** **12**

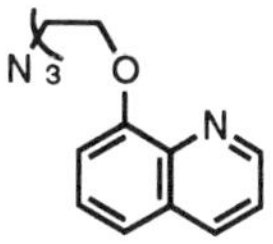

attenuated in ethanol and acetonitrile. These results suggest that in the non-polar solvent Li^+ coordinates with the ethereal oxygens and the quinolyl nitrogen, which induces a helical structure and causes a stacking interaction of the terminal aromatic groups of the podand. In agreement, the strongest fluorescence quenching is observed for the lithium salt in dichloromethane.

Table 10.2 Stability Constants of Alkali Metal Complexes with Podands 10–12 in Methanol at 25°C[26]

	log *K*				
Podand	Li^+	Na^+	K^+	Rb^+	Cs^+
10	2.37	3.22	3.51	3.06	2.6
11	3.45	3.41	3.24	3.19	<1.7
12	2.18	3.69	2.58	2.05	1.57

Stability constants for the complexation of alkali metal cations by podands **10–12** in methanol are presented in Table 10.2.[26] Although Li^+ selectivity was noted for bis-quinolyl ionophores **4** and **5** (vide supra), the bis-quinolyl and tris-quinolyl polyethers **10** and **12**, respectively, do not exhibit selective Li^+ complexation. Similarly podand **11**, which has two terminal benzoic acid groups is not selective for Li^+.

The results presented above demonstrate that selectivity for Li^+ complexation in homogeneous solution can be achieved only with specially designed podands.

2.2 Coronands

In 1967 the seminal publication of C. J. Pedersen on the synthesis of macrocyclic polyethers (crown ethers) and the metal ion complexing properties of these cyclic compounds appeared.[27] In this landmark paper, Pedersen reported that the ultraviolet absorption at 275 nm for a solution of dibenzo-14-crown-4 **(13)** in methanol was altered somewhat when lithium bromide was added. Furthermore, he noted that the solubility of this crown ether in aqueous methanol was greater in the presence of lithium hydroxide than in the presence of sodium or potassium hydroxide. When an equimolar solution of **13** and lithium thiocyanate in methanol was evaporated, a solid was formed which had a different melting point than that of the crown ether itself and gave correct combustion analysis for a 1 : 1 complex of the lithium salt and the crown ether. These results suggested that small-ring crown ethers might be good complexing agents for Li^+.

In 1974, Liotta and co-workers[28] reported that the first viable preparation of 12-crown-4 **(14)**. Conductance studies on acetonitrile solutions containing 12-crown-4, tetramethyl-12-crown-4 (**15**, formed by oligomerization of propylene oxide induced by boron trifluoride[29]), 15-crown-5,[28] and 18-crown-6[30,31] and lithium iodide, sodium tetraphenylborate, and potassium tetraphenylborate were performed by Hopkins and Norman.[32] The stability constant for complexation of Li^+ by 12-crown-4 was about 20% higher than that for Na^+ (Table 10.3). On the other hand, for 15-crown-5 the stability constant for Na^+ complexation was much higher than that for Li^+. The stability constant for complexation of Li^+ by tetramethyl-12-crown-4 was slightly higher than with 12-

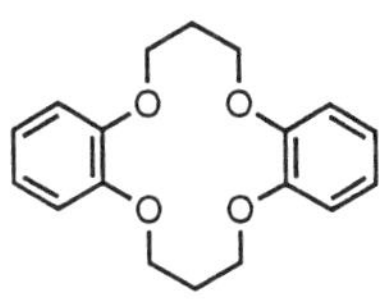

13 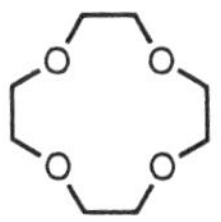 **14** 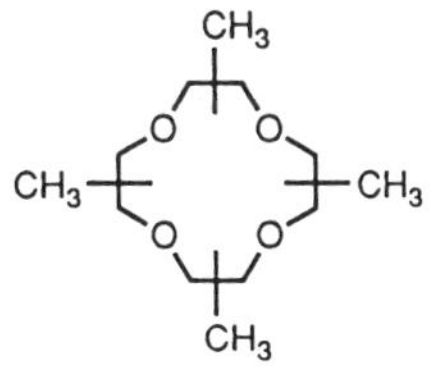 **15**

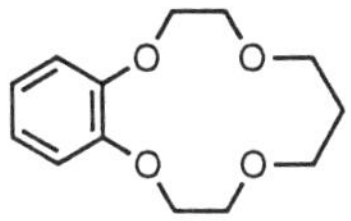

16 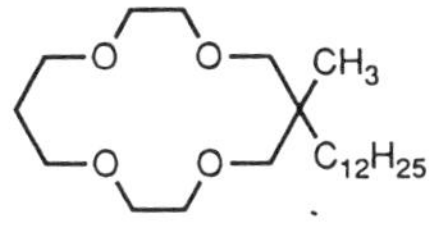 **17** **18–20**

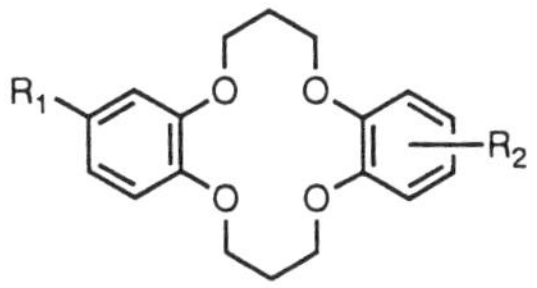

	R_1	R_2
18	CH_3	H
19	*t*-Bu	H
20	*t*-Bu	*t*-Bu

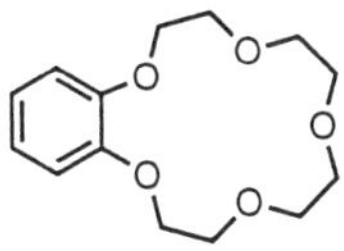

21 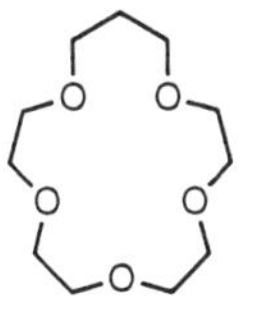 **22**

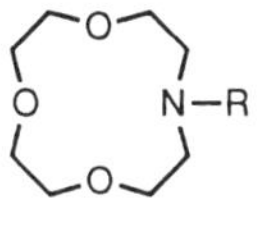

	R
23	CH_3
24	$CH_2CH_2OCH_3$

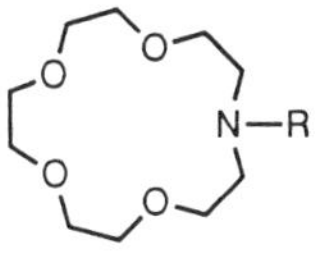

	R
25	CH_3
26	$CH_2CH_2OCH_3$
27	CH_2CH_2OH

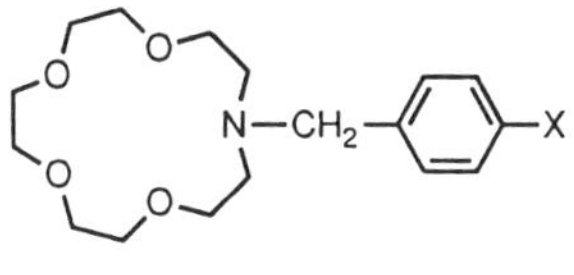

	X
28	OCH_3
29	H
30	Cl
31	CN
32	NO_2

crown-4. Infrared studies on acetonitrile solutions of lithium perchlorate showed an intense band at about 400 cm^{-1}, attributed to the vibration of Li^+ against its solvent cage, being weakened by the addition of increasing amounts of 12-crown-4.

Stability constants for complexation of Li^+ and, in some cases, Na^+ for comparison by a variety of coronands **2**, **14–42**, are recorded in Table 10.3.

Table 10.3 Stability Constants for Formation of 1:1 Complexes of Lithium and Sodium Ions with Coronands[a]

Ionophore	Cation	log *K*	Method	*t*(°C)	Solvent	Ref.
		Crown ethers				
12-crown-4 (**14**)	Li	3.40	cond	25	Acetonitrile	32
	Na	3.32	cond	25	Acetonitrile	32
	Li	2.93	pot	25	PC	33
	Li	4.25	NMR	27	Acetonitrile	34
	Li	~0	NMR	27	Water	34
	Li	−0.1	NMR	30	DOH	35
	Li	1.62	NMR	27	Acetone	34
	Li	~0	NMR	27	Methanol	34
	Li	0.70	NMR	27	Pyridine	34
	Li	~0	NMR	27	DMSO	34
	Li	1.87	NMR	rt	Melt[b]	36
Tetramethyl-12-crown-4 (**15**)	Li	3.46	cond	25	Acetonitrile	32
Benzo-13-crown-4 (**16**)	Li	2.4	NMR	25	Acetonitrile	37
	Li	~5	NMR	25	Dichloromethane	37
	Li	~3	NMR	25	Nitromethane	37
3-Dodecyl-3-methyl-14-crown-4 (**17**)	Li	6.0	cond	25	Acetonitrile	38
	Na	3.7	cond	25	Acetonitrile	38
Dibenzo-14-crown-4[c] (**13**)	Li	~4.5	NMR	26	Acetonitrile	39
	Li	3.60	NMR	26	PC	39
	Li	3.15	NMR	26	Acetone	39
	Li	1.85	NMR	26	THF	39
	Li	1.97	NMR	26	Pyridine	39
4-Methyldibenzo-14-crown-4[c] (**18**)	Li	~5	NMR	26	Acetonitrile	39
	Li	4.40	NMR	26	PC	39
	Li	4.06	NMR	26	Acetone	39
	Li	2.28	NMR	26	THF	39
	Li	2.16	NMR	26	Pyridine	39
4-(*tert*-Butyl)-dibenzo-14-crown-4[c] (**19**)	Li	~5	NMR	26	Acetonitrile	39
	Li	4.25	NMR	26	PC	39
	Li	3.97	NMR	26	Acetone	39
	Li	2.18	NMR	26	THF	39
	Li	2.04	NMR	26	Pyridine	39
4,4′(5′)-di-(*tert*-Butylbenzo)-14-crown-4[c] (**20**)	Li	4.70	NMR	26	Acetonitrile	39
	Li	4.25	NMR	26	PC	39
	Li	3.72	NMR	26	Acetone	39
	Li	2.00	NMR	26	THF	39
	Li	2.02	NMR	26	Pyridine	39
15-Crown-5 (**2**)	Li	3.60	cond	25	Acetonitrile	32
	Na	5.28	cond	25	Acetonitrile	32
	Li	5.34	cond	25	Acetonitrile	40
	Na	5.38	cond	25	Acetonitrile	40
	Li	4.26	cond	25	PC	41
	Na	3.7	cond	25	PC	41
	Li	2.39	NMR	rt	Melt[b]	36

Table 10.3 (*Continued*)

Ionophore	Cation	log *K*	Method	*t*(°C)	Solvent	Ref.
		Crown ethers (Cont.)				
Benzo-15-crown-5 (**21**)	Li	4.46	cond	25	Acetonitrile	42
	Na	4.25	cond	25	Acetonitrile	42
	Li	3.77	cond	25	PC	41
	Na	4.35	cond	25	PC	41
	Li	2.31	cond	25	Methanol	41
	Na	2.99	cond	25	Methanol	41
	Li	2.33	NMR	rt	Melt[b]	36
16-Crown-5 (**22**)	Li	4.48	cond	25	Acetonitrile	40
	Na	5.39	cond	25	Acetonitrile	40
	Li	3.25	cond	25	PC	40
	Na	5.7	cond	25	PC	40
		Azacrown ethers				
N-Methyl monoaza-12-crown-4 (**23**)	Li	<2.0	pot	25	Aq. methanol[d]	43
	Na	2.10	pot	25	Aq. methanol[d]	43
N-(2-Methoxyethyl) monoaza-12-crown-4 (**24**)	Li	<2.0	pot	25	Aq. methanol[d]	43
	Na	3.10	pot	25	Aq. methanol[d]	43
N-Methyl monoaza-15-crown-5 (**25**)	Li	<2.0	pot	25	Aq. methanol[d]	43
	Na	3.41	pot	25	Aq. methanol[d]	43
N-(2-Ethoxyethyl) monoaza-15-crown-5 (**26**)	Li	1.96	pot	25	Aq. methanol[d]	43
	Na	3.81	pot	25	Aq. methanol[d]	43
N-(2-Hydroxyethyl)-monoaza-15-crown-5 (**27**)	Li	2.34	pot	25	Aq. methanol[d]	43
	Na	3.95	pot	25	Aq. methanol[d]	43
N-(*p*-Methoxy-benzyl) monoaza-15-crown-5 (**28**)	Li	4.82	pot	25	Acetonitrile	44
	Na	5.74	pot	25	Acetonitrile	44
N-Benzyl monoaza-15-crown-5 (**29**)	Li	4.63	pot	25	Acetonitrile	44
	Na	5.68	pot	25	Acetonitrile	44
N-(*p*-Chlorobenzyl) monoaza-15-crown-5 (**30**)	Li	4.46	pot	25	Acetonitrile	44
	Na	4.70	pot	25	Acetonitrile	44
N-(*p*-Cyanobenzyl) monoaza-15-crown-5 (**31**)	Li	4.31	pot	25	Acetonitrile	44
	Na	4.52	pot	25	Acetonitrile	44
N-(*p*-Nitrobenzyl) monoaza-15-crown-5 (**32**)	Li	3.70	pot	25	Acetonitrile	44
	Na	3.97	pot	25	Acetonitrile	44
1,10-Diaza-18-crown-6 (**33**)	Li	4.39	NMR	rt	Acetonitrile	45
	Li	3.67	NMR	rt	PC	45
	Li	>5	NMR	rt	Nitromethane	45
	Na	3.37	NMR	rt	Nitromethane	45
	Li	2.13	NMR	rt	Acetone	45

Table 10.3 (***Continued***)

Ionophore	Cation	log *K*	Method	*t*(°C)	Solvent	Ref.
		Azacrown ethers (Cont.)				
	Na	1.96	NMR	rt	Acetone	45
	Li	~0	NMR	rt	DMSO	45
	Na	1.19	NMR	rt	DMSO	45
	Li	0.43	NMR	rt	Pyridine	45
	Na	4.12	NMR	rt	Pyridine	45
N,N′-Di(2-hydroxyethyl)-1,7-diaza-12-crown-4 (**34**)	Li	2.4	pot	25	Aq. methanol[d]	46
	Na	3.6	pot	25	Aq. methanol[d]	46
N,N′-Di(*N,N*-dimethyl oxyacetamido)-1,7-diaza-12-crown-4 (**35**)	Li	5.38	calor	25	Methanol	47
	Na	4.77	calor	25	Methanol	47
N,N′-Di(*N,N*-dimethyl oxypropanamido)-1,7-diaza-12-crown-4 (**36**)	Li	2.99	calor	25	Methanol	47
	Na	3.01	calor	25	Methanol	47
N,N′,N″-Tris-(diphenylphosphinyl-methyl)-triaza-9-crown-3 (**37**)	Li	5.6	cond	25	THF-Ch[e]	48
	Na	4.5	cond	25	THF-Ch[e]	48
Tetraaza-12-crown-4 (**38**)	Li	~0	pot	25	0.5 M aq. KNO_3	49
Tetraaza-14-crown-4 (**39**)	Li	~6	polar	25	Acetonitrile	50
trans-Tetraaza macrocycle **40**	Li	3.9	spec	25	Methanol	51
	Na	2.1	spec	25	Methanol	51
cis-Tetraaza macrocycle **41**	Li	6.3	spec	25	Methanol	51
	Na	2.6	spec	25	Methanol	51
		Other ligands				
Dibenzomethyl-phosphonyl-20-crown-7 (**42**)	Li	3.36	spec	23	THF	52
	Na	>4	spec	23	THF	52

[a]Abbreviations: cal = titration calorimetry; cond = conductivity; pot = potentiometry; spec = spectroscopy; rt = room temperature; PC = propylene carbonate; DMSO = dimethyl sulfoxide; THF = tetrahydrofuran.

[b]Melt = room temperature melt of aluminum chloride-*N*-butylpyridinium chloride (0.8 : 1).

[c]log $K \simeq 0$ in DMSO and DMF.

[d]Methanol–water, 95/5 (v/v).

[e]Tetrahydrofuran–chloroform, 4/1 (v/v).

Ionophores which exhibit approximately equal stability constants for complexation of Li^+ and Na^+ in at least one solvent include: 12-crown-4 **(14)**, 15-crown-5 **(2)**, benzo-15-crown-5 **(21)**, 1,10-diaza-18-crown-6 **(33)**, and the diaza-12-crown-4 compound **36** which has two amide-containing side arms. Stronger complexation of Li^+ than Na^+ is found for the 14-crown-4 compound **17**, the *N,N′*-disubstituted diaza-12-crown-4 compound **35**, and the *N,N′,N″*-

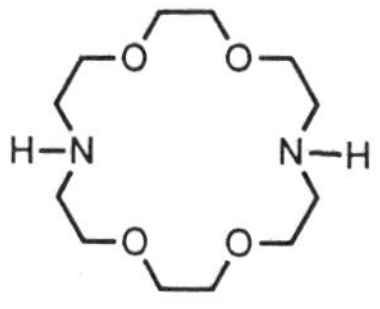

33

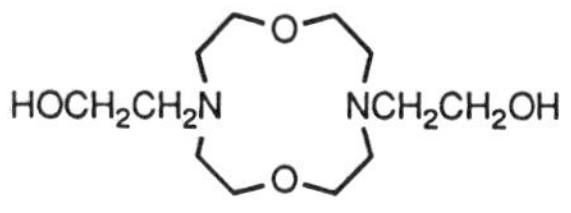

34

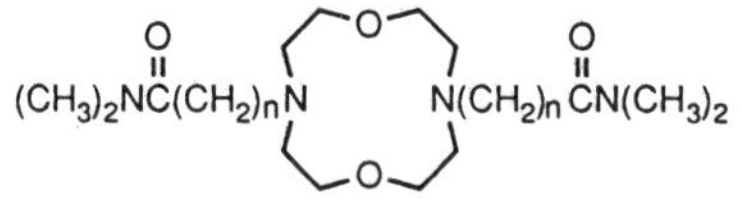

	n
35	1
36	2

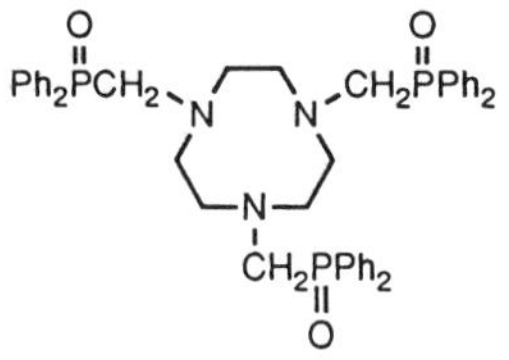

37

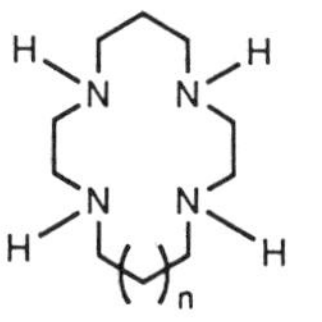

	n
38	0
39	1

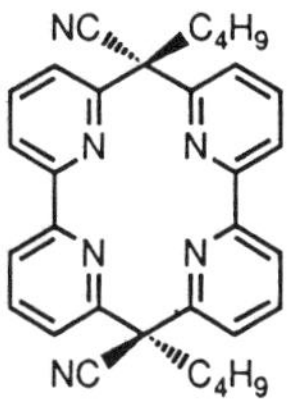

40

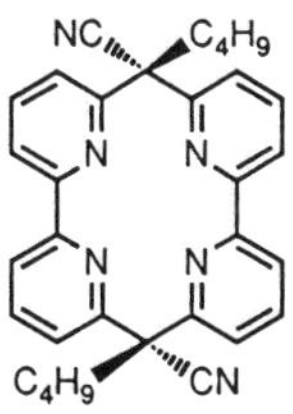

41

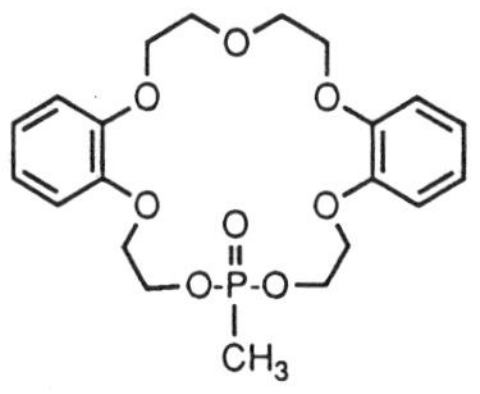

42

trisubstituted triaza-9-crown-3 compound **37** and the tetraaza macrocycles **40** and **41**.

Diameters of the alkali metal cations,[53] as determined from X-ray crystal structures, and potential cavity diameters for various crown ether rings,[54] as estimated from CPK (Corey–Pauling–Kortum) space-filling models, and, in some cases, as determined from X-ray crystal structures are compared in Table 10.4. (It should be noted that the concept of "hole size" for crown ethers is simplistic, since the "hole" does not exist in the absence of the cation.) It has been found that the strongest complexation of an alkali metal cation by a crown ether takes place when the metal ion diameter is somewhat smaller than that of the cavity.[10] Therefore Li^+ is too large to be accommodated within the cavity of an ionophore with a 12-crown-4 ring and too small for strong complexation within the cavity of a 15-crown-5 ring. Thus ionophores with 13- and 14-membered rings might be expected to afford strong and selective complexation of Li^+.

For the various coronands listed in Table 10.3, the largest ratios of log $K(Li^+)$/log $K(Na^+)$ are found with 3-dodecyl-3-methyl-14-crown-4 **(17)** and the *cis*-tetraaza macrocycle **41**, which suggests an excellent fit of Li^+ within the cavities of these coronands. However, considerably stronger complexation of Li^+ than Na^+ is also observed for the *N,N′*-disubstituted diaza-12-crown-4 compound **35** and the *N,N′,N″*-trisubstituted triaza-9-crown-3 compound **37**. Since Li^+ is too large to be accommodated with the macrocyclic cavities of **35** and **37**, it must "perch" on the nitrogen and/or oxygen heteroatoms which comprise the macrocyclic ring. In such a position, Li^+ can interact with the coordinating functions in one or more of the side arms. For such interactions, the length of the side arm(s) would be a critical factor. Thus stronger Li^+ than Na^+ complexation by ionophore **35**, which has acetamido side arms but equal binding of the two ions by ionophore **36** with propanamido side arms, may be rationalized.

Recently Dale and co-workers[55] reported the synthesis of 12-crown-3 **(43)**, hexamethyl-12-crown-3 **(44)**, 16-crown-4 **(45)**, and octamethyl-16-crown-4 **(46)**. The stability constant for 1 : 1 complex formation with lithium perchlorate as determined by NMR spectroscopy in deuterioacetonitrile–deuteriomethanol (19 : 1) was larger for 12-crown-4 ($K > 1000\ M^{-1}$) than for 16-crown-

Table 10.4 Diameters of Alkali Metal Cations[53] and Crown Ether Cavities[54]

Cation	Diameter (Å)	Crown Ether	Cavity Diameter	
			X-Ray Data (Å)	CPK Models (Å)
Li^+	1.48	12C4		1.2–1.5
Na^+	2.04	15C5	1.72–1.84	1.70
K^+	2.76	18C6	2.68–2.86	2.86
Rb^+	2.98	21C7		3.40
Cs^+	3.40	24C8		4.00

4 ($K \sim 100\ M^{-1}$). However, 12-crown-3 formed a strong 2 : 1 complex with sodium perchlorate, whereas 16-crown-4 did not. Hexamethyl-12-crown-3 gave weaker complexation of lithium perchlorate ($K \sim 5\ M^{-1}$) than did 12-crown-3, but with no detectable complexation of sodium perchlorate. It was proposed that the methyl substituents in hexamethyl-12-crown-3 prevented the formation of a 2 : 1 complex with Na^+.

The stability constants for association of Li^+ with 12-crown-4 **(14)** are markedly affected by the solvent (Table 10.3). From the listing of data for 12-crown-4 obtained by various researchers using several different techniques, the stability constant is seen to decrease with solvent variation in the order: acetonitrile > propylene carbonate > acetone > pyridine > water, methanol, DMSO. The influence of the solvent upon Li^+ complexation by small ring crown ethers has been more completely examined by Chen, Wang, and Wu.[39] For the series of dibenzo-14-crown-4-compounds **13**, **18–20**, stability constants were determined in five organic solvents. Stability constants were found to be strongly influenced by solvent variation and decreased in the order: acetonitrile > propylene carbonate > acetone > THF, pyridine. With the exception of pyridine, this order is the inverse of the Gutman donor numbers[56] for the solvents. The stability constant for Li^+ association was also found to be enhanced by the addition of an alkyl group to one or both of the benzo groups of dibenzo-14-crown-4 (Table 10.3).

A quantitative evaluation of the influence of electronic effects of substituents upon stability constants for complexation of Li^+ by lariat ethers has been conducted by Echegoyen, Gokel, and co-workers.[57] For a series of *N*-benzyl monoaza-15-crown-5 compounds **28–32**, stability constants for Li^+ association were determined by potentiometry in acetonitrile (Table 10.3). The measured stability constants were found to correlate with Taft inductive constants (σ^0) for the para substituents on the benzyl group with $\rho = -1.0$. Electron-donating substituents enhanced the stability constant and electron-withdrawing groups diminished it. Even though the aromatic ring and the nitrogen atom of the macrocyclic ring are separated by an insulating methylene group, change of the para substituent from a methoxy group to nitro produced a 13-fold decrease in the stability constant for Li^+ association.

Electroactive substituents can exert a pronounced influence upon Li^+ association by lariat ethers. Echegoyen, Gokel, and co-workers[24,57] have prepared lariat ethers **47** and **48** and studied their electrochemical behavior by cyclic voltammetry in the presence of Li^+. In acetonitrile with tetrabutylammonium perchlorate as the supporting electrolyte, electrochemical reduction of the side arms in **47** and **48** gave Li^+ binding enhancements of 2000 and 800,000, respectively. These very large enhancements in Li^+ binding were attributed to intramolecular ion pairing.

Intramolecular ion pairs are readily formed by ion-exchange reactions of lariat ethers which bear an acidic group on the side arm, for example, **49**. Using carbon-13 NMR, Echegoyen, Gokel, and co-workers[59] have demonstrated that the lithium salt of **48** has a much more rigid structure in deuter-

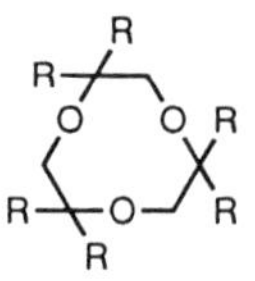

	R
43	H
44	CH_3

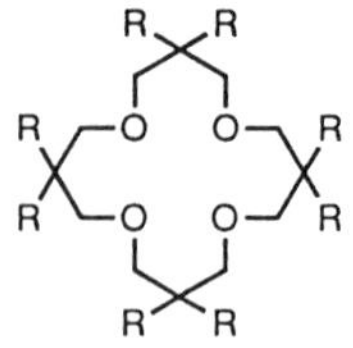

	R
45	H
46	CH_3

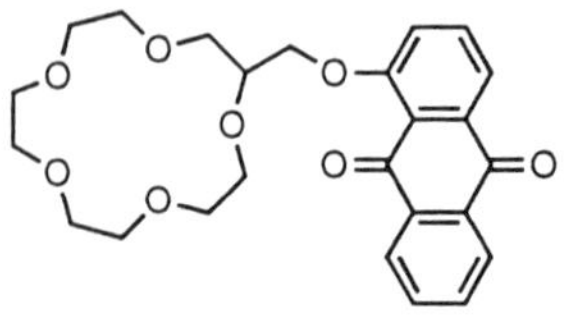

47 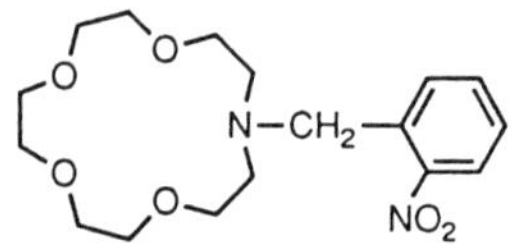 **48**

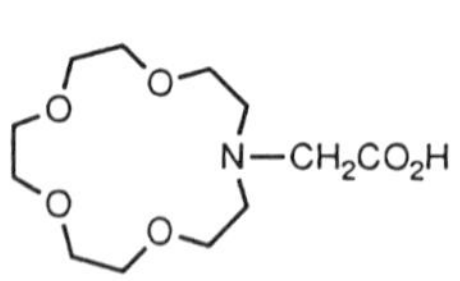

49 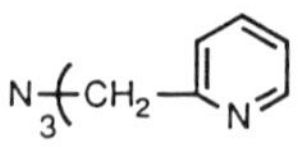 **50** 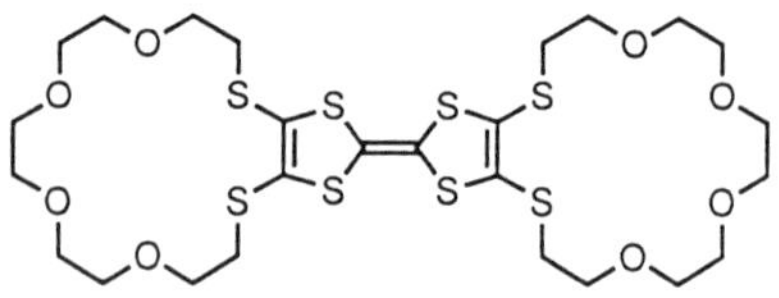 **51**

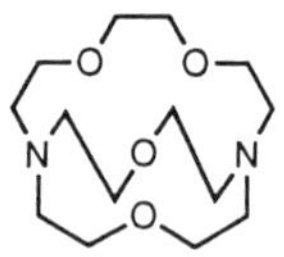

52  **53**  **54**

ioacetonitrile then does the carboxylic acid form or the carboxylic acid form in the presence of lithium perchlorate.

Complexation of Li^+ by cyclic oligopeptides has been studied by Shimizu and co-workers[59] and by others[60,61] using NMR spectroscopy. The stability constant for 1 : 1 complexation of Li^+ by the synthetic cyclic octopeptide cyclo[Gly-L-Lys(Z)-Sar-L-Pro]$_2$ is $2.4 \times 10^3\ M^{-1}$ at 25°C in acetonitrile.

Jasplakinolide, a novel polyketide–depsipeptide isolated from a marine sponge, binds Li^+ weakly ($K = 60\ M^{-1}$) at 20°C in acetonitrile.[62]

Very recently several papers have appeared which report studies of alkali metal cation complexation by acyclic and cyclic multidentate ligands in the gas phase. Such studies are designed to assess the importance of the solvent in metal ion–ionophore complexation reactions.

By Californium-257 plasma desorption spectrometry, Becher and co-workers[63] studied the competitive formation of gas phase complexes of Li^+, Na^+, and K^+ with several ionophores. The selectivity for 12-crown-4 **(14)** was $Na^+ > Li^+ > K^+$. With podand **50** and bis-coranand **51**, formation of the Li^+ complex was favored over that with Na^+ and the K^+ complex was not detected. With fast atom bombardment–mass spectrometry (FAB-MS), Takahashi[64] observed a selectivity of $Li^+ > Na^+ > K^+$ for formation of complexes with dibenzo-14-crown-4 **(13)**.

Maleknia and Brodbelt[65] utilized a tandem mass spectrometer to generate ion complexes of two different alkali metals with crown ethers by liquid secondary ion mass spectrometry. Upon gas phase isolation and high energy dissociation, these complexes predominantly produced crown ether/alkali metal adduct ions. Although 12-crown-4 **(14)** was not examined, 15-crown-5 **(2)** gave a selectivity order of $Li^+ \gg Na^+ > K^+ > Cs^+$. The concept of "maximum contact point" was developed to rationalize the binding preference for a smaller metal ion diameter than is predicted by the "best fit" concept.

Deardon and co-workers[66,67] employed Fourier transform ion cyclotron resonance mass spectrometry to study interactions of alkali metal cations with 12-crown-4 **(14)**, 15-crown-5 **(2)**, and 18-crown-6 in the gas phase. Alkali metal cations initially complexed to a smaller crown ether were observed to transfer to a larger crown ether in every case. Also it was noted that the initially formed metal ion–ionophore complex underwent reaction with a second crown ether molecule to form a metal ion-bound dimer species.

2.3 Cryptands

In 1975 Lehn and Sauvage[68] reported stability constants and selectivities for complexation of alkali metal and alkaline earth cations by a new class of bicyclic ionophores, the cryptands. In these multidentate ligands, two nitrogen atoms were linked by three oxygen atom containing bridges (e.g., **52–54**). The ionophores are commonly identified by the number of oxygen atoms in the bridges. Thus **52**, **53**, and **54** are called [2.1.1]-cryptand, [2.2.1]-cryptand, and [2.2.2]-cryptand, respectively.

Diameters of the alkali metal cations[53] and the cryptand cavities[69] are recorded in Table 10.5. The cavity size of [2.1.1]-cryptand **(52)** is appropriate for strong complexation of Li^+, but too small to accommodate Na^+ or any of the other alkali metal cations. Hence much stronger complexation of Li^+ than Na^+ ion is anticipated.

Stability constants for complexation of Li^+, Na^+, and K^+ by cryptands **52–54** in water[70] are presented in Table 10.6. Contrary to the situation with coronands (Table 10.3), very appreciable association of alkali metal cations with

Table 10.5 Diameters of Alkali Metal Cations[53] and Cryptand Cavities[69]

Cation	Diameter (Å)	Cryptand	Diameter (Å)
Li^+	1.48	[2.1.1]	1.6
Na^+	2.04	[2.2.1]	2.2
K^+	2.76	[2.2.2]	2.8
Rb^+	2.98		
Cs^+	3.40		

Table 10.6 Stability Constants for Formation of Alkali Metal Ion Complexes with Cryptands 50–52 in Water at 25°C[70]

	log *K*				
Cryptand	Li^+	Na^+	K^+	Rb^+	Cs^+
[2.1.1] (**50**)	5.5	3.2	<2	<2	
[2.2.1] (**51**)	2.5	5.4	4.0	2.6	<2.0
[2.2.2] (**52**)	1.0	4.0	5.5	4.2	1.5

the bicyclic ionophores is evident even in water. The stability constants demonstrate that [2.1.1]-cryptand is a very selective ionophore for Li^+.

The energetics for complex formation with [2.1.1]-cryptand are similar to those obtained for complexes of aniline diacetic acid sulfonates and carboxylates.[71] However, the alkaline earth cations interfere strongly with these simple complexones, whereas [2.1.1]-cryptand is very selective for Li^+ over the alkaline earth cations.

Stability constants for complexation of Li^+ and Na^+ with [2.1.1]-cryptand in a variety of solvents are compared in Table 10.7.[70] Although association with Li^+ is preferred over Na^+ in all solvents, the difference is stability constants varies from a low of 16 to dimethyl sulfoxide to 5000 in propylene

Table 10.7 Stability Constants for Formation of Lithium and Sodium Ion Complexes with [2.1.1]-Cryptand (52) in Various Solvents at 25°C[70]

	Dielectric	log *K*		
Solvent	Constant	Li^+	Na^+	log $K(Li^+)$–log $K(Na^+)$
Water	79	5.5	3.2	2.3
Methanol	33	8.0	6.1	1.9
Ethanol	24	8.5	7.1	1.4
Acetonitrile	36	>10	>9	
Propylene carbonate	65	12.5	8.8	3.7
Dimethylformamide	37	7.0	5.2	1.8
Dimethyl sulfoxide	49	5.8	4.6	1.2

carbonate. The difference is stability constants for Li^+ and Na^+ association is largest in propylene carbonate and water. Of the solvents listed on Table 10.7, these two have the highest dielective constants. Based upon the stability constants for association with [2.1.1]-cryptand, the Li^+/Na^+ selectivity is calculated to be 5000 in propylene carbonate and 200 in water.

Evidence that [2.1.1]-cryptand totally encloses the complexed Li^+ was provided by lithium-7 NMR studies.[72] The chemical shift of the Li^+ complex of [2.1.1]-cryptand was independent of the nature of the solvent used. Thus the cryptand effectively shields the complexed Li^+.

The stability constant, a thermodynamic parameter, is the ratio of rate constants, kinetic parameters, for the complexation (formation) and decomplexation (dissociation) steps (Eq. 3). Formation and dissociation rate constants for Li^+ and Na^+ ion complexes of [2.1.1]-cryptand in water and methanol are presented in Table 10.8.[73] The rates of formation for the Na^+ complex with [2.1.1]-cryptand are greater than those of the Li^+ complex by factors of 28 and 6 in water and methanol, respectively. On the other hand, the rates of dissociation of the Na^+ complex are 5600 and 570 times faster than those of the Li^+ complex with [2.1.1]-cryptand in water and methanol, respectively. Thus even though the rate of formation of the Li^+ complex is slower than that of the Na^+ complex, its dissociation rate is much slower which produces higher stability constants for the Li^+ complex with [2.1.1]-cryptand.

An investigation of the kinetics of formation of the lithium complex of [2.1.1]-cryptand using stopped-flow calorimetry suggests that the complexation occurs initially at one face of the cryptand, such that the metal is only partially enclosed (to form an "exclusive complex").[74] Rearrangement of this species follows to yield the more stable complex (the "inclusive" complex) which contains the metal ions inside the cryptand.

Lincoln and co-workers[75] have prepared cryptand **55**, in which the oxygen in one of the five-atom bridges between the two nitrogens in [2.1.1]-cryptand **(52)** has been replaced by a methylene group. Stability constants (log K) for the Li^+ and Na^+ complexes of **55** in dimethylformamide at 25°C are 2.80 and 2.87, respectively. Thus the strength of Li^+ complexation by **55** is comparable with that for Na^+ complexation and there is no apparent selectively between Li^+ and Na^+ binding by this cryptand. The Li^+ complex is inclusive, whereas the Na^+ complex is exclusive.[75]

Table 10.8 Formation (k_f) and Dissociation (k_d) Rate Constants for Lithium and Sodium Ion Complexes of [2.1.1]-Cryptand (52) in Water and Methanol at 25°C[73]

Solvent	Cation	k_f ($M^{-1}s^{-1}$)	k_d (s^{-1})
Water	Li^+	8.0×10^3	2.5×10^{-2}
Water	Na^+	2.2×10^5	1.4×10^2
Methanol	Li^+	4.8×10^5	4.4×10^{-3}
Methanol	Na^+	3.1×10^6	2.5

Cryptand **56** is a structural isomer of [2.1.1]-cryptand **(52)** and has a clam-like structure in which the two $-(CH_2)_2O(CH_2)_2O(CH_2)_2-$ jaws are hinged about a $>N(CH_2)_2N<$ moiety. Stability constants (log K) as determined by Lincoln and co-workers[76,77] for the Li^+ and Na^+ complexes of **56** in dimethylformamide are 3.5 and 6.1, respectively, which shows strong selectivity for Na^+ complexation. Thus the pronounced Li^+ selectivity of [2.1.1]-cryptand is lost for the structurally modified cryptands **55** and **56**.

High Li^+ selectivity with respect to Na^+ and K^+ has been reported by Ciampolini and Micheloni[78,79] for cryptands **57** and **58** which have nitrogen hetero-

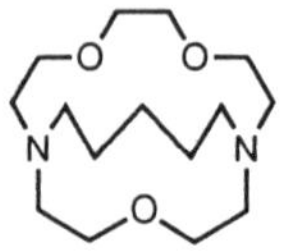

55 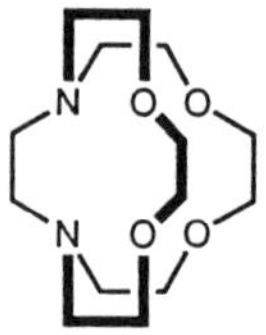 **56** 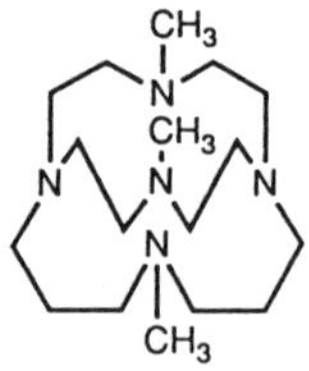 **57**

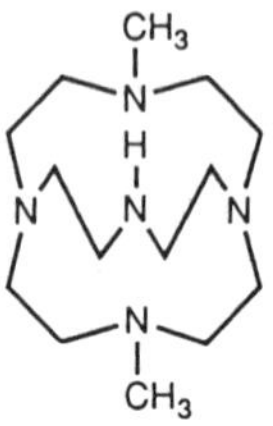

58 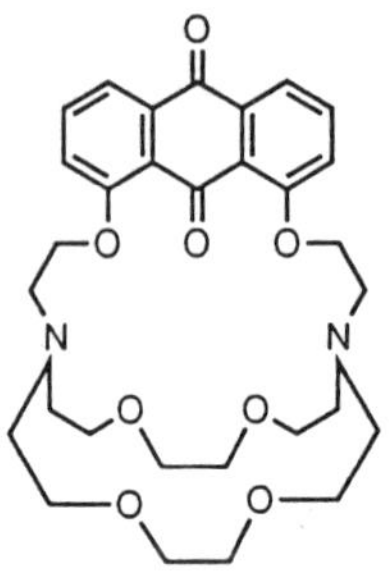 **59**

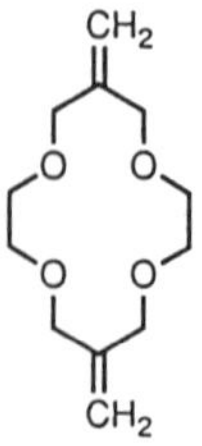

60 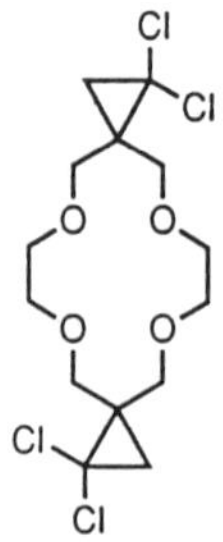 **61** 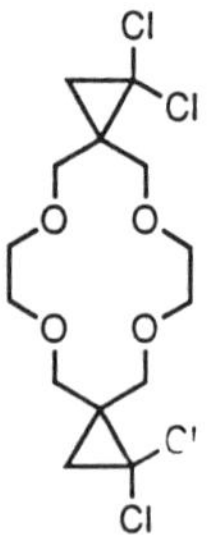 **62**

atoms in the sidearms as well as at the bridgehead positions. By carbon-13 and lithium-7 NMR spectroscopy, strong Li^+ complexation by **57** was clearly demonstrated in highly alkaline aqueous solutions. Using carbon-13 NMR spectroscopy as the diagnostic technique, no evidence was found for complexation of Na^+ or K^+ by the macrobicyclic polyaza ligand **57**.

For the macrobicyclic polyaza compound **58**, complete discrimination for Li^+ complexation over that for Na^+ was shown by a combination of potentiometric and NMR spectroscopic measurements.[80] For the Li^+ complex of **58**, a high log K value of 4.8 was obtained in very alkaline aqueous solution. Although bicyclic polyaza compounds **57** and **58** exhibit high Li^+ selectivity, strongly alkaline conditions are required to fully deprotonate the ionophores.

Echegoyen, Gokel, and co-workers[81] detected a very unusual 1:2 ionophore:lithium ion complex of anthraquinone cryptand **59** by cyclic voltammetry. The existence of this complex was confirmed by ESR spectroscopy and by a binding study using lithium-7 NMR spectroscopy. The formation of 1:2 complexes of **59** with Na^+ and K^+ was also detected by cyclic voltametry and ESR spectroscopy.

3 SOLID STATE STRUCTURES OF IONOPHORE-COMPLEXED LITHIUM IONS

No naturally occurring molecules are known which exhibit preferential binding of Li^+. Synthetic ionophores that would selectively complex Li^+ in its physiological concentration have been the objective of preparative endeavors in several different research laboratories. Elucidation of the coordination properties of Li^+ should lead to the design of ionophores with enhanced Li^+ selectivity and to improved understanding of the biological activity of Li^+.

Several summaries of structural chemistry for natural and synthetic ionophores and their complexes with cations are available.[11–13, 82–88] The recent review by Olsher, Izatt, Bradshaw, and Dalley[13] is devoted to the coordination chemistry of Li^+.

Coordination numbers from 2 to 8 are exhibited by Li^+ compounds and complexes.[2, 5, 89, 90] In organolithium compounds[89] and Li^+ complexes,[69, 91] the binding is primarily electrostatic. Since coulombic forces are nondirectional, the radius of Li^+ plays an important role in determining the structure of a complex. In a complex, the coordination number of Li^+ is determined primarily by the number of binding sites, usually ions (ion–ion interaction) and negative poles of the neutral ligand and solvent (ion–dipole interaction), that can be packed around Li^+. The small radius of Li^+ provides a possibility for a variety of coordination numbers and polyhedra.

The 1:1 complexes of Li^+ with podands and coronands are 4- and 5-coordinated.[13] The most Li^+ selective of these ionophores exhibit fourfold tetrahedral (Td) coordination and fivefold square pyramidal (SP) coordination (Fig. 10.2).

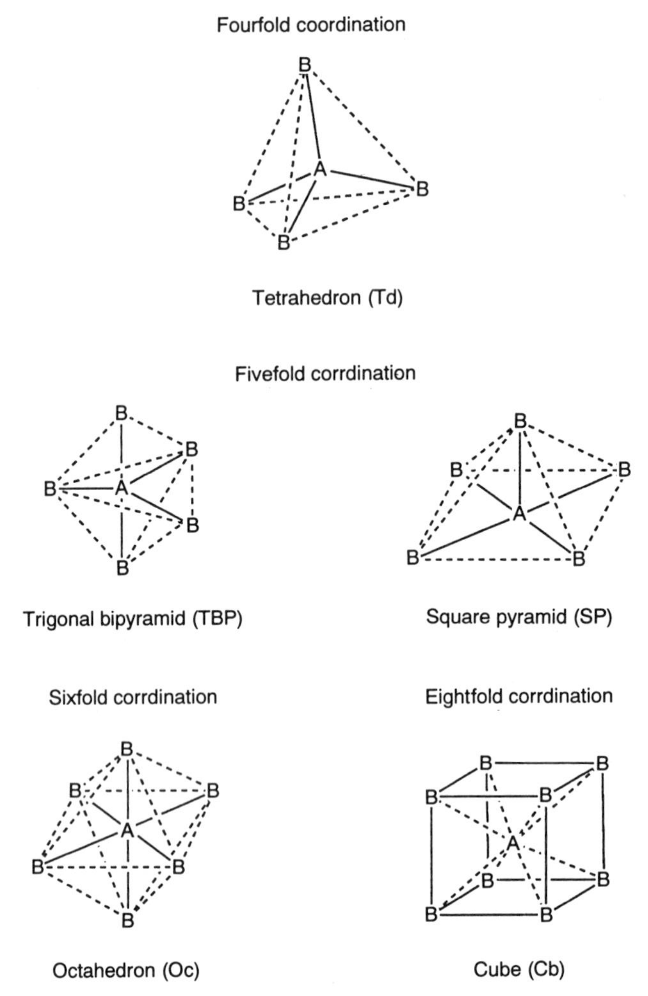

Figure 10.2 Coordination complexes observed in solid-state, lithium ion–ionophore complexes.

Hexadentate Li^+-selective ligands include rigid-cage molecules with very small radii, such as cryptands and spherands.[13]

Eightfold coordination is found in the sandwich complex of Li^+ with two 12-crown-4 **(14)** molecules in aprotic media.[92,93] Trace amounts of water caused the complexes to decompose.

In view of the excellent coverage for solid-state structures of lithium ion–ionophore complexes which is provided in the recent review by Olsher, Izatt, bradshaw, and Dalley,[13] only results published since 1989 will be presented here. These include structures in which the ionophores are substituted 14-crown-4[94–96] and dibenzo-14-crown-4[96–100] compounds, three Group 6 phosphinite crown ethers,[101] a polyaza bicyclic ligand,[79] a spherand,[102] and a torand.[103]

In homogeneous solution, it has been shown that substituted 14-crown-4 compounds may exhibit high selectivity for Li^+. Therefore the structures of substituted 14-crown-4 compounds and their complexes are of interest.

For the lithium thiocyanate complex of *sym*-bismethylene-14-crown-4 **(60)**, Olsher, Krakowiak, Dalley, and Bradshaw report a square pyramidal coordination geometry for Li^+ (Fig. 10.2) with the four ether oxygens forming the base of the pyramid.[94] Czech, Bartsch, and co-workers reacted **60** with dichlorocarbene to give a mixture of *trans*- and *cis*-6,13-bis(spiro-2,2-dichlorocyclopropyl)-14-crown-4 compounds (**61** and **62**, respectively) for which structure determination of the lithium thiocyanate complexes was utilized to differentiate the geometrical isomers.[95] Once again the coordination geometry for Li^+ was square pyramidal with four ether oxygens, providing the base of the pyramid and the nitrogen of the thiocyanate ion at the apex. In the lithium thiocyanate complex of 2,2,3,3,6,9,9,10,10-nonamethyl-14-crown-4 **(63)**, Sachleben and Burns found a coordination geometry for Li^+ of square planar also.[96]

Crystal structures of the four uncomplexed, substituted 14-crown-4 compounds **63–66** were reported recently.[94–96] In the uncomplexed ionophores **63** and **64**, Sachleben and Burns note that two of the oxygen atoms are turned outward and two carbon atoms of the polyether ring are pointed inward which fills the interior of the rings.[96] Thus marked reorganization of the crown ether structures would be required to form a square pyramidal complex with Li^+.

For uncomplexed *cis*-bis(hydroxymethyl)-14-crown-4 **(65)**, a least-squares plane was calculated by Olsher, Dalley, and co-workers for the four ring oxygen atoms.[94] Each oxygen atom deviates by 0.2 Å from the plane in an alternating manner with one atom above the plane and the next below the plane. Thus some structural reorganization would be necessary to produce the planar geometry for a square pyramidal Li^+ complex. In contrast in the crystal structure for uncomplexed 6,6,13,13-tetra(benzyloxymethyl)-14-crown-4 **(66)**, Czech, Olsher, Shoham, Dalley, Bartsch, and co-workers observed that the four ring oxygen atoms form a perfect plane.[95] Thus the four oxygen atoms in **66** are "preorganized" for complexation of Li^+. As first recognized by Cram,[104] such "preorganization of the binding site" can produce strong and selective cation binding.

From the solid-state structures of the uncomplexed, substituted 14-crown-4 compounds **63–66** it is seen that the level of preorganization varies markedly with changes in the nature and attachment sites of the substituents. The situation is very different with dibenzo-14-crown-4 compounds. In the crystal structure of dibenzo-14-crown-4 **(13)** as reported in 1991 by two groups, the four ring oxygen atoms form a plane which is a preorganized base for square pyramidal coordination of Li^+.[96,97] The molecule is butterfly-shaped with an angle of 123° between the two major planes which are defined by each benzene ring with the two attached oxygen atoms.

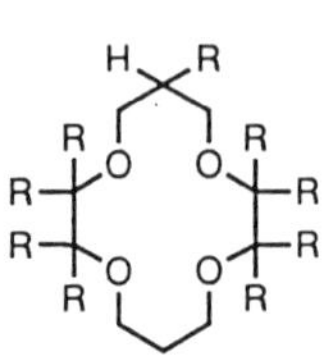

63 (R = CH_3) 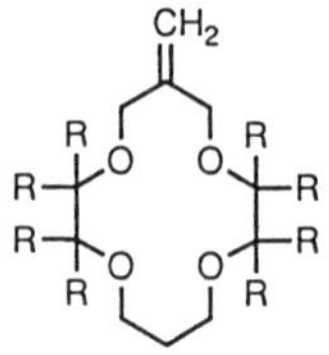 **64** (R = CH_3)  **65**

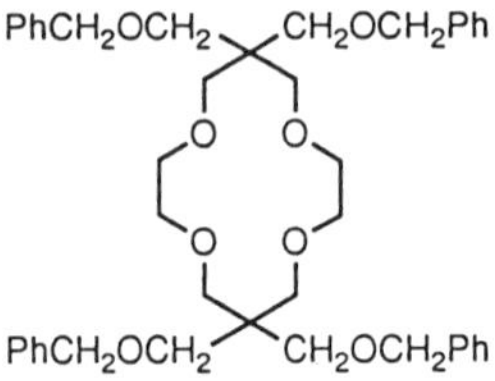

66

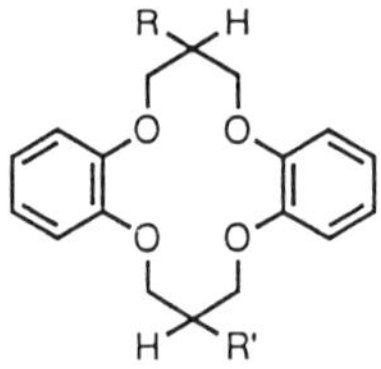

	R	R'
67	OH	H
68	OH	OH
69	CH_2OH	OH (trans)
70	CH_2CO_2H	H
71	$CH_2P(O)(Ph)(OH)$	H
72	OCH_2CO_2H	H

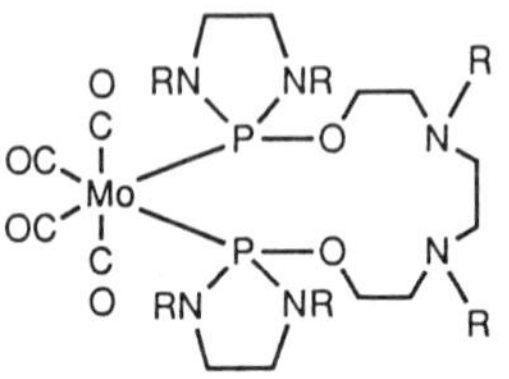

73 (R = CH_3)

Structural modifications of dibenzo-14-crown-4 by attachment of one or more pendent hydroxyl groups offers the potential for anion solvation as well as Li^+ complexation by the crown ether.[98] To assess this possibility, molecular structures were determined by Olsher, Frolow, Dalley, Bartsch, and co-workers for lithium thiocyanate complexes of dibenzo-14-crown-4 alcohols **67** and **69** and diol **68**. In addition the solid-state structure for the lithium nitrate complex of crown ether alcohol **69** was determined. Crystal structures for these four complexes show intracomplex "scorpion-like" and intercomplex "head-to-tail" hydrogen bonding of the crown ether alcohols with the anions. Thus these structurally modified crown ethers serve as bifunctional ligands for concomitant anion solvation and cation complexation.

Structures for two internal salts of dibenzo-14-crown-4 compounds with proton-ionizable side arms have been reported by Burns and Sachbelen.[99] For the lithium carboxylate and phosphinate salts of lariat ethers **70** and **71**, respectively, the coordination geometry for Li^+ is square pyramidal with the four ring oxygens forming the base of the pyramid. The apical coordination site is occupied by an oxygen atom of the ionized group of a second complex to form dimeric structures. This arrangement is quite different from that reported by Shoham, Bartsch, Olsher, Lipscomb, and co-workers[100] for the lithium carboxylate salt of the closely related lariat ether carboxylic acid **72**. In this complex, the Li^+ coordination is once again square pyramidal with the four ring oxygens forming the base of the pyramid. However, the apical coordination site is occupied by the oxygen of a water molecule which bridges Li^+ and the carboxylate group in the side arm of the same molecule. Thus the lithium carboxylate complex of **72** is monomeric while the lithium salts of the ionized forms of lariat ethers **70** and **71** are dimeric. Presumably the longer, more flexible side arm in lariat ether **72** facilitates formation of an intramolecular complex.

Powell and co-workers have reported crystal structures for Li^+ complexes of the molybdaphosphinito crown ethers **73–75**.[101] These ditopic ionophores combine a subunit containing a ''soft'' binding site with one bearing a ''hard'' site. The coordination geometry of the encapsulated Li^+ is square pyramidal in the complex with **73** and distorted tetrahedral in the complexes with **74** and **75**.

The solid state structure of the lithium tetraphenylborate complex of the bicyclic polyaza ionophore **58** has been determined by Ciampolini, Micheloni, and co-workers.[79] The Li^+ is wholly enclosed within the cryptand cavity and has a five-coordinate trigonal–bipyramidal structure.

Trueblood, Maverick, and Knobler have reported the solid-state structures for spherand **76** and its complex with lithium chloride.[102] The oxygen atoms of the inward-facing methoxy groups in the ionophore and the Li^+ complex have the same symmetry. The closest O· · ·O distance shortens by 0.13 Å when the small Li^+, with its strong force field, enters the cavity. In the complex, Li^+ is six-coordinate with an octahedral geometry.

Despite the fact that Li^+ is much smaller than the central cavity of torand **77**, Bell and co-workers observed that stable 1 : 1 complexes with lithium picrate are formed.[103] Structure determination has revealed that the complex is a supramolecular assembly of two torands, two Li^+, and three water molecules.

4 DETERMINATION OF LITHIUM IONS USING IONOPHORES

Applications of synthetic multidentate ligands in analytical chemistry for the determination and separation of desired metal ion species has been the subject of several reviews.[105–117] A variety of techniques are described, including solvent extraction, liquid membrane transport, chromatographic separations, spectrophotometric determination, and ion—selective electrodes.

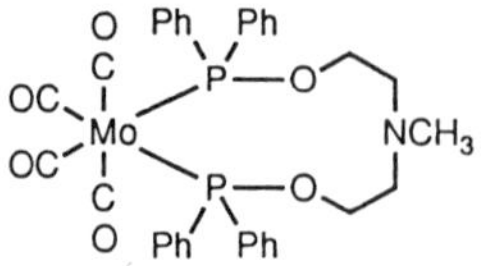

74 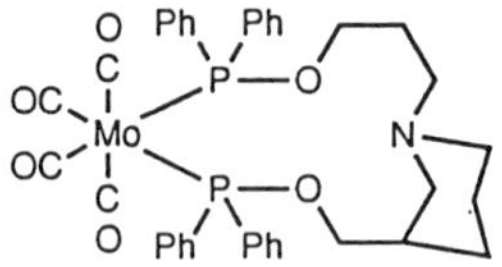 **75**

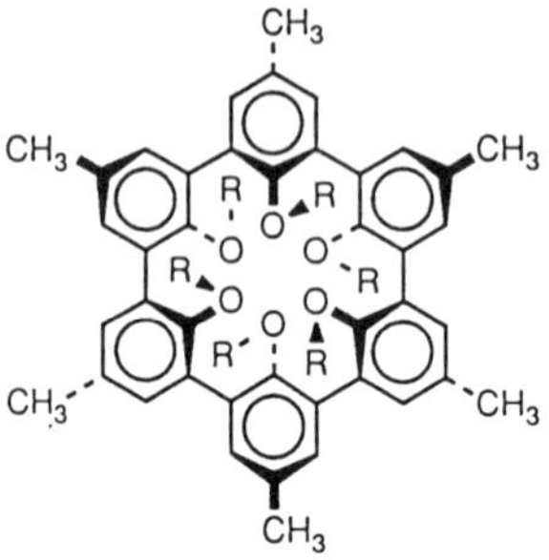

76 ($R = CH_3$) 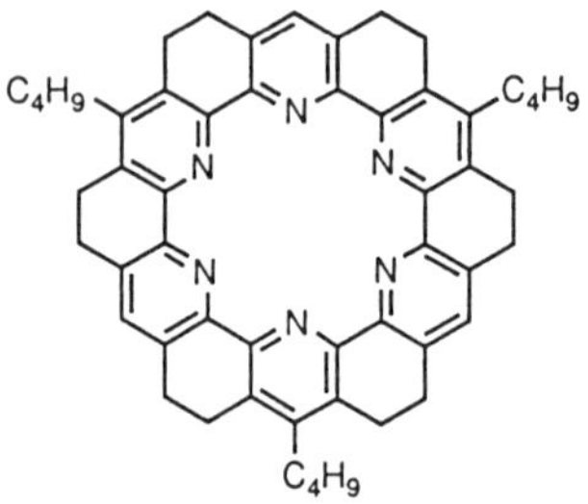 **77**

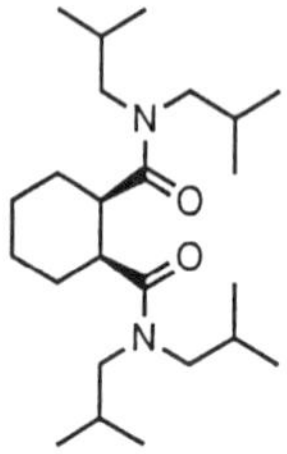

78 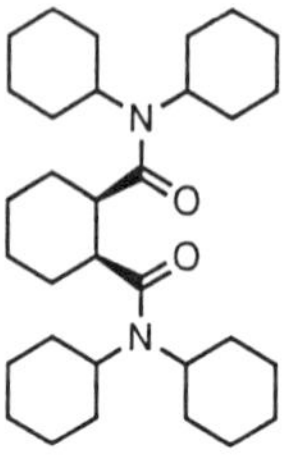 **79** 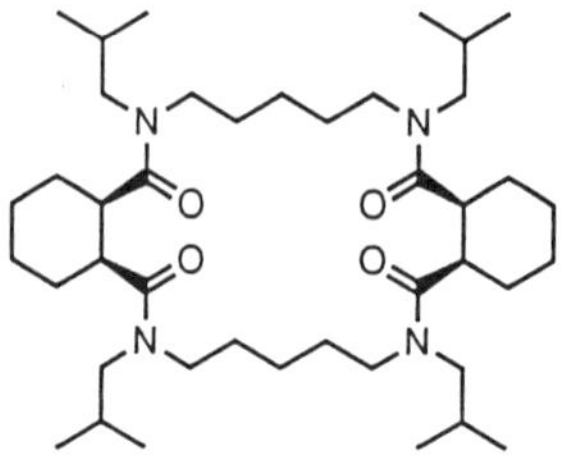 **80**

Under the current heading, the determination of Li^+ by ion-selective electrodes using multidentate ligands and by spectrophotometry with chromogenic and fluorogenic ionophores will be examined. Separations of Li^+ by ionophore-based extraction, transport, and sorption processes will be explored under the next major heading.

Lithium salts are effective therapeutic agents in disorders, such as maniac depressive illness (MDI). Lithium carbonate tablets are the prevalent form of administration. The dosage is critical because of toxicity and undesirable side effects. Strict monitoring of the Li^+ concentration in body fluids is required during therapy. Therefore considerable attention has been focused upon methodology for Li^+ determination in body fluids, particularly in plasma and whole blood.

For such determinations, Na^+ interference is the greatest problem. Blood contains about 140 mM Na^+, whereas effective Li^+ concentrations for MDI treatment are 0.7–1.5 mM. An effective assay for Li^+ in human blood demands a very high selectivity for Li^+ over Na^+. Another significant problem is interference by the proteins in blood.

4.1 Ion-Selective Electrodes

4.1.1 History of Lithium Ion-Selective Electrodes. An early attempt for the determination of Li^+ utilized lithium containing glass electrodes.[18] Membranes of Li_2O-containing glasses, called LAS glass (15% Li_2O; 25% Al_2O_3; 60% SiO_2), showed some Li^+ selectivity when used in electrodes. Although regular glass electrodes exhibit high preference for H^+ and also Na^+, LAS-based glass electrodes gave a Li^+/Na^+ selectivity of about 3. This is unsatisfactory for the determination of Li^+ in blood. In addition, deterioration of these electrodes by blood proteins was severe. Very frequently the electrodes had to be cleaned to remove films of serum proteins.

A so-called lithium-bronze is the active inorganic component of a Li^+ sensitive electrode. An epoxy resin containing lithium vanadium oxide bronze[119] is employed to provide a Li^+/Na^+ selectivity ratio of about 50, which is still insufficient for clinical use.

The systems described above utilize charge carriers as the active constituents. Neutral ion-exchanging materials were considered to be potentially more effective. Therefore, the design, synthesis, and evaluation of neutral carrier-type electrodes for determination of Li^+ in body fluids has received the attention of several research groups. In the following discussion neutral carrier membrane electrodes are characterized by descriptions of the requisite membrane potentials, the design of Li^+-selective membranes in accord with the Nernst law, and the chemical characterization of neutral ionophore carriers for Li^+-selective electrodes.

4.1.2 Membrane Potential. An electrochemical membrane is a thin sheet that is selectively permeable for certain ions. When two different electrolyte solutions are separated by such an electrochemical membrane, an electrical potential difference is generated across the sheet. The membrane potential varies according to the difference in ionic activity of the ions. The Nernst equation describes the potential in terms of the log of the activity coefficient. If a membrane is highly specific to a certain ion, the membrane potential obeys the Nernst equation irrespective of various coexisting ions. Such ion-selective membranes, when incorporated into electrodes, become devices which can be used to determine the ionic activity in the sample solution. Since the measured potential is an expression of the activity coefficient and the ion activity in the membrane is a constant, the potential provides a direct measure of the ionic activity in the sample solution.

Membrane potentials for neutral-carrier-type, ion-selective electrodes also can be interpreted in terms of ion transfer by the neutral carrier in the membrane.[120]

For a system in which an electrochemical membrane is in contact with two electrolyte solutions, several equilibria hold with regard to distribution and complex formation of the cation (I^+), anion (X^-), neutral carrier (S), and neutral carrier–cation complex (IS^+) at one of the interfaces.[114] This is illustrated in Scheme 1, where the asterisk identifies species in the organic phase (i.e., the membrane phase).

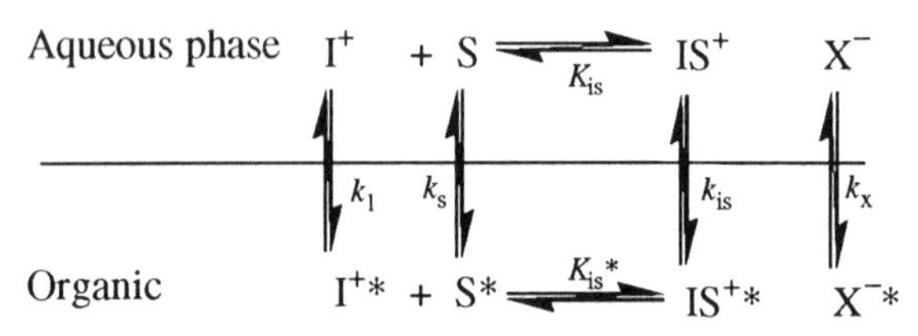

Scheme 1. Equilibria at an Electrochemical Membrane between Two Electrolyte Solutions

The equilibrium constants are defined as follows, where *a* stands for activity and *C* for concentration:

Complex formation constant in one of the aqueous phases: $K_{IS} = C_{IS^+}/(a_I \times C_S)$

Complex formation constant in the membrane phase: $K_{IS*} = C_{IS^{+*}}/(a_{I*} \times C_{S*})$

Distribution constant for I^+: $k_I = a_{I*}/a_I$

Distribution constant for S: $k_s = C_{S*}/C_S$

Distribution constant for IS^+: $k_{IS} = C_{IS*}/C_{IS}$

Distribution constant for X^-: $k_X = a_{X*}/a_X$

The same equilibria hold for cation J^+, also at both interfaces. If the neutral carrier molecule complexes both cations with a 1 : 1 stoichiometry, the potential difference between aqueous phases (i) and (ii) can be described as follows:

$$E = \frac{RT}{F} \ln \frac{a_I(\text{i}) + \left[\dfrac{U_{JS*}k_{JS}K_{JS}}{U_{IS*}k_{IS}K_{IS}}\right] a_J(\text{i})}{a_I(\text{ii}) + \left[\dfrac{U_{JS*}k_{JS}K_{JS}}{U_{IS*}k_{IS}K_{IS}}\right] a_J(\text{ii})} \tag{6}$$

where R, T, and F are the gas constant, temperature, and Faraday constant respectively, and U_{IS*} or u_{JS*} is the mobility of the neutral carrier-complexed ion in the membrane (organic phase).

4.1.3 Selectivity Coefficient. The membrane potential in an ion-selective electrode is based on the concentration difference of the ion in the sample solution and in the internal solution within the electrode body. Generally the concentration of internal solution of the primary ion (I^+) in the electrode is constant, if I^+ is the ion for which the electrode is selective. If aqueous phase (ii) is the internal filling solution of the electrode, Eq. 6 can be rewritten in the form of Eq. 7, the so-called Nicolsky–Eisenman equation:

$$E = E_0 + \frac{RT}{F} \ln \left[a_I + \sum_J K_{IJ}^{Pot} a_J \right] \tag{7}$$

where

$$K_{IJ}^{Pot} = \frac{U_{JS*} k_{JS} K_{JS}}{U_{IS*} k_{IS} K_{IS}} \tag{8}$$

The quantity K_{IJ}^{Pot} is called the selectivity coefficient, which is often used to characterize the ion selectivity of a given ion-selective electrode. A small value of this coefficient for I^+ with respect to J^+ means high selectivity for I^+ over the interfering ion J^+. (The reciprocal of the selectivity coefficient, $1/K_{IJ}^{Pot}$, is the I/J selectivity.)

Since the mobilities of the neutral carrier-complexed ions in the membrane (organic phase) usually do not depend upon the identity of the cation (i.e., $U_{IS*} = U_{JS*}$), the selectivity coefficient can be defined in the simplified form shown in Eq. 9. Thus the selectivity coefficient for a

$$K_{IJ}^{Pot} = \frac{k_{JS} K_{JS}}{k_{IS} K_{IS}} = \frac{K_J}{K_I} \tag{9}$$

neutral carrier-type ion-selective electrode is usually equal to the ratio of extraction equilibrium constants for I^+ and J^+. Such extraction equilibria are defined as shown in Eq. 10.

$$I^+ \text{ (or } J^+) + S^* + X^- \rightleftharpoons IS^{+}* \text{ (or } JS^{+}*) + X^- \tag{10}$$

If the distribution constant for the neutral carrier-complexed ion is the same for both cations (i.e., $k_{IS} = k_{JS}$), the selectivity coefficient can be further simplified to become just the ratio of the complex formation constants, as shown in Eq. 11. Equation 11 can be used only in very

$$K_{IJ}^{Pot} = \frac{K_{JS}}{K_{IS}} \tag{11}$$

special cases.[121] More commonly, Eq. 9 is employed.

Selectivity coefficients are customarily determined by the fixed interference method (FIM), which is a mixed solution method, or the separate solution method (SSM). In the FIM the potential is measured with solutions containing a constant activity of the interfering ion (a_J) and varying activities of the primary ion I using the ion-selective electrode and an appropriate reference electrode. The potential values are plotted against the activity of the primary ion. The intersection of the extrapolated linear portions of the pilot affords a value for the primary ion's activity (a_I), which is used to compute the K_{IJ}^{Pot} value from Eq. 12, where Z_I and Z_J are the charges on the I and J ions, respectively.

$$K_{IJ}^{Pot} = \left[\frac{a_I}{a_J}\right]^{Z_I/Z_J} \tag{12}$$

The SSM is based on potential measurements made with two separate solutions and is valid only in cases where the electrode inhibits a Nerstian response. The FIM is generally more desirable than the SSM since the mixed solution represents the actual condition under which the electrodes are employed.

4.1.4 Prerequisites for Neutral Carriers for Lithium Ion-Selective Electrodes. A neutral carrier-based, ion-selective electrode is usually equipped with a solvent-polymeric membrane that contains an appropriate neutral ionophore as the cation carrier. There are three main requirements for synthetic ionophores to be used in Li^+-selective electrodes.

First, the carrier should be highly selective for Li^+. In particular, high selectivity for Li^+ over Na^+ is desired if the electrodes are to be used for Li^+ assay in blood or serum. More specifically, they should have high ratios of cation extraction equilibrium constants for Li^+ over Na^+, as expressed in Eq. 9. It should be recalled that Li^+ is more strongly hydrated than Na^+, which biases extraction into an organic medium against Li^+. The complex formation constants for Li^+ with such neutral carriers do not have to be especially high—the ratio of $K(Li^+)/K(Na^+)$ is the important factor.

Second, the molecular structure of the ionophore must be sufficiently flexible to allow for a high rate of cation exchange. A rigid carrier structure results in slow response in establishing the potential and the resulting Li^+-selective electrodes would show low sensitivity. Thus although [2.1.1]-cryptand **(52)** exhibits excellent Li^+/Na^+ selectivity for complexation in homogeneous solution, the ionophore is too rigid for use in a Li^+-selective electrode.

Third, the neutral ionophore should be quite lipophilic, since it must be dissolved in the solvent-polymeric membrane matrix. Lipophilicity also prevents loss of the carrier from the sensing membrane by dissolution into the aqueous sample and internal solutions, which would cause rapid deterioration of the electrode. Attachment of long aliphatic chains to the neutral carrier is an effective way to increase the lipophilicity when plasticizers or membrane solvents are used for the sensing membranes.[121] On the other hand, a carrier with too high a molecular weight will cause poor mobility and produce sluggish cation exchange, resulting in slow electrode response.

4.1.5 Podands in Lithium-Ion Selective Electrodes. Shortly before Pedersen's first publication on crown ether chemistry,[27] a note on the complexation of alkali metal salts by an acyclic aromatic diamine appeared.[123] The stoichiometry of the sodium salt–diamine complex was 1:3 which confirmed a hexacoordinated Na^+. Subsequently it was hinted that different alkali metal ions were complexed by 2,3-di(*p*-aminophenyl)butane with varying stabilities, all of which were higher than those for salts of alkaline earth metal ions.[124]

In 1975, Simon and co-workers[125] described a Li^+-selective electrode based upon the lipophilic acyclic dioxadimide **3** incorporated into poly(vinyl chloride) (PVC) plasticized by tris(2-ethylhexyl)phosphate (TEHP) in a standard type ion-selective electrode (Fig. 10.3). A Li^+/Na^+ selectivity of 20 was achieved which markedly surpassed the Li^+-selectivity of a LAS glass electrode. However interference by H^+ was severe.[125,126]

Considerable effort was expended to design lipophilic acyclic diamide-type ionophores with higher Li^+/Na^+ selectivities. A K^{Pot}_{LiNa} value of less than 1×10^{-2} (a selectivity of >100) was realized with electrodes containing lipophilic diamide **78** in PVC/NPOE (*o*-nitrophenyl octyl ether) membranes.[127,128] An even better Li^+ selectivity against Na^+ ($K^{Pot}_{LiNa} = 3.5 \times 10^{-3}$) was attained with lipophilic diamide **79**.[129]

Since complexation of Li^+ is thought to involve more than one molecule of the ionophore, dimeric diamide **80** was prepared and evaluated. When employed in plasticized PVC membrane electrodes **80** gave K^{Pot}_{LiNa} 1×10^{-2}.[130,131]

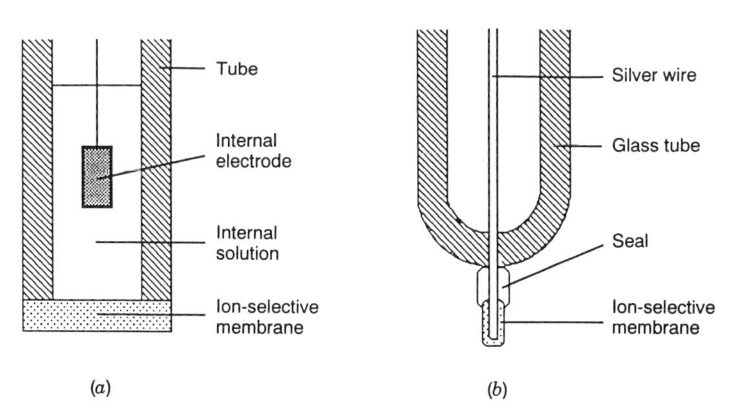

Figure 10.3 Construction of ion-selective electrodes: (*a*) standard type; (*b*) coated-wire type.

Solvent polymeric membrane electrodes based upon these lipophilic acyclic diamide-type ionophores have been applied to Li^+ assay in snail neurons by a microelectrode technique[132] and in blood serum.[129]

A variety of other lipophilic acyclic diamides were synthesized by Simon and co-workers,[133, 134] by Christian and co-workers[135, 136] (e.g., **81** and **82**), and by Sugihara, Okada, and Hiratani.[137] However their Li^+/Na^+ selectivities do not surpass those of the abovementioned acyclic diamide ionophores.

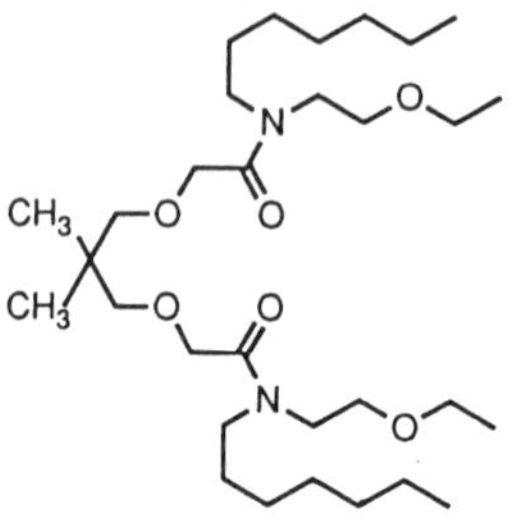

81

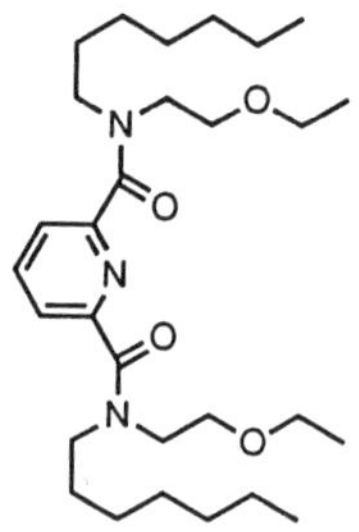

82

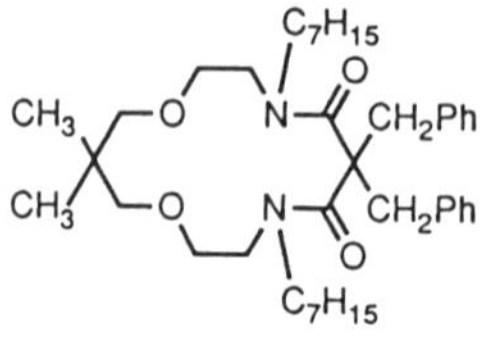

83

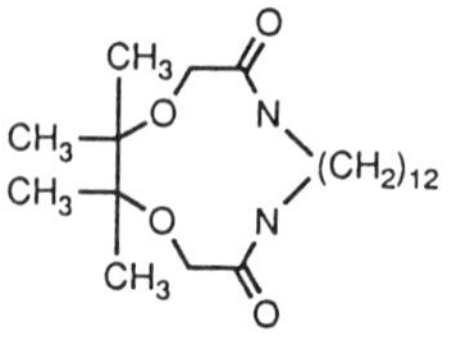

84

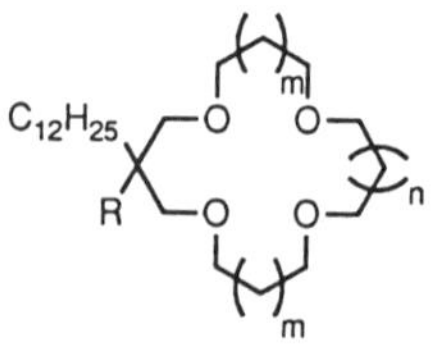

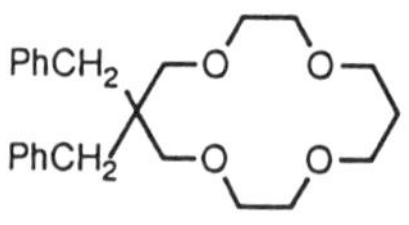

	m	n	R
85	0	0	H, CH_3
86	0	1	H, CH_3
87	1	0	H, CH_3
88	1	1	H, CH_3

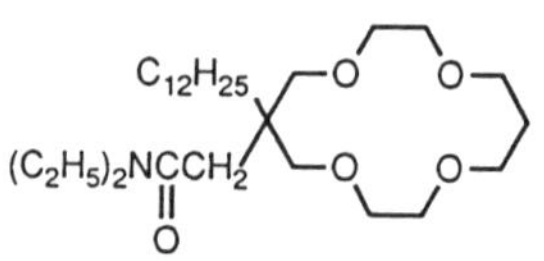

89

90

It was thought that macrocyclization of the acyclic diamide ionophores might enhance Li^+ selectivity. Two types of macrocyclic polyether diamides, **83** and **84**, were synthesized and evaluated, but no improvement in Li^+ selectivity was discerned.[138–140]

Acyclic diether **4**, which has a quinoline moiety at each end of the chain, is also a good candidate as a Li^+ ionophore in ion-selective electrodes. Plasticized PVC membranes containing ionophore **4** gave a K_{LiNa}^{Pot} value of 6.3×10^{-3} in an optimized formulation.[141,142] Owing to the basicity of the quinoline units, the electrodes employing ionophore **4** exhibit a greater response to H^+ than Li^+.

The lipophilic phenanthroline derivative **6** is highly selective for Li^+ in solvent polymeric membrane electrodes. A membrane containing **6** in NPOE-plasticized PVC gave a K_{LiNa}^{Pot} value below 1×10^{-3}.[143] However interference by H^+ may be a complicating factor.

Slight structural modification of monensin and ionomycin shifted the natural preference of these acyclic polyether antibiotics from Na^+ for the former and Ca^{2+} for the latter to a pronounced selectivity for Li^+.[144–146] Lactonized monensin monoisobutyrate gives a Li^+/Na^+ selectivity of 80 in a solvent polymeric membrane electrode. Acyclic, structurally modified derivatives of the two antibiotics exhibited much lower Li^+ selectivity.

4.1.6 Coronands in Lithium-Ion Selective Electrodes. Crown ethers (cyclic polyethers) are good candidates for Li^+ ionophores in ion-selective electrodes. Ion selectivities of crown ethers vary considerably with ring size and small-ring crown ethers are promising Li^+ ionophores. Crown ethers containing four ring oxygens (crown-4 compounds) were anticipated to provide the basic skeletons for the Li^+ ionophores.

The first crown ethers to be studied as Li^+ ionophores in ion-selective electrodes were 16-crown-4 **(45)** and its octamethyl derivative **46**. Ion-selective electrodes with **45** or **46** incorporated into PVC membranes plasticized with tris(2-ethylhexyl)phosphate (TEHP) were found by Aalmo and Krane[147] to give modest Li^+ selectivities with K_{LiNa}^{Pot} values of about 0.1. Shortly thereafter Olsher[148] reported a similar Li^+/Na^+ selectivity for dibenzo-14-crown-4 **(13)** incorporated in a PVC/NPOE (*o*-nitrophenyl octyl ether) membrane system. Use of 12-crown-4 **(14)** in PVC membrane electrodes by Gadzekpo and Christian[149] gave a lower Li^+ selectivity.

To systematically probe the influence of ring size variation within crown-4 compounds upon Li^+ selectivity in solvent polymeric membrane electrodes Kimura, Shono, and co-workers[150,151] examined crown-4 compounds **85–88**, which have ring sizes of 13- to 16-members, in PVC/NPOE membrane electrodes. The 13-crown-4 compounds **85** exhibited greater affinity for Na^+ than Li^+. With the 14-crown-4 compound **86** ($R{=}CH_3$), a K_{LiNa}^{Pot} value of less than 1×10^{-2} was obtained which showed high Li^+ selectivity. For the 15-crown-4 derivative **87**, the Li^+/Na^+ selectivities were almost the same as those noted with the 14-crown-4 compounds. Further expansion of the ring size gave diminished Li^+/Na^+ selectivity for the 16-crown-4 compounds **88**. Thus the 14-crown-4 and 15-crown-4 ring sizes appeared to be the best for producing high Li^+ selectivities in this series.

The Li^+/Na^+ selectivity can be increased by introducing geminal alkyl groups on the 14-crown-4 ring.[150–152] Usually Li^+ forms only 1 : 1 complexes with the 14-crown-4 ring; whereas Na^+ forms both 2 : 1 (crown ether : cation) sandwich complexes as well as 1 : 1 complexes. Since the 2 : 1 Na^+ complex is more lipophilic than the 1 : 1 Na^+ complex, its formation in the hydropholic medium of the membrane is accentuated which reduces the Li^+ selectivity in the membrane electrode. Thus prevention of formation of the 2 : 1 Na^+ complex by the presence of bulky geminal alkyl groups on the 14-crown-4 compound augments the Li^+ selectivity. Thus 6-dodecyl-6-methyl-14-crown-4 (**86**, R=CH_3) exhibits higher Li^+ selectivity in solvent polymeric membrane electrodes than does 6-dodecyl-14-crown-4 (**86**, R=H).[150,151] Kimura, Shono, and co-workers[152] have also found that 6,6-dibenzyl-14-crown-4 (**89**) exhibits high Li^+/Na^+ selectivity. Crown ether **89** is now commercially marketed as an Li^+ ionophore for use in ion-selective electrodes.

Incorporation of an additional binding site into the side arm of substituted 14-crown-4 compounds may also produce an increase in Li^+ selectivity. For example, the presence of an amide group, which possesses high affinity for cations with high charge density,[13] in ionophore **90** provides enhanced Li^+ selectivity ($K^{Pot}_{LiNa} = 4.4 \times 10^{-3}$) as compared with that of **86** (R=CH_3) with $K^{Pot}_{LiNa} = 6.9 \times 10^{-3}$.[152] This also seems to be the case for **91**, a chiral 14-crown-4 compound with two pendent amide functions, which Parker and co-workers[153–155] have found to give an excellent K^{Pot}_{LiNa} value of 1.6×10^{-3} when incorporated into PVC/NPOE membrane electrodes.

The Li^+ selectivity of an ionophore in a solvent polymeric membrane is also influenced by the membrane formulation. Addition of a small amount of an organophosphorus compound, such as trioctylphosphine oxide (TOPO) or tris(2-ethylhexyl)phosphate (TEHP) to liquid membranes or plasticized PVC membranes based on 14-crown-4 derivatives may promote their Li^+ selectivities with respect to Na^+, presumably because the Li^+-crown ether complexes are stabilized by axial coordination with the phosphorous compound.[156,157] For example, addition of 1×10^{-2} M TOPO to a dibenzo-14-crown-4 in nitrobenzene liquid membrane changed the K^{Pot}_{LiNa} value from 1×10^{-1} to 2.7×10^{-2}.[156] Also, an electrode based on PVC/NPOE-TEHP (98 : 2)/**89** gave an excellent K^{Pot}_{LiNa} value of 1.3×10^{-3}.[158] Li^+-selective electrodes based on ionophores **89** and **91** have been applied to the determination of Li^+ in serum[155,158] and, with the former compound, combination of the Li^+-selective membrane with a dialysis membrane can eliminate protein interference.[158]

In a side-by-side comparison of the lipophilic acyclic diamide ionophore **92** and the crown ether ionophore **86** (R=CH_3) in PVC/NPOE membrane electrodes, Christian and co-workers[136] found somewhat higher Li^+ selectivity for the former, but faster response at low Li^+ concentrations and stability for five months with the latter.

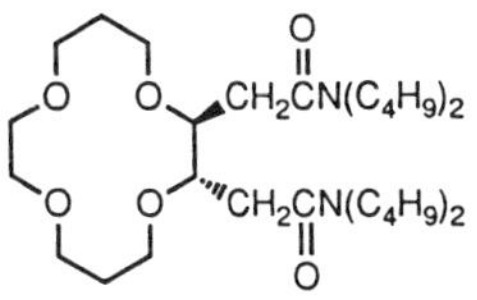

$CH_2C(O)N(C_4H_9)_2$

$CH_2C(O)N(C_4H_9)_2$

91

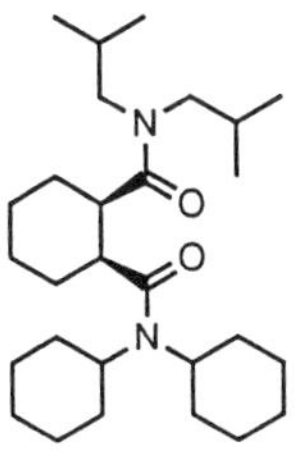

92

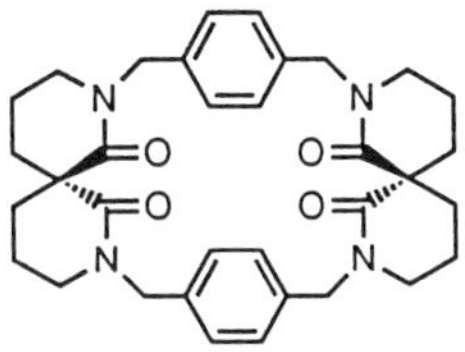

93

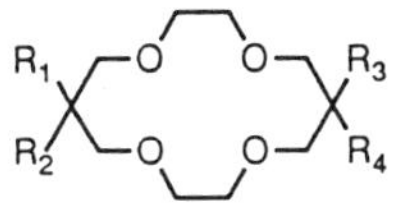

	R_1	R_2	R_3	R_4
94	CH_2Ph	CH_2Ph	CH_2Ph	H
95	C_2H_5	C_2H_5	C_2H_5	C_2H_5

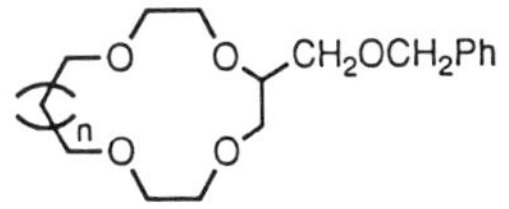

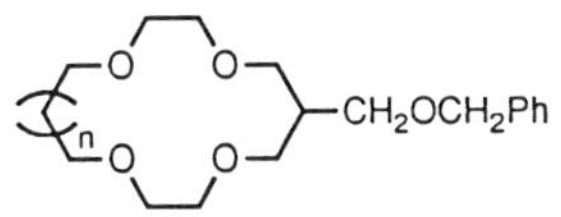

CH_2OCH_2Ph

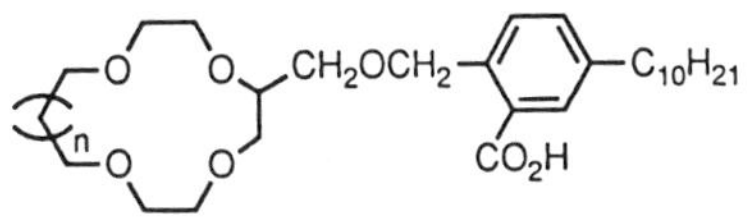

	n
96	0
97	1

CH_2OCH_2Ph

	n
98	0
99	1

CH_2OCH_2 $C_{10}H_{21}$ CO_2H

	n
100	0
101	1

CH_2OCH_2 $C_{10}H_{21}$ CO_2H

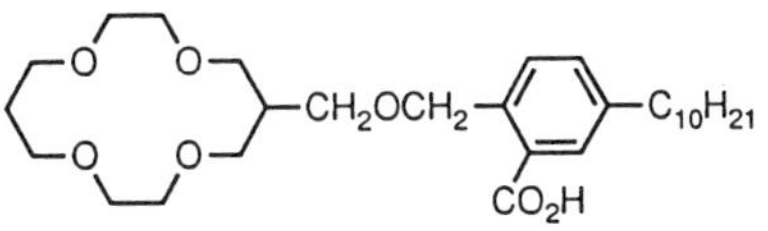

102

Several other types of macrocycles have been tested for their selectivities as Li^+ ionophores for ion-selective electrodes. These include a cyclic oligosiloxane,[159] a variety of macrocyclic lactones and lactone lactams,[160] 12-crown-3 derivatives,[161] dibenzo-14-crown-4 derivatives,[162,163] and cyclic tetraamide **93**.[164] All of these ionophores give inferior Li^+ selectivity when compared with high-performance electrodes based on 14-crown-4 derivatives.

4.1.7 Flow Injection Analysis (FIA) Systems. Flow through-type analytical systems are often more convenient and efficient than corresponding batch-type analytical methods. In some cases, the selectivity and sensitivity in analysis are enhanced in flow-through systems. FIA systems with flow through-type Li^+ electrodes give rapid response time, good reproducibility, and rapid assay time, and require very small sample volumes as compared with batch-type analysis using conventional ion-selective electrodes.

Several different FIA systems have been utilized.[165–175] Christian and co-workers[166, 168, 169, 171–175] developed a flow cell containing a set of silver wire microelectrodes (Fig. 10.3). The electrode membrane is prepared by dip coating of the silver wire with a solution of PVC, NPOE, the ionophore, and potassium tetrakis(*p*-chlorophenyl)borate in tetrahydrofuran. With this FIA system, Li^+ selectivities of several new cyclic diaamides,[169] a series of benzo-13-crown-4 compounds,[175] and a group of new substituted 14-crown-4 compounds[175] were determined. With the 14-crown-4 compounds it was discovered that 6,6,13-tribenzyl-14-crown-4 **(94)** and 6,6,13,13-tetraethyl-14-crown-4 **(95)** gave considerably higher Li^+/Na^+ selectivities than did the commercially available Li^+ ionophore **86**.

Christian and co-workers[171–174] investigated the responses of a series of (benzyloxy)methyl crown ethers **96–99** in the FIA system with coated silver wire electrodes and the fixed interference method (140 mM sodium chloride). The effect of incorporating 1% trioctylphophone oxide (TOPO) into the PVC/NPOE/ionophore/potassium tetrakis(*p*-chlorophenyl)borate matrix was also probed. In the absence of TOPO, the Li^+/Na^+ selectivity increased as the crown ether ring size was varied in the order 13-crown-4 < 12-crown-4 < 14-crown-4. In the presence of TOPO, the Li^+/Na^+ selectivity order changed to 13-crown-4 < 14-crown-4 < 12-crown-4. For the series of lipophilic crown ether carboxylic acids **100–102**, the Li^+/Na^+ selectivity was found to increase with crown ether ring size in the order 13-crown-4 < 14-crown-4 < 12-crown-4. In the presence of TOPO, the ordering changed to 13-crown-4 < 12-crown-4 < 14-crown-4. Also it was found that the weak Li^+ selectivity observed for *sym*-dibenzo-14-crown-4-oxyacetic acid **(72)** changed to mild Na^+ selectivity in the presence of TOPO. Thus the composition of the membrane matrix is once again shown to have an important influence on the selectivity of Li^+ ionophores.

4.1.8 Improvements in Ion Selectivity and Protein Interference. To the present time, the best selectivity coefficients for Li^+ with respect to Na^+ for neutral carrier-type solvent polymeric membrane electrodes are about 1×10^{-3}. Under these circumstances, a potentiometric Li^+ assay in human blood is just marginally possible.[122, 129, 155, 158, 167] It seems that K^{Pot}_{LiNa} values of about 10^{-4} are necessary for Li^+-selective electrodes with which a determination of approximately 1 mM Li^+ can be made with only 1% Na^+ interference. Additional research is necessary to provide new neutral synthetic ionophores with even higher Li^+/Na^+ selectivities. Also, optimization of the membrane matrix may enhance the Li^+/Na^+ selectivities of currently available and new Li^+ ionophores.[156,157]

In diluted blood samples, interference by proteins in the assay for Li^+ with ion-selective electrodes can be neglected. However, the interference is severe in undiluted samples. To avoid this problem, removal of proteins by pretreatment is necessary. Dialysis and ultracentrifugation have been used with undiluted blood to remove the proteins, especially those of high molecular weight. Since the pretreatment is quite time consuming, alternatives are under investigation.

Two convenient methods for the elimination of protein interference have been proposed. The method of Xie and Christian[168] utilizes a flow injection analysis (FIA) system containing a dialysis membrane. In this system samples are dialyzed automatically before the determination of Li^+ using a flow-through-type ion-selective electrode. The alternative method of Kimura, Oishi, Miura, and Shono[158] employs a Li^+-selective electrode with a composite dialysis membrane. With both of these systems, a whole blood sample can be analyzed for Li^+ without dilution in the important clinical range with an error of only a few percent.

4.2 Spectrophotometry with Chromoionophores and Fluoroionophores

In classical colorimetric determination methods, molar absorptivities of metal ions are enhanced by complex formation with chelate reagents.[177] If an ionophore bearing a chromophore or fluorophore in the vicinity of the metal ion binding site complexes a metal ion, the chromophore may be perturbed with significant changes in the absorption or emission spectra.[178] Although the subject of chromo- and fluoroionophores for metal ion determination was reviewed,[109,110,112,115,117,178] the present treatment focuses upon Li^+-selective chromogenic and fluorogenic ionophores. The coverage will be separated into sections for chromogenic and fluoroionophores which respond to Li^+ in homogeneous media and those that are utilized in extraction–spectrophotometric systems.

4.2.1 Lithium Ion Responsive Agents in Homogeneous Solution. Acyclic Li^+ ionophores **4–6**, which are good neutral carriers in Li^+-selective electrodes, may also be applied to Li^+ fluorimetry.[20-22] Hiratani[20,21] reported that the fluorescence emission intensity of podands **4** and **5** in acetonitrile or methanol was significantly enhanced by the addition of lithium perchlorate. The presence of sodium perchlorate increased the fluorescence slightly, but potassium perchlorate had no effect. Quantitative determination of lithium perchlorate in acetonitrile at concentrations of 10^{-4}–10^{-6} M by fluorescence spectrophotometry with podand **5** was achieved.[21] With 2,9-dibutyl-1,10-phenanthroline (**6**), Hiratani and co-workers[22] observed a very strong increase in fluorescence intensity in acetonitrile when lithium perchlorate was added.

The presence of sodium or potassium perchlorate had no influence on the fluorescence intensity of **5**.

Various monoazacrown ethers with a chromophore bound to the ring nitrogen, such as **103**, were prepared by Vögtle and co-workers[178,179] and spectral changes upon addition of alkali and alkaline-earth metal salts to their acetonitrile solutions were monitored. Although marked color changes were observed, none of these chromogenic coronands exhibited Li^+ selectivity.

It was surmised that the selectivity and strength of chromogenic response upon metal ion complexation by the ionophore would be enhanced if a chromogenic unit with an ionizable proton were employed.[178] This would alter the metal ion-ionophore interaction from an ion-dipole type to a stronger ion-pair interaction.

Crown ether azaphenol **104**, in which the acidic phenol group of the chromophore points into the crown ether cavity (i.e., is intraannular), under-

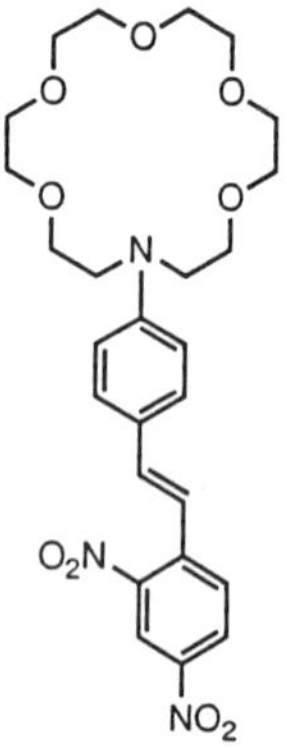

103 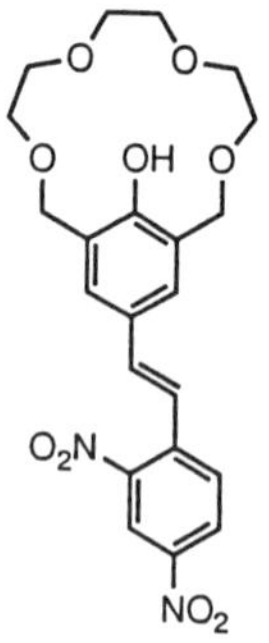 **104** 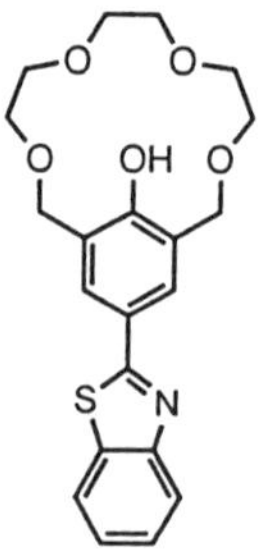 **105**

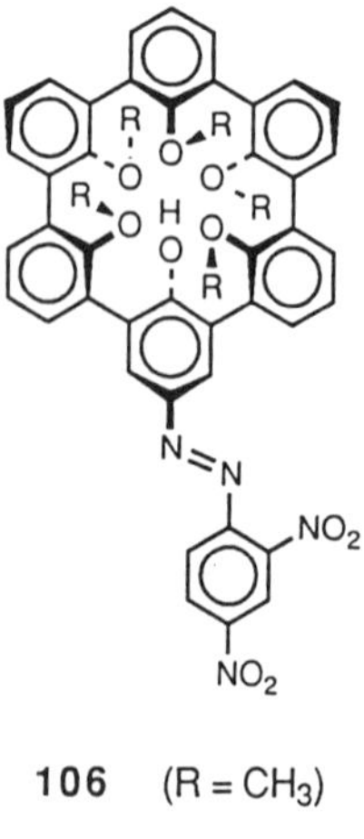

106 ($R = CH_3$) 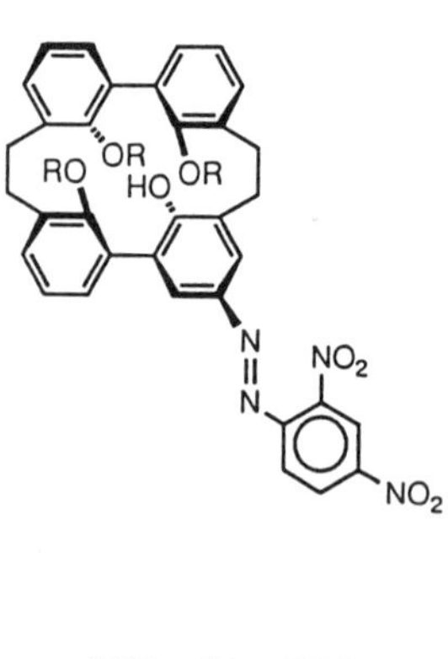 **107** ($R = CH_3$)

goes highly Li^+-selective coloration in the presence of an appropriate amine. Misumi and co-workers[180–183] found that addition of lithium salts to a basic organic solution of **104** changed its absorption spectrum with a red shift of 150 nm. The presence of other alkali metal salts produced only very minor spectral changes. Chromomionophore **104** in chloroform–dimethyl sulfoxide–triethylamine is a sensitive colorimetric assay for Li^+ in the range of 25 to 250 ppb.[183] This method was applied to the analysis of a commercial pharmaceutical preparation (a solid lithium carbonate tablet). When a fluoroionophore is incorporated into the coronand (e.g., in **105**) instead of a chromophore, the resultant fluoroionophore can be applied to Li^+ fluorimetry.[184–186]

Cram and co-workers[187] observed that yellow solutions of chromogenic spherand **106** containing an equivalent of pyridine in chloroform, ethanol, dioxane, or aqueous dioxane became deep blue-violet when lithium or sodium perchlorate was added. The presence of potassium perchlorate produced no change from the yellow color of **106**. Although the absorption maximum for the sodium phenolate complex of **106** occurs at a longer wavelength than that of the lithium phenolate complex, the separation is insufficient to allow Li^+ or Na^+ to be determined in the presence of the other ion.

Spherand azophenol dye **107** was designed by Kaneda, Misumi, and co-workers[188] to be a Li^+-specific chromoionophore. Addition of lithium salts to a solution of **107** in chloroform caused an immediate color change from yellow (λ_{max} = 430 nm) to violet (λ_{max} = 580 nm). Addition of a wide variety of alkali, alkaline-earth, and other metal salts produced no change in the yellow color of this chromogenic spherand.

Spirobenzopyran derivative **108** and **109** with monoazacrown moieties incorporated at different positions have been prepared recently by Inoue and co-workers[189] and by Kimura and co-workers,[190] respectively. These ionophores gave cation-induced coloration in organic solvents which was quite selective for Li^+.[189,190] Spirobenzopyran derivatives generally isomerize photochemically to their corresponding zwitterionic merocyanine forms which absorb in the visible region of the spectrum. Interestingly for **108** and **109**, metal ion complexation by the monoazacrown ether moieties induced the isomerization to their merocyanine forms even in the dark. Intramolecular interaction of a macrocycle-complexed Li^+ with the phenolate anion in the merocyanine forms enhanced the isomerization due to its high charge density.

Ogwa and co-workers[191] report that treatment of bis(bipyridyl) macrocycles **40** and **41** with sulfuric acid produces the red macrocyclic compound **110**. As lithium chloride was added to a dichloromethane solution of **110**, the red color changed to colorless. Addition of Na^+ or K^+ to a solution of **110** did not induce a color change. It was proposed that the bleaching of the red color by Li^+ is caused by the formation of complex **111**. In agreement, the ultraviolet spectra of **111** and the Li^+ complexes of **40** and **41** are very similar.

A weakness of chromogenic and fluorogenic ionophores **103–105**, **107–110** is that their Li^+ selective responses are limited to organic solvent systems.

The 14-membered formazane **112** exhibited high Li^+ selectivity in aqueous

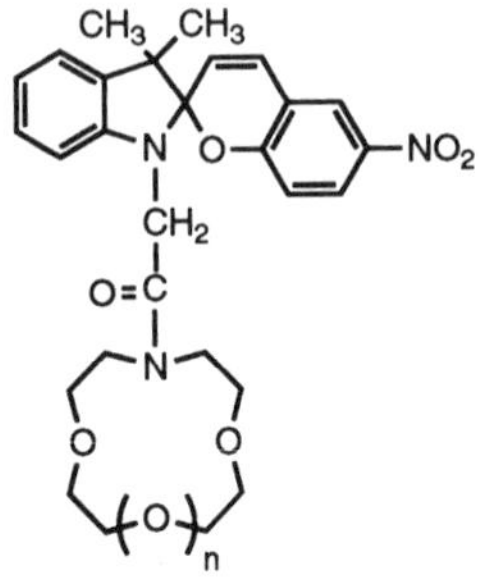

108 (n = 1, 2)

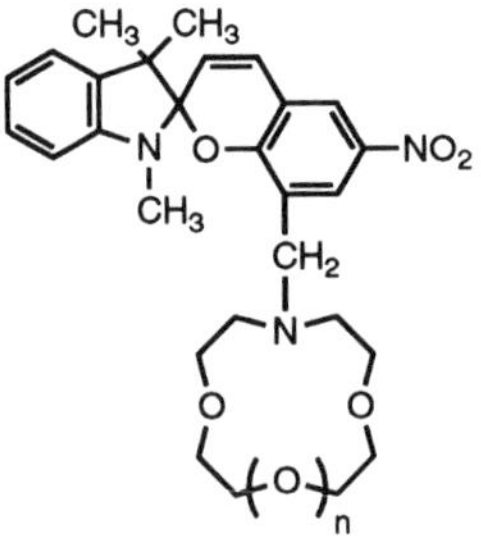

109 (n = 1, 2)

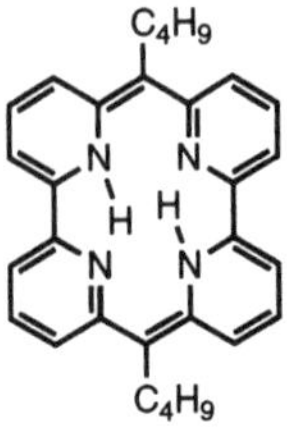

110

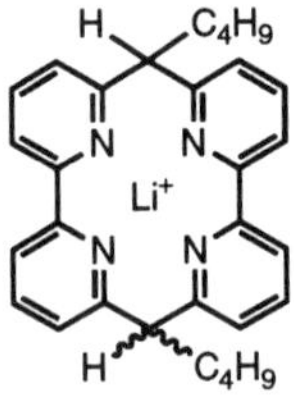

111

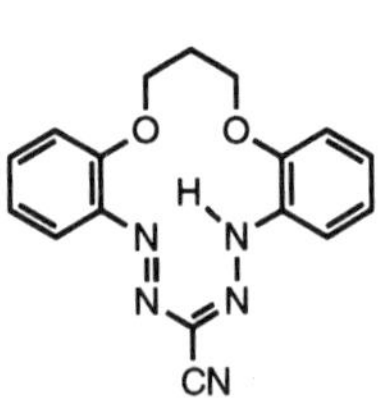

112

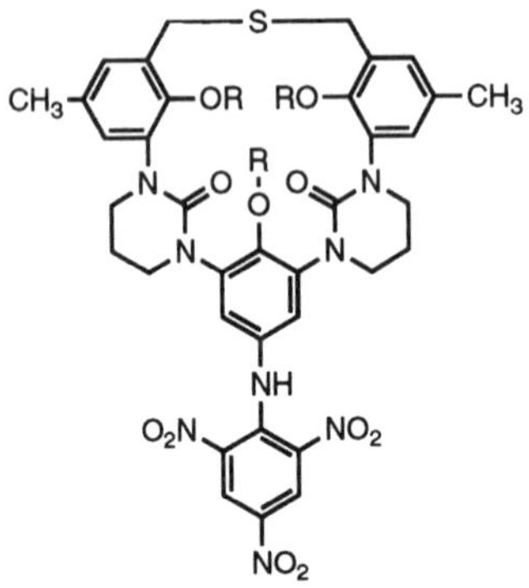

113 (R = CH_3)

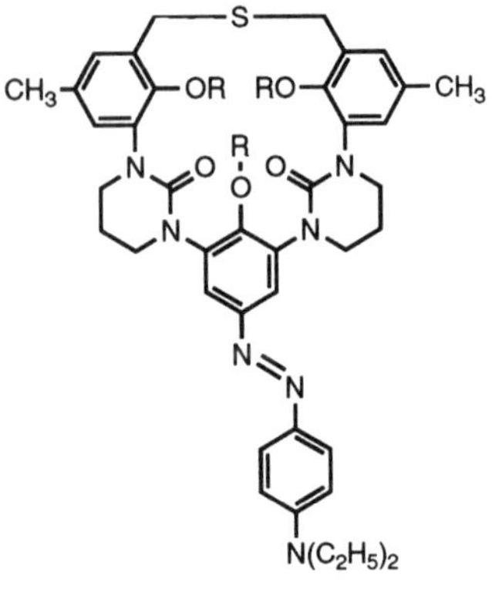

114 (R = CH_3)

acetone. However Attiyat and Christian[192] noted that the ionophore had poor stability and measurements must be made within 15 minutes.

Very recently chromogenic ionophores with high Li^+ selectivity which function in homogeneous aqueous media were reported by Czech, Chapoteau, Kumar, and co-workers.[193–195] In a 9:1 water–diethylene glycol monoethyl ether (DEGMEE) solution, chromogenic bis(urea) spherands **113** and **114** bound Li^+ better than Na^+.[192] At an optimal pH of 9.0, λ_{max} values for the absorbances of **113** and the Na^+ and Li^+ complexes of the anionic form of **113** are 406, 424, and 435 nm, respectively. Smaller spectral shifts were observed with the nonionizable chromogen **114**. Spectral differences between the Li^+ and Na^+ complexes of ionized **113** are insufficient for the determination of Li^+ in the presence of Na^+ in homogeneous aqueous media.

Azophenol cryptand **115** was prepared by Zazulak, Chapoteau, Czech, and Kumar[193] and found to exhibit extremely high selectivity for Li^+ over Na^+ in 10% aqueous DEGMEE. By a spectrophotometric method, an association constant value of 3200 M^{-1} was obtained for the lithium phenolate complex in 10% aqueous DEGMEE in the presence of tetramethylammonium hydroxide. Under these conditions there was no spectrophotometric response for Na^+ or K^+. Chapoteau, Czech, Zazulak, and Kumar[195] have reported the first practical colormetric method for determining Li^+ in blood serum without sample pre-

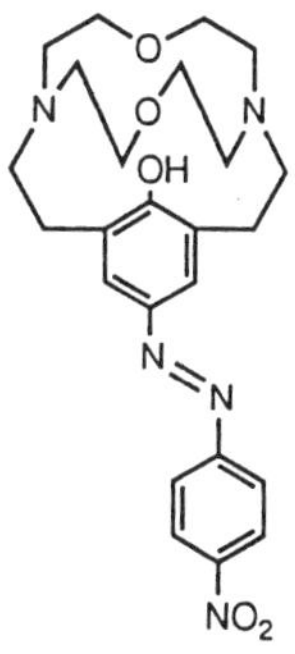

115 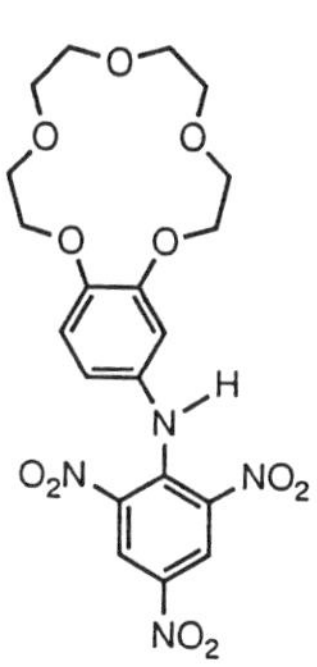 **116** 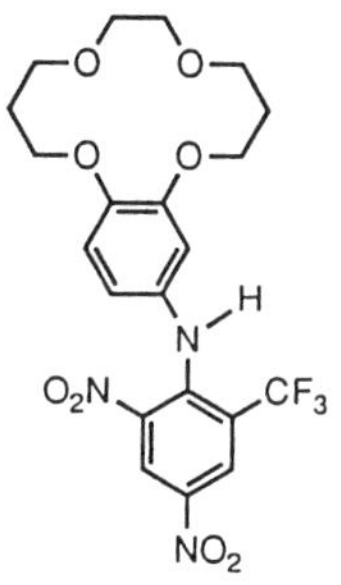 **117**

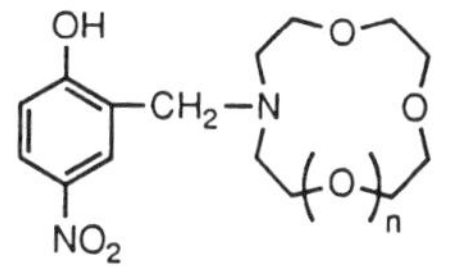

	n
118	1
119	2

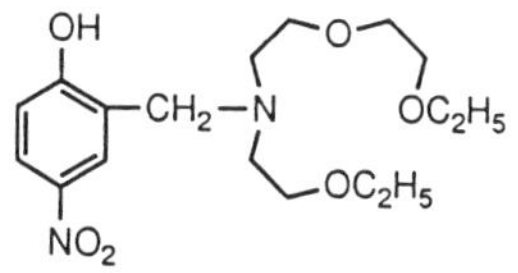

120

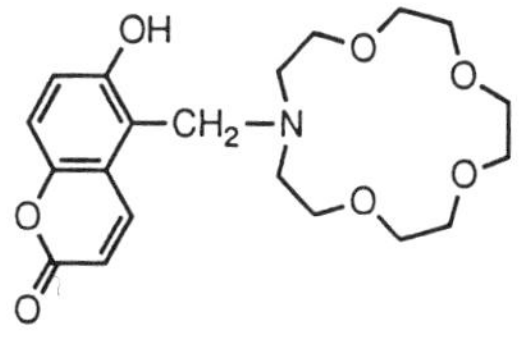

121 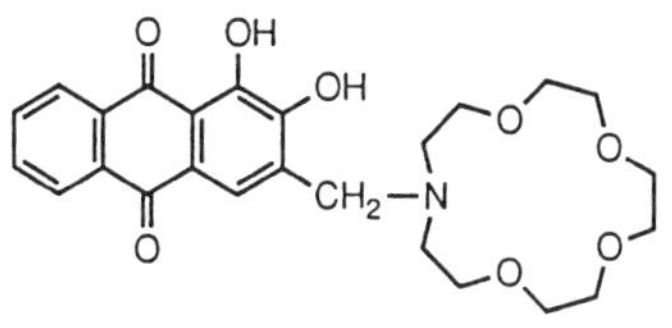 **122**

treatment or solvent extraction steps. The method is based on chromogenic cryptand **115**, which exhibits exceptionally high selectivity (>>4000 : 1) for Li^+ over Na^+ under the analysis conditions. The standard curve for the method is linear up to 3.5 mM, which exceeds the therapeutic range for Li^+. Results for blood serum samples from 57 patients correlated well with those obtained by flame photometry.

4.2.2 Lithium Ion Responsive Agents for Extraction Spectrophotometry. The mechanism of metal ion extraction from an aqueous solution into an organic solution of a proton-ionizable (i.e., proton-dissociable) crown ether is shown in Figure 10.4. In this process, a proton of the chromogenic ionophore in the organic phase is exchanged for a metal ion from the aqueous phase. If the ionophore is selective for a given metal ion, preferential extraction of that cation from a metal ion mixture may be achieved. Following the extraction, spectrophotometry on the organic phase determines the concentration of the metal ion complex(es).[110, 115, 117]

Assuming the formation of only 1 : 1 complexes for the extraction of alkali metal cations by a proton-ionizable ionophore (HL), the equilibrium and extraction equilibrium constant expressions are shown in Eqs. 13 and 14, respectively, where ML stands for a complex of L with M and the subscripts "o" and "a" denote the organic and aqueous phases,

$$(HL)_o + (M^+)_a \rightleftharpoons (ML)_o + (H^+)_a \tag{13}$$

$$K_{ex} = \frac{[ML]_o\,[H^+]_a}{[HL]_o\,[M^+]_a} \tag{14}$$

respectively. If the following assumptions are made, where "t" means total, $[HL]_t \simeq [HL]_o + [ML]_o + [L^-]_a$, $[M^+]_t \simeq [M^+]_a$, $[HL]_a \simeq 0$, $[ML]_a \simeq 0$, and $[L^-] \simeq 0$, Eq. 14 can be rewritten as shown in Eq. 15, where ϵ_{L^-} and A_a stand for the molar absorptivity and absorbance for L^- in the aqueous phase.

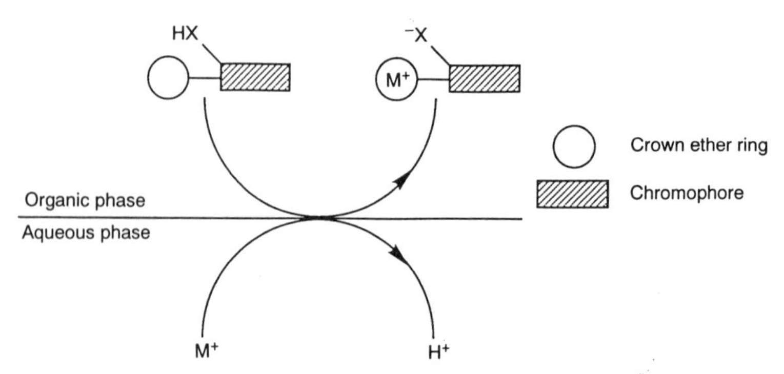

Figure 10.4 Extraction of a metal ion by a proton-ionizable crown ether.

Also, ϵ_{ML} and A_o signify the molar absorptivity and absorbance for ML in the

$$\frac{[H^+]_a}{[M^+]_t} = \left[\frac{K_{ex}\epsilon_{ML}(\epsilon_L - [HL]_t - A_a)}{\epsilon_L - A_o}\right] - K_{ex} \tag{15}$$

organic phase at a wavelength where HL does not absorb. Furthermore, if distribution of the chromogenic ionophore is negligible due to its high lipophilicity, Eq. 15 can be simplified to Eq. 16. In systems where significant cation extraction is realized, plots of $[H^+]_a/[M^+]_t$ against $(\epsilon_L\text{-}[HL]_t\text{-}A_L)/(\epsilon_L\text{-}A_o)$ or $1/A_o$ give straight lines, which show that the assumptions made above are reasonable. The K_{ex} value is obtained from the intercept of the line. Most frequently the extraction equilibrium constant is expressed in terms of pK_{ex}, as shown in Eq. 17. A larger value of pK_{ex} signifies a lower level

$$\frac{[H^+]_a}{[M^+]_t} = \frac{K_{ex}\epsilon_{ML}[HL]_t}{A_o} - K_{ex} \tag{16}$$

$$-\log pK_{ex} = pK_{ex} \tag{17}$$

of metal ion extraction.

In the first report of extraction spectrophotometry with a chromogenic crown ether, Takagi, Ueno, and co-workers[196] used an organic solution of ionophore **116** to extract alkali metal cations from basic aqueous solutions. Upon ionization of the picrylamino group of **116**, the alkali metal cation is transferred into the organic phase forming an electroneutral, intramolecular ion-pair complex. When employed in the extraction pectrophotometry method, ionophore **116** was selective for K^+.

For Li^+ determination, Wu and Pacey[197] designed and synthesized the closely related benzo-14-crown-4 compound **117** and utilized this chromogenic crown ether to extract Li^+ from aqueous solutions into chloroform. Although the level of extraction was low ($pK_{ex} = 11.0$), the Li^+ selectivity exhibited by **117** was applied to the determination of Li^+ in blood serum.[197]

Macrocyclic polyethers with side arms containing coordination sites are termed lariat ethers.[7a] If the side arm is attached to a carbon atom of a crown ether ring, the ionophore is identified as a C-pivot lariat ether. If the attachment site is the nitrogen atom of a monoazacrown ether, the compound is called a N-pivot lariat ether.

In Table 10.9 are collected extraction constant data for transfer of Li^+ and Na^+ from aqueous solutions into 1,2-dichloroethane by chromogenic and fluorogenic lariat ethers **118**, **119**, **121–136**, and model compound **120**. From the extraction constants for Li^+ and Na^+, the Li^+/Na^+ selectivity may be calculated. The calculated selectivities are also given in Table 10.9.

Takagi and co-workers[199–202, 204] and Pacey and co-workers[198, 203] prepared chromogenic and fluorogenic N-pivot lariat ethers **118**, **119**, and **121–126** by attaching a chromogenic or fluorogenic group to the nitrogen atom of a mono-

Table 10.9 Extraction Constants for Transfer of Li^+ and Na^+ from Aqueous Solution into 1,2-Dichloroethane at 25°C by Proton-Ionizable Chromogenic and Fluorogenic Lariat Ethers

	pK_{ex}		$K_{ex}(Li^+)$	
Ionophore	Li^+	Na^+	$K_{ex}(Na^+)$	Ref.
118[a]	10.2	12.5	209	198
118	9.7	12.4	500	199
119	9.2	9.8	4.0	200, 201
120	11.0	13.1	130	201
121	9.7	10.3	4.0	202
122	8.3	8.7	2.5	201
123	11.7	12.8	13	199
124[a]	7.1	7.4	2.0	203
125	9.0	9.3	2.0	199, 201, 204
126	9.8	10.3	3.7	199, 201, 204
127	10.1	9.9	0.8	199, 201, 204
128	9.8	10.3	3.7	199, 201, 204
129[a]	7.1	7.4	2.0	203
130	11.3	13.7	240	205, 206
131	6.9	8.9	87	205, 206
132	10.9	12.9	91	206
133	8.6	10.4	68	206
134	7.0	8.7	49	206
135	6.9	8.8	76	206
136	6.8	8.8	100	206
137	11.1	13.4	200	207

[a]For extraction into chloroform.

azacrown ether. Of these compounds the Li^+/Na^+ selectivity for extraction into 1,2-dichloroethane is the highest for compounds **118** and **123**, which incorporate a monoaza-12-crown-4 unit. The Li^+/Na^+ selectivity of **120** which is an acyclic analogue of **118** is also high, but the level of Li^+ extraction with this proton-ionizable chromogenic podand is very low. The C-pivot chromogenic lariat ethers **128** and **129** provide poor Li^+/Na^+ selectivity.

Kimura, Shono, and co-workers[205,206] prepared and evaluated the series of chromogenic C-pivot lariat ethers **130–136** which are based on 14-crown-4. In general the Li^+/Na^+ selectivities for extraction into 1,2-dichloroethane are in the range of 50–240. Chromogenic ionophore **131** has been applied to the analysis of Li^+ in blood serum using an extraction spectroscopy flow injection analysis method.[207,208] Fluorogenic **137** and other compounds have been used in Li^+ extraction fluorimetry.[209,210]

Crowned spirobenzopyrans **109** are Li^+ selective for extractions from aqueous solution into 1,2-dichloroethane.[190]

Chromogenic bis(urea) spherands **113** exhibited selectivity for Na^+ over

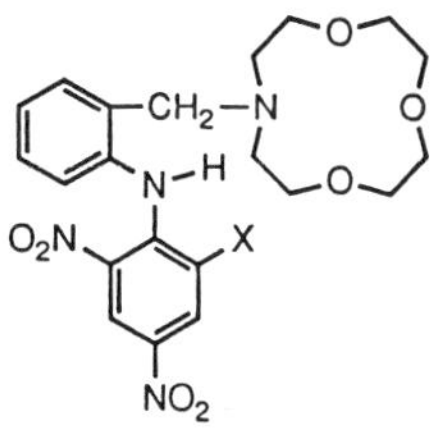

	X
123	NO_2
124	CF_3

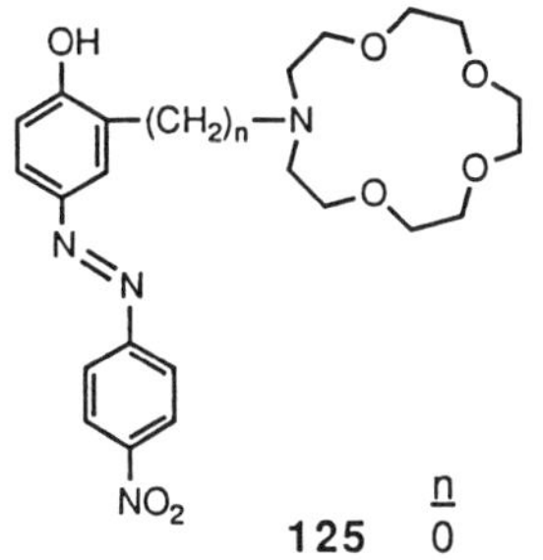

	n
125	0
126	1
127	2

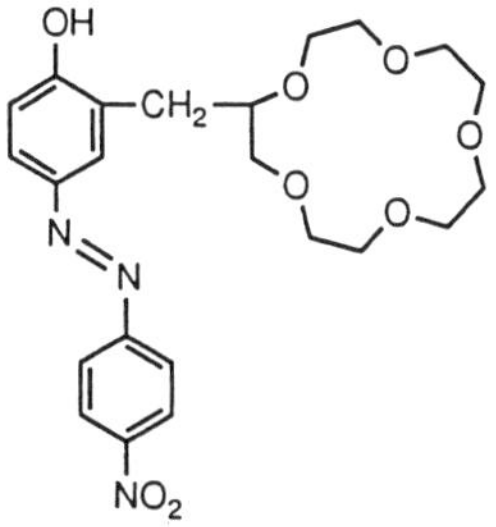

128

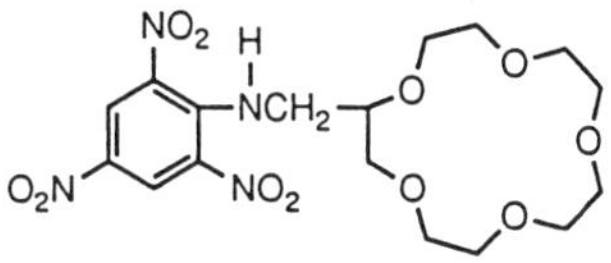

129

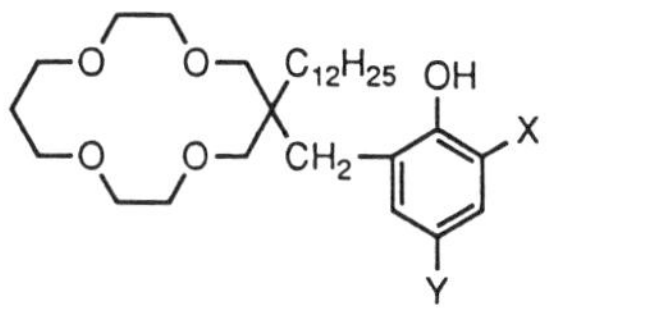

	X	Y
130	H	NO_2
131	NO_2	NO_2
132	N=N-C_6H_4-NO_2	N=N-C_6H_4-NO_2
133	NO_2	N=N-C_6H_4-NO_2

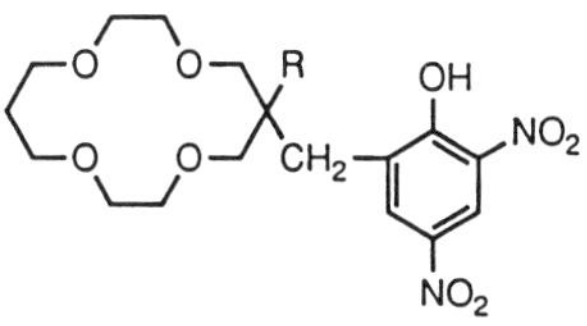

	R
134	H
135	CH_3
136	CH_2Ph

Li^+ in chloroform–water extractions, even though Li^+ was bound better than Na^+ in homogeneous aqueous media.[193]

Very recently Sholl and Southerland[211] reported the preparation of aza-phenol cryptand **138**. Chloroform solutions of **138** were found to extract Li^+ from aqueous solutions at pH 7–8 with very high selectivity over Na^+, K^+, Mg^{2+}, and Ca^{2+}. Based upon the extraction constants, the Li^+/Na^+ selectivity was about 400.

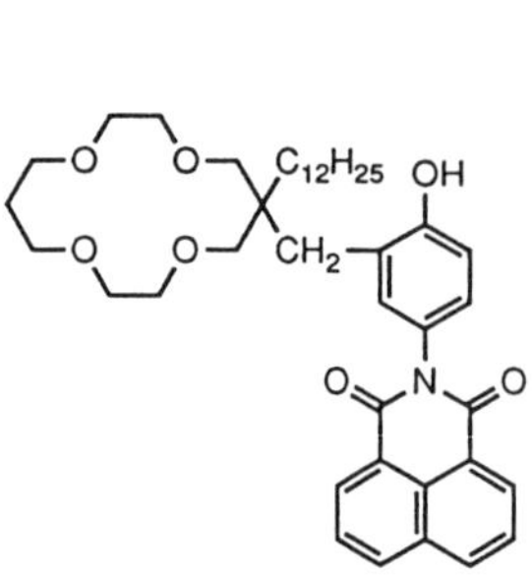

137 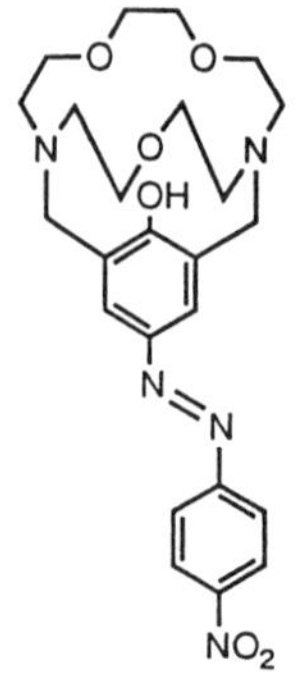 138 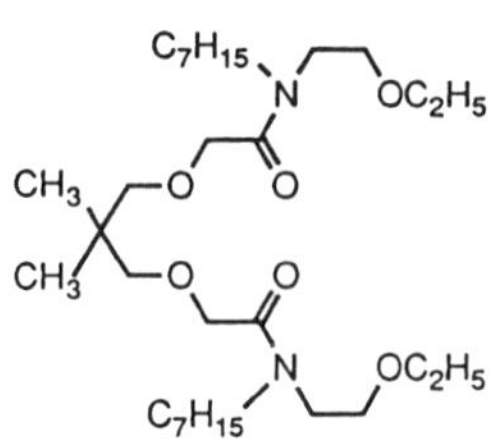 139

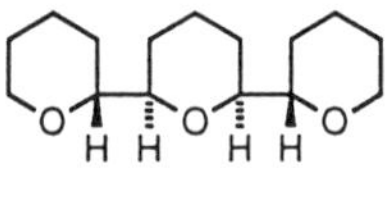

140 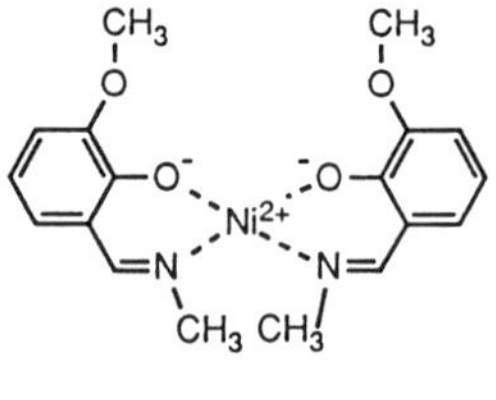 141

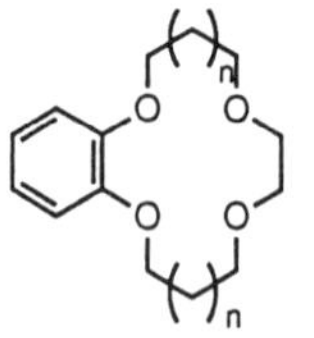

	n
142	1
143	0

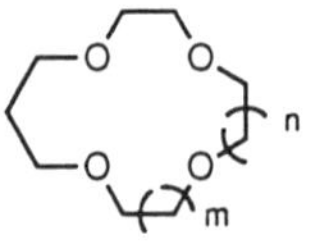

	n	m
144	1	1
145	1	2
146	2	2

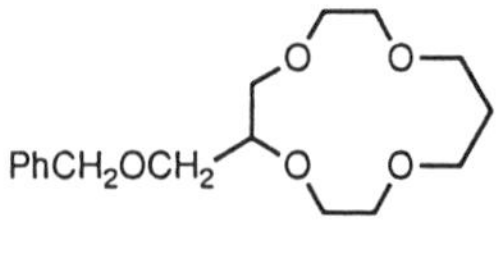

147

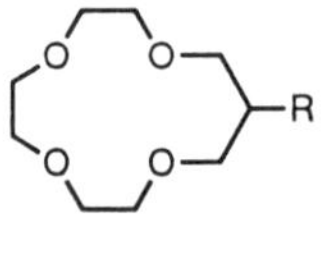

	R
148	$=CH_2$
149	CH_2OCH_2Ph

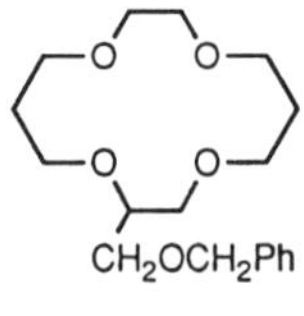

150

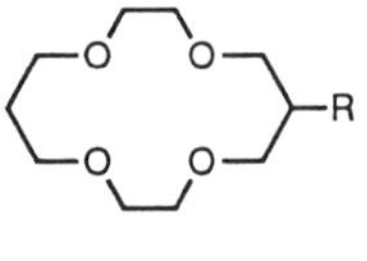

	R
151	$=CH_2$
152	CH_2OCH_2Ph

5 SEPARATION OF LITHIUM IONS USING IONOPHORES

Lithium has been recovered from minerals such as spodumene, lepidolite, petalite, and amblygonite. To recover lithium from lake or ocean brines, attempts to precipitate lithium carbonate to separate Li^+ from other alkali and alkaline earth metal ions have been made. This method is ineffective except for Li^+-rich brines, such as in the Salar de Atacama in Chile.

With the advent of crown ethers,[27] a potential for the separation of a particular alkali metal cation from other metal ions appeared. In the early years of this chemistry, no synthetic ionophore with the capability to selectively extract Li^+ from aqueous solution into an organic phase was known. This is due to the strong hydration energy of Li^+ and the similarities of Li^+ and Na^+ in their chemical behavior. Through design and synthesis efforts of many researchers in laboratories worldwide, several synthetic ionophores are now available which exhibit good Li^+ selectivity in separation processes.

In this section, ionophores for the separation of Li^+ by solvent extraction[108] and transport across liquid membranes[106,111] will be discussed. The Li^+ complexing agents will be separated into two categories: (1) neutral ligands which form ternary complexes with a metal cation and its counteranion; and (2) proton-ionizable (i.e., proton-dissociable) ligands[212,213] which form neutral complexes with the metal cation. For the latter, the anionic site results from proton dissociation.

5.1 Separation of Lithium Ions by Solvent Extraction

5.1.1 Neutral Ionophores

Extraction Constant. For solvent extraction with a neutral ionophore, the extraction equilibrium is given in Eq. 18. Subscripts "o" and "a" identify species

$$(L)_o + (M^+)_a + (A^-)_a \rightleftharpoons (MLA)_o \qquad (18)$$

in the organic and aqueous phases, respectively, and MLA designates an ion pair between, for instance, a crown ether complexed with the metal ion (ML^+) and the counterion (A^-). Thus the extraction may be designated as an "ion-pair" extraction.

A popular method for determining cation-binding strengths of synthetic ionophores was originally developed by Frensdorff[214] and subsequently refined by others.[215,216] In this technique, an aqueous solution is prepared of the metal ion of interest and a colored anion, typically picrate. When shaken with an immiscible organic solvent, typically dichloromethane, 1,2-dichloroethane, or chloroform, no extraction occurs and the organic phase remains colorless. When a synthetic ionophore is present in the organic phase, this phase becomes colored. Using Beer's law, the amount of anion in the organic phase can be assessed under the assumption that the amounts of anion and metal ion in the organic phase are the same. Then through a series of equilibrium expressions, the extraction constant, K_{ex}, is calculated. The technique must be used with care, since the values obtained depend on several variables. These include the organic solvent, the concentrations of cation and anion in the aqueous phase, the volumes of the aqueous and organic solutions, the ratio of the ligand to the salt, the ionic strength of the aqueous phase, the temperature, and even the mixing procedure. When all of these variables are controlled, the resultant extraction constants provide useful information regarding the relative extraction efficiencies of different synthetic ionophores.

Recently Inoue and co-workers[217] reported that dicationic complexes may be formed using the picrate extraction method with unsubstituted crown ethers if the metal ion diameter is considerably smaller than that of the polyether cavity.

Podands. Because of the high hydration energy of Li^+ and close similarity in the complexation behaviors of Li^+ and Na^+, extraction selectivity for Li^+ by podands is rare.

For picrate extractions with acyclic tetraether diamide **139**, Shanzer and co-workers[218] observed that Li^+ was extracted from an aqueous solution into dichloromethane 40 times better than Na^+. Presumably 2 : 1 (podand : metal ion) extraction complexes were formed.

Still and co-workers[219] report that podand **140** extracts lithium picrate into chloroform seven times better than sodium picrate. Molecular mechanics calculations reveal a low-energy conformation of **140** which is preorganized for Li^+ complexation.

A self-assembling podand has been described by Scheparta and McDevitt.[220] A complex of two molecules of *N*-methyl 2-hydroxy-3-methoxybenzaldimine with Ni^{2+} gave complex **141** in which a four-oxygen binding site has been assembled. Complex **141** extracts alkali metal picrates into chloroform with the efficiency ordering $Na^+ > Li^+ > K^+, Rb^+, Cs^+$.

Coronands. As described in Section 4.1.6, crown-4 compounds which contain four oxygens in the crown ether rings may complex Li^+ effectively. Selectivity for Li^+ extraction with a crown-4 compound was first reported by Olsher and Jagur-Grodzinski.[37] Benzene and dichloromethane solutions of benzo-13-crown-4 **(16)** were found to selectively extract Li^+ from aqueous solutions containing the other alkali metal cations. Selectivity ratios for Li^+ over the other metal ions (Li^+/M^+) for extraction into dichloromethane were 2.5, 44, 216, and 355 for Na^+, K^+, Rb^+, and Cs^+, respectively.

For picrate extraction into dichloromethane, a Li^+/Na^+ selectivity of 4.0 is calculated for benzo-14-crown-4 **(142)**.[198] The effect of organic solvent variation upon the extraction of lithium picrate by dibenzo-14-crown-4 **(13)** and its analogues has been reported by Wang and co-workers.[221]

Inoue and co-workers[222] extracted alkali metal picrates with dichloromethane solutions of benzo-12-crown-4 **(143)**, 12-crown-4 **(14)**, 13-crown-4 **(144)**, 14-crown-4 **(145)**, 15-crown-4 **(146)**, and 16-crown-4 **(45)**. Only with 14-crown-4 **(145)** was significantly greater extractability of lithium picrate than sodium picrate observed.

Bartsch and co-workers[223] synthesized a series of novel crown-4 compounds

147–157 in which the crown ether ring size is systematically varied from 13 to 16 atoms. Extraction constants for extraction of lithium and sodium picrates into chloroform by these crown-4 compounds are shown in Table 10.10. The highest level Li^+ extraction was achieved with the 14-crown-4 compounds **150–152**. The methylene-substituted 14-crown-4 compound **151** exhibited very high Li^+ selectivity with $Li^+/Na^+ = 44$. In a very recent study of substituted 14-crown-4 compounds, Bartsch, Czech and co-workers[175] observed that the bis-methylene-substituted 14-crown-4 compound **60** has the same Li^+/Na^+ extraction selectivity as **152**. For the series of benzyloxymethyl-substituted crown-4 compounds **147**, **149**, **150**, **152**, **153**, **155**, and **157**, the Li^+ extracting ability decreased as the crown ether ring size was varied 14-crown-4 > 13-crown-4 ≥ 15-crown-4 >> 16-crown-4. This ordering may be rationalized in terms of the size-fit concept.[38]

Recently, Okahara and co-workers[224] prepared a series of 12-, 13-, and 14-crown-4 derivatives which bear geminal methyl and lariat groups on the same ring carbon atom (e.g., **158** and **159**). The side arm of the lariat ether contains potential coordination sites. The Li^+/Na^+ extraction selectivity was greater for the 14-crown-4 compound **159** than for 13-crown-4 derivative **158**.

Lipophilic 14-crown-4 derivatives **160** and **162**, which are preorganized by decalin walls, were synthesized by Koribo and co-workers.[225,226] These ionophores combine high selectivity and extractability for Li^+. Very recently, Sachleben and co-workers[227] prepared closely related di-decalin compound **161** and the synthetically more accessible nonamethyl 14-crown-4 **63**. Both compounds give high selectivity and efficiency for Li^+ in competitive alkali metal chloride and nitrate extractions into 1-octanol.

Macroheterocycles **163–166** which contain tetrahydrofuran units are reported by Kobuke and co-workers.[228] The coronands discriminated between

Table 10.10 Extraction Constants for Extraction of Lithium and Sodium Picrates from Aqueous Solution into Chloroform by Crown-4 Compounds[223]

Ionophore	Ring Size	K_{ex} Li$^+$	K_{ex} Na$^+$	$K_{ex}(Li^+)/K_{ex}(Na^+)$
147	13-crown-4	39	8	5
148	13-crown-4	22	2	11
149	13-crown-4	9	3	3
150	14-crown-4	119	12	10
151	14-crown-4	445	10	44
152	14-crown-4	200	8	25
153	15-crown-4	6	6	1
154	15-crown-4	5	<1	
155	15-crown-4	11	7	2
156	16-crown-4	2	<1	
157	16-crown-4	<1	<1	

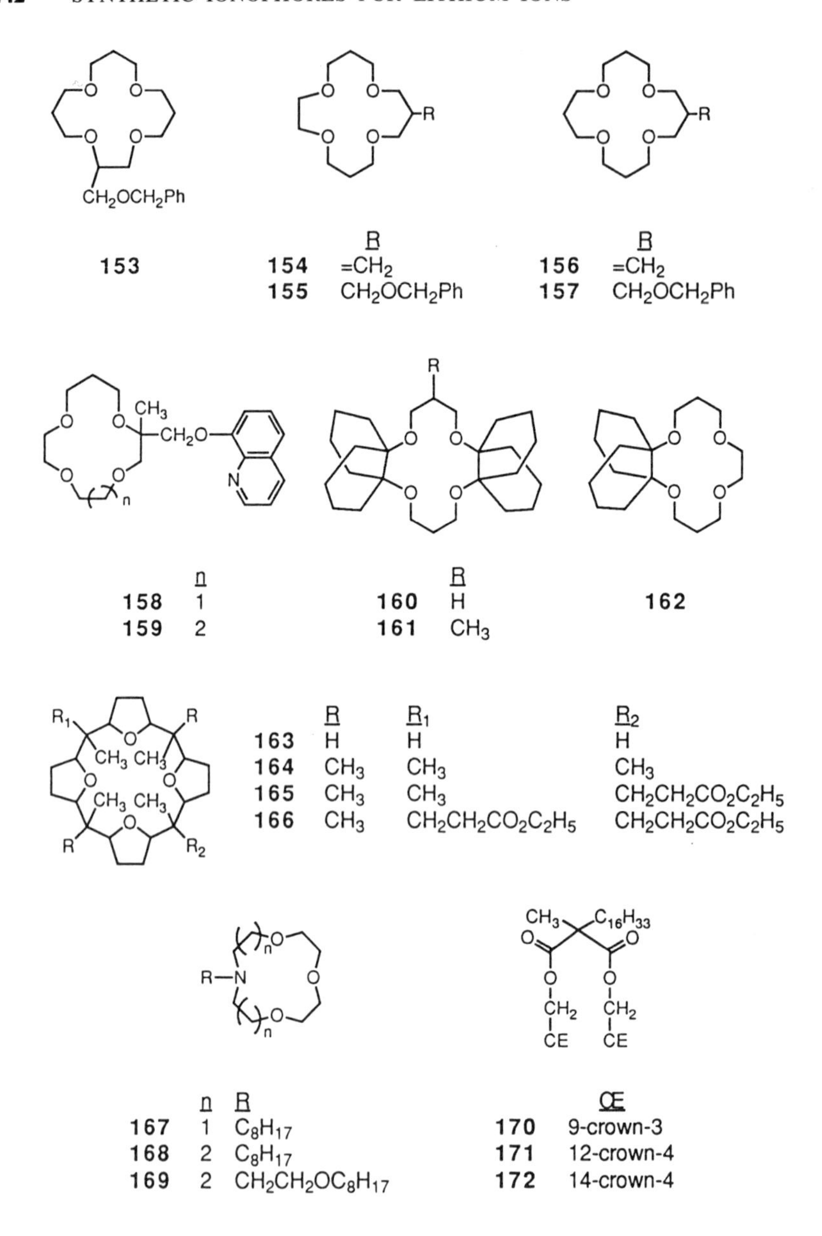

Li^+ and the other alkali metal cations for picrate extractions into chloroform. The simplest compound **163** was the most efficient extraction agent, but offered poor Li^+/Na^+ selectivity. Compound **165** with a pendent ethyl carboethoxy group was the most selective for Li^+.

Okahara and co-workers[229] observed high Li^+/Na^+ selectivities for picrate extractions into dichloromethane by small-ring azacrown ethers **167** and **168** and N-pivot lariat ether **169**.

For the bis(crown ether) compounds **170–172**, a cation may interact cooperatively with two crown ether rings to form an intramolecular sandwich-type complex. For picrate extractions into chloroform, Bartsch and co-workers[230] observed a cation extraction selectivity order of $Li^+ > Na^+ > K^+ > Rb^+$ with a calculated selectivity ratio $Li^+/Na^+ = 7$ for the bis(9-crown-3) compound **170**. For the bis(14-crown-4) compound **172** the selectivity ratio was the same, but with the bis(12-crown-4) compound **171** the ratio was 37.

Nishimura and co-workers[231] have communicated recently the synthesis of crownophores **173**. In competitive extractions of Li^+, Na^+, and K^+ from alkaline aqueous media by chloroform solutions of the crownophore and dodecanoic acid, Li^+/Na^+ selectivities of 15–25 were achieved.

Synergistic Li^+ extraction by a combination of a crown ether and bis(2-ethylhexyl)phosphate **(174)** in chloroform has been reported by Takahashi and co-workers.[232] The combination of the crown ether and a lipophilic acidic component gave a higher level of Li^+ extraction than when either the crown ether or phosphate ester were used independently.

Surprisingly the cyclophane-type macrocyclic ligand **175** was found by Saigo and co-workers[233] to exhibit pronounced Li^+/Na^+ selectivity for alkali metal picrate extractions into chloroform. Since the cavity size in **175** is much larger than that of Li^+, it was proposed that $Li(OH_2)_6^+$ was complexed.

Spherands. Preorganization of the binding site is an effective method for enhancing the ion selectivity and complexing ability of a synthetic ionophore.[104,234] The spherands prepared by Cram and co-workers are highly preorganized compounds.[235] Spherands **76, 176–181** are highly selective for Li^+.[87,234–238] Association constants for complexation of alkali metal picrates in chloroform saturated with water were determined by ion-pair extraction of deuterium oxide solutions of the metal picrate with a deuteriochloroform solution of the ionophore. Calculated values of $K(Li^+)/K(Na^+)$ for spherands **76, 176–180** are 580, 28, 790, 840, 36, and 90, respectively. Although these spherands have excellent selectivity for Li^+, the sophisticated structures and slow extraction of Li^+ salts from aqueous solution into immiscible organic solvents[234] reduce their potential as practical extractants.

5.1.2 Proton-Ionizable Ionophores. For the solvent extraction results described above, neutral ligands were employed which form ternary complexes with the cation and counterion. A ligand containing a proton-ionizable (proton-dissociable) group can complex with a metal cation to form a neutral complex (Eq. 13, Section 4.2.2). Thus proton-ionizable ionophores possess the distinct advantage over neutral ligands in that extraction of the metal ion from the aqueous phase into the organic phase does not require concomitant transport of an aqueous phase anion. This is especially important for potential practical applications in which hard aqueous phase anions, such as chloride, nitrate, and sulfate, would be involved.[113]

Although the proton-ionizable chromoionophores and fluoroionophores are

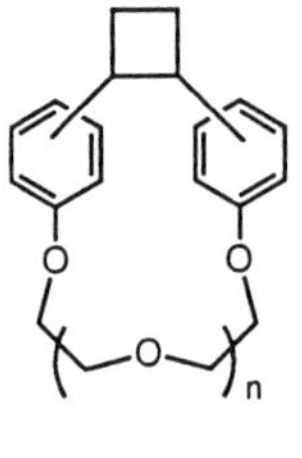

(n = 1 - 4)

173 **174** 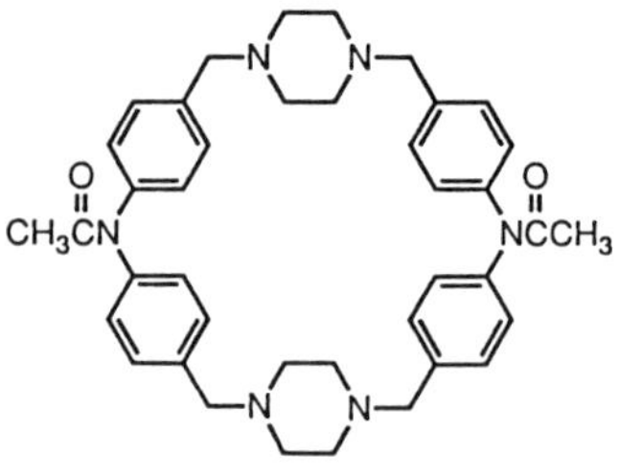 **175**

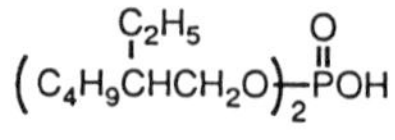

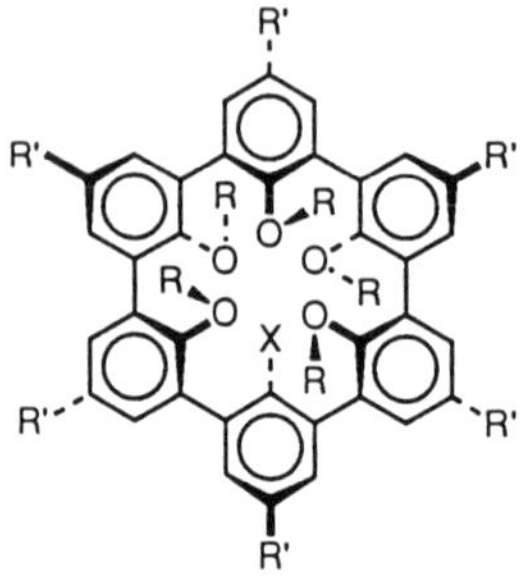

	X	R	R'
176	OH	CH_3	CH_3
177	OH	CH_3	H
178	H	CH_3	CH_3

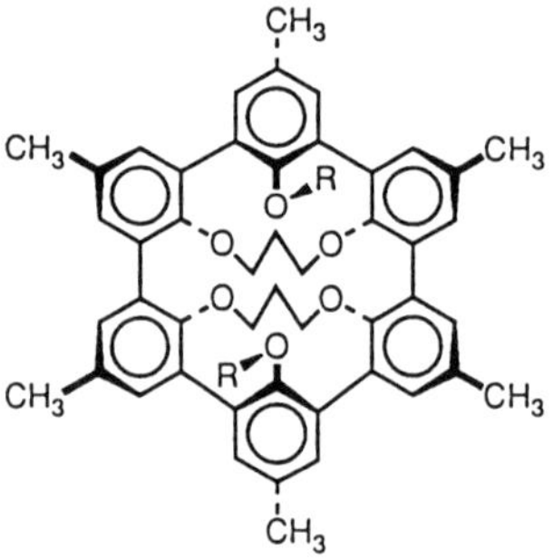

179 (R = CH_3)

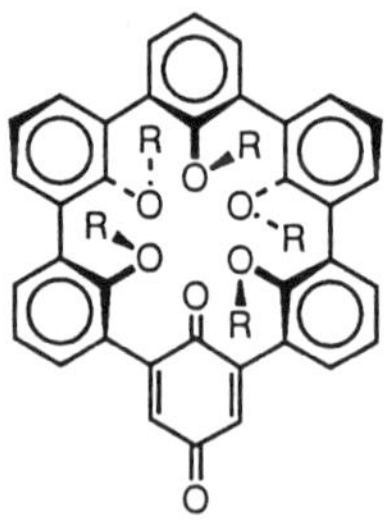

180 (R = CH_3)

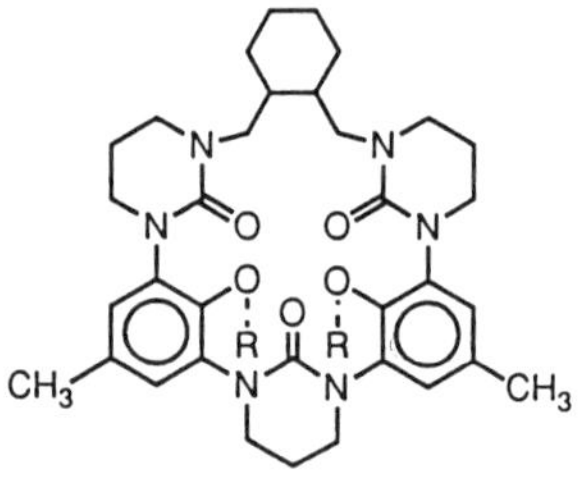

181 (R = CH_3)

included in this category, they have already been surveyed in Section 4.2.2 on the determination of Li^+ by extraction spectrophotometry. In this section nonchromophoric podand and coronand extractants which have pendent carboxylic or phosphorous acid functions are described.

Podands. Only very limited information is available concerning proton-ionizable podands with extraction selectivity for Li^+.

Recently, Bartsch and co-workers[239] reported results for competitive solvent extractions of alkali metal cations from aqueous solutions into chloroform solutions of lipophilic polyether carboxylic acids **182–189**. Li^+ extractivity was more pronounced when the polyether chain terminated in an ether linkage ($R{=}CH_2Ph$) rather than an alcohol group. For **186–189** the Li^+/Na^+ extraction diminished as the polyether chain was lengthened. Noteworthy was the extraction selectivity order of $Li^+ \gg Na^+ > K^+, Rb^+ > Cs^+$ observed for the rather simple acyclic ligand **186** which gave a Li^+/Na^+ selectivity of 4.9.

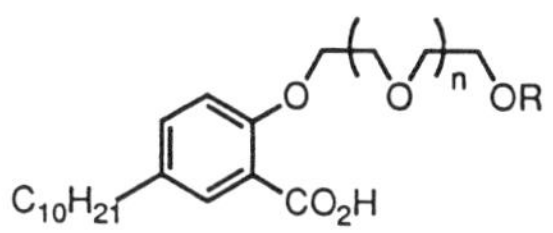

	n	R
182	0	H
183	1	H
184	2	H
185	3	H
186	0	CH_2Ph
187	1	CH_2Ph
188	2	CH_2Ph
189	3	CH_2Ph

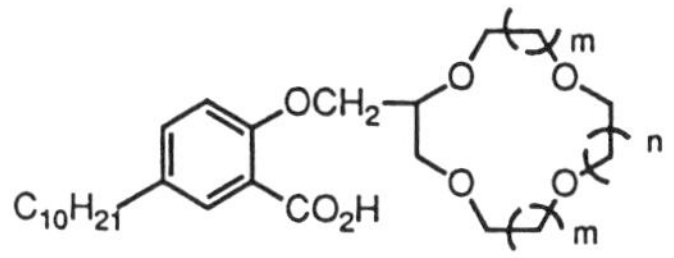

	n	m
190	1	1
191	2	1
192	1	2

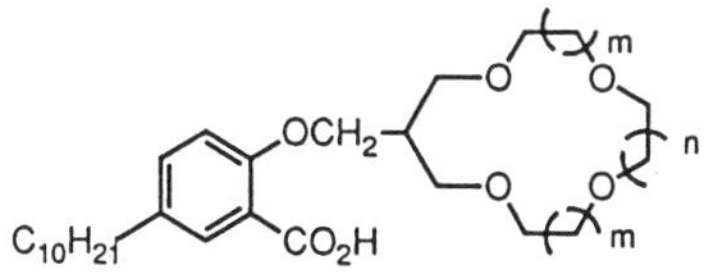

	n	m
193	1	1
194	2	1
195	1	2

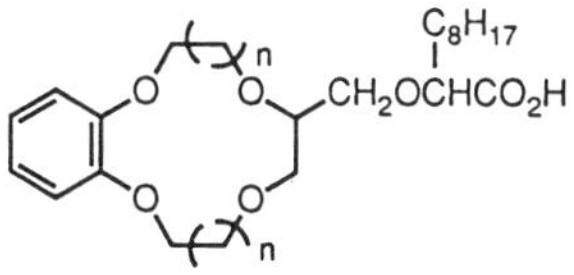

	n
196	1
197	2

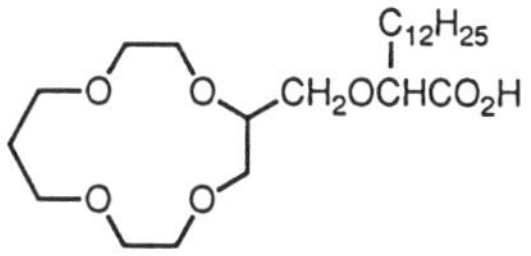

198

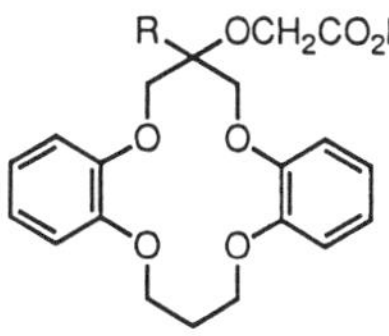

	R
199	$C_{10}H_{21}$
200	H

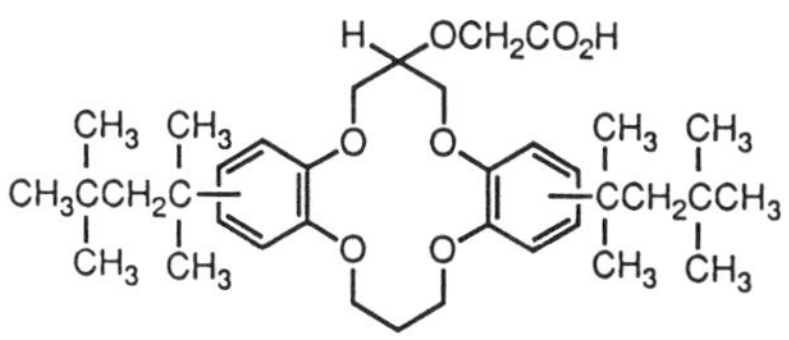

201

Coronands. Bartsch and co-workers have studied the extraction of alkali and alkaline earth metal ions using lariat ether carboxylic acids.[113] Specific compounds have been identified which exhibit high selectivity and efficiency for Li^+ in competitive extractions from aqueous solution of alkali metal chloride and hydroxides into chloroform.[240–243]

A series of lipophilic lariat ether carboxylic acids **190–195** with CE = crown-4 were synthesized.[244,245] The crown ether ring size was systematically varied from 12 to 15 members. For competitive alkali metal cation extraction into chloroform, the extraction selectivity order for the 12-crown-4, 13-crown-4, and 15-crown-4 compounds was $Li^+ > Na^+ > K^+ > Rb^+, Cs^+$ with Li^+/Na^+ selectivities of 2.4–3.7. In sharp contrast, the 14-crown-4 derivatives **192** and **194** extracted only Li^+ and Na^+ into the chloroform phase with Li^+/Na^+ selectivities of 19 and 20, respectively.[239,240] Metal ion loadings were 90–100% for competitive alkali metal cation extractions into chloroform by **190–195**. The high Li^+ selectivity and efficiency for extractions with lipophilic lariat ether carboxylic acids **192** and **194** are of magnitudes suitable for application in the recovery of lithium from natural sources.

Other lipophilic lariat ether carboxylic acids prepared by Bartsch and co-workers[240,241,244] which showed Li^+ selectivity in competitive alkali metal cation extractions into chloroform are compounds **196–199**. Of these extractants the Li^+/Na^+ selectivity of 4.7 for **197** was the highest.

Uhlman and co-workers[246,247] synthesized *sym*-(dibenzo-14-crown-4)-oxy-acetic acid **(200)** and investigated its propensity for extraction from a mixture of alkali and alkaline earth metal cations as a function of the pH of the aqueous solutions. The extractant exhibited good selectivity for Li^+ at pH 5.5–6.5, but gave preferential extraction of alkaline earth cations at higher pH.

Very recently, Sachleben, Moyer and co-workers[248] compared the extraction of alkali metal cations from aqueous solutions of their chlorides into *o*-xylene by bis-(*t*-butylbenzo)-14-crown-4-acetic acid **(201)** and 2-methyl-2-heptylnonanoic acid. Lipophilic lariat ether carboxylic acid **201** extracted Li^+ and Na^+ at significantly lower pH than did the aliphatic carboxylic acid. Preferential extraction of Li^+ by **201** at low loading, but little or no selectivity at high loading, was observed. This suggested an aggregated organic-phase species for the Li^+-selective extraction which involved both ionized and unionized forms of **201**.

Lariat ether phosphates **202–205** were reported recently by Habata, Akabori, and co-workers.[249,250] For extractions from basic aqueous solutions into chloroform, Li^+/Na^+ selectivities of 5.4 and 3.4 were observed for ionophores **202** and **205**, respectively, with moderate extraction efficiencies.

5.2 Separation of Lithium Isotopes

Metal isotopes have different physical properties. For example, the radius of $^6Li^+$ is greater by 0.04 pm than that of $^7Li^+$. Therefore, the potential exists for isotope separation by solvent extraction or chromatographic separations

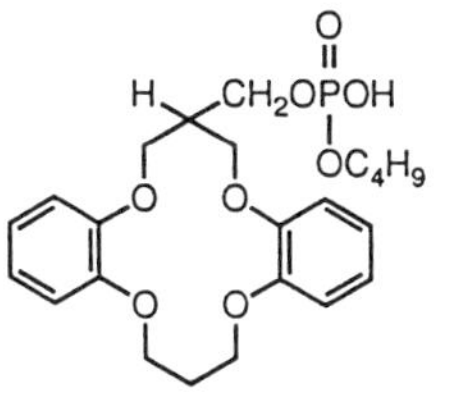

202

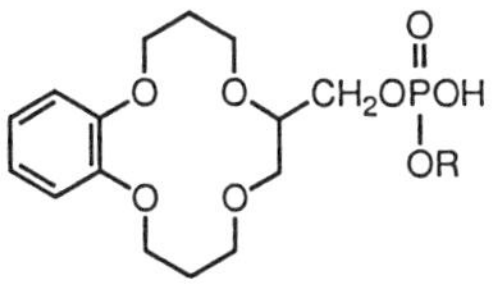

	R
203	C_4H_9
204	C_8H_{17}
205	$C_{12}H_{25}$

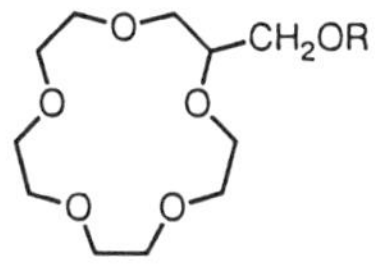

	R
206	$C_{12}H_{25}$
207	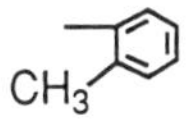

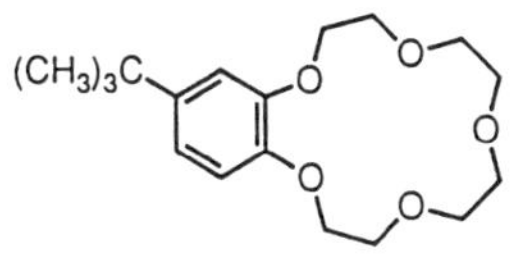

208

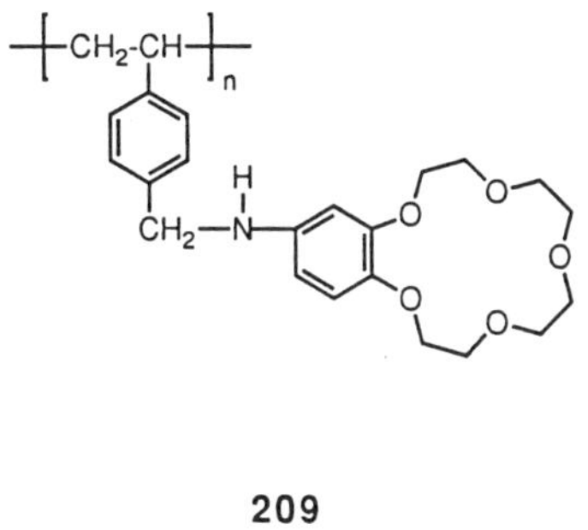

209

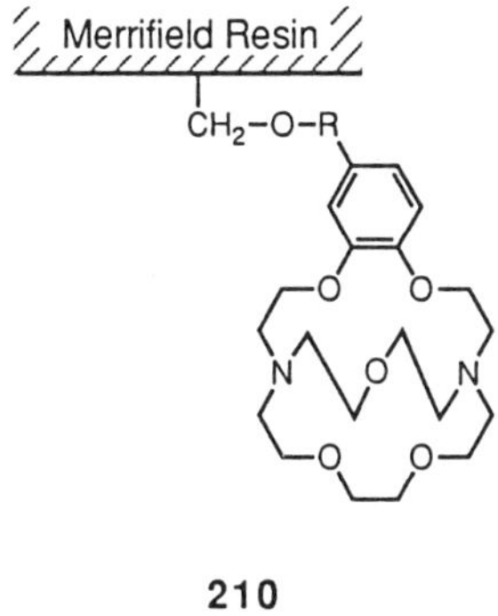

210

involving ionophores. In the pioneering research in this area, Jepsen and DeWitt[251] applied 18-crown-6 derivatives in a solvent extraction system to the separation of $^{40}Ca^{2+}$ and $^{44}Ca^{2+}$. The use of crown ethers and cryptands in isotope separations has been reviewed.[252,253]

Lithium possesses two isotopes, ^{6}Li and ^{7}Li, with natural abundances of 7.5 and 92.5%, respectively. Each lithium isotope plays an important role in the nuclear industry. Although naturally abundant lithium hydroxide is utilized for corrosion control in CANDU pressurized fission reactors, highly enriched ^{7}Li lithium hydroxide is used to adjust the cooling water pH for light water reactors.[254] Highly enriched ^6Li in the form of lithium hydride its employed as a shield against thermal neutrons.

Several different crown ethers and lariat ethers have been applied to the separation of lithium isotopes in solvent extraction processes by Chinese, Japanese, and Russian researchers.[255–262] These include 12-crown-4 **(14)**,[259] 15-crown-4 **(2)**,[262] lauryloxymethyl-15-crown-5 **(206)**,[259] tolyloxymethyl-15-crown-5 **(207)**,[259] benzo-15-crown-5 **(21)**,[256,257,259,261,262] 4-*t*-butylbenzo-15-crown-5 **(208)**,[58,259] dicyclohexano-18-crown-6,[257,262] dibenzo-18-crown-6,[262] and "ionophore C_{401}."[255] For lithium isotope separation by solvent extraction, the separation factor α is defined as $\{[^6Li^+]/[^7Li^+]\}_{org}/\{[^6Li^+]/[^7Li^+]\}_{aq}$. The extraction of lithium iodide with 12-crown-4 **(14)** in chloroform afforded a maximal single-stage isotope separation factor of 1.057 at 0°C.[259] Maximal separation factors for crown ethers **21**, **206**, and **207** were 1.042, 1.041, and 1.043, respectively.[259]

Chinese researchers have measured infrared (IR) spectra for ^6Li and ^7Li lithium salts complexed with a variety of crown ethers.[263–265] Isotopic shifts were noted in both the mid-IR[264] and far-IR[263–265] spectral regions.

The earliest application of ionophores to lithium isotope separation by solvent extraction was made by Jepsen and Cairns.[266] Separation factors of 1.026–1.041 were obtained for solvent extractions with [2.2.1]-cryptand **(53)**.

Another lithium isotope separation method is chromatography with ionophore-containing stationary phases. By reaction of 4-aminobenzo-15-crown-5 with a chloromethylated copolymer of styrene and divinylbenzene, Korean researchers prepared monobenzo-15-crown-5 resin **209**.[267] Use of the resin in column chromatography with water–acetonitrile (5:95, v/v) as eluent gave a single-stage separation factor of 1.053.

Two different Japanese research groups have investigated lithium isotope separations with commercially available (Merck) benzo[2.2.1]-cryptand resin **210** in both column chromatographic[268,269] and batch sorption [270] methods. A maximum single-stage separation factor of 1.047 was obtained for lithium chloride in methanol at 0°C with $^6Li^+$ being concentrated on the cryptand resin.[269] For aqueous methanol as the mobile phase, enhancing the water content decreased the interaction between the cryptand moiety and Li^+ which diminished the separation factor.[270] Thus continuous lithium isotope separation by column chromatography with crown ether and cryptand-containing stationary phases appears to be feasible.

Very recently Echegoyen and Chen[271] reported the use of redox-active macrocyclic and macrobicyclic ligands and complexes to separate lithium isotopes. Anthraquinone cryptand **59**, crown ethers **211** and **212**, and lariat ether **213** were quantitatively reduced via controlled potential electrolysis to their corresponding anion radicals in dichloromethane and allowed to react with a solid ^6Li and ^7Li lithium perchlorate mixture of known composition. The separation factor, defined as $\{[^6Li^+]/[^7Li^+]\}_{org}/\{[^6Li^+]/[^7Li^+]\}_{solid}$, was determined by atomic absorption spectrophotometry. The values were 1.18 ± 0.07, 1.04 ± 0.06, 1.15 ± 0.07, and 1.16 ± 0.07 for ionophores **59** and **211–213**,

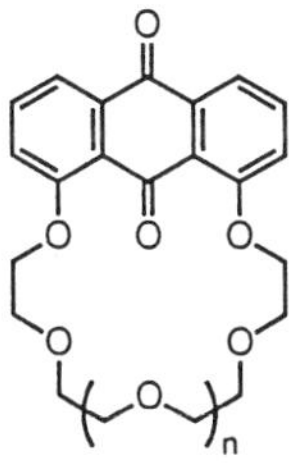

	n
211	1
212	2

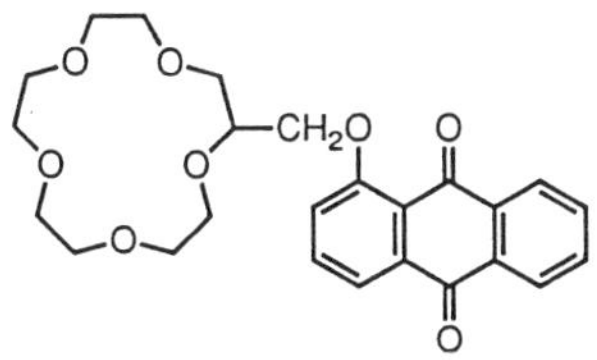

213

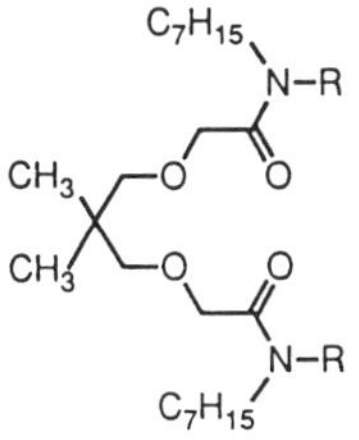

	R
214	CH_3
215	$CH_2CH_2CO_2C_2H_5$
216	$CH_2CH_2C(O)N(C_2H_5)_2$
217	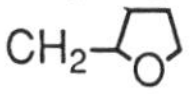 CH_2-(tetrahydrofuranyl)
218	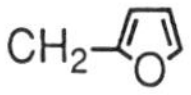 CH_2-(furanyl)

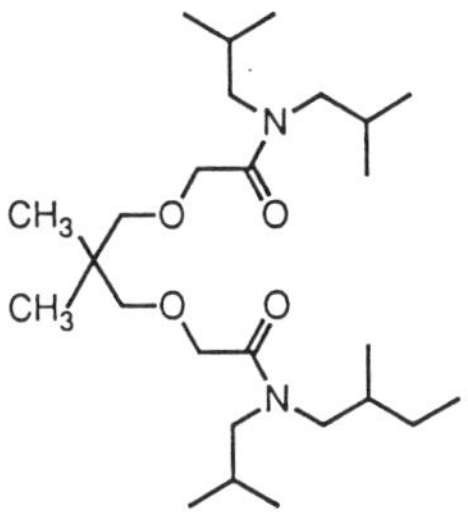

219

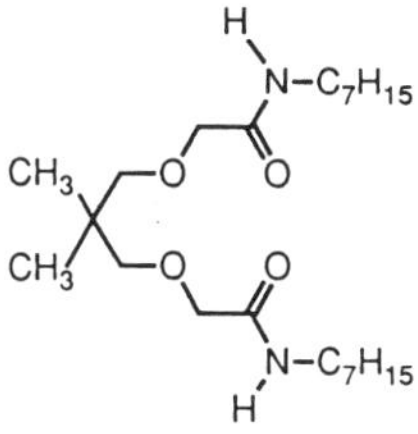

220

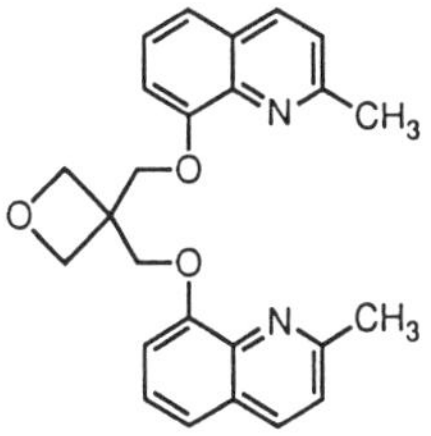

221

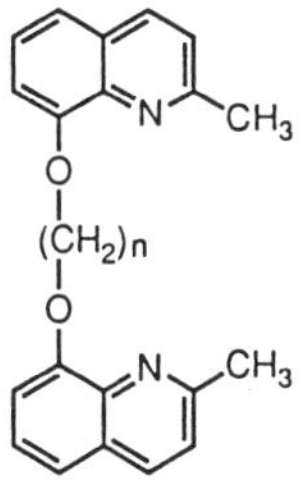

	n
222	3
223	4

respectively. Thus for ionophores **59**, **212**, and **213**, very large separation factors were observed.

5.3 Separation of Lithium Ions by Transport across Liquid Membranes

Although there are a variety of natural membrane systems that transport Na^+ and K^+ selectivity, this is not the case for Li^+. However, artificial membrane systems that utilize synthetic ionophores were developed which exhibit Li^+ selectivity. Transport of metal ions across liquid membranes facilitated by synthetic ionophores has been reviewed.[106,111,272–275]

5.3.1 Passive and Active Mechanisms for Cation Transport. Metal ion transport across artificial membranes may be separated into the two general classifications of passive or ionophore-facilitated passive transport and of active transport. In passive transport, the metal ion is transferred by diffusion due to a concentration gradient between an aqueous phase on one side of the membrane and a second aqueous phase on the other side. Ionophore-facilitated passive transport of the metal ion across the membrane is promoted by an ionophore complex (Fig. 10.5). The neutral ionophore (represented by the circle) forms a ternary complex with the metal ion and its counterion in the membrane phase. Transport abilities in either the passive or ionophore-facilitated passive transport systems depend on the hydration of both the metal ion and its counteranion. Soft anions with low hydration energies aid the metal ion transport. Ionophore-facilitated passive transport is also dependent upon the metal ion complexing ability of the ionophore.

In an active transport system, the cation is transferred by coupling with movement of mobile protons or electrons. Proton-driven transport is the most familiar system and is based on a large change in the cation-complexing ability of the ionophore induced by contact with acidic and basic aqueous phases.

Two types of ligands are employed for proton-driven transport. The more commonly encountered type involves proton-ionizable (i.e., proton-dissociable) ionophores (Fig. 10.6). At the interface between the basic aqueous source phase and the membrane, the ionophore loses a proton and concurrently binds a metal ion to form a complex. This electroneutral complex diffuses through

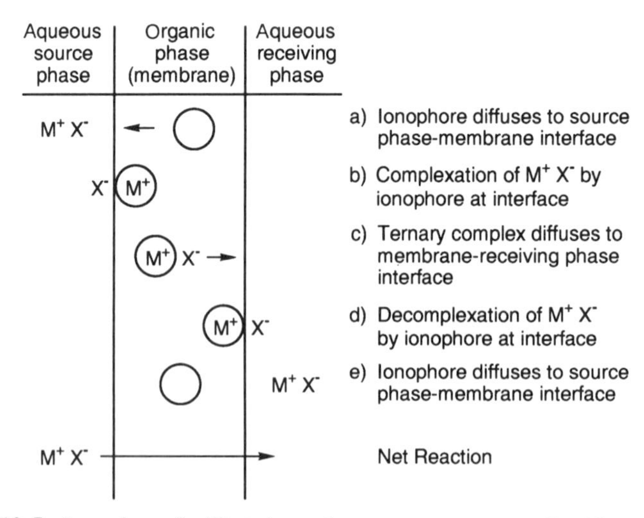

Figure 10.5 Ionophore-facilitated passive transport across a liquid membrane.

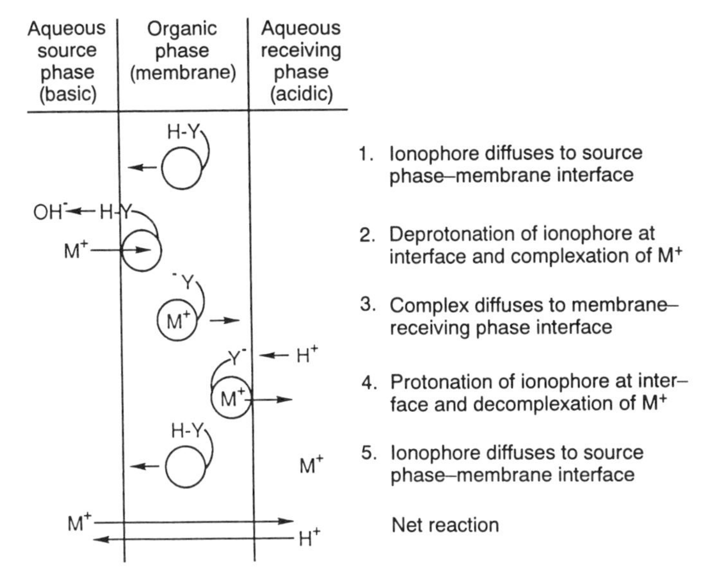

Figure 10.6 Proton-driven active transport across a liquid membrane by a proton-ionizable ionophore.

the membrane to its interface with the acidic receiving phase, where the metal ion is released by reprotonation of the anionic site. The overall reaction is coupled movement of the metal ion and a proton across the membrane in opposite directions. Thus a pH gradient can be utilized to transport the metal ion against its concentration gradient. Since concomitant transfer of an aqueous phase anion is not required for movement of the metal ion across the membrane, active transport with proton-ionizable ionophores has the greatest potential for practical applications in which hard aqueous phase anions, such as chloride, nitrate, and sulfate, would be involved.

In the second type of proton-driven transport, the neutral ionophore contains a basic nitrogen atom which is contact with an acidic aqueous solution produces positively a charged ammonium ion (Fig. 10.7). At the interface between the basic aqueous source phase and the membrane, the ionophore is deprotonated and forms a ternary complex with the metal ion and the counter-anion. This complex diffuses through the membrane to its interface with the acidic aqueous receiving phase where the nitrogen atom of the ionophore is reprotonated which releases the metal ion into the receiving phase. Once again there is coupled movement of the metal ion and a proton across the membrane in opposite directions. Since the anion Y^- is shuttled back and forth across the membrane, there is no need for an anion from the aqueous source phase to accompany the transported metal ion. Usually Y^- is a lipophilic organic anion which has been added to the membrane.

In this section, passive and ionophore-facilitated passive transport across liquid membranes are considered first. Next published results for Li^+ separa-

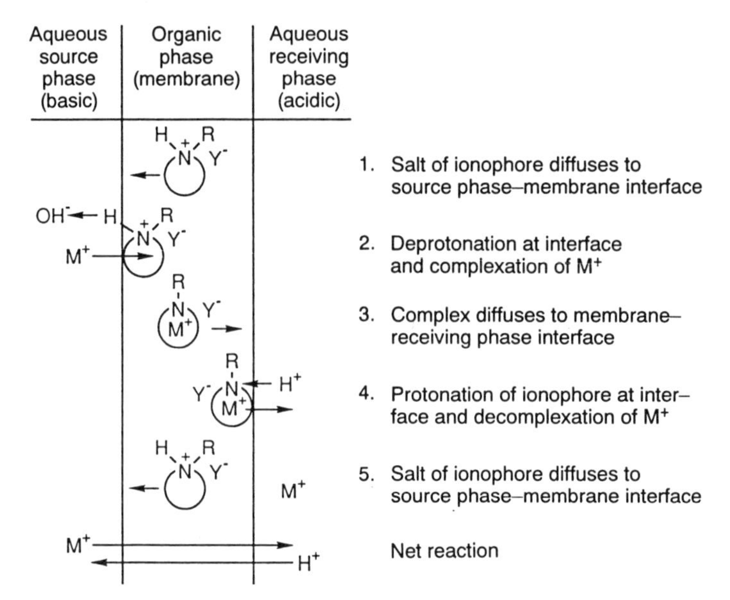

Figure 10.7 Proton-driven active transport across a liquid membrane by an ionophore with a basic nitrogen atom.

tion and enrichment by proton-driven transport through membranes containing proton-ionizable or proton-associable ionophores are summarized.

5.3.2 Passive or Ionophore-Facilitated Transport. Attempts to recover Li^+ selectively from Dead Sea brines were made using solvent-polymeric membranes prepared from poly(vinyl chloride) with phosphates, phosphonates, and/or carboxylates as plasticizers in a passive membrane transport system.[276] The membranes exhibited high selectivities and efficiencies for Li^+ transport when tribromide was present as a lipophilic counteranion. Among the membranes investigated, one containing diphenyl(2-ethylhexyl)phosphate gave the best Li^+ selectivity. With this membrane a Li^+/Na^+ transport selectivity of 20 was achieved when the bromine concentration in the source phase was 1.2×10^{-3} M.

Ionophore-facilitated passive transport of metal ions across solvent-polymeric membrane electrodes is the central feature of the Li^+-selective electrodes discussed in Section 4.1. However, only minute quantities of metal ions are transported in this analytical technique which makes it inappropriate as a method for Li^+ separation on a practical scale.

In general there are far fewer neutral ionophores which provide Li^+-selective membrane transport than those that exhibit Li^+ selectivity for metal ion complexation in homogeneous media or in solvent extraction. Very high stability of the cation–ionophore complex is a disadvantage for movement of metal ions across a liquid membrane since it retards metal ion release into the receiving phase.[106,273] Cryptands, spherands, and macrocyclic ligands which complex Li^+ strongly give poor transport efficiency and selectivity. Thus appropriate kinetics for both metal ion–ionophore complex formation and dissociation are very important factors.[106,273]

Podands. Acyclic amide and ether-containing ionophores were developed as neutral carriers for Li^+-selective electrodes by Simon and co-workers.[127] Lipid bilayer membranes containing acyclic ligands **139**, **214–219** exhibited highly Li^+-selective permeability as determined by measurement of steady-state electrical properties.[277-280] It was assumed that the ligands form 2:1 (ionophore: monovalent cation) complexes as the major permeate species and operate as "equilibrium-domain" carriers of monovalent cations. The Li^+/Na^+ selectivity for membrane permeation, $P(Li^+)/P(Na^+)$, decreased in the order **219** > **139** > **217** > **216** > **214** > **215** > **218**. The Li^+/Na^+ and Li^+/K^+ permeation selectivities with ionophore **219** were both 59.

Shanzer and co-workers studied the Li^+ permeability of compounds **139**, **215**, **216**, and **220** by in vitro experiments with liposomes using a fluorescence assay.[218] The most efficient carrier for Li^+ was **139**, while **215** and **216** were less effective and **220** was ineffective.

To evaluate the metal ion transporting capabilities of synthetic ionophores, transport across a bulk liquid membrane in a U-tube (Pressman) cell (Fig. 10.8) is frequently employed. The aqueous source phase of the metal salt (A) is separated from the aqueous receiving phase (C) by an immiscible organic liquid membrane (B) which contains the ionophore. Periodically the aqueous receiving phase is sampled to determine the concentration of the transported metal ion. Each phase is slowly stirred (ca 200 rpm) mechanically.

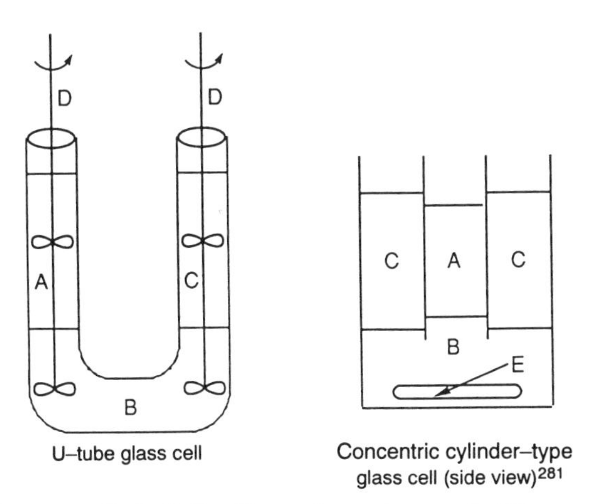

Figure 10.8 Apparatus for bulk liquid membrane transport: (A) aqueous source phase containing the metal salt; (B) membrane phase, usually chloroform, containing the ionophore; (C) aqueous receiving phase, usually distilled water; (D) glass stirrer; (E) teflon-coated magnetic stirring bar.

An alternative is the concentric cylinder-type cell (Fig. 10.8) which consists of an open-ended glass tube suspended in a glass vial or beaker. One end of the tube is immersed in the membrane phase (B) which separates the aqueous source phase (A) inside the tube from the aqueous receiving phase (C) outside the tube.[281]

Competitive transport of alkali and alkaline earth metal thiocyanates across bulk chloroform membranes by acyclic, 8-quinolyl terminated compounds **5**, **221–223**, in a U-tube cell was investigated by Hiratani and co-workers.[282] The level of Li^+ transport selectivity was strongly influenced by the structure of the podand. Ionophore **5** gave a high Li^+/Na^+ transport selectivity of 110. Evidence that the ionophore forms a pseudo cavity to accommodate Li^+ has been obtained by proton NMR spectroscopy.

Very recently Hirose, Hiratani, and co-workers[283] reported the results of competitive lithium, sodium, and potassium thiocyanate transport across chloroform membranes by acyclic carriers **224** which have two terminal benzoquinolyl units and **225** with one quinolyl and one benzoquinolyl unit. For **224** the transport selectivity and efficiency were found to be markedly influenced by variation of the number of methylene units in the spacer which joins the two benzoquinolyl groups. The fastest and most selective ($Li^+/Na^+ = 84$; $Li^+/K^+ = 274$) transport was noted with $n = 2$. Podand **225** was somewhat less selective and effective for Li^+ transport.

The Li^+ selectivity noted for 2,9-dibutyl-1,10-phenanthroline **(6)** for fluorescence measurements in homogeneous solution[22] and in solvent polymer membrane electrodes[143] was proposed to arise from 2:1 (podand:Li^+) complexation. Sugihara, Hiratani, and co-workers[284] studied competitive transport of lithium, sodium, and potassium thiocyanates across a bulk chloroform membrane with **6** and the bisphenanthrolines **226–230**. The amount of Li^+ transported by the bisphenanthrolines in a specified time period was 4–160 times that of phenanthroline **6**. The greatest level of transport and selectivity ($Li^+/Na^+ = 1850$) was obtained with bisphenanthroline **228** in which the spacer has sufficient flexibility to allow the nitrogens of both phenanthroline units to simultaneously coordinate with one Li^+.

Coronands. Competitive alkali metal picrate transport across bulk chloroform membranes with 14-crown-4 **(145)**, decalino-14-crown-4 **(162)**, and the unsaturated decalino-14-crown-4 compound **(231)** in a concentric tube cell (Fig. 10.8) was investigated by Kobiro and co-workers.[225] The rate of Li^+ transport by decalino-14-crown-4 **(162)** was about tenfold faster than that for **145** and **231**. For ionophore **162**, transport of K^+, Rb^+, and Cs^+ was undetectable and a Li^+/Na^+ transport selectivity of 20 was observed. Thus the asymmetrical nature of **162** with lipophilicity on only one side of the crown ether ring was found to facilitate transport.

By comparison of single species alkali metal picrate transport across bulk chloroform membranes, Tsukube and co-workers[285] estimated Li^+/Na^+ transport selectivities of about 2 for ester side-armed polyazacrown compounds **232** and **233**.

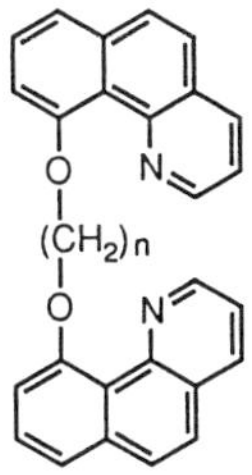

(n = 2 - 6)

224

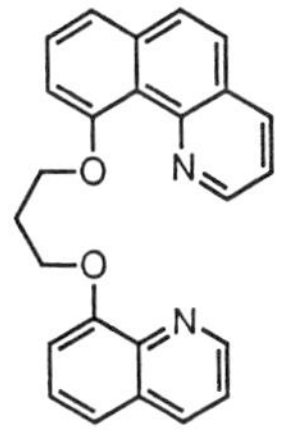

225

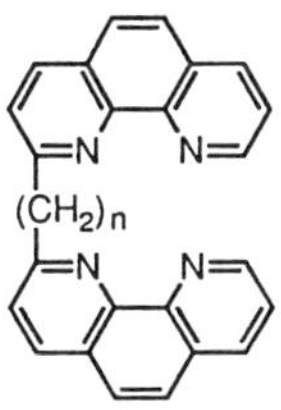

	n		n
226	4	**229**	7
227	5	**230**	8
228	6		

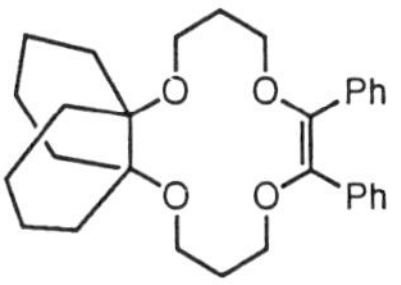

231

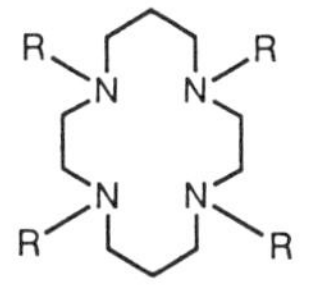

(R = $CH_2C(O)N(C_2H_5)_2$)

232

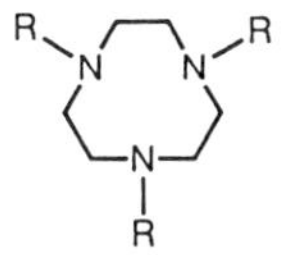

(R = $CH_2C(O)N(C_2H_5)_2$)

233

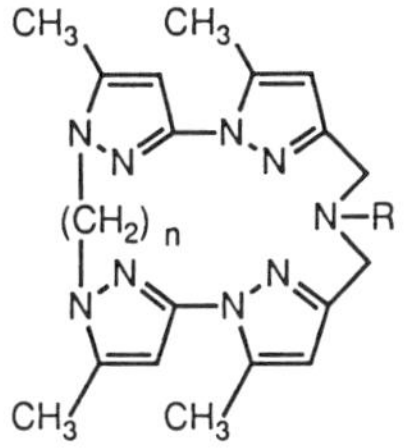

	n	R
234	2	CH_3
235	2	CH_2CH_2Pyridine
236	2	CH_2CH_2OPyrazine
237	3	CH_3

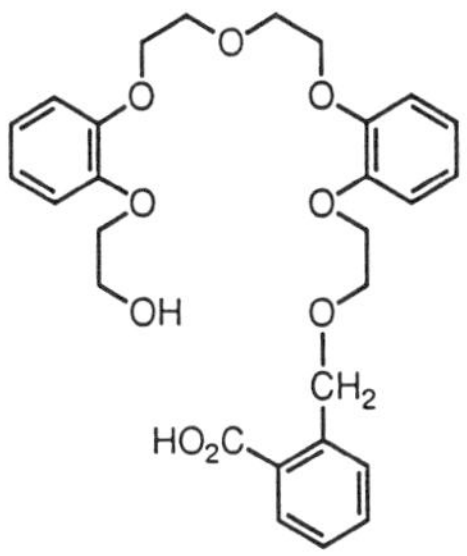

238

For the tetrazolic macrocycles **234–237** Tarrago and co-workers[286] recorded Li^+/Na^+ selectivities of about 2 for **234**, **236**, and **237**, but a higher value of 8.3 for **235**, in competitive transport of alkali metal picrates across bulk dichloromethane membranes. With less lipophilic nitrate salts, the transport rate was markedly reduced, but the Li^+/Na^+ selectivities increased substantially to 10.0, 21.6, 5.9, and 5.0 for ionophores **234–237**, respectively. Thus the identity of the counteranion affects both the rate and selectivity in competitive transport of metal ions by neutral ionophores. Tetrazolic macrocycles **234–237** also exhibited Li^+ selectivity for picrate extractions into dichloromethane.

5.3.3 Active Transport. As mentioned above for a proton-ionizable (proton-dissociable) ionophore, metal ion transport across an artificial membrane does not require concomitant transfer of an anion from the aqueous source phase. Therefore, the presence of lipophilic anions like thiocyanate or picrate in the aqueous source phase is not needed for efficient metal ion transport.

Podands. Proton-coupled, Li^+ selective competitive transport from a basic aqueous source phase through a bulk organic membrane containing the carrier into an acidic aqueous receiving phase has been reported for acyclic polyethers with carboxylic acid functions at one terminus and a variety of groups at the other.[287–292]

Using an H-type cell to accommodate an organic membrane solvent which was lighter than water, Yamazaki and co-workers[287] observed much faster transport of Li^+ than Na^+ by proton-ionizable podand **238**.

The influence of structural variation within the podand upon Li^+ transport selectivity has been carefully probed by Hiratani and co-workers.[288–291] Acylic polyethers **9**, **239–241**, have a benzoic acid group at one terminus and a quinolyl group at the other. For competitive transport across bulk chloroform membranes, the Li^+/Na^+ transport selectivities were: **9** (3.4) > **241** (2.2) > **240** (2.0) > **239** (1.3). It was proposed that in ionophore **9**, Li^+ coordinates with the ethereal oxygens and the quinolyl nitrogen to induce a helical structure **242** and causes a stacking interaction of the terminal aromatic groups of the podand.[25,288,289] Support for this contention was obtained by proton NMR and electronic spectral studies and by molecular modeling.

The effect of structural variation within the terminal aromatic group was examined with compounds **243–249**, all of which gave Li^+/Na^+ transport selectivities in the range of 3.2–4.6 except **243**, for which a much higher selectivity of 14 was observed. It should be noted that the nonaromatic terminal hydrocarbon group in **249** also gave good transport selectivity.

Acyclic polyether dicarboxylic acids **250** gave a Li^+/Na^+ transport selectivity of 3.3 for $n = 2$ and lower selectivities as the polyether chain was lengthened.[291] It was postulated that podands **250** with $n = 3$ and 5 have chains long enough to encapsulate two Li^+ inside the pseudocavity.

Hiratani and co-workers[290] also developed an acyclic polyether ligand with high transport selectivity for K^+. With a dual transport cell, Li^+ and K^+ were simultaneously separated using Li^+- and K^+-selective carrier molecules.[291]

A different variety of acyclic polyether carboxylic acid ionophores **251–255** was synthesized by Okahara and co-workers.[292] For competitive transport across bulk dichloromethane membranes, the Li^+/Na^+ selectivities decreased in the order **252** (4.8) > **255** (4.2) > **251** (3.6) > **253** (2.4) > **254** (1.1). For this series of carriers, elongation of the polyether chain was detrimental to the transport selectivity for Li^+.

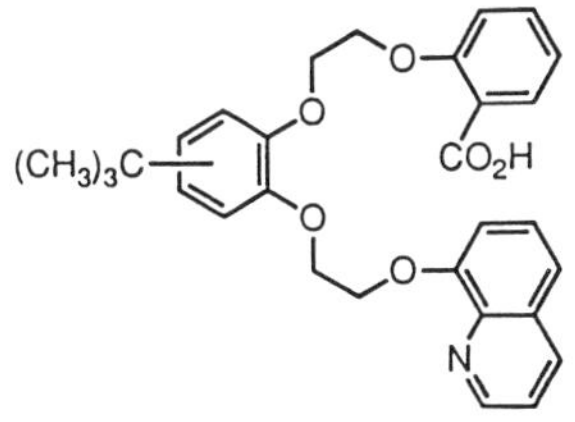

$(CH_3)_3C$ O O CO_2H O O N

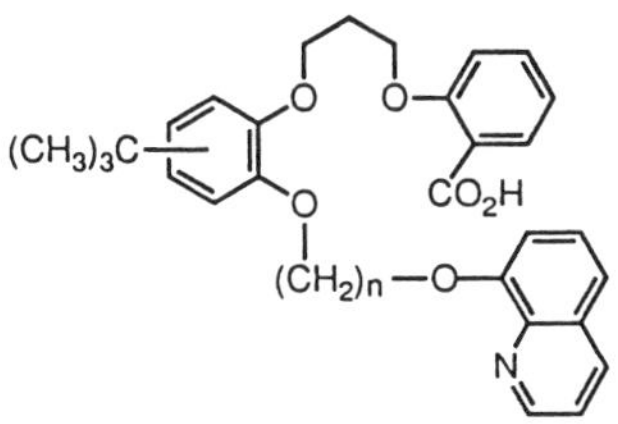

239

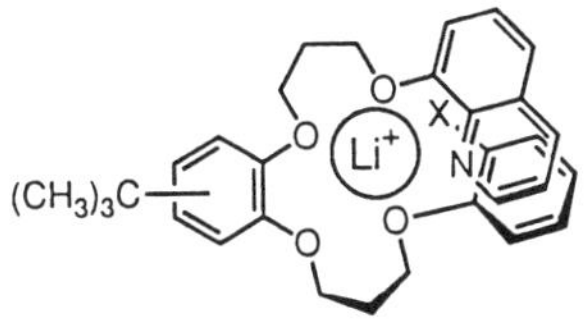

$(CH_3)_3C$ O O CO_2H O $(CH_2)_n$—O N

	n
240	2
9	3
241	4

$(CH_3)_3C$ O O Li^+ X N O O

$(X = CO_2^-)$

242

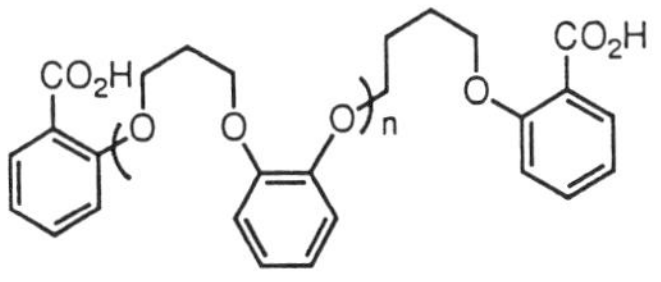

$(CH_3)_3C$ O O CO_2H O O—R

	R
243	N CH_3
244	N
245	
246	
247	
248	OC_2H_5
249	C_2H_5

CO_2H O O O n O CO_2H

(n = 2, 3, 5)

250

HO_2C N OR

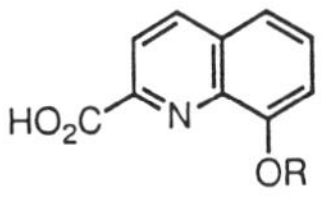

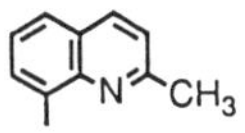

	R
251	C_8H_{17}
252	$CH_2CH_2OC_8H_{17}$
253	$CH_2CH_2OCH_2CH_2OC_8H_{17}$
254	$CH_2CH_2(OCH_2CH_2)_2OC_8H_{17}$
255	$CH_2CH_2CH_2OC_8H_{17}$

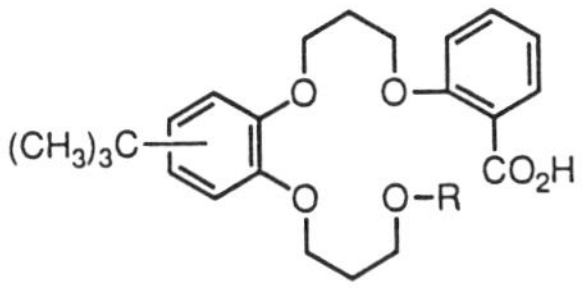

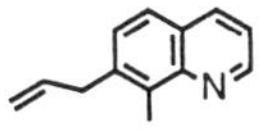

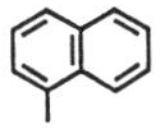

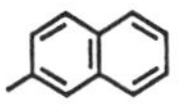

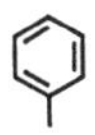

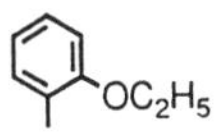

Coronands. Takahashi and co-workers[232] recently reported results for Li^+ transport by a crown ether–alkylphosphoric acid system which serves as a model for proton-ionizable coronand carriers. For transport of Li^+ and Na^+ from a basic aqueous source phase through a chloroform membrane into an acidic aqueous receiving phase, dibenzo-14-crown-4 **(13)** alone was ineffective. Although bis(2-ethylhexyl)phosphoric acid **(174)** alone transported Li^+ and Na^+, the Li^+ selectivity was very low ($Li^+/Na^+ = 1.3$). For a mixed-carrier system of **13** and **174**, the transport rate for Na^+ remained the same as that found with just **174** as the carrier. On the other hand, Li^+ transport increased by a factor of 2.4 with the mixed carrier system, which enhanced the Li^+/Na^+ transport selectivity. Thus the presence of both a Li^+-selective ionophore and a proton-ionizable compound in the membrane was demonstrated to enhance both the selectivity and efficiency of Li^+ transport.

In another recent publication, Nishimura and co-workers[231] observed that the mixed-carrier systems of crownophanes **173** and dodecanoic acid provided competitive Li^+/Na^+ transport selectivities as high as 8.6 for bulk chloroform membranes.

In the following discussion, a Li^+ selective coronand and a proton-ionizable group are incorporated into a single lipophilic carrier molecule. For example, Inokuma and co-workers[293] utilized lipophilic lariat 14-crown-4 carboxylic acid **256** to obtain a Li^+/Na^+ transport selectivity of 1.8 in a chloroform membrane system.

In coronands **257** and **258**, which were prepared and studied by Bradshaw, Izatt, and co-workers,[294–296] the proton-ionizable group is part of the macrocyclic ring. In competitive proton-driven transport across a bulk dichloromethane membrane, the lipophilic pyridono 15-crown-5 compound **257**,[294,295] and 1-hydroxy-4-pyridino-14-crown-4 compound **258**,[296] transported Li^+ selectively. Under optimized condition, Li^+ transport by **257** is more rapid than by **258**. It seems reasonable that the resultant cavity when the former is ionized can accommodate Li^+, whereas in ionized **258** Li^+ would be oriented above the ring cavity due to interference with the intraannular *N*-oxide group. A disadvantage of carriers **257** and **258** is the extremely alkaline source phase pH required for transport.

Bartsch and co-workers[297] investigated competitive alkali metal cation transport from nearly neutral aqueous solution across bulk chloroform membranes by lipophilic lariat ether phosphonic acid monoethyl esters, including **259** and **260**, and model compound **261**. For the 14-crown-4 carrier **260**, the transport selectivity order was $Li^+ > Na^+ > K^+, Rb^+, Cs^+$ with considerably higher Li^+/Na^+ selectivity than that observed with the 12-crown-4 carrier **259**. Although the transport rate for model compound **261** was slower than that observed with the crown ether-containing carriers, **261** still exhibited good Li^+/Na^+ transport selectivity. To rationalize this unanticipated selectivity for transport of Li^+, it was proposed that a phosphonate–phosphonic acid complex, such as **262**, may be responsible.

Competitive, proton-coupled transport of alkali metal cation transport through bulk chloroform membranes by lariat ether phosphonic acid diesters **202–205** was reported very recently by Habata, Akabori, and co-workers.[249,250] Highly Li^+-selective transport ($Li^+/Na^+ = 14.4$) was obtained with the dibenzo-14-crown-4 compound **202**. For the benzo-14-crown-4 carriers **203–205**, the Li^+/Na^+ transport selectivities were 5–6.

The use of lipophilic 14-crown-4 nitrophenol derivatives **130**, **131**, **263**, and **264** as carriers for transport of alkali metal cations across polymer-supported liquid membranes was investigated by Kimura and co-workers.[298,299] The membrane employed for this proton-driven cation transport was a disk of microporous polypropylene film which had been impregnated with an *o*-nitrophenyl octyl ether (NPOE) solution of the ionophore. These well-designed carriers combine a crown ether ring which has specificity for Li^+ with an easily ionizable nitrophenol or dinitrophenol function and a lipophilic group. The membranes of **130**, **131**, and **263**, which have 2-phenol groups, exhibited extremely high Li^+ selectivities. When **130** was the carrier in the membrane, the Li^+/Na^+ transport selectivity was 50. Compared with **130**, **131**, and **263**, the transport ability for carrier **264** decreased significantly. In **264** the 4-phenoxide anion cannot interact intramolecularly with Li^+ coupled by the polyether ring.

A polymer-supported liquid membrane with the dinitrophenol compound **131** as carrier was utilized for Li^+ enrichment.[299] Proton-driven transport was conducted under conditions where the basic aqueous source phase containing Li^+ and Na^+ had eight times the volume of the acidic receiving phase. After transport for 30 h, the Li^+ concentration in the receiving phase was about five times greater than its initial source phase concentration.

Lithium separation in a flow-through transport system has been achieved with hollow fiber membranes impregnated with an NPOE solution of ionophore **131**.[300] The polymer-supported, hollow-fiber membrane system permitted continuous metal ion transport for 25 h with a Li^+/Na^+ selectivity of approximately 30. This type of flow-through transport system should be applicable to Li^+ separation and Li^+ analysis.

For the proton-ionizable podands and coronands described above, the proton-driven transport mechanism is that shown in Figure 10.6. Okaha and coworkers[229] reported competitive proton-coupled, liquid-membrane transport of Li^+, Na^+, and K^+ by monoaza-12-crown-4 and monoaza-14-crown-4 carriers **167–169** and **265–268**, respectively, for which the alternative active transport mechanism depicted in Figure 10.7 is utilized. For transport across bulk dichloromethane membranes containing both the ionophore and picric acid, all carriers except **168** exhibited Li^+/Na^+ selectivity with both the Li^+ transport rate and selectivity being influenced by the structural variation in the N-pivot side arm. The most effective and selective ($Li^+/Na^+ = 24$) transport for Li^+ was observed with lariat ether **266**, which provides both an appropriate 14-crown-4 ring size for Li^+ complexation and a side arm which contains a fourth hard oxygen donor atom for Li^+ binding.

5.4 Separation by Sorption

Selective sorption of Li^+ from aqueous solutions by an insoluble, ionophore-containing, polymeric resin would have many advantages for large-scale Li^+ separation from an engineering viewpoint. Thus the solvent inventory required for solvent extraction or the limited lifetimes of liquid membrane transport

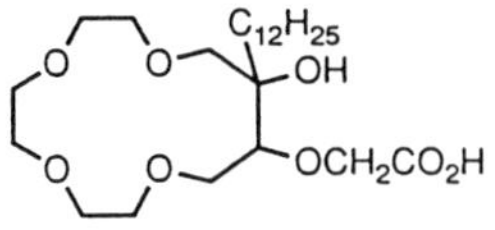

256 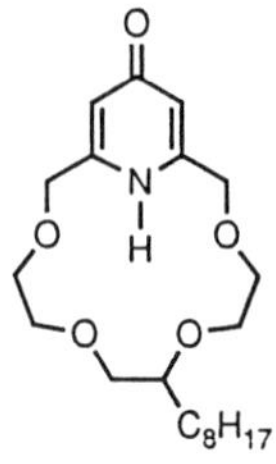 **257** 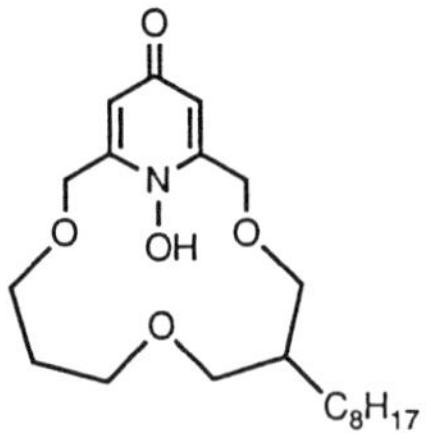 **258**

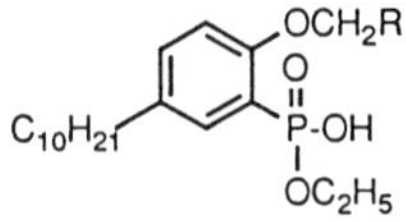

	R
259	12-crown-4
260	14-crown-4
261	cyclohexyl

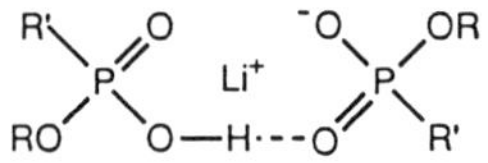

262

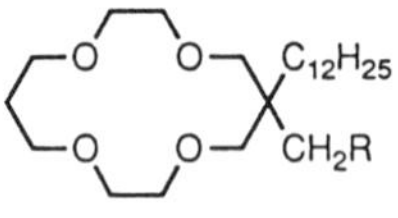

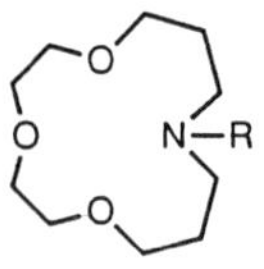

	R
265	C_8H_{17}
266	$CH_2CH_2OC_8H_{17}$
267	$CH_2CH_2CH_2OC_8H_{17}$
268	$CH_2CH_2OCH_2CH_2OC_8H_{17}$

130

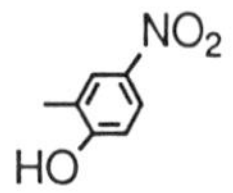

263

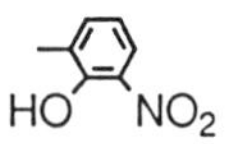

131

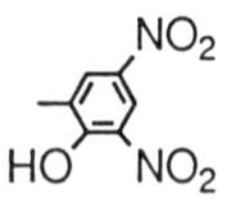

264

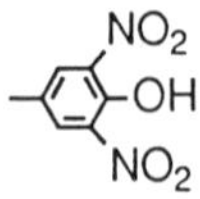

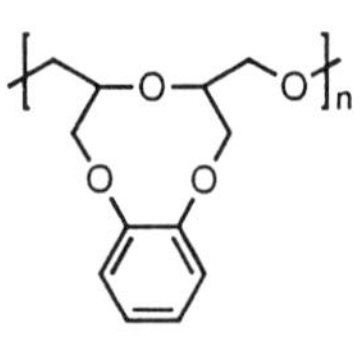

269

systems would no longer be an important factor. Also, loss of the expensive and potentially physiologically active ionophore would be eliminated by covalent bonding of the chelating agent within the polymeric matrix.

Presumably podands and coronands that exhibit high Li^+ selectivity in homogeneous solution or in solvent extraction or liquid membrane transport pro-

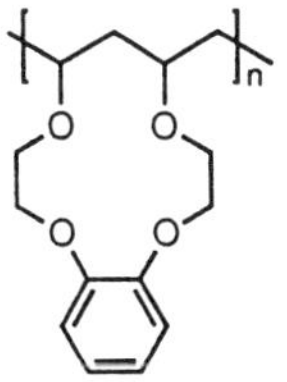

270 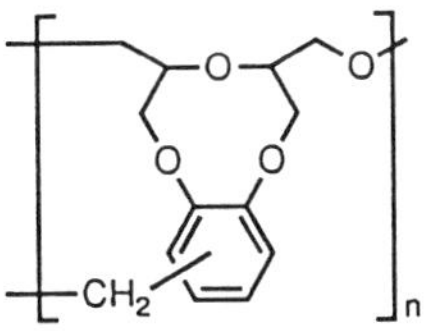 **271** 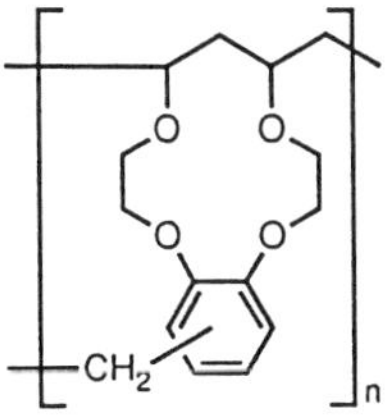 **272**

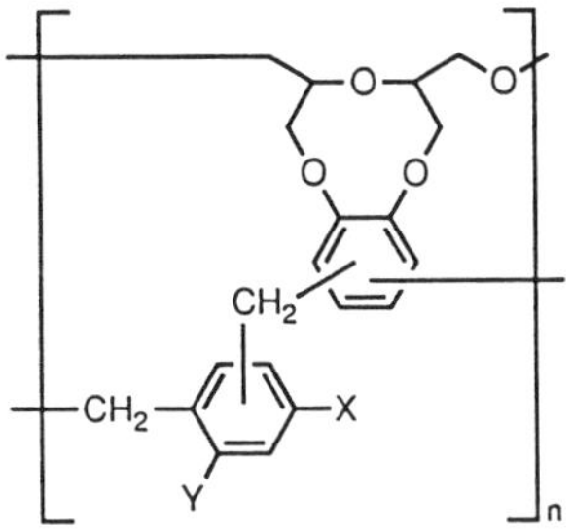

	X	Y
273	OH	H
274	OH	OH

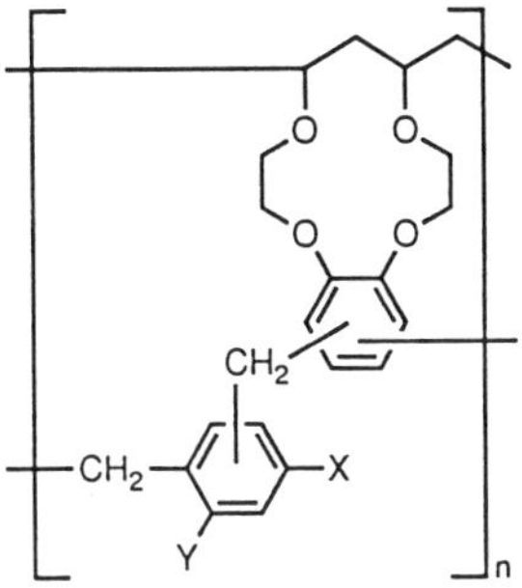

	X	Y
275	OH	H
276	OH	OH

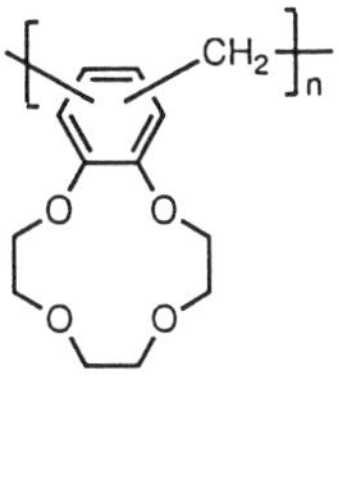

277

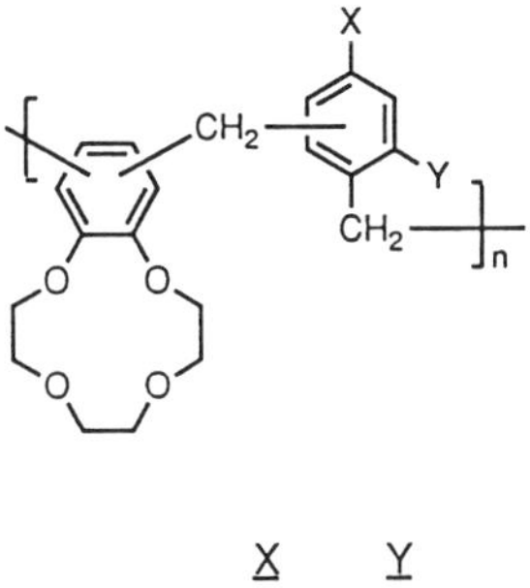

	X	Y
278	OH	H
279	OH	OH

cesses would be good candidates for incorporation into polymers. In a 1984 review of macrocyclic ligands on polymers by Smid and Sinta,[301] no polymers with Li^+ sorption selectivity are mentioned.

With the objective of developing resins for use with the lithium-rich brines in Chile, Bartulin and co-workers[302,303] synthesized polymers **269** and **270**, which contain benzo-9-crown-3 and benzo-13-crown-4 units. Subsequently these crown ether polymers were crosslinked under two sets of conditions to form polymers **271–276**. Also, condensation polymerization of benzo-12-crown-4 **(143)** with formaldehyde, phenol and formaldehyde, and resorcinol and formaldehyde gave polymers **277–279**, respectively.[304] Sorption of Li^+ and Na^+ from neutral and basic (pH = 12) aqueous solutions of polymeric resins **271–279** was examined. All of the polymers exhibited sorption of Li^+, but not Na^+. The degree of Li^+ sorption was found to be strongly influenced by the structure of the crown ether polymer and the pH of the aqueous solution. Polymers **274** and **276** showed the highest efficiency with 70 and 78% Li^+ sorption, respectively. Unfortunately, stripping of the sorbed Li^+ from resins **271–276** by contact with water or methanol gave only poor recoveries (6–23%).

Weak Li^+/Na^+ sorption selectivity by gel-type polymer **280**, which contains benzo[2.2.2]-cryptand units, was reported by Manecke and Reuter.[305] When the *N*-vinyl pyrrolidinone monomer was absent, the copolymer showed much stronger sorption of Na^+ than Li^+.

Although polymer **281**, which is prepared by condensation polymerization of *sym*-(propyl)-dibenzo-14-crown-4-oxyacetic acid with formaldehyde, gave good sorption selectivity for Li^+ and Na^+ over the other alkali metal cations, no differentiation between Li^+ and Na^+ was noted by Bartsch and co-workers.[306] This is a new type of ion-exchange resin which incorporates both cyclic polyether and ion-exchange sites for metal ion complexation.[307]

New ionomeric polymers with high Li^+ selectivity and ready subsequent release of sorbed Li^+ or its salts is an attractive area for further investigation.

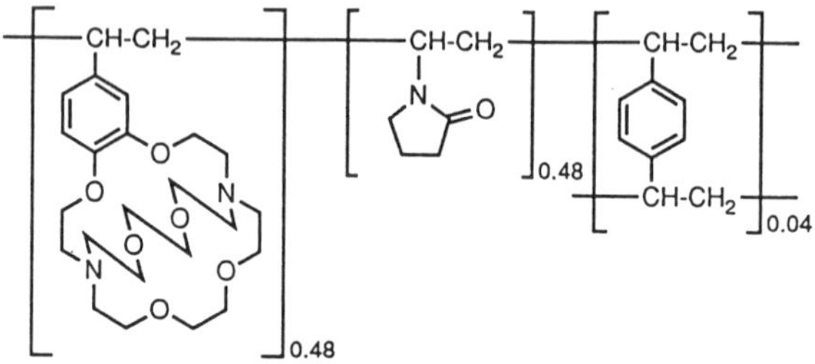

280

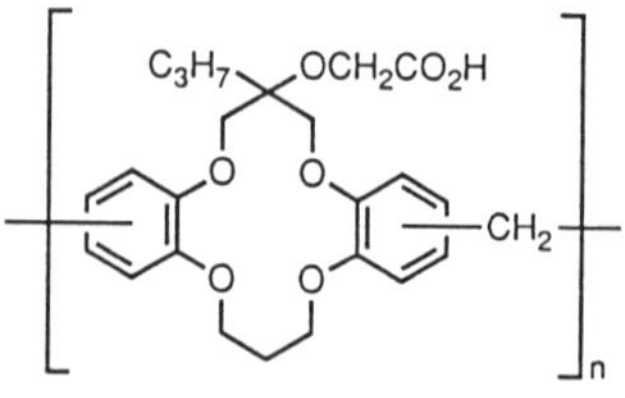

281

6 CONCLUDING REMARKS

The use of synthetic Li^+-selective or specific ionophores is applicable in several fields. In the analytical area, these ionophores allow Li^+ to be determined precisely in the presence of other alkali or alkaline earth metal cations. Ionophore-containing, ion-selective electrodes can be utilized in industry for process control. They can also be applied to monitoring of industrial waste streams with the dual purpose of maintaining the Li^+ concentration at an environmentally safe level for discharge and facilitating the operation of Li^+ recycle systems.

Determination of Li^+ in the blood of patients treated with lithium compounds is of the utmost importance. There is a sharp limit, varying on an individual basis, between beneficial concentrations and toxic levels for treatment of manic depressive illness. Because of the presence of much higher concentrations of Na^+ in the blood, very high Li^+/Na^+ selectivity is required for Li^+ to be determined. At present the Li^+/Na^+ selectivity obtainable for ionophore-containing, ion-selective electrodes is marginally adequate for measurement of Li^+ concentrations in blood serum.[155,158] Therefore, the search for new ionophores with even higher Li^+ selectivity, as well as properties suitable for use in the solvent-polymeric membranes of ion-selective electrodes, continues.

In the clinical laboratory, spectrophotometric determination with Li^+-specific chromoionophores is now possible.[193,207,208] This can replace the traditional flame photometric determination with a method that involves less expensive instrumentation which can be utilized by less skilled technicians. Such factors may be an important advantage for Third World countries. Noteworthy is the very recent development of a chromogenic cryptand which allows Li^+ in serum to be determined by colorimetry without extraction.[193]

New methodology for the separation of lithium from other elements is the key to success for the lithium industry. It is conceivable that Li^+-specific ionophores could be applied to extractive metallurgy. Leachates obtained from the processing of spodumene could provide purified lithium-containing streams. More important would be their application in the separation of Li^+ or its salts from brine sources. It would seem that polymeric resins with incorporated ionophores which would specifically retain Li^+ or its salts possess the greatest potential for success in such applications.

The role of lithium in living systems is an area of active investigation.[308] A new journal, *Lithium*, publishes articles on biochemical, physiological, pharmacological, psychopharmacological, and clinical aspects of lithium together with papers on the physical and chemical aspects of lithium which have a bearing on its biological properties. The potential for application of synthetic ionophores with high Li^+ selectivity in such studies is very high. For example, ionophore-induced Li^+ transport across human erythrocyte membranes by 3,3-dibenzyl-14-crown-4 **(89)** was reported recently by Abraha and Mota de Freitas.[309]

Synthetic Li^+-specific ionophores also hold potential for use as adjuvants in the lithium treatment of manic depressive illness and of viral infections, such as Herpes and Varicela Zoster (Shingles).[310] Cautious addition of such an ionophore might reduce the amount of lithium required for successful treatment. This is based on the premise that ionophore-complexed Li^+ could more easily cross the mostly lipid phase of biological membranes.[311,312] Research must first establish the nontoxic character of the synthetic ionophore. If suitable ionophores are found, it would be of some importance to correlate the thermodynamic stability of the Li^+-ionophore complex with the desired Li^+ concentration in the target microenvironment. It is entirely possible that complexes of high stability may be less effective owing to their inability to release Li^+. The kinetics of Li^+ complexation and decomplexation in the biological environment will play an important role.

We hope that this survey of synthetic ionophores for lithium will serve as a catalyst for further research and development.

ACKNOWLEDGMENTS

RAB wishes to acknowledge support by the Division of Chemical Sciences of the Office of Basic Energy Sciences of the U.S. Department of Energy for his research on metal ion complexation by proton-ionizable podands and coronands.

ROB thanks FMC Lithium Division for access to their library, help in obtaining references, and facilitation of communication with the other authors. Special appreciation is expressed to Mrs. Joann Trull, Librarian, for her sustained assistance in securing accurate information and to Mr. Michael Yarbrough for help in communications. Both are with FMC Lithium Division.

TS expresses appreciation to Dr. Hidefumi Sakamoto for his help in writing a portion of the chapter.

REFERENCES

1. *Lithium—Current Applications in Science, Medicine, and Technology*, Bach, R. O., Ed.; Wiley: New York, 1985.
2. Tosteson, D. *Sci. Am.* **1981**, *244*, 164.
3. Lazarus, J. H.; Collard, K. J. *Endocrine and Metabolic Effects of Lithium*; Plenum: New York, 1986.
4. *Basic Mechanisms in the Action of Lithium*, Proceedings of a symposium held at Schloss Ringberb, Bavaria, FRG, October 4–6, 1981; Emrich, H. M.; Aldenhoff, J. B.; Lux, H. D., Eds.; Elsevier: New York, 1982.
5. Bach, R. O. *Med. Hypotheses* **1987**, *23*, 157.
6. Wakefield, B. J. *Organolithium Methods*; Academic Press: New York, 1988.
7. Gokel, G. W. *Crown Ethers and Cryptands*; The Royal Society of Chemistry: Cambridge, England, 1991, (a) p. 21, (b) p. 71.

8. Christensen, J. J.; Eatough, D. J.; Izatt, R. M. *Chem. Rev.* **1974**, *74*, 351.
9. Hubberstey, P. *Coord. Chem. Rev.* **1985**, *66*, 1.
10. Izatt, R. M.; Bradshaw, J. S.; Nielson, S. A.; Lamb, J. D.; Christensen, J. J.; Sen, D. *Chem. Rev.* **1985**, *85*, 271.
11. Hubberstey, P. *Coord. Chem. Rev.* **1988**, *85*, 1.
12. Bajaj, A. V.; Poonia, N. S. *Coord. Chem. Rev.* **1988**, *87*, 55.
13. Olsher, U.; Izatt, R. M.; Bradshaw, J. S.; Dalley, N. K. *Chem. Rev.* **1991**, *91*, 137.
14. Izatt, R. M.; Pawlak, K.; Bradshaw, J. S.; Bruenig, R. L. *Chem. Rev.* **1991**, *91*, 1721.
15. Buncel, E.; Shin, H. S.; van Truong, Ng.; Bannard, R. A. B.; Purdon, J. C. *J. Phys. Chem.* **1988**, *92*, 4176.
16. Davidson, R. B.; Izatt, R. M.; Christensen, J. J.; Schultz, R.; Dishong, D. M.; Gokel, G. M. *J. Org. Chem.* **1984**, *49*, 5080.
17. Yatsimirskii, K. B.; Sinyavskaya, E. I.; Tsymbal, L. V.; Tsvetkov, E. N.; Kron, T. E. *Zh. Neorg. Khim.* **1983**, *28*, 1410.
18. Evreinov, V. I.; Baulin, V. E.; Vostroknutova, Z. N.; Bondarenko, N. A.; Syundyukova, T. Kh.; Tsvetkov, E. N. *Izv. Akad. Nauk SSSR, Ser. Khim.* **1989**, 1990.
19. Olsher, U.; Elgavish, G. A.; Jagur-Grodzinski, J. *J. Am. Chem. Soc.* **1980**, *102*, 3338.
20. Hiratani, K. *J. Chem. Soc., Chem. Commun.* **1987**, 960.
21. Hiratani, K. *Analyst* **1988**, *113*, 1067.
22. Hiratani, K.; Nomoto, M.; Sugihara, H.; Okada, T. *Chem. Lett.* **1990**, *43*.
23. Beer, P. D.; Sikanayika, H.; Blackburn, C.; McAleer, J. F. *J. Organometal. Chem.* **1988**, *350*, C15.
24. Gutowski, D. A.; Delgado, M.; Gatto, V. J.; Echegoyen, L.; Gokel, G. W. *J. Am. Chem. Soc.* **1986**, *108*, 7553.
25. Hiratani, K. *Bull. Chem. Soc. Jpn.* **1985**, *58*, 420.
26. Tümmler, B.; Maass, G.; Vögtle, F.; Sieger, H.; Heimann, U.; Weber, E. *J. Am. Chem. Soc.* **1979**, *101*, 2588.
27. Pedersen, C. J. *J. Am. Chem. Soc.* **1967**, *89*, 7017.
28. Cook, F. L.; Caruso, T. C.; Byrne, M. P.; Bowers, C. W.; Speck, D. H.; Liotta, C. L. *Tetrahedron Lett.* **1974**, 4029.
29. Kem, R. J. *J. Org. Chem.* **1968**, *33*, 388.
30. Greene, G. N. *Tetrahedron Lett.* **1972**, 1793.
31. Gokel, G. W.; Cram, D. J.; Liotta, C. L.; Harris, H. P.; Cook, F. L. *J. Org. Chem.* **1974**, *39*, 2445.
32. Hopkins, Jr., H. P.; Norman, A. B. *J. Phys. Chem.* **1980**, *84*, 309.
33. Massaux, J.; Desreux, J.; Duyckaerts, G. *J. Chem. Soc., Dalton Trans.* **1980**, 865.
34. Smetana, A. J.; Popov, A. I. *J. Solut. Chem.* **1980**, *9*, 183.
35. Erk, C. *Spectrosc. Lett.* **1985**, *18*, 723.
36. Rhinebarger, R. R.; Rovang, J. W.; Popov, A. I. *Inorg. Chem.* **1984**, *23*, 2557.
37. Olsher, U.; Grodzinski, J. J. *J. Chem. Soc., Dalton Trans.* **1981**, 501.

38. Kitazawa, S.; Kimura, K.; Yano, H.; Shono, T. *J. Am. Chem. Soc.* **1984**, *106*, 6978.
39. Chen, S.-S.; Wang, S.-J.; Wu, S.-C. *Inorg. Chem.* **1984**, *23*, 3901.
40. Takeda, Y.; Katsuta, K.; Inoue, Y.; Hakushi, T. *Bull. Chem. Soc. Jpn.* **1988**, *61*, 627.
41. Takeda, Y. *Bull. Chem. Soc. Jpn.* **1982**, *55*, 2040.
42. Takeda, Y.; Kumazawa, T. *Bull. Chem. Soc. Jpn.* **1988**, *61*, 655.
43. Wickstrom, T.; Dale, J.; Lund, W. *Anal. Chim. Acta* **1988**, *211*, 233.
44. Gutowski, D. A.; Gatto, V. J.; Maller, J.; Echegoyen, L.; Gokel, G. W. *J. Org. Chem.* **1987**, *52*, 5172.
45. Shamsipur, M.; Popov, A. I. *Inorg. Chim. Acta* **1980**, *43*, 243.
46. Matthes, K. E.; Parker, D.; Buschmann, H. J.; Ferguson, G. *Tetrahedron Lett.* **1987**, *28*, 5503.
47. Amble, E.; Dale, J. *Acta Chem. Scand. B* **1979**, *33*, 698.
48. Yatsimirshii, K. B.; Kabachnik, M. I.; Sinyavskaya, E. I.; Medved, T. Ya; Polikarpov, Yu. M.; Shcherbakov, B. K. *Russ. J. Inorg. Chem.* **1984**, *29*, 510.
49. Ruangpornvisuti, V. W.; Probst, M. M.; Rode, B. M. *Inorg. Chim. Acta* **1988**, *144*, 21.
50. Hojo, M.; Imai, Y. *Anal. Sci.* **1985**, *1*, 185.
51. Ogawa, S.; Uchida, T.; Hirano, T.; Sabuni, M.; Uchida, Y. *J. Chem. Soc., Perkin Trans. 1* **1990**, 1649.
52. Bidzilya, V. A.; Golovkova, L. P.; Yatsimirskii, K. B. *Russ. J. Inorg. Chem.* **1981**, *26*, 664.
53. Shannon, R. D.; Prewitt, C. T. *Acta Crystallogr., Sect. B: Struct. Sci.* **1969**, *B25*, 925.
54. Lamb, J. D.; Izatt, R. M.; Christensen, J. J. In *Progress in Macrocyclic Chemistry*, Vol. 2; Izatt, R. M.; Christensen, J. J., Eds.; Wiley: New York, 1981; p. 54.
55. Dale, J.; Eggestad, J.; Fredriksen, S. B.; Groth, P. *J. Chem. Soc.* **1987**, 1391.
56. Gutmann, V. *Coordination Chemistry in Nonaqueous Solvents*; Springer-Verlag: Vienna, 1968.
57. Kaifer, A.; Gutowski, D. A.; Echegoyen, L.; Gatto, V. J.; Schultz, R. A.; Cleary, T. P.; Morgan, C. R.; Goli, D. M.; Rios, A. M.; Gokel, G. W. *J. Am. Chem. Soc.* **1985**, *107*, 1958.
58. Echegoyen, L.; Parra, D.; Bartolotti, L. J.; Hanlon, C.; Gokel, G. W. *J. Coord. Chem.* **1988**, *18*, 85.
59. Shimizu, T.; Tanaka, Y.; Tsuda, K. *J. Chem. Soc. Jpn.* **1985**, *58*, 3436.
60. Kimura, S.; Imanishi, Y. *Int. J. Biol. Macromol.* **1981**, *3*, 225.
61. Kessler, H.; Hehlein, W.; Schuck, R. *J. Am. Chem. Soc.* **1982**, *104*, 4534.
62. Inman, W.; Crews, P.; McDowell, R. *J. Org. Chem.* **1989**, *54*, 2523.
63. Malhotra, N.; Roestorff, P.; Hansen, T. K.; Becher, J. *J. Am. Chem. Soc.* **1990**, *112*, 3709.
64. Takahashi, T. "Analysis of Cation-Macrocycle Complexation by Fast Atom Bombardment Mass Spectrometry;" Presentation at the 7th International Symposium on Molecular Recognition and Inclusion: Kyoto, Japan, July 1992.

65. Maleknia, S.; Brodbelt, J. *J. Am. Chem. Soc.* **1992**, *114*, 4295.
66. Zhang, H.; Chu, I.-H.; Leming, S.; Dearden, D. V. *J. Am. Chem. Soc.* **1991**, *113*, 7415.
67. Zhang, H.; Dearden, D. V. *J. Am. Chem. Soc.* **1992**, *114*, 2754.
68. Lehn, J.-M.; Sauvage, J. P. *J. Am. Chem. Soc.* **1975**, *97*, 6700.
69. Lehn, J.-M. *Struct. Bonding (Berlin)* **1973**, *16*, 1.
70. Cox, B. J.; Garcia-Rosas, J.; Schneider, H. *J. Am. Chem. Soc.* **1981**, *103*, 1384.
71. Schwarzenbach, G.; Willi, A.; Bach, R. O. *Helv. Chim Acta* **1947**, 1303.
72. Cahen, Y. M.; Dye, J. L.; Popov, A. I. *J. Phys. Chem.* **1975**, *79*, 1289.
73. Cox, B. G.; Garcia-Rosas, J.; Schneider, H. *J. Am. Chem. Soc.* **1981**, *103*, 1054.
74. Liesegang, G. W. *J. Am. Chem. Soc.* **1981**, *103*, 953.
75. Abou-Hamdan, A.; Brereton, I. M.; Hounslow, A. M.; Lincoln, S. F.; Spotswood, T. M. *J. Inclusion Phenom.* **1987**, *5*, 137.
76. Abou-Hamdan, A.; Lincoln, S. F. *Inorg. Chem.* **1991**, *30*, 462.
77. Lincoln, S.; Stephens, A. K. W. *Inorg. Chem.* **1991**, *30*, 3529.
78. Bencini, A.; Bianchi, A.; Borselli, A.; Ciampolini, M.; Garcia-España, E.; Dapporto, P.; Micheloni, M.; Paoli, P.; Ramirez, J. A.; Valtancoli, B. *Inorg. Chem.* **1989**, *28*, 4279.
79. Bencini, A.; Bianchi, A.; Chimichi, S.; Ciampolini, M.; Dapporto, P.; Garcia-España, E.; Micheloni, M.; Nardi, N.; Paoli, P.; Valtancoli, B. *Inorg. Chem.* **1991**, *30*, 3687.
80. Bencini, A.; Bianchi, A.; Borselli, A.; Campolini, M.; Micheloni, M.; Nardi, N.; Paoli, P.; Valtancoli, B.; Chimichi, S.; Dapporto, P. *J. Chem. Soc., Chem. Commun.* **1990**, 174.
81. Chen, Z.; Schall, O. F.; Alcala, M.; Li, Y.; Gokel, G. W.; Echegoyen, L. *J. Am. Chem. Soc.* **1992**, *114*, 444.
82. Truter, M. R. *Struct. Bonding (Berlin)* **1973**, *16*, 71.
83. Dalley, N. K. In *Synthetic Multidentate Macrocyclic Compounds*; Izatt, R. M.; Christensen, J. J., Eds.; Academic Press: New York, 1978; pp. 207–243.
84. Goldberg, I. *The Chemistry of Ethers, Crown Ethers, Hydroxyl Groups and Their Sulfur Analogues*, Supplement E1; Patai, S., Ed.; Wiley: New York, 1980; pp. 175–214.
85. Hilgenfeld, R.; Saenger, W. *Top. Curr. Chem.* **1982**, *101*, 1.
86. Goldberg, I. In *Inclusion Compounds*, Vol. 2; Atwood, J. L.; Davies, J. E. D.; MacNicol, D. D., Eds.; Academic Press: New York, 1984; pp. 261–335.
87. Goldberg, I. In *Crown Ethers and Analogs*; Patai, S.; Rappoport, Z., Eds.; Wiley: New York, 1989; pp. 359–398, 399–476.
88. Fronczek, F. R.; Gandour, R. D. In *Cation Binding by Macrocycles: Complexation of Cationic Species by Crown Ethers*; Inoue, Y.; Gokel, G. W., Eds.; Marcel Dekker: New York, 1990; pp. 311–361.
89. Setzer, W. N.; Schleyer, P. v. R. *Adv. Organomet. Chem.* **1985**, *24*, 353.
90. Seebach, D. *Angew. Chem., Int. Ed. Engl.* **1988,** *27*, 1624.
91. Simon, W.; Morf, W. E.; Meier, P. C. *Struct. Bonding (Berlin)* **1973**, *16*, 113.

92. Hope, H.; Olmstead, M. M.; Power, P. P.; Xu, X. *J. Am. Chem. Soc.* **1984**, *106*, 819.
93. Power, P. P. *Acc. Chem. Res.* **1988**, *21*, 147.
94. Olsher, U.; Krakowiak, K. E.; Dalley, N. K.; Bradshaw, J. S. *Tetrahedron* **1991**, *47*, 2947.
95. Czech, B. P.; Zazulak, W.; Kumar, A.; Olsher, U.; Feinberg, H.; Cohen, S.; Shoham, G.; Dalley, N. K.; Bartsch, R. A. *J. Heterocycl. Chem.* **1992**, *29*, 1389.
96. Sachleben, R. A.; Burns, J. H. *J. Chem. Soc., Perkin Trans. 2* **1992**, 1971.
97. Dalley, N. K.; Jiang, W.; Olsher, U. *J. Inclusion Phenom. Mol. Recognit. Chem.* **1992**, *12*, 305.
98. Olsher, U.; Frolow, F.; Dalley, N. K.; Weiming, J.; Yu, Z.-Y.; Knobeloch, J. M.; Bartsch, R. A. *J. Am. Chem. Soc.* **1991**, *113*, 6570.
99. Sachleben, R. A.; Burns, J. H. *Acta Crystallogr., Sec. C* **1991**, *C47*, 1968, 2339.
100. Shoham, G.; Christianson, D. W.; Bartsch, R. A.; Heo, G. S.; Olsher, U.; Lipscomb, W. N. *J. Am. Chem. Soc.* **1984**, *106*, 1280.
101. Powell, J.; Sawyer, J. F.; Meindl, P.; Smith, S. J. *Acta Crystallogr. Sect. B: Struct. Sci.* **1990**, *B46*, 753.
102. Trueblood, K. H.; Maverick, E. F.; Knobler, C. B. *Acta Crystallogr., Sect. B: Struct. Sci.* **1991**, *B47*, 389.
103. Bell, T. W.; Cragg, P. J.; Drew, M. G. B.; Firestone, A.; Kwok, D.-I. A. *Angew. Chem., Int. Ed. Engl.* **1992**, *31*, 348.
104. Cram, D. J. *Angew Chem., Int. Ed. Engl.* **1986**, *25*, 1039.
105. Blasius, E.; Janzen, K.-P. *Top. Curr. Chem.* **1981**, *98*, 163.
106. Lamb, J. D.; Izatt, R. M.; Christensen, J. J. In *Progress in Macrocyclic Chemistry*, Vol. 2; Izatt, R. M.; Christensen, J. J., Eds.; Wiley-Interscience: New York, 1981; pp. 41–90.
107. Hiraoka, M. *Crown Compounds: Their Characteristics and Applications*; Elsevier: New York, 1982.
108. Takeda, Y. *Top. Curr. Chem.* **1984**, *121*, 1.
109. Takagi, M.; Ueno, K. *Top. Curr. Chem.* **1984**, *121*, 39.
110. Takagi, M.; Nakamura, H. *J. Coord. Chem.* **1986**, *15*, 53.
111. Bartsch, R. A.; Charewicz, W. A.; Kang, S. I.; Walkowiak, W. In *Liquid Membranes: Theory and Applications—ACS Symposium Series 347*; American Chemical Society: Washington, DC, 1987; pp. 86–97.
112. Blanco, G. D.; Arias, A. P.; Sanz, M. A. *Quimica Anal.* **1988**, *7*, 371.
113. Bartsch, R. A. *Solv. Extn. Ion Exch.* **1989**, *7*, 829.
114. Kimura, K.; Shono, T. In *Cation Binding by Macrocycles: Complexation of Cationic Species by Crown Ethers*; Inoue, Y.; Gokel, G. W., Eds.; Marcel Dekker: New York, 1990; pp. 429–463.
115. Takagi, M. In *Cation Binding by Macrocycles: Complexation of Cationic Species by Crown Ethers*; Inoue, Y.; Gokel, G. W., Eds.; Marcel Dekker: New York, 1990; pp. 465–495.

116. Lockhart, J. C. In *Inclusion Compounds, Volume 5—Inorganic and Physical Aspects of Inclusion*; Atwood, J. L.; Davies, J. E. D.; MacNicol, D. A., Eds.; Oxford University Press: New York, 1991; pp. 345–363.

117. Kaneda, T. In *Crown Ethers and Analogous Compounds*; Hiraoka, M., Ed.; Elsevier: New York, 1992; pp. 311–334.

118. Rechnitz, G. A. *Chem. Eng. News* **1967**, 146.

119. Manakova, L. I. *Z. Anal. Khim.* **1982**, *37*, 539.

120. Ciani, S.; Eisenman, G.; Szabo, G. *J. Membr. Biol.* **1969**, *1*, 1.

121. Morf, W. E.; Ammann, D.; Pretsch, D.; Simon, W. *Pure Appl. Chem.* **1973**, *36*, 421.

122. Dinten, O.; Spichinger, O. E.; Chaniotakis, N.; Gehrig, P.; Rusterholz, B.; Morf, W. E.; Simon, W. *Anal. Chem.* **1991**, *63*, 596.

123. Marullo, N. P.; Lloyd, R. A. *J. Am. Chem. Soc.* **1966**, *88*, 1076.

124. Marullo, N. P. U. S. Patent 3,461,166, August 12, 1969.

125. Güggi, M.; Fiedler, U.; Pretsch, E.; Simon, W. *Anal. Chim. Acta* **1981**, *131*, 117.

126. Kirsch, N. L.; Funck, R. J.; Pretsch, E.; Simon, W. *Helv. Chim. Acta* **1977**, *60*, 2326.

127. Zhukov, A. F.; Erne, D.; Ammann, D.; Güggi, M.; Pretsch, E.; Simon, W. *Anal. Chim. Acta* **1981**, *131*, 117.

128. Metzger, E.; Aeschimann, R.; Egli, M.; Suter, G.; Dohner, R.; Ammann, D.; Dobler, M.; Simon, W. *Helv. Chim. Acta* **1986**, *69*, 1821.

129. Metzger, E.; Ammann, D.; Aspen, R.; Simon, W. *Anal. Chem.* **1986**, *58*, 132.

130. Bliggensdorfer, R.; Suter, G.; Simon, W. *Helv. Chim. Acta* **1989**, *72*, 1164.

131. Morf, W. E.; Bliggensdorfer, R.; Simon, W. *Anal. Sci.* **1989**, *5*, 453.

132. Thomas, R. C.; Simon, W.; Oehme, M. *Nature* **1975**, *258*, 754.

133. Metzger, E.; Ammann, D.; Shefer, U.; Pretsch, E.; Simon, W. *Chimia* **1984**, *36*, 440.

134. Bochenska, M.; Simon, W. *Mikrochim. Acta (Wien)* **1990**, *3*, 277.

135. Gadzekpo, V. P. Y.; Hungerford, J. M.; Kadry, A. M.; Ibrahim, Y. A.; Christian, G. D. *Anal. Chem.* **1985**, *57*, 493.

136. Gadzekpo, V. P. Y.; Hungerford, J. M.; Kadry, A. M.; Ibrahim, Y. A.; Xie, R. Y.; Christian, G. D. *Anal. Chem.* **1986**, *58*, 1948.

137. Sugihara, H.; Okada, T.; Hiratani, K. *Anal. Chim. Acta* **1986**, *182*, 275.

138. Hruska, Z.; Petranek, J. *Polym. Bull.* **1987**, *17*, 103.

139. Hruska, Z.; Petranek, J. *Coll. Czech. Chem. Commun.* **1988**, *53*, 68.

140. Attiyat, A. S.; Kadry, A. M.; Hanna, H. R.; Ibrahim, Y. A.; Christian, G. D. *Anal. Sci.* **1990**, *6*, 233.

141. Hiratani, K.; Okada, T.; Sugihara, H. *Anal. Chem.* **1987**, *59*, 766.

142. Okada, T.; Hiratani, K.; Sugihara, H. *Analyst* **1987**, *112*, 587.

143. Sugihara, H.; Okada, T.; Hiratani, K. *Chem. Lett.* **1987**, 2391.

144. Suzuki, K.; Tohda, K.; Sasakura, H.; Inoue, H.; Tatsuta, K.; Shirai, T. *J. Chem. Soc., Chem. Commun.* **1987**, 932.

145. Suzuki, K.; Tohda, K.; Tominaga, M.; Tatsuta, K.; Shirai, T. *Anal. Lett.* **1987**, *20*, 927.
146. Todha, K.; Suzuki, K.; Kosuge, N.; Watanabe, K.; Nagashima, H.; Inoue, H.; Shirai, T. *Anal. Chem.* **1990**, *62*, 936.
147. Aalmo, K. M.; Krane, J. *Acta Chem. Scand., Ser. A.* **1982**, *36*, 227.
148. Olsher, U. *J. Am. Chem. Soc.* **1982**, *104*, 4006.
149. Gadzekpo, V. P. Y.; Christian, G. D. *Anal. Lett.* **1983**, *16*, 1371.
150. Kimura, K.; Kitazawa, S.; Shono, T. *Chem. Lett.* **1984**, 639.
151. Kitazawa, S.; Kimura, K.; Yano, H.; Shono, T. *J. Am. Chem. Soc.* **1984**, *106*, 6978.
152. Kimura, K.; Yano, H.; Kitazawa, S.; Shono, T. *J. Chem. Soc., Perkin Trans. 2* **1986**, 1945.
153. Kataky, R.; Nicholson. P. E.; Parker, D. *Tetrahedron Lett.* **1989**, *30*, 4559.
154. Kataky, R.; Nicholson, P. E.; Parker, D. *J. Chem. Soc., Perkin Trans.* **1990**, *2*, 321.
155. Kataky, R.; Nicholson, P. E.; Parker, D. *Analyst* **1991**, *116*, 135.
156. Imato, T.; Katahira, M.; Ishibashi, N. *Anal. Chim. Acta* **1984**, *165*, 285.
157. Kitazawa, S.; Kimura, K.; Yano, H.; Shono, T. *Analyst* **1985**, *110*, 295.
158. Kimura, K.; Oishi, H.; Miura, T.; Shono, T. *Anal. Chem.* **1987**, *59*, 2331.
159. Schindler, J. G.; Stork, G.; Struh, H. J.; Schal, W. *Fresenius Z. Anal. Chem.* **1978**, *45*, 290.
160. Bogatski, A. V.; Lukyanenko, N. G.; Golubev, V. N.; Nozarova, N. Yu.; Karkenko, L. P.; Popkov, Yu. A.; Shapkin, V. A. *Anal. Chim. Acta* **1984**, *157*, 151.
161. Tanaka, M.; Miura, T.; Sakamoto, H.; Shono, T. *Chem. Exp.* **1990**, *5*, 713.
162. Olsher, U.; Frolow, F.; Shoham, G.; Heo, G.-S.; Bartsch, R. A. *Anal. Chem.* **1989**, *61*, 1618.
163. Olsher, U.; Grolow, F.; Shoham, G.; Luboch, E.; Yu, Z.-Y.; Bartsch, R. A. *J. Inclusion Phenom. Molec. Recognit. Chem.* **1990**, *9*, 125.
164. Bell, T. W.; Choi, H.-J.; Heil, G. *Tetrahedron Lett.* **1993**, *34*, 971.
165. Gadzekpo, V. P. Y.; Moody, G. J.; Thomas, J. D. *Anal. Proc.* **1986**, *23*, 62.
166. Xie, R. Y.; Gadzekpo, V. P. Y.; Kadry, A. M.; Ibrahim, Y. A.; Ruzicka, J.; Christian, J. D. *Anal. Chim. Acta* **1986**, *184*, 259.
167. Metzger, E.; Dohner, R.; Simon, W.; Vonderschmitt, D. J.; Gantschi, K. *Anal. Chem.* **1987**, *59*, 1600.
168. Xie, R. Y.; Christian, G. D. *Anal. Chem.* **1986**, *58*, 1806.
169. Attiyat, A. S.; Ibrahim, Y. A.; Kadry, A. M.; Xie, R. Y.; Christian, G. D. *Fresenius Z. Anal. Chem.* **1987**, *329*, 12.
170. Pacey, G. E.; Wu, Y. P.; Sasaki, K. *Anal. Biochem.* **1987**, *160*, 243.
171. Attiyat, A. S.; Christian, G. D. *Flow Injection Anal.* **1987**, *4*, 103.
172. Attiyat, A. S.; Christian, G. D. *Anal. Sci.* **1988**, *4*, 13.
173. Attiyat, A. S.; Christian, G. D.; Xie, R. Y.; Wen, X.; Bartsch, R. A. *Anal. Chem.* **1988**, *60*, 2561.
174. Attiyat, A. S.; Christian, G. D.; Bartsch, R. A. *Electroanalysis* **1989**, *1*, 63.

175. Bartsch, R. A.; Goo, M.-J.; Christian, G. D.; Wen, X.; Czech, B. P.; Chapoteau, E.; Kumar, A. *Anal. Chim. Acta* **1993**, *272*, 285.
176. Eugster, R.; Gehrig, P. M.; Morf, W. E.; Spichiger, O. E.; Simon, W. *Anal. Chem.* **1991**, *63*, 2285.
177. Zoltov, Yu. A. *Extraction of Chelate Compounds*; Ann Arbor—Humphrey Science Publishers: Ann Arbor, MI, 1970; pp. 177–191.
178. Löhr, H. G.; Vögtle, F. *Acc. Chem. Res.* **1985**, *18*, 65.
179. Dix, J. P.; Vögtle, F. *Angew. Chem., Int. Ed. Engl.* **1978**, *17*, 857.
180. Kaneda, T.; Sugihara, K.; Kamiya, H.; Misumi, S. *Tetrahedron Lett.* **1981**, *22*, 4407.
181. Nakashima, K.; Nakatsuji, S.; Akiyama, S.; Kaneda, T.; Misumi, S. *Chem. Lett.* **1982**, 1781.
182. Sugihara, K.; Kaneda, T.; Misumi, S. *Heterocycles* **1982**, *18*, 57.
183. Nakashima, K.; Nakatsuji, S.; Akiyama, S.; Kaneda, T.; Misumi, S. *Chem. Pharm. Bull.* **1986**, *34*, 168.
184. Tanigawa, I.; Tsuemoto, K.; Kaneda, T.; Misumi, S. *Tetrahedron Lett.* **1984**, *25*, 5327.
185. Nakashima, K.; Nakatsuji, S.; Akiyama, S.; Tanigawa, I.; Kaneda, T.; Misumi, S. *Talanta* **1984**, *31*, 749.
186. Nakashima, K.; Nagaoka, Y.; Nakatsuji, S.; Kaneda, T.; Tanigawa, I.; Hirose, K.; Misumi, S.; Akiyama, S. *Bull. Chem. Soc. Jpn.* **1987**, *60*, 3219.
187. Cram, D. J.; Carmack, R. A.; Helgeson, R. C. *J. Am. Chem. Soc.* **1988**, *110*, 571.
188. Kaneda, T.; Umeda, S.; Tanigawa, H.; Misumi, S. *J. Am. Chem. Soc.* **1985**, *107*, 4802.
189. Inouye, M.; Ueno, M.; Kitao, T.; Tsuchiya, K. *J. Am. Chem. Soc.* **1990**, *112*, 8977.
190. Kimura, K.; Yamashita, T.; Yokoyama, M. *J. Chem. Soc., Chem. Commun.* **1991**, 147.
191. Ogawa, S.; Narushima, R.; Arai, Y. *J. Am. Chem. Soc.* **1984**, *106*, 5760.
192. Attiyat, A. S.; Christian, G. D. *Microchem. J.* **1988**, *37*, 114.
193. Chapoteau, E.; Chowdhary, M. S.; Czech, B. P.; Kumar A.; Zazulak, W. *J. Org. Chem.* **1992**, *57*, 2804.
194. Zazulak, W.; Chapoteau, E.; Czech, B. P.; Kumar, A. *J. Org. Chem.* **1992**, *57*, 6720.
195. Chapoteau, E.; Czech, B. P.; Zazulak, W.; Kumar, A. *Clin. Chem.* **1992**, *38*, 1654.
196. Takagi, M.; Nakamura, H.; Ueno, K. *Anal. Lett.* **1977**, *10*, 1115.
197. Wu, Y. P.; Pacey, G. E. *Anal. Chem. Acta* **1984**, *162*, 285.
198. Sasaki, K.; Pacey, G. *Anal. Chem. Acta* **1985**, *174*, 141.
199. Katayama, Y.; Fukuda, R.; Hiwatario, K.; Takagi, M. *Rep. Asahi Glass Found. Ind. Technol.* **1986**, *48*, 193.
200. Nakamura, H.; Sakka, H.; Takagi, M.; Ueno, U. *Chem. Lett.* **1981**, 1305.
201. Katayama, Y.; Nida, K.; Ueda, M.; Nakamura, H.; Takagi, M. *Anal. Chim. Acta* **1985**, *173*, 193.

202. Nishida, H.; Katayama, Y.; Katsuki, H.; Nakamura, H.; Takagi, M.; Ueno, K. *Chem. Lett.* **1982**, 1853.
203. Bubnis, B. P.; Pacey, G. E. *Tetrahedron Lett.* **1984**, *25*, 1107.
204. Katayama, Y.; Nakamura, H.; Takagi, M. *Anal. Sci.* **1985**, *1*, 393.
205. Kimura, K.; Tanaka, M.; Kitazawa, S.; Shono, T. *Chem. Lett.* **1985**, 1239.
206. Kimura, K.; Tanaka, M.; Iketani, S.; Shono, T. *J. Org. Chem.* **1987**, *52*, 836.
207. Kimura, K.; Iketani, S.; Shono, T. *Anal. Chim. Acta* **1987**, *203*, 85.
208. Kimura, K.; Iketani, S.; Sakamoto, H.; Shono, T. *Anal. Sci.* **1988**, *4*, 221.
209. Kimura, K.; Iketani, S.; Sakamoto, H.; Shono, T. *Analyst* **1990**, *115*, 1251.
210. Forrest, H.; Pacey, G. *Talanta* **1989**, *36*, 335.
211. Scholl, A. F.; Sutherland, I. O. *J. Chem. Soc., Chem. Commun.* **1992**, 1716.
212. McDaniel, C. W.; Bradshaw, J. S.; Izatt, R. M. *Heterocycles* **1990**, *30*, 665.
213. Brown, P. R.; Bartsch, R. A. In *Inclusion Aspects of Membrane Chemistry*; Osa, T.; Atwood, J. L., Eds.; Kluwer Academic Publishers: Boston, 1991; pp. 1–57.
214. Frensdorff, H. K. *J. Am. Chem. Soc.* **1971**, *93*, 4684.
215. Sadakane, A.; Iwachico, T.; Toei, K. *Bull. Soc. Chem. Jpn.* **1975**, *48*, 60.
216. Moore, S. S.; Tarnowski, T. L.; Newcomb, M.; Cram, D. J. *J. Am. Chem. Soc.* **1977**, *99*, 6398.
217. Inoue, Y.; Liu, Y.; Amano, F.; Ouchi, M.; Tai, A.; Hakushi, T. *J. Chem. Soc., Dalton Trans.* **1988**, 2735.
218. Shanzer, A.; Samuel, D.; Korenstein, D. *J. Am. Chem. Soc.* **1983**, *105*, 3815.
219. Iimori, T.; Still, W. C.; Rheingold, A. L.; Stalley, D. L. *J. Am. Chem. Soc.* **1989**, *111*, 3439.
220. Schepartz, A.; McDevitt, J. P. *J. Am. Chem. Soc.* **1989**, *111*, 5976.
221. Chen, C.-S.; Chao, H.-E.; Wang, S.-J.; Wu, S.-C. *Inorg. Chim. Acta* **1988**, *145*, 85.
222. Liu, Y.; Inoue, Y.; Hakushi, T. *Bull. Chem. Soc. Jpn.* **1990**, *63*, 3044.
223. Czech, B. P.; Babb, D. A.; Son, B.; Bartsch, R. A. *J. Org. Chem.* **1984**, *49*, 4805.
224. Wakita, R.; Yonetani, M.; Nakatsuji, Y.; Okahara, M. *J. Org. Chem.* **1990**, *55*, 2752.
225. Kobiro, K.; Matsuoka, T.; Takada, S.; Kakiuchi, K.; Tobe, Y.; Odaira, Y. *Chem. Lett.* **1986**, 713.
226. Kobiro, K.; Hiro, T.; Matsuoka, T.; Kakiuchi, K.; Tobe, Y.; Odaira, Y. *Bull. Chem. Soc. Jpn.* **1988**, *61*, 4164.
227. Sachleben, R. A.; Burns, J. H.; Davis, M. C.; Driver, J. L.; Chen, Z.; Deng, Y.; Moyer, B. A. *Recognition of Lithium Cation by Substituted 14-Crown-4 Macrocycles*; 205th National ACS Meeting; Denver, CO; April 1993; Abstract I&EC 104.
228. Kobuke, Y.; Hanji, K.; Horguchi, K.; Asada, M.; Nakayama, Y.; Furukawa, J. *J. Am. Chem. Soc.* **1976**, *98*, 7414.
229. Nakatsuji, Y.; Wakita, R.; Horada, Y.; Okahara, M. *J. Org. Chem.* **1989**, *54*, 2988.
230. Pugia, M. J.; Knudsen, B. E.; Bartsch, R. A. *J. Org. Chem.* **1987**, *52*, 2617.

231. Inokuma, S.; Katoh, R.; Yamamoto, T.; Nishimura, J. *Chem. Lett.* **1991**, 1751.
232. Takahashi, T.; Habata, Y.; Iri, Y. *J. Inclusion Phenom. Mol. Recognit. Chem.* **1991**, *11*, 379.
233. Kihara, N.; Saigo, K.; Habata, Y.; Ohno, M.; Hasegawa, M. *Chem. Lett.* **1989**, 1289.
234. Cram, D. J.; Lein, G. M.; Kaneda, T.; Helgeson, R. C.; Knobler, C. B.; Maverick, E.; Trueblood, K. N. *J. Am. Chem. Soc.* **1981**, *103*, 6228.
235. Cram, D. J.; Kaneda, T.; Helgeson, R. C.; Lein, G. M. *J. Am. Chem. Soc.* **1979**, *101*, 6752.
236. Cram, D. J.; Dicker, I. B.; Knobler, C. B.; Trueblood, K. N. *J. Am. Chem. Soc.* **1982**, *104*, 6828.
237. Cram, D. J.; Lein, G. M. *J. Am. Chem. Soc.* **1985**, *107*, 3657.
238. Koenig, K. E.; Helgeson, R. C.; Cram, D. J. *J. Am. Chem. Soc.* **1976**, *98*, 4018.
239. Walkowiak, W.; Ndip, G. M.; Desai, D. H.; Lee, H. K.; Bartsch, R. A. *Anal. Chem.* **1992**, *64*, 1685.
240. Bartsch, R. A.; Czech, B. P.; Kang, S. I.; Stewart, L. E.; Walkowiak, W.; Charewicz, W. A.; Heo, G. S.; Son, B. *J. Am. Chem. Soc.* **1985**, *107*, 4997.
241. Walkowiak, W.; Charewicz, W. A.; Kang, S. I.; Pugia, M. J.; Bartsch, R. A. *Anal. Chem.* **1990**, *62*, 2018.
242. Walkowiak, W.; Kang, S. I.; Stewart, L. E.; Ndip, G.; Bartsch, R. A. *Anal. Chem.* **1990**, *62*, 2022.
243. Bartsch, R. A.; Yang, I.-W.; Jeon, E.-G.; Walkowiak, W.; Charewicz, W. A. *J. Coord. Chem.* **1992**, *27*, 75.
244. Czech, B.; Son, B.; Bartsch, R. A. *Tetrahedron Lett.* **1983**, *24*, 2923.
245. Czech, B. P.; Czech, A.; Son, B.; Lee, H. K.; Bartsch, R. A. *J. Heterocycl. Chem.* **1986**, *23*, 465.
246. Uhlemann, E.; Geyer, H.; Gloe, K.; Mühl, P. *Anal. Chim. Acta* **1986**, *185*, 279.
247. Uhlemann, E.; Bukowsky, H.; Dietrich, F.; Gloe, K.; Mühl, P.; Mosler, H. *Anal. Chim. Acta* **1989**, *224*, 47.
248. Sachleben, R. A.; Moyer, B. A.; Case, F. I.; Garmon, S. A. *Sep. Sci. Technol.* **1993**, *28*, 1.
249. Habata, Y.; Ikeda, M.; Akabori, S. *Tetrahedron Lett.* **1992**, *33*, 3157.
250. Habata, Y.; Akabori, S. *Tetrahedron Lett.* **1992**, *33*, 5815.
251. Jepsen, B. E.; Dewitt, R. *J. Inorg. Nucl. Chem.* **1976**, *38*, 1175.
252. Heumann, K. G. *Top. Curr. Chem.* **1985**, *127*, 77.
253. Katal'nikov, S. G.; Myshletsov, I. A. *Tr. Inst.-Mosk. Khim.-Teknol. Inst. im. D. I. Mendeleeva* **1989**, *156*, 3.
254. Nishizawa, K.; Watanabe, H.; Ishino, S.; Shinagawa, M. *J. Nucl. Sci. Technol.* **1984**, *21*, 133.
255. Zhi, K.; Fu, L.; Yao, Z.; Gao, Z.; Zhang, F. *Lanzhou Daxue Xuebao, Ziran Kexueban* **1982**, *18*, 187.
256. Nishizawa, K.; Ishino, S.; Watanabe, H. *J. Nucl. Sci. Technol.* **1984**, *21*, 694.

257. Sheng, H.; Li, S.; Chen, Y.; Jin, P.; Cai, Q. *Youji Huaxue* **1984**, *281*.
258. Fang, S.; Zhi, K.; Fu, L.; Gao, Z.; Yao, Z.; Zhang, F. *He Huaxue Yu Fangshe Huaxue* **1987**, *9*, 142.
259. Nishizawa, K.; Takano, T.; Ikeda, I.; Okahara, M. *Sep. Sci. Technol.* **1988**, *23*, 333.
260. Nishizawa, K.; Takano, T. *Sep. Sci. Technol.* **1988**, *23*, 751.
261. Fu, L.; Fang, S.; Yao, Z.; Gao, Z.; Tan, G. *He Huaxue Yu Fangshe Huaxue* **1989**, *11*, 142.
262. Tsivadze, A. Yu.; Demin, S. V.; Levkin, A. V.; Zhilov, V. I.; Nikol'skii, S. F.; Knyazev, D. A. *Zh. Neorg. Khim.* **1990**, *35*, 2158.
263. He, S.; Wu, J.; Chang, T. *Rev. Chem. Miner.* **1983**, *20*, 737.
264. He, S.; Wu, J.; Zhang, Q. *Huaxue Xuebao* **1984**, *42*, 1183.
265. Li, S.; Luo, W.; Wang, D. *He Huaxue Yu Fangshe Huaxue* **1987**, *9*, 236.
266. Jepsen, B. E.; Cairns, G. A. *MLM*-2622 **1979**.
267. Kim, D. W.; Jeon, J. S.; Eon, T. Y.; Suh, M. Y.; Lee, C. H. *J. Radioanal. Nucl. Chem.* **1991**, *150*, 417.
268. Fujine, S.; Saito, K.; Shiba, K. *J. Nucl. Sci. Technol.* **1983**, *20*, 439.
269. Nishizawa, K.; Watanabe, H.; Ishino, S.; Shinagawa, M. *J. Nucl. Sci. Technol.* **1984**, *21*, 133.
270. Nishizawa, K.; Watanabe, H. *J. Nucl. Sci. Technol.* **1986**, *23*, 843.
271. Echegoyen, L.; Chen, Z. *Redox-Active Crowns and Cryptands for Isotope Separations*; 48th Southwest Regional Meeting of the American Chemical Society; Lubbock: Texas, October 1992; Abstract 203.
272. Christensen, J. J.; Lamb, J. D.; Brown, P. R.; Oscarson, J. L.; Izatt, R. M. *Sep. Sci. Technol.* **1981**, *16*, 1193.
273. McBride, Jr., D. W.; Izatt, R. M.; Lamb, J. D.; Christensen, J. J. In *Inclusion Compounds, Volume 3—Physical Properties and Applications*; Atwood, J. L.; Davies, J. E. D.; MacNicol, D. D., Eds.; Academic Press: New York, 1984; pp. 571–628.
274. Izatt, R. M.; Clark, G. A.; Bradshaw, J. S.; Lamb, J. D.; Christensen, J. J. *Sep. Purif. Methods* **1986**, *15*, 21.
275. Izatt, R. M.; LindH, G. C.; Bruening, R. L.; Bradshaw, J. S.; Lamb, J. D.; Christensen, J. J. *Pure Appl. Chem.* **1986**, *58*, 1453.
276. Jagur-Grodzinski, J.; Shori, E. *Isr. J. Chem.* **1985**, *26*, 65.
277. Margalit, R.; Eisenman, G. *J. Membrane Biol.* **1981**, *61*, 209.
278. Margalit, R.; Shanzer, A. *Biochem. Biophys. Acta* **1981**, *649*, 441.
279. Margalit, R.; Shanzer, A. *Pflügers Arch.* **1982**, *395*, 87.
280. Zeevi, A.; Margalit, R. *J. Membrane Biol.* **1985**, *86*, 61.
281. Lamb, J. D.; Christensen, J. J.; Izatt, S. R.; Bedke, K.; Austin, M. S.; Izatt, R. M. *J. Am. Chem. Soc.* **1978**, *100*, 3219.
282. Hiratani, K.; Taguchi, K.; Sugihara, H.; Okada, T. *Chem. Lett.* **1986**, 197.
283. Hirose, T.; Hiratani, K.; Fujiwara, K.; Kasuga, K. *Chem. Lett.* **1993**, 369.
284. Collin, J.-P.; Sugihara, H.; Hiratani, K.; Okada, T.; Takahaski, T. *Polymethylene-Bridged Bisphenanthrolines as Highly Selective Complexing Agents for*

Lithium; 7th International Symposium on Molecular Recognition and Inclusion, July 1992; Kyoto, Japan; Abstract PA41.

285. Tsukube, H.; Adachi, H.; Morosawa, S. *J. Chem. Soc., Perkin Trans. 1* **1989**, 1537.
286. Tarrago, G.; Zidane, I.; Marzin, C.; Tep, A. *Tetrahedron* **1988**, *44*, 41.
287. Yamazaki, N.; Nakahama, S.; Hirao, A.; Negi, S. *Tetrahedron Lett.* **1978**, 2429.
288. Hiratani, K. *Chem. Lett.* **1982**, 1021.
289. Hiratani, K.; Taguchi, K.; Sugihara, H.; Iio, K. *Bull. Chem. Soc. Jpn.* **1984**, *57*, 1976.
290. Hiratani, K.; Taguchi, K.; Sugihara, H.; Iio, K. *J. Membrane Sci.* **1987**, *35*, 91.
291. Hiratani, K.; Taguchi, K. *Bull. Chem. Soc. Jpn.* **1987**, *60*, 3827.
292. Wakita, R.; Matsumoto, M.; Nakatsuji, Y.; Okahara, M. *J. Membrane Sci.* **1991**, *57*, 297.
293. Inokuma, S.; Kohno, T.; Inoue, K.; Yabusa, K.; Kuwamura, T. *Nippon Kagaku Kaishi* **1985**, 1585.
294. Bradshaw, J. S.; Izatt, R. M.; Huszthy, P.; Nakatsuji, Y.; Biernat, J. F.; Koyama, H.; McDaniel, C. W.; Wood, S. A.; Nielsen, R. B.; LindH, G. C.; Bruening, R. L.; Lamb, J. D.; Christensen, J. J. *Stud. Org. Chem. (Amsterdam)* **1987**, *31*, 553.
295. Izatt, R. M.; LindH, G. C.; Bruening, R. L.; Huszthy, P.; Lamb, J. D.; Bradshaw, J. S.; Christensen, J. J. *J. Inclusion Phenom.* **1987**, *5*, 739.
296. Bradshaw, J. S.; Guynn, J. M.; Wood, S. G.; Krakowiak, K. E.; Izatt, R. M.; McDaniel, C. W.; Wilson, B. E.; Dalley, N. K.; LindH, G. C. *J. Org. Chem.* **1988**, *53*, 2811.
297. Walkowiak, W.; Brown, P. R.; Shukla, J. P.; Bartsch, R. A. *J. Membrane Sci.* **1987**, *32*, 59.
298. Kimura, K.; Sakamoto, H.; Kitazawa, S.; Shono, T. *J. Chem. Soc., Chem. Commun.* **1985**, 669.
299. Sakamoto, H.; Kimura, K.; Shono, T. *Anal. Chem.* **1987**, *59*, 1513.
300. Sakamoto, H.; Kimura, K.; Tanaka, M.; Shono, T. *Bull. Chem. Soc. Jpn.* **1989**, *62*, 3394.
301. Smid, J.; Sinta, R. *Top. Curr. Chem.* **1984**, *121*, 105.
302. Bartulin, J.; Parra, M.; Ramirez, A.; Zunza, H. *Polym. Bull. (Berlin)* **1989**, *22*, 33.
303. Bartulin, J.; Parra, M.; Ramirez, A.; Zunza, H. *Polym. Bull. (Berlin)* **1990**, *24*, 129.
304. Bartulin, J.; Parra, M.; Ramirez, A.; Zunza, H. *Bol. Soc. Chil. Quim.* **1989**, *34*, 191.
305. Manecke, G.; Reuter, P. *Makromol. Chem.* **1981**, *182*, 1973.
306. Hayashita, T.; Lee, J. H.; Hankins, M. G.; Lee, J. C.; Kim J. S.; Knobeloch, J. M.; Bartsch, R. A. *Anal. Chem.* **1992**, *64*, 815.
307. Hayashita, T.; Goo, M.-J.; Lee, J. C.; Kim, J. S.; Krzykawski, J.; Bartsch, R. A. *Anal. Chem.* **1990**, *62*, 2283.

308. *Lithium and Cell Physiology*; Bach, R. O.; Gallicchio, V. S., Eds.; Springer-Verlag: New York, 1990.

309. Abraha, A.; Mota De Freitas, D. *Lithium* **1992**, *3*, 203.

310. Bach, R. O.; Specter, S. In *Lithium: Inorganic Pharmacology and Psychiatric Use*; Birch, N. J., Ed.; IRL Press: New York, 1988; pp. 91–92.

311. Riddell, F. G.; Arumugan, S.; Cox, B. G. In *Lithium: Inorganic Pharmacology and Psychiatric Use*; Birch, N. J., Ed.; IRL Press: New York, 1988; pp. 293–295.

312. Riddell, F. G.; Arumugan, S. *Biochim. Biophys. Acta* **1989**, 6.

11

PREPARATION AND REACTIONS OF POLYLITHIUMORGANIC COMPOUNDS

A. MAERCKER
Institute of Organic Chemistry
University of Siegen
Siegen, Germany

1 INTRODUCTION

As discussed in Chapter 1, the lithium–carbon bond is highly polarized. Extremely stable organolithium compounds are ionized completely, so that contact or solvent separated ion pairs are formed. However, dissociation usually does not occur since protic solvents, which might solvate anions by hydrogen

Lithium Chemistry: A Theoretical and Experimental Overview, Edited by Anne-Marie Sapse and Paul von Ragué Schleyer.
ISBN 0-471-54930-4

bonding, generally cannot be used. The highly basic carbanions would deprotonate such solvents immediately.

Lithiumorganic compounds are usually not monomeric but are associated (aggregated) through lithium bridges. In dilithium compounds, the intramolecular equivalent—double lithium bridges between the carbanionic centers—can help to stabilize the systems.[1,2] Several reviews describe the remarkable variety of polylithiumorganic structures.[1-4] The preparation as well as some reactions of these species will be treated in this chapter.

2 SYNTHESES

2.1 Halogen–Metal Exchange Reactions

2.1.1 with Lithium Metal. The direct replacement of halogen in organic molecules by treatment with lithium metal, the most direct method for the synthesis of alkyllithium compounds, often is only of limited value for the synthesis of polylithiumorganic compounds. After the first step, α, β, or gamma elimination of lithium halide is faster than the second halogen–metal exchange. Thus, only 1,4-dilithiobutane and higher 1,n-dilithioalkanes ($n > 4$) can be prepared straightforwardly starting from the corresponding 1,n-dibromoalkanes:[5]

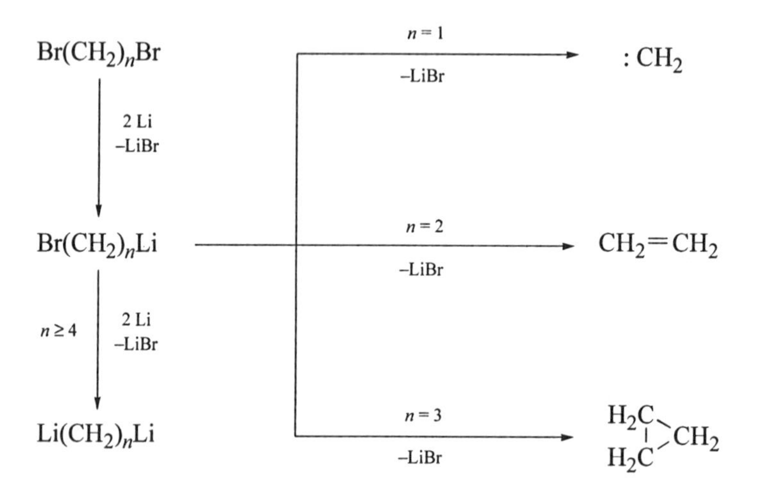

Early reports of the preparation of 1,2-dilithioethane from 1,2-dichloro- or 1,2-dibromoethane and lithium[6] as well as 1,3-dilithiopropane from 1,3-dichloropropane[7] proved to be incorrect. Both of these dilithium compounds are unstable and decompose by lithium hydride elimination.[8-13]

$$\begin{array}{ccccc}
CH_2Br_2 & \xrightarrow[-2\,LiBr]{4\,Li} \!\!\!\!\!/\;\; & CH_2Li_2 & \xrightarrow[-2\,LiCl]{2\,Me_3SiCl} & CH_2(SiMe_3)_2 \\
\downarrow {\scriptstyle 2\,Li \mid -LiBr} & & & & \uparrow {\scriptstyle Me_3SiCl \mid -LiCl} \\
BrCH_2Li & \xrightarrow[-LiCl]{Me_3SiCl} & Me_3SiCH_2Br & \xrightarrow[-LiBr]{2\,Li} & Me_3SiCH_2Li
\end{array}$$

Bis(trimethylsilyl)methane, obtained in 6% yield by treating dibromomethane with lithium in the presence of chlorotrimethylsilane,[5] is not formed from dilithiomethane but arises in a stepwise reaction via bromomethyltrimethylsilane. Instead of lithium metal, only the highly reactive lithium-4,4′-di-tert-butylbiphenyl (LDBB)[14] and diiodomethane in THF/diethyl ether (5 : 1) at −110°C gave dilithiomethane in 93% purity. The side products were 1% 1,2-dilithioethane (which decomposed to vinyllithium at > −70°C) as well as 5% 1,3-dilithiopropane and 1% 1,4-dilithiobutane.[12] Bis(trimethylsilyl)-[15] and diphenyldilithiomethane[16] could be prepared analogously starting from the corresponding geminal dichloro compounds.

$$R_2CCl_2 \xrightarrow[THF,\,-95°C]{LDBB} R_2CLi_2$$

$R = Me_3Si;\ Ph$

The success of such reactions depends critically on the stability of the carbenoid, R_2CLiX, formed initially. Starting from dibromodiphenylmethane, for instance, the yield was only 4–8%.[16] When di-tert-butyldibromomethane was reacted with LDBB under similar conditions, no gem-dilithium compound formed at all:[17]

$$Ph_2CBr_2 \xrightarrow{LDBB} Ph_2C\genfrac{}{}{0pt}{}{Li}{Br} \xrightarrow[-2\,LiBr]{2\,x}$$

$$Ph_2C{=}CPh_2 \xrightarrow{LDBB} Ph_2\bar{C}{-}\bar{C}Ph_2 \quad 2\,Li^+$$

$$(Me_3C)_2CBr_2 \xrightarrow{LDBB} (Me_3C)_2C\genfrac{}{}{0pt}{}{Li}{Br} \xrightarrow{-LiBr}$$

$$(Me_3C)_2C: \longrightarrow \begin{array}{l} Me_3CCH \\ \quad\;| \;\rangle CH_2 \\ Me_2C \end{array}$$

On the other hand, 7,7-dilithionorbornane[17] and 1,1-dilithio-2,2-diphenylethylene[18] could be prepared successfully from the corresponding geminal dibromo compounds. Both the resulting dilithio compounds undergo interesting rearrangements on warming: ring opening and intramolecular 1,4-proton shift, respectively.

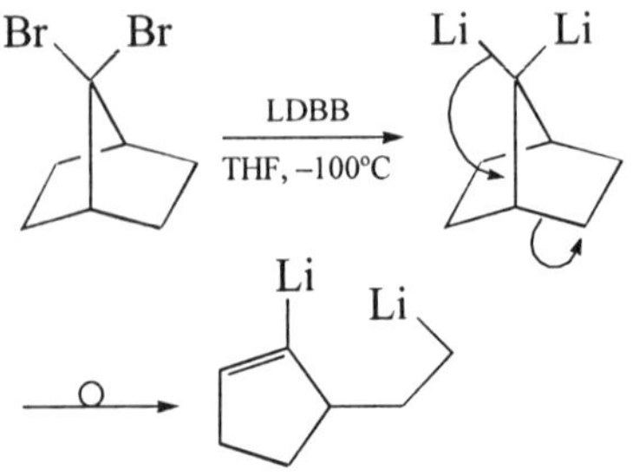
Br
Br
LDBB
THF, −100°C
Li
Li
Li
Li
O

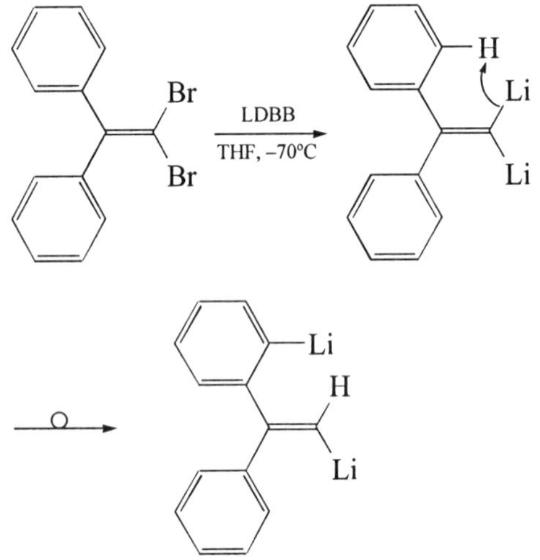
Br
Br
LDBB
THF, −70°C
H
Li
Li
Li
H
Li
O

To obtain bis(trimethylsilyl)dilithiomethane,[15] LDBB could be replaced by a lithium suspension prepared in a special apparatus.[19] Solvents were THF, diethyl ether or diethyl ether/pentane (1 : 1) at −100°C. In contrast, a commercial lithium dispersion did not work.[15]

Even lithium powder does not react with 1,4- and 1,5-dichloro-2-alkoxides. However, lithium naphthalene was a good catalyst in these cases:[20]

$$Cl(CH_2)_n\overset{OLi}{\underset{CH_2CH=CH_2}{C}}CH_2Cl \xrightarrow[THF,\ -78^\circ C]{Li\text{-}Naph} Li(CH_2)_n\overset{OLi}{\underset{CH_2CH=CH_2}{C}}CH_2Li$$

$n = 2;3$

Moreover, 1,4-dichloro-bicyclo[2.2.1]heptane does react with lithium metal, although a 1% sodium alloy is needed:[21]

$$\text{1,4-dichlorobicyclo[2.2.1]heptane} \xrightarrow[-2\,LiCl]{4\,Li} \text{1,4-dilithiobicyclo[2.2.1]heptane}$$

A 1,3-dilithio-2-silapropane is the only open-chain 1,3-dilithio derivative available from the corresponding dichloro compound:[22,23]

$$ClCH_2SiMe_2CH_2Cl \xrightarrow[-2\,LiCl]{4\,Li} LiCH_2SiMe_2CH_2Li$$

The direct replacement of halogen by lithium metal can also be used to prepare dilithiumaromatic compounds (other than *o*-dilithioarenes). Thus 2,2′-dilithiobiphenyl was obtained from 2,2′-diiodo-[24–27] or 2,2′-dibromobiphenyl.[28]

$$\text{2,2'-}X_2\text{-biphenyl} \xrightarrow[-2\,LiX]{4\,Li} \text{2,2'-}Li_2\text{-biphenyl}$$

X = Br; I

Nevertheless, the homogeneous halogen–metal exchange reaction with alkyllithium compounds (2.1.2) gives better results.

2.1.2 with Alkyllithium Compounds. Since halogen–metal exchange is an equilibrium reaction, only polylithium organic compounds which are more stable than the starting alkyllithium compounds can be prepared. Thus, merely an 18% yield of 1,4-dilithiooctafluorobutane was obtained from 1,4-diiodooctafluorobutane and *n*-butyllithium:[29]

$$I(CF_2)_4I + 2\,C_4H_9Li \underset{-80°C}{\overset{Et_2O}{\rightleftharpoons}} Li(CF_2)_4Li + 2\,C_4H_9I$$

It is better to use the thermodynamically less stable tert-butyllithium for such purposes. With alkyl iodides, this offers the additional advantage that the other exchange product, tert-butyl iodide, is destroyed by reaction with a second mole of tert-butyllithium:[30,31]

$$I(CH_2)_nI \; + \; 4\,Me_3CLi \xrightarrow[-78^\circ C \,\to\, 25^\circ C]{Et_2O}$$

$$Li(CH_2)_nLi \; + \; 2\,Me_3CH \; + \; 2\,Me_2C{=}CH_2 \; + \; 2\,LiI$$

$n = 4;5;6$

I, I; 2 t-BuLi, THF/pentane, −78°C; I, Li; 2 t-BuLi; Li, Li

t-BuLi

Li; 2 t-BuLi; t-Bu, Li, Li

However, even with tert-butyllithium, the reaction with isocentric (i.e., on the same carbon atom) polyhalogenated alkanes,[32,33] as well as with geminal dibromo-[33-35] and dichlorocyclopropanes,[36] usually stops after the replacement of only one halogen. Literature reports[37] of reaction products from ''dilithio compounds'' probably arise instead via reaction of the initially formed carbenoid with the electrophile, followed by halogen–lithium exchange and quenching with a second equivalent of the electrophile. Such two-stage mechanisms have been demonstrated, e.g., by the reaction of dibromomethane with lithium in the presence of chlorotrimethylsilane.[5]

Even with tert-butyllithium at −110°C, benzylic geminal chlorines are not replaced satisfactorily.[38] This can be achieved more successfully by lithium-4,4′-di-tert-butylbiphenyl (LDBB).[16]

On the other hand, halogen–metal exchanges with alkyllithium compounds are ideal for the introduction of one or more lithium atoms into aromatic and heteroaromatic ring systems as well as for the synthesis of vinyllithium derivatives. It even has been claimed[38a] that all six chlorines of hexachlorobenzene can be replaced by lithium by using a large excess of tert-butyllithium in pentane/1,4-dioxane at −125°C. The yield of hexalithiobenzene is reported to be 53%.

Since dilithioarenes have long been synthesized by this method, mostly from the corresponding bromo compounds,[39] only recent, interesting examples will be mentioned here.

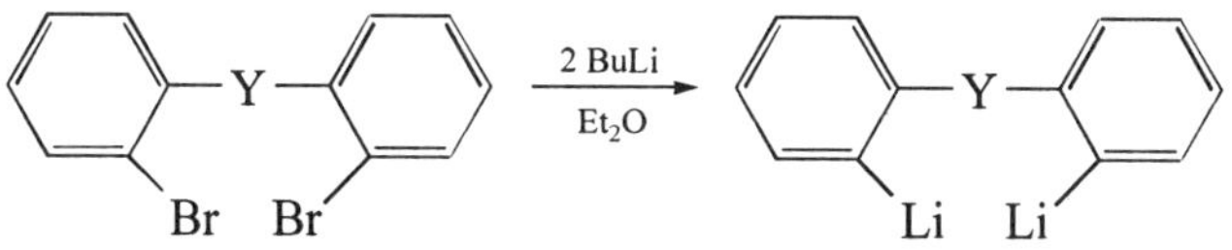

Y Br Br → 2 BuLi, Et_2O → Y Li Li

Y = CH_2[40]; CHPh[40]; CH_2CH_2[41,42]; O[43]; CH_2O[43]; CH_2NLi[44]

Tris(*o*-lithiophenyl)amine was prepared in an analogous manner:[45]

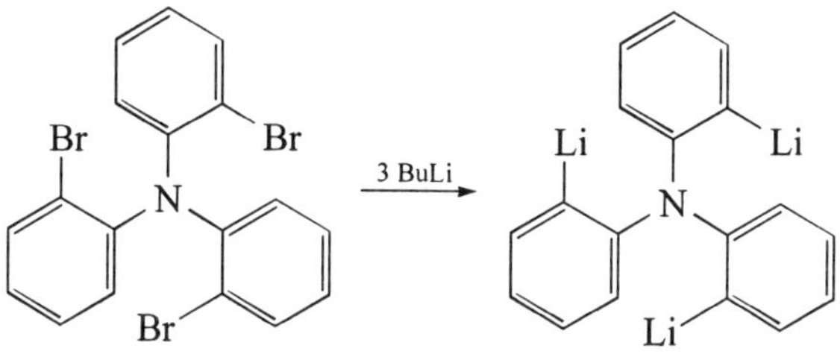

Br Br N Br → 3 BuLi → Li Li N Li

Even with excess *n*-butyllithium, only two of the bromines of 2,4,6-tribromoanisole are replaced by lithium in a pentane suspension at −20°C:[46]

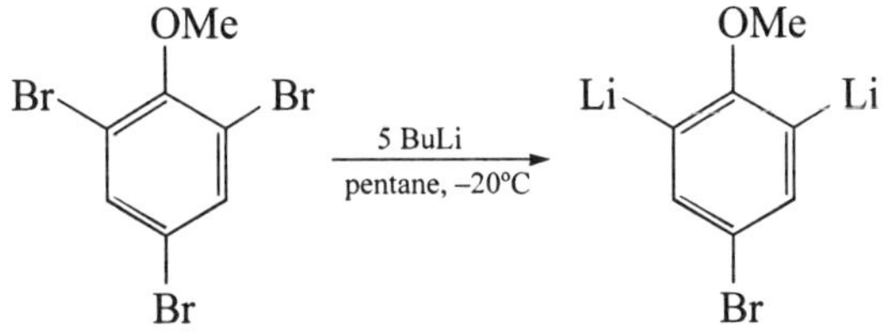

OMe Br Br Br → 5 BuLi, pentane, −20°C → OMe Li Li Br

THF solvent is necessary for the preparation of 2,7-dilithionaphthalene and 2,7-dilithioanthracene; in diethyl ether only one bromine is replaced:[47]

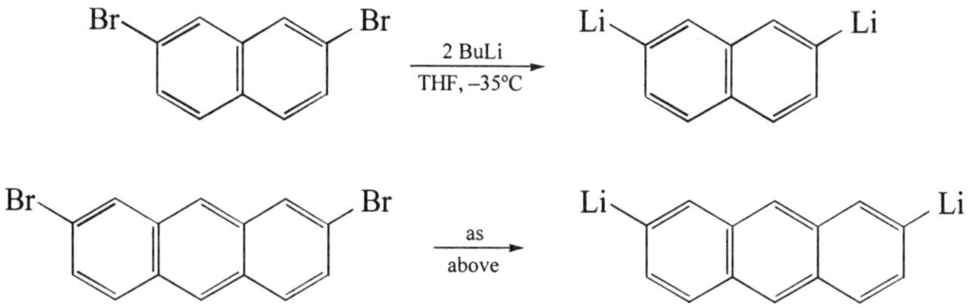

Br Br → 2 BuLi, THF, −35°C → Li Li

Br Br → as above → Li Li

However, 9,10-dilithioanthracene is obtained in diethyl ether solution:[48]

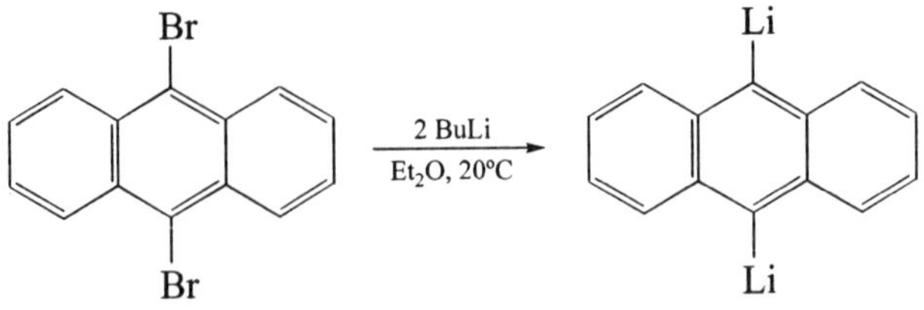

Br Li

2 BuLi

Et_2O, 20°C

Br Li

The formation of 5,6-dilithioacenaphthene is accelerated by the addition of *N,N,N′,N′*-tetramethylethylenediamine (TMEDA); only 15–30 min at −10° to 0°C are necessary instead of one hour at reflux:[49]

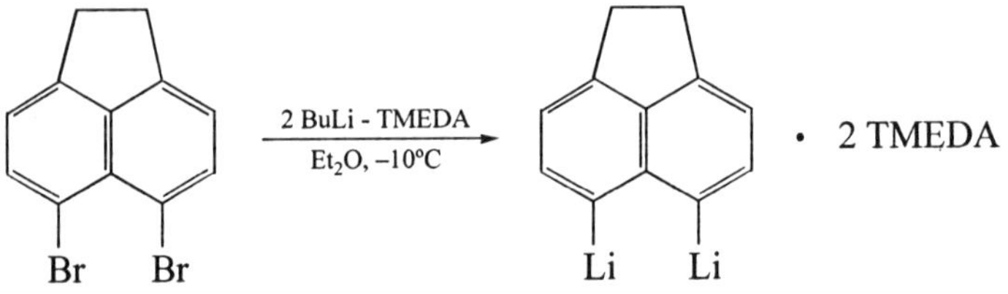

2 BuLi - TMEDA

Et_2O, −10°C

· 2 TMEDA

Br Br Li Li

Iodine–lithium exchange of 1,5-diiodonaphthalene takes place at −78°C using tert-butyllithium in diethyl ether:[50]

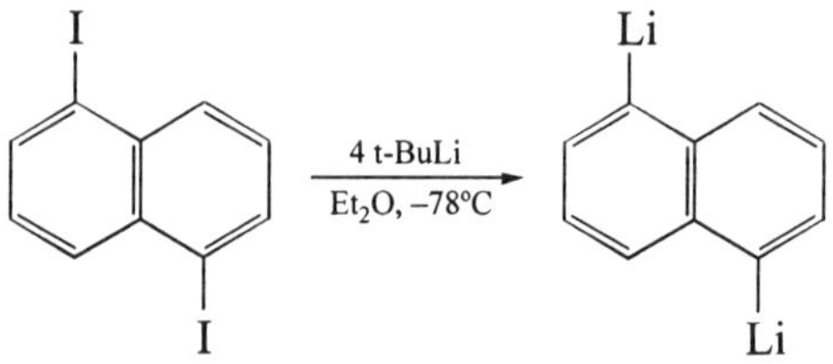

I Li

4 t-BuLi

Et_2O, −78°C

I Li

The bromine–lithium exchange reaction can even be performed in the presence of tertiary amide functional groups with tert-butyllithium in THF at −78°C:[51]

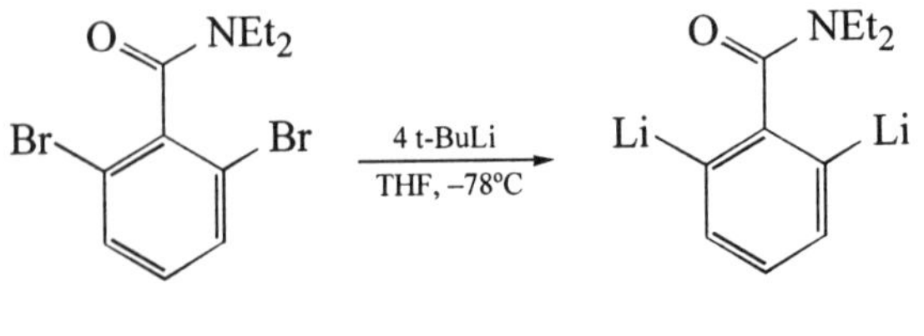

O NEt_2 O NEt_2

Br Br 4 t-BuLi Li Li

THF, −78°C

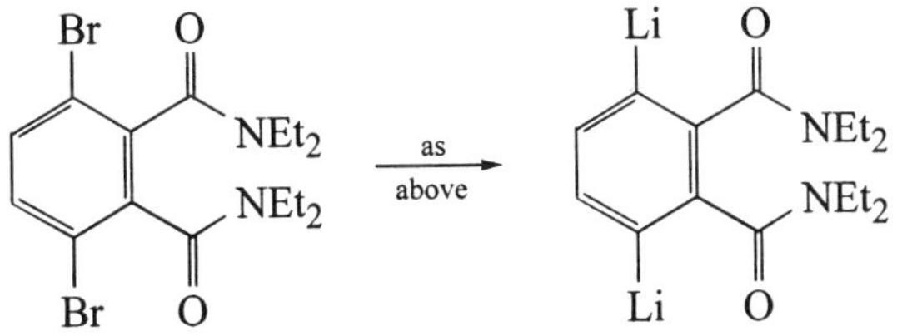

Perfluoralkyl ethers also are not attacked by *n*-butyllithium at −78 °C:[52]

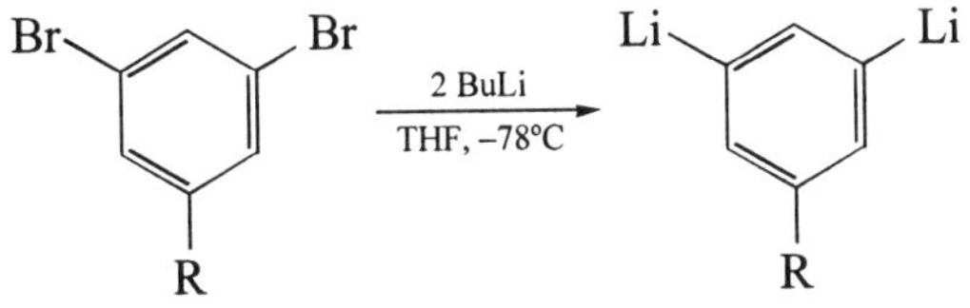

$R = CF_2CF_2O(CF_2CF_2O)_4CF_2CF_3$

The para-selectivity shown by hexahalobenzenes in halogen–metal exchange reactions is noteworthy:[53]

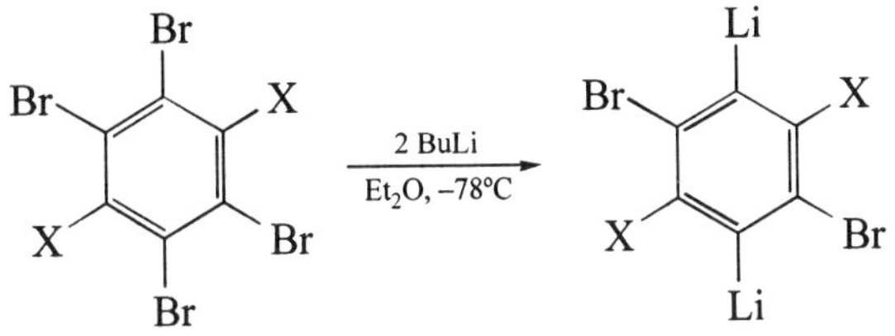

X = Cl; Br

As expected, bromine–lithium exchange is favored over chlorine–lithium exchange; on warming, benzyne is formed via elimination of lithium bromide rather than lithium chloride [53]. Lithium fluoride elimination is even slower; consequently, the following dilithium aromatic compounds are stable at low temperatures:[53,54]

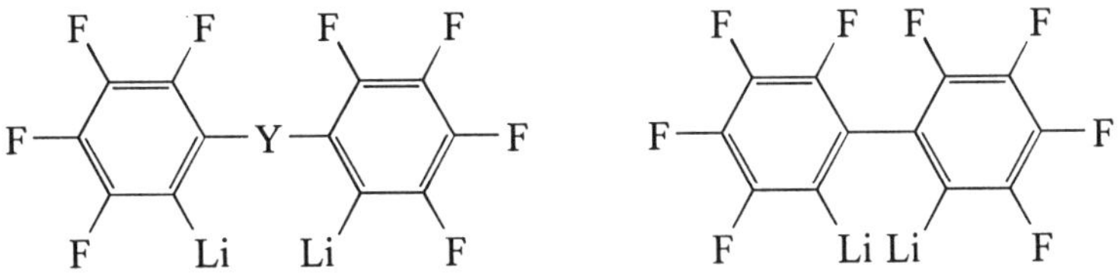

Y = S; $GePh_2$

It is not known whether the latter favors a planar conformation, like the unsubstituted 2,2′-dilithiobiphenyl with its symmetric double lithium bridge above and below the π system.[56] However, both 2,2′-dilithio-6,6′-dimethylbiphenyl[57] and 2,2′-dilithio-1,1′-binaphthyl[58] are chiral, enantiomerically stable compounds, and obviously cannot be planar:

Me Me I I

2 BuLi, Et_2O, –10°C ⇌ I_2

Me Me Li Li

X X

2 BuLi, Me_2O, –44°C →

Li Li

X = Br; I

The stabilization energy of the lithium bridging in a planar molecule—according to calculations 16.8 kcal/mole in the gas phase[59]—is estimated in solution to be reduced to no more than 5–10 kcal/mole.[60]

The reported synthesis of 1,2-dilithiotetrafluorobenzene by bromine–lithium exchange is highly questionable; good yields of derivatization products were only obtained with excess butyllithium. A stepwise reaction probably took place:[61]

F F F F Br Br

2 BuLi, Et_2O, –78°C →

F F F F Li Li

BuLi | CO_2

2 CO_2

F F F F CO_2Li Br

BuLi, CO_2 →

F F F F CO_2Li CO_2Li

The stepwise reaction of the following heterocycle could be shown unequivocally; the corresponding dilithiumorganic compound is not involved:[62]

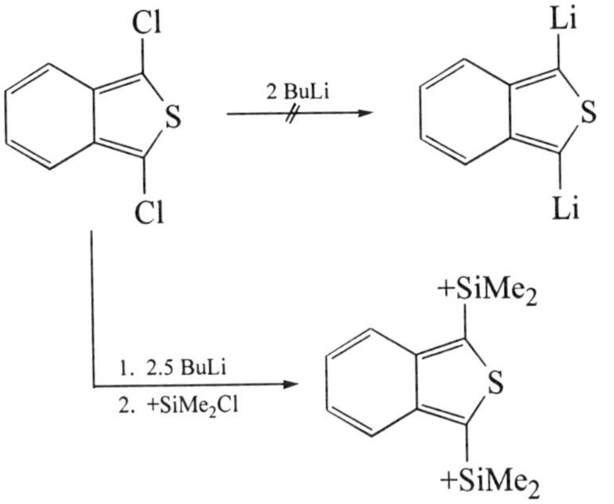

On the other hand, dibromo- and diiodohetarenes react like dihaloarenes,[63–65] e.g.,

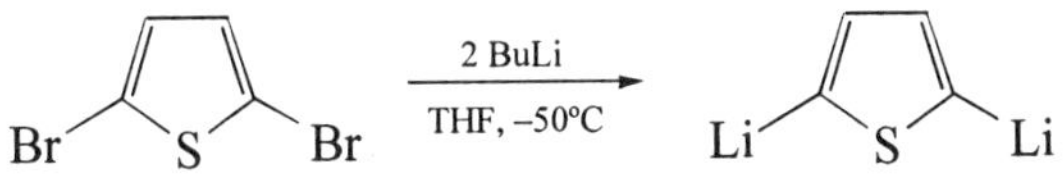

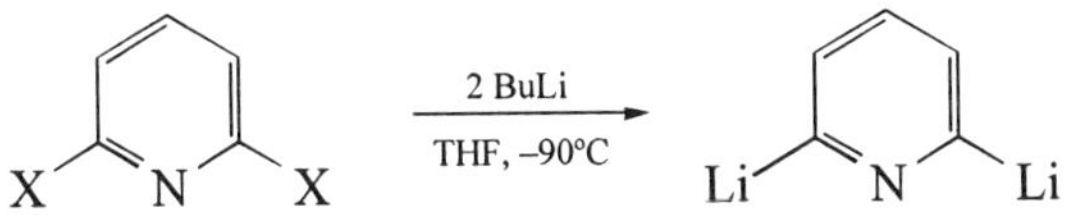

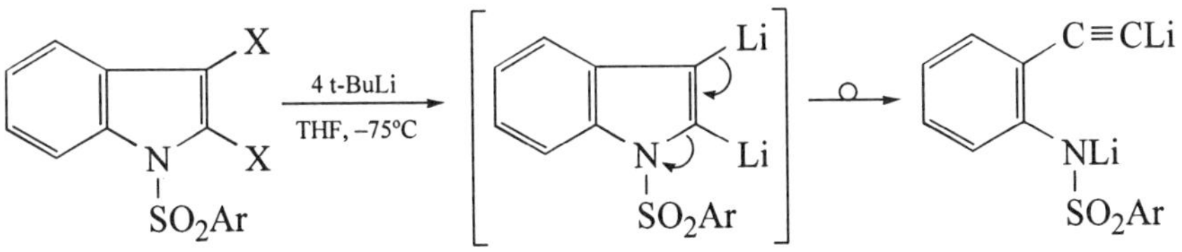

X = Br; I • Ar = C_6H_5; C_6H_4OMe-p

In the last case, rearrangement could only be prevented by using Ar = p-anisyl.[65] A trilithioheterocycle has also been reported:[66]

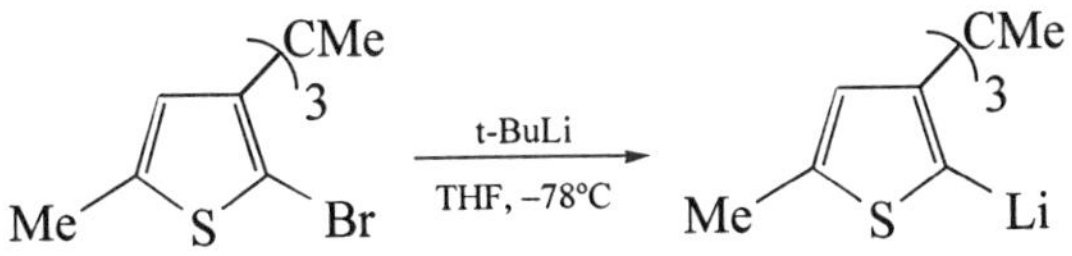

However, even a tenfold excess of butyllithium in diethyl ether only replaces two iodine atoms of 2,3,4-triiodofuran, and does so quite selectively (at the 2 and 4 positions):[67]

$$\text{2,3,4-triiodofuran} \xrightarrow[\text{Et}_2\text{O, }-70^\circ\text{C}]{\text{2 BuLi}} \text{2,4-dilithio-3-iodofuran}$$

The selectivity of these halogen–metal exchange reactions also is illustrated impressively by the reaction of octachloro-4,4′-bipyridyl with butyllithium. Only the 5 and 5′ chlorines are replaced by lithium.[68,69] The preparation of hexachloro-6,6′-dilithio-4,4′-bipyridyl requires the corresponding dibromo compound:[70]

$$\text{octachloro-4,4'-bipyridyl} \xrightarrow[\text{Et}_2\text{O, }-75^\circ\text{C}]{\text{2 BuLi}} \text{5,5'-dilithio-hexachloro-4,4'-bipyridyl}$$

$$\text{6,6'-dibromo-hexachloro-4,4'-bipyridyl} \xrightarrow[\text{above}]{\text{as}} \text{hexachloro-6,6'-dilithio-4,4'-bipyridyl}$$

The introduction of more than one vinyl lithium by halogen–metal exchange is possible only if the corresponding monolithium intermediate is rather stable and does not readily eliminate lithium halide. Thus, (Z,Z)-1,5-dilithio-1,4-pentadiene can be prepared from the corresponding diiodo compound in good yields with butyllithium either in pentane at −50°C or in diethyl ether at −30°C as well as with methyllithium in diethyl ether at −40°C:[71]

$$\text{I–CH=CH–CH}_2\text{–CH=CH–I} \xrightarrow[\text{pentane, }-50^\circ\text{C or Et}_2\text{O, }-30^\circ\text{C}]{\text{2 BuLi}} \text{Li–CH=CH–CH}_2\text{–CH=CH–Li}$$

While (Z,Z)-1,4-dilithio-1,3-butadiene might be synthesized by iodine-[72] or by bromine–lithium exchange[73] with tert-butyllithium at −78°C in diethyl ether, much better yields are achieved by tin–lithium transmetallation[74] (Section 2.3.2.2). According to theoretical computations,[75] symmetrical bridging similar to that in 2,2′-dilithiobiphenyl is to be expected:[56]

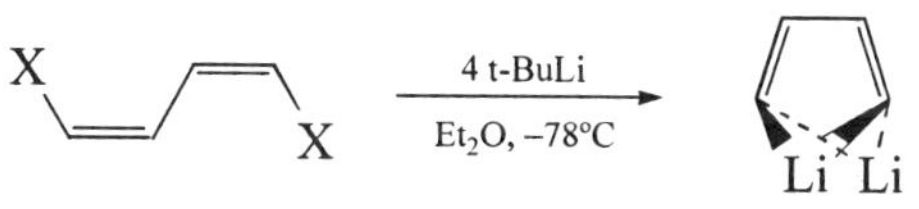

X = Br; I

Both the *Z,Z*- as well as the *E,E*-derivative of the corresponding 1,4-diphenyl compound could be prepared with *n*-butyllithium in THF at −78°:[76]

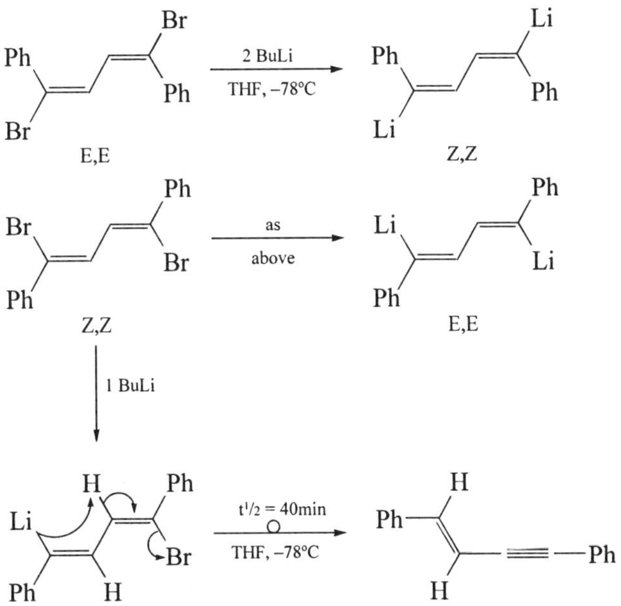

However, even at −78°C in THF (but not in diethyl ether), the monolithium intermediate of the *E,E*-compound decomposes slowly by LiBr loss.[76] Although intramolecular 1,3-proton shifts normally are not observed during organolithium rearrangements,[77–79] the stability of the monolithium intermediate formed during the preparation of the *Z,Z*-isomer does suggest (but certainly does not prove) that the subsequent reaction is intramolecular.

Another 1,4-dilithium compound, 2,2′-dilithio-3,3′-bicyclohexenyl, was obtained similarly from *n*-butyllithium and the corresponding bromine compound in diethyl ether at room temperature:[80]

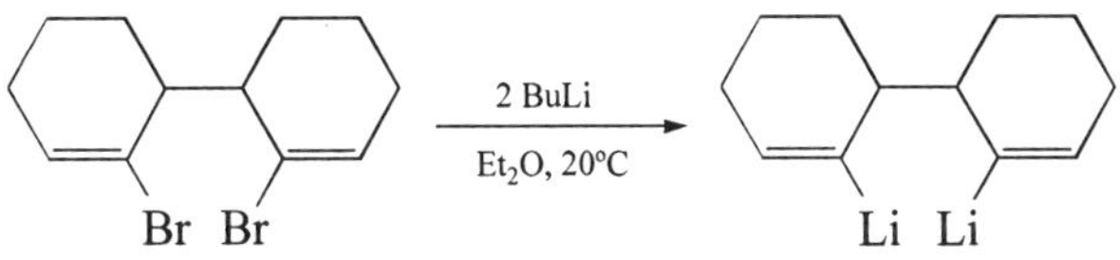

However, only one halogen atom of 1,1-dihalo-1-alkenes can be replaced by lithium even with tert-butyllithium. This was shown for aromatic as well as for aliphatic substituted geminal dihaloalkenes.[81] With 1,1-dibromo-2,2-diphenylethylene, a carbenoid complex with the alkyllithium compound is formed. However, this simulates the behavior expected of a geminal dilithioalkene, e.g., by the reaction with methyl iodide or with dimethyl sulfate:[81]

$PhC{\equiv}CPh$ (−LiBr)

$Ph_2C{=}CBr_2 \xrightarrow[-RBr]{2\ RLi} Ph_2C{=}C(Li)Br \cdot RLi \not\xrightarrow[-RBr]{} Ph_2C{=}CLi_2$

$Ph_2C{=}CLi_2 \xrightarrow{1,4\text{-}H}$ (2-lithiophenyl)(Ph)C=CH(Li)

$Ph_2C{=}CLi_2 \xrightarrow[-2\ LiI]{2\ MeI} Ph_2C{=}CMe_2$

$Ph_2C{=}C(Li)Br \cdot RLi \xrightarrow[-LiI]{MeI} Ph_2C{=}C(Me)Br \cdot RLi \xrightarrow[-RBr]{} Ph_2C{=}C(Me)Li \xrightarrow[-LiI]{MeI} Ph_2C{=}CMe_2$

The name "Quasi Dianion Complex" (QUADAC) was coined[82] for such species which react as if they were "dianions." QUADACs can be distinguished from real dilithium compounds by warming up the reaction mixture: diphenylacetylene, the product of the well known Fritsch–Buttenberg–Wiechell rearrangement of carbenoids[83] was obtained. A true dilithium intermediate should give products from a 1,4-proton shift;[18] this happened when lithium-4,4′-di-tert-butylbiphenyl (LDBB)[14] was employed.

The alleged 1,1-dilithio-2-isopropoxy-2-phenylethene,[84] in reality, is a carbenoid complexed with methyllithium:[18,85]

(Ph)(i-PrO)C=CI₂ —1. s-BuLi, THF, −70°C; 2. MeLi→ (Ph)(i-PrO)C=C(Li)I · MeLi —✗→ [(Ph)(i-PrO)C=CLi₂]

(Ph)(i-PrO)C=C(Li)I · MeLi —MeI, −LiI→ (Ph)(i-PrO)C=CMe₂

(Ph)(i-PrO)C=C(Li)I · MeLi —(−LiI)→ i-PrO—≡—Ph

[(Ph)(i-PrO)C=CLi₂] —(−LiO-i-Pr)→ Ph—≡—Li

When the geminal dilithioalkene was prepared with LDBB, β-elimination of lithium isopropoxide took place spontaneously, yielding lithium phenylacetylide even at $-70°C$.[18,85]

In the analogous aliphatic example, the main reaction was the introduction of the alkyl group of the organolithium compound used:[81]

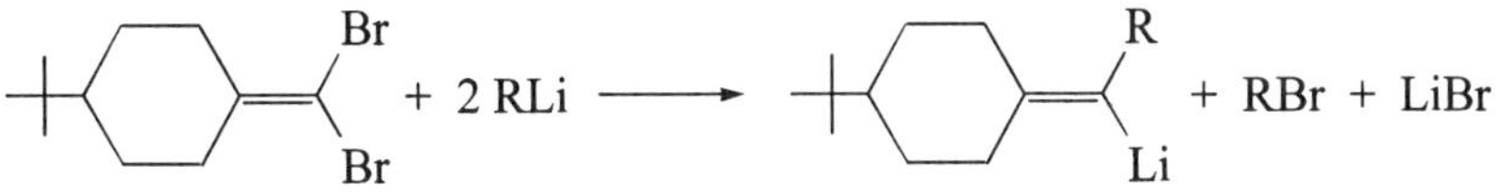

Two mechanistic alternatives, A and B, require consideration:

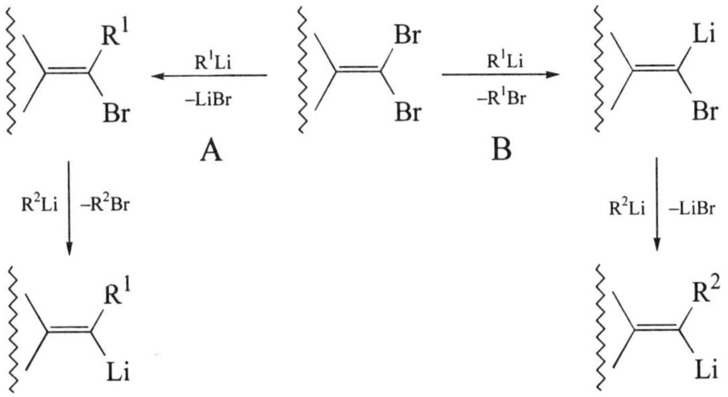

By using two different alkyllithium compounds sequentially, first R^1 = *t*-Bu and then R^2 = Et, mechanism A could be excluded since the main product had an ethyl group.[81] While mechanism A had been proposed[86] for the coupling of tert-butyllithium with 2-(dibromomethylene)adamantane, mechanism B may be a more likely alternative.

At low concentrations of RLi, the QUADAC couples with the carbenoid to yield a vicinal dilithium dimer.[81] This 2,3-dilithio-1,3-butadiene derivative also can be prepared by the addition of lithium metal to the corresponding 1,2,3-butatriene[87] (Section 2.4.1):

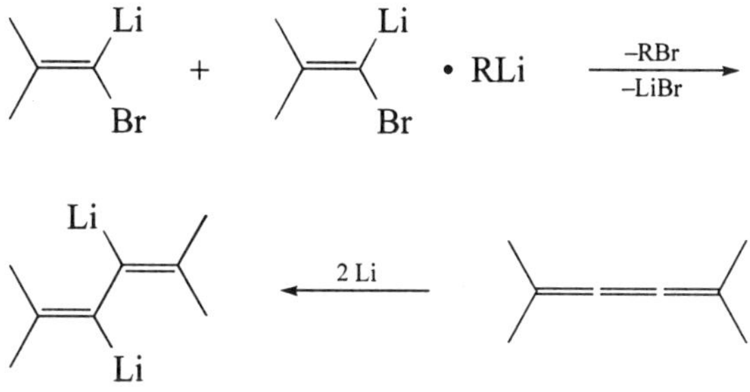

2.2 Ether Cleavage Reactions

The cleavage of ethers with lithium metal[88] are especially useful for the synthesis of allyl- and benzyllithium compounds. Up to 10 lithiums have been introduced by this method,[89] cf., e.g.,

OEt OEt OEt OEt

8 Li
THF
−4 LiOEt

Li$^+$ Li$^+$ Li$^+$ Li$^+$

Thioethers are cleaved more easily than ethers.[88] Thio functions react selectively in the presence of ether bonds. To accelerate the reaction, naphthalene may be added or a dispersion of lithium may be employed. Numerous dilithium compounds have been synthesized in this manner,[90,91] e.g.,

$$PhS(CH_2)_3O(CH_2)_3SPh \xrightarrow[-2\ LiSPh]{4\ Li} Li(CH_2)_3O(CH_2)_3Li$$

$$CH_3CH(SPh)-C_6H_4-CH(SPh)CH_3 \xrightarrow[above]{as} CH_3\bar{C}H(Li^+)-C_6H_4-\bar{C}H(Li^+)CH_3$$

$$PhCH(SPh)-Y-CH(SPh)Ph \xrightarrow[above]{as} Ph\bar{C}H(Li^+)-Y-\bar{C}H(Li^+)Ph$$

$Y = (CH_2)_{3-6,10}$; $CH_2-C_6H_4-CH_2$

Both 1,3-dilithio-2-sila- and 2-germapropane derivatives could be prepared:[92]

$$PhCH(SPh)-MMe_2-CH(SPh)Ph \xrightarrow[-2\ LiSPh]{4\ Li} Ph\bar{C}H(Li^+)-MMe_2-\bar{C}H(Li^+)Ph$$

M = Si; Ge

Lithium-4,4′-di-tert-butylbiphenyl (LDBB)[14] is especially effective[93] for the cleavage of tris(phenylthiomethyl)cyclohexane derivatives:

$$\text{MeO, MeO, OMe-cyclohexane}(CH_2SPh)_3 \xrightarrow[\text{THF, -78°C; -3 LiSPh}]{LDBB} \text{MeO, MeO, OMe-cyclohexane}(CH_2Li)_3$$

Analogous to the halogen–lithium exchanges, sulfur–lithium replacements also succeed with alkyllithium reagents.[94] In the following example, sec-butylphenyl sulfide instead of lithium thiophenolate is formed, i.e., the trithio orthoester group is attacked thiophilically by sec-butyllithium.

$$(PhS)_3C(CH_2)_4C(SPh)_3 \xrightarrow[\text{THF, }-45°C \atop -2\text{ s-BuSPh}]{2\text{ s-BuLi}} (PhS)_2\bar{C}(Li^+)(CH_2)_4\bar{C}(Li^+)(SPh)_2$$

2.3 Metal–Metal Exchange Reactions (Transmetallations)

2.3.1 with Lithium Metal. Transmetallation reactions often succeed when halogen–metal exchanges fail, e.g., for the preparation of 1,2-dilithiobenzene by mercury–lithium replacement:[95]

1,2-dibromobenzene $\xrightarrow[-LiBr]{2\text{ Li}}$ 1-bromo-2-lithiobenzene $\xrightarrow{-LiBr}$ benzyne

1-bromo-2-lithiobenzene $\not\xrightarrow[-LiBr]{2\text{ Li}}$ 1,2-dilithiobenzene

$[o\text{-}C_6H_4Hg-]_3 \xrightarrow[-Li/Hg]{Li}$ 1,2-dilithiobenzene

Transmetallation has an additional advantage over halogen–metal exchanges in giving salt-free organolithium compounds. For this purpose, halogen-free organomercury reagents often are used as starting materials, e.g., for the synthesis of 2,2′-dilithiobiphenyl from biphenylenemercury.[25,96]

Although not salt free, the first heterosubstituted 1,3-dilithiopropane and 1,4-dilithiobutane derivatives were prepared by direct mercury–lithium exchange reactions.[97,98]

$$Hg(CH_2CH(NLiPh)CH_2HgBr)_2 \xrightarrow[-2\text{ LiBr} \atop -Li/Hg]{Li \atop \text{THF, }-78°C} 2\ LiCH_2CH(NLiPh)CH_2Li$$

$$BrHgCH_2CH(NLiPh)CH(NLiPh)CH_2HgBr \xrightarrow[\text{above}]{\text{as}} LiCH_2CH(NLiPh)CH(NLiPh)CH_2Li$$

However, lithium halides do not generally interfere with subsequent reactions. The same is true for the lithium amalgam which is always present after Hg-Li transmetallations.

Hg-Li exchanges are of special value for the synthesis of isocentric polylithiated hydrocarbons, i.e., compounds bearing two (geminal) or more than two lithium atoms on the same carbon atom. These are often not available via other methods. Diethylether at room temperature turned out to be the best medium; reactions with lithium powder (2% sodium) are complete after two to three hours,[99] e.g.,

$$CH_3CH(HgCl)_2 \xrightarrow[\substack{-2\,LiCl \\ -Li/Hg}]{Li} CH_3CHLi_2 \xrightarrow[-LiH]{8\,h\ 20°C} CH_2{=}CHLi$$

$$C_6H_{13}CH(HgCl)_2 \xrightarrow[above]{as} \left[C_6H_{13}CHLi_2\right] \xrightarrow[-LiH]{} C_5H_{11}CH{=}CHLi$$

The following isocentric polylithiated compounds could be prepared similarly: CH_2Li_2,[99] $CHLi_3$,[100] CLi_4,[101] $CH_2{=}CHCH_2CHLi_2$,[4,102] $(CH_3)_2$-$C{=}CLi_2$,[103,104] and other 1,1-dilithio-1-alkenes.[103,104]

While 1,1-dilithioethane is rather stable (it takes eight hours to decompose at room temperature to give vinyllithium and lithium hydride),[99] higher 1,1-dilithioalkanes are much less stable. Thus, 1,1-dilithioheptane decomposes during its preparation and only 1-lithio-1-heptene is obtained.[4,102] The fair stability of 4,4-dilithio-1-butene is noteworthy in this context. According to MNDO calculations, this may be due to intramolecular coordination of the lithium atoms with the π system.[4,102]

These transmetallation reactions may proceed via radicals and the mechanisms obviously are rather complicated. Dimers also are implicated as intermediates. Thus, up to 7% 2,3-dilithio-1,3-butadienes have been obtained as side products during synthesis of 1,1-dilithio-1-alkenes.[103] The butadienes probably arise from the corresponding mercury intermediates. This could be shown by interrupting the reaction shown below after 10 min: the dimeric mercury compound was formed in 95% yield. However, subsequent reaction with additional lithium dust resulted in C,C bond cleavage, but gave up to 8% of 2,3-dilithio-1,3-butadiene again as a side product,[87] e.g.,

2 HgCl HgCl —Li, −4 LiCl, −Li/Hg→ 2 Li Li

HgCl ClHg → Li Li

A similar mechanism might operate during 1,1-dilithioalkane preparations. Besides CH_2Li_2, traces of dilithioacetylene (lithium carbide, C_2Li_2) are found when $CH_2(HgI)_2$ is treated with lithium dust in diethyl ether.[4,102]

$C(HgCl)_4$ gave hexalithioethane (C_2Li_6) and tetralithioethylene (C_2Li_4) (in a 4:3 ratio) predominantly, evidently due to the dimerization of radical intermediates.[4,102] Consequently, the direct mercury–lithium exchange reaction is not very suitable for the synthesis of tetralithiomethane.[101]

2.3.2 *with Alkyllithium Compounds*

2.3.2.1 Starting with Organomercury Compounds. As with halogen–lithium exchange, mercury replacement can also be achieved by reaction with alkyllithium compounds instead of lithium metal. The products are salt and amalgam free. The first polylithiumorganic compounds prepared by this variant were 1,3-dilithiopropanes.[13] According to ab initio calculations,[1,105] symmetrical doubly lithium bridged (C_{2v} symmetry) or related dimeric structures can be expected.

$$BrHgCH_2CR_2CH_2HgBr \xrightarrow[-2\ LiBr]{2\ t\text{-}BuLi}$$

$$t\text{-}BuHgCH_2CR_2CH_2Hg\text{-}t\text{-}Bu \xrightarrow[-2\ t\text{-}Bu_2Hg]{2\ t\text{-}BuLi} LiCH_2CR_2CH_2Li$$

a : R = H ; b : R = CH_3

The first step is rapid with two equivalents of tert-butyllithium in pentane at 0°C; the precipitated lithium bromide was removed by filtration. Two further equivalents of tert-butyllithium resulted in a slow reaction at room temperature (several hours for R = H, one week for R = CH_3). The 1,3-dilithiopropanes, which had precipitated as white powders, could be freed from di-tert-butylmercury by washing with pentane.

This was the method of choice for the synthesis of tetralithiomethane. However, from $C(HgCl)_4$ it was not possible to separate the lithium chloride after the first step.[101,106]

$$C(HgCl)_4 \xrightarrow[-4\ LiCl,\ -4\ t\text{-}Bu_2Hg]{8\ t\text{-}BuLi} CLi_4 \xleftarrow[-4\ EtLi,\ -4\ t\text{-}Bu_2Hg]{8\ t\text{-}BuLi} C(HgEt)_4$$

We therefore employed $C(HgEt)_4$, hoping to be able to wash out the di-tert-butylmercury as well as the ethyllithium, since both are soluble in cyclopentane. However, to our surprise, stirring $C(HgEt)_4$ in cyclopentane with tert-butyllithium for one day at room temperature gave a deep red-brown, very light-sensitive solution containing a complex of CLi_4 with di-tert-butylmercury and ethyllithium. These accompanying products had to be removed by sublimation. Interestingly, the pure, extremely pyrophoric tetralithiomethane which remained no longer dissolved in hydrocarbon solvents.[101]

Cis- and *trans*-1,2-dilithioethylene also became available for the first time via this transmetallation reaction with tert-butyllithium:[107]

$$\begin{array}{ccc} cis\text{-}(ClHg)HC{=}CH(HgCl) & & trans\text{-}(ClHg)HC{=}CH(HgCl) \\ \downarrow & \text{4 t-BuLi, }-75^\circ\text{C};\ -2\ \text{t-Bu}_2\text{Hg};\ -2\ \text{LiCl} & \downarrow \\ cis\text{-}LiHC{=}CHLi & \not\rightleftharpoons & trans\text{-}LiHC{=}CHLi \end{array}$$

As both compounds are rather unstable—lithium hydride eliminates very easily, especially from the cis compound—one has to work at $-75°C$ by introducing a pentane solution of tert-butyllithium into a suspension of the mercury compound in diethyl ether or THF. As shown by monitoring the di-tert-butylmercury gas chromatographically using an internal standard, the reaction is complete in only half an hour. The lithioacetylene formed by lithium hydride elimination is metallated immediately to give dilithioacetylene (lithium carbide). Not only the excess tert-butyllithium, but also the dilithioethylene products, function as bases. As a consequence, vinyllithium forms as well.[107]

$$LiCH{=}CHLi \xrightarrow{-LiH} HC{\equiv}CLi$$
$$LiCH{=}CHLi + HC{\equiv}CLi \longrightarrow CH_2{=}CHLi + LiC{\equiv}CLi$$

In contrast, no reaction of the dimercury percursors—*cis* or *trans*—could be observed with lithium metal, in cyclopentane, in diethyl either, or in THF (presumably because of poor solubility).

However, transmetallation with tert-butyllithium is not always the best way to achieve mercury–lithium exchange reactions. Although we had no problems with the synthesis of dilithiomethane,[99] the corresponding reaction of 1,1-*bis*-(chloromercurio)ethane with tert-butyllithium—although faster than with lithium metal—only yielded vinyllithium instead of 1,1-dilithoethane.[4] Evidently, the ate-complex intermediate eliminates lithium hydride faster than the geminal dilithium compound. The allyllithium product in the 1,3-dilithiopropane reaction[13] may well have the same origin.

$$CH_3CH(HgCl)_2 \xrightarrow[-2\,LiCl,\ -t\text{-}Bu_2Hg]{4\ t\text{-}BuLi} CH_2(H\ Li^+)\text{—}CHLi(\overline{Hg}\text{-}t\text{-}Bu_2) \xrightarrow{-t\text{-}Bu_2Hg}$$

$$CH_3CH(HgCl)_2 \xrightarrow[-2\,LiCl,\ -Li/Hg]{Li} CH_3CHLi_2 \xrightarrow{8\,h\ 20°C} CH_2{=}CHLi + LiH$$

The mercury–lithium exchange reaction with alkyllithium reagents also was used successfully for the preparation of dilithiocubanes.[108] The neighboring aminocarbonyl groups are so strongly stabilizing that methyllithium suffices as the reagent. However, tin–lithium exchange (Section 2.3.2.2) generally is preferable for such cases.

ClHg, A, A, HgCl cubane $\xrightarrow[-2\,LiCl,\ -2\,Me_2Hg]{4\,MeLi,\ THF,\ -78°C}$ Li, A, A, Li cubane

A = $CON(CHMe_2)_2$

2.3.2.2 Starting with Tin- or Other Organometallic Compounds. Transmetallation using tin–lithium exchange is an excellent method—even better than mercury–lithium exchange—for the preparation of organolithium compounds. However, it is only applicable to compounds more stable than *n*-butyllithium. This also is true for the synthesis of polylithiumorganic compounds. Examples are α-metallated ethers,[109] (*E*,*E*)-1,5-dilithio-1,4-pentadiene,[110] (Z,Z)-1,5-dilithio-1,4-pentadiene derivatives[111,112] (surprisingly the unsubstituted (Z,Z)-1,5-dilithio-1,4-pentadiene could not be prepared in this way),[110] and a compound tentatively formulated as 1,4-dilithio-2-butyne.[113]

Bu_3SnCH_2O / OCH_2SnBu_3 bicyclic diene $\xrightarrow[-2\,Bu_4Sn]{2\,BuLi,\ THF,\ -78°C}$ $LiCH_2O$ / OCH_2Li bicyclic diene

Me_3Sn–pentadiene–$SnMe_3$ $\xrightarrow[-2\,Me_3SnBu]{2\,BuLi,\ Et_2O,\ -78°C}$ Li–pentadiene–Li

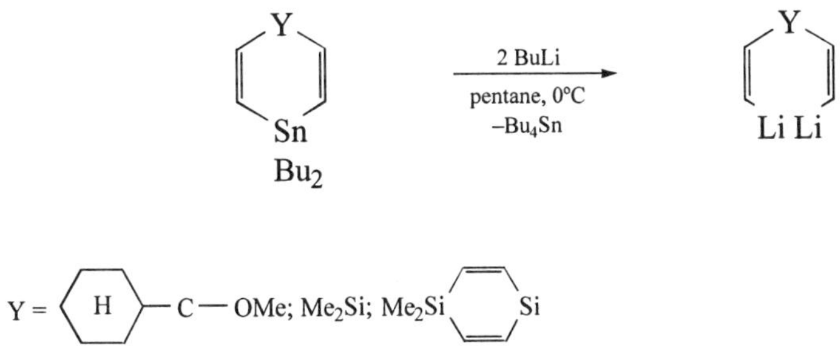

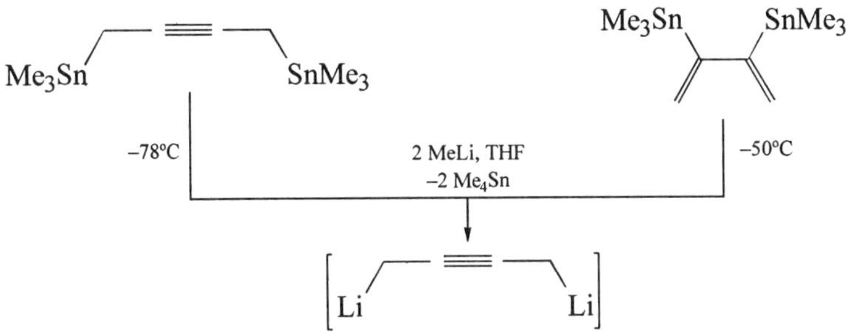

Tin–lithium exchanges also have been used recently to prepare 1,4-dilithio-1,3-butadienes,[74] 1,3- and 1,4-dilithiobenzene-chromium tricarbonyl complexes[114] as well as poly-[4-(lithiomethyl)-styrene]:[115]

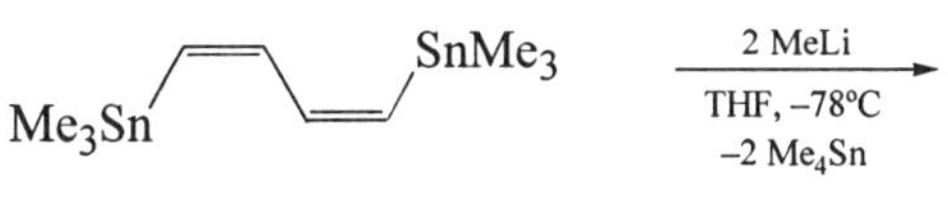

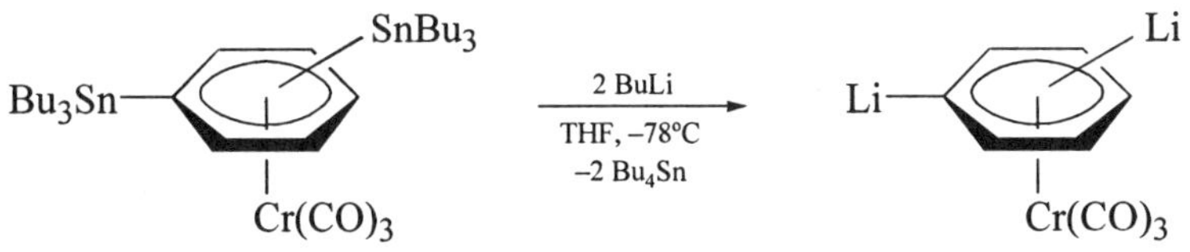

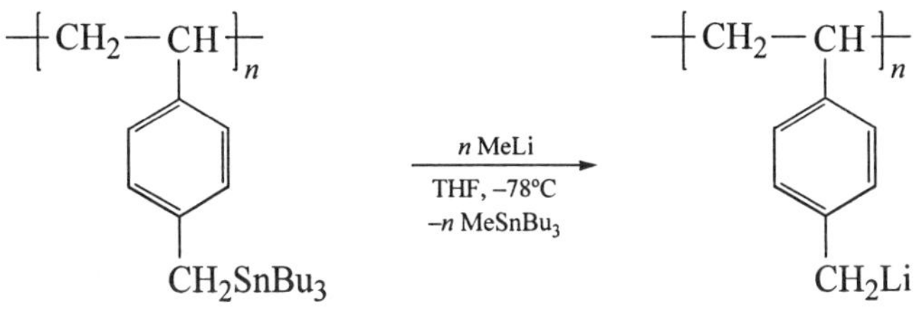

The corresponding ate-complex intermediates could be detected in the following examples. In agreement with ab initio calculations it was shown that (2*E*,4*E*)-2,5-dilithio-2,4-hexadiene is more stable than the isomeric 2,5-dilithio-1,5-hexadiene:[116]

$$\text{Me}_3\text{Sn(CH}_3\text{)C=CH–CH=C(CH}_3\text{)SnMe}_3 \xrightarrow[\text{THF, }-78^\circ\text{C}; -\text{Me}_4\text{Sn}]{\text{MeLi}} \text{[stannacyclopentadiene ate complex, CH}_3\text{, CH}_3\text{] Me}_3\text{Li}^+$$

$$\xrightarrow[-\text{Me}_4\text{Sn}; -\text{MeLi}]{2\text{ x}; > -30^\circ\text{C}} \text{[spiro Sn(CH}_3\text{) Li}^+\text{ complex]} \xrightarrow[-\text{Me}_4\text{Sn}]{3\text{ MeLi}} 2\ \text{[CH}_3\text{, CH}_3\text{, Li Li]}$$

$$\text{Me}_3\text{Sn–C(=CH}_2\text{)–CH}_2\text{–CH}_2\text{–C(=CH}_2\text{)–SnMe}_3 \xrightarrow[\text{THF, }-78^\circ\text{C}; -\text{Me}_4\text{Sn}]{\text{MeLi}} \text{[Sn ate complex] Me}_3\text{Li}^+$$

$$\xrightarrow[-\text{Me}_4\text{Sn}; -\text{MeLi}]{2\text{ x}; > -40^\circ\text{C}} \text{[spiro Sn(CH}_3\text{) Li}^+\text{ complex]} \xrightarrow[-\text{Me}_4\text{Sn}]{3\text{ MeLi}} 2\ \text{[Li Li]}$$

However, neither trans-1,2-dilithioethylene[117] nor 1,1-dilithio-1-alkenes[118] could be prepared by tin–lithium exchange; the reaction stops after the replacement of only one stannyl group:

$$\text{Bu}_3\text{Sn(H)C=C(H)SnBu}_3 \xrightarrow[-\text{Bu}_4\text{Sn}]{\text{BuLi}} \text{Bu}_3\text{Sn(H)C=C(H)Li} \xrightarrow{\text{BuLi}}\!\!\!\!\!\!\not\ \ \ \ $$

$$\text{RCH=C(SnMe}_3\text{)}_2 \xrightarrow[-\text{Me}_4\text{Sn}]{\text{MeLi}} \text{RCH=C(Li)(SnMe}_3\text{)} \xrightarrow{\text{MeLi}}\!\!\!\!\!\!\not\ \ \ \ $$

This suggests that dilithioethylenes may be less stable than *n*-butyllithium in contrast to theoretical[119] conclusions (based on the unsolvated monomers) that trans-1,2-dilithioethylene should be more stable than methyllithium and even vinyllithium.

The successful preparation of *extremely unstable* polylithiumorganic compounds by metal–metal exchange requires the use of tert-butyllithium as the reagent.[101,103,106,107] However, tin–lithium transmetallation cannot be employed with *t*-BuLi, because of excessive steric crowding in the ate-complex intermediate.[120,121] Four tert-butyl groups cannot be accommodated around the pentacoordinate tin atom. Indeed, when tetramethylstannane is treated with tert-butyllithium only two methyl groups are replaced by tert-butyl groups.[122] This problem has been solved by using dicoordinate mercury rather than tetracoordinate tin, i.e., mercury–lithium exchange does proceed with *t*-BuLi[101,103,106,107] (Section 2.3.2.1).

$$\text{t-Bu}_3\text{SnR} \xrightarrow[\quad]{\text{t-BuLi}\ (\not\ )} \left[\text{t-Bu}_4\text{SnR}\right]^- \text{Li}^+ \longrightarrow \text{RLi} + \text{t-Bu}_4\text{Sn}$$

$$\text{t-BuHgR} \xrightarrow{\text{t-BuLi}} \left[\text{t-Bu}_2\text{HgR}\right]^- \text{Li}^+ \longrightarrow \text{RLi} + \text{t-Bu}_2\text{Hg}$$

Transmetallation reactions with several other elements of the periodic table can also be used for the synthesis of lithiumorganic compounds[123,124] but polylithiated hydrocarbons have only rarely been prepared in this manner. An exception is the synthesis of dilithiodiacetylene by silicon–lithium exchange:[125]

$$\text{Me}_3\text{Si}-\text{C}\equiv\text{C}-\text{C}\equiv\text{C}-\text{SiMe}_3 \xrightarrow[\text{Et}_2\text{O}\,|\,\text{THF, 0°C};\ -2\ \text{SiMe}_4]{2\ \text{MeLi}} \text{Li}-\text{C}\equiv\text{C}-\text{C}\equiv\text{C}-\text{Li}$$

Tellurium–lithium exchange is extremely valuable in special cases and enabled the synthesis of an interesting dilithiostyrene which could not be prepared by tin–lithium exchange:[126]

The choice of solvent is critical; α-metallation took place exclusively in THF.[126]

Like the corresponding organotin examples, the reaction of gem-organodiboron compounds with butyllithium stops after replacement of only one boron group. The product is an ate-complex involving the remaining boron atom.[127–129]

$$RCH_2CH(BR'_2)_2 \xrightarrow[-BuBR'_2]{2\ BuLi} RCH_2CH(Li)(\overline{B}R'_2(Bu))\ Li^+$$

The replacement of potassium by lithium in a dianion is the final example:[130]

$$KC{\equiv}C{-}C(CH_2)_2^{-}\ K^+ \xrightarrow[-2\ KBr]{2\ LiBr} LiC{\equiv}C{-}C(CH_2)_2^{-}\ Li^+$$

2.4 Addition Reactions

2.4.1 Reductive Metallations with Lithium Metal. It has been known since 1928 that phenyl substituted C,C-double bonds react readily with lithium metal to yield stable, π-delocalized benzylic carbanions (Schlenk-addition). Examples are the addition of lithium to stilbene[131–133] and to 1,4-diphenyl-1,3-butadiene:[132,134]

$$PhCH{=}CHPh \xrightarrow[THF]{2\ Li} Ph\overline{C}H{-}\overline{C}HPh\ \ 2\ Li^+$$

$$PhCH{=}CH{-}CH{=}CHPh \xrightarrow[above]{as} Ph\overline{C}H{-}CH{=}CH{-}\overline{C}HPh\ \ Li^+\ \ Li^+$$

Depending on the solvent contact or solvent-separated ion pairs may be formed. The dark red solutions (THF-d_8) from *cis* as well as from *trans* stilbene give identical ^1H and ^{13}C NMR spectra. Hence, only one species exists even at −70°C.[135] 1,4-Dilithio-1,4-diphenyl-2-butene in THF-d_8 prefers the *cis* configuration (NMR data).[136] X-ray analysis of crystals of the same compound (prepared by another route with two moles of TMEDA) shows a somewhat disordered doubly lithium bridged *cis* structure (C_2 rather than C_{2v} symmetry) in agreement with MNDO calculations.[137] In contrast, NMR data show the more bulky 1,1,4,4-tetra-phenylbutadiene dianion to favor the *trans* geometry in solution even with lithium counterions.[136]

The bis(tricarbonylchromium)-1,4-diphenylbutadiene[138] and higher 1,*n*-diphenyl conjugated polyenes[139] also have been reduced by lithium metal. The stereodynamic behavior as a function of chain length in the latter dianions has been studied.[139]

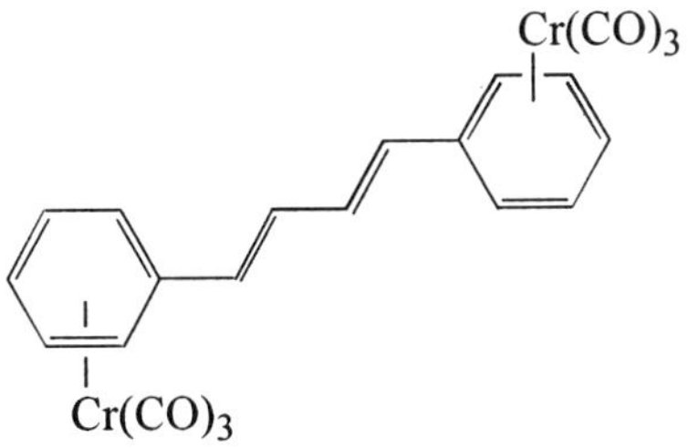

$Cr(CO)_3$

$Cr(CO)_3$

$Ph(CH{=}CH)_nPh$

$n = 2; 3; 4$

The reacting double bond may also be part of a fulvene system,[140] e.g.,

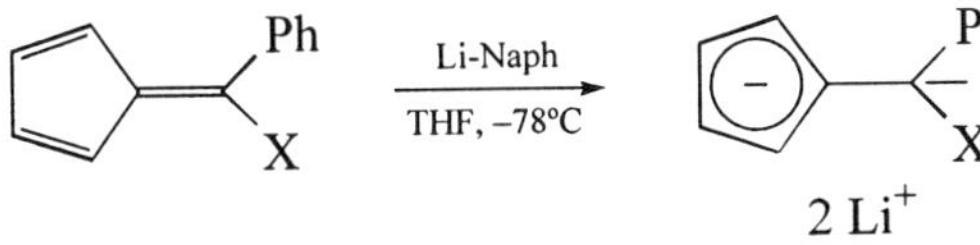

Ph X Li-Naph THF, −78°C Ph X 2 Li^+

X = Ph; NMe_2

Several hydrocarbons with more than one stilbene unit have been reduced to tetraanion salts:[141]

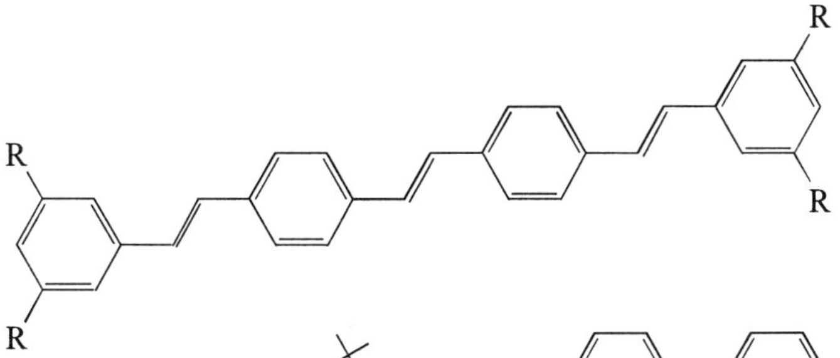

R R R R

R = H; t-Bu

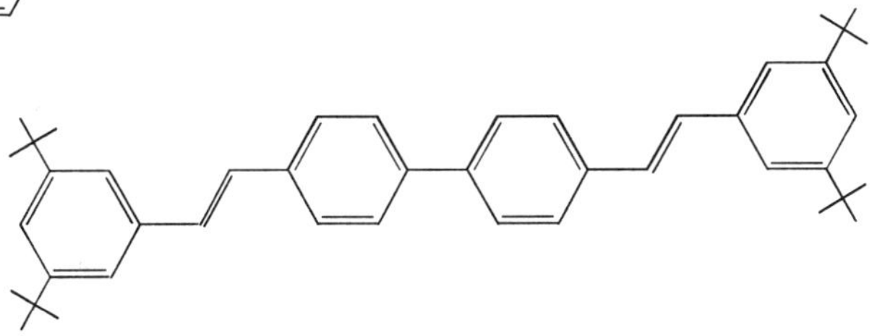

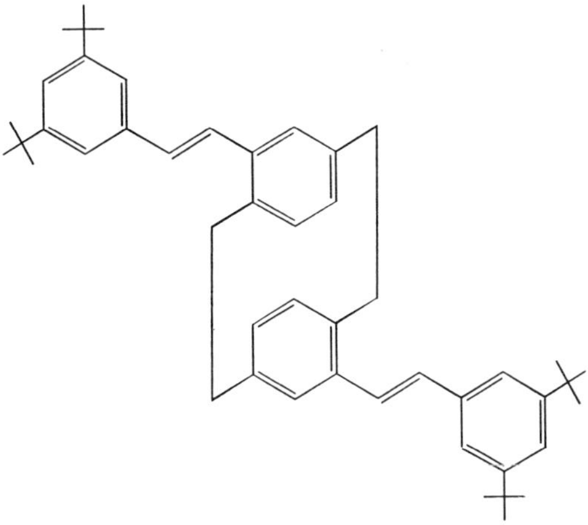

Although multiply charged anions from molecules with extended π-systems[142] will not be discussed in this chapter, we note that even octaanion salts have been obtained by the reduction of tetra(stilbenyl)ethylene and related hydrocarbons.[143]

Four lithiums can be introduced into *cis,cis*-1,2,3,4-tetraphenyl-1,3-butadiene[144] and the related 1,1-dimethyl-2,3,4,5-tetraphenyl-1-silacyclopentadiene[132] to form the tetraanion salts:

$$PhCH{=}CPh{-}CPh{=}CHPh \xrightarrow[THF]{4\,Li} Ph\bar{C}H{-}\bar{C}Ph{-}\bar{C}Ph{-}\bar{C}HPh \quad 4\,Li^+$$

$$\text{2,3,4,5-}Ph_4C_4SiMe_2 \xrightarrow[above]{as} [Ph_4C_4SiMe_2]^{4-}\ 4\,Li^+$$

Unsymmetrically phenyl-substituted ethylenes, like styrene[145] or 1,1-diphenylethylene,[146,147] dimerize via the intermediate radical anions. This reaction is known as "dimerizing addition." According to ab initio calculations the unsubstituted 1,4-dilithiobutane monomer favors a doubly bridged structure (C_2 symmetry).[105]

$$2\,PhCH{=}CH_2 \xrightarrow{2\,Li} Ph\bar{C}HCH_2CH_2\bar{C}HPh \quad Li^+ \quad Li^+$$

$$2\,Ph_2C{=}CH_2 \xrightarrow{2\,Li} Ph_2\bar{C}CH_2CH_2\bar{C}Ph_2 \quad Li^+ \quad Li^+$$

CH2 CH2 CH2 CH2 Li Li

The 1,4-dilithiobutanes formed by the reaction of β-alkylstyrenes with lithium have been thoroughly investigated by cryoscopic measurements and by ^1H, ^{13}C, and ^6Li NMR spectroscopy.[148] The dilithio compounds are monomers in THF. The butane chain does adopt a cisoid conformation, but the two benzyl subunits chelate only one lithium cation while the other Li^+ is solvent-separated.[148]

1,4-Dilithio-1,1,4,4-tetraphenyl butane, the dimer of 1,1-diphenylethylene, was shown by X-ray analysis to form a monomeric contact triple ion with an anti-periplanar butane chain and nearly trigonal planar carbanionic centers.[149]

Depending on the conditions, the addition of lithium to diphenylacetylene (tolane) gives either the unstable *cis*-1,2-dilithio-1,2-diphenylethylene (*cis*-dilithiostilbene)[150] or the dimer 1,4-dilithio-1,2,3,4-tetraphenyl-1,3-butadiene.[151-157] According to calculations[75,158] on the phenyl-free parent species, both the compounds will prefer a doubly bridged conformation.[159]

$$\mathrm{PhC{\equiv}CPh} \xrightarrow[\mathrm{THF,\ -80^{\circ}C}]{\mathrm{2\ Li}} \mathrm{(Ph)(Li)C{=}C(Li)(Ph)}\ \text{(doubly Li-bridged)}$$

$$\mathrm{2\ PhC{\equiv}CPh} \xrightarrow[\mathrm{Et_2O}]{\mathrm{2\ Li}} \mathrm{Ph(Li)C{=}C(Ph){-}C(Ph){=}C(Li)Ph}\ \text{(doubly Li-bridged)}$$

The claimed[160] addition of two further lithium atoms to dilithiostilbene to form tetralithiobibenzyl (sonication with excess lithium in the presence of 4,4′-di-tert-butylbiphenyl in dry THF) was disproven.[161]

Phenylcyclopropylacetylene gives a monomeric vicinal *cis* dilithioalkene derivative even in diethyl ether.[162] However, above −30°C a series of reactions occur. The cyclopropylmethyl moiety first undergoes ring opening. The resulting open-chain dilithium compound then loses ethylene to give a "sesquiacetylene"[163,164] which undergoes deprotonation to yield a known trilithiumorganic compound.[165] The fragmentation is reminiscent of the loss of ethylene from 3,3-diphenylpropyllithium to give diphenylmethyllithium.[166]

$$\mathrm{Ph{-}C{\equiv}C{-}CH(CH_2)_2} \xrightarrow[\mathrm{Et_2O,\ -30^{\circ}C}]{\mathrm{2\ Li}} \mathrm{Ph(Li)C{=}C(Li){-}CH(CH_2)_2} \xrightarrow[> -30^{\circ}\mathrm{C}]{} $$

$$\mathrm{Ph(Li)C{=}C{=}CH{-}CH_2{-}CH_2{-}Li} \xrightarrow[-\mathrm{CH_2{=}CH_2}]{} \mathrm{[Ph{-}C{\equiv}C{\equiv}CH]^{2-}\ 2\,Li^+}$$

$$\longrightarrow \mathrm{[Ph{-}C{\equiv}C{\equiv}CH{-}CH_2{-}CH_3]^{-}\ Li^+} + \mathrm{[Ph{-}C{\equiv}C{\equiv}C{-}Li]^{2-}\ 2\,Li^+}$$

A tetralithium four-membered ring compound was obtained upon treatment of 1,6-diphenyl-1,5-hexadiyne with lithium in THF:[167]

$$\mathrm{Ph{-}C{\equiv}C{-}CH_2{-}CH_2{-}C{\equiv}C{-}Ph} \xrightarrow[\text{THF}]{4\,\text{Li}}$$

$$\mathrm{PhCLi_2{-}C{=}C{-}CLi_2Ph}\ \ (\text{ring closed by } \mathrm{CH_2{-}CH_2})$$

Another interesting lithium-induced cyclization of a tribenzo cyclotriyne leads to the formation of a fulvalene (helicene) dianion derivative in high yield, as confirmed by X-ray crystallography.[168] Lithium–hydrogen exchange at the 1 and 4 positions involves the solvent, THF:

Lithium metal can also be added to double bonds activated by silyl groups, e.g., tetrakis(trimethylsilyl)ethylene,[169,170] 1,1-*bis*(trimethylsilyl)ethylene,[171] and triphenylvinylsilane.[172] The last two are further examples of dimerizing additions:

$$\mathrm{(Me_3Si)_2C{=}C(SiMe_3)_2} \xrightarrow[\text{THF, 20°C}]{2\,\text{Li}} \mathrm{(Me_3Si)_2\bar{C}{-}\bar{C}(SiMe_3)_2}\ \ 2\,\mathrm{Li^+}$$

$$2\,\mathrm{(Me_3Si)_2C{=}CH_2} \xrightarrow[\text{above}]{\text{as}} \mathrm{(Me_3Si)_2\bar{C}CH_2CH_2\bar{C}(SiMe_3)_2}\ \ \mathrm{Li^+}\ \ \mathrm{Li^+}$$

$$2\,\mathrm{Ph_3SiCH{=}CH_2} \xrightarrow[\text{THF, −75°C}]{2\,\text{Li}} \mathrm{Ph_3Si\bar{C}HCH_2CH_2\bar{C}HSiPh_3}\ \ \mathrm{Li^+}\ \ \mathrm{Li^+}$$

The dark green radical anion intermediate could be detected on the way to 1,2-dilithio[tetrakis(trimethylsilyl)]ethane, the first stable nonconjugated 1,2-dilithioethane. After removal of the solvent and addition of hexane, the dark red THF solution of the dilithiumorganic compound gives yellow crystals of an adduct with two moles of the THF. X-ray analysis showed that the structure of the starting material is essentially retained after the lithiation. Each tricoordinated lithium atom is bound to THF and to both carbanionic centers; a *trans* doubly bridged structure results.[170] 1,2-Dilithioethane, according to ab initio calculations,[173] prefers a *trans* conformation (C_{2h}) with a partially bridged geometry, although the symmetrically *trans* doubly bridged (D_{2h}) structure is only 1.9 kcal/mol higher in energy.

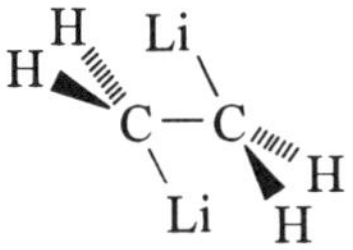
H
H
Li
C
C
H
Li
H

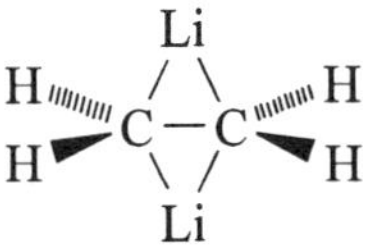
Li
H
H
C
C
H
H
Li

Trimethylsilyl groups also were used to activate benzene for reduction with lithium metal:[174,175]

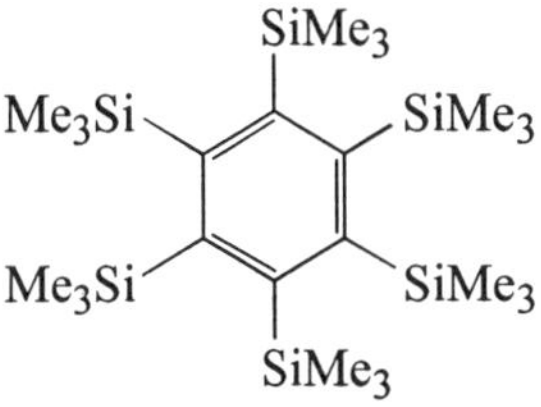
SiMe3
Me3Si
SiMe3
Me3Si
SiMe3
SiMe3

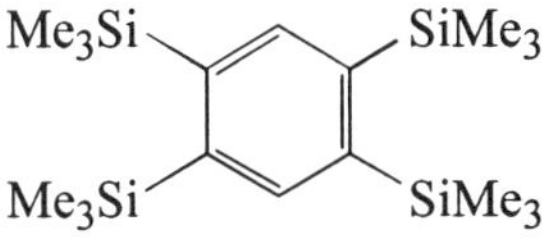
Me3Si
SiMe3
Me3Si
SiMe3

The X-ray structure of the *hexasubstituted* benzene dianion[174] reveals a boat form with both lithium(THF) cations on the same side of the benzene ring. However, the *tetrasubstituted* dianion derivative[175]—prepared in dimethoxyethane (DME) instead of THF—is nearly planar, i.e., truly antiaromatic, with the two lithium(DME) cations located quite symmetrically on opposite faces of the benzene ring. The 8 π-electron antiaromaticity was responsible for a uniquely large downfield ^{7}Li NMR shift. This shows that the molecular structure in the crystal is retained in solution. According to ab initio calculations,[176] the experimentally unknown unsubstituted dilithiobenzenide has a folded structure similar to the corresponding hexakis(trimethylsilyl)benzenide with synfacial lithium cations.

Another dimerizing addition of lithium was observed starting with a silyl substituted methyleneborane in toluene.[177]

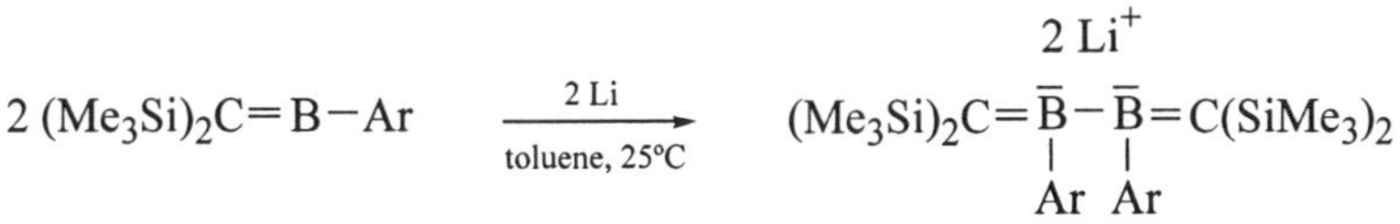
2 (Me3Si)2C=B−Ar
2 Li
toluene, 25°C
2 Li+
(Me3Si)2C=B−B=C(SiMe3)2
Ar Ar

Ar = Mesityl

An interesting organodilithium compound with one σ-bonded lithium is obtained by the addition of lithium metal to tetraphenylallene.[178] The structure deduced from NMR spectroscopy was controversial at first.[179,180] However, the equivalence of both the lithium atoms can be explained by fast equilibration on the NMR time scale.[181] This also is reasonable according to ab initio calculations on the parent compound, 2,3-dilithiopropene.[182]

$$Ph_2C{=}C{=}CPh_2 \xrightarrow[\text{THF, }-78^\circ C]{2\,Li} Ph(Ph)C{\cdots}C(Li){\cdots}C(Ph)Ph\ \ Li^+$$

The hydrolysis product, 1,1,3,3-tetraphenylpropene, also adds lithium, but the primarily formed dianion undergoes a phenyl rearrangement from the 3 to the 2 position:[179,181]

$$Ph_2C{=}CH{-}CHPh_2 \xrightarrow[\text{THF, }-20^\circ C]{2\,Li} Ph_2\bar{C}{-}CH(Ph){-}\bar{C}HPh_2\ \ Li^+\ \ Li^+$$

No phenyl migration takes place in the corresponding allyl anion.[181] As with 1,2,3-triphenylallyllithium, an allyl trianion is formed:[183]

$$PhCH{\cdots}CPh{\cdots}CHPh^{-}\ Li^+ \xrightarrow[\text{THF}]{2\,Li} Ph\bar{C}H{-}\bar{C}Ph{-}\bar{C}HPh\ \ 3\,Li^+$$

Even nonactivated aliphatic alkenes and alkynes can be forced to react with metallic lithium—catalyzed or even uncatalyzed. Vicinal dilithiated hydrocarbons are formed, although sometimes only as intermediates. While 1,2-dilithioethane can be expected on the basis of ab initio calculations[173] to be reasonably stable, the alleged[184] "1,2-dilithioethane," prepared by the addition of lithium metal to ethylene in dioxane solvent, was shown to be dilithioacetylene (lithium carbide). The following sequence of reactions explains this result:[11]

$$CH_2{=}CH_2 \xrightarrow[-4\,LiH]{6\,Li} LiC{\equiv}CLi$$

$$CH_2{=}CH_2 \xrightarrow{2\,Li} LiCH_2{-}CH_2Li \xrightarrow{-LiH} CH_2{=}CHLi \xrightarrow{-CH_2{=}CH_2} LiC{\equiv}CLi$$

$$CH_2{=}CHLi \xrightarrow{2\,Li} LiCH_2{-}CHLi_2 \xrightarrow{-LiH} \begin{bmatrix} LiCH{=}CHLi \\ CH_2{=}CLi_2 \end{bmatrix} \xrightarrow{-LiH} HC{\equiv}CLi \longrightarrow LiC{\equiv}CLi$$

The solvent plays a decisive role: 1,4-dioxane can be replaced by 1,2-dimethoxyethane (DME), but no reaction takes place in diethyl ether or in THF.[11] According to a theoretical study,[185] the solvent effect on these electron transfer reactions of lithium atoms should be small. The easy elimination of lithium hydride from 1,2-dilithioethane has also been confirmed computationally.[105] Vinyllithium forms a very stable complex (mixed dimer) with lithium hydride. Only reaction of highly active lithium at −120°C allowed 1,2-dilithioethane to be trapped in 8% yield together with the same amount of 1,4-dilithiobutane.[186] Several mechanisms are plausible for the formation of the latter:[186]

$$CH_2{=}CH_2 \xrightarrow{Li} LiCH_2\dot{C}H_2 \xrightarrow{Li} LiCH_2CH_2Li$$

$$CH_2{=}CH_2 + LiCH_2\dot{C}H_2 \longrightarrow LiCH_2CH_2CH_2\dot{C}H_2 \xrightarrow{Li} LiCH_2CH_2CH_2CH_2Li$$

$$2\times\ LiCH_2\dot{C}H_2 \longrightarrow LiCH_2CH_2CH_2CH_2Li; \quad LiCH_2CH_2Li \xrightarrow{CH_2=CH_2} LiCH_2CH_2CH_2CH_2Li$$

Vinyllithium, the second intermediate leading from ethylene to dilithioacetylene, reacts with lithium as well.[11] The trilithioethane and dilithioethylene intermediates, however, are only postulated and could not be trapped. On the other hand, the reaction can be stopped at the vinyllithium step by using biphenyl and naphthalene as the catalyst in dimethoxymethane solvent.[8] However, considerable amounts of 1,4-dilithiobutane, 1,6-dilithiohexane as well as 4-lithio-1-butene and butyllithium are obtained as by-products.[8] The vinyllithium yields are much better (70–75%) by using lithium sand in THF at 0°C along with the trithiapentalenes Ia and Ib in combination with water-free zinc or iron(III) chloride catalysts.[9,10]

$$CH_3CH{=}CH_2 \xrightarrow[\text{THF, 0°C}]{\text{2 Li, catalyst}} CH_3CHLiCH_2Li \xrightarrow{-LiH}$$

$$(E)\text{-}H_3C(H)C{=}C(H)Li + (Z)\text{-}H_3C(H)C{=}C(Li)H + H_3C(Li)C{=}CH_2 + [CH_2{\cdots}CH{\cdots}CH_2]^- Li^+$$

Structure I: 1,6,6a-trithiapentalene (S—S—S) bearing Ph, H, R^1, R^2

I a: R^1 = H; R^2 = Ph

b: R^1 = Ph; R^2 = H

The four isomeric lithiopropenes are formed from propene in a total yield of 75–85%; the mixture contains 85–90% of (*E*)-1-lithiopropene as the main reaction product. Interestingly, the reaction gives allyllithium as the main product by simply adding $PtCl_2$ or $PdCl_2$ to the catalyst I.[9] However, not even traces of the postulated 1,2-dilithioalkanes could be detected.

Like propene, higher 1-alkenes can been lithiated catalytically—even more selectively—to yield (*E*)-1-lithio-1-alkenes. (However, isobutene and (*Z*)-2-butene are exceptions.) While higher 1,*n*-dienes like 1,5-hexadiene and 1,7-octadiene react similarly at both double bonds, an interesting trilithio compound was obtained from 1,4-pentadiene:[9]

$$CH_2{=}CHCH_2CH{=}CH_2 \xrightarrow[-3\,LiH]{6\,Li} \overbrace{CH_2{\cdots}CLi{\cdots}CH{\cdots}CH{\cdots}CHLi}^{-}\ Li^+$$

Some 1-alkenes react with lithium dispersions under relatively mild conditions without neither a catalyst nor a solvent to give the 1-lithio-1-alkynes and lithium hydride as the major products.[187] Thus, 1-hexene was converted to 1-lithio-1-hexyne in 65% yield in one hour at the reflux temperature of the olefin (64°C). It is reasonable to assume that the first-formed 1,2-dilithiohexane loses lithium hydride to give 1-lithio-1-hexene. This was detected as a side-product in 4% yield:[187]

$$BuCH{=}CH_2 \xrightarrow{2\,Li} BuCHLiCH_2Li \xrightarrow[-LiH]{}$$

$$BuCH{=}CHLi \xrightarrow[-2\,LiH]{2\,Li} BuC{\equiv}CLi$$

A 1 : 1-mixture of the corresponding vinyllithium and lithioalkyne were obtained from trimethylvinylsilane and lithium in hexane at 0°C—again without a catalyst:[188]

$$2\,Me_3SiCH{=}CH_2 \xrightarrow[\text{hexane, 0°C};\ -4\,LiH]{6\,Li} (Me_3Si)(H)C{=}C(H)(Li) + Me_3SiC{\equiv}CLi$$

In THF solution, however, dimerizing addition occurs as with triphenylvinly-silane[172] (see above).

While isolated triple bonds normally react rather slowly with lithium dust, strained rings are exceptions. Lithium adds to the triple bond of cyclooctyne even at −35°C in diethylether to give a 70% yield of *cis*-1,2-dilithiocyclooctene. The yellow solution also contains up to 20% of the dimerizing addition product, a 1,4-dilithio-1,3-butadiene derivative:[189–191]

Li, Et_2O, −35°C; Li; 2x; Li

One of the three mechanisms which might lead to dimerization could be excluded here—1,2-dilithiocyclooctene does not add to cyclooctyne.[191] As expected, the reactivity decreases on going to the less strained, higher cycloalkynes. Moreover, increasing amounts of the *trans* compounds are obtained besides the *cis*-1,2-dilithiocycloalkenes, e.g., 5% *trans*-1,2-dilithiocyclodecene and 73% *trans*-1,2-dilithiocyclododecene.[191] The reactivity of the 12-membered ring is already nearly as low as that of open-chain alkynes: 36 hours at room temperature are necessary.

The reaction of open-chain alkynes with lithium dust at room temperature in diethyl ether generally takes 48 hours. As with cyclododecyne, only very small amounts of dimeric side products could be detected. These additions of lithium are truly *trans*. The open-chain vicinal *trans*-dilithioalkenes—in contrast to *cis*-1,2-dilithiocyclooctene—are not soluble in diethyl ether.[189–191]

$$R^1{-}C{\equiv}C{-}R^2 \xrightarrow[Et_2O,\ 20°C]{2\ Li} (R^1)(Li)C{=}C(Li)(R^2)$$

a: $R^1 = C_2H_5$; $R^2 = C_4H_9$

b: $R^1 = R^2 = C_3H_7$

c: $R^1 = R^2 = C_4H_9$

It is not known if *cis*-addition occurs first followed by *cis-trans* isomerization. As this isomerization could be excluded with the parent compound 1,2-dilithioethylene,[107] inversion may only be possible with the radical anion, i.e., before introduction of the second lithium atom.[191]

As expected, two conjugated triple bonds react readily with lithium metal. However, the product depends on the substituents and the solvent used. A *cis*-1,4-dilithiobutatriene derivative was obtained as a crystalline *bis*-THF adduct by conjugate addition of lithium to di-tert-butyl diacetylene in THF:[192]

$$\text{t-Bu}-C\equiv C-C\equiv C-\text{t-Bu} \xrightarrow[\text{THF, }-10^{\circ}\text{C}]{2\,\text{Li}} (\text{t-Bu})(\text{Li})C=C=C=C(\text{t-Bu})(\text{Li})$$

X-ray analysis revealed the product to have a dimeric structure with two different kinds of lithium atoms. This agrees with model MNDO calculations.[192] The tert-butyl groups are very important for the success of the reaction. Replacement by trimethylsilyl[193] or phenyl groups[194] leads mainly to polymerization or to cleavage with the formation of two molecules of the corresponding lithioacetylene. The latter is also the main reaction starting with di-tert-butyl-diacetylene when diethyl ether is used instead of THF as the solvent.[192]

On the other hand, phenyl substituted butatrienes react with alkali metals without any problems to yield butyne derivatives again by 1,4-addition.[195,196] Even the corresponding dilithium compound (with 2 TMEDA), prepared by a different route with the one and four positions bearing only one phenyl group, has a true butyne structure.[197]

Interestingly enough, alkyl substituents prevent 1,4-addition; the lithium atoms are steered to the central double bond to give the 2,3-dilithio-1,3-butadiene derivatives I.[87,198]

$$Ph_2C=C=C=CPh_2 \xrightarrow{2\,M} Ph_2\bar{C}-C\equiv C-\bar{C}Ph_2 \quad M^+ \quad M^+$$

$$(R^1)(R^2)C=C=C=C(R^3)(R^4) \underset{I_2}{\overset{2\,\text{Li}}{\rightleftharpoons}} (R^1)(R^2)C=C(\text{Li})-C(\text{Li})=C(R^3)(R^4)$$

I

a: $R^1 = R^2 = R^3 = R^4 = CH_3$

b: $R^1 = R^3 = CH_3$; $R^2 = R^4 =$ t-Bu

c: $R^1 = R^2 = CH_3$; $R^3R^4 = (CH_2)_5$

d: $R^1R^2 = R^3R^4 = (CH_2)_5$

The tetramethyl compound Ia—3,4-dilithio-2,5-dimethyl-2,4-hexadiene—could be obtained in crystalline form; X-ray analysis showed it to be a tetramer with two different kinds of lithium atoms and two crystallographically-different diene residues in a nearly orthogonal gauche conformation. The C,C bond lengths (C1—C2 = C3—C4 = 1.40 Å and C2—C3 = 1.50 Å on average) show Ia to be a true butadiene derivative.[4,199]

As revealed by ^{6}Li NMR spectroscopy (especially a ^{6}Li,^{6}Li-COSY experiment), the tetrameric structure is retained even in a THF solution.[200,201] Obviously, in diethyl ether solution, other aggregates coexist as well.

In special cases, even a single bond can be reductively cleaved by lithium metal to yield dilithiumorganic compounds. The first example was the reaction of 1,1-diphenyl-2-cyclopropylethylene with lithium in diethyl ether or THF:[202]

$$Ph_2C{=}CH{-}CH\langle CH_2{-}CH_2 \rangle \xrightarrow{2\,Li} Ph_2C{\overset{-}{=}}CH{=}CHCH_2CH_2Li \quad Li^+$$

Other strained molecules can be cleaved, expecially if both carbanionic centers formed are resonance stabilized by π delocalization of the two negative charges. Thus, on treating semibullvalene with lithium in THF or dimethyl ether at $-78\,°C$ two diastereoisomeric dimers of "dilithium semibullvalenide" have been obtained,[203,204] and the same is true for barbaralane and 2,6-diphenylbarbaralane yielding the corresponding *bis*-allyl-anions:[205]

2 Li, THF, –78°C → Li^+ Li^+

as above → Li^+ Li^+

R = H; Ph

On the other hand, the related 3,4-homotropilidene did not react in the same manner; a monoanion formed instead:[206]

2 Li, –LiH → Li^+

An interesting dilithiumorganic compound with two different benzylic centers has been obtained by the reductive cleavage of the following polycyclic compound, a photoadduct of diphenylacetylene (tolane) with naphthalene.[207] Using only ^1H NMR spectroscopy, the cleavage of the σ bond was not recognized at first.[208]

2 Li, THF, –78°C

During the reduction of cyclic conjugated hydrocarbons with lithium metal (a subject not discussed in full here) [see reference 142], σ bonds are cleaved occasionally even when the carbanionic centers formed are not resonance stabilized. Examples are the reductive cleavage of benzocyclobutene[209] and of biphenylene:[210,211]

2 Li
THF
CH_2CH_2Li
Li

as
above
Li Li

The following is an aliphatic analogue:[212]

2 Li
Et_2O, 30°C
Li Li

Remarkably—as shown by the last example[212] and by benzocyclobutene[209]—the weakest σ bond is not always the one that cleaves. Notably, the reaction of methylenecyclopropane with lithium metal does not yield the extremely stable Y-delocalized trimethylenemethane dianion,[213–215] but rather 2,4-dilithio-1-butene by scission of the stronger cyclopropane σ bond.[216]

$CH_2{=}C(CH_2)_2$ $\xrightarrow[Et_2O]{2\,Li}$ $\not\rightarrow$ $[CH_2{\cdots}C(CH_2)_2]^{2-}$ 2 Li^+

$\rightarrow$ $CH_2{=}C(Li)CH_2CH_2Li$

Therefore, we argued that the electron transfer does not take place to the σ^* orbital but rather to the π^* orbital of the neighboring double bond. This is followed by a ring–chain rearrangement.[162,202,216] The initially formed vicinal dilithiumorganic intermediate could even be trapped in one case.[162] However, as this was not possible with methylenecyclopropane, ring opening of the radical anion intermediate had to be considered as an alternative mechanism. This could be excluded, however, by introducing two geminal methyl groups into the cyclopropane ring of methylenecyclopropane.[216] In addition, ab initio calculations[217] showed the cyclic radical anion to be 6.6 kcal/mol more stable than the corresponding open chain alternative.

$$\dot{C}H_2-C(Li)<(CMe_2)(CH_2) \xrightarrow{Li} Li-CH_2-C(Li)<(CMe_2)(CH_2)$$

$$\dot{C}H_2-C(Li)<(CMe_2)(CH_2) \not\rightarrow CH_2{=}C(Li)CH_2\dot{C}Me_2 \qquad Li-CH_2-C(Li)<(CMe_2)(CH_2) \rightarrow CH_2{=}C(Li)CMe_2CH_2Li$$

$$\dot{C}H_2-C(Li)<(CMe_2)(CH_2) \xrightarrow{Li} CH_2{=}C(Li)CMe_2CH_2Li$$

Direct ring–chain rearrangement during attack of the second lithium at the cyclopropane ring of the radical anion also is consistent with the experimental results.[217]

Compared with the unsubstituted 1,3-dilithiopropane, which loses lithium hydride with a half life of one hour at room temperature,[13] 2,4-dilithio-1-butene is remarkably stable: not even a trace of the expected monolithio tautomers could be detected.[216] Interestingly, the tetramethyl substituted derivative does eliminate lithium hydride, although in the other orientation.[216]

$$CH_2{=}C(Li)CH_2CH_2Li \underset{-LiH}{\not\rightarrow} CH_2{=}C(Li)-CH{=}CH_2 \quad \text{or} \quad CH_2{=}C{=}CH-CH_2Li$$

$$CH_2{=}C(Li)CMe_2CMe_2Li \xrightarrow[-LiH]{} CH_2{=}C(Li)CMe_2CMe{=}CH_2$$

While two lithium atoms usually are necessary for the reductive cleavage of a cyclopropane σ bond, **two** σ bonds of the "butterfly olefin" 1,1′-bicyclopropylidene are cleaved by two lithiums. In diethyl ether solution—but not in THF—the primarily formed (1,3-dilithiopropylidene)cyclopropane can be isolated before it rearranges to 1,6-dilithio-3-hexyne (less than two and one-half hours at room temperature):[218]

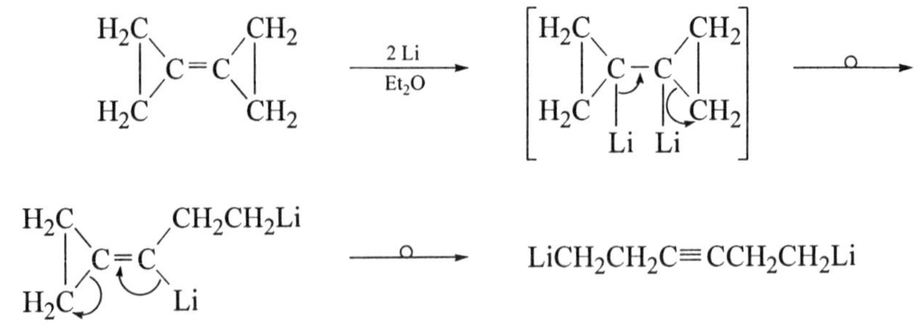

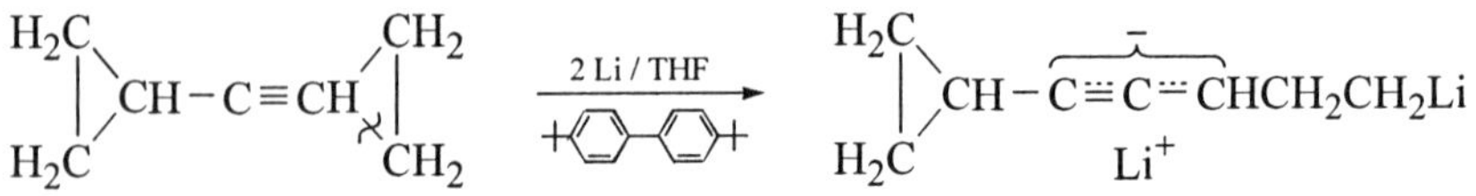

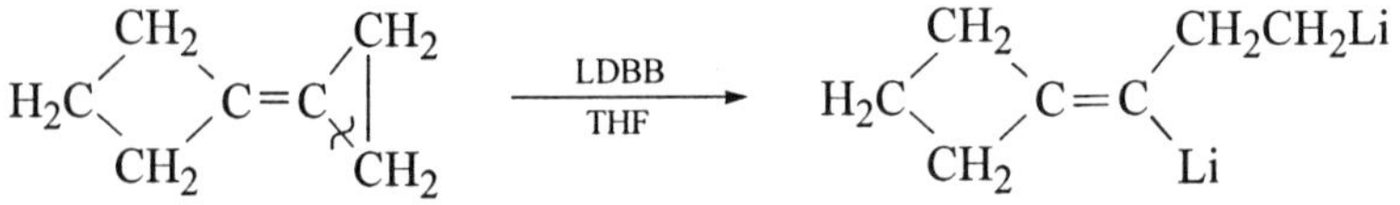

Catalytic amounts of di-tert-butylbiphenyl are necessary for the reaction of dicyclopropylacetylene with lithium. Even in THF as the solvent, this stops after the cleavage of only one cyclopropane ring.[162] LDBB was also used for the reductive metallation of the next higher homolog of 1,1′-bicyclopropylidene, cyclopropylidenecyclobutane. The cyclobutane ring remains intact.[217]

Compared with methylenecyclopropane, unsubstituted methylenecyclobutane shows the expected very low reactivity towards lithium metal.[219] The yield was only 16% even with LDBB after eight hours in THF at −20°C.[217]

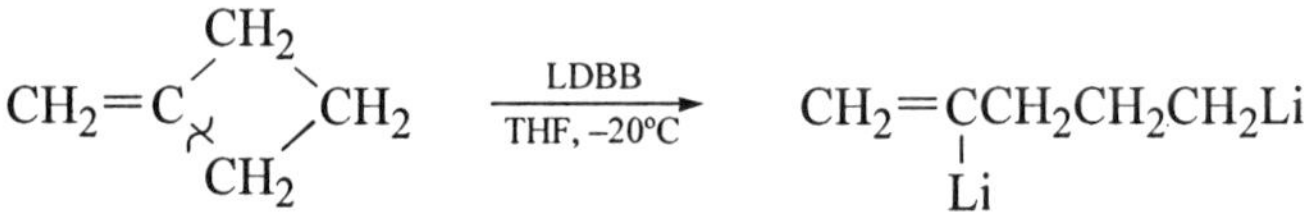

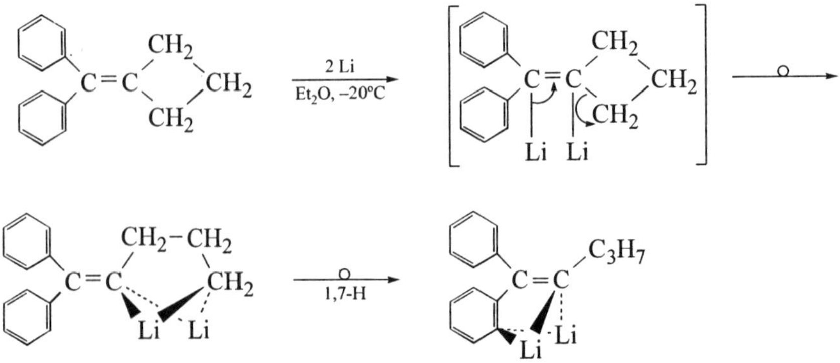

On the other hand, diphenylmethylenecyclobutane reacts smoothly with lithium powder in diethyl ether even without a catalyst. The initially formed

1,4-dilithio product, however, slowly rearranges via a 1,7-proton shift into another 1,4-dilithio compound (presumably also doubly-bridged) which is rather stable in diethyl ether.[219] As shown by homonuclear scalar ^{6}Li, ^{6}Li spin–spin coupling, the final compound is dimeric. This coupling was detected for the first time via the ^{6}Li, ^{6}Li INADEQUATE experiment.[220,221] This very sensitive technique is a promising alternative to the COSY experiment[200,201,222] and provides an additional tool for structural studies of organolithium compounds in solution.

A similar rearrangement was shown also to take place even faster with diphenylmethylenecyclopropane, this time via a 1,6-proton shift:[217]

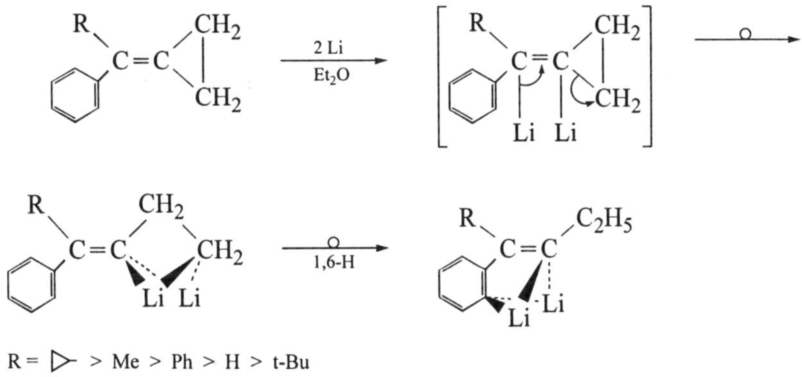

R = ▷– > Me > Ph > H > t-Bu

Benzylidenecyclopropane derivatives with only one phenyl group behave similarly: after ring opening the lithium is forced to occupy the position *cis* to the phenyl group. The reactivity depends on the substituent R, and decreases in the sequence given. The intermolecular character of the rearrangement could be proven with R = Me:[217]

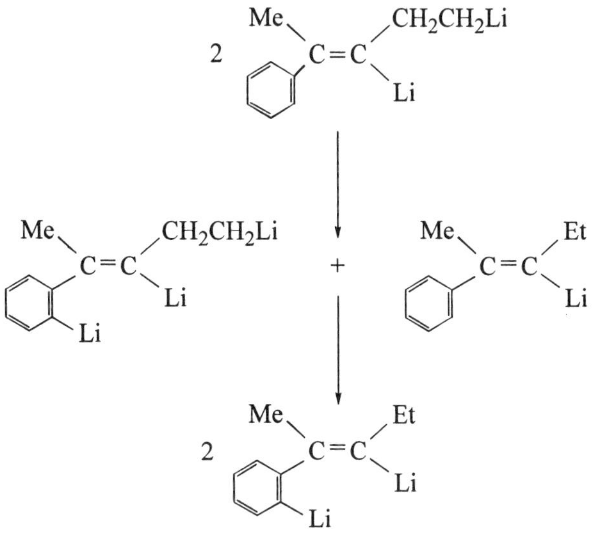

Not only is α-cyclopropylbenzylidenecyclopropane highly reactive, but it also shows an extremely interesting reversible ring–chain rearrangement. Additional ring opening of the cyclopropyl substituent occurs at higher temperatures:[217]

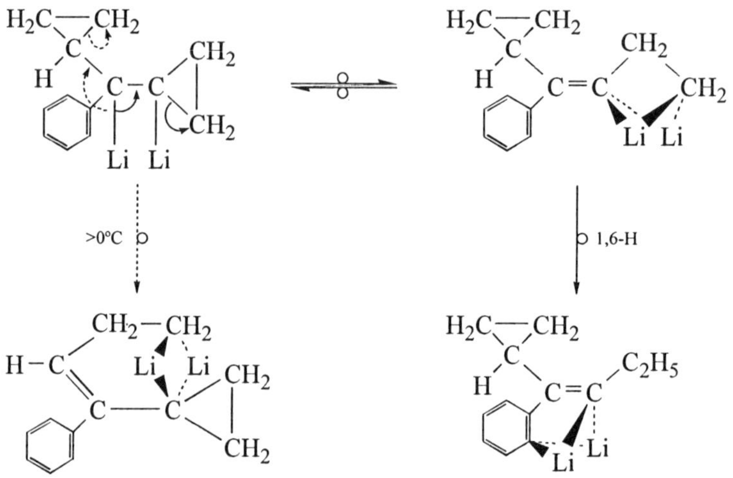

Again only the (*E*)-derivative is formed stereoselectively; a doubly bridged 1,5-dilithio compound is the most stable end product. However, the 1,6-proton shift (via the initially formed 1,3 doubly bridged kinetic controlled intermediate) is an irreversible side reaction which competes at temperatures above 0°C.

Replacement of the phenyl group by a second cyclopropyl substituent leads to very low reactivity. Even with LDBB, only the methylenecyclopropane ring is cleaved irreversibly.[217] Dialkylmethylenecyclopropanes as well as cyclopropylidene-cyclopentane and -cyclohexane do not react at all, neither with lithium powder upon sonication nor with LDBB.[217,219] The same is true for bicyclobutane and spiropentane despite their high ring strain.

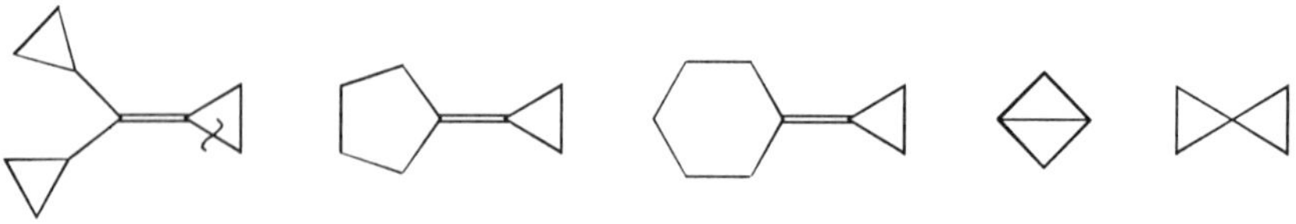

In contrast, methylenespiropentane, an isomer of the butterfly olefin, does react with lithium metal, although only after the addition of di-tert-butylbiphenyl as a catalyst. The reaction takes place in THF even at −60°C. Over

80% of a dilithio-butadiene derivative is produced, but at 0°C this adds two further lithium atoms to yield an interesting tetralithium organic compound with different kinds of σ- and π-bonded lithiums.[223]

$$CH_2{=}C(\text{spiro-}C_5H_8)\xrightarrow[\text{4,4'-di-}t\text{-butylbiphenyl}]{2\ Li/THF}\left[CH_2{=}C(Li)-C(CH_2Li)(CH_2)_2\right]\longrightarrow$$

$$CH_2{=}C(Li)-C(CH_2CH_2Li){=}CH_2 \xrightarrow{2\ Li} \overbrace{CH_2{\cdots}C(Li){\cdots}C(CH_2CH_2Li){\cdots}CH_2}^{2-}\ 2\ Li^+$$

No catalyst is necessary for the corresponding reaction of phenylspiropentane with lithium dust (2% sodium). In THF at −30°C, a dark red dilithiumorganic product results:[224]

$$PhCH(\text{spiro-}C_4H_6)\xrightarrow[THF,\ -30°C]{2\ Li}\left[PhCH(Li)-C(CH_2Li)(CH_2)_2\right]\longrightarrow$$

$$PhCH(Li)-C(CH_2CH_2Li){=}CH_2$$

The complete regioselectivity of the last two reactions is remarkable. Under the same conditions, 1,1-diphenylcyclopropane is cleaved as well.[224] However, more forcing conditions (14 hours sonication at room temperature) are necessary for the corresponding cleavage of phenylcyclopropane. The initially formed 1-phenyl-1,3-dilithiopropane eliminates lithium hydride; the end product once again is a tetralithiumorganic compound with two different kinds of σ- and π-bonded lithiums:[224]

$$Ph_2C(CH_2)_2 \xrightarrow[\text{THF, }-30^\circ C]{2\,Li} Ph_2C(Li)CH_2CH_2Li$$

$$PhCH(CH_2)_2 \xrightarrow[\text{THF, }20^\circ C\text{))))))}]{2\,Li} PhCH(Li)CH_2CH_2Li \xrightarrow{-LiH}$$

$$PhCH{=}CHCH_2Li \xrightarrow{Li} PhCH(Li)\text{—}\overset{*}{C}H(CH_2Li)\text{—}\overset{*}{C}H(CH_2Li)\text{—}CH(Li)Ph$$

meso + d,l

In contrast to the parent bicyclobutane, phenyl-substituted bicyclobutanes can be cleaved with lithium metal even at $-70^\circ C$ in THF solution:[224]

$$Ph\text{–}C(CH_2)_2C\text{–}R \xrightarrow[\text{THF, }-70^\circ C]{2\,Li} (Ph)(Li)C(CH_2)_2C(R)(Li)$$

R = H; CH_3

At room temperature in diethyl ether, elimination of lithium hydride was observed, but not dimerization.[224]

[1.1.1]Propellane is the only hydrocarbon without a neighboring π system known to react with lithium 4,4′-di-tert-butylbiphenyl; obviously the energy of the σ^* LUMO is especially low. The solvent is important; the yield is 61% in dimethyl ether at its boiling point but only 6% in THF:[225]

$$\xrightarrow[Me_2O,\ -23^\circ C]{2\,Li} \text{Li–(bicyclo[1.1.1]pentyl)–Li}$$

2.4.2 Addition of Lithiumorganic Compounds. The kinetics of the addition of two molecules of butyllithium to meta- and para-diisopropenylbenzene have been investigated thoroughly in cyclohexane and in benzene solvents.[226] The reactivity of the two double bonds of the meta-derivative are almost the same and are comparable to that of α-methylstyrene. However, the first addition of the para isomer is faster than the second and the overall reaction is more rapid.[226]

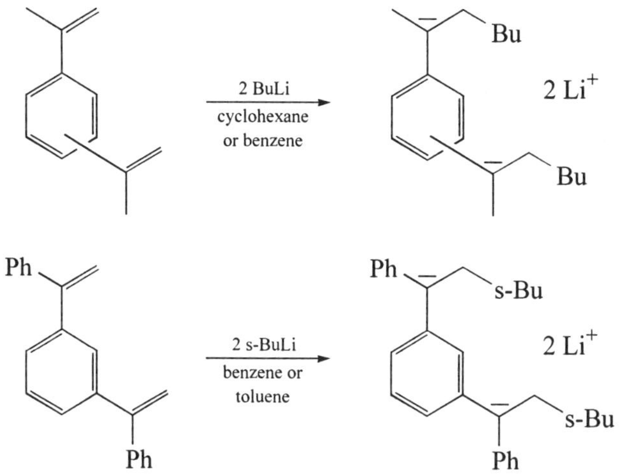

Of the corresponding α-phenylvinyl compounds, only the meta-derivative gives satisfactory yields with sec-butyllithium.[227] The diadducts behave as efficient bifunctional anionic initiators. These can be used in nonpolar solvents, e.g., for the polymerization of 1,3-dienes.[226,227]

A *bis*-cyclopentadienyllithium compound has been prepared elegantly by nucleophilic attack of methyllithium at the exocyclic double bonds of the fulvene units of a difulvene:[228]

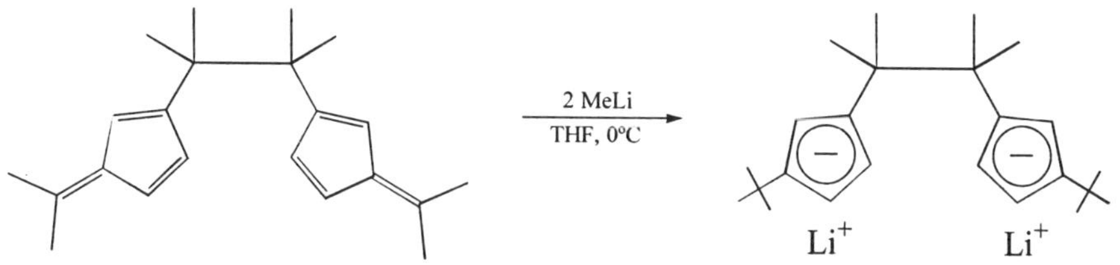

The addition of tert-butyllithium to a B=C double bond gives an interesting diboryldilithiomethane derivative; X-ray crystallography reveals a monomeric structure:[229]

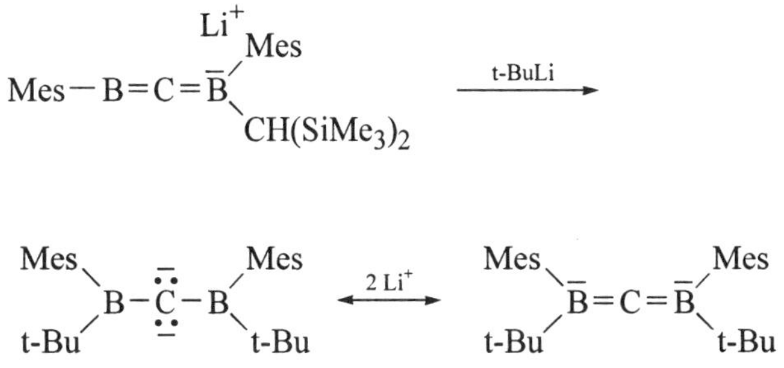

2.5 Polymetallation of Acidic Hydrocarbons (Hydrogen–Metal Exchange)

When the CH-acidic compounds are activated, acid–base reactions can often lead to the replacement of more than one proton by lithium. Such methods are especially suitable for the preparation of π-delocalized carbanions or highly inductively stabilized polylithiumorganic compounds.

2.5.1 Arenes and Heteroarenes as well as Metallocenes. Tertiary amino groups and ether oxygens are known to acidify ortho positions of arenes; indeed, *bis*(*N*,*N*-dimethyl-4-aminophenyl)methane[230] and diphenyl ether[2,3,231–234] can be lithiated twice:

Me_2N–C_6H_4–CH_2–C_6H_4–NMe_2 —(2 BuLi; hexane, 60°C)→ Me_2N–$C_6H_3(Li)$–CH_2–$C_6H_3(Li)$–NMe_2

C_6H_5–O–C_6H_5 —(2 BuLi; THF / Et_2O, reflux)→ Li, Li (ortho-dilithiated diphenyl ether)

When two different heteroatom substituents are present, bulky groups at nitrogen force the lithium substitution to proceed ortho to the oxygen:[235]

Phenoxazine (N–R) —(BuLi; THF, 0°C)→ 1,9-dilithio derivative (Li, Li; O; N–R)

R = PhCHMe; t-$BuSiMe_2$

The introduction of two lithium atoms into the same ring, however, is much more difficult; the activation by at least two methoxy groups is necessary.

Depending on the arrangement of the groups, the double metallation can work well, badly, or not at all.[236,237] Anisole and the isomeric dimethoxy- and trimethoxybenzenes give with *n*-butyllithium/TMEDA the following dilithiated products (judging from the yields of the silylation products):[236]

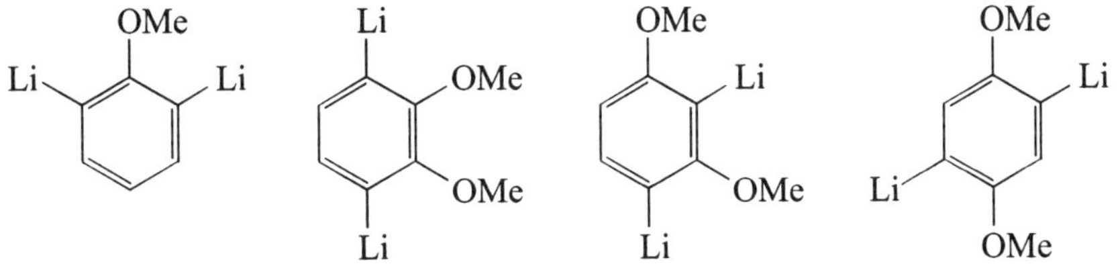

Similarly, 1,2,4,5-tetraalkoxybenzenes can be lithiated twice. However, some monolithio compounds below 0°C are poorly soluble both in diethyl ether and in THF so that bismetallation can only be achieved by using large quantities of solvent at 40°C.[238]

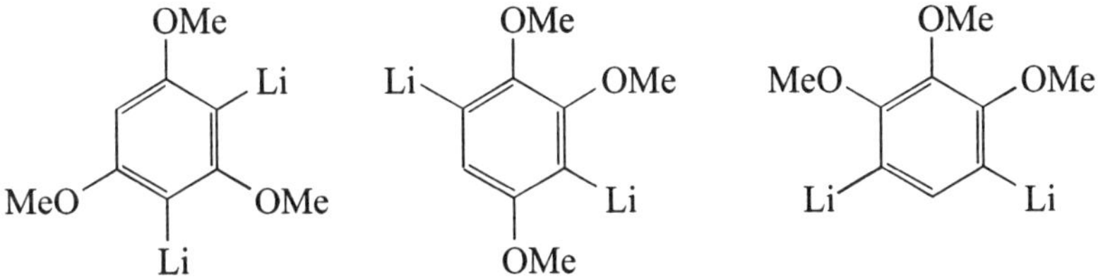

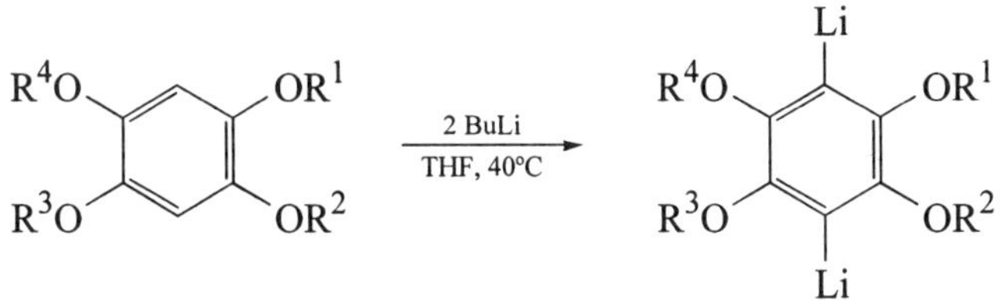

$R^1 = R^2 = R^3 = R^4 = CH_3$
$R^1 = R^3 = CH_3$; $R^2 = R^4 = CH_3OCH_2$
$R^1 = R^2 = CH_3$; $R^3R^4 = CH_2$
$R^1R^2 = R^3R^4 = CH_2$

Alkylthio groups activate nearly as well, but methylthio compounds are metallated preferably in the side chain α to sulfur,[239,240] e.g.,

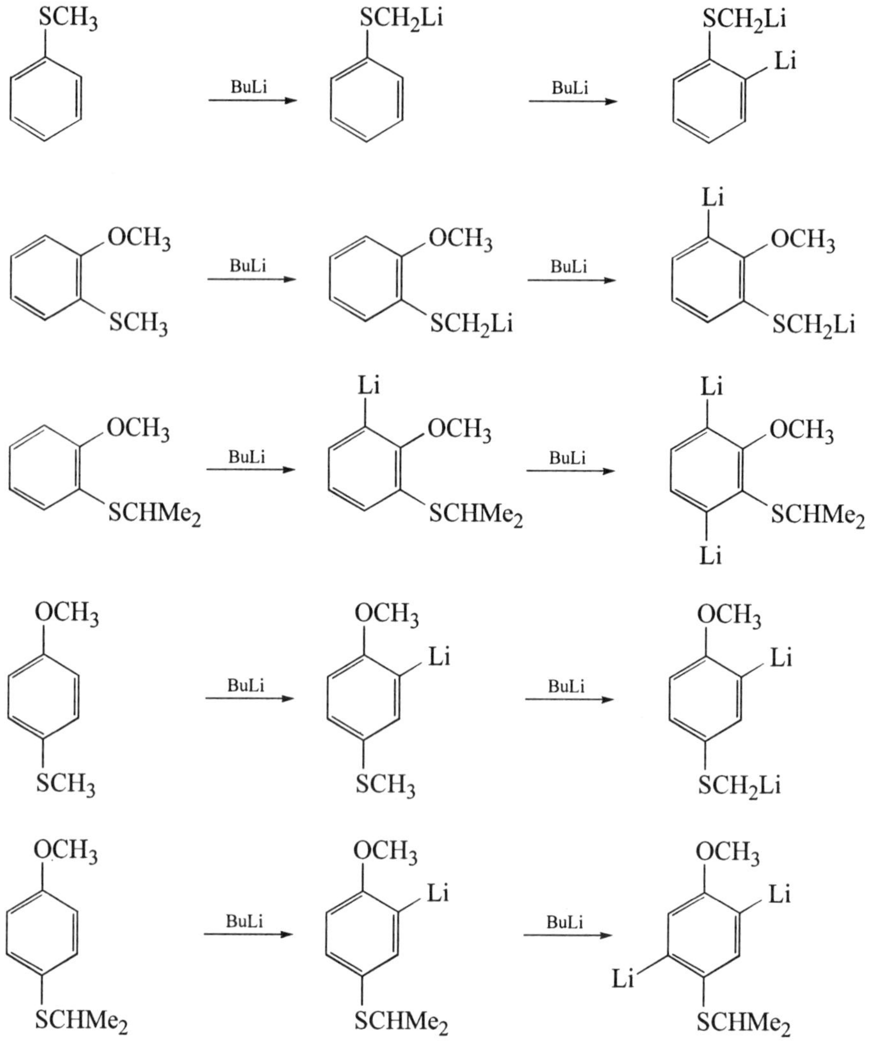

Fluorines acidify even more than alkoxy groups. Thus in 1,3,5-trifluorobenzene, all three hydrogens can be replaced by lithium with tert-butyllithium even at $-115°C$:[241]

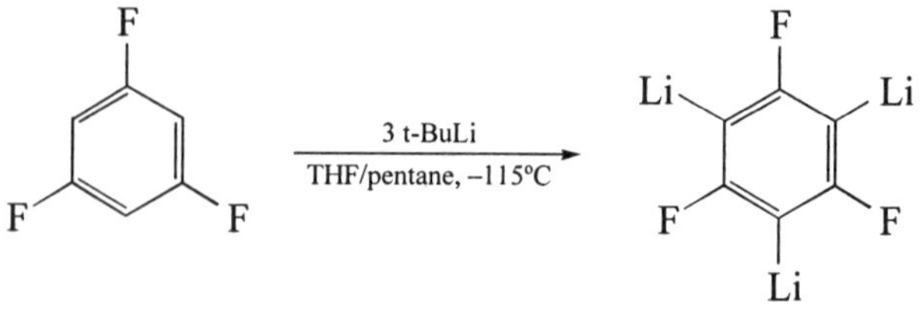

At higher temperatures, elimination of lithium fluoride takes place with aryne formation. The following sequence of reactions is similar:[242]

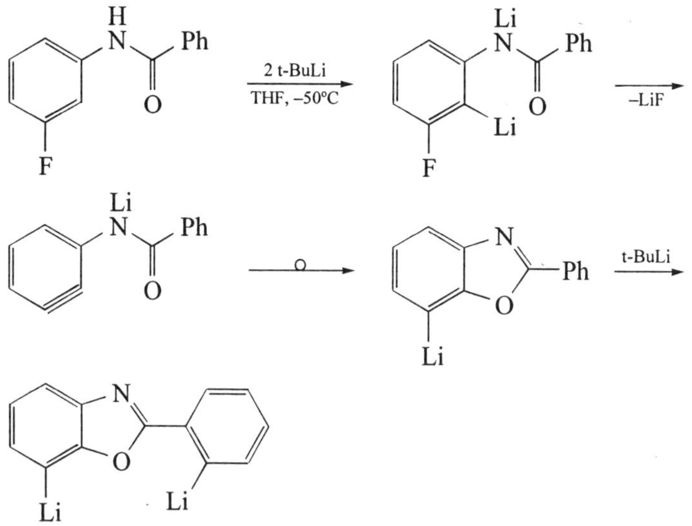

2 t-BuLi
THF, −50°C

−LiF

t-BuLi

As with the corresponding hydroquinone ethers, dicarbamates can be dilithiated:[51]

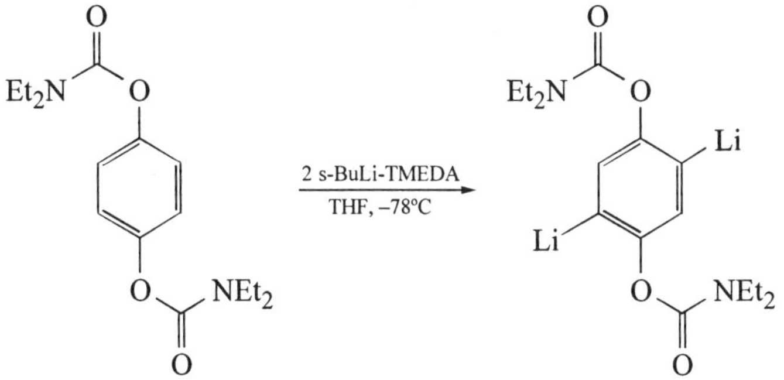

2 s-BuLi-TMEDA
THF, −78°C

With *N*,*N*-diethylbenzamide, -phthalamide, and -terephthalamide, only monolithiated products were obtained under the same conditions. A tertiary carbonamide functionality directly bonded to the aromatic ring obviously is not sufficiently activating for double lithiation.[51] However, in the presence of chlorotrimethylsilane, which reacts only slowly with sec-butyllithium, the disilylated compounds are formed by a stepwise reaction sequence.[51] After replacement of the silyl groups by bromine, the corresponding dilithio compounds can finally be obtained by a separate double halogen–lithium exchange reaction.[51]

This goal can be achieved, however, much more easily by the newer, "reverse transmetallation'' method.[108,243] Lithium tetramethylpiperidide (LiTMP) is used as the base in the presence of mercuric chloride. Even though only traces of lithiated products are present at equilibrium, these are trapped and the equilibrium is displaced. Eventually, a 60% yield of *o,o'*-dimercurated *N,N*-diethylbenzamide is produced. The latter can then be treated with *n*-butyllithium to yield 2,6-dilithio-*N,N*-diethylbenzamide by mercury–lithium exchange:[108,243]

O NEt2 → (LiTMP / $HgCl_2$) → ClHg, HgCl, O NEt2 → (4 BuLi / THF, –60°C) → Li, Li, O NEt2

LiTMP = Lithiumtetramethylpiperidide

Interestingly, the reaction of biphenyl with *n*-butyllithium in the presence of TMEDA yields a dilithiumorganic compound even though activating groups are absent. The two lithium atoms are introduced solely into the 2 and 2′ positions:[59]

→ (2 BuLi • TMEDA / hexane, 20°C) → TMEDA • Li Li • TMEDA

The symmetrical doubly lithium-bridged structure, predicted theoretically,[56] could be confirmed by X-ray crystallography. It is reasonable to suppose that 2,2′-dilithio-1-phenylpyrrole, first prepared analogously in 1979,[244] has a similar Li-bridged structure.

N → (2 BuLi • TMEDA / Et_2O) → N, Li Li

Other polycyclic aromatic hydrocarbons like triphenylene, phenanthrene, and chrysene can be dilithiated as well, although—especially the latter two—less selectively.[245]

On the other hand, the doubly lithiated dimer of 6-dimethylamino-1-azafulvene is formed in excellent yields, obviously due to chelation (intramolecular solvation):[246]

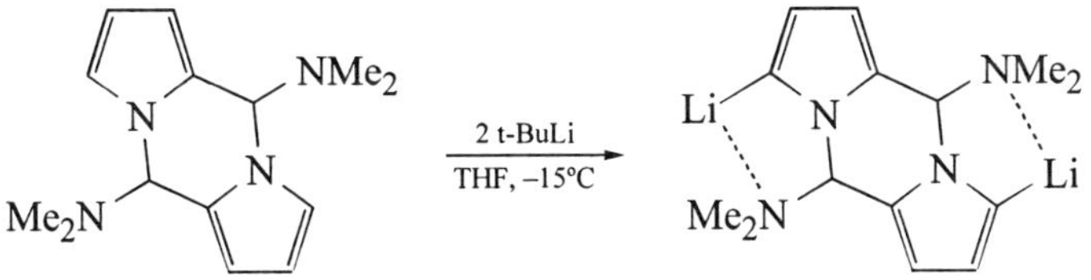

The double metallation of 2-benzofuran[247] and 3,4-dichlorothiophene[248] introduces, as expected, the two lithium atoms into the only free positions, i.e., α to the hetero atoms:

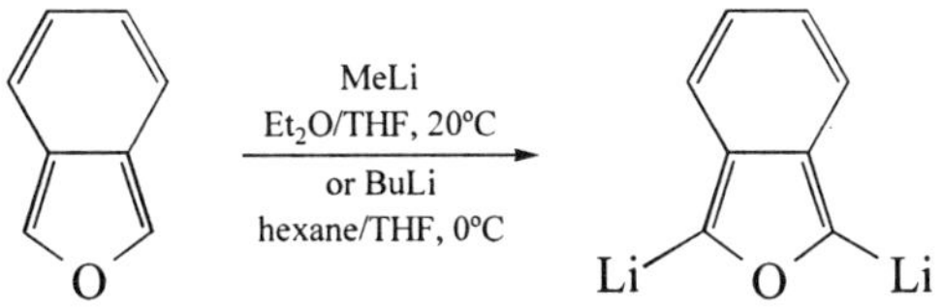

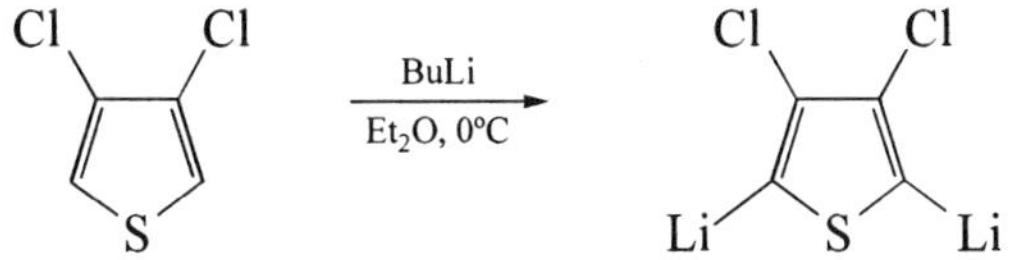

However, even the unsubstituted furan[249,250] and thiophene[249-251] as well as 3-tert-butylthiothiophene[252,253] are 2,5-dilithiated exclusively:

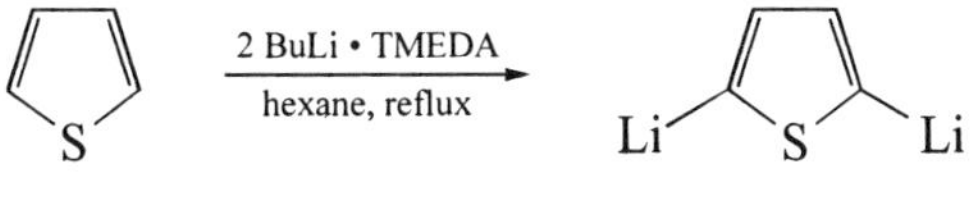

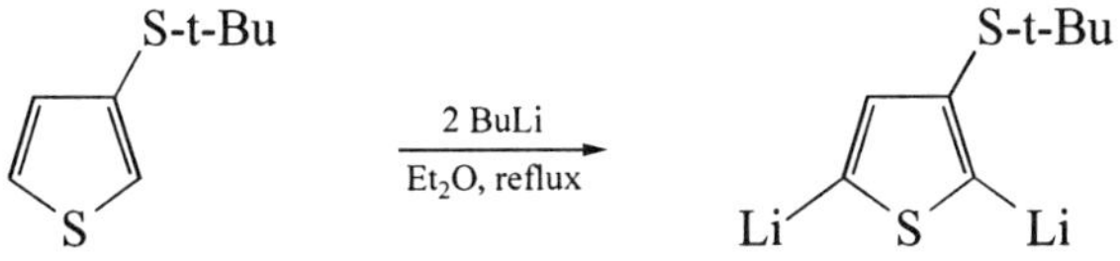

When the 2 position is occupied, i.e., by a carbonamide group,[254] a sulfonamide group,[255] or an oxazoline ring,[256] 3,5-metallation takes place stepwise, first in the 5 position and then in the 3 position, e.g.,

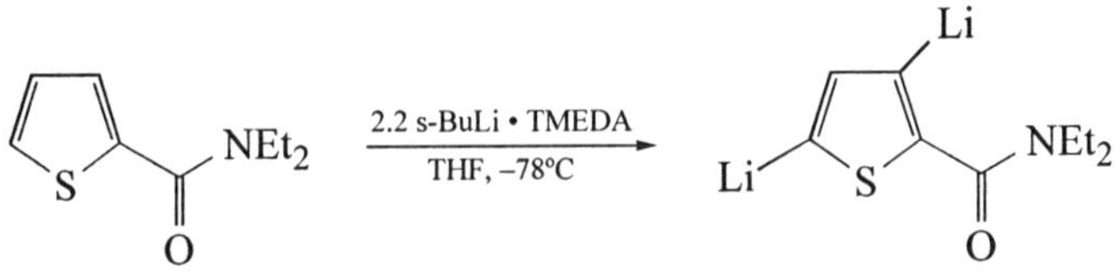

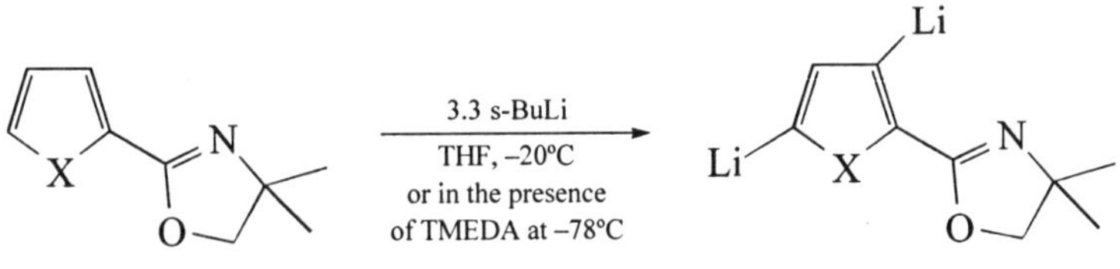

X = O; S

The two carbanionic centers can react stepwise in a tandem reaction with two different electrophiles; the more reactive lithium in the 3 position undergoes substitution first.[254]

The metallation of 3-(ethoxymethyl)-2-phenylimidazole first takes place at the heterocyclic ring and then at the ortho position of the phenyl ring:[257]

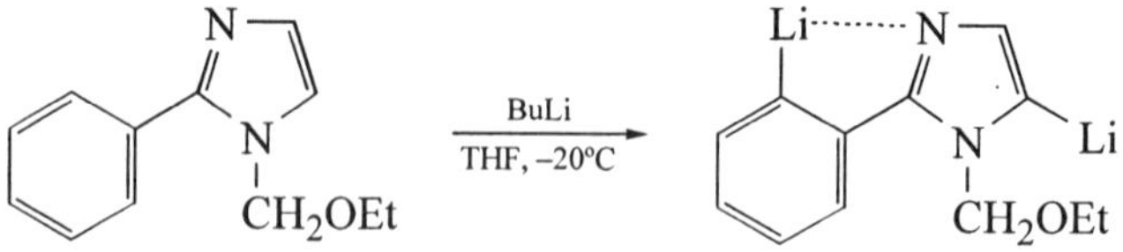

N-Methylpyrrole yields a mixture of both the 2,4- and 2,5-dilithio compounds.[249] When the *N*-methyl is replaced by a trimethylsilyl or triethylsilyl group, migration of the silyl group into the 2 position has been observed in addition:[258]

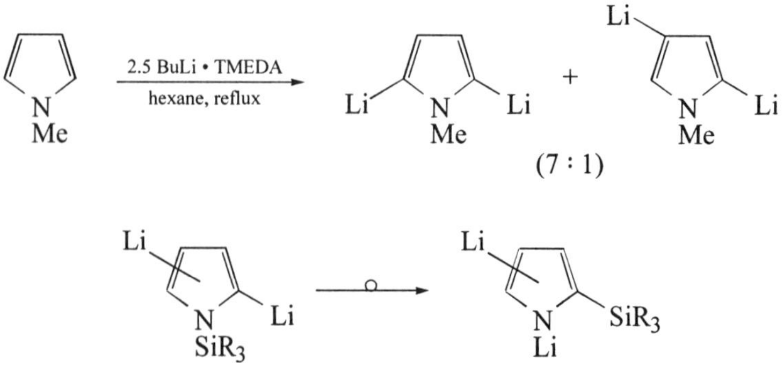

(7 : 1)

R = Me; Et

Metallocenes[259] e.g., ferrocene,[260-266] 1,1″-bi-ferrocene,[267] and other sandwich compounds,[268-271] can be polymetallated:

2.5 BuLi • TMEDA
hexane, 20°C

5 BuLi • TMEDA
cyclohexane, 80°C

With a tenfold excess of butyllithium/TMEDA at 70°C, up to seven lithium atoms can be introduced into ferrocene.[272] Substituted ferrocenes have been dilithiated as well,[273-276] e.g.,

BuLi • TMEDA
hexane, 20°C

With 1,1′-disubstituted ferrocenes, much more dl dilithio products are formed than the corresponding meso isomers.[277,278]

2.5.2 ***Vinyl Compounds, Bicyclobutanes, and Cubanes.*** Strongly activating groups are essential for the double lithiation of strainless alkenes. Examples are the *cis* and *trans* cinnamic nitriles[279] and tetrathiafulvalene:[280-282]

2 $i\text{-}Pr_2NLi$

2 $i\text{-}Pr_2NLi$

4 $i\text{-}Pr_2NLi$

The corresponding tetraselenafulvalene,[283] as well as the related vinylene trithiocarbonate (1,3-dithiole-2-thione)[284,285] and selena analogs,[285] react similarly. In contrast to the unsubstituted 1,2-dilithioethylene,[107] the *cis/trans* isomers of α,β-dilithiocinnamic nitrile equilibrate.[279]

Due to the high ring strain leading to higher s-character of the C—H bonds, the acidity of the vinyl protons in cyclopropene is increased so that dilithiation is possible even in the absence of electron withdrawing groups.[286] The same is true for certain bicyclobutane derivatives:[287–291]

2 PhLi, Et_2O, 0°C

Li Li

2 BuLi, Et_2O, 20°C

Li Li

Me_2 Si

as above

Me_2 Si

Li Li

1,2-Dilithio-3,3-diphenylcyclopropene shows an interesting rearrangement to 1,1-dilithio-3,3-diphenylallene:[292]

Ph Ph

Li Li

Li Ph

C=C=C

Li Ph

Cubane can be lithiated only when activating groups, e.g., carbonamides, are present. "Reverse transmetallation" (p. 525)[108] is the best method; the traces of lithiated products at equilibrium are trapped by mercuric chloride. This displaces the equilibrium and the process continues. A separate mercury–lithium exchange reaction follows:

A A

LiTMP, $HgCl_2$

ClHg A A HgCl

MeLi

Li A A Li

A = $CON(CHMe_2)_2$

LiTMP = Lithiumtetramethylpiperidide

Polylithiation takes place with the bis-cyclopropyl carbinol, but the reaction with the corresponding methyl ether only introduces one lithium:[293]

$$\xrightarrow[\text{Et}_2\text{O, 30°C}]{\text{6 s-BuLi}}$$

On the other hand, the icosahedral carborane 1,2-$C_2B_{10}H_{12}$ can readily be dilithiated at both the carbon atoms:[294]

$$\text{HC}\!-\!\text{CH}\,(B_{10}H_{10}) \xrightarrow[\text{Et}_2\text{O, 20°C}]{\text{2 BuLi}} \text{LiC}\!-\!\text{CLi}\,(B_{10}H_{10})$$

2.5.3 *Benzyl and Allyl Compounds.* Benzyl and allyl groups can be readily deprotonated due to the formation of π-delocalized carbanions. No wonder that molecules bearing two independent benzyl positions, like 2,2′-dimethylbiphenyl,[295] 2,2′-dimethyl-1,1′-binaphthyl,[296,297] and 4,4′-dimethyl-2,2′-bipyridyl[298,299] are lithiated twice, the latter even with lithium diisopropylamide as the base:

$$\xrightarrow[\text{hexane, 20°C}]{\text{BuLi • TMEDA}}$$

$$\xrightarrow[\text{hexane, –20°C}]{\text{BuLi • TMEDA}}$$

$$\xrightarrow[\text{THF, 0°C}]{\text{i-Pr}_2\text{NLi}}$$

The metallation site of 1,1-di-p-tolylethane depends on the base. One of the methyl protons is substituted by tert-butyllithium (kinetic control) whereas with benzyllithium the thermodynamically more stable benzhydryl derivative, 1,1-di-p-tolylethyllithium, is formed instead:[300]

H_3C–C_6H_4–$CH(CH_3)$–C_6H_4–CH_3 $\xrightarrow[\text{THF, 20°C}]{\text{t-BuLi}}$ $Li^+H_2\bar{C}$–C_6H_4–$CH(CH_3)$–C_6H_4–$\bar{C}H_2Li^+$

$\xrightarrow{\text{PhCH}_2\text{Li | THF, 20°C}}$ H_3C–C_6H_4–$\bar{C}(CH_3)Li^+$–C_6H_4–CH_3

The xylenes as well as mesitylene and other polymethyl benzenes can be metallated several times. The reactivity decreases in the order meta > ortho > para.[301,302] With butyllithium/TMEDA as the base, *bis*-1,4-lithiomethylbenzenes do not form; geminal dimetallated compounds are observed instead. These were found only in minor amounts under other conditions. The effects influencing the reactivity and selectivity of the metallation of polymethyl benzenes have been studied thoroughly.[303]

o-xylene ($C_6H_4(CH_3)_2$) $\xrightarrow[\text{hexane, 20°C}]{\text{BuLi • TMEDA}}$ $C_6H_4(\bar{C}H_2Li^+)_2$

H_3C–C_6H_4–CH_3 $\xrightarrow[\text{above}]{\text{as}}$ H_3C–C_6H_4–$CHLi_2$

mesitylene (H_3C, CH_3, CH_3) $\xrightarrow[\text{above}]{\text{as}}$ $C_6H_3(\bar{C}H_2Li^+)_3$ ($Li^+H_2\bar{C}$, $\bar{C}H_2Li^+$, $\bar{C}H_2Li^+$)

1-Methyl-2-(methylthio)benzene also gives a dilithiated product; further reactions serve to introduce substituents into the thiomethyl and methyl groups:[304]

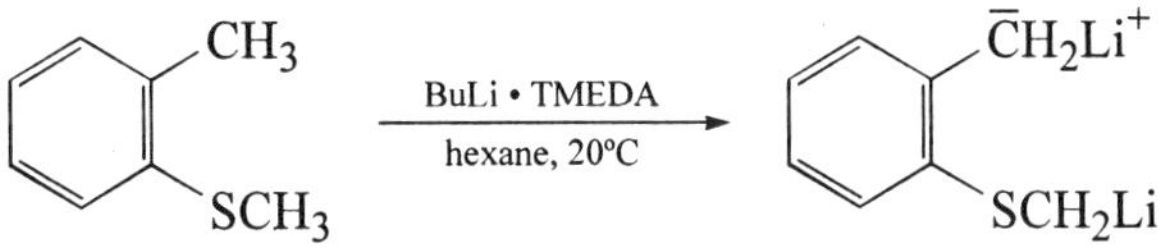

Surprisingly, polymethyl naphthalenes only are monometallated by excess *n*-butyllithium/TMEDA in THF/hexane. Substitution at the C2 methyl group is preferred.[305,306] In contrast, both the benzyl protons in m-diisoheptyl-benzene are metallated even in the absence of TMEDA.[307]

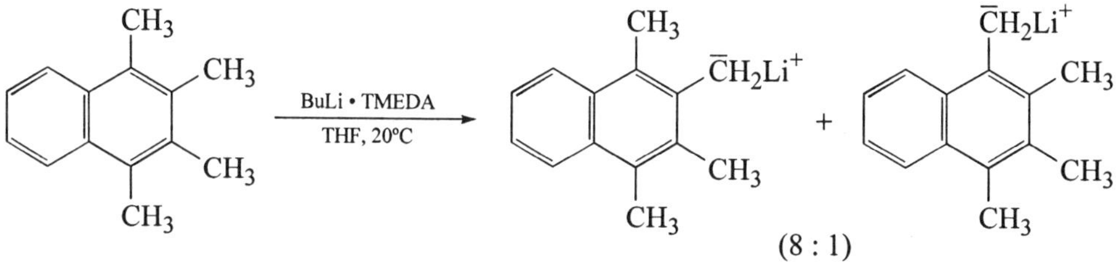

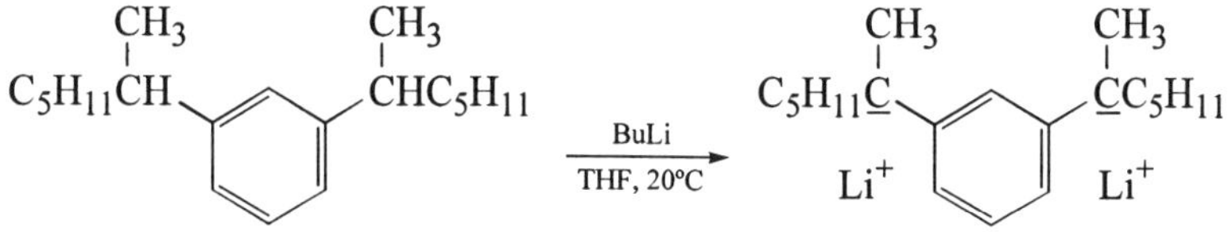

(8 : 1)

Lithium di-isopropylamide suffices for the dilithiation of 2,3-dimethyl-chinoxaline:[308]

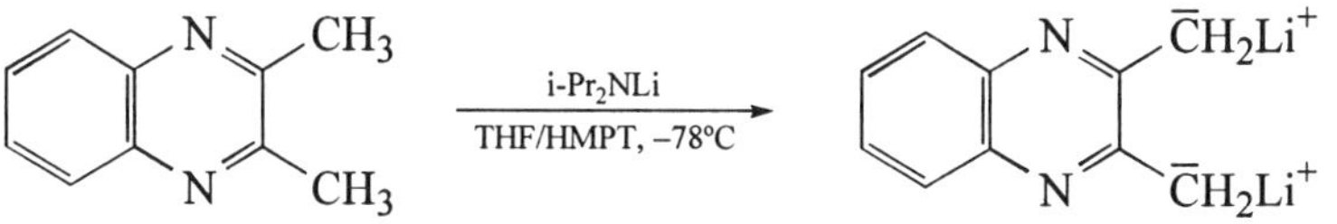

As expected, silylated side-chains show higher reactivity. The nicely crystalline dilithium products were investigated by X-ray analysis.[295,309-314] A second trimethylsilyl group hinders benzyl proton attack sterically so that the reactivity actually is lower than that of an unsubstituted methyl group; the order is $ArCH_2SiMe_3 > ArCH_3 > ArCH(SiMe_3)_2$.[310]

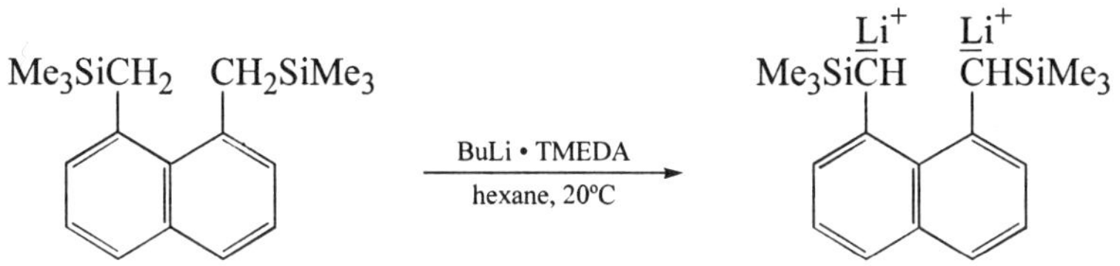

CH_2SiMe_3 / CH_2SiMe_3 (o, m, p) —as above→ $Li^+\bar{C}HSiMe_3$ / $\bar{C}HSiMe_3$ Li^+

Me_3SiCH_2–(pyridine)–CH_2SiMe_3 —as above→ $Me_3Si\bar{C}H$ Li^+ –(pyridine)– $\bar{C}HSiMe_3$ Li^+

However, replacement of the trimethylsilyl groups in *o*-xylene by phenyl groups leads to a dramatic change in structure of the dilithium derivatives. The lithiums in doubly metallated *o-bis*-(trimethylsilylmethyl)benzene show the usual symmetrical doubly bridged (C_{2v} symmetry).[310] In contrast, only one lithium atom in the corresponding *o*-dibenzylbenzene derivative bridges; the other bonds in benzyl-like fashion.[315,316] Polymorphy has even been observed in one case: the same compound had two crystalline modifications with completely different lithium positions.[309]

An excess of *n*-butyllithium/TMEDA introduces up to three lithium atoms in toluene; benzyllithium is assumed to be the intermediate.[317,318]

CH_3 (toluene) —BuLi • TMEDA, hexane, 20°C→ CH_2Li (23%) + CH_3 / Li (2%) +

CH_3 / Li (1%) + CH_2Li / Li (7%) + CH_2Li / Li (11%) + $CHLi_2$ / Li (46%)

Of course, electron withdrawing side-chain substituents in toluene facilitate the removal of a benzyl proton; weaker bases like lithium *bis*-trimethylsilylamide are effective. These form ''Quasi Dianion Complexes [QUADACs]'' which *simulate* a second metallation,[82] e.g.,

$$PhCH_2CN \xrightarrow{2\ LiNR_2} \text{Ph(H)C=C=N}\langle Li_2 \rangle NR_2$$

$$PhCH_2CN \not\xrightarrow{2\ LiNR_2} PhCLi_2CN \xrightarrow{2\ D_2O} PhCD_2CN$$

(the lithiated ketenimine also gives $PhCD_2CN$ with D_2O, D_2O)

R = Me_3Si; i-Pr

The corresponding QUADAC with lithium di-isopropylamide (LDA) as the base could be crystallized (with 2 TMEDAs) and characterized by X-ray analysis.[319] However, both the α hydrogens of trimethylsilylacetonitrile[320] could be replaced by lithium with butyllithium or LDA in diethyl ether/hexane. The solid-state structure of the crystalline product, $\{Li_2(Me_3SiCCN)\}_{12}$-$(Et_2O)_6(C_6H_{14})$[321] differs significantly from the solid-state structures of monolithionitriles.

Two vicinal benzyl protons of dibenzyl can be removed.[322] Interestingly, the stilbene dianion salt, lithium bridged according to X-ray analysis, is in equilibrium with the corresponding radical anion;[323]

$$PhCH_2CH_2Ph \xrightarrow[\text{benzene/hexane, } 20°C]{BuLi \cdot TMEDA} Ph\bar{C}H-\bar{C}HPh \quad 2\ Li^+$$

$$\text{(Ph)(H)C=C(H)(Ph) bridged by 2 Li} \rightleftharpoons [PhCH{=}CHPh]^{\cdot -}\ Li^+ + Li^{\circ}$$

The same compound had been obtained earlier by the addition of lithium metal to stilbene (see p. 502). The acenaphthylene dianion salt, too, is available from acenaphthene as well as from acenaphthylene:[324,325]

$$\text{acenaphthene} \xrightarrow[\text{THF, 0°C}]{2\ BuLi} [\text{acenaphthylene}]^{2-}\ 2\ Li^+ \xleftarrow[\text{THF}]{2\ Li} \text{acenaphthylene}$$

The reaction of *n*-butyllithium/TMEDA with 1,4-dihydronaphthalene[326] and 9,10-dihydroanthracene[327–329] gave dianion salts. However, this did not occur with 9,10-dihydrophenanthrene;[328] uncharged phenanthrene results from butyllithium assisted lithium hydride elimination of the initial monolithiated intermediate.[328]

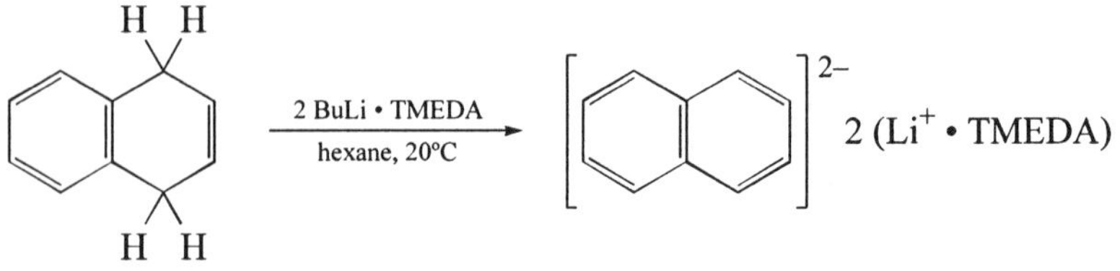
H H
2 BuLi • TMEDA
hexane, 20°C
2–
2 (Li+ • TMEDA)
H H

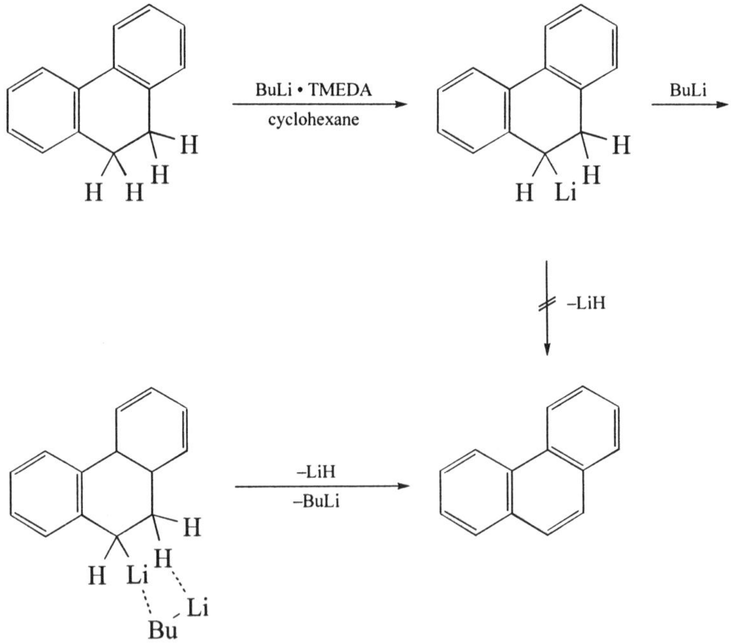
BuLi • TMEDA
cyclohexane
H
H H H
BuLi
H
H Li H
–LiH
–LiH
–BuLi
H
H Li H
Li
Bu

In the presence of cadmium(II) chloride, dianion salts of anthracene and anthracene derivatives can be converted to the corresponding neutral hydrocarbons.[328,329] As with naphthalene, elimination of lithium hydride from the monoanion does not occur spontaneously:

R, R' = H; Me; Et; $SiMe_3$

In agreement with MNDO calculations,[330] X-ray analysis[3,325–327] shows the lithium atoms to be located on different sides of the ring systems of these dianion salts:

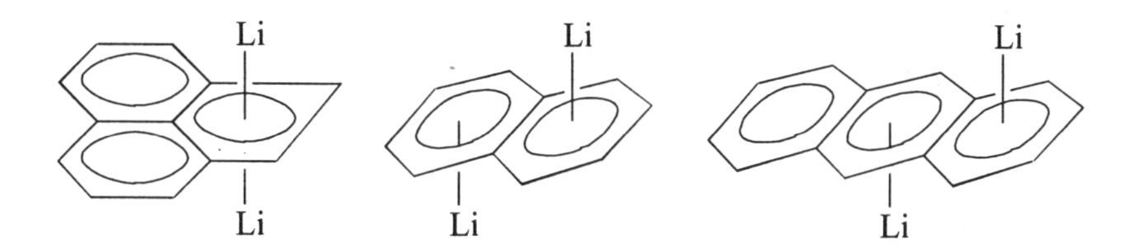

The nonsymmetric crystal structures are not the result of crystal-packing forces, but reflect the preferred electrostatics and perhaps the partial covalency of the carbon–lithium bonding.[330]

As mentioned above (p. 507), the structure of benzene dianion salts depend on the number of trimethylsilyl substituents. While the 1,2,4,5-tetra-substituted benzene dianion moiety[175] is nearly planar with the two lithium cations located above and below the center of the ring, the hexakis-(trimethylsilyl)benzenide dianion[174] favors a boat form with both lithium cations located on the same side of the benzene ring. Ab initio calculations[176] predict the same arrangement for the unsubstituted dilithiobenzenide.

Another case of polymorphy was observed with the product of deprotonation of *trans*-1,2-diphenylbenzocyclobutene using *n*-butyllithium/TMEDA.[331] The two crystalline modifications show different positions for the lithium atoms and the TMEDA molecules (''lithiotropy'').[332,333] Additional stabilization by 6 π Hückel cyclobutadienyl dianion delocalization could not be detected.[331]

$$\text{(trans-1,2-diphenylbenzocyclobutene)} \xrightarrow[\text{THF, 0°C}]{\text{BuLi} \cdot \text{TMEDA}} [\text{1,2-diphenylbenzocyclobutadiene}]^{2-} \; 2\,(\text{Li}^+ \cdot \text{TMEDA})$$

On the other hand, branched *bis*-allylic dianion salts are distinctly more stable than the corresponding linear systems. While propene[334] and 2-butene[213] can be lithiated twice with *n*-butyllithium/TMEDA, the dilithium salt of the cross-conjugated trimethylenemethane dianion,[213–215,335–340] is formed especially readily:

$$CH_2{=}CHCH_3 \xrightarrow[\text{hexane, 20°C}]{\text{BuLi} \cdot \text{TMEDA}} [CH_2{\cdots}CH{\cdots}CH]^{2-} \; 2\,Li^+$$

$$CH_3CH{=}CHCH_3 \xrightarrow[\text{above}]{\text{as}} [CH_2{\cdots}CH{\cdots}CH{\cdots}CH_2]^{2-} \; 2\,Li^+$$

$$CH_2{=}C(CH_3)_2 \xrightarrow[\text{above}]{\text{as}} [CH_2{\cdots}C(CH_2)(CH_2)]^{2-} \; 2\,Li^+$$

The ''Y-aromaticity'' concept has been employed to explain the higher stability; however, this rationalization is highly controversal.[303,315,341–352] Stable Y-delocalized dianions bearing 4*n*, e.g., 8 π electrons,[345] are known. According to a recent ab initio study,[353] Y-aromaticity does not exist. The isolated trimethylenemethane dianion is a nonplanar system with pyramidal CH_2 groups. At higher levels of theory, the relative energy of the trimethylenemethane dianion becomes less than that of the butadiene dianion.[354] However, this result on the isolated dianions does not agree with experimental results on the lithium salts which indicate that the cross-conjugated trimethylenemethane system is indeed more stable than the isomeric linear butadiene dianion derivatives. The importance of the counter ions is obvious.

Crystalline derivatives of the trimethylenemethane dianion have only been obtained with phenyl substituted derivatives.[349,351,355] Of course, these systems exhibit benzyl anion-like character. The X-ray structures[355] also show the lithiums to be on different sides of the cross-conjugated system.

$$PhCH{=}C(CH_2Ph)_2 \xrightarrow[\text{hexane, 25°C}]{2\,\text{BuLi} \cdot \text{TMEDA}} [PhCH{\cdots}C(CHPh)(CHPh)]^{2-} \; 2\,Li^+$$

The conformers of an interesting phosphino-stabilized trimethylenemethane dianion interconvert rapidly:[356]

$$Ph_2P\text{–}CH_2C(=CH_2)CH_2\text{–}PPh_2 \xrightarrow[\text{THF, –40°C}]{\text{neo-pentyllithium}}$$

With sulfur and nitrogen-containing substituents, trilithiation of the isobutene skeleton is allowed:[357]

$$CH_2{=}C(CH_3)CH_2X \xrightarrow[\text{hexane, 20°C}]{\text{BuLi • TMEDA}} \left[CH_2{\cdots}C(CH)(CHX)\right]^{3-}\ 3\,Li^+$$

X = SPh; NMePh

More extended linear and cross-conjugated systems also have been prepared, e.g.,[213,345,346,358–362]

$$CH_2{=}C(CH_2CH{=}CH_2)_2 \xrightarrow[\text{THF}]{\text{BuLi}} \left[CH_2{\cdots}C(CH{\cdots}CH{\cdots}CH_2)_2\right]^{2-}\ 2\,Li^+$$

$$CH_2{=}CHCH_2CH_2CH{=}CH_2 \xrightarrow[\text{hexane, 20°C}]{\text{BuLi • TMEDA}}$$

$$\left[CH_2{\cdots}CH{\cdots}CH{\cdots}CH{\cdots}CH{\cdots}CH_2\right]^{2-}\ 2\,Li^+ \equiv$$

Li • TMEDA

Li • TMEDA

Dimetallation of 1,4-*bis*(methylene)cyclohexane results in *p*-xylene dianion formation. According to MNDO calculations, the linearly conjugated dianion intermediate is effectively planar, with a symmetrical charge distribution.[363] In contrast, 2-methyl-1,5-hexadiene, an acyclic analogue of 1,4-*bis*-

(methylene)cyclohexane, undergoes isomerization rather than hydride elimination. Due to steric effects, the linearly conjugated dianion intermediate must be nonplanar, with a strongly bent π system. Two orthogonal allyl anion systems are present.[363]

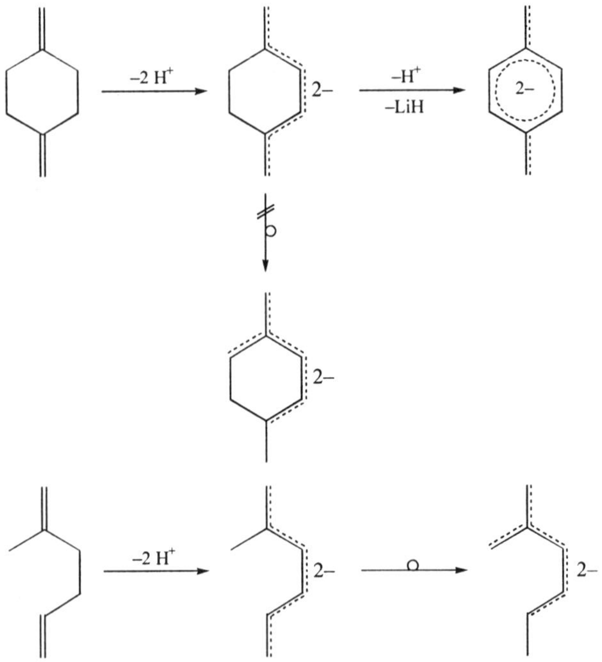

A slow, stepwise rearrangement of 3,4-dilithio-2,5-dimethyl-2,4-hexadiene[364] into the "Y-delocalized" 2,5-dimethyl-2,4-hexadiene dianion has been observed:

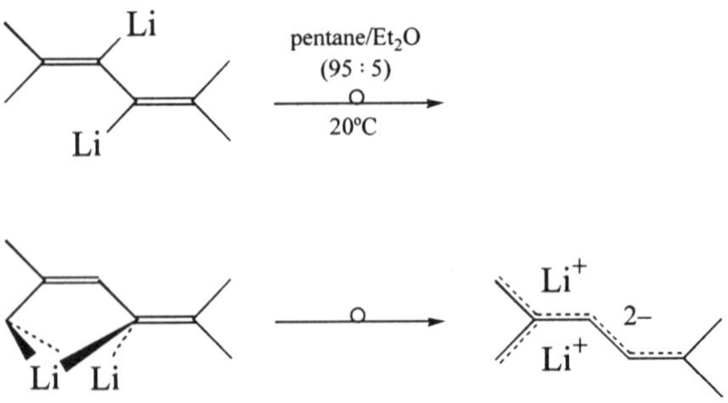

Upon metallation, 1,3,6-heptatriene cyclizes to the cycloheptatrienyl trianion[365] which, of course, owes its existence to the close contact with the three lithium cations.[366] A similar example is the conversion of (*Z,Z,Z*)-1,3,6-cyclononatriene into the homocyclooctatetraene dianion.[367]

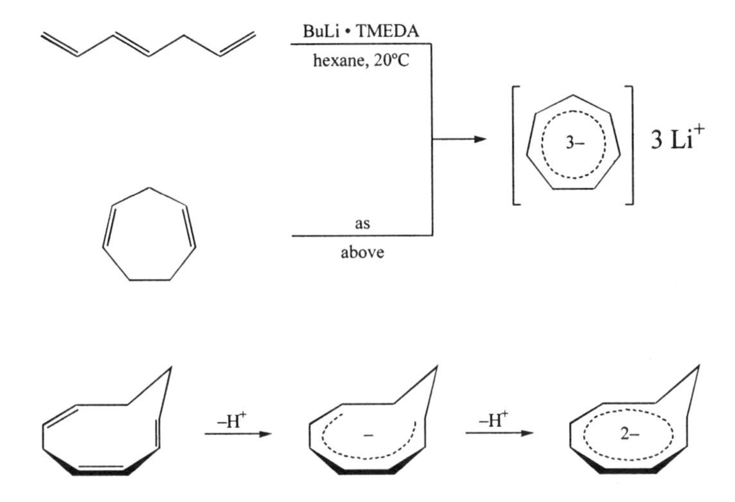

The parent cyclooctatetraene dianion is formed when 1,5-cyclooctadiene is treated with *n*-butyllithium/TMEDA. (Interestingly, neither 1,3-cycloocta-diene nor 1,3,5-cyclootatriene undergo this reaction.) Oxidation with cad-mium(II) chloride gives cyclooctatetraene in very good yields:[368]

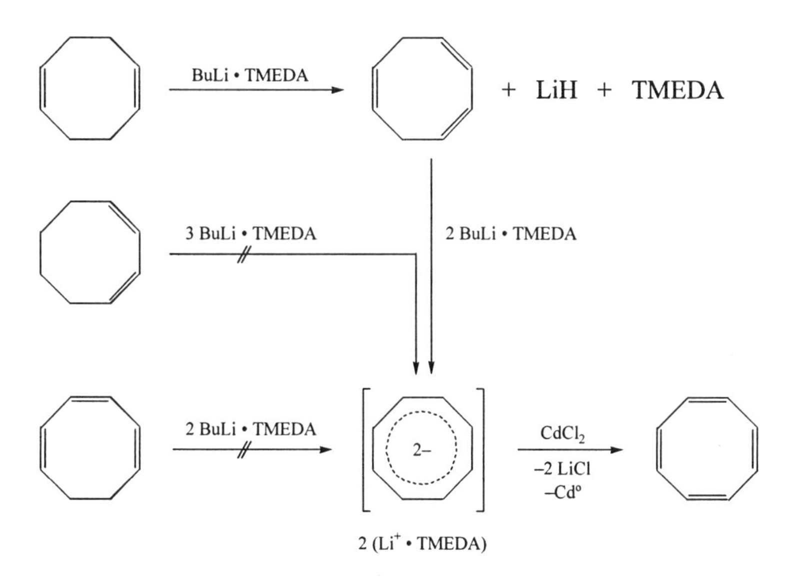

2.5.4 Fluorene and Cyclopentadiene Compounds. As is well known, flu-orene and cyclopentadiene show especially high CH acidity. Indeed, hydro-carbons containing these structural units twice can be dilithiated. Examples are *bis*(cyclopentadienyl)methane[369,370] and its octamethyl derivative[371] as well as the 1,*n*-*bis*(9-fluorenyl)alkanes,[372,373] the methano-bridged tetraphenylene,[374] and also certain fluorenophanes.[375]

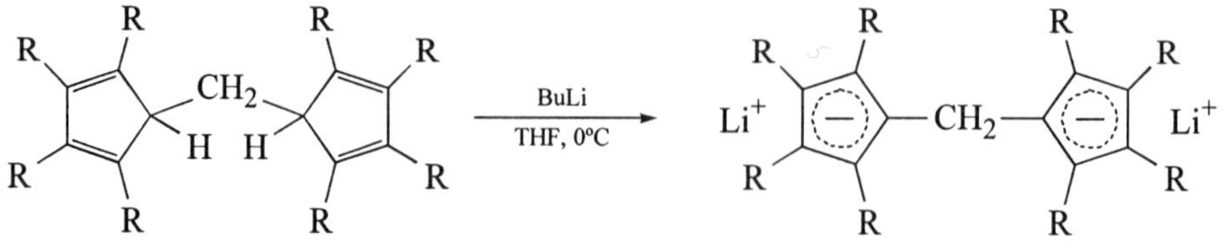

BuLi
THF, 0°C

R = H; CH_3

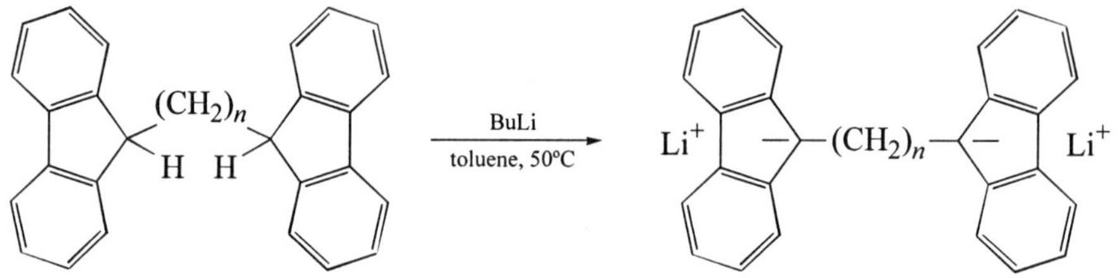

BuLi
toluene, 50°C

n = 2; 4; 6

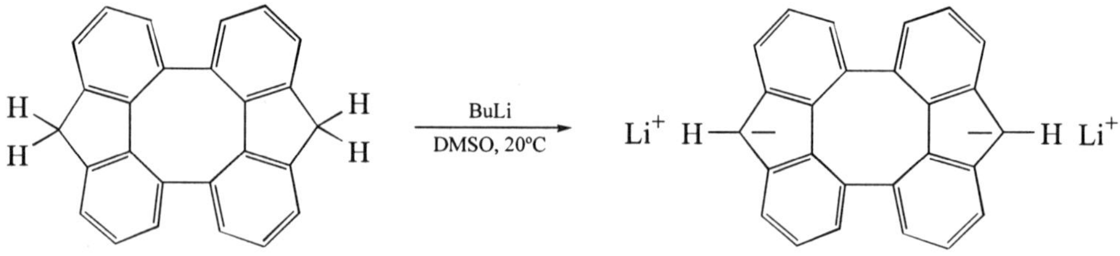

BuLi
DMSO, 20°C

Some more recent examples[228,264,370,376–385] follow:

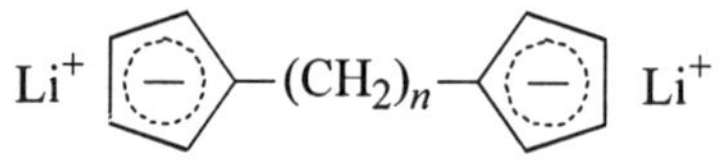

n = 2–5

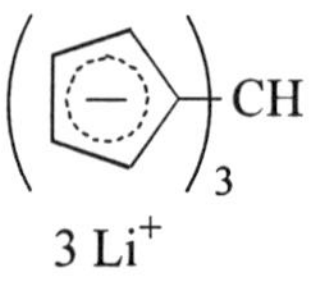

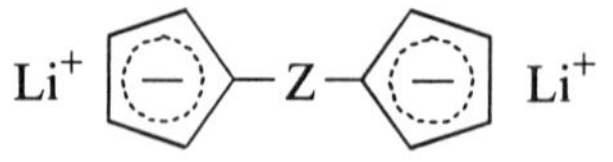

3 Li^+

Z = CMe_2; $SiMe_2$; $MeSiCH_2CH_2SiMe_2$

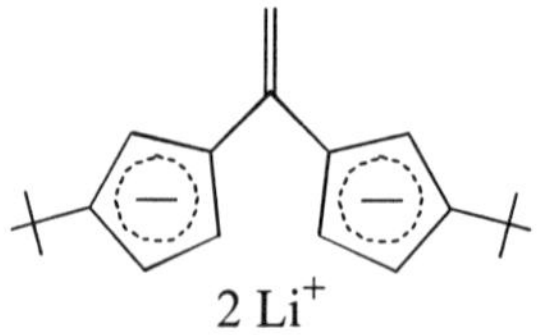

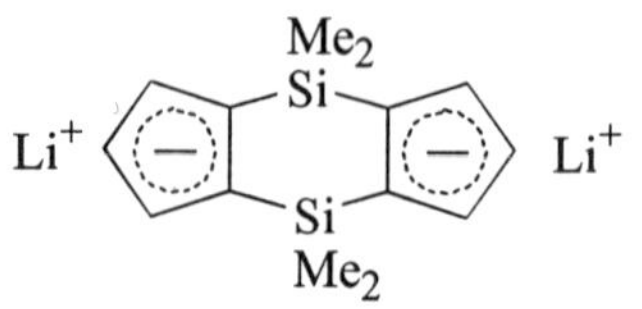

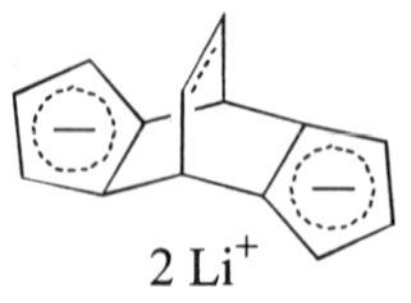

2 Li^+

2 Li^+

According to dynamic ^{1}H NMR spectroscopy, the 7.6 ($\pm$ 0.5) kcal/mol barrier to internal rotation in the *tris*(cyclopentadienyl)methane trianion is almost identical with that for the triphenylmethyl anion (8 kcal/mol).[264]

When dissolved in toluene, the 1,*n*-*bis*(9-lithio-9-fluorenyl)alkanes (n = 2, 4 or 6), are present as intramolecular aggregates. These can be broken up by addition of tetrahydrofuran or tetrahydropyran to form solvated tight ion pairs as well as loose ion pairs; the intramolecular aggregates are more stable for n = 2.[372] The systems with n = 2 and 6 could even been crystallized with eight THFs per complex.[373] X-ray analysis reveals solvent separated ion pairs with each lithium solvated tetrahedrally by four THF molecules. The planar fluorenyl units are parallel to each other.[373]

There is NMR-spectroscopic evidence for a ferrocene-bridged lithium *bis*(vinylcyclopentadienyl)lithiate complex in THF solution with two differently bonded lithiums.[380] Two different dynamic processes were detected. The "ate-bound" lithium ion exchanged with the solvent-separated lithium species. The hydrocarbon decks rotated intramolecularly; this involved cleavage of the "ate complex" in the transition state.[380]

Even a Li-Cp-Li-Cp super sandwich has been synthesized and characterized structurally by X-ray crystallography.[385] Again there are two modes of coordination for lithium ions. One lithium ion is $\eta^5:\eta^5$ coordinated, sandwich-fashion, to two cyclopentadienyl rings of the substituted fulvalene ligands. The second lithium ion is η^5-bonded to a ligand cyclopentadienyl ring and to two THF molecules.[385] A recent paper supporting the existence of carbanion triplet structures of dilithium 2,2′- and 1,2′-biindenide (as evidenced by NMR spectroscopy and MNDO/AMI calculations[386]) is pertinent.

The dilithium salt of 9,9′-bifluorene[387] can be oxidized to *bis*-fluorenylidene:[388]

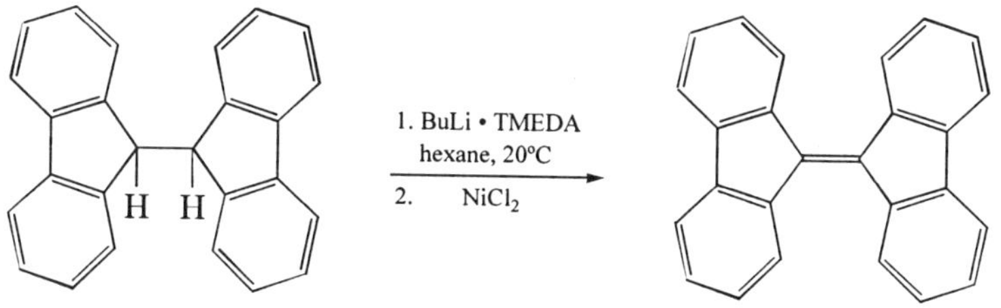

H H

1. BuLi • TMEDA
hexane, 20°C
2. $NiCl_2$

Oxygen gave similar results:[376]

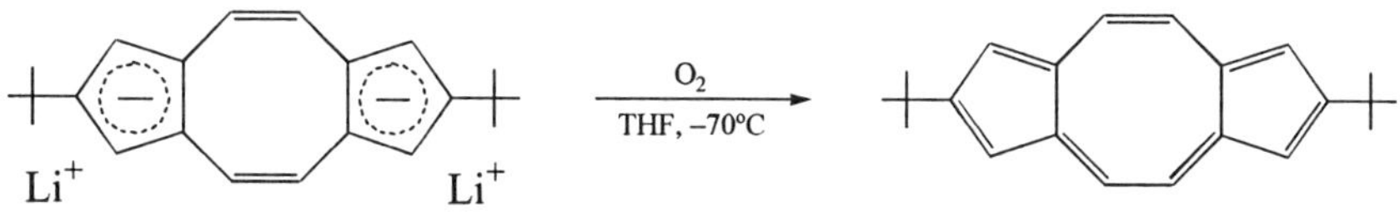

Li^+ Li^+

O_2
THF, –70°C

The Cu(II)-induced oxidation of 1,*n*-di(cyclopentadienyl)alkanediides strongly favors intermolecular over intramolecular coupling. The yields of intramolecular coupling products decrease strongly from 7% (n = 2) to 1% (n = 3) to traces (n = 4, 5). Intramolecular coupling is enhanced considerably when the H atoms of the CH_2CH_2 bridge are replaced by methyl groups:[228]

Li^+ ⊖–$(CH_2)_n$–⊖ Li^+ —$CuCl_2$→

n = 2: 92% 7%
n = 3: 82% 1%
n = 4: 82% traces
n = 5: 87% traces

Li^+ Li^+ —$CuCl_2$→

The preparation of dilithium pentalenide from dihydropentalene is noteworthy:[389,390]

—BuLi • TMEDA / hexane, 20°C→ 2 (Li^+ • TMEDA)

Li

Li

X-ray analyses confirmed the MNDO predictions[391] that the two lithium atoms occupy different sides of the aromatic 10 π-electron system. The dianion can be converted into the radical anion of pentalene photolytically.[392]

2.5.5 Alkynes and Allenes. The reactivity of alkynes towards butyllithium depends critically on the position of the triple bond.[303b,393] After the initial lithiation, terminal acetylenes can be metallated further quite readily. Propyne[394,395] and 1,3-pentadiyne[396] could even be perlithiated to C_3Li_4 and C_5Li_4. According to MNDO calculations,[2] both these "lithiocarbons"[318] favor bridged structures.

$$CH_3C{\equiv}CH \xrightarrow[\text{hexane, 20°C}]{\text{BuLi}} C_3Li_4$$

$$CH_3C{\equiv}C{-}C{\equiv}CH \xrightarrow[\text{hexane, 20°C}]{\text{BuLi}\cdot\text{TMEDA}} C_5Li_4$$

Propyne reacts with *n*-butyllithium in stages. The propargylide (with one σ- and one π-bonded lithium) formed in the second step also can be prepared from allene in hexane/THF (1:1) at −50°C:[397]

$$CH_3C{\equiv}CH \longrightarrow CH_3C{\equiv}CLi \longrightarrow H_2C{\cdots}C{\cdots}CLi\ Li^+ \longleftarrow CH_2{=}C{=}CH_2$$

$$\longrightarrow HC{\cdots}C{\cdots}CLi\ 2\,Li^+ \longrightarrow LiC{\cdots}C{\cdots}CLi\ 2\,Li^+$$

According to computational results,[398,399] all the polylithioalkynes have bridging lithiums.

Starting with propyne, the lithiation can be halted at the dilithio stage by using two equivalents of *n*-butyllithium and one equivalent of TMEDA in hexane/diethyl ether solution at −60'°C.[400] The longer chain and substituted[401] 1-alkynes react similarly; e.g., with 3-methyl-1-butyne[402] the metallation stops at the propargylide step as expected:

$$(CH_3)_2CHC{\equiv}CH \xrightarrow[\text{hexane, reflux}]{\text{BuLi}} (CH_3)_2C{\cdots}C{\cdots}CLi\ Li^+$$

Under these conditions, only dimetallation takes place even with 1-butyne[403] and higher 1-alkynes.[404–407] The same dilithiobutyne also was obtained from 1,2-butadiene:[408]

$$CH_3CH_2C{\equiv}CH \xrightarrow[\text{hexane, reflux}]{\text{BuLi}} CH_3CH{\cdots}C{\cdots}CLi\ Li^+ \xleftarrow[\text{hexane}]{\text{BuLi}} CH_3CH{=}C{=}CH_2$$

Only the more reactive tert-butyllithium introduces a third lithium:[395]

$$CH_3CH_2C\equiv CH \xrightarrow[\text{pentane, reflux}]{\text{t-BuLi}} \overbrace{CH_3C\cdots C\cdots CLi}^{2-} \quad 2\,Li^+$$

The two well-separated C,C-triple bonds of 1,6-heptadiyne react independently to yield the following products:[409]

$$HC\equiv C(CH_2)_3C\equiv CH \xrightarrow[Et_2O]{BuLi} LiC\equiv C(CH_2)_3C\equiv CLi \; +$$

$$\overbrace{LiC\cdots C\cdots CH}^{-}(CH_2)_2C\equiv CLi \;\; Li^+ \; + \; \overbrace{LiC\cdots C\cdots CH}^{-}CH_2\overbrace{CH\cdots C\cdots CLi}^{-} \;\; Li^+ \;\; Li^+$$

The reaction can be stopped at the terminal dilithium stage in THF solution at $-78\,°C$:[410–412]

$$(H)(HC\equiv C)C{=}C(H)(C\equiv CH) \xrightarrow[\text{THF, }-78°C]{\text{2 BuLi}} (H)(LiC\equiv C)C{=}C(H)(C\equiv CLi)$$

$$(H)(HC\equiv CCH_2)C{=}C(H)(CH_2C\equiv CH) \xrightarrow[\text{above}]{\text{as}} (H)(LiC\equiv CCH_2)C{=}C(H)(CH_2C\equiv CLi)$$

$$CH_2{=}C(CH_2C\equiv CH)_2 \xrightarrow[\text{above}]{\text{as}} CH_2{=}C(CH_2C\equiv CLi)_2$$

The preparation of the following terminal dilithiodiynes[413,414] is straightforward:

$$o\text{-}C_6H_4(C\equiv CH)_2 \xrightarrow[\text{THF, }-10°C]{\text{2 BuLi}} o\text{-}C_6H_4(C\equiv CLi)_2$$

$$HC\equiv C-SiR_2-C\equiv CH \xrightarrow[\text{THF, 20°C}]{\text{2 BuLi}} LiC\equiv C-SiR_2-C\equiv CLi$$

R = Me; Ph

Starting with 4-methoxy-3-butene-1-yne, elimination of methanol takes place under the metallation conditions: the dilithium salt of butadiyne is the final product:[415,416]

$$\mathrm{MeOCH{=}CHC{\equiv}CH} \xrightarrow[\text{THF, }-25^\circ\text{C}]{\text{2 BuLi}} \mathrm{HC{\equiv}C{-}C{\equiv}CLi}$$

$$\xrightarrow{\text{BuLi}} \mathrm{LiC{\equiv}C{-}C{\equiv}CLi}$$

Dimerization occurs when 3-pentene-1-yne and its derivatives are treated with *n*-butyllithium in a mixture of THF/hexane:[417]

$$\mathrm{CH_3CH{=}CHC{\equiv}CH} \xrightarrow[\text{THF, hexane, }0^\circ\text{C}]{\text{BuLi}} \mathrm{LiCH_2CH{=}CHC{\equiv}CLi}$$

$$\downarrow$$

$$\mathrm{LiC{\equiv}CCH_2\underset{\displaystyle CH_3}{\underset{|}{C}}HCH_2CH{=}CHC{\equiv}CLi}$$

According to the following mechanism, the dimerization only requires catalytic amounts of butyllithium:[418]

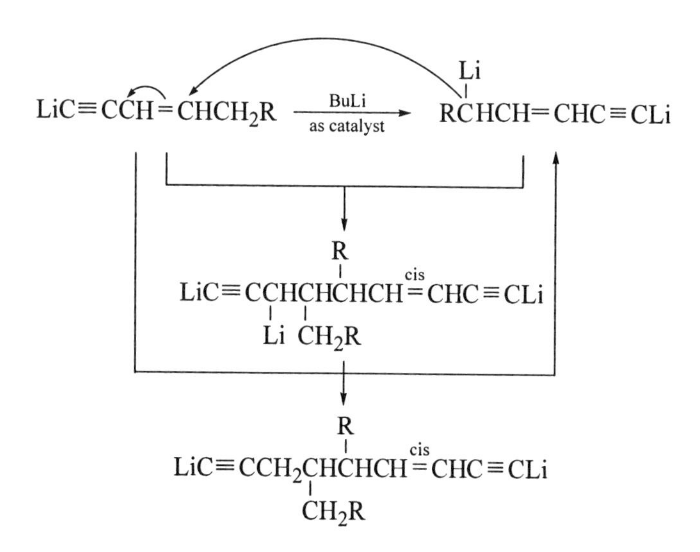

R = H; Me; OMe; SMe; NMe_2

The *Z*-configuration predominates in the dimers obtained after aqueous work-up. Lithiated acetylenes add analogously to the double bond of lithiated enynes:[418]

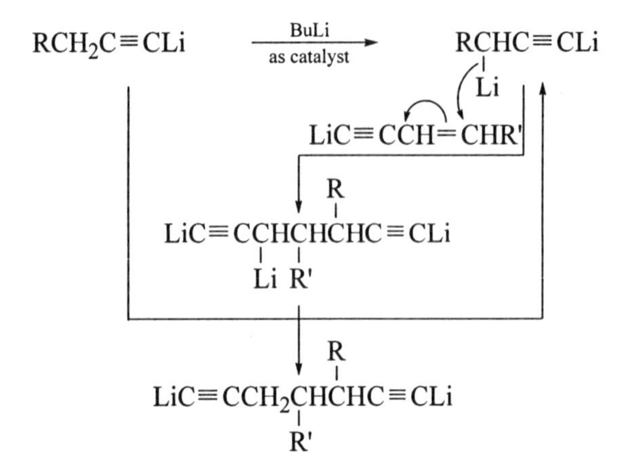

R = Ph; Pr; OMe; SMe; NMe_2
R' = H; Et

Alkynes with internal triple bonds also can be polylithiated, but with more difficulty. Interestingly, detachment of two propargylic protons of a single methylene group is favored even when propargylic protons are present on the other side of the triple bond. This results in 2-propynyl dianions with "sesquiacetylene" structures:[163,164]

$$RC\equiv CCH_2R' \longrightarrow \overbrace{RC\equiv C = CHR'}^{-}\ Li^+ \longrightarrow \overbrace{RC \equiv C \equiv CR'}^{2-}\ 2\,Li^+$$

The dilithiation of 1-phenyl-1-butyne[419,420] and the introduction of four lithium atoms into 1,8-cyclotetradecadiyne with two isolated triple bonds[421] provide examples:

$$PhC\equiv CCH_2CH_3 \xrightarrow[Et_2O,\ 25°C]{BuLi} \overbrace{PhC \equiv C \equiv CCH_3}^{2-}\ 2\,Li^+$$

$$\text{cyclo-}[CH_2C\equiv C(CH_2)_4CH_2C\equiv C(CH_2)_4] \xrightarrow[THF,\ 0°C]{BuLi} \text{cyclo-}[\overbrace{C \equiv C \equiv C}^{2-}(CH_2)_4\overbrace{C \equiv C \equiv C}^{2-}(CH_2)_4]\ 4\,Li^+$$

The especially stable sesquiacetylene structure is comprised of two equivalent orthogonal four-electron three-center bonds with symmetrical overlap.

The localized negative charges of the C_3 fragment only interact weakly with other groups capable of resonance;[163] the doubly bridged lithiums undoubtedly contribute strongly to this behavior.

$$PhC{\equiv}CCH_2CR{=}CHR' \xrightarrow[Et_2O]{BuLi} \overbrace{PhC{\cdots}C{\cdots}C}^{2-}CR{=}CHR' \quad 2\,Li^+$$

R = R' = H
R = H; R' = Me
R = Me; R' = H
R = H; R' = Ph

In contrast, the sesquiacetylenes from 1-alkynes with a propargylic methyl group—contrary to the literature[422,423]—rearrange via 1,3-hydrogen shift to the even more stable propargylide. The final products are the same as those from the corresponding 1-alkynes.[167]

$$RC{\equiv}CCH_3 \longrightarrow \overbrace{RC{\cdots}C{\cdots}CH_2}^{-}\ Li^+ \longrightarrow$$

$$\overbrace{RC{\cdots}C{\cdots}CH}^{2-}\ 2\,Li^+ \longrightarrow \overbrace{RC{\cdots}C{\cdots}CLi}^{2-}\ 2\,Li^+$$

$$\overbrace{RC{\cdots}C{\cdots}CH}^{2-}\ 2\,Li^+ \xrightarrow{1,3\text{-H}} \overbrace{RCH{\cdots}C{\cdots}CLi}^{-}\ Li^+ \longrightarrow \overbrace{RC{\cdots}C{\cdots}CLi}^{2-}\ 2\,Li^+$$

$$RCH_2C{\equiv}CH \longrightarrow \overbrace{RCH{\cdots}C{\cdots}CLi}^{-}\ Li^+$$

The reverse reaction, i.e., the rearrangement of the simplest propargylide (R=H) into the corresponding sesquiacetylene, could be excluded by isotopic labeling[424] (according to ab initio calculations,[398] the energy difference is 12.9 kcal/mol):

$$^{13}CH_3C{\equiv}CH \xrightarrow{BuLi} \overbrace{H_2{}^{13}C{\cdots}C{\cdots}CLi}^{-}\ Li^+ \xrightarrow[\not\ ]{1,3\text{-H}} \overbrace{H\overset{*}{C}{\cdots}C{\cdots}\overset{*}{C}H}^{2-}\ 2\,Li^+$$

The reaction of butyllithium with 2-butyne[402] and with 1-phenyl-1-propyne[318,419,423] exemplify the polymetallation of 2-alkynes:

$$CH_3C{\equiv}CCH_3 \xrightarrow[TMEDA]{BuLi} \overbrace{CH_3C{\cdots}C{\cdots}CLi}^{2-} \; 2\,Li^+$$

$$PhC{\equiv}CCH_3 \xrightarrow[Et_2O]{BuLi} \overbrace{PhC{\cdots}C{\cdots}CLi}^{2-} \; 2\,Li^+$$

According to the literature,[423] dimetallation of 1-phenyl-1-propyne yields a sesquiacetylene (I) different from the propargylide (II) derived from 3-phenyl-1-propyne. Both did not interconvert (two different ^{1}H NMR spectra). Since the reactions with electrophiles from each yield the same reaction products (III), it was assumed initially that (I) underwent a 1,3-hydrogen shift during derivatization:

$$PhC{\equiv}CCH_3 \xrightarrow[Et_2O{-}d_{10}]{2\,BuLi} \overbrace{PhC{\cdots}C{\cdots}CH}^{2-} \; 2\,Li^+ \; (\textbf{I}, \delta = 2.96) \xrightarrow{E^+} \overbrace{PhC(E){\cdots}C{\cdots}CH}^{-} \; Li^+ \xrightarrow{1,3\text{-}H} PhCH(E)C{\equiv}CLi \xrightarrow{E^+} PhCH(E)C{\equiv}CE \; (\textbf{III})$$

$$\textbf{I} \not\rightleftharpoons \textbf{II}$$

$$PhCH_2C{\equiv}CH \xrightarrow[Et_2O{-}d_{10}]{2\,BuLi} \overbrace{PhCH{\cdots}C{\cdots}CLi}^{-} \; Li^+ \; (\textbf{II}, \delta = 3.30) \xrightarrow{2\,E^+} PhCH(E)C{\equiv}CE \; (\textbf{III})$$

$E = D; Me_3Si$

However, our reinvestigation[167] showed that the 1,3-hydrogen shift really occurs during dianion formation. The ^{1}H NMR signals of the propargylide (II) could hardly be detected due to low solubility. The two different ^{1}H NMR spectra assigned[423] to (I) and (II) correspond instead to the two corresponding monoanion salts:

$$PhC{\equiv}CCH_3 \xrightarrow[Et_2O\text{-}d_{10}]{BuLi} \overbrace{PhC{\equiv}C{=}CH_2}^{-}\ Li^+\ (\delta = 2.92) \xrightarrow[Et_2O\text{-}d_{10}]{BuLi} \overbrace{PhC{\equiv}C{\equiv}CH}^{2-}\ 2\,Li^+\ \mathbf{I}$$

$$\mathbf{I} \xrightarrow{1,3\text{-}H} \mathbf{II}$$

$$PhCH_2C{\equiv}CH \xrightarrow[Et_2O\text{-}d_{10}]{BuLi} PhCH_2C{\equiv}CLi\ (\delta = 3.31) \xrightarrow[Et_2O\text{-}d_{10}]{BuLi} \overbrace{PhCH{=}C{\equiv}CLi}^{-}\ Li^+\ (\delta = 3.05\ \text{(very broad)})$$

II (badly soluble)

$$\mathbf{II} \xrightarrow[Et_2O\text{-}d_{10}]{BuLi} \overbrace{PhC{\equiv}C{\equiv}CLi}^{2-}\ 2\,Li^+$$

$$PhCH{=}C{=}CH_2 \xrightarrow[Et_2O\text{-}d_{10}]{BuLi} \overbrace{PhCH{=}C{\equiv}CH}^{-}\ Li^+ \xrightarrow[Et_2O\text{-}d_{10}]{BuLi} \mathbf{II}$$

(could not be observed)

According to a recent paper,[425] a similar rearrangement involving a 1,3-silyl shift takes place upon dilithiation of 1,3-bis(trimethylsilyl)propyne:

$$Me_3SiC{\equiv}CCH_2SiMe_3 \xrightarrow[THF]{2\,BuLi}$$

$$\overbrace{Me_3SiC{\equiv}C{\equiv}CSiMe_3}^{2-}\ 2\,Li^+ \rightleftharpoons \overbrace{(Me_3Si)_2C{=}C{\equiv}CLi}^{-}\ Li^+$$

Additional ring metallation takes place with phenyl-substituted acetylenes. Thus, the ortho- and para-phenyl hydrogens of 1-phenyl-1-propyne are replaced by lithium.[318] The second metallation of phenylacetylene takes place primarily ($>80\%$) in the ortho position; comparable amounts of the meta- and para-compounds are formed but only 15% together.[426] Ortho metallation occurs exclusively in the absence of TMEDA, but good yields are obtained only with sec-butyllithium.[81]

$$C_6H_5C{\equiv}CH \xrightarrow[pentane]{s\text{-}BuLi} o\text{-}Li\text{-}C_6H_4C{\equiv}CLi$$

Three or four lithium atoms can be introduced into 2,4-hexadiene, depending on the conditions:[409,427]

$$CH_3C{\equiv}C-C{\equiv}CCH_3 \xrightarrow[\text{hexane}]{\text{BuLi}} CH_3C_5Li_3$$

$$CH_3C{\equiv}C-C{\equiv}CCH_3 \xrightarrow[\text{Et}_2\text{O or THF}]{\text{BuLi}} C_6H_2Li_4$$

As expected, dicyclopropylacetylene[428] and 2,2,8,8-tetramethyl-3,6-nonadiyne[3] can only be lithiated twice:

$$\text{c-C}_3\text{H}_5-C{\equiv}C-\text{c-C}_3\text{H}_5 \xrightarrow[\text{hexane, 20°C}]{\text{BuLi} \cdot \text{TMEDA}} \text{c-C}_3\text{H}_4(Li)-C{\equiv}C-\text{c-C}_3\text{H}_4(Li)$$

$$\text{t-BuC}{\equiv}\text{CCH}_2\text{C}{\equiv}\text{C-t-Bu} \xrightarrow{\text{BuLi}} \text{t-BuC}{\equiv}\text{CC(Li)}_2\text{C}{\equiv}\text{C-t-Bu}$$

The latter dilithium compound exists in the crystal as a complicated tetrameric complex.[3] The following 1,2,6,7-octatetraene-4-ynes and 1,2,8,9-decatetraene-4,6-diynes are only dilithiated with n-butyllithium in THF at $-78°C$:[429]

$$R^1R^2C{=}C{=}CH-(C{\equiv}C)_n-CH{=}C{=}CR^1R^2 \xrightarrow[\text{THF, }-78°\text{C}]{\text{2 BuLi}}$$

$$R^1R^2C{=}C{=}C(Li)-(C{\equiv}C)_n-C(Li){=}C{=}CR^1R^2$$

$n = 1; 2.$ $R^1 = R^2 = \text{Et}.$ $R^1 = \text{Me}; R^2 = \text{Et}$

These structural representations showing only one tautomer, of course, are oversimplified. In reality, the lithiums are not σ-bonded, and cannot be assigned to a single carbon atom. A better but more complicated formulation would be a resonance stabilized carbanion with the lithium cations bridging between (or among) the centers with the highest negative charges. Although tautomers (or tautomeric mixtures) are often formulated in the literature, there is not a single example where two tautomers, presumably in equilibrium, could be detected individually.

2.6 Combined Methods

2.6.1 Halogen–Metal Exchange Followed by Metal–Metal-Exchange. Although 1-bromo-2-[(trimethylstannyl)methyl]benzene only reacted with *n*-butyllithium in THF at $-70°C$ to give tin–lithium transmetallation, halogen–lithium exchange occurred with tert-butyllithium in diethyl ether at $-80°C$. α,2-Dilithiotoluene could be prepared by continuing the reaction in diethyl ether at room temperature or in THF solution at $-80°C$:[430]

SnMe$_3$ Br — t-BuLi, Et$_2$O, –80°C → SnMe$_3$ Li — t-BuLi, Et$_2$O, 25°C or THF, –80°C → Li Li

BuLi, THF, –70°C → Li Br

SnMe$_3$ Li$^+$ → Li SnMe$_3$

2.6.2 Halogen–Metal Exchange Followed by Metallation. Halogen–lithium exchange is often followed by the introduction of a second lithium atom by hydrogen exchange. This is steered by the "agostic" interaction of the lithium already present[59,244,405,431,432] e.g.,

N Br — 2 BuLi, Et$_2$O, 0°C → N Li Li

Br — BuLi, Et$_2$O/hexane, –20° → 10°C → Li + BuBr

Li — BuLi • TMEDA, hexane, reflux → Li Li

The butyl bromide formed in the first step should be removed before proceeding, in order to avoid Wurtz-coupling side reactions.

2.6.3 Metallation Followed by Halogen–Metal Exchange. Metallation probably precedes bromine–lithium exchange in the following example:[433,434]

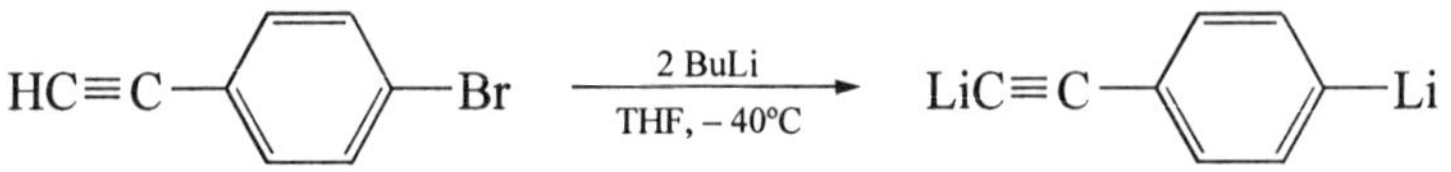

$$HC{\equiv}C{-}C_6H_4{-}Br \xrightarrow[\text{THF, }-40^\circ C]{\text{2 BuLi}} LiC{\equiv}C{-}C_6H_4{-}Li$$

2.6.4 Metallation Followed by Metal–Metal Exchange. A tinorganic intermediate is recommended for the regio- and stereoselective lithiation of diallyl amine via the reaction sequence below:[435]

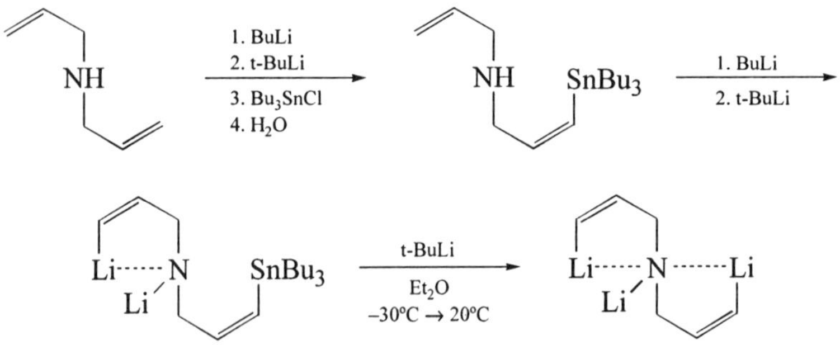

1. BuLi 2. t-BuLi 3. Bu_3SnCl 4. H_2O → NH ... $SnBu_3$; 1. BuLi 2. t-BuLi →

Li······N (Li) ... $SnBu_3$ —t-BuLi, Et_2O, −30°C → 20°C→ Li······N······Li (Li)

2.6.5 Metallation Followed by Addition of a Lithiumorganic Compound. The sequential reaction of diallyl amines with alkyllithium reagents involves, as the last step, regioselective addition of a lithiumorganic compound to one of the C,C-double bonds:[436–439]

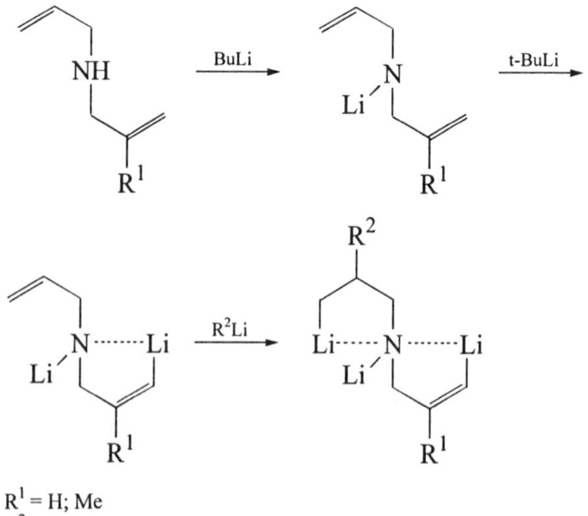

NH (R^1) —BuLi→ N–Li (R^1) —t-BuLi→

N·······Li (Li) (R^1) —R^2Li→ (R^2) Li······N······Li (Li) (R^1)

R^1 = H; Me
R^2 = Et; Bu; t-Bu

In the next examples, the addition also follows the metallation step:[286,440]

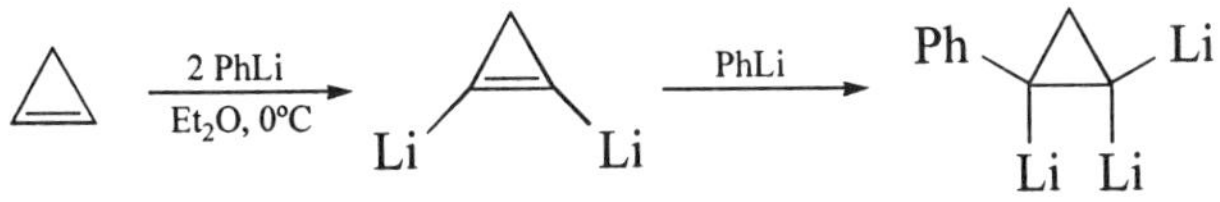
2 PhLi
Et2O, 0°C
Li
Li
PhLi
Ph
Li
Li Li

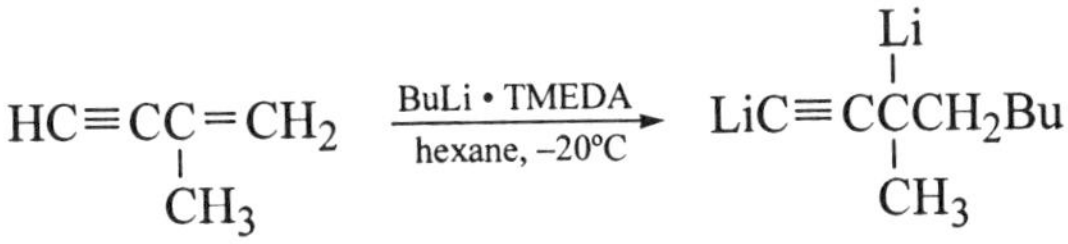
HC≡CC=CH2
CH3
BuLi • TMEDA
hexane, –20°C
Li
LiC≡CCCH2Bu
CH3

2.6.6 Addition of a Lithiumorganic Compound Followed by Metallation. The reaction of *n*-butyllithium with diphenylacetylene (tolane) yields a dilithio product via a stepwise addition-metallation sequence.[441–443] The yield is especially good in the presence of TMEDA.[222,444,445]

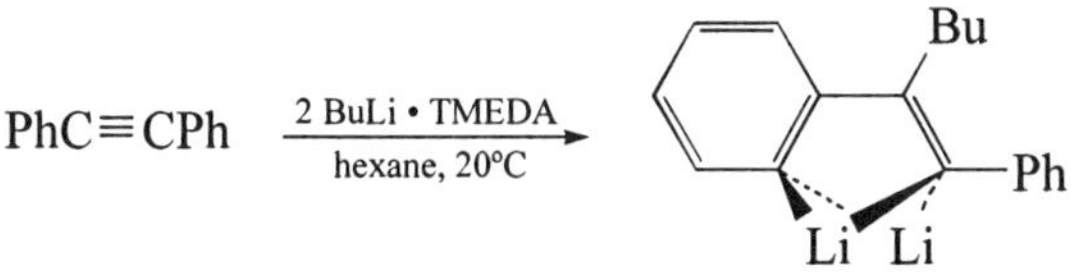
PhC≡CPh
2 BuLi • TMEDA
hexane, 20°C
Bu
Ph
Li Li

In THF, one equivalent of *n*-butyllithium adds to tolane to give exclusively the *trans* monolithio compound (monomeric in THF).[222] Thorough characterization by two-dimensional NMR spectroscopy also revealed, in agreement with MNDO predictions, the agostic activation of the second metallation site.[222] The X-ray analysis structure of the dilithio product shows a doubly lithium bridged monomer chelated with one TMEDA ligand per lithium atom. This structure is retained in benzene-d_6 solution, while in THF-d_8 a monomer–dimer equilibrium is observed.[222]

The addition of tert-butyllithium to phenylisonitrile also is followed by ortho-metallation:[446,447]

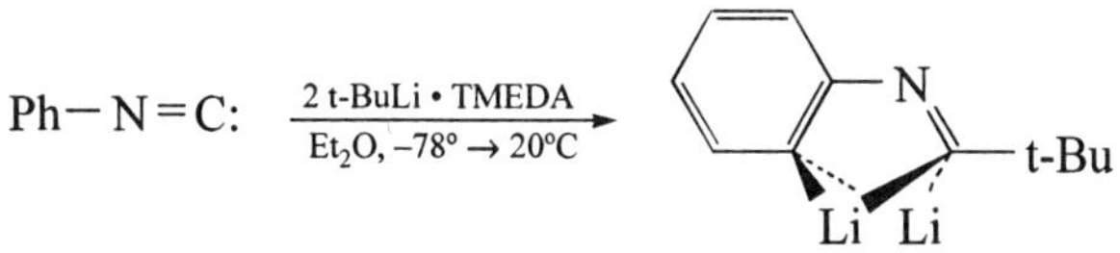
Ph−N=C:
2 t-BuLi • TMEDA
Et2O, –78° → 20°C
N
t-Bu
Li Li

The dianions obtained from cyclooctatetraene and organolithiums[448,449] can be oxidized by air[448] or by cadmium chloride[449] to the corresponding monosubstituted cyclooctatetraene derivatives:

$$\text{C}_8\text{H}_8 \xrightarrow[\text{Et}_2\text{O, 20°C}]{\text{RLi}} [\text{C}_8\text{H}_8\text{R}]^- \text{Li}^+ \xrightarrow[-\text{RH}]{\text{RLi}} [\text{C}_8\text{H}_7\text{R}]^{2-}\ 2\,\text{Li}^+ \xrightarrow[\text{CdCl}_2]{\text{O}_2 \text{ or}} \text{C}_8\text{H}_7\text{R}$$

R = Me; Bu; s-Bu; t-Bu; Ph; $PhCH_2$

n-Butyllithium and phenyllithium also react with cyclooctatetraene in the presence of TMEDA,[449] and TMEDA activation is required for the reaction of methyllithium.[448]

2.7 Special Methods

2.7.1 Pyrolysis Reactions. In 1955 Ziegler and co-workers[450] discovered that pyrolysis of *halide-free* methyllithium gives dilithiomethane (and methane) in excellent yields:

$$2\,CH_3Li \xrightarrow{225°C} CH_4 + CH_2Li_2$$

This disproportionation reaction, according to model ab initio computations, proceeds endothermically through a polar hydrogen-transfer mechanism within a tetrameric methyllithium aggregate:[451]

$$(CH_3Li)_4 \xrightarrow{225°C} CH_4 + CH_2Li_2 \bullet (CH_3Li)_2$$

The free energy of reaction, ΔG = 19.6 kcal/mol, and an activation energy, ΔG = 53.8 kcal/mol at 500 K are calculated.[451] However, pyrolysis of an equal molar mixture of $(CD_3Li)_4$ and $(CH_3Li)_4$ ruled out an entirely intramolecular mechanism.[452]

The Ziegler procedure[450]—improved by Lagow[453–455]—still is the best method for preparing dilithiomethane, although temperature control is important. At temperatures higher than 230°C decomposition to hydrogen (H_2), lithium hydride (LiH) and dilithioacetylene (lithium carbide, C_2Li_2) occurs. Pyrolysis at different temperatures and reaction times revealed perlithiopropyne (C_3Li_4) to be an intermediate:[453,454]

$$CH_2Li_2 \xrightarrow[6\ min]{350°C} \underset{20\%}{C_3Li_4} + \underset{80\%}{C_2Li_2}$$

$$CH_2Li_2 \xrightarrow[10\ min]{400°C} \underset{99\%}{C_2Li_2}$$

$$C_3Li_4 \xrightarrow[3\ min]{300°C} \underset{40\%}{C_3Li_4} + \underset{60\%}{C_2Li_2}$$

$$C_3Li_4 \xrightarrow[10\ min]{350°C} \underset{100\%}{C_2Li_2}$$

Dilithioacetylene (C_2Li_2) also is the end product of the thermal decomposition of tetralithiomethane (CLi_4). In addition to perlithiopropyne (C_3Li_4), perlithioethylene (C_2Li_4) also was indicated to be an intermediate by trapping experiments:[454]

$$CLi_4 \xrightarrow[2\ min]{225°C} \underset{60\%}{CLi_4} + \underset{20\%}{C_2Li_4} + \underset{20\%}{C_3Li_4}$$

$$CLi_4 \xrightarrow[8\ min]{225°C} \underset{20\%}{CLi_4} + \underset{30\%}{C_2Li_4} + \underset{40\%}{C_3Li_4} + \underset{10\%}{C_2Li_2}$$

$$CLi_4 \xrightarrow[10\ min]{225°C} \underset{100\%}{C_2Li_2}$$

Nevertheless, polylithioalkanes are stable in the gas phase for a short period of time prior to pyrolytic decomposition. By using special flash-vaporization techniques (''Flash Vaporization Mass Spectrometry''), mass spectra have been obtained even of compounds which have no observable vapor pressure below 650°C in very high vacuum.[453,454] Temperatures of 1500°C are reached in less than three seconds, and e.g., dilithiomethane could be transported over a distance of 10 cm with less than 10% decomposition[453,454]

$$CH_2Li_2 \xrightarrow[2\ s]{1500°C} \underset{90\%}{CH_2Li_2} + \underset{2\%}{C_3Li_4} + \underset{8\%}{C_2Li_2}$$

On the other hand, attempted mass spectroscopy of CLi_4 using the FAB-technique, failed.[456] Nevertheless, CLi_4 has been detected in the gas phase along with CLi_3 by heating crystalline dilithioacetylene (C_2Li_2) in a Knudsen cell to 725°C.[457] Reinvestigation in 1992[458,459] revealed, in addition, traces of hexalithiomethane (CLi_6). The considerable stability of this hyperlithiated methane had been prophesied by ab initio computations.[460] While CLi_5 could not be detected in these experiments,[458] it also was predicted to be a viable species in isolation.[460] These results cast some doubt on Lagow's assignments of stoichiometry based, e.g., on the hydrolysis products (see below). Both CLi_6 and CLi_5, as well as CLi_4, can be expected to give CD_4 with D_2O.

The Ziegler procedure has a major disadvantage. It cannot be used for the preparation of sensitive polylithiumorganic compounds, especially when elimination of lithium hydride is possible. Thus, this technique was not successful with cyclopropyllithium, but it worked excellently to give 1,1-dilithio-2,2,3,3-tetramethylcyclopropane.[455]

$$2\ (Me_2C)_2CHLi \xrightarrow[12\ h]{170^\circ C} (Me_2C)_2CH_2 + (Me_2C)_2CLi_2$$

Three additional geminal dilithiumorganic compounds have been prepared this way:[461]

$$2\ Me_3CCH_2Li \xrightarrow{180^\circ C} Me_4C + Me_3CCHLi_2$$

$$2\ Me_3SiCH_2Li \xrightarrow{150^\circ C} Me_4Si + Me_3SiCHLi_2$$

$$2\ (Me_3Si)_2CHLi \xrightarrow{160^\circ C} (Me_3Si)_2CH_2 + (Me_3Si)_2CLi_2$$

However, 9-lithiofluorene gave considerable amounts of dimer and trimer products along with 9,9-dilithiofluorene:[462]

180°C
72 h

Li H → H H + Li Li (15%) + Li Li (70%) + Li Li (15%)

The following reactions did not succeed. Either the ''wrong'' dilithiumorganic compound was formed[463,464] or dimerization with an unexplained loss of lithium took place:[463]

$$(CH_3)_2C{=}CHLi \not\longrightarrow (CH_3)_2C{=}CLi_2$$

$$\xrightarrow[24\ h]{140°C} \left[CH_2{=}C(CH_2)_2\right]^{2-}\ 2\ Li^+$$

$$H_2C{=}C(CH_3)CH_2Li \not\longrightarrow H_2C{=}C(CH_3)CHLi_2$$

$$Ph_2C{=}CHLi \xrightarrow[24\ h]{130°C} Ph_2C{=}CH{-}CH{=}CPh_2$$

$$Ph_2C{=}CHLi \not\longrightarrow Ph_2C{=}CLi_2$$

Intermolecular elimination of lithium hydride also may be a further complication. An example is the 4 h pyrolysis of halide-free benzyllithium at 150–170°C:[4,102]

$$2\ C_6H_5CH_2Li \not\longrightarrow C_6H_5CH_3 + C_6H_5CHLi_2$$

$$2\ C_6H_5CH_2Li \xrightarrow{-LiH} m\text{-}LiCH_2C_6H_4CH_2C_6H_5 \quad 75\%$$

$$2\ C_6H_5CH_2Li \xrightarrow{-LiH} o\text{-}LiCH_2C_6H_4CH_2C_6H_5 \xrightarrow{-LiH} \text{9,10-dihydroanthracene} \quad 10\%$$

While the final product of the pyrolysis of salt-free allyllithium is tetralithiopropyne (C_3Li_4), dilithiopropene can be detected as an intermediate under milder conditions (100–120°C). Hence, lithium hydride elimination takes place here only after the disproportionation:[4,102]

$$2\,CH_2{=}CHCH_2Li \longrightarrow CH_2{=}CHCH_3 + [CH_2{\cdots}CH{\cdots}CH]\,Li_2$$

$$[CH_2{\cdots}CH{\cdots}CH]\,Li_2 \xrightarrow{-LiH} [LiCH_2C{\equiv}CH] \longrightarrow CH_3C{\equiv}CLi \xrightarrow{-CH_2{=}CHCH_3} C_3Li_4$$

2.7.2 ***Reactions with Lithium Vapor.*** Lithium vapor is much more reactive than even the finest lithium dust. This was first shown in 1955 when dilithioacetylene (lithium carbide) was obtained from graphite in quantitative yield.[465]

In 1972 Lagow and co-workers[466,467] developed a general technique for preparing polylithiumorganic compounds by reacting organic substrates with lithium vapor. Lithium atoms at high temperature react with e.g., a hydrocarbon or a halocarbon substrate in vacuum. A special stainless steel reactor is employed. This consists of a Knudsen cell containing the lithium metal heated in a furnace, an inlet tube for the organic reactant, and a liquid nitrogen cold finger on which the product (along with unreacted lithium metal) condenses. Besides the high reactivity of lithium vapor, this nonsolution method has the advantages of reducing rearrangements and secondary reactions due to the rapid quenching of the products on the cold finger.

Tetralithiomethane (CLi_4) and hexalithioethane (C_2Li_6), the first perlithiated alkanes, were prepared by reacting lithium vapor with the halocarbon precursors between 800 and 1000°C:[466]

$$CCl_4 + 8\,Li\,(g) \xrightarrow{850°C} CLi_4 + 4\,LiCl$$

The main side products were reported to be C_2Li_4 and C_2Li_2. With partially halogenated hydrocarbons, not just the halogen atoms, but some and perhaps even all the hydrogens are replaced by lithium.[468,469] The selectivity could be improved by working at 750°C instead of 850°C.[470,471] In this manner, CH_2Li_2 and CLi_4 could be synthesized in 65.7 and 40.5% yields, respectively.[471] However, even at 750°C, the chloroform hydrogen was labile toward substitution by a lithium: $CHLi_3$ was obtained in only 15.5% yield; the main by-products were 19.1% CH_2Li_2, 20.1% CLi_4, and 39.7% C_2Li_2.[470,471] The more highly chlorinated chloromethanes give greater amounts of C_2 molecules (mainly lithium carbide). Thus even the "purest" CLi_4 contained 58.5% C_2Li_2.[471]

The most homogeneous C_2Li_6 was obtained by using diethylmercury instead of C_2Cl_6 as the starting material. Interestingly, $(C_2H_5)_4Sn$ and $(C_2H_5)_4Pb$ gave ethyllithium (C_2H_5Li) predominately.[472] The atomic lithium–carbon vapor reaction yielded C_3Li_4 as the main product.[473]

Both substitution of lithium for hydrogen and addition of lithium to the double bonds occurred when lithium vapor reacted with alkenes and benzene. Polylithiated alkanes and alkenes resulted.[474,475]

The trapping of the lithiocarbon product in the lithium–metal matrix is a major limitation of the Lagow procedure for preparative purposes. However, techniques have been developed to minimize this problem.[471] Unfortunately, it is neither possible—at least yet—to separate the mixtures of lithium-substituted products nor to characterize individual polylithiated species by X-ray analysis or spectroscopically. The stoichiometries of solid products (which contain excess lithium) have been assigned on the basis of the derivatives resulting from quenching experiments. But the C_1, C_2, or C_3 carbon fragments are likely to be embedded in a lithium ''matrix,'' and it is not clear what kinds of products such ''hyperlithiated'' species can be expected to give. While interesting enough, the observation of C_2H_4 (or C_2D_4) on hydrolysis, for example, does not necessarily mean that ''C_2Li_4'' was the true precursor.

REFERENCES

1. Schleyer, P. v. R. *Pure Appl. Chem.* **1983**, *55*, 355.
2. Schleyer, P. v. R. *Pure Appl. Chem.* **1984**, *56*, 151.
3. Setzer, W. N.; Schleyer, P. v. R. *Adv Organomet. Chem.* **1985**, *24*, 353.
4. Maercker, A.; Theis, M. *Top. Curr. Chem.* **1987**, *138*, 1.
5. West, R.; Rochow, E. G. *Naturwissenschaften* **1953**, *40*, 142; *J. Org. Chem.* **1953**, *18*, 1739.
6. Kuus, H. *Tartu Riikliku Ulik. Toim.* **1968**, *219*, 245; *C. A.* **1969**, *71*, 49155.
7. U.S. Patent 2 947 793 **1960**, Firestone Tire & Rubber Co., Inv.: Eberly, K. C. *C. A.* **1961**, *55*, 382.
8. Rautenstrauch, V. *Angew. Chem.* **1975**, *87*, 254; *Angew. Chem. Int. Ed. Engl.* **1975**, *14*, 259.
9. Bogdanović, B.; Wermeckes, B., *Angew. Chem.* **1981**, *93*, 691; *Angew. Chem. Int. Ed. Engl.* **1981**, *20*, 684.
10. Bogdanović, B. *Angew. Chem.* **1985**, *97*, 253; *Angew. Chem. Int. Ed. Engl.* **1985**, *24*, 262.
11. Maercker, A.; Grebe, B. *J. Organomet. Chem.* **1987**, *334*, C21.
12. van Eikema Hommes, N. J. R.; Bickelhaupt, F.; Klumpp, G. W. *Recl. Trav. Chim. Pays-Bas* **1987**, *106*, 514.
13. Seetz, J. W. F. L.; Schat, G.; Akkerman, O. S.; Bickelhaupt, F. *J. Am. Chem. Soc.* **1982**, *104*, 6848.
14. Freeman, P. K.; Hutchinson, L. L. *J. Org. Chem.* **1980**, *45*, 1924; **1983**, *48*, 4705.
15. van Eikema Hommes, N. J. R.; Bickelhaupt, F.; Klumpp, G. W. *Tetrahedron Lett.* **1988**, *29*, 5237.
16. van Eikema Hommes, N. J. R.; Bickelhaupt, F.; Klumpp, G. W. *J. Chem. Soc., Chem Commun.* **1991**, 438.

17. Vlaar, C. P.; Klumpp, G. W. *Tetrahedron Lett.* **1991**, *32*, 2951.
18. Maercker, A.; Bös, B. Sixth IUPAC Symposium on Organometallic Chemistry directed towards Organic Synthesis, August 25–29, 1991, Utrecht, The Netherlands, Abstract A 11.
19. Kündig, E. P.; Perret, C. *Helv. Chim. Acta.* **1981**, *64*, 2606.
20. Barluenga, J.; Fernandez, J. R.; Yus, M. *Synthesis* **1985**, 977.
21. Wilcox, Jr., C. F.; Leung, C. *J. Org. Chem.* **1968**, *33*, 877.
22. Seyferth, D.; Rochow, E. G. *J. Am. Chem. Soc.* **1955**, *77*, 907.
23. Seyferth, D.; Attridge, C. J. *J. Organomet. Chem.* **1970**, *21*, 103.
24. Wittig, G.; Geissler, G. *Justus Liebigs Ann. Chem.* **1953**, *580*, 44.
25. Wittig, G.; Herwig, W., *Chem. Ber.* **1954**, *87*, 1511.
26. Heaney, H.; Heinekey, D. M.; Mann, F. G.; Millar, I. T. *J. Chem. Soc.* **1958**, 3838.
27. Wittig, G.; Hellwinkel, D. *Chem. Ber.* **1964**, *97*, 789.
28. Gelius, R. *Chem. Ber.* **1960**, *93*, 1759.
29. Johncock, P. *J. Organomet. Chem.* **1966**, *6*, 433.
30. Negishi, E.; Swanson, D. R.; Rousset, C. J. *J. Org. Chem.* **1990**, *55*, 5406.
31. Eaton, P. E.; Tsanaktsidis, J. *J. Am. Chem. Soc.* **1990**, *112*, 876.
32. Siegel, H.; Hiltbrunner, K.; Seebach, D. *Angew. Chem.* **1979**, *91*, 845; *Angew. Chem. Int. Ed. Engl.* **1979**, *18*, 785.
33. Seebach, D.; Siegel, H.; Gabriel, J., Hässig, R. *Helv. Chim. Acta* **1980**, *63*, 2046.
34. Seebach, D.; Siegel, H.; Müllen, K.; Hiltbrunner, K. *Angew. Chem.* **1979**, *91*, 844; *Angew. Chem. Int. Ed. Engl.* **1979**, *18*, 784.
35. Warner, P. M.; Herold, R. D. *J. Org. Chem.* **1983**, *48*, 5411.
36. Warner, P. M.; Chang, S.-C.; Koszewski, N. J. *J. Org. Chem.* **1985**, *50*, 2605.
37. Hiyama, T.; Takehara, S.; Kitatani, K.; Nozaki, H. *Tetrahedron Lett.* **1974**, 3295.
38. Baran, Jr., J. R.; Lagow, R. J. *J. Am. Chem. Soc.* **1990**, *112*, 9415.
38a. Baran, Jr., J. R.; Hendrickson, C.; Laude, Jr., D. A.; Lagow, R. J. *J. Org. Chem.* **1992**, *57*, 3759.
39. Schöllkopf, U. "Methoden zur Herstellung und Umwandlung von lithium-organischen Verbindungen." In Houben-Weyl-Müller, *Methoden der Organischen Chemie, 4.* Auflage, Band 13/1, pp. 87–253. Thieme, Stuttgart, 1970.
40. Winkel, Y. v. d.; Baar, B. L. M. v.; Bastiaans, H. M. M.; Bickelhaupt, F.; Schenkel, M.; Stegmann, H. B. *Tetrahedron.* **1990**, *46*, 1009.
41. Corey, J. Y.; Dueber, M.; Malaidza, M. *J. Organomet. Chem.* **1972**, *36*, 49.
42. Lange, L. D.; Corey, J. Y.; Rath, N. P. *Organometallics* **1991**, *10*, 3189.
43. Chang, V. H. T.; Corey, J. Y. *J. Organomet. Chem.* **1980**, *190*, 217.
44. Early, R. A.; Gallagher, M. J. *J. Chem. Soc. (C)* **1970**, 158.
45. Hellwinkel, D.; Schenk, W. *Angew. Chem.* **1969**, *81*, 1049; *Angew. Chem. Int. Ed. Engl.* **1969**, *8*, 987.
46. Green, K. *J. Org. Chem.* **1991**, *56*, 4325.

47. Porzi, G.; Concilio, C. *J. Organomet. Chem.* **1977**, *128*, 95.
48. Duerr, B. F.; Chung, Y.-S.; Czarnik, A. W. *J. Org. Chem.* **1988**, *53*, 2120.
49. Tanaka, N.; Kasai, T. *Bull. Chem. Soc. Jpn.* **1981**, *54*, 3020.
50. Wang, W.; D'Andrea, S. V.; Freeman, J. P.; Szmuszkovicz, J. *J. Org. Chem.* **1991**, *56*, 2914.
51. Mills, R. J.; Horvath, R. F.; Sibi, M. P.; Snieckus, V. *Tetrahedron Lett.* **1985**, *26*, 1145.
52. Eapen, K. C.; Saba, C. S.; Tamborski, C. *J. Fluorine Chem.* **1987**, *35*, 571.
53. Hart, H.; Nwokogu, G. C. *Tetrahedron Lett.* **1983**, *24*, 5721.
54. Cohen, S. C.; Massey, A. G. *J. Organomet. Chem.* **1968**, *12*, 341.
55. Cohen, S. C.; Massey, A. G. *J. Organomet. Chem.* **1967**, *10*, 471.
56. Schubert, U.; Neugebauer, W.; Schleyer, P. v R. *J. Chem. Soc., Chem. Commun.* **1982**, 1184.
57. Frejd, T.; Klingstedt, T. *J. Chem. Soc., Chem. Commum.* **1983**, 1021.
58. Brown, K. J.; Berry, M. S.; Waterman, K. C.; Lingenfelter, D.; Murdoch, J. R. *J. Am. Chem. Soc.* **1984**, *106*, 4717.
59. Neugebauer, W.; Kos, A. J.; Schleyer, P. v. R. *J. Organomet. Chem.* **1982**, *228*, 107.
60. Brown, K. J.; Murdoch, J. R. *J. Am. Chem. Soc.* **1984**, *106*, 7843.
61. Tamborski, C.; Soloski, E. J. *J. Organomet. Chem.* **1969**, *20*, 245; Prabhu, U. D. G.; Eapen, K. C.; Tamborski, C. *J. Org. Chem.* **1984**, *49*, 2792.
62. Okuda, Y.; Lakshmikantham, M. V.; Cava, M. P. *J. Org. Chem.* **1991**, *56*, 6024.
63. Pham, C. v.; Macomber, R. S.; Mark, Jr., H. B.; Zimmer, H. *J. Org. Chem.* **1984**, *49*, 5250.
64. Newkome, G. R.; Roper, J. M. *J. Organomet. Chem.* **1980**, *186*, 147.
65. Gribble, G. W. *Synlett* **1991**, *289*, 295.
66. Ishii, A.; Kodachi, M.; Nakayama, J.; Hoshino, M. *J. Chem. Soc., Chem. Commun.* **1991**, 751.
67. Šrogl, J.; Janda, M.; Stibor, I.; Procházková, H. *Z. Chem.* **1971**, *11*, 464.
68. Foulger, N. J.; Wakefield, B. J. *J. Organomet. Chem.* **1974**, *69*, 161.
69. Foulger, N. J.; Wakefield, B. J. *J. Organomet. Synth.* **1986**, *3*, 369.
70. Wakefield, B. J. *J. Organomet. Chem.* **1975**, *99*, 191.
71. Jutzi, P.; Baumgärtner, J.; Schraut, W. *J. Organomet. Chem.* **1977**, *132*, 333.
72. Ashe, A. J., III; Drone, F. J. *Organometallics* **1985**, *4*, 1478.
73. Ferede, R.; Noble, M.; Cordes, A. W.; Allison, N. T. Lay, Jr., J. *J. Organomet. Chem.* **1988**, *339*, 1.
74. Ashe, A. J., III; Mahmoud, S. *Organometallics* **1988**, *7*, 1878.
75. Kos, A. J.; Stein, P.; Schleyer, P. v. R. *J. Organomet. Chem.* **1985**, *280*, C1.
76. Reich, H. J.; Reich. I. L. *J. Org. Chem.* **1975**, *40*, 2248.
77. Maercker, A.; Passlack, M. *Chem. Ber.* **1982**, *115*, 540.
78. Maercker, A.; Passlack, M. *Chem. Ber.* **1983**, *116*, 710.
79. Maercker, A.; Stötzel, R. *J. Organomet. Chem.* **1983**, *254*, 1.

80. Moore, W. R.; Bell, L. N.; Daumit, G. P. *J. Org. Chem.* **1971**, *36*, 1694.
81. Maercker, A.; Bös B., unpublished results; B. Bös, doctorate thesis, University of Siegen, 1992.
82. Crowley, P. J.; Leach, M. R.; Meth-Cohn, O.; Wakefield, B. J. *Tetrahedron Lett.* **1986**, *27*, 2909.
83. Köbrich, G.; Reitz, G.; Schumacher, U. *Chem. Ber.* **1972**, *105*, 1674; Review: Köbrich, G. *Angew. Chem. Int. Ed. Engl.* **1972**, *11*, 473.
84. Barluenga, J.; Rodriguez, M. A.; Campos, P. J.; Asensio, G. *J. Am. Chem. Soc.* **1988**, *110*, 5567.
85. Maercker, A.; Bös, B. *Main Group Metal Chemistry* **1991**, *14*, 67; we were not permitted to publish our work in the journal where the errors originated.
86. Olah, G. A.; Wu, A. *Synthesis* **1990**, 885.
87. Maercker, A.; Dujardin, R. *Angew. Chem.* **1985**, *97*, 612; *Angew. Chem. Int. Ed. Engl.* **1985**, *24*, 571.
88. Maercker, A. *Angew. Chem.* **1987**, *99*, 1002; *Angew. Chem. Int. Ed. Engl.* **1987**, *26*, 972.
89. (a) Rajca, A. *J. Am. Chem. Soc.* **1990**, *112*, 5889, 5890; (b) Rajca, A. *J. Org. Chem.* **1991**, *56*, 2557; (c) Rajca, A.; Utamapanya, S.; Xu, J. *J. Am. Chem. Soc.* **1991**, *113*, 9235; (d) Utamapanya, S.; Rajca, A. *J. Am. Chem. Soc.* **1991**, *113*, 9242; (e) Rajca, A.; Utamapanya, S. *J. Org. Chem.* **1992**, *57*, 1760.
90. Screttas, C. G.; Micha-Screttas, M. *J. Org. Chem.* **1978**, *43*, 1064.
91. Screttas, C. G.; Micha-Screttas, M. *J. Org. Chem.* **1979**, *44*, 713.
92. Akkerman, O. S.; Bickelhaupt, F. *J. Organomet. Chem.* **1988**, *338*, 159.
93. Rücker, C. *J. Organomet. Chem.* **1986**, *310*, 135.
94. Cohen, T.; Ritter, R. H.; Ouellette, D. *J. Am. Chem. Soc.* **1982**, *104*, 7142.
95. Wittig, G.; Bickelhaupt, F. *Chem. Ber.* **1958**, *91*, 883.
96. Wittig, G.; Herwig, W. *Chem. Ber.* **1955**, *88*, 962.
97. Barluenga, J.; Villamaña, J.; Fañanás, F. J.; Yus, M. *J. Chem. Soc., Chem. Commun.* **1982**, 355.
98. Barluenga, J.; Fañanás, F. J.; Villamaña, J.; Yus, M. *J. Chem. Soc., Perkin Trans.* **1984**, *1*, 2685.
99. Maercker, A.; Theis, M.; Kos, A. J.; Schleyer, P. v. R. *Angew. Chem.* **1983**, *95*, 755; *Angew. Chem. Int. Ed. Engl.* **1983**, *22*, 733.
100. Bleshinskii, V. S.; Bleshinskii, S. V. *Izv. Akad. Nauk Kirg. SSR* **1982**, *47; C. A.* **1982**, *97*, 216264q.
101. Maercker, A.; Theis, M. *Angew Chem.* **1984**, *96*, 990; *Angew. Chem. Int. Ed. Engl.* **1984**, *23*, 995.
102. Theis, M. doctorate thesis, University of Siegen, 1985.
103. Maercker, A.; Dujardin, R. *Angew. Chem.* **1984**, *96*, 222; *Angew Chem. Int. Ed. Engl.* **1984**, *23*, 224.
104. Maercker, A.; Dujardin, R. *Organomet. Synth.* **1986**, *3*, 356.
105. Schleyer, P. v. R.; Kos, A. J.; Kaufmann, E. *J. Am. Chem. Soc.* **1983**, *105*, 7617.
106. Maercker, A.; Theis, M. *Organomet. Synth.* **1986**, *3*, 378.

107. Maercker, A.; Graule, T.; Demuth, W. *Angew. Chem.* **1987**, *99*, 1075; *Angew. Chem. Int. Ed. Engl.* **1987**, *26*, 1032.
108. Eaton, P. E.; Cunkle, G. T.; Marchioro, G.; Martin, R. M. *J. Am. Chem. Soc.* **1987**, *109*, 948.
109. Farnum, D. G.; Monego, T. *Tetrahedron Lett.* **1983**, *24*, 1361.
110. Jutzi, P.; Baumgärtner, J. *J. Organomet. Chem.* **1978**, *148*, 257.
111. Märkl, G.; Hofmeister, P. *Tetrahedron Lett.* **1976**, 3419.
112. Sakurai, H.; Koyama, T.; Kira, M.; Hosomi, A.; Nakadaira, Y. *Tetrahedron Lett.* **1982**, *23*, 543.
113. Reich, H. J.; Yelm, K. E.; Reich, I. L. *J. Org. Chem.* **1984**, *49*, 3438.
114. Wright, M. E. *Organometallics* **1989**, *8*, 407.
115. Brix, B.; Clark, T. *J. Org. Chem.* **1988**, *53*, 3365.
116. Ashe, A. J., III; Lohr, L. L.; Al-Taweel, S. M. *Organometallics* **1991**, *10*, 2424.
117. Seyferth, D.; Vick, S. C. *J. Organomet. Chem.* **1978**, *144*, 1.
118. Mitchell, T. N.; Amamria, A. *J. Organomet. Chem.* **1983**, *252*, 47.
119. Apeloig, Y.; Clark, T.; Kos, A. J.; Jemmis, E. D.; Schleyer, P. v. R. *Israel J. Chem.* **1980**, *20*, 43.
120. Reich, H. J.; Phillips, N. H. *J. Am. Chem. Soc.* **1986**, *108*, 2102.
121. Reich, H. J.; Phillips, N. H. *Pure Appl. Chem.* **1987**, *59*, 1021.
122. Farah, D.; Karol, T. J.; Kuivila, H. G. *Organometallics* **1985**, *4*, 662.
123. Kauffmann, T. *Top. Curr. Chem.* **1980**, *92*, 109.
124. Kauffman,.T. *Angew. Chem.* **1982**, *94*, 401; *Angew. Chem. Int. Ed. Engl.* **1982**, *21*, 410.
125. Ishikawa, M.; Hasegawa, Y.; Kunai, A.; Yamanaka, T. *J. Org. Chem.* **1990**, *381*, C57.
126. Maercker, A.; Bodenstedt, H.; Brandsma, L. *Angew. Chem.*, **1992**, *104*, 1387; *Angew. Chem. Int. Ed. Engl.* **1992**, *31*, 1339.
127. Cainelli, G.; Dal Bello, G.; Zubiani, G. *Tetrahedron Lett.* **1965**, 3429.
128. Cainelli, G.; Dal Bello, G.; Zubiani, G. *Tetrahedron Lett.* **1966**, 4315.
129. Zweifel, G.; Arzoumanian, H. *Tetrahedron Lett.* **1966**, 2535.
130. Kulik, W.; Verkruijsse, H. D.; deJong, R. L. P.; Hommes, H.; Brandsma, L. *Tetrahedron Lett.* **1983**, *24*, 2203.
131. Schlenk, W.; Bergmann, E. *Justus Liebigs Ann. Chem.* **1928**, *463*, 106.
132. O'Brien, D. H.; Breeden, D. L. *J. Am. Chem. Soc.* **1981**, *103*, 3237.
133. Smith, J. G.; Oliver, E.; Boettger, T. J. *Organometallics* **1983**, *2*, 1577.
134. Yokoyama, Y.; Takahashi, K. *Chem. Lett.* **1987**, 589.
135. Yokoyama, Y.; Koizumi, T.; Kikuchi, O. *Chem. Lett.* **1991**, 2205.
136. Yokoyama, Y.; Takahura, Y.; Swasaki, T.; Takahashi, K. *Bull. Chem. Soc. Jpn.* **1989**, *62*, 682.
137. Wilhelm, D.; Clark, T.; Schleyer, P. v. R.; Dietrich, H.; Mahdi, W. *J. Organomet. Chem.* **1985**, *280*, C6.
138. Rieke, R. D.; Darnwala, K. P.; Forkner, M. W. *Organometallics* **1991**, *10*, 2946.

139. Schenk, R.; Huber, W.; Schade, P.; Müllen, K. *Chem. Ber.* **1988**, *121*, 2201.
140. Kawase, T.; Fujino, S.; Oda, M. *Chem. Lett.* **1990**, 1683.
141. Schenk, R.; Hucker, J.; Hopf, H.; Räder, H.-J.; Müllen, K. *Angew. Chem.* **1989**, *101*, 942; *Angew. Chem. Int. Ed. Engl.* **1989**, *28*, 904.
142. Reviews: Müllen, K. *Angew. Chem.* **1987**, *99*, 192; *Angew. Chem. Int. Ed. Engl.* **1987**, *26*, 204; Müllen, K. *Chem. Rev.* **1984**, *84*, 603; Staley, S. W. *React. Intermed.* **1983**, *3*, 19.
143. Schenk, R.; Müllen, K.; Wennerström, O. *Tetrahedron Lett.* **1990**, *31*, 7367.
144. Sandel, V. R.; Belinky, B.; Stefaniak, T.; Kreil, D. *J. Org. Chem.* **1975**, *40*, 2116.
145. Weyenberg, D. R.; Toporcer, L. H.; Bey, A. E. *J. Org. Chem.* **1965**, *30*, 4096.
146. Schlenk, W.; Bergmann, E. *Justus Liebigs Ann. Chem.* **1928**; *463*, 1; **1930**, *479*, 58 and 78.
147. Reed, P. J.; Urwin, J. R. *J. Organomet. Chem.* **1972**, *39*, 1.
148. Schade, P.; Schäfer, T.; Müllen, K.; Bender, D.; Knoll, K.; Bronstert, K. *Chem. Ber.* **1991**, *124*, 2833.
149. Bock, H.; Ruppert, K.; Havlas, Z.; Bensch, W.; Hönle, W.; Schnering, H. G. v. *Angew. Chem.* **1991**, *103*, 1197; *Angew. Chem. Int. Ed. Engl.* **1991**, *30*, 1183.
150. Levin, G.; Jagur-Grodzinski, J.; Szwarc, M. *J. Am. Chem. Soc.* **1970**, *92*, 2268.
151. Schlenk, W.; Bergmann, E. *Justus Liebigs Ann. Chem.* **1928**, *463*, 71.
152. Smith, L. I.; Hoehn, H. H. *J. Am. Chem. Soc.* **1941**, *63*, 1184.
153. Leavitt, F. C.; Manuel, T. A.; Johnson, F. *J. Am. Chem. Soc.* **1959**, *81*, 3163.
154. Leavitt, F. C.; Manuel, T. A.; Johnson, F.; Matternas, L. U.; Lehman, D. S. *J. Am. Chem. Soc.* **1960**, *82*, 5099.
155. Braye, E. H.; Hübel, W.; Caplier, I. *J. Am. Chem. Soc.* **1961**, *83*, 4406.
156. Eisch, J. J. *Organomet. Synth.* **1981**, *2*, 98 and 176.
157. Sekiguchi, A.; Zigler, S. S.; Haller, K. J.; West, R. *Recl. Trav. Chim. Pays-Bas* **1988**, *107*, 197.
158. Kos, A. J.; Schleyer, P. v. R. *J. Am. Chem. Soc.* **1980**, *102*, 7928.
159. In the meantime, the doubly bridged structure of the dimer compound has been confirmed by X-ray analysis (P. v. R. Schleyer, personal communication).
160. Karaman, R.; Fry, J. L. *Tetrahedron Lett.* **1989**, *30*, 4931.
161. Maercker, A.; Girreser, U. *Tetrahedron Lett.* **1990**, *31*, 7595.
162. Maercker, A.; Girreser, U. *Angew. Chem.* **1990**, *102*, 718; *Angew. Chem. Int. Ed. Engl.* **1990**, *29*, 667.
163. Klein, J.; Brenner, J. *J. Am. Chem. Soc.* **1969**, *91*, 3094.
164. Klein, J.; Becker, J. Y. *J. Chem. Soc., Chem. Commun.* **1973**, 576.
165. Mulvaney, J. E.; Folk, T. L.; Newton, D. J. *J. Org. Chem.* **1967**, *32*, 1674.
166. Maercker, A.; Passlack, M. *Chem. Ber.* **1983**, *116*, 710.
167. Maercker, A.; Fischenich, J., unpublished results; Fischenich, J., doctorate thesis, University of Siegen, 1994.
168. Youngs, W. J.; Djebli, A.; Tessier, C. A. *Organometallics* **1991**, *10*, 2089.
169. Sakurai, H.; Nakadaira, Y.; Tobita, H. *Chem. Lett.* **1982**, 771.

170. Sekiguchi, A.; Nakanishi, T.; Kabuto, C.; Sakurai, H. *J. Am. Chem. Soc.* **1989**, *111*, 3748.

171. Kira, M.; Hino, T.; Kubota, Y.; Matsuyama, N.; Sakurai, H. *Tetrahedtron Lett.* **1988**, *29*, 6939.

172. Eisch, J. J.; Beuhler, R. J. *J. Org. Chem.* **1963**, *28*, 2876.

173. Kos, A. J.; Jemmis, E. D.; Schleyer, P. v. R.; Gleiter, R.; Fischbach, U.; Pople, J. A. *J. Am. Chem. Soc.* **1981**, *103*, 4996.

174. Sekiguchi, A.; Ebata, K.; Kabuto, C.; Sakurai, H. *J. Am. Chem. Soc.* **1991**, *113*, 1464.

175. Sekiguchi, A.; Ebata, K.; Kabuto, C.; Sakurai, H. *J. Am. Chem. Soc.* **1991**, *113*, 7081.

176. Sygula, A.; Rabideau, P. W. *J. Am. Chem. Soc.* **1991**, *113*, 7797.

177. Pilz, M.; Allwohn, J.; Willershausen, P.; Massa, W.; Berndt, A. *Angew. Chem.* **1990**, *102*, 1085; *Angew. Chem. Int. Ed. Engl.* **1990**, *29*, 1030.

178. Dowd, P. *J. Chem. Soc., Chem. Commun.* **1965**, 568.

179. Bernard, J.; Schnieders, C.; Müllen, K. *J. Chem. Soc., Chem. Commun.* **1985**, 12.

180. Rajca, A.; Tolbert, L. M. *J. Am. Chem. Soc.* **1985**, *107*, 2969.

181. Rajca, A.; Tolbert, L. M. *J. Am. Chem. Soc.* **1987**, *109*, 1782.

182. Rajca, A.; Streitwieser, Jr., A.; Tolbert, L. M. *J. Am. Chem. Soc.* **1987**, *109*, 1790.

183. Boche, G.; Buckl, K. *Angew. Chem.* **1978**, *90*, 291; *Angew. Chem. Int. Ed. Engl.* **1978**, *17*, 284.

184. Kuus, H. *Uch. Zap. Tartu. Gos. Univ.* **1966**, *193*, 130; *C. A.* **1968**, *69*, 67443.

185. Hänsele, E.; Clark, T. *Z. Phys. Chem.* **1991**, *171*, 21.

186. van Eikema Hommes, N. J. R.; Bickelhaupt, F.; Klumpp, G. W. *Angew. Chem.* **1988**, *100*, 1100; *Angew. Chem. Int. Ed. Eng.* **1988**, *27*, 1083.

187. Skinner, D. L.; Peterson, D. J.; Logan, T. J. *J. Org. Chem.* **1967**, *32*, 105.

188. Khotimskii, V. S.; Bryantseva, I. S.; Durgar'yan, S. G.; Petrovskii, P. V. *Izv. Akad. Nauk SSSR, Ser. Khim.* **1984**, 470; *C. A.* **1984**, *100*, 209971.

189. Maercker, A.; Graule, T.; Girreser, U. *Angew. Chem.* **1986**; *98*, 174; *Angew. Chem. Int. Ed. Engl.* **1986**, *25*, 167.

190. Maercker, A.; Graule, T.; Girreser, U. *Organomet. Synth.* **1988**, *4*, 366.

191. Maercker, A.; Girreser, U., *Tetrahedron* **1994**, *50*, 8019.

192. Neugebauer, W.; Geiger, G. A. P.; Kos, A. J.; Stezowski, J. J.; Schleyer, P. v. R. *Chem. Ber.* **1985**, *118*, 1504.

193. Walton, D. R. M.; Waugh, F. *J. Organomet. Chem.* **1972**, *37*, 45.

194. Schlenk, W.; Bergmann, E. *Justus Liebigs Ann. Chem.* **1928**, *463*, 82.

195. Zweig, A.; Hoffmann, A. K. *J. Am. Chem. Soc.* **1962**, *84*, 3278.

196. Nahon, R.; Day, A. R. *J. Org. Chem.* **1965**, *30*, 1973.

197. Setzer, W. N.; Schleyer, P. v. R. *Adv. Organomet. Chem.* **1985**, *24*, 390.

198. Maercker, A.; Dujardin, R.; Brauers, F. *Organomet. Synth.* **1988**, *4*, 362.

199. Maercker, A.; Dujardin, R.; Engelen, B.; Buchmeier, W.; Jung, M., unpublished results.

200. Günther, H.; Moskau, D.; Dujardin, R.; Maercker, A. *Tetrahedron Lett.* **1986**, *27*, 2251.
201. Moskau, D.; Brauers, F.; Günther, H.; Maercker, A. *J. Am. Chem. Soc.* **1987**, *109*, 5532.
202. Maercker, A. *Justus Liebigs Ann. Chem.* **1970**, *732*, 151.
203. Goldstein, M. J.; Wenzel, T. T.; Whittaker, G.; Yates, S. F. *J. Am. Chem. Soc.* **1982**, *104*, 2669.
204. Goldstein, M. J.; Wenzel, T. T. *J. Chem. Soc., Chem. Commun.* **1984**, 1654; 1655.
205. Trinks, R.; Müllen, K. *Chem. Ber.* **1987**, *120*, 1481.
206. Trinks, R.; Müllen, K. *Tetrahedtron Lett.* **1988**, *29*, 3929.
207. Hoell, D.; Lex, J.; Müllen, K. *J. Am. Chem. Soc.* **1986**, *108*, 5983.
208. Müllen, K. *Helv. Chim. Acta* **1978**, *61*, 1305.
209. Maercker, A.; Berkulin, W.; Schiess, P. *Angew Chem.* **1983**, *95*, 248; *Angew. Chem. Int. Ed. Engl.* **1983**, *22*, 246.
210. Benken, R.; Finneiser, K.; Puttkamer, H. v.; Günther, H.; Eliasson, B.; Edlund, U. *Helv. Chim. Acta* **1986**, *69*, 955.
211. Bausch, J. W.; Gregory, P. S.; Olah, G. A.; Prakash, G. K. S.; Schleyer, P. v. R.; Segal, G. A. *J. Am. Chem. Soc.* **1989**, *111*, 3633.
212. Moore, W. R.; Bell, L. N.; Daumit, G. P. *J. Am. Chem. Soc.* **1970**, *92*, 6680.
213. Bates, R. B.; Beavers, W. A.; Greene, M. G.; Klein, J. H. *J. Am. Chem. Soc.* **1974**, *96*, 5640.
214. Klein, J.; Medlik, A. *J. Chem. Soc., Chem. Commun.* **1973**, 275.
215. Klein, J.; Medlik-Bala, A.; Meyer, A. Y.; Chorev, M. *Tetrahedron* **1976**, *32*, 1839.
216. Maercker, A.; Klein, K.-D. *Angew. Chem.* **1989**, *101*, 63; *Angew. Chem. Int. Ed. Engl.* **1989**, *28*, 83.
217. Maercker, A.; Daub, V. E. E., *Tetrahedron* **1994**, *50*, 2439.
218. Maercker, A.; Klein, K.-D. *J. Organomet. Chem.* **1991**, *410*, C35.
219. Maercker, A.; Klein, K.-D. *J. Organomet. Chem.* **1991**, *401*, C1.
220. Eppers, O.; Günther, H.; Klein, K.-D.; Maercker, A. *Magn. Reson. Chem.* **1991**, *29*, 1065.
221. Eppers, O.; Fox, T.; Günther, H. *Helv. Chim. Act* **1992**, *75*, 883.
222. Bauer, W.; Feigel, M.; Müller, G.; Schleyer, P. v. R. *J. Am. Chem. Soc.* **1988**, *110*, 6033.
223. Maercker, A.; Daub, V. E. E.; Deskowski, A. *Third International Symposium on Carbanion Chemistry*, June 14–18, 1992, Gallipoli (Lecce), Italy, Abstract PS-6.
224. Maercker, A.; Oeffner, K., unpublished results, Oeffner, K., doctorate thesis, University of Siegen, 1993.
225. Bunz, U.; Szeimies, G. *Tetrahedron Lett.* **1990**, *31*, 651.
226. Lutz, P.; Beinert, G.; Franta, E.; Rempp, P. *Eur. Polym. J.* **1979**, *15*, 1111.
227. Schulz, G.; Höcker, H. *Angew. Chem.* **1980**, *92*, 212; *Angew. Chem. Int. Ed. Engl.* **1980**, *19*, 219.

228. You, S.; Gubler, M.; Neuenschwander, M. *Chimia* **1991**, *45*, 387.
229. Pilz, M.; Allwohn, J.; Hunold, R.; Massa, W.; Berndt, A. *Angew. Chem.* **1988**, *100*, 1421; *Angew. Chem. Int. Ed. Engl.* **1988**, *27*, 1370.
230. Friedmann, G.; Linder, P.; Brini, M.; Cheminat, A. *J. Org. Chem.* **1979**, *44*, 237.
231. Gilman, H.; Trepka, W. J. *J. Org. Chem.* **1961**, *26*, 5202.
232. Granoth, I.; Levy, J. B.; Symmes, Jr., C.; *J. Chem. Soc., Perkin Trans. 2* **1972**, 697.
233. Kostenko, N. L.; Nesterova, S. V.; Toldov, S. V.; Skvortsov, N. K.; Reikhsfel'd, V. O. *Zh. Obshch. Khim.* **1987**, *57*, 716.
234. Toldov, S. V.; Kostenko, N. L.; Bel'skii, V. K.; Skvortsov, N. K.; Reikhsfel'd, V. O. *Zh. Obshch. Khim.* **1990**, *60*, 1257; engl. 1120.
235. Antonio, Y.; Barrera, P.; Contreras, O.; Franco, F.; Galeazzi, E.; Garcia, J.; Greenhouse, R.; Guzmán, A.; Yelarde, E.; Muchowski, J. M. *J. Org. Chem.* **1989**, *54*, 2159.
236. Crowther, G. P.; Sundberg, R. J.; Sarpeshkar, A. M. *J. Org. Chem.* **1984**, *49*, 4657.
237. Cabiddu, S.; Contini, L.; Fattuoni, C.; Floris, C.; Gelli, G. *Tetrahedron* **1991**, *47*, 9279.
238. Dallacker, F.; Sanders, G. *Chem.-Ztg.* **1986**, *110*, 369.
239. Cabiddu, C.; Floris, C.; Melis, S. *Tetrahedron Lett.* **1986**, *27*, 4625.
240. Cabiddu, S.; Fattuoni, C.; Floris, C.; Gelli, G.; Melis, S.; Sotgiu, F. *Tetrahedron* **1990**, *46*, 861.
241. Howells, R. D.; Gilman, H. *Tetrahedron Lett.* **1974**, 1319.
242. Clark, R. D.; Caroon, J. M. *J. Org. Chem.* **1982**, *47*, 2804.
243. Eaton, P. E.; Martin, R. M. *J. Org. Chem.* **1988**, *53*, 2728.
244. Cheeseman, G. W. H.; Greenberg, S. G. *J. Organomet. Chem.* **1979**, *166*, 139.
245. Ashe, A. J., III; Kampf, J. W.; Savla, P. M. *J. Org. Chem.* **1990**, *55*, 5558.
246. Muchowski, J. M.; Hess, P. *Tetrahedron Lett.* **1988**, *29*, 777.
247. Crump, S. L.; Rickborn, B. *J. Org. Chem.* **1984**, *49*, 304.
248. Rahman, M. T.; Gilman, H. *J. Indian Chem. Soc.* **1976**, *53*, 582.
249. Chadwick, D. J.; Willbe, C. *J. Chem. Soc., Perkin Trans. 1* **1977**, 887.
250. Ishii, A.; Horikawa, Y.; Takaki, I.; Shibata, J.; Nakayama, J.; Hoshino, M. *Tetrahedron Lett.* **1991**, *32*, 4313.
251. Brandsma, L.; Verkruijsse, H. D. *Preparative Polar Organometallic Chemistry*, Vol. 1, p. 164, Springer, Heidelberg 1987.
252. Gol'dfarb, Ya. L.; Stoyanovich, F. M.; Chermanova, G. B. *Izv. Akad. Nauk SSSR, Ser. Khim.* **1973**, 2290; *C. A.* **1974**, *80*, 47744.
253. Gol'dfarb, Ya. L.; Stoyanovich, F. M.; Chermanova, G. B.; Lubuzh, E. D. *Izv. Akad. Nauk SSSR, Ser. Khim.* **1978**, 2760; *C. A.* **1979**, *90*, 151900.
254. Doadt, E. G.; Snieckus, V. *Tetrahedron Lett.* **1985**, *26*, 1149.
255. Graham, S. L.; Scholz, T. H. *J. Org. Chem.* **1991**, *56*, 4260.
256. Carpenter, A. J.; Chadwick, D. J. *Tetrahedron Lett.* **1985**, *26*, 5335.
257. Demuth, Jr., T. P.; Lever, D. C.; Gorgos, L. M.; Hogan, C. M.; Chu, J. *J. Org. Chem.* **1992**, *57*, 2963.

258. Chadwick, D. J.; Hodgson, S. T. *J. Chem. Soc., Perkin Trans. 1* **1982**, 1833.
259. Review: Slocum, D. W.; Engelmann, T. R.; Ernst, C.; Jennings, C. A.; Jones, W.; Koonssvitsky, B.; Lewis, J.; Shenkin, P. *J. Chem. Educ.* **1969**, *46*, 144.
260. Rausch, M. D.; Ciappenelli, D. J. *J. Organomet. Chem.* **1967**, *10*, 127.
261. Bishop, J. J.; Davison, A.; Katcher, M. L.; Lichtenberg, D. W.; Merrill, R. E.; Smart, J. C. *J. Organomet. Chem.* **1971,** *27*, 241.
262. Walczak, M.; Walczak, K.; Mink, R.; Rausch, M. D.; Stucky, G. *J. Am. Chem. Soc.* **1978**, *100*, 6382.
263. Butler, I. R.; Cullen, W. R.; Ni, J.; Rettig, S. J. *Organometallics* **1985**, *4*, 2196.
264. Kinoshita, T.; Tasumi, S.; Zanka, Y.; Tsuji, S.; Takamuki, Y.; Fukumasa, M.; Takeuchi, K.; Okamoto, K. *Tetrahedron Lett.* **1990**, *31*, 6673.
265. Herberhold, M.; Leitner, P. *J. Organomet. Chem.* **1991**, *411*, 233.
266. Herberhold, M.; Brendel, H.-D.; Nuyken, O.; Pöhlmann, T. *J. Organometal. Chem.* **1991**, *413*, 65.
267. Herberhold, M.; Brendel, H.-D.; Thewalt, U. *Angew. Chem.* **1991**, *103*, 1664; *Angew. Chem. Int. Ed. Engl.* **1991**, *30*, 1652.
268. Elschenbroich, C. *J. Organomet. Chem.* **1968**, *14*, 157.
269. Elschenbroich, C.; Heikenfeld, G.; Wünsch, M.; Massa, W.; Baum, G. *Angew. Chem.* **1988**, *199*, 397; *Angew. Chem. Int. Ed. Engl.* **1988**, *27*, 414.
270. Crocco, G. L.; Gladysz, *Chem. Ber.* **1988**, *121*, 375.
271. Elschenbroich, C.; Sebbach, J.; Metz, B. *Helv. Chim. Acta* **1991**, *74*, 1718.
272. Halasa, A. F.; Tate, D. P. *J. Organomet. Chem.* **1970**, *24*, 769.
273. Hedberg, F. L.; Rosenberg, H. *J. Am. Chem. Soc.* **1970**, *92*, 3239; **1973**, *95*, 870.
274. Huffman, J. W.; Cope, J. F. *J. Org. Chem.* **1971**, *36*, 4068.
275. Butler, I. R.; Cullen, W. R.; Reglinski, J.; Rettig, S. J. *J. Organomet. Chem.* **1983**, *249*, 183.
276. Butleer, I. R.; Cullen, W. R.; Rettig, S. J. *Can. J. Chem.* **1987**, *65*, 1452.
277. Hayashi, T.; Yamamoto, A.; Hojo, M.; Ito, Y. *J. Chem. Soc., Chem. Commun.* **1989**, 495.
278. Petter, R. C.; Milberg, C. I. *Tetrahedron Lett.* **1989**, *30*, 5085.
279. Feit, B. A.; Haag, B.; Schmidt, R. R. *J. Org. Chem.* **1987**, *52*, 3825.
280. Aharon-Shalem, E.; Becker, J. Y.; Bernstein, J.; Bittner, S.; Shaik, S. *Tetrahedron Lett.* **1985**, *26*, 2783.
281. Jϕrgensen, M.; Bechgaard, K. *Synthesis* **1989**, 207.
282. Bryce, M. R.; Cooke, G. *Synthesis* **1991**, 263.
283. Rajeswari, S.; Jackson, Y. A.; Cava, M. P. *J. Chem. Soc., Chem. Commun.* **1988**, 1089.
284. Nigrey, P. J. *Synth. Metals* **1988**, *27*, B 365.
285. Poleschner, H.; Radeglia, R.; Fuchs, J. *J. Organomet. Chem.* **1992**, *427*, 213.
286. Applequist, D. E.; Saurborn, E. G. *J. Org. Chem.* **1972**, *37*, 1676.
287. Murata, I.; Nakazawa, T.; Kato, M.; Tatsuoka, T.; Sugihara, Y. *Tetrahedron Lett.* **1975**, 1647.
288. Schlüter, A.-D.; Huber, H.; Szeimies, G. *Angew. Chem.* **1985**, *97*, 406; *Angew. Chem. Int. Ed. Engl.* **1985**, *24*, 404.

289. Butkowskyj-Walkiv, T.; Szeimies, G. *Tetrahedron* **1986**, *42*, 1845.
290. Ando, W.; Igarashi, Y.; Kabe, Y.; Tokitoh, N. *Tetrahedron Lett.* **1990**, *31*, 4185.
291. Igarashi, Y.; Kabe, Y.; Hagiwara, T.; Ando, W. *Tetrahedron* **1992**, *48*, 89.
292. Binger, P.; Müller, P.; Wenz, R.; Mynott, R. *Angew. Chem.* **1990**, *102*, 1070; *Angew. Chem. Int. Ed. Engl.* **1990**, *29*, 1037.
293. Gallagher, D. J.; Garrett, C. G.; Lemieux, R. P.; Beak, P. *J. Org. Chem.* **1991**, *56*, 853.
294. Yang, X.; Knobler, C. B.; Hawthorne, M. F. *Angew. Chem.* **1991**, *103*, 1519; *Angew. Chem. Int. Ed. Engl.* **1991**, *30*, 1507.
295. Engelhardt, L. M.; Leung, W.-P.; Raston, C. L.; Twiss, P.; White, A. H. *J. Chem. Soc., Dalton Trans.* **1984**, 321.
296. Noyori, R; Sano, N.; Murata, S.; Okamoto, Y.; Yuki, H.; Ito, T. *Tetrahedron Lett.* **1982**, *23*, 2969.
297. Gross, U.-M.; Bartels, M.; Kaufmann, D. *J. Organomet. Chem.* **1988**, *344*, 277.
298. Griggs, C. G.; Smith, D. J. H. *J. Chem. Soc., Perkin Trans 1* **1982**, 3041.
299. Garelli, N.; Vierling, P. *J. Org. Chem.* **1992**, *57*, 3046.
300. Bank, S.; Sturges, J. S. *J. Organomet. Chem.* **1978**, *156*, 5.
301. Klein, J.; Medlik, A.; Meyer, A. Y. *Tetrahedron* **1976**, *32*, 51.
302. Klein, J.; Medlik-Balan, A. *J. Am. Chem. Soc.* **1977**, *99*, 1473.
303. Reviews: a) Klein, J. *Tetrahedron* **1983**, *39*, 2733; b) Klein, J. *Tetrahedron* **1988**, *44*, 503.
304. Cabiddu, S.; Floris, C.; Gelli, G.; Melis, S. *J. Organomet. Chem.* **1989**, *366*, 1.
305. Dunkelblum, E.; Hart, H. *J. Org. Chem.* **1979**, *44*, 3482.
306. Inagaki, S.; Imai, T.; Mori, Y. *Bull. Chem. Soc. Jpn.* **1989**, *62*, 79.
307. Murav'eva, L. S.; Sergutin, V. M.; Mushina, E. A.; Zgonnik, V. N.; Kreutsel', B. A. *Izv. Akad. Nauk SSSR, Ser. Khim.* **1982**, 705; *C. A.* **1982**, *97*, 23846.
308. Kaiser, E. M.; Petty, J. D. *J. Organomet. Chem.* **1976**, *108*, 139.
309. Engelhardt, L. M.; Papasergio, R. I.; Raston, C. L.; White, A. H. *J. Chem. Soc., Dalton Trans.* **1984**, 311.
310. Lappert, M. F.; Raston, C. L.; Skelton, B. W.; White, A. H. *J. Chem. Soc., Chem. Commun.* **1982**, 14.
311. Engelhardt, L. M.; Leung, W.-P.; Raston, C. L.; White, A. H. *J. Chem. Soc., Dalton Trans.* **1985**, 337.
312. Leung, W.-P.; Raston, C. L.; Skelton, B. W.; White, A. H. *J. Chem. Soc., Dalton Trans.* **1984**, 1801.
313. Hacker, R.; Schleyer, P. v. R.; Reber, G.; Müller, G.; Brandsma, L. *J. Organomet. Chem.* **1986**, *316*, C4.
314. Lappert, M. F.; Martin, T. R.; Raston, C. L. *Inorg. Synth.* **1989**, *26*, 144.
315. Schleyer, P. v. R.; Kos, A. J.; Wilhelm, D.; Clark, T.; Boche, G.; Decher, G.; Etzrodt, H.; Dietrich, H.; Mahdi, W. *J. Chem. Soc., Chem. Commun.* **1984**, 1495.
316. Boche, G.; Decher, G.; Etzrodt, H.; Dietrich, H.; Mahdi, W.; Kos, A. J.; Schleyer, P. v. R. *J. Chem. Soc., Chem. Commun.* **1984**, 1493.

317. West, R.; Jones, P. C. *J. Am. Chem. Soc.* **1968**, *90*, 2656.
318. West, R. *Adv. Chem. Ser.* **1974**, *130*, 211.
319. Zarges, W.; Marsch, M.; Harms, K.; Boche, G. *Angew. Chem.* **1989**, *101*, 1424; *Angew. Chem. Int. Ed. Engl.* **1989**, *28*, 1392.
320. Gornowicz, G. A.; West, R. *J. Am. Chem. Soc.* **1971**, *93*, 1714.
321. Zarges, W.; Marsch, M.; Harms, K.; Boche, G. *Chem. Ber.* **1989**, *122*, 1307.
322. Walczak, M.; Stucky, G. *J. Am. Chem. Soc.* **1976**, *98*, 5531.
323. Wilhelm, D.; Clark, T.; Courtneidge, J. L.; Davies, A. G. *J. Chem. Soc., Chem. Commun.* **1983**, 213.
324. Kershner, L. D.; Gaidis, J. M.; Freedman, H. H. *J. Am. Chem. Soc.* **1972**, *94*, 985.
325. Rhine, W. E.; Davis, J. H.; Stucky, G. *J. Organomet. Chem.* **1977**, *134*, 139.
326. Brooks, J. J.; Rhine, W.; Stucky, G. D. *J. Am. Chem. Soc.* **1972**, *94*, 7346.
327. Rhine, W. E.; Davis, J.; Stucky, G. *J. Am. Chem. Soc.*, **1975**, *97*, 2079.
328. Harvey, R. G.; Nazareno, L.; Cho, H. *J. Am. Chem. Soc.* **1973**, *95*, 2376.
329. Cho, H.; Harvey, R. G. *J. Org. Chem.* **1975**, *40*, 3097.
330. Sygula, A.; Lipkowitz, K.; Rabideau, P. W. *J. Am. Chem. Soc.* **1987**, *109*, 6602.
331. Boche, G.; Etzrodt, H.; Marsch, M.; Thiel, W. *Angew. Chem.* **1982**, *94*, 141; *Angew. Chem. Int. Ed. Engl.* **1982**, *21*, 132; *Angew. Chem. Suppl.* **1982**, 345.
332. Boche, G.; Etzrodt, H.; Massa, W.; Baum, G. *Angew. Chem.* **1985**, *97*, 858; *Angew. Chem. Int. Ed. Engl.* **1985**, *24*, 863.
333. Review: Boche, G. *Top. Curr. Chem.* **1988**, *146*, 1.
334. Klein, J.; Medlik-Balan, A. *J. Chem. Soc., Chem. Commun.* **1975**, 877.
335. Lokey, R. S.; Mills, N. S.; Rheingold, A. L. *Organometallics* **1989**, *8*, 1803.
336. Herberich, G. E.; Ganter, C.; Wesemann, L.; Boese, R. *Angew. Chem.* **1990**, *102*, 914; *Angew. Chem. Int. Ed. Engl.* **1990**, *29*, 912.
337. Herberich, G. E.; Eigendorf, U.; Ganter, C. *J. Organomet. Chem.* **1991**, *402*, C17.
338. Herberich, G. E.; Spaniol, T. P. *J. Chem. Soc., Chem. Commun.* **1991**, 1457.
339. Herberich, G. E.; Englert, U.; Ganter, C.; Wesemann, L. *Chem. Ber.* **1992**, *125*, 23.
340. Guyot, B.; Pornet, J.; Miginiac, L. *Tetrahedron* **1991**, *47*, 3981.
341. Gund, P. *J. Chem. Educ.* **1972**, *49*, 100.
342. Bates, R. B.; Hess, Jr., B. A.; Ogle, C. A.; Schaad, L. J. *J. Am. Chem. Soc.* **1981**, *103*, 5052.
343. Inagaki, S.; Hirabayashi, Y. *Chem. Lett.* **1982**, 709.
344. Mills, N. S.; Shapiro, J.; Hollingsworth, M. *J. Am. Chem. Soc.* **1981**, *103*, 1263.
345. Rusinko, A., III; Mills, N. S.; Morse, P. *J. Org. Chem.* **1982**, *47*, 5198.
346. Mills, N. S. *J. Am. Chem. Soc.* **1982**, *104*, 5689.
347. Mills, N. S.; Rusinko, A. R., III. *J. Org. Chem.* **1986**, *51*, 2567.
348. Clark, T.; Wilhelm, D.; Schleyer, P. v. R. *Tetrahedron Lett.* **1982**, *23*, 3547.

349. Wilhelm, D.; Clar, T.; Schleyer, P. v. R.; Buckl, K.; Boche, G. *Chem. Ber.* **1983**, *116*, 1669.
350. Wilhelm, D.; Clark, T.; Schleyer, P. v. R. *Tetrahedron Lett.* **1983**, *24*, 3985.
351. Wilhelm, D.; Clark, T.; Schleyer, P. v. R. *J. Chem. Soc., Perkin Trans. 2* **1984**, 915.
352. Agranat, I.; Skancke, A. *J. Am. Chem. Soc.* **1985**, *107*, 867.
353. Gobbi, A.; MacDougall, P. J.; Frenking, G. *Angew. Chem.* **1991**, *103*, 1038; *Angew. Chem. Int. Ed. Engl.* **1991**, *30*, 1001.
354. Agranat, I.; Radhakrishnan, T. P.; Herndon, W. C.; Skancke, A. *Chem. Phys. Lett.* **1991**, *181*, 117.
355. Wilhelm, D.; Dietrich, H.; Clark, T.; Mahdi, W., Kos, A. J.; Schleyer, P. v. R. *J. Am. Chem. Soc.* **1984**, *106*, 7279.
356. van Doorn, J. A.; van der Heijden, H. *Recl. Trav. Chim. Pays-Bas* **1990**, *109*, 302.
357. Klein, J.; Weiler, L. *Israel J. Chem.* **1984**, *24*, 69.
358. Klein, J.; Medlik-Balan, A. *Tetrahedron Lett.* **1978**, 279.
359. Arora, S. K.; Bates, R. B.; Beavers, W. A.; Cutler, R. S. *J. Am. Chem. Soc.* **1975**, *97*, 6271.
360. Bahl, J. J.; Bates, R. B.; Beavers, W. A.; Mills, N. S. *J. Org. Chem.* **1976**, *41*, 1620.
361. Review: Bates, R. B. *Dianions and Polyanions*, in Buncel, E.; Dust, T. *Comprehensive Carbanion Chemistry*, Part A, p. 1-53, Elsevier, Amsterdam 1980.
362. Review: Barry, C. E., III; Bates, R. B.; Beavers, W. A.; Camou, F. A.; Gordon, B., III; Hsu, H. F.-J.; Mills, N. S.; Ogle, C. A.; Siahaan, T. J.; Suvannachut, K.; Taylor, S. R.; White, J. J.; Yager, K. M. *Synlett* **1991**, 207.
363. Meyer, S. D.; Mills, N. S.; Runnels, J. B.; de la Torre, B.; Ruud, C. C.; Johnson, D. K. *J. Org. Chem.* **1991**, *56*, 947.
364. Maercker, A.; Wunderlich, H., unpublished results, Wunderlich, H., doctorate thesis, University of Siegen, 1990.
365. Bahl, J. J.; Bates, R. B.; Beavers, W. A.; Launer, C. R. *J. Am. Chem. Soc.* **1977**, *99*, 6126.
366. Schleyer, P. v. R.; Wilhelm, D.; Clark, T. *J. Organomet. Chem.* **1985**, *281*, C17.
367. Barfield, M.; Bates, R. B.; Beavers, W. A.; Blacksberg, I. R.; Brenner, S.; Mayall, B. I.; McCulloch, C. S. *J. Am. Chem. Soc.* **1975**, *97*, 900.
368. Gausing, W.; Wilke, G. *Angew. Chem.* **1978**, *90*, 380; *Angew. Chem. Int. Ed. Engl.* **1978**, *17*, 371.
369. Katz, T. J.; Acton, N.; Martin, G. *J. Am. Chem. Soc.* **1969**, *91*, 2804.
370. Nifant'ev, I. E.; Shestakova, A. K.; Lemenovskii, D. A.; Slovokhotov, Yu. L.; Struchkov, Yu. T. *Metallorg. Khim.* **1991**, *4*, 292.
371. Scholz, H. J.; Werner, H. *J. Organomet. Chem.* **1986**, *303*, C8.
372. Takaki, U.; Collins, G. L.; Smid, J. *J. Organomet. Chem.* **1978**, *145*, 139.
373. Becker, B.; Engelmann, V.; Müllen, K. *Angew. Chem.* **1989**, *101*, 501; *Angew. Chem. Int. Ed. Engl.* **1989**, *28*, 458.
374. Hellwinkel, D.; Haas, G. *Justus Liebigs Ann. Chem.* **1979**, 145.

375. Haenel, M. W. *Tetrahedron Lett.* **1977**, 1273.
376. Hafner, K.; Thiele, G. F.; Mink, C. *Angew. Chem.* **1988**, *100*, 1213; *Angew. Chem. Int. Ed. Engl.* **1988**, *27*, 1191.
377. Nifant'ev, I. E.; Yarnykh, V. L.; Borzov, M. V.; Mazurchik, B. A.; Mstyslavsky, V. I.; Roznyatovsky, V. A.; Ustynyuk, Y. A. *Organometallics* **1991**, *10*, 3739.
378. Burger, P.; Hortmann, K.; Diebold, J.; Brintzinger, H.-H. *J. Organomet. Chem.* **1991**, *417*, 9.
379. Lang, H.; Seyferth, D. *Organometallics* **1991**, *10*, 347.
380. Eiermann, M.; Hafner, K. *J. Am. Chem. Soc,* **1992**, *114*, 135.
381. Hiermeier, J.; Köhler, F. H.; Müller, G. *Organometallics* **1991**, *10*, 1787.
382. Atzkern, H.; Hiermeier, J.; Kanellakopulos, B.; Köhler, F. H.; Müller, G.; Steigelmann, O. *J. Chem. Soc., Chem. Commun.* **1991**, 997.
383. Siemeling, U.; Jutzi, P.; Neumann, B.; Stammler, H.-G.; Hursthouse, M. B. *Organometallics* **1992**, *11*, 1328.
384. Atzkern, H.; Köhler, F. H.; Müller, R. *Z. Naturforsch.* **1990**, *45b*, 329.
385. Malaba, D.; Chen, L.; Tessier, C. A.; Youngs, M. J. *Organometallics* **1992**, *11*, 1007.
386. Sethson, I.; Johnels, D.; Lejon, T.; Edlund, U.; Wind, B.; Sygula, A.; Rabideau, P. W. *J. Am. Chem. Soc.* **1992**, *114*, 953.
387. Walczak, M.; Stucky, G. D. *J. Organomet. Chem.* **1975**, *97*, 313.
388. Yasuda, H.; Walczak, M.; Rhine, W.; Stucky, G. *J. Organomet. Chem.* **1975**, *90*, 123.
389. Katz, T. J.; Rosenberg, M.; O'Hara, R. K. *J. Ann. Chem. Soc.* **1964**, *86*, 249.
390. Wilhelm, D.; Clark, T.; Schleyer, P. v. R. *Organomet. Synth.* **1986**, *3*, 358.
391. Stezowski, J. J.; Hoier, H.; Wilhelm, D.; Clark, T.; Schleyer, P. v. R. *J. Chem. Soc., Chem. Commun.* **1985**, 1263.
392. Wilhelm, D.; Courtneidge, J. L.; Clark, T.; Davies, A. G. *J. Chem. Soc., Chem. Commun.* **1984**, 810.
393. Review: Klein, J. *Propargylic metalation*, in Patai, S. *The Chemistry of the Carbon–Carbon Triple Bond*, p. 343–379, Wiley, New York 1978.
394. West, R.; Carney, P. A.; Mineo, I. C. *J. Am. Chem. Soc.* **1965**, *87*, 3788.
395. West, R.; Jones, P. C. *J. Am. Chem. Soc.* **1969**, *91*, 6156.
396. Chwang, T. L.; West, R. *J. Am. Chem. Soc.* **1973**, *95*, 3324.
397. Jaffé, F. *J. Organomet. Chem.* **1970**, *23*, 53.
398. Jemmis, E. D.; Chandrasekhar, J.; Schleyer, P. v. R. *J. Am. Chem. Soc.* **1979**, *101*, 2848.
399. Jemmis, E. D.; Poppinger, D.; Schleyer, P. v. R.; Pople, J. A. *J. Am. Chem. Soc.* **1977**, *99*, 5796.
400. Bhanu, S.; Scheinmann, F. *J. Chem. Soc., Chem. Commun.* **1975**, 817.
401. West, R. *Adv. Chem. Ser.* **1974**, *130*, 211.
402. Priester, W.; West, R.; Chwang, T. L. *J. Am. Chem. Soc.* **1976**, *98*, 8413.
403. Eberly, K. C.; Adams, H. E. *J. Organomet. Chem.* **1965**, *3*, 165.
404. Sagar, A. J. G.; Scheinmann, F. *Synthesis* **1976**, 321.

405. Bhanu, S.; Khan, E. A.; Scheinmann, F. *J. Chem. Soc., Perkin Trans 1*, **1976**, 1609.
406. Kahn, G. R.; Power, K. A.; Scheinmann, F. *J. Chem. Soc., Chem. Commun.* **1979**, 215.
407. Pover, K. A.; Scheinmann, F. *J. Chem. Soc., Perkin Trans. 1* **1980**, 2338.
408. Phenylallene and 3-phenyl-1,2-butadiene yield analog dilithio compounds: Medlik-Balan, A.; Klein, J. *Tetrahedron* **1980**, *36*, 299; Brandsma, L.; Verkruijsse, H. D. *Preparative Polar Organometallic Chemistry*, Vol. 1, p. 45, Springer, Heidelberg 1987; Brandsma, L.; Mugge, E. *Recl. Trav. Chim. Pays-Bas* **1973**, *92*, 628; Meijer, J.; Ruitenberg, K.; Westmijze, H.; Vermeer, P. *Synthesis* **1981**, 551.
409. Klein, J.; Becker, J. Y.; *J. Chem. Soc., Perkin Trans. 2* **1973**, 599.
410. Danishefsky, S. J.; Yamashita, D. S.; Matlo, N. B. *Tetrahedron Lett.* **1988**, *29*, 4681.
411. Gleiter, R.; Merger, R. *Tetrahedron Lett.* **1989**, *30*, 7183.
412. Gleiter, R.; Merger, R. *Tetrahedron Lett.* **1990**, *31*, 1845.
413. Bortolin, R.; Parbhoo, B.; Brown, S. S. D. *J. Chem. Soc., Chem. Commun.* **1988**, 1079.
414. Wong, H. N. C.; Sondheimer, F. *Tetrahedron Lett.* **1980**, *21*, 217.
415. Zweifel, G.; Rajagopalan, S. *J. Am. Chem. Soc.* **1985**, *107*, 700.
416. Rajagopalan, S.; Zweifel, G. *Organomet. Synth.* **1988**, *4*, 496.
417. Brandsma, L.; Verkruijsse, H. D.; Hommes, H. *J. Chem. Soc., Chem. Commun.* **1982**, 1214.
418. Klusener, P. A. A.; Hommes, H.; Hanekamp, J. C.; van der Kerk, A. C. H. T. M.; Brandsma, L. *J. Organomet. Chem.* **1991**, *409*, 67.
419. Becker, J. Y.; Klein, J. *J. Organomet. Chem.* **1978**, *157*, 1.
420. Medlik-Balan, A.; Klein, J. *Tetrahedron* **1980**, *36*, 299.
421. Becker, J. Y. *Tetrahedron Lett.* **1976**, 2159.
422. Klein, J.; Becker, J. Y. *Tetrahedron* **1972**, *28*, 5385.
423. Klein, J.; Brenner, S. *Tetrahedron* **1970**, *26*, 2345.
424. Hooz, J.; Calzada, J. G.; McMaster, D. *Tetrahedron Lett.* **1985**, *26*, 271.
425. Märkl, G.; Pflaum, S.; Maack, A. *Tetrahedron Lett.* **1992**, *33*, 1981.
426. Brandsma, L.; Hommes, H.; Verkruijsse, H. D.; de Jong, R. L. P. *Recl. Trav. Chim. Pays-Bas* **1985**, *104*, 226.
427. Priester, W.; West, R. *J. Am. Chem. Soc.* **1976**, *98*, 8426.
428. Köbrich, G.; Merkel, D. *Justus Liebigs Ann. Chem.* **1972**, *761*, 50.
429. Märkl, G.; Attenberger, P.; Kellner, J. *Tetrahedron Lett.* **1988**, *29*, 3651.
430. de Boer, H. J. R.; Akkermann, O. S.; Bickelhaupt, F. *Organometallics* **1990**, *9*, 2898.
431. Neugebauer, W.; Clark, T.; Schleyer, P. v. R. *Chem. Ber.* **1983**, *116*, 3283.
432. Brandsma, L.; Verkruijsse, H. D. *Preparative Polar Organometallic Chemistry*, Vol. 1, p. 195, Springer, Heidelberg 1987.
433. Furber, M.; Taylor, R. J. K.; Burford, S. C. *J. Organomet. Chem.* **1986**, *311*, C35.

434. Furber, M.; Taylor, R. J. K.; Burford, S. C. *J. Chem. Soc., Perkin Trans. 1* **1987**, 1573.
435. Barluenga, J.; Ganzález, R.; Fañanás, F. J. *Tetrahedron Lett.* **1992**, *33*, 831.
436. Barluenga, J.; Foubelo, F.; González, R.; Fañanás, F. J.; Yus, M. *J. Chem. Soc., Chem. Commun.* **1990**, 587.
437. Barluenga, J.; Foubelo, F.; González, R.; Fañanás, F. J., Yus, M. *J. Chem. Soc., Chem. Commun.* **1990**, 1521.
438. Barluenga, J.; Foubelo, F.; Gonzáles, R.; Fañanás, F. J.; Yus, M. *J. Chem. Soc., Chem. Commun.* **1991**, 1001.
439. Review: Barluenga, J. *Pure Appl. Chem.* **1990**, *62*, 595.
440. Klusener, P. A. A.; Kulik, W.; Brandsma, L. *J. Org. Chem.* **1987**, *52*, 5261.
441. Mulvaney, J. E.; Gardlund, Z. G.; Gardlund, S. L. *J. Am. Chem. Soc.* **1963**, *85*, 3897.
442. Mulvaney, J. E.; Gardlund, Z. G.; Gardlund, S. L.; Newton, D. J. *J. Am. Chem. Soc.* **1966**, *88*, 476.
443. Mulvaney, J. E.; Carr, L. J. *J. Org. Chem.* **1968**, *33*, 3286.
444. Rausch, M. D.; Klemann, L. P. *J. Am. Chem. Soc.* **1967**, *89*, 5732.
445. Mulvaney, J. E.; Newton, D. J. *J. Org. Chem.* **1969**, *34*, 1936.
446. Walborsky, H. M.; Ronman, P. *J. Org. Chem.* **1978**, *43*, 731.
447. Review: Periasamy, M. P.; Walborsky, H. M. *Org. Prep. Proc. Int.* **1979**, *11*, 293.
448. Miller, J. T.; DeKock, C. W.; Brault, M. A. *J. Org. Chem.* **979**, *44*, 3508.
449. Gausing, W.; Wilke, G.; Krüger, C.; Liu, L. K. *J. Organomet. Chem.* **1980**, *199*, 137.
450. Ziegler, K.; Nagel, K.; Patheiger, M. *Z. Anorg. Allg. Chem.* **1955**, *282*, 345.
451. van Eikema Hommes, N. J. R.; Schleyer, P. v. R.; Wu, Y.-D. *J. Am. Chem. Soc.* **1992**, *114*, 1146.
452. Baran, Jr., J. R.; Lagow, R. J. *J. Organomet. Chem.* **1992**, *427*, 1.
453. Gurak, J. A.; Chinn, Jr., J. W.; Lagow, R. J. *J. Am. Chem. Soc.* **1982**, *104*, 2637.
454. Shimp, L. A.; Morrison, J. A.; Gurak, J. A.; Chinn, Jr., J. W.; Lagow, R. J. *J. Am. Chem. Soc.* **1981**, *103*, 5951.
455. Kawa, H.; Manley, B. C.; Lagow, R. *J. Am. Chem. Soc.* **1985**, *107*, 5313.
456. Guilhaus, M.; Brenten, A. G.; Beynon, J. H.; Theis, M.; Maercker, A. *Org. Mass. Spectrom.* **1985**, *20*, 592.
457. Kudo, H. *Chem. Lett.* **1989**, 1611.
458. Kudo, H. *Nature* **1992**, *355*, 432.
459. Review: Maercker, A. *Angew. Chem.* **1992**, *104*, 598; *Angew. Chem. Int. Ed. Engl.* **1992**, *31*, 584.
460. Schleyer, P. v. R.; Würthwein, E.-U.; Kaufmann, E.; Clark, T.; Pople, J. A. *J. Am. Chem. Soc.* **1983**, *105*, 5930.
461. Kawa, H.; Manley, B. C.; Lagow, R. J. *Polyhedron* **1988**, *7*, 2023.
462. Chauhan, H. P. S.; Kawa, H.; Lagow, R. J. *J. Org. Chem.* **1986**, *51*, 1632.
463. Ludvig, M. M.; Lagow, R. J. *J. Org. Chem.* **1990**, *55*, 4880.

464. Ludvig, M. M.; Kawa, H.; Lagow, R. J. *Synth. Commun.* **1990**, *20*, 1657.

465. Hérold, A. *Bull. Soc. Chim Fr.* **1955**, 999.

466. Chung, C.; Lagow, R. J. *J. Chem. Soc., Chem. Commun.* **1972**, 1078.

467. Reviews: Lagow, R. J.; Gurak, J. A. *Polylithium Organic Chemistry: A Research of Emerging Significance*, in Grünewald, H. *Chem. Future, Proc. 29th IUPAC Congr. 1983*, pp. 107–113, Pergamon, Oxford 1984; Chinn, Jr., J. W.; Gurak, J. A.; Lagow, R. J., *New Developments in the Chemistry of Polylithium Compounds*, in Bach, R. O., *Lithium: Curr. Appl. Sci., Med., Technol.*, pp. 291–305, Wiley, New York 1985; Lagow, J. R. *Met.-Met. Bonds Clusters Chem. Catal.*, [*Proc. Ind.-Univ. Coop. Chem. Program*], 7th 1989 (Pub. 1990), 171.

468. Sneddon, L. G.; Lagow, R. J. *J. Chem. Soc., Chem. Commun.* **1975**, 302.

469. Shimp, L. A.; Lagow, R. J. *J. Org. Chem.* **1979**, *44*, 2311.

470. Landro, F. J.; Gurak, J. A.; Chinn, Jr., J. W.; Newman, R. M.; Lagow, R. J. *J. Am. Chem. Soc.* **1982**, *104*, 7345.

471. Landro, F. J.; Gurak, J. A.; Chinn, Jr., J. W.; Lagow, R. J. *J. Organomet. Chem.* **1983**, *249*, 1.

472. Shimp, L. A.; Lagow, R. J. *J. Am. Chem. Soc.* **1979**, *101*, 2214.

473. Shimp, L. A.; Lagow, R. J. *J. Am. Chem. Soc.* **1973**, *95*, 1343.

474. Morrison, J. A.; Chung, C.; Lagow, R. J. *J. Am. Chem. Soc.* **1975**, *97*, 5015.

475. Shimp, L. A.; Chung, C.; Lagow, R. J. *Inorg. Chim. Acta* **1978**, *29*, 77.

INDEX